LUDWIG FADDEEV
MEMORIAL VOLUME

A Life in Mathematical Physics

LUDWIG FADDEEV
MEMORIAL VOLUME
A Life in Mathematical Physics

Editors

Mo-Lin Ge
Chern Institute of Mathematics, China

Antti J. Niemi
Stockholm University and Uppsala University, Sweden

Kok Khoo Phua
Nanyang Technological University, Singapore

Leon A. Takhtajan
Stony Brook University, USA and Euler International Mathematical Institute, Russia

World Scientific

NEW JERSEY • LONDON • SINGAPORE • BEIJING • SHANGHAI • HONG KONG • TAIPEI • CHENNAI • TOKYO

Published by

World Scientific Publishing Co. Pte. Ltd.
5 Toh Tuck Link, Singapore 596224
USA office: 27 Warren Street, Suite 401-402, Hackensack, NJ 07601
UK office: 57 Shelton Street, Covent Garden, London WC2H 9HE

Library of Congress Cataloging-in-Publication Data
Names: Ge, M. L. (Mo-Lin), editor. | Niemi, Antti J., editor. | Phua, K. K., editor. |
Takhtadzhian, L. A. (Leon Armenovich), editor.
Title: Ludwig Faddeev memorial volume : a life in mathematical physics /
edited by Molin Ge (Nankai University, China), Antti Niemi (Stockholm University, Sweden),
Kok Khoo Phua (Nanyang Technological University, Singapore),
Leon A. Takhtajan (Stony Brook University, USA).
Other titles: Faddeev memorial volume
Description: New Jersey : World Scientific, 2018. | Includes bibliographical references.
Identifiers: LCCN 2018012481 | ISBN 9789813233768 (hardcover : alk. paper)
Subjects: LCSH: Faddeev, L. D. | Mathematical physics. | Festschriften.
Classification: LCC QC20 .L81975 | DDC 530.15--dc23
LC record available at https://lccn.loc.gov/2018012481

British Library Cataloguing-in-Publication Data
A catalogue record for this book is available from the British Library.

Photo on back cover taken by Dr. Arthur Budagov (1978)

For any available supplementary material, please visit
http://www.worldscientific.com/worldscibooks/10.1142/10811#t=suppl

Printed in Singapore

Preface

Ludwig D. Faddeev (1934–2017) was a scientific giant whose impact significantly extends from pure mathematics to theoretical physics. He had made enormous contributions in filling the gaps between mathematics and physics, and his research will continue to both influence and inspire generations of mathematical physicists.

Faddeev was born in Leningrad. His parents were both mathematicians, and his father was also an excellent pianist. When Ludwig was young, his parents aspired a musical career for their son. Unfortunately, these plans could not be realized, as the Second World War put a stop to them. However, a strong passion for music stayed with Ludwig throughout his whole life.

At the Leningrad State University, Ludwig decided to enrol at the Faculty of Physics, instead of the Faculty of Mathematics, of which his father was Dean at the time. The decision reflected both his non-conformist character and his desire to always find his own ways, both in life and in science. Ludwig prepared his PhD thesis under the guidance of Olga Ladyzhenskaya at the Leningrad Department of the Steklov Mathematical Institute of Russian Academy of Sciences (LOMI). The subject of his thesis was a complete solution of the inverse scattering problem for the one-dimensional Schrödinger operator. LOMI, that subsequently became POMI, remained his scientific home during his entire career.

Ludwig Faddeev was strongly attracted to mathematical beauty and the intricacies of quantum field theories. He believed that mathematical, and in particular geometrical beauty must be considered an essential feature of any theoretical description of Nature that is correct. As he kept on challenging himself with technically difficult questions, beauty and elegance were two guiding principles throughout his scientific career.

Among Ludwig's numerous achievements were the discovery of the Faddeev equation of the quantum mechanical three-body problem, several aspects of automorphic functions, quantization of gauge theories and the development of Feynman path integral techniques into a universal tool of quantization. Together with his students and colleagues, Faddeev developed the quantum inverse scattering method into a general approach to quantum integrability. He was a pioneer in the investigation of solitons in various field theory models and analyzed the structure of the Yang–Baxter equation, paving the road to the discovery of quantum groups. Ludwig Faddeev made several groundbreaking discoveries on gauge anomalies, and his fascination with complex topology led him to discover field theory models that support knots as solitons.

Ludwig was a modest person whose life always centered around his family and students, who formed an extended family to him. Together with his wife Anna, Ludwig enjoyed the solitude of the forests on Karelian Isthmus and around Lake Ladoga, where the two often escaped hectic city life to live quietly in their beloved *dacha*.

As editors, we take great pride in presenting our readers this memorial volume that we have prepared in collaboration with World Scientific Publishing. We have collected both personal reminiscences by friends and scientific contributions by colleagues, students and former associates. These articles celebrate the rich and eventful life and science of Ludwig Faddeev, our teacher, colleague and a good friend. We all miss you.

The Editors
M.-L. Ge, A. J. Niemi, K. K. Phua and L. A. Takhtajan

Contents

Part 1
Personal Memories

Some Photographs with Ludwig Faddeev

C. N. Yang

C. N. Yang Institute for Theoretical Physics, USA
and
Tsinghua University, P. R. China

I am surprised and saddened to hear that Ludwig Faddeev had passed away. I had last seen him at a conference in Singapore in 2015. He was then in good health and spirits.

Ludwig and I had been good friends since the early 1970s. We had overlapping interests in several areas of physics. He was very powerful mathematically. I had written in several places that he should have shared the 1999 Nobel Prize in Physics with 't Hooft and Veltman.

I found four old photographs with Faddeev which I reprint below:

Fig. 1. 1997 in Seoul. Faddeev with Baxter and Yang.

Fig. 2. 2005 in Tsinghua University. Left to right: Faddeev, Yang, Niemi and Ge.

Fig. 3. 2008 in Hong Kong. Faddeev with Yang.

Fig. 4. 2008 in Hong Kong. Faddeev was awarded a Shaw Prize in Mathematical Sciences. In the photograph, Faddeev was shaking hands with Sir R. R. Shaw.

Faddeev

Roman Jackiw*,‡ and So-Young Pi†,§

*Department of Physics, Massachusetts Institute of Technology,
Cambridge, MA 02139-4307, USA

†Department of Physics, Boston University,
Boston, Massachusetts 02115, USA

‡jackiw@mit.edu

§soyoung@bu.edu

Today's scholars who pursue several unrelated topics in their research rarely attain important results in more than one area. This observation, due to M. Bos, is well exemplified by the duo of mathematical physics: physicists who establish fundamental natural phenomena hardly ever attain results that are significant to mathematics, and vice versa: frontline mathematicians are not known for their successful physical ideas.

Of course, every rule has its exceptions. Indeed a very small cohort of doubly preeminent mathematical physicists can be readily identified. Ludwig Faddeev belongs to this select company. He moved physics forward when he developed quantization procedures using functional integration for a variety of models. He introduced "ghosts" (with Popov) for non-Abelian theories, which made possible the standard particle physics model. Additionally, he simplified (with Jackiw and Senjanovic) quantization of constrained dynamical systems, and gave a prescription (with Kulish) for regulating infrared singular theories.

Simultaneously, within mathematics with his students, Faddeev explored the quantum inverse scattering method, which led to the emergence of a new field in mathematics (viz. Drinfeld's Quantum Groups).

Faddeev's papers give ample evidence of his contributions at the forefront of mathematics and physics. The breadth, vigor and beauty of Ludwig's permanent accomplishments fully justifies placing him in the company of his namesake.

Ludvig Faddeev

Olli Lehto
University of Helsinki, Finland

I am not sure when I first met Ludvig Faddeev. He was an invited speaker in the International Congress of Mathematicians (ICM) 1978 in Helsinki, where I was the President of the Congress.

Faddeev and I were members of the Executive Committee of the International Mathematical Union (IMU) for many years. Faddeev was the Vice-President of the IMU in 1983–1986 and the President in 1987–1990. During these eight years, I was the Secretary of the IMU. Our cooperation was smooth, although in those days it was difficult to make a phone call to Leningrad.

In the discussion of the election of the new IMU President in 1985 it was agreed that Faddeev was a mathematician of high standing and had excellent personal qualities. In favor of Faddeev was also the fact that of the mathematical superpowers, the USA had had three IMU Presidents, while the USSR none. Yet some opposition was expressed from the United States because Faddeev was from the USSR. After long negotiations, however, it was agreed that taking a stand in favor of Faddeev did not mean the approval of the political system of the USSR. In the final election in 1986, Faddeev was the only candidate and he said, "An election like in my country."

After 1990, Faddeev received many invitations abroad. However, the Faddeevs did not want to move to the US permanently. Helsinki is close to St. Petersburg, and Faddeev asked if I could help him to arrange a part-time job at the University of Helsinki. This soon succeeded and the Faddeevs had a home in Helsinki.

In 1988, Faddeev founded the Euler International Mathematical Institute in St. Petersburg. Their scientific activity began in 1990, and the Finns had great hopes for cooperation. Later, the institute became a part of the Russian Academy of Sciences.

In the summer of 1993, Ludvig and Anya invited my wife Eva and me to their summerhouse in the Karelian Isthmus. We visited Jukkola, formerly a part of Finland, where my mother had her childhood home, which I had last visited 60 years earlier.

For the last time, we met Faddeev in 2015. He was one of the three invited speakers at my 90th birthday celebration at Helsinki University.

Fig. 1. Ludvig Faddeev and Olli Lehto in Jukkola, Karelian Isthmus in 1993. *Photo: Courtesy of Eva Lehto.*

Fig. 2. Ludvig Faddeev speaking at Olli Lehto's 90th birthday celebration at Helsinki University in 2015. *Photo: Courtesy of Erkki Lehto.*

Fig. 3. Olli Lehto (right) speaking to Ludvig Faddeev during Lehto's 90th birthday celebration dinner in Helsinki in 2015. Next to Faddeev is David Drasin. *Photo: Courtesy of Erkki Lehto.*

Ludvig and Helsinki — A Personal Recollection

Antti J. Niemi
Nordita, Stockholm University, Roslagstullsbacken 23,
SE-106 91 Stockholm, Sweden
and
Department of Physics and Astronomy, Uppsala University,
P.O. Box 803, S-75108, Uppsala, Sweden
Antti.Niemi@physics.uu.se

Starting from 1989 and lasting until he passed away, Ludvig Faddeev had a very close and special relationship with Helsinki and Finland that was build on our personal friendship and professional collaboration. Here is my recollection on how it all started and how it then evolved.

My 30 years of friendship with Ludvig Faddeev began during the International Congress of Mathematicians that was held at Berkeley in 1986. During the Congress, Ludvig was elected as the new President of the International Mathematical Union. We met a few times in between the various sessions, to talk about anomalies in gauge theories which at the time interested both of us. I was not a registered participant at the ICM, but at that time, I worked at UC Berkeley; so Ludvig was also very kind and helped me passed through the security to participate in Sasha Polyakov's talk. Later on, I learned that during the election of the IMU President, the General Secretary Olli Lehto had forcefully argued in favour of Ludvig against those who did not want to see a President of IMU to come from the Soviet Union. Three years later, Olli played again an important role in both of our lives, when he as the President of Helsinki University worked out the circumstances, for Ludvig to work part-time in Helsinki.

In 1989, I returned to my native Finland, as a group leader of theoretical high energy physics at the now-defunct Research Institute for Theoretical Physics at Helsinki University. After several years in the United States, one of my dreams in coming back to Finland was to establish scientific contacts with the excellent groups in the Soviet Union, in particular with the Faddeev School at Steklov Mathematical Institute in Leningrad. Ludvig was a giant of quantum field theory and mathematical physics. For me, his style and his research on anomalies, co-cycles and topology in gauge theories were the most exciting things ever. My Grand Plan was to try and build a research group in Helsinki, grounded on an active scientific collaboration with the prime research institutions of the Soviet Union at the time. Besides Faddeev's school at Steklov, I was also keen in a collaboration with Alexei Morozov and his colleagues at ITEP in Moscow. The year before, during a workshop in Warsaw, I had befriended both Alusha and Ludvig's student, Samson Shatashvili. At that workshop, the most interesting lectures, by far, were those that took place informally and without a time limit at the guest house of the Banach Institute where all three of us stayed.

In the summer of 1989, Samson came to visit me in Helsinki, for an extended period, together with his family. During a walk, he started to tell that Ludvig would like to have contact with me, and he would like to come and visit me in Finland. In fact, Samson then went on, Ludvig would like to have some kind of long-term arrangement that would allow him to come and visit Helsinki University on a regular basis, and Ludvig wanted to build a collaboration between his school and my little group. For me, this sounded like a dream come true: the Soviet Union was still viewed as a stable superpower, but despite some cracks, the Iron Curtain seemed steady. During Samson's visit, we wondered a lot about the world development, what will happen and what it meant for life and science. I remembered in particular, a weekend trip to my summer house in August 1989. Matthias Blau who was then also visiting me, was very adamant that there will never be any kind of unification between East and West Germany. Samson and I wondered, I think Samson even doubted, and it took only a few months until the Wall collapsed.

After my discussion with Samson about Ludvig's interests, I immediately started to prepare an application for funding, to set up an exchange project between Faddeev and Helsinki University. My colleagues in theoretical physics at Helsinki University lacked both ability and perspective to help and understand, so instead I contacted Olli Lehto who was then President of Helsinki University, in addition of being General Secretary of IMU. Olli got really excited and he advised me to contact Tuulikki Mäkeläinen for practical help. It turned out that the European Mathematical Society had an office at the Mathematics Department of the University. Tuulikki was running the office, and she had regular contacts with Ludvig on EMS/IMU matters. She helped me to find effective ways to communicate with Ludvig. This was at a time when email was still a novelty and not even available in the Soviet Union. The only good way to communicate was by telephone. But direct lines to Leningrad were few and mostly one received an answer from someone who was only speaking in Russian. However, Tuulikki had a phone number that I could use to get in touch with Ludvig, directly. This number came into very active use during the years that followed. Tuulikki also gave me a host of practical advice that helped me to prepare the proposal and run the project.

After some contacts with Ludvig, I presented an outline of the project plan to Olli Lehto: Ludvig should be able to come and go as he wishes, he should be able to stay in Finland whenever he wants. There should be resources for him to bring a couple of scientists from Steklov to visit Helsinki, on a long-term basis. A few days later, I received a phone call from Elisabeth Helander who was then the research director at the Academy of Finland, the main funding agency for fundamental research in Finland. She told me that Olli Lehto had contacted the Academy in a most persuasive manner and funding for a five-year mathematical physics project will be made available along the lines I had proposed. The so-called Faddeev project then started, with a generous budget that gave us more or less what Ludvig wanted. The project became a flagship for the Academy: At that time, Finland was collapsing into a deep recession. Several excellent scientists were leaving the country. Many of them went to Sweden, eventually including myself. Ludvig went the opposite direction: He came to Finland. After the contract was announced and signed, a major Finnish magazine had a cover article on Ludvig, declaring him "the biggest capture of all times".

Ludvig and I agreed that the ambition should be for him to stay at least three full months each year in Helsinki. He will also bring at least two persons from his group to Helsinki, full-time and on a long-term basis. The first to come were Sklyanin and Volkov. Anton Alekseev also came very early, for one of his many visits to Helsinki. Then followed by Reiman, Bytsko, Kashaev, Kulish, Isergin, Tarasov, Stroganoff, Mnev and many, many others. Both Bytsko and Volkov spent several years in Helsinki, and so did Kashaev who even worked a few years at the Nokia Research Center before continuing to Geneva. Helsinki University provided us with very comfortable office space, specially designed and renovated for the needs of Ludvig and his group members. The offices were located adjacent to the Research Institute for Theoretical Physics, even though the project functioned quite independently. Helsinki University even arranged for a very nice apartment for Ludvig and his wife Anya to stay, a place that they both called home for several years.

Together with Anton Alekseev, we also started to attract students into mathematical physics. Anton came for several visits often with more than a handful of young students. For a few years, we then organised a series of schools that we called "The Baltic Rim Student Seminars". In 1991 and 1993, the school was in Helsinki, and in 1992, it took place in Leningrad. In this activity, we were joined by Alexei Morozov who brought several students from ITEP and elsewhere, but mainly from Moscow. We also organised joint meetings in a little more exotic places including eastern Georgia, southern Ukraine and elsewhere. When I went to Uppsala in 1992 and started to share my time between Sweden and Finland, many of the activities with students also moved to Uppsala even though Helsinki stayed as a focal point. Later on, when Anton joined me in Uppsala on a permanent basis, the remaining student activities followed.

I also started to visit Leningrad, then St. Petersburg, on a regular basis. There, Ludvig often drove me to his *dacha* near Kivennapa on the Karelian Isthmus. Now and then, I went to pick mushrooms in the nearby forests, with both Ludvig and Anya. Afterwards we enjoyed what we found, either at their *dacha* or at their home in St. Petersburg. Both Ludvig and Anya were like professional mushroom hunters. In the forest, they simply opened the car window, now and then, and apparently there was some smell in the forest that told them when to stop. They both knew instinctly where to find mushrooms. They also exercised their skills in Finland where the competition in the forest was less fierce. There was never a shortage of mushrooms in those years, fresh during the season and preserved otherwise. When I was unsure of what I ate, I just followed Ludvig's advice that a glass of vodka always provides a sufficient amount of self-assurance.

For the forest trips and for driving between St. Petersburg and Helsinki, Ludvig bought a strong and roomy four-wheel drive Mitsubishi Pajero. I helped him to buy the car tax-free in Finland, where he then exported it to the Soviet Union. On the day he was due to make the payment, we met on the street in front of a bank in Helsinki. Ludvig was standing there, waiting for me with a worn-down plastic shopping bag in his hand. In the bank, he then opened the bag and out came a huge pile of French francs, more than enough for a full cash payment of the car. The money was from a Prize he had received in Paris, a little earlier.

A few times I had also drove myself to Leningrad. A particularly memorable event took place in 1990 when I visited Samson Shatashvili and his family, over a weekend. On the way back to Helsinki, I recognised the side road to Ludvig and Anya's *dacha* and made a quick decision to go and check if they were there. When I parked, Anya came out from the *dacha*, totally surprised. Later on, she explained to me that that was the moment she understood the world has changed: I had driven with my Finnish-registered car through an area that was supposedly closed to all foreigners.

The new world order then converted Leningrad into St. Petersburg and the Soviet Union became largely replaced by Russia. This brought enormous challenges to science that very much affected Ludvig's group at Steklov. Many of his best performing students got positions abroad, mainly in France and USA and things were moving very fast: When I made my first visit to Steklov Institute, I met Nicolai Reshetikhin, on his last day at office and a day before he departed to Harvard. Others were also packing and soon Samson Shatashvili, then Anton Alekseev, then others started to tell me about their ambitions and plans for a future abroad. Those who stayed had to endure a very hard life and our joint project played a key role in the very survival of the Faddeev School by providing support at a critical moment. But it was a close call: In the spring of 1992, Ludvig phoned me and explained that he had received a letter from Chen-Ning Yang who apparently planned to retire and proposed Ludvig to take over as the Director of the Institute for Theoretical Physics at Stony Brook. The offer must have been very attractive to Ludvig, in particular, since two of his closest former students Takhtajan and Korepin were already professors at Stony Brook. I was certain that the letter spelled the end of our project. However, to my happy and very glad surprise, Ludvig told me a few weeks later that he and Anya will not leave Russia. They have their hearts set in the beautiful St. Petersburg and deep in the forests of Karelian Isthmus. With the financial stability that was provided to them by our joint project, Ludvig decided to say "no" to the proposal Yang made. The offer must have been much better than what Helsinki could provide and Ludvig's quick and firm refusal must have been a big surprise: A few months later, C. N. Yang came to visit me in Helsinki, together with his wife. We spent a few very enjoyable days touring Helsinki and he gave both a wonderful talk and a beautiful public lecture at the University. He also made it known what he thought of Ludvig: During a dinner with Stig Stenholm and Pekka Jauho, Stig asked him about his opinion on Ludvig. The answer left no doubts: Ludvig is the best ever Russian physicist; the only one who comes close is Landau. But what Ludvig does is very demanding, much more difficult for people to understand, that's why Landau gets more credit.

We also had a meeting with the new President of Helsinki University, Risto Ihamuotila, during which Yang described the exceptional economical progress that is transforming China. He told us that Finland should hurry up to take advantage of the development. Shortly afterwards, I received an invitation from Mo-Lin Ge to come and visit Nankai Institute of Mathematics, now Chern Institute of Mathematics, in Tianjin. Like me, Mo-Lin is a great admirer and a very good friend of Ludvig. In the subsequent years, I occasionally joined Ludvig to visit Mo-Lin in Tianjin including car drives that we all three took to Beijing to a meeting with C. N. Yang. Thus, I have Ludvig to thank, also for the wonderful and continuing friendship that I have with Mo-Lin. Our ongoing collaboration is now starting to contribute to the new generation of professors in China.

Initially, the Faddeev project functioned as an independent activity but in a close association with the Research Institute for Theoretical Physics. Early on, I made a few failed efforts to integrate it with the wider activities of theoretical physics in Helsinki. For example, I wanted to set up an advisory committee which was headed by Olli Lehto and included Elisabeth Helander and a theoretical physics professor from the University as initial members. In our first and only meeting, this professor proposed that the entire Faddeev project should be closed. After that, Olli Lehto and Elisabeth Helander took the project directly under the protective wings of the President's office. Then they arranged for further stability by having Ludvig nominated as an Academician in Finland. This is very rare, and it is the highest level of honour that Finland can give to a scientist. Ludvig received the title from Riitta Uosukainen, then the Minister of Science and Education, at an event in the Round Room of Kalastajatorppa where celebrities from Brezhnev to Putin, from Reagan to Clinton have been honoured. After the formal ceremony, we had a dinner, with no limits on the amount of vodka consumed.

An additional element of stability to the Faddeev project came when Finland became a member state of CERN. As a consequence, the Research Institute for Theoretical Physics was joined together with some other institutions in physics, into a national institute for theoretical and experimental physics HIP. The founding director of HIP was Eero Byckling, my teacher from the undergraduate years who fully understood the value of the Faddeev collaboration. He provided the project an environment in which it could continue, free from intrigues.

For Ludvig, the early years of the project certainly played a pivotal role in terms of providing him both with economical stability and a sense of security, when everything around him seemed to collapse with that of the Soviet Union: The life of a scientist in the early days of Russia was pretty void of economical rewards. Ludvig received a substantial salary in Finland and with Anya, they had a very modest lifestyle. Thus, I presume that most of the income was used to support their children: Early on, Ludvig invested in particular to the business endeavours of his son-in-law who eventually succeeded on a truly grand scale. As a thank-you, he then rewarded Ludvig for the early help he had received and, Ludvig and Anya's life became economically secure, and apparently, even very comfortable.

The project in Helsinki had an impact and influenced also developments elsewhere: In the early 1990s, when I accepted the chair of theoretical physics at Uppsala University and started to commute between Finland and Sweden, the project had a pivotal role in my success to build up the Department at Uppsala. The first person whom I offered employment was Samson Shatashvili, but he immediately left for Yale. Then came Anton Alekseev and he brought with him a huge number of visitors and students. Anton started as a postdoc but eventually became a full professor and stayed until he moved permanently to Geneva. I suppose that to some extent, Konstantin Zarembo can be seen as part of this tradition, a replacement of Anton Alekseev who secured the continuity of the Steklov-style mathematical physics in Sweden. Ludvig became then an honourary doctor at Uppsala, and he took a keen interest in supporting the University. For example, when the professorship in mathematics after Lennart Carleson became vacant, I asked Ludvig for help to find a good candidate and this is the way how Oleg Viro ended up in Uppsala. With Oleg and Jouko Mickelsson and the support from Ludvig, we organised a full-year program in mathematical physics at the Mittag-Leffler Institute. It was a dramatic year, in one of our joint articles, Ludvig thanked the cardiology unit of Danderyd Hospital where he spent time recuperating from a heart attack while we finished our calculations.

All in all, the project in Helsinki lasted around ten years at its full capacity, then continued with decreasing level of activity until Ludvig passed away: There was no formal end, the formal part of the project simply ran its course until there was no longer a need, when both Ludvig and I in a somehow natural fashion ended up spending much more time elsewhere. Towards the end, Ludvig's visits to Helsinki became increasingly rare. He no longer came by car but took the train instead. HIP with Byckling as the Director supported Ludvig and his group after the Academy funding ran out, for as long as Ludvig had the strength to come for extended periods. At the end, Ludvig and Anya opted to come only for a few weeks, short visits during the summer. These visits were always generously hosted by Antti Kupiainen and the Department of Mathematics of Helsinki University. Occasionally, I also came to Helsinki, to meet Ludvig and Anya in memory of the good old days.

Scientifically the 1980s, in particular, the years immediately preceding the start of our project, must be considered as the most impressive ones of the Faddeev School at the Steklov Institute. In Helsinki, the project also produced the most impressive list of articles. A few of them have collected over 500 citations and there are probably ten or so with more than 100 citations. Many are continuing to grow. Among the noteworthy articles are those that came from the collaboration between Faddeev, Kashaev and Volkov on quantum dilogarithms, on discrete quantum integrable models and on lattice Virasoro algebras. Sklyanin, partly with Kulish, made beautiful work on integrable quantum spin chains. Faddeev had a collaboration with Korchemsky on high-energy QCD as an integrable model. Many others also produced articles with high impact. Worth mentioning was a joint article by Faddeev and my former student, Olav Tirkkonen, on connections between Liouville model and spin chains. Subsequently, Kashaev and Tirkkonen teamed up and collaborated extensively at the Nokia Research Center in Helsinki, where they built the foundations for the future of mobile technology.

Perhaps the most extensive — and probably also the most visible — research outcome from the Faddeev project was the collaboration that developed between Ludvig and me. This collaboration was the core for both of our scientific activity, for about ten years between 1996 and 2006. We started our joint research during a workshop at the Schrödinger Institute in Vienna. During a tram trip to the hotel, Ludvig described to me about his dream of knots as solitons. Subsequently, we produced something like 15 joint articles, all the way from knots and such in Yang–Mills theories to the structure of Higgs and the Standard Model, topology of two-condensate superconductors, separation between spin and charge and even aspects of plasma physics. A milestone in our collaboration was Ludvig's Plenary Talk at the 2002 ICM in Beijing, in which he described our results on knotted solitons in Yang–Mills theories and elsewhere.

For Ludvig, the family was always the most important aspect of his life. Anya was both his wife and his best friend. The extended family consisted of his many students and I think I got as close to becoming a family member as one can, without being either a relative in blood or a student from Steklov. One of Ludvig's former students once told me that "you are like one of us". Ludvig was my very good friend, a great collaborator and the memory of our joint research and friendship, the time we spent together, continues to live with me.

Remembering Ludwig Dmitrievich Faddeev, Our Lifelong Partner in Mathematical Physics*

Daniel Sternheimer†

Department of Mathematics, Rikkyo University,
3-34-1 Nishi-Ikebukuro, Toshima-ku, Tokyo 171-8501, Japan
and
Institut de Mathématiques de Bourgogne, Université de Bourgogne,
BP 47870, F-21078 Dijon Cedex, France
Daniel.Sternheimer@u-bourgogne.fr

We briefly recount the long friendship that developed between Ludwig and us (Moshé Flato and I), since we first met at ICM 1966 in Moscow. That friendship extended to his school and family, and persists to this day. Its strong personal impact and main scientific components are sketched, including reflections on what mathematical physics is (or should be).

1. In the Beginning

Since there are, and will certainly be (including in this volume) many accounts of the seminal works of Ludwig Faddeev (23 March 1934–26 February 2017) and their impact on modern science, especially modern mathematical physics and mathematical physicists, I shall concentrate this short contribution on events, sometimes anecdotical, that are characteristic and not so well known, and on reflections on mathematical physics. Most involve my friend and coworker for almost 35 years, Moshé Flato (17 September 1937–27 November 1998). Inevitably, these include also a number of major scientific developments and prominent Russian scientists. Though (at the start) younger than many of the latter, Ludwig quickly rose to their level. Like them (maybe even more than them) he created and developed his own school, a scientific family, where appeared and flourished a number of leading mathematical physicists, now present in a variety of academic institutions in leading scientific countries. At the level attained, that was possible only in the USSR of that time, and only with a scientist with a vision.

†Honorary Professor, St. Petersburg State University, Russia.

1.1. *ICM 1966 and around*

1.1.1. *First meeting and background*

Moshé Flato and I first met Ludwig Faddeev at the International Congress of Mathematicians that was held in Moscow (August 16–26, 1966), where the acronym ICM came into wide usage, and later that year when he spent two months at IHES (cf. Ludwig's recollection in [10]). It was "love at first sight" between the young mathematical physicists that we were then (he was born in 1934, Moshé in 1937 and I in 1938). Though we sometimes had to refer to him as "Faddeev", between us we always called each other by first names. So for us he was always "Ludwig" and not "Ludwig Dmitrievich" (Russian style) or "Professor Faddeev", a formal Western way to address colleagues.

His first name is unusual for a Russian. As one may guess, it was given to him in homage to Ludwig van Beethoven. Music (classical of course) was important in his family, and for him as a corollary. In my mind, I can still hear him humming "in petto" some tune. Music was an interest that Ludwig shared with Moshé, who could play piano "by ear" (from classical music to Gershwin) and at some point hesitated between a career in piano and in science.

Ludwig decided to get higher education at the Department of Physics of Leningrad University, mostly to be independent of his father who was a renowned algebraist and a Professor at the Department of Mathematics. Yet the leopard can't change its spots (in French, "chassez le naturel, il revient au galop") and Ludwig's starting point in research was more in mathematics, due to the influence of his tutor from the third year of undergraduate studies, Olga Alexandrovna Ladyzhenskaya, a renowned specialist in PDE, who however gave him full freedom to follow his own path.

Similarly (see [9] for more details), after an undergraduate education in both mathematics and physics, and some hesitation between Shimshon Amitsur (ring theory), Nathan Rosen (relativity) and Giulio Racah (spectroscopy), Moshé settled for theoretical physics with Racah, starting by working in applications of group theory to physics. Another parallel is that a main reason that (in October 1963) brought Moshé to France, where he soon made a career in the mathematical community, was his desire to become independent of Racah, his "scientific father".

For different reasons, both did not choose between mathematics and physics, but became original mathematical physicists, at a time when that was not so fashionable (at least, in the way they pursued it). Thus they had many common points in spite of seemingly different backgrounds. Given their strong personalities, they could not remain indifferent to each other. As happened most of the time with these dedicated scientists, "love at first sight" was the obvious outcome — and naturally extended to me.

Of course, Moshé's chemistry with the Russians owed a lot to his "Russian genes", including the fact he spoke Russian (and Polish, incidentally) very well, thanks to his parents and grandmothers. Moshé's musical ear was excellent, so his Russian accent was perfect, though Russians could not quite figure out from which Russian city he originated!

1.1.2. *Mathematical Physics?*

An even more important factor for empathy was the fact that we were all (especially Ludwig and Moshé) true mathematical physicists, able (as Ludwig once told us) to "speak mathematics" to mathematicians and "speak physics" to physicists (the opposite is more common). Not many are truly bilingual in this way. As Sir Michael Atiyah said in London at the 2000 International Congress of Mathematical Physics [1] "mathematics and physics are two communities separated by a common language". The language is mathematics of course, but it is spoken with very different accents and grammars.

Though the terminology may seem self-explanatory, it is not. To be slightly pedantic, one ought to distinguish between *mathematical physics*, i.e. (hopefully rigorous) mathematical treatment of physical problems, and *physical mathematics*, i.e. parts of mathematics inspired by physics. [There is also a somewhat old-fashioned notion of mathematical physics, mainly in the UK, referring essentially to differential equations coming from physics and engineering.] What is commonly called mathematical physics covers, in fact, all these notions.

Ludwig excelled in all aspects, and knew what he was doing. I remember a talk he gave in Moshé's seminar in Dijon in which (in part for pedagogical reasons) he went a bit too fast in his derivation. His co-author and son-in-law Sasha Volkov, a very honest and precise scientist who was then for a couple of years postdoc in Dijon, remarked during the talk that they had not yet proved a statement in that generality (it was eventually).

There is an important difference between rigorous mathematical physics and pure mathematics. The purpose is often different. In mathematics, one tends to study a problem in as general a context as possible (even if one does not go as far as Bourbaki). While in mathematical physics, where the aim is usually to tackle specific physical problems, and though generalizations may turn out to have unexpected consequences, it is often enough to use (and develop) mathematical tools adapted to the desired applications.

The main difference between mathematical and theoretical physics is that mathematical rigor is a concern in the former but not in the latter. Some famous theoretical physicists do not quite understand why mathematical physicists are developing such an ingenious machinery to prove results they know for a long time to be true, and are often verified experimentally. Many others do not even realize that there may be a problem. The issue is not purely academic, even for physicists, because it may happen that a rigorous mathematical treatment points out to an effect that is real and potentially important.

Such distinctions developed with time. In the Antiquity, there were practically none. Until a couple of centuries ago, most leading scientists were both mathematicians and physicists, as is attested by the variety of theorems and laws that bear their name. Gradually, science became a kind of Babel tower, even within a domain. One sometimes distinguishes between "specialists", who know almost everything on a very restricted area, and "generalists" who know a lot (albeit often superficially) on many domains. [With humor, going to the limit, that can be expressed by saying that specialists know everything on nothing while generalists know nothing on everything.] Ludwig was a rare example of someone who was both, having a huge knowledge and being able to go deep in many domains.

1.1.3. *ICM 1966*

ICM 1966 was a major event for the development of mathematics. Until then the scientific exchanges between USSR and "the West" had been very limited. A record number of mathematicians attended (4,282 according to official statistics), of which 1,479 came from the USSR, 672 from other "Socialist countries" in Europe, while over 1,200 came from "Western countries", including 280 from France: Moshé (then still only citizen of Israel but working in France since October 1963) and I were among the latter. I remember that we and many from the French delegation traveled to Moscow on a Tupolev plane, organized as a train with compartments seating eight.

In Moscow, we were accommodated, together with many "ordinary" participants (some VIPs were accommodated in "smaller" hotels closer to the Kremlin), in the huge hotel Ukraina (opened in 1957, the largest hotel in Europe), one of the seven Stalinist skyscrapers in Moscow with a height of 206 meters (including the spire, 73 meters long) and total floor area ca. 88,000 m^2, a small city in itself. On every floor, there were "etazhniks" supposed to help but in fact checking on the guests. (We encountered the same system 10 years later in Taipei ...).

The opening ceremony of ICM 1966, as it is now known, was held in the Kremlin. As far as I remember, it is during the long party which followed that we met, and instantly became friends with, Ludwig and other leading Soviet scientists like Israel Moiseevich Gelfand and Nikolay Nikolayevich Bogolyubov. The latter (N.N.) who turned 57 during the Congress was then, *inter alia*, Director and main founder of the Joint Institute of Nuclear Sciences in Dubna (the "Eastern CERN"), a new city which is now "Science City Dubna". He invited us to spend a week there after ICM, which we did (that is another story), staying in Hotel Dubna, which was then a few months old; the building has not changed since that time, but the rooms were of course eventually modernized.

N.N. was also Academician Secretary of the Division of Mathematics of the Soviet Academy of Sciences, a capacity in which Ludwig became his successor. N.N. was not affiliated with the Division of Physics, mainly due to local scientific policy reasons. One of the factors was probably a conflict with Landau, who was a kind of "Russian Pauli" and in particular strongly objected Quantum Field Theory. In a centralized scientific community, several "crocodiles" have a hard time to coexist in the same pond, however large it may be. The same is true for France, especially in physics, and Moshé [9], like N.N. and Ludwig, quickly found "scientific hospitality" in the more open mathematical community.

The bulk of the Congress sessions was located in another of the seven Stalinistic-style skyscrapers, that of the Moscow State University.

1.2. *ICMP 1972 and some mathematical physics*

1.2.1. *ICMP 1972 and before*

The next occasion we could meet was, if I remember correctly, in Moscow in December 1972, where the logo "$M \cap \Phi$" was introduced. We may have met also in Warsaw, earlier that month, in the "Franco–Polish–Swedish" meeting on "Fundamental problems in elementary particle physics", which evolved from the Franco–Swedish "Gunnar Källén Colloquia", initiated in 1967 by Moshé and Gunnar Källén (who, like Moshé, was not mincing his words) and so-named after Källén's accidental death at 42 in 1968. That may also have contributed to inspire Bogolyubov to organize "$M \cap \Phi$" just after our conference in Warsaw.

After IAMP (the International Association of Mathematical Physics) was created in 1976, that large "$M \cap \Phi$" meeting was *a posteriori* defined as the First International Congress of Mathematical Physics (ICMP). The following meeting in Warsaw in March 1974 (also an avatar of the Gunnar Källén colloquia), where we met again and where Moshé and I suggested the creation of IAMP (again that is another story, told in part by Rudolf Haag in [11] and by Elliott Lieb in [10]), was then defined as the Second ICMP. ICMP 1972 was organized by N. N. Bogolyubov. There was again a grand party in the Kremlin, during which Moshé had further extensive discussions with Ludwig and Bogolyubov.

1.2.2. *Integrable systems, quantum groups and deformation quantization*

When we met with Ludwig for the first time, Ludwig and Moshé concurred that physics is based on a 4-dimensional space-time, so that 2-dimensional models were not true physics. But in 1970, a crucial "phase transition" occurred in Ludwig's scientific interests [7], albeit in a different scientific context. As he writes in [15]:

"In 1970 I was introduced by V. Zakharov to the inverse scattering method of solving the nonlinear evolution equation on 2-dimensional space-time. Our first joint result — the Hamiltonian interpretation and complete integrability of the Korteveg–de Vries equation — defined my activity for 20 years. The main achievements here, made together with a large group of excellent students (now called "Leningrad School"), are the unravelling of the algebraic structure of quantum integrable models (the Yang–Baxter equation) and formulation of the Algebraic Bethe Ansatz. This development eventually became a base of construction of quantum groups by V. Drinfeld."

Incidentally, as he told us, the terminology "Yang–Baxter" is due to Ludwig and Leon Takhtajan [8] who showed that, implicit in Baxter's work, is the machinery which relates the eight-vertex model to a linear problem, and gave a clear identification of its purely algebraic aspects. It seems that Baxter did not realize the paramount importance of the structure he had introduced, until Ludwig and Leon made it clear.

Around the same time (in the 70's), building on Moshé's ideas about the role of deformations in physics (see e.g., [16] and references therein) we started to develop what is now called "deformation quantization". One of the factors leading to "the founding 1978 papers" [2] was the fact that in 1975 at a meeting in Austin, Texas where I lectured on our works, Jerzy Plebański told me that it reminded him of Moyal, on whose work he had based lecture notes he had written (in Polish), a copy of which he eventually sent us. I had not heard of Moyal's work, and on *Mathematical Reviews* what is now called the Moyal bracket was not even mentioned, but Moshé had heard of it, so we ordered a copy from CNRS. Together with Jacques Vey's 1975 paper (who, inspired by our works on deformations of the Poisson brackets, rediscovered the Moyal bracket), and a lot of further study we made, that developed to [2].

Interestingly, Ludwig told us much later that already in the late 60's he had lectured to his students in Leningrad on the passage from classical to quantum mechanics and on Weyl (and Moyal) quantization. That was also around the time (late 60's to mid-70's) when Felix Berezin was developing his approach to quantization (which roughly speaking, he eventually defined as a functor from classical to quantum observables). But the lack of enough interactions in those days between various schools, especially on both sides of the "iron curtain", prevented what would have been a fruitful convergence to develop. Only a decade later, after quantum groups were introduced in Leningrad in the context of integrable systems, and the term coined by Drinfeld (together with the relation to our deformation quantization), it became clear that Ludwig and we were developing various aspects of essentially the same physical idea.

And there is more, which surprisingly relates to our unconventional works in the 60's around symmetries of elementary particles, for which Moshé's approach followed from his education in the Racah school in Jerusalem (see a short account in [16]). As Ludwig wrote in [6]:

"An unfinished physical fundamental problem in a narrow sense is physics of elementary particles. This puts this part of physics into a special position. And it is here where modern mathematical physics has the most probable chances for a breakthrough."

That is in part what I am promoting lately [16], at least for the symmetries part of it. Indeed (in a nutshell) I am advocating deforming the Poincaré group of special relativity, first to Anti de Sitter (with a tiny negative curvature), explaining the photons and leptons as composites (of "singletons") and then for hadrons deforming it further to some quantum group version. Of course, field theory needs to be built on that basis. We miss the impact Ludwig could give to such a project, helping solve the "unfinished physical fundamental problem" he mentioned in 2000.

1.2.3. *ICM 1990 in Kyoto and before*

During the period of 1986–1990, Ludwig was President of the International Mathematical Union and then (it ended after him, but that is another story) ex-officio headed the Fields Medals committee for the Kyoto ICM 1990. In March 1990, we met him in London around the time of the final meeting of the Fields committee in Oxford. At some point, he told us that he had a problem. He did not know how to break the news to one of the laureates, whom incidentally we had known already when he was a graduate student. (In those days, email was not so widespread, especially when people were on the move.) Moshé told Ludwig: "No problem, in two days he is coming for dinner at my place in Paris."

So Ludwig wrote the announcement, in his own very characteristic handwriting. Two days later, when the guest came for dinner, Moshé congratulated him. He was sure Moshé was "pulling his leg" but when he saw Ludwig's letter he realized that this was true and asked Moshé: "Can I call my folks?" and jumped on the phone in Moshé's office. He was one of the "three quantum Fields medals" at ICM 1990 in Kyoto.

2. Some Scientific Achievements and Their Recognition

2.1. *Prizes and around*

It is often said that the ways of the Lord are mysterious. So may seem to be the ways of the Nobel committees and of many prizes committees, unless of course you are in the confidence, but then you are not supposed to tell.

Moshé did know something (in great part indirectly, and I through him) because every year from 1971 to his death, since his independent and original opinions were valued, he was asked (under Chap. 6 of the regulations) to nominate candidates for the Prize in physics. He also had "off the record" discussions with Swedish friends, of course not explicitly about the Prize. The last request to nominate arrived after his death in November 1998. Most of his nominations were handwritten by himself on the form he received, and some parts of these eventually made it through. It would be interesting for a historian of science, when these become available (not in our lifetime, which is the purpose of the cap of at least 50 years) to study his very original nominations in the light of what happened afterward.

A few things may however be told, since they are in the public domain. One is that (at least for quite some time last century), accredited journalists who were waiting for the announcement of the Nobel prize were given three closed envelopes containing the names of the laureates and the motivation for the decision. When the announcement came, they were told to open one of the three, and discard the other two without looking into these. [All did as they were requested.] So there must be at least twice more scientists who are of "Nobel level" than there are actual laureates. [It can be reasonably conjectured that, for different reasons, Ludwig and Bogolyubov belong to that category.] A number of factors may tip the balance, including a part of luck as in any human context and extra-scientific factors, including living long enough for making room for a proper occasion.

Several reasons make the Nobel prize more prominent than others. For a long time, it was the prize carrying the highest award, though for every individual that depended on the number of laureates sharing it (at most three, not always in equal parts) and for some time on how successful were the investments of the endowment. The main reasons are probably its ancientness and the grandiose Prize ceremonies (which Moshé attended a couple of times, including with me in 1991 for its 90th anniversary), that have no equivalent.

Incidentally, being awarded such a socially important recognition may become a curse for a scientist, in particular by encouraging complacency and in leading to that by many social duties. (Ludwig was spared that "curse", and he remained active to his last day.) It is not easy for a Nobel laureate to continue and pursue research at a significant level, but there are exceptions. The Nobel medal has two sides, while that is in general not true for the Fields medal which must be awarded before the age of 40.

In the Nobel 1999 physics prize announcements [14], one can see a filiation of ideas from QED (Quantum Electrodynamics) [12] with its Abelian gauge $U(1)$ to the 1954 construction by C. N. Yang and R. L. Mills of the first example of a nonabelian gauge theory, to the electroweak model introduced in the 60's by Weinberg, Salam and Glashow (in separate publications) which got [13] the Nobel prize in 1979 after consequences were observed experimentally and to the works recognized in the 1999 Prize, for which works by "R. P. Feynman, B. S. DeWitt, L. Ḋ. Faddeev and V. N. Popov" "made significant contributions." Indeed it was the 1967 paper and Kiev preprint of Faddeev and Popov which enabled 't Hooft to prove in 1971 that the electroweak model is renormalizable, hence of physical relevance.

There is a cute anecdote, which experts can understand. In December 1979, after the Prize ceremony, a program was aired on Swedish radio, with Salam and Weinberg. At some point, Weinberg quoted some phrase from the Bible. Salam (an immense scientist who well deserved the Nobel prize) remarked that it exists also in the Qur'an, to which Weinberg quipped: "Yes, but we published it before!"

One may wonder why Ludwig wrote such a short (two pages) *Physics Letters* paper, very hard to understand, and gave the (important) details only in a Kiev preprint. There was a strict page limit in *Physics Letters*, and in Soviet Union in those days, longer papers could be sent for publication only after getting a permission from the Division of Nuclear Physics in the Academy, which was controlled by the Landau school who stood against Quantum Field Theory. Indeed in those days [6] "Quantum Field Theory was virtually forbidden, especially in the Soviet Union, due to the influence of Landau." Still a number of scientists, who understood Russian, soon managed to get their hands on the Kiev preprint, and use its results. [The important Kiev preprint was translated into English only after five years, but preprints are not taken into consideration for the Nobel prize.]

More precisely, the MSN review MR1773036 (2001j:81001) of [6] reads: "The natural set of quantization rules for Yang–Mills fields, valid to all orders of perturbation theory, were obtained by L. D. Faddeev and V. N. Popov in 1966 and published in a short communication to *Physics Letters* [*Phys. Lett. B* **25** (1967), 29–30] and in an extended version as a preprint of the Kiev Institute of Theoretical Physics ["Perturbation theory for gauge-invariant fields", Preprint No. FERMILAB-PUB-72-057-T/NAL-THY-57, Fermi Natl. Accel. Lab., Batavia, IL, 1972; translated from Preprint No. ITF-67-036, Inst. Theor. Phys., Acad. Sci. Ukraine, Kiev, 1967]."

And the first part of Ludwig's citation for the 2008 Shaw prize (which carries a comparable amount) reads [15]: "Ludwig Faddeev has made many important contributions to quantum physics. Together with Victor Popov he showed the right way to quantize the famous Yang–Mills equations which underlie all contemporary work on sub-atomic physics. This led in particular to the work of 't Hooft and Veltman which was recognized by the Nobel Prize for Physics of 1999."

2.2. *Ludwig, an unusual group leader and great friend*

Victor Popov was one of Ludwig's first collaborators and, unlike most of his collaborators, he was not technically his student since [4] he started to work in physics with Yuri Novozhilov, with whom we became close friends when he spent eight years at UNESCO in Paris. But Popov can be considered as the first member of his school, the famous "Leningrad school" of mathematical physics.

The second member was Ludwig's student Petr Petrovich Kulish (1944–2016), a good scientist and a fine person. His membership in the group was precious in many ways, both scientific and administrative.

On the administrative part, a typical attitude of Ludwig towards the scientific and political establishments is that he always was guided by the pursue of scientific excellence and, in case of problems, knew up to which point he could stretch the limit.

Soon other members joined the group, who are still alive, many of whom must have contributed to this volume. I will give only one name because of an anecdote that is characteristic of the closely knit group it was.

Indeed, after Alexander Rudolfowitsch Its (Ph.D. 1977) got his Russian Doctor degree, there has been a party during our visit to Leningrad in 1989. The party was held in Its' small apartment, which essentially became one huge table. Though it was hard then to get good products, the many guests brought an impressive variety of goodies. That was probably the best party we ever had in Russia, in great part due to the exceptionally friendly and warm atmosphere. The material conditions eventually (and fortunately) improved in Russia, but the sense of solidarity that was felt then, typical of Ludwig's school, was unique.

In addition to the above, the long overdue improvement of material conditions in Russia had another "perverse effect" related to Ludwig and us. In Soviet times, Ludwig was one of the few who had a car. That permitted him to go to the forests near Leningrad and pick a large variety of excellent wild mushrooms of which (like many Russians) he was an expert, and which he would preserve in salt. Every time he was visiting us in Paris, he would bring a jar or two. After the demise of the USSR, ownership of cars was gradually democratized and Ludwig had to go further and further to find less and less mushrooms. To the point that a dozen years ago, when my late wife Noriko Sakurai and I visited his wife Anya and him in their house of Komarovo and went looking for mushrooms in the woods, Noriko was the only one who found some, small yellow chanterelles. At that time, when visiting Paris, Ludwig was instead bringing as presents excellent tea produced by his very successful son-in-law.

2.3. *Some other important works*

Ludwig had many important contributions, besides his seminal work with Popov. These are certainly detailed in the present volume, by persons more qualified than me to do so. And there are still some works "in progress" as we say. I shall mention here only a few.

As he wrote in his 2008 autobiography for the Shaw prize [15], his first scientific paper was published in 1956, so he has been involved in active scientific work for 60 years. "[He] began by treating the mathematical questions of the quantum scattering theory, both direct and inverse problems. The treatment of the quantum scattering theory for the system of three particles, based on the integral equations, now bearing [his] name, brought [him his] first success. The work was highly appreciated by the specialists in nuclear physics. The attention of mathematicians came later and now the theory of many body quantum scattering is an active subject of modern mathematical physics." At some point it even seemed that this work might lead him to (at least part of) a Nobel prize in nuclear physics.

In another part of his citation for the Shaw prize (in 2008), one can read: "Faddeev also developed the quantum version of the beautiful theory of integrable systems in two dimensions which has important applications in solid state physics as well as in recent models of string theory." That is an area in which his "Leningrad school" excelled, and one of the reasons why he moved to physics in two dimensions. That brought to his school very strong cooperation with important groups in Japan (especially the school of Mikio Sato in Kyoto) and in France (in addition to our strong interactions with him).

The Shaw citation continues: "In another application of the scattering theory of differential operators, Faddeev (jointly with Boris Pavlov) discovered a surprising link with number theory and the famous Riemann Hypothesis."

One of his more recent prophetic works can be found in his contribution to the conference held in Dijon in 1999 after Moshé's untimely death in November 1998 (at the age of 61). In this apparently simple and elegant paper [5] he introduced "The modular double [which] is just the double which defines the hidden symmetry in conformal field theory."

I can conclude with one short sentence: Ludwig was a phenomenon.

References

[1] M. Atiyah, On the unreasonable effectiveness of physics in mathematics, in *Highlights of Mathematical Physics* (*London, 2000*) (Amer. Math. Soc., Providence, RI, 2002), pp. 25–38.

[2] F. Bayen, M. Flato, C. Fronsdal, A. Lichnerowicz and D. Sternheimer, Deformation theory and quantization. I. Deformations of symplectic structures, *Ann. Phys.* **111** (1978) 61–110;
Deformation theory and quantization. II. Physical applications, *Ann. Phys.* **111** (1978) 111–151.

[3] L. D. Faddeev and V. N. Popov, Feynman diagrams for the Yang–Mills field, *Phys. Lett. B* **25**(1) (1967) 29–30;
V. N. Popov and L. D. Faddeev, Kiev Inst. Theor. Phys. Acad. Sci. preprint ITP 67–36 (1967), English translation: Perturbation theory for gauge-invariant fields (preprint NAL-THY-57, (1972));
Reprinted e.g. in *50 Years of Yang–Mills Theory*, ed. G. 't Hooft (World Scientific, 2005);
http://www.scholarpedia.org/article/Faddeev-Popov_ghosts

[4] L. D. Faddeev, How I came to work with Victor Popov, *J. Math. Sci.* **88**(2) (1998) 111–113; [Translated from *Zap. Nauchn. Sem. POMI* **224** (1995) 5–9.]

[5] L. Faddeev, Modular double of a quantum group, in *Conférence Moshé Flato 1999, Vol. I* (*Dijon*), Math. Phys. Stud., Vol. 21 (Kluwer Acad. Publ., Dordrecht, 2000), pp. 149–156.

[6] L. D. Faddeev, Modern mathematical physics: What it should be, in *Mathematical Physics 2000* (Imperial College Press, London, 2000), pp. 1–8; arXiv:math-ph/0002018.

[7] L. D. Faddeev and V. E. Zakharov, The Korteweg–de Vries equation is a fully integrable Hamiltonian system, *Funkcional. Anal. i Prilozen.* **5**(4) (1971) 18–27; English translation: *Funct. Anal. Appl.* **5** (1971) 280–287.

[8] L. D. Faddeev and L. A. Takhtajan, The quantum method for the inverse problem and the XYZ Heisenberg model, *Uspekhi Mat. Nauk* **34**(5)(209) (1979) 13–63, 256 (Russian).

[9] M. Flato, The man and the scientist, in *Conférence Moshé Flato 1999, Vol. I* (*Dijon*), Math. Phys. Stud., Vol. 21 (Kluwer Acad. Publ., Dordrecht, 2000), pp. 3–54; MR1805881 (2001m:01061).

[10] M. Flato, Personal recollections, *Lett. Math. Phys.* **48**(1) (1999) 5–34.

[11] R. Haag, Some people and some problems met in half a century of commitment to mathematical physics, *Eur. Phys. J. H* **35**(3) (2010) 263–307.

[12] Link for the 1965 Nobel prize in physics:
https://www.nobelprize.org/nobel_prizes/physics/laureates/1965/

[13] Link for the 1979 Nobel prize in physics:
https://www.nobelprize.org/nobel_prizes/physics/laureates/1979/press.html

[14] References from Nobel 1999 Advanced Material:
M. Veltman, *Nucl. Phys. B* **7** (1968) 637;
G. 't Hooft, *Nucl. Phys. B* **35** (1971) 167;
G. 't Hooft and M. Veltman, *Nucl. Phys. B* **44** (1972) 189; *ibid.* **50** (1972) 318, https://www.nobelprize.org/nobel_prizes/physics/laureates/1999/

[15] http://www.shawprize.org/en/shaw.php?tmp=3&twoid=49&threeid=57&fourid=75

[16] D. Sternheimer, "The important thing is not to stop questioning", including the symmetries on which is based the standard model, in *Geometric Methods in Physics*, Trends Math. (Birkhäuser/Springer, Cham, 2014), pp. 7–37.

A Friendly Ghost Story

Frank Wilczek
Department of Physics, MIT, Cambridge MA 02139, USA
wilczek@mit.edu

I never got to know Ludvig Faddeev well, but his work had a big impact on my career. I'm very happy to have this opportunity to contribute a small personal tribute in thanks.

In 1972, I was a graduate student in mathematics at Princeton, thinking about converting to physics. There was a lot of excitement about gauge theories, inspired by the then-recent discovery of their renormalizability, and by their promising application to modeling the weak interaction. There was also a lot of excitement about the renormalization group, inspired by Wilson's theory of critical phenomena and his vision of how to analyze short-distance behavior in quantum field theory. I was extremely intrigued by the possibility of putting those two ideas together. I had the great good fortune to begin work with David Gross, who was a world-leading expert on the renormalization group side, and a force of nature in taking on intellectual adventures.

Today it is completely routine to combine gauge theory and the renormalization group, but in those days, it was not quite clear that it even made sense. After all, the renormalization group centers on scale transformations, while gauge theories center on local gauge transformations and — as Sidney Coleman (who was visiting Princeton at the time) kept reminding me — those two sorts of transformations don't commute. At a more down-to-earth level, practical calculation of renormalization group coefficients involves off-shell Green's functions, and in such calculations gauge symmetry is far from obvious.

I tried to read the breakthrough papers of 't Hooft and Veltman on renormalization in gauge theories, but I found it slow going, because they used an idiosyncratic diagram technique, and because they assumed a level of sophistication which, as a newcomer, I just didn't have.

To get some traction, I retreated to the earlier work of Faddeev and Popov, in which they introduced their famous ghosts. Those papers were so clean and lucid that I understood them right away. Their central result, a simple supplement to standard Feynman diagrams which enforces gauge invariance off-shell, at least in loops, was encouraging in a general way. Also, the very concrete, algorithmic nature of their answer invited me to just go ahead and use it in calculations, trusting that "the work will teach you how to do it".

David and I were very keen to get answers for the beta function and other quantities which we believed, on physical grounds, should be gauge invariant. Since the foundations of our intuition were shaky, it seemed important to do the calculations in as many logically independent ways as possible. That would give a potential failure of gauge invariance opportunities to reveal itself. Conversely, if all the methods agreed, we could be pretty confident of the gauge invariance — and also that we'd got the right answer!

There were two obvious paths to pursue. One was to fix the gauge in different ways, most simply using "Fermi-type" conditions, which roughly speaking amount to mixing different amounts of longitudinal modes into the overall gluon propagator. Their physical effects are supposed to cancel, but of course that will only occur if the Faddeev–Popov ghost is included. The other was to calculate renormalization of gluon couplings to different kinds of particles — fermions, scalars, (transverse) gluons ... and Faddeev–Popov ghosts, whose couplings are especially simple. Since symmetry requires that the gauge coupling is universal, symmetry requires that all those couplings should get renormalized the same way.

These (presumably) highly redundant calculations diagnosed several mistakes in the complicated algebra. Once we fixed those, everything harmonized. I vividly remember the moment of joy, relief, and satisfaction when I first saw that happening, late one night in Jadwin Hall.

In that way, the Faddeev–Popov work played an important inspirational and technical role in the genesis of asymptotic freedom, and changed my life.

In a more general way, it was instrumental in changing attitudes toward quantum field theory and path integrals, from the dark ages of distrust and neglect in the 1960s to their secure, central positions in our world-vision today.

The decades-long failure of quantum field theory to foster progress in understanding the strong interaction, together with principled objections to renormalization theory, had led many — including such luminaries as Dirac, Landau, Chew, and Gell-Mann — to express grave reservations about it. Path integrals were widely regarded as a clever curiosity, with only niche applications. You find that attitude reflected even in the book of Feynman and Hibbs, which was long the only full-scale treatment of the subject.

In that atmosphere, the fact that Faddeev and Popov discovered a very nontrivial, concrete, check-able consequence of taking path integrals, at face value, as providing the definition of a quantum field theory came as a revelation. It made an indelible impression on a whole generation of theoretical physicists, and we've never looked back.

In the late 1940s.

On summer vacation in the late 1940s.

Over the piano, in the early 1950s.

Ski vacation in Dombai, North Caucasus, Russia, 1960.

During a conference in Uzhgorod, Western Ukraine, 1961.

Together with young theoretician D. Dyakonov.

The First Conference on Quantum Solitons, Steklov Mathematical Institute, Leningrad, Russia, October 1978.

With Samson L. Shatashvili and Sir Michael Atiyah during the official launch of the Hamilton Mathematics Institute TCD and the Dublin meeting of the British Association for the Advancement of Science (BA), Trinity College Dublin Front Square, Ireland, September 5, 2005.

During the Shaw Prize award ceremony, Hong Kong, 2008.

Ludwigfest. During the conference in honor of his 75th birthday, Geneva, Switzerland, March 2009. Professor Faddeev with his pupils Professors A. B. Venkov and S. L. Shatashvili.

L. D. Faddeev and V. E. Zakharov in Dijon, France, June 2009.

Together with his wife Anna Mikhailovna.

Ski resort at Lech, Austria, 2010.

During a talk in Aarhus University, Denmark, August 2010.

L. D. Faddeev and his students (from left to right): I. Ya. Aref'eva, S. L. Shatashvili, P. P. Kulish, L. A. Takhtajan and A. Yu. Volkov at Steklov Mathematical Institute in Saint Petersburg, September 2010.

During the RAQIS conference in Dijon, France, September 2014.

L. D. Faddeev at Steklov Mathematical Institute in Saint Petersburg.

During a conference in Chamonix, France, March 2015.

From left to right: Mrs. Semenov-Tian-Shansky, Mrs. Takhtajan, Elena Volkova (Faddeev's granddaughter), Maria Volkova (Faddeev's daughter), A. Yu. Volkov, L. A. Takhtajan and M. A. Semenov-Tian-Shansky, during a conference in Chamonix, France, March 2015.

At Aiguille de Midi with an altitude of 3,842 m, France, March 2015.

Part 2
Scientific Papers

Coadjoint Orbits, Cocycles and Gravitational Wess–Zumino

Anton Alekseev[*,‡] and Samson L. Shatashvili[†,§,¶]

[*] *Department of Mathematics, University of Geneva, Geneva, Switzerland*

[†] *The Hamilton Mathematics Institute and the School of Mathematics, Trinity College Dublin, Dublin 2, Ireland*

and

Simons Center for Geometry and Physics, Stony Brook, USA

and

Israel Gelfand Chair, IHES, Bures-sur-Yvette, France

[‡] *Anton.Alekseev@unige.ch*

[§] *samson@maths.tcd.ie*

About 30 years ago, in a joint work with L. Faddeev we introduced a geometric action on coadjoint orbits. This action, in particular, gives rise to a path integral formula for characters of the corresponding group G. In this paper, we revisit this topic and observe that the geometric action is a 1-cocycle for the loop group LG. In the case of G being a central extension, we construct Wess–Zumino (WZ) type terms and show that the cocycle property of the geometric action gives rise to a Polyakov–Wiegmann (PW) formula with boundary term given by the 2-cocycle which defines the central extension. In particular, we obtain a PW type formula for Polyakov's gravitational WZ action. After quantization, this formula leads to an interesting bulk-boundary decoupling phenomenon previously observed in the WZW model. We explain that this decoupling is a general feature of the Wess–Zumino terms obtained from geometric actions, and that in this case, the path integral is expressed in terms of the 2-cocycle which defines the central extension.

In memory of our teacher Ludwig Faddeev

1. Introduction

When we were invited to contribute a paper in the Ludwig Faddeev memorial volume, we turned to a project that we started together with him more than 30 years ago. And when we started working on the paper, it was a pleasant surprise that many interesting things can be added after so many years.

Cocycles and geometric actions were the topics of our joint papers with Faddeev [1–3], and this theme continued in [4–6]. Recent developments in a wide variety

[¶]On leave of absence from Institute for Information Transmission Problems, Moscow, Russia.

of topics in modern theoretical physics turned out to be connected to those old ideas. To list just a few interesting connections, we mention the SYK model [8, 9] as studied in [10–17] and the vacuum decay in CFT of [18].

Faddeev advocated that one should think of quantization as a path integral in the Hamiltonian formalism

$$Z = \int \mathcal{D}p\mathcal{D}q\, e^{\frac{i}{\hbar}S}.$$

Here the action is written as

$$S = \int pdq - H(p,q)dt = \int d^{-1}\omega - Hdt,$$

ω is the symplectic form on phase space and d^{-1} is understood in the sense of Novikov. The integration measure in the path integral is induced by the Liouville measure on the phase space:

$$\mathcal{D}p\mathcal{D}q = \prod_t dp(t)dq(t).$$

In [3], we considered a coadjoint orbit of a compact Lie group G as a phase space. For $G = SU(n), SO(n)$ we chose (following Guillemin–Sternberg [19]) a Darboux chart and imposed boundary conditions on the coordinates $q(t)$ at $t = t_1; t = t_2$ such that the path integral reproduced the correct character formula.

In the present paper, we revisit the construction of geometric actions on coadjoint orbits. It turns out that the geometric action for the group G is a group 1-cocycle for the loop group LG. This cocycle property gives rise to an interesting identity for path integrals of the geometric action relating matrix elements before and after the gauge transformation.

We then follow [4] and focus on the case when the group G is a central extension

$$1 \to S^1 \to G \to G' \to 1$$

of some other group G' by the circle S^1. The main examples are G being the standard central extension of a loop group of a compact connected Lie group and G being the Virasoro–Bott central extension of the diffeomorphism group of the circle. We observe that a linear combination of geometric actions for the three groups (G, G' and S^1) gives rise to the Wess–Zumino (WZ) type term:

$$\mathrm{WZ} = S_G - S_{G'} - S_{S^1}.$$

In our examples, this term is given by the action of the WZW model and by the action of the Polyakov gravitational WZ model, respectively.

Using the cocycle property of the geometric action, we prove a Polyakov–Wiegmann (PW) type formula for the WZ terms. The boundary contribution in this formula is given by the 2-cocycle α which defines the central extension.[a]

[a]When this paper was completed, we learnt that the theory of PW formulas for WZ terms was developed by Aratyn–Nessimov–Pacheva and their collaborators in a series of papers, see e.g., [20, 21].

We focus our attention on the gravitational Wess–Zumino action. In quantum theory, the corresponding PW formula gives rise to an interesting bulk-boundary decoupling phenomenon where the path integral is completely determined by the classical action with given boundary conditions. This phenomenon was first observed in [22] for the WZW model (it is motivated by the background independent open string field theory [23–26] and the open/closed string correspondence). We use our formalism to establish the bulk-boundary decoupling for the gravitational WZ action. Our calculation is inspired by the recent studies in [18]. Among other things, we observe that the path integral is given by the 2-cocycle which defines the central extension.

2. Coadjoint Orbits and Cocycles

In this section, we recall the construction of geometric actions on coadjoint orbits of Lie groups (orbit actions), show that they satisfy the 1-cocycle property and compare the Kirillov character formula and the path integral character formula which makes use of an orbit action.

2.1. *Orbit actions and the cocycle property*

Let G be a connected Lie group, $\mathfrak{g}$ be the Lie algebra of G and $\mathfrak{g}^*$ its dual space. The group G acts on $\mathfrak{g}$ by the adjoint action $g\colon x \to \mathrm{Ad}_g\, x$ and on $\mathfrak{g}^*$ by the coadjoint action $g\colon \xi \to \mathrm{Ad}^*_g\, \xi$. To simplify notation, we pretend that $G, \mathfrak{g}$ and $\mathfrak{g}^*$ are realized as spaces of matrices (this is the case in many examples) and write $\mathrm{Ad}_g\, x = g^{-1}xg$ and $\mathrm{Ad}^*_g\, \xi = g^{-1}\xi g$. For $\lambda \in \mathfrak{g}^*$, we define its coadjoint orbit

$$\mathcal{O}_\lambda = \{\xi = g^{-1}\lambda g; g \in G\}.$$

There is a canonical projection $\pi_\lambda\colon G \to \mathcal{O}_\lambda$ given by $\pi_\lambda(g) = g^{-1}\lambda g$. Coadjoint orbits carry canonical Kirillov–Kostant–Souriau (KKS) symplectic forms. It is easy to describe their pull-backs to G under projection π_λ:

$$\pi^*_\lambda \omega_\lambda = \frac{1}{2}\langle \lambda, [\mathrm{d}gg^{-1}, \mathrm{d}gg^{-1}]\rangle = \langle \lambda, (\mathrm{d}gg^{-1})^2\rangle = \mathrm{d}\langle \lambda, \mathrm{d}gg^{-1}\rangle.$$

Here d is the de Rham differential and $\mathrm{d}gg^{-1} \in \Omega^1(G, \mathfrak{g})$ is the right-invariant Maurer–Cartan form on G. The 2-form ω_λ defines a non-trivial cohomology class in $H^2(\mathcal{O}_\lambda, \mathbb{R})$. The Liouville form $\omega^N_\lambda/N!$, where $N = \dim(\mathcal{O}_\lambda)/2$, is a G-invariant volume form on the orbit $\mathcal{O}_\lambda$.

The interest in the study of coadjoint orbits originates in the Borel–Weil–Bott Theorem and in the Kirillov orbit method which establish a correspondence between irreducible unitary representations of G and coadjoint orbits. One of the manifestations of this correspondence is the Kirillov character formula [27]. We write it in the case of G compact and connected: let $x \in \mathfrak{g}$ and $\exp(x) \in G$ be an element of the Lie algebra and its image under the exponential map, λ be the highest weight of the irreducible representation σ_λ: $G \to \mathrm{End}(V_\lambda)$ and $\chi_\lambda(g) = \mathrm{Tr}_{V_\lambda}\sigma_\lambda(g)$ be its character. Then, the Kirillov character formula states

$$\chi_\lambda(\exp(x)) = j^{-1/2}(x) \int_{\mathcal{O}_{\lambda+\rho}} \frac{\omega_{\lambda+\rho}^N}{N!}\, e^{i\langle \xi, x\rangle}, \tag{2.1}$$

where ρ is the half sum of positive roots of G and $j(x)$ is the Jacobian of the exponential map $x \mapsto \exp(x)$ with respect to the Lebesgue measure on $\mathfrak{g}$ and the Haar measure on G.

The KKS symplectic form allows to define an action of a Hamiltonian system on a coadjoint orbit. Fix $\lambda \in \mathfrak{g}^*$ and consider maps g: $I \to G$ for I a 1-dimensional manifold (a segment or a circle). The orbit action is given by the following expression:

$$S(\lambda, g) = \int_I \langle \lambda, dgg^{-1}\rangle = \int_I \langle \xi, g^{-1}dg\rangle, \tag{2.2}$$

where $\xi = g^{-1}\lambda g$. Observe that $S(\lambda, g)$ is a 1-cocycle for the group of paths (or loops if $I = S^1$) $\{g$: $I \to G\}$ with point-wise multiplication. That is, $\delta S = 0$ where

$$\delta S(\lambda, g, h) = S(\xi = g^{-1}\lambda g, h) - S(\lambda, gh) + S(\lambda, g). \tag{2.3}$$

This observation plays a key role in the rest of the paper.

2.2. *The path integral character formula*

The main feature of the orbit action is the following path integral identity. Let $A \in \Omega^1(I, \mathfrak{g})$ be a connection on I (all the G-bundles on I are trivial). Then, for $I = S^1$ we have the following character formula

$$\chi_\lambda\left(P\exp\int_I A\right) = \int \mathcal{D}g\, e^{i\left(S(\lambda,g)+\int_I\langle\xi,A\rangle\right)}. \tag{2.4}$$

Here $P\exp\int_I A$ is the holonomy of A along I (with some base point chosen on the circle). The path integral is loosely written with the integration measure $\mathcal{D}g$ even though $\mathcal{D}\xi$ would be more appropriate (there is a gauge symmetry preserving ξ). In (2.4), we are using the coadjoint orbit $\mathcal{O}_\lambda$ but choosing a different regularization of the path integral allows to use the coadjoint orbit $\mathcal{O}_{\lambda+\rho}$ as in the Kirillov character formula. The significant difference of the path integral formula (2.4) from the finite-dimensional formula (2.1) is the absence of the normalization factor $j^{-1/2}(x)$ which is automatically taken into account by the path integral.

In the case of $I = [0, T]$, the path integral defines matrix elements of the representation σ_λ:

$$\langle a|\sigma_\lambda\left(P\exp\int_I A\right)|b\rangle = \int_{g(0)\in\mathcal{L}_b, g(T)\in\mathcal{L}_a} \mathcal{D}g\, e^{i\left(S(\lambda,g)+\int_I\langle\xi,A\rangle\right)}. \tag{2.5}$$

Here $\langle a|$ and $|b\rangle$ label two vectors in V_λ, and $\mathcal{L}_a, \mathcal{L}_b$ are appropriate boundary conditions which correspond to these vectors. For $G = SU(n)$ and $G = SO(n)$ such boundary conditions for elements of the Gelfand–Zeitlin basis were constructed in [3].

Note that under the gauge transformations

$$A \mapsto A^h = hAh^{-1} + dhh^{-1}$$

the orbit action transforms as follows:

$$\begin{aligned} S(\lambda, g) + \int_I \langle g^{-1}\lambda g, A^h\rangle &= S(\lambda, g) + S(g^{-1}\lambda g, h) + \int_I \langle g^{-1}\lambda g, hAh^{-1}\rangle \\ &= S(\lambda, gh) + \int_I \langle (gh)^{-1}\lambda(gh), A\rangle. \end{aligned}$$

Here we have used the cocycle property (2.3). This leads to the following transformation property for the path integral on the right-hand side of (2.5) (to make notation shorter, we drop the boundary conditions):

$$\begin{aligned} \int \mathcal{D}g\, e^{i\left(S(\lambda,g)+\int_I\langle\xi(g),A^h\rangle\right)} &= \int \mathcal{D}g\, e^{i\left(S(\lambda,gh)+\int_I\langle(gh)^{-1}\lambda(gh),A\rangle\right)} \\ &= \int \mathcal{D}v\, e^{i\left(S(\lambda,v)+\int_I\langle\xi(v),A\rangle\right)}, \end{aligned}$$

where we have made a substitution $v = gh$. For $I = S^1$, this change of variables does not affect the result. For $I = [0, T]$, the boundary conditions $\mathcal{L}_a, \mathcal{L}_b$ get affected and this is represented by the factors $h(0)$ and $h(T)$ in the gauge transformation of the matrix elements of the holonomy:

$$\langle a|P\exp\int_I A^h|b\rangle = \langle a|h(T)\left(P\exp\int_I A\right)h(0)^{-1}|b\rangle. \tag{2.6}$$

In particular, for $A = 0$ we obtain the following formula for the path integral:

$$\int_{g(0)\in\mathcal{L}_b, g(T)\in\mathcal{L}_a} \mathcal{D}g\, e^{i(S(\lambda,g)+\int_I\langle\xi,dhh^{-1}\rangle)} = \langle a|\sigma_\lambda(h(T))\sigma_\lambda(h(0))^{-1}|b\rangle. \tag{2.7}$$

In what follows, we will see several applications of this formula in Conformal Field Theory.

3. Central Extensions

In this section, we study in detail orbit actions for central extensions, introduce the notion of a Wess–Zumino action and derive a general Polyakov–Wiegmann (PW) type formula. Our main examples are the Wess–Zumino–Witten (WZW) action which corresponds to the central extensions of loop groups and the Polyakov gravitational Wess–Zumino action which corresponds to the central extension of the diffeomorphism group of the circle.

3.1. *Orbit actions and the Polyakov–Wiegmann formula*

Let $\widehat{G}$ be a central extension of the group G by the circle S^1 defined by an exact sequence of group homomorphisms:

$$1 \to S^1 \to \widehat{G} \to G \to 1. \tag{3.1}$$

The corresponding central extension of the Lie algebra $\mathfrak{g}$ is given by the exact sequence of Lie algebra homomorphisms:

$$0 \to \mathbb{R} \to \widehat{\mathfrak{g}} \to \mathfrak{g} \to 0. \tag{3.2}$$

In what follows, we will pretend that the S^1-bundle $\widehat{G} \to G$ is trivial, and that it admits a section s: $G \to \widehat{G}$. We will need the following objects defined by this section. The group law in $\widehat{G}$ defines a 2-cocycle α: $G \times G \to S^1$ such that for $g, h \in G$ one has

$$s(g)s(h) = s(gh)e^{i\alpha(g,h)}.$$

The section s induces a splitting $\mathfrak{g} \to \widehat{\mathfrak{g}}$ of the exact sequence (3.2) (denoted by the same letter) and a Lie algebra 2-cocycle $\omega(x, y)$ via the formula

$$[s(x), s(y)]_{\widehat{\mathfrak{g}}} = s([x, y]_{\mathfrak{g}}) + \omega(x, y).$$

We introduce the notation

$$\beta(g) = ds(g)s(g)^{-1} - dgg^{-1} \in \Omega^1(G)$$

for the 1-form $\beta(g)$ which represents the central part of the Maurer–Cartan form $ds(g)s(g)^{-1}$.

The splitting s also defines an isomorphism $\widehat{\mathfrak{g}}^* \cong \mathfrak{g}^* \oplus \mathbb{R}$. The coadjoint action of $\widehat{G}$ descends to G since the central circle acts trivially. We will use the notation

$$\mathrm{Ad}^*_g(\lambda, c) = (g^{-1}\lambda g + c\gamma(g), c),$$

where $\lambda \in \mathfrak{g}^*$, $c \in \mathbb{R}$, $g^{-1}\lambda g$ stands for the coadjoint action of G on $\mathfrak{g}^*$ and $\gamma(g) \in \mathfrak{g}^*$ is the contribution of the central extension.

For an element $(\lambda, c) \in \widehat{\mathfrak{g}}^*$, we consider the orbit action for $\widehat{g} = s(g)e^{i\theta}$:

$$S_{\widehat{G}}((\lambda, c), \widehat{g}) = \int_I \langle (\lambda, c), d\widehat{g}\,\widehat{g}^{-1} \rangle = \int_I (\langle \lambda, dgg^{-1} \rangle + c(d\theta + \beta(g))).$$

Note that the first term is the orbit action for the group G corresponding to the orbit of λ and that the term $S_c(\theta) = c \int_I d\theta$ is the orbit action for the circle S^1. The expression

$$\mathrm{WZ}(g) = \int_I \beta(g) \tag{3.3}$$

is independent of λ and c and is called the Wess–Zumino action. To summarize, we have

$$S_{\widehat{G}}((\lambda, c), \widehat{g}) = S_G(\lambda, g) + S_{S^1}(c, \theta) + c\,\mathrm{WZ}(g). \tag{3.4}$$

The orbit actions $S_{\widehat{G}}((\lambda, c), \widehat{g}), S_G(\lambda, g)$ and $S_{S^1}(c, \theta)$ satisfy their cocycle conditions. The comparison of these cocycle conditions gives rise to the Polyakov–Wiegmann (PW) type formula of the Wess–Zumino action $\mathrm{WZ}(g)$. This formula is one of our main results:

$$\mathrm{WZ}(g) - \mathrm{WZ}(gh) + \mathrm{WZ}(h) = \int_I \langle \gamma(g), dhh^{-1} \rangle + \int_I \alpha_2(g, h). \tag{3.5}$$

Note that the definitions of $\mathrm{WZ}(g), \gamma(g)$ and $\alpha_2(g, h)$ depend on the choice of the section s. If one chooses another section, all these quantities change so as to preserve Eq. (3.5). Formula (3.5) (in the case without boundary) was first obtained in [20] using the 1-cocycle property of the expression $\gamma(g)$.

In what follows, we will give two interesting examples orbit actions for central extensions and of the Wess–Zumino actions and their PW formulas.

3.2. *Example: Current algebras and WZW action*

Let G be a connected and simply connected semisimple Lie group, $\mathfrak{g}$ be its Lie algebra and $L\mathfrak{g}$ be the space of loops x: $S^1 \to \mathfrak{g}$. Given an invariant scalar product B: $\mathfrak{g} \times \mathfrak{g} \to \mathbb{R}$, there is a central extension of $L\mathfrak{g}$ defined by the cocycle

$$\kappa(\alpha, \beta) = \int_{S^1} B(\alpha, d\beta).$$

As a vector space, the central extension $\widehat{L\mathfrak{g}}$ is a direct sum of the loop algebra $L\mathfrak{g}$ and the real line, $\widehat{L\mathfrak{g}} = L\mathfrak{g} \oplus \mathbb{R}$. The corresponding group is given by the short exact sequence

$$1 \to S^1 \to \widehat{LG} \to LG \to 1,$$

where $LG = \{g: S^1 \to G\}$ is the loop group of G. A convenient way to think of the dual space to $\widehat{L\mathfrak{g}}$ is to identify it with pairs (A, c), where $c \in \mathbb{R}$ is a real number and $A \in \Omega^1(S^1, \mathfrak{g})$ is a connection on the circle. The pairing with $\widehat{L\mathfrak{g}}$ is given by

$$\langle (A, c), (x, \mu) \rangle = \int_{S^1} B(A, x) + c\mu.$$

The coadjoint action of $\widehat{LG}$ descends to LG, and it is given by gauge transformations

$$g: (A, c) \mapsto (g^{-1}Ag + c\, g^{-1}\partial_x g, c).$$

Hence, we identify the map $\gamma(g) = g^{-1}\partial_x g$. For $c \neq 0$, the coadjoint orbits are in one-to-one correspondence with conjugacy classes in G. The correspondence is given by mapping a pair (c, A) to the conjugacy class of the holonomy $P\exp\left(\frac{1}{c}\int_{S^1} A\right)$.

The orbit action is naturally defined on $M = I \times S^1$, where $I = [a, b]$ is a segment or a circle $I = S^1$. The variable on I is denoted by t and plays the role of time, the variable on S^1 is denoted by x and is interpreted as a space variable. The orbit action is given by the expression

$$S((A, c), (g, \theta)) = \frac{1}{4\pi} \int_M \langle a(x), \partial_t g g^{-1} \rangle dt dx + c \int_I \theta + c\, \mathrm{WZ}(g) \tag{3.6}$$

where $A = a(x)\mathrm{d}x$. In this case, the expression $\mathrm{WZ}(g)$ is the chiral Wess–Zumino–Witten given by the following formula:

$$\mathrm{WZ}(g) = \frac{1}{4\pi}\int_M \langle \partial_t g g^{-1}, \partial_x g g^{-1}\rangle dt dx + \frac{1}{12\pi}\int_N \langle \mathrm{d}g g^{-1}, (\mathrm{d}g g^{-1})^2\rangle. \tag{3.7}$$

In the case of $M = S^1 \times S^1$, the manifold N is a handlebody with $\partial N = M$, the field $g\colon M \to G$ extends to N (since $\pi_2(G)$ is trivial) and the second term is defined modulo $2\pi\mathbb{Z}$. Hence, in this case $\mathrm{WZ}(g)$ is a multi-valued functional in the sense of Novikov [28]. The coefficients in front of the terms on the right-hand side of (3.7) are on the one hand traditional in the literature, and on the other hand they ensure the integrality of the action if the scalar product is the standard invariant scalar product on the Lie algebra $\mathfrak{g}$.

In the case when $M = [0,1] \times S^1$, the action $\mathrm{WZ}(g)$ should be supplemented with boundary conditions at $t = 0, 1$ in order to make the second term in (3.7) well defined. The ambiguity in the definition of the WZW action arises because the bundle $\widehat{LG} \to LG$ is nontrivial and it does not admit a globally defined section s.

The PW formula for the action $\mathrm{WZ}(g)$ reads

$$\mathrm{WZ}(g) - \mathrm{WZ}(gh) + \mathrm{WZ}(h) = \frac{1}{4\pi}\int_M \langle g^{-1}\partial_x g, \partial_t h h^{-1}\rangle + \int_I d\alpha_2(g,h).$$

The first term on the right-hand side is given by the classical PW formula of [29]. The second term was observed in [22] and it is the 2-cocycle which defines the central extension $\widehat{LG}$. It depends on the choice of a (local) section s. For local sections defined on the neighborhood of the group unit, the universal expressions for $\alpha(g,h)$ were classified in [31].

3.3. *Example: Virasoro algebra and gravitational WZ*

One more example of an interesting Lie algebra which gives rise to a useful orbit action is the Virasoro algebra. One considers the Lie algebra $\mathrm{Vect}(S^1)$ of vector fields on the circle with elements of the form $u(x)\partial_x$, where $u(x)$ is a 2π-periodic function. The central extension of $\mathrm{Vect}(S^1)$ is defined by the Lie algebra Virasoro–Bott 2-cocycle

$$\omega(u(x)\partial_x, v(x)\partial_x) = \int_{S^1} u(x)v'''(x)dx.$$

Elements of the dual space are pairs (B, c), where c is a real number (the central charge) and $B = b(x)dx^2$ is a quadratic differential. The coadjoint action of the diffeomorphism group $\mathrm{Diff}(S^1)$ is given by

$$F\colon (B = b(x)dx^2, c) \mapsto (B^F = b(F(x))F'(x)^2 + cS(F))dx^2, c).$$

Here

$$S(F) = \frac{F'''}{F'} - \frac{3(F'')^2}{2(F')^2}$$

is the Schwarzian derivative of F. Hence, we identify the element $\gamma(F) = S(F)$.

The orbit action acquires the form

$$S((B,c),(F,\theta)) = \int_M b(F)F'^2\,\frac{\dot F}{F'}\,dtdx + c\int_I d\theta + c\,\mathrm{WZ}_{\mathrm{grav}}(F), \tag{3.8}$$

where $b(F) = b(F(x,t))$, the expression $\dot F/F'$ represents the Maurer–Cartan form and the Wess–Zumino term (the gravitational Wess–Zumino action) is given by

$$\mathrm{WZ}_{\mathrm{grav}} = \int_M \frac{\dot F}{F'}\left(\frac{F'''}{F'} - 2\left(\frac{F''}{F'}\right)^2\right) dtdx. \tag{3.9}$$

Note that using the inverse diffeomorphism $f(F(x,t),t) = x$, this expression can be re-written as the Polyakov action for induced 2-dimensional gravity [7]:

$$\int_M \frac{\dot F}{F'}\left(\frac{F'''}{F'} - 2\frac{F''^2}{F'^2}\right) dtdx = \int_M \left(\frac{\dot f' f''}{(f')^2} - \frac{\dot f (f'')^2}{(f')^3}\right) dtdy, \tag{3.10}$$

where $y = F(t,x)$.

Our second main result is the PW type formula for the gravitational Wess–Zumino action:

$$\mathrm{WZ}_{\mathrm{grav}}(F) - \mathrm{WZ}_{\mathrm{grav}}(F\circ G) + \mathrm{WZ}_{\mathrm{grav}}(G) = \int_M S(F)\,\frac{\dot g}{g'}\,dxdt + \int_I d\alpha_2^{VB}(F,G). \tag{3.11}$$

Here $g(G(x,t),t) = x$ is the inverse diffeomorphism to G and $\alpha(F,G)$ is the group Virasoro–Bott cocycle

$$\alpha_2^{VB}(F,G) = \int_{S^1} \log(F'(G(x)))' \log(G'(x))dx.$$

Note that the structure of the gravitational PW formula (3.11) resembles the standard PW formula. The right-hand side is a sum of the bulk term and a boundary term. The boundary term is a cocycle which defines the central extension of the diffeomorphism group. The bulk term is an integral of a product of two factors defined by the two group elements. Note also that there are important new features in the formula (3.11): the factors in the bulk term are no longer symmetric. One of them is the Schwarzian derivative which is a contribution in the coadjoint action which comes from the central extension. The other one is the Maurer–Cartan form on the diffeomorphism group. In the case of loop groups, these factors resemble each other, and they turn out to be quite different in the case of the diffeomorphism group.

4. Application: The Bulk-Boundary Decoupling for WZ

In this section, we apply formula (2.7) to describe certain 2-dimensional path integrals for WZW and gravitational WZ models. Recall that for $\lambda = 0$, the orbit action essentially coincides with the WZ term

$$S_{\widehat G}((0,c),\widehat g) = c\int d\theta + c\mathrm{WZ}(g).$$

In what follows, we consistently drop the first term and consider the path integral for the WZ term[b]:

$$\int_{g(0)\in\mathcal{L}_b, g(T)\in\mathcal{L}_a} \mathcal{D}g e^{i(c\mathrm{WZ}(g))+\int_I\langle\xi, dhh^{-1}\rangle)} = \langle a|\sigma_0(h(T))\sigma_0(h(0))^{-1}|b\rangle. \tag{4.1}$$

We will focus on the situation when the right-hand side yields 1 trivializing the path integral.

4.1. *Example: Decoupling in WZW*

Here we recall the instance of quantum decoupling observed in the standard WZW model in [22] (a slightly different argument was given in [30]). To match notation, we replace the real variables x, t with complex variables $z, \bar{z}$. This implies that the fields $g(z, \bar{z})$ are complex valued (otherwise, the WZW model does not admit solutions of equations of motion).

The boundary condition at one end of the cylinder will be the vacuum state $|b\rangle = |0\rangle$ which makes the cylinder into a disc D. This state has the following property: it is annihilated by all non-positive components of the holomorphic and anti-holomorphic WZW currents:

$$J^a_n|0\rangle = \bar{J}^a_n|0\rangle = 0$$

for all $n \geq 0$.

The idea is to compute the path integral as a function of values of the field g on the other boundary:

$$Z(u) = \int_{v|_{\partial D}=u} \mathcal{D}v\, e^{ic\mathrm{WZ}(v)}.$$

The strategy of [22] is to make a substitution $v = gh$, where h is the unique solution of WZW equations of motion with boundary values given by u and $g|_{\partial D} = 1$. Using the PW formula, one obtains

$$Z(u) = e^{ic\mathrm{WZ}(h)} \int_{g|_{\partial D}=1} \mathcal{D}g\, e^{i(c\mathrm{WZ}(g)+c\int_\Sigma(g^{-1}\partial g, \bar{\partial}hh^{-1}))}.$$

Note that the 2-cocycle term $\alpha(g, h) = 0$ since $g = 1$ on the boundary of the disc. The state $\langle a|$ corresponding to the boundary conditions $g|_{\partial D} = 1$ is a maximally symmetric Cardy D-brane state (see [32, 33] for Lagrangian description) which satisfies the boundary conditions

$$\langle a|(J_n - \bar{J}_{-n}) = 0 \tag{4.2}$$

for all $n \in \mathbb{Z}$.

Classical solutions of equations of motion of the WZW model are of the form $h = h_1(\bar{z})h_2(z)$. Hence,

$$\bar{\partial}hh^{-1} = \bar{\partial}h_1(\bar{z})h_1^{-1}(\bar{z})$$

[b]Path integrals of the type (4.1) (in the case without boundary) were studied in [21].

which is an entire function of $\bar{z}$ on the disc D. Note also that $J(z) = g^{-1}\partial g$ is the holomorphic current of the quantum WZW model. Using Eq. (4.1), we obtain

$$\int_{g|_{\partial D}=1} \mathcal{D}g\, e^{i(c\mathrm{WZ}(g))+\int_I \langle \xi, dhh^{-1}\rangle)} = \langle a|\sigma_0(h_1(\bar{z})|_{|z|=1})|0\rangle.$$

Here the contribution of $h(0)$ at the center of the disc disappears because of the invariance of the vacuum state $|0\rangle$. The group element $h_1(\bar{z})$ is of the form

$$h_1(\bar{z}) = \exp\left(\sum_{n<0,a} c_{n,a} J^a_n\right),$$

where n's are all negative (non-negative n's correspond to a singularity at $\bar{z} = 0$). This yields

$$\langle u|\sigma_0\left(\exp\left(\sum_{n,a} c_{n<0,a} J^a_n\right)\right)|0\rangle = \langle u|\sigma_0\left(\exp\left(\sum_{n<0,a} c_{n,a} \bar{J}^a_{-n}\right)\right)|0\rangle = \langle u|0\rangle = 1,$$

where we have used the properties of the D-brane state $\langle a|$ and of the vacuum state $|0\rangle$.

In fact, the argument above applies to any maximally symmetric Cardy state $\langle a|$ which satisfies Eq. (4.2). For all such states, the quantum decoupling phenomenon occurs. However, the boundary conditions for the field $v = gh$ turn out to be rather involved:

$$(h(v^{-1}\partial v)h - \partial hh^{-1} - \bar{\partial} vv^{-1} + vh^{-1}\bar{\partial} hv^{-1})|_{\partial D} = 0.$$

Another observation is that the 2-cocycle $\alpha_2(g, h)$ is in general nonvanishing and it contributes to the answer.

In the original setup, we can use the PW formula one more time to we obtain (following [22]) a compact and beautiful answer:

$$Z(u) = e^{ic\mathrm{WZ}(h)} = e^{ic\alpha_2(h_1|_{\partial D}, h_2|_{\partial D})}. \tag{4.3}$$

That is, the partition function is completely described in terms of the Riemann–Hilbert decomposition $h = h_1(\bar{z})h_2(z)$ and of the 2-cocycle α_2.

4.2. *Decoupling in gravitational WZ*

In this section, we apply the example of the WZW calculation of the previous section to the case of the gravitational WZ action. As before, we impose a vacuum state $|b\rangle = |0\rangle$ at one end of the cylinder making it into a disc D. Recall that the vacuum state is annihilated by Virasoro generators

$$L_n|0\rangle = \bar{L}_n|0\rangle = 0$$

for $n \geq -1$. In order to address the decoupling phenomenon, we consider the path integral of the type (4.1) for gravitational WZ action:

$$I(H) = \int_{G|_{\partial D}\in\mathcal{L}} \mathcal{D}G\, e^{i(c\mathrm{WZ}_{\mathrm{grav}}(G)+\int_D S(G)\frac{\bar{\partial} h}{\partial h} d^2z)},$$

where $H(z,\bar{z})$ is a classical solution of equations of motion of the action $\mathrm{WZ}_{\mathrm{grav}}$, $h(H(z,\bar{z}),\bar{z}) = z$ is the inverse diffeomorphism and $\mathcal{L}$ stands for a conformal boundary condition $\langle a|$ which satisfies the equation

$$\langle a|(L_n - \bar{L}_{-n}) = 0$$

for all n. In particular, we assume $G|_{\partial D}(z) = z$.

Solutions of equations of motion $H(z,\bar{z})$ are given by formula

$$H(z,\bar{z}) = \frac{A(\bar{z})u(z) + B(\bar{z})}{C(\bar{z})u(z) + D(\bar{z})},$$

where $AD - BC = 1$. Hence, it can be represented as $H = H_2 \circ H_1$, where

$$H_1\colon z \mapsto u(z), \quad H_2\colon z \mapsto \frac{A(\bar{z})z + B(\bar{z})}{C(\bar{z})z + D(\bar{z})}. \tag{4.4}$$

Recall (see [4]) that the combination $\bar{\partial}h/\partial h$ admits the following presentation

$$\frac{\bar{\partial}h}{\partial h} = z^2 j_+(\bar{z}) - 2zj_0(\bar{z}) + j_-(\bar{z}),$$

where $j_{+,0,-}$ are currents discovered in [7] (they are functions of the parameters A, B, C, D and of their derivatives).

Equation (4.1) suggests the following answer for the path integral $I(H)$:

$$I(H) = \langle a|\sigma_0(H_2)\sigma_0(H_1)|0\rangle.$$

Note that H_2 is a Möbius transformation generated by $L_{1,0,-1}$. Using the property of the conformal invariant state $\langle a|$, we transform it into a Möbius transformation generated by $\bar{L}_{1,0,-1}$ which then preserves the vacuum $|0\rangle$. Hence, we have

$$I(H) = \langle a|\sigma_0(H_1)|0\rangle.$$

Similar to the arguments in the previous section, we write

$$H_1 = \exp\left(\sum_{n\le 0} c_n L_n\right)$$

and compute

$$I(H) = \langle a|\sigma_0\left(\exp\left(\sum_{n\le 0} L_n\right)\right)|0\rangle = \langle a|\sigma_0\left(\exp\left(\sum_{n\le 0} \bar{L}_{-n}\right)\right)|0\rangle = \langle a|0\rangle = 1.$$

Here we have used conformal invariance of the state $\langle a|$ and the properties of the vacuum.

We are now in the position to compute a more general path integral for the gravitational WZ action:

$$Z = \int_{V|_{\partial D}\in\mathcal{L}'} \mathcal{D}V e^{ic\mathrm{WZ}_{\mathrm{grav}}(V)} = e^{ic\mathrm{WZ}_{\mathrm{grav}}(H)} I(H) = e^{ic\mathrm{WZ}_{\mathrm{grav}}(H)}.$$

Here $V = G \circ H$, where G satisfies boundary conditions $\mathcal{L}$ and H is a solution of equations of motion for $\mathrm{WZ}_{\mathrm{grav}}$. Note that the Virasoro–Bott cocycle vanishes:

$\alpha_2(G, H) = 0$ since $\log(\partial G) = \log(1) = 0$ on the boundary of the disc. At the moment, the description of the boundary conditions $\mathcal{L}'$ is implicit. Giving a more explicit description, similar to the WZW case, remains an interesting question (see also [30]).

Using the gravitational PW formula, we can further simplify the answer for the partition function Z:

$$Z = e^{ic\mathrm{WZ}_{grav}(H)} = e^{ic\alpha_2^{VB}(H_1|_{\partial D}, H_2|_{\partial D})}.$$

This is an analogue of Eq. (4.3) in the WZW case. We observe that the partition function is completely described in terms of the decomposition (4.4). We interpret this decomposition as a gravitational analogue of the Riemann–Hilbert problem.

Acknowledgments

We are grateful to N. Nekrasov, G. Pimentel and A. M. Polyakov for important discussions. We would like to thank F. Valach for useful comments. We are indebted to E. Nissimov and S. Pacheva, former students of L. Faddeev, who attracted our attention to their work on geometric actions and in particular to [20, 21]. The research of A.A. was supported in part by the ERC project MODFLAT, by the grant numbers 165666 and 159581 of the Swiss National Science Foundation and by the National Center of Competence in Research SwissMAP. The research of S.Sh. was supported by Simons Foundation under the program "Targeted Grants to Institutes".

References

[1] L. D. Faddeev and S. L. Shatashvili, Algebraic and Hamiltonian methods in the theory of nonabelian anomalies, *Theor. Math. Phys.* **60** (1985) 770–778; *Teor. Mat. Fiz.* **60** (1984) 206–217.

[2] L. D. Faddeev and S. L. Shatashvili, Realization of the Schwinger term in the Gauss law and the possibility of correct quantization of a theory with anomalies, *Phys. Lett. B* **167** (1986) 225–228.

[3] A. Alekseev, L. D. Faddeev and S. L. Shatashvili, Quantization of symplectic orbits of compact Lie groups by means of the functional integral, *J. Geom. Phys.* **5** (1988) 391–406.

[4] A. Alekseev and S. L. Shatashvili, Path integral quantization of the coadjoint orbits of the Virasoro group and 2D gravity, *Nucl. Phys. B* **323** (1989) 719–733.

[5] A. Alekseev and S. L. Shatashvili, From geometric quantization to conformal field theory, *Comm. Math. Phys.* **128** (1990) 197–212.

[6] A. Alekseev and S. L. Shatashvili, Quantum groups and WZW models, *Comm. Math. Phys.* **133** (1990) 353–368.

[7] A. M. Polyakov, Quantum gravity in two dimensions, *Mod. Phys. Lett. A* **2** (1987) 893–898.

[8] S. Sachdev and J.-W. Ye, Gapless spin fluid ground state in a random, quantum Heisenberg magnet, *Phys. Rev. Lett.* **70** (1993) 3339–3342; arXiv:cond-mat/9212030.

[9] A. Kitaev, Talk given at the Fundamental Physics Prize Symposium (November 10, 2014); A. Kitaev, KITP seminar (February 12, 2015); A simple model of quantum holography, talks at KITP (April 7, 2015 and May 27, 2015).

[10] G. Turiaci and H. Verlinde, On CFT and quantum chaos, *JHEP* **2016** (2016) No. 12, 110; arXiv:1603.03020.

[11] J. Maldacena and D. Stanford, Remarks on the Sachdev–Ye–Kitaev model, *Phys. Rev. D* **94**(10) (2016) 106002; arXiv:1604.07818.

[12] J. Maldacena, The SYK model and near extremal black holes, http://scgp.stonybrook.edu/video/video.php?id=2733.

[13] J. Maldacena, D. Stanford and Z. Yang, Conformal symmetry and its breaking in two dimensional nearly anti-de-Sitter space, arXiv:1606.01857.

[14] J. Engelsoy, T. G. Mertens and H. Verlinde, An investigation of AdS2 backreaction and holography, *JHEP* **2016** (2016) No. 7, 139; arXiv:1606.03438.

[15] G. Mandal, P. Nayak and S. R. Wadia, Coadjoint orbit action of Virasoro group and two-dimensional quantum gravity dual to SYK/tensor models, *JHEP* **2017** (2017) No. 11, 046; arXiv:1702.04266.

[16] D. Gaiotto and M. Rapcak, Vertex algebras at the corner, arXiv:1703.00982.

[17] D. Stanford and E. Witten, Fermionic localization of the Schwarzian theory, *JHEP* **2017** (2017) No. 10, 008; arXiv:1703.04612.

[18] G. L. Pimentel, A. M. Polyakov and G. M. Tarnopolsky, Vacuum decay in CFT and the Riemann–Hilbert problem, *Nucl. Phys. B* **907** (2016) 617–632; arXiv:1512.06721.

[19] V. Guillemin and S. Sternberg, The Gel'fand–Cetlin system and quantisation of complex flag manifolds, *J. Funct. Anal.* **52**(1) (1983) 106–128.

[20] H. Aratyn, E. Nissimov and S. Pacheva, On the group theoretical meaning of the conformal field theories in the framework of coadjoint orbits, *Phys. Lett. B* **251** (1991) 401–405.

[21] H. Aratyn, E. Nissimov and S. Pacheva, Infinite-dimensional Noether symmetry groups and quantum effective actions from geometry, *Phys. Lett. B* **255** (1991) 359–366.

[22] M. Baumgartl, I. Sachs and S. L. Shatashvili, Factorization conjecture and the open/closed string correspondence, *JHEP* **2005** (2005) No. 05, 040; hep-th/0412266.

[23] E. Witten, On background independent open-string field theory, *Phys. Rev. D* **46** (1992) 5467–5473; arXiv:hep-th/9208027.

[24] E. Witten, Some computations in background independent open-string field theory, *Phys. Rev. D* **47** (1993) 3405–3410; arXiv:hep-th/9210065.

[25] S. L. Shatashvili, Comment on the background independent open string theory, *Phys. Lett. B* **311** (1993) 83–86; hep-th/9303143.

[26] S. L. Shatashvili, On the problems with background independence in string theory, *Alg. Anal.* **6** (1994) 215–226; hep-th/9311177.

[27] A. A. Kirillov, *Elements of the Theory of Representations*, Grundlehren der Mathematischen Wissenschaften, Band 220 (Springer-Verlag, Berlin-New York, 1976).

[28] S. P. Novikov, Multivalued functions and functionals. An analogue of the Morse theory, *Soviet Math. Dokl.* **24**(2) (1981) 222–226.

[29] A. M. Polyakov and P. B. Wiegmann, Goldstone fields in two dimensions with multivalued actions, *Phys. Lett. B* **141** (1984) 223–228.

[30] S. L. Shatashvili, Factorisation and string field theory, EuroStrings 2006, Cambridge, UK (2006); http://www.damtp.cam.ac.uk/events/estg06/talks/shatashvili/index.html.

[31] A. Alekseev, F. Naef, X. Xu and C. Zhu, Chern–Simons, Wess–Zumino and other

cocycles from Kashiwara–Vergne and associators, to appear in *Lett. Math. Phys.*; arxiv:1702.08857.

[32] J. L. Cardy, Boundary conditions, fusion rules and the Verlinde formula, *Nucl. Phys. B* **324** (1989) 581–596.

[33] A. Alekseev and V. Schomerus, D-branes in the WZW model, *Phys. Rev. D* **60** (1999) 061901; hep-th/9812193.

K-Theory and the Jones Polynomial

Michael F. Atiyah[*,‡] and Carlos Zapata-Carratala[†,§]

[*] *The University of Edinburgh, Office 6319, James Clerk Maxwell Building, Peter Guthrie Tait Road, Edinburgh EH9 3FD, UK*

[†] *The University of Edinburgh, Office 6324, James Clerk Maxwell Building, Peter Guthrie Tait Road, Edinburgh EH9 3FD, UK*

[‡] *m.atiyah@ed.ac.uk*

[§] *c.zapata.carratala@gmail.com*

In this paper, we present a new approach to the definition of the Jones polynomial using equivariant K-theory.

Dedicated to Ludwig Faddeev

1. Links and the Jones Polynomial

A **link**, is a smooth embedding of an oriented 1-dimensional manifold L into $\mathbb{R}^3$. The number of components will be denoted by m. When $m = 1$, L becomes a knot. In [4], Jones defined a polynomial in the variable t that was a topological invariant associated to any link L, it will be denoted by $J_L(t)$. The precise definition is done inductively using invariance under isotopy, the normalization $J_{\text{unknot}}(t) = 1$ and the skein formula

$$\frac{1}{t}J_{L_+}(t) - tJ_{L_-}(t) = \left(\sqrt{t} - \frac{1}{\sqrt{t}}\right) J_{L_0}, \tag{1.1}$$

where L_+, L_- and L_0 are links with identical planar diagrams except near one crossing, where the branches of L_+ cross in the opposite order to those of L_-, and where L_0 has non-crossing branches. It is then easy to see that the **Jones polynomial** is a Laurent polynomial in $\sqrt{t}$ with integer coefficients. Some explicit examples are shown in Fig. 1.

2. Tubular Neighborhood

Given a link L, we can find a ball $B \subset \mathbb{R}^3$ centered at an origin o such that $o \notin L$ and $L \subset B$. Such a choice is unique up to isotopy. If we construct the

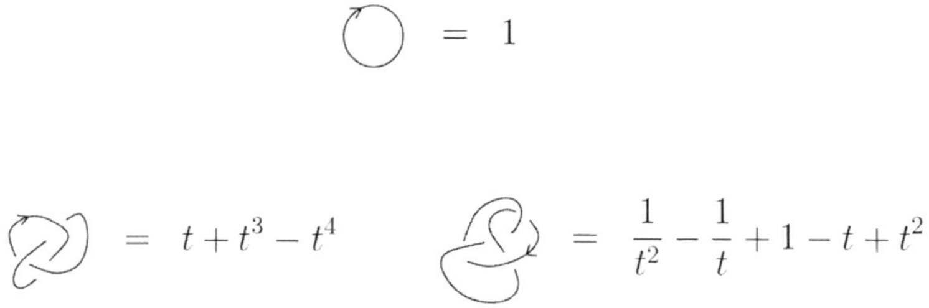

Fig. 1. The Jones polynomials of some knots.

one-point-compactification of $\mathbb{R}^3$, $S^3 = \mathbb{R}^3 \cup \{\infty\}$, the link embedded in S^3 does not contain the point at infinity.

The Euclidean metric of $\mathbb{R}^3$ allows us to construct an open tubular neighborhood L' by taking suitably small open disks in each of the fibers of the normal bundle. Let $D \subset \mathbb{C}$ be the open unit disk in the complex plane, then the tubular neighborhood is diffeomorphic to $L \times D$.

An orientation of $\mathbb{R}^3$, together with the orientation of L, induces a natural orientation on the normal bundle that, in turn, induces an orientation for the tubular neighborhood.

3. Group Actions

First note that the circle $S^1 \cong U(1)$ acts on the link L by rotating each component and thus also on L'. On the other hand, the unit disk in the complex plane D has a natural $U(1)$ action given by phase rotation. These two actions combined give the action of the torus $T^2 = S^1 \times S^1$ on each component of $L' = L \times D$ which in particular restricts to the obvious action on the boundary.

Recall that the link was contained in a ball B centered around the origin. We shall consider the natural $SU(2)$ action on the complement $\mathbb{R}^3 - B$ whose deformation retract is simply S^2. This $SU(2)$ action extends to an action of $U(2)$ on $L \times S^3$, by adding the action of its center.

4. K-Theory

Our aim is to derive the Jones polynomial of a link from G-equivariant K-theory. In Sec. 3, we saw that $G = U(2)$ was the appropriate symmetry group and we will use G-equivariant K-theory with compact supports as in [1]. This involves, as in [2], the representation ring $R(G)$.

We recall that, for any compact connected Lie group G with maximal torus T and Weyl group $W = N(T)/T$, with $N(T)$ its normalizer, we have the process of *holomorphic induction*. In the Borel–Weil construction this assigns, to any positive weight λ of T (taken to be maximal), a holomorphic line bundle over G/T whose sections give the irreducible representation V_λ of G. This extends to a surjective homomorphism Ind: $R(T) \to R(G)$ where $\mathrm{Ind}(\lambda) = \mathrm{Ind}(\mu) \Leftrightarrow \mu = w(\lambda)$ with $w \in W$ and $\mathrm{sign}(w) = 1$.

In the case of interest to us, we have $G = U(2)$, $T = U(1) \times U(1)$, $W = \mathbb{Z}_2$. Consider the embedding $U(1) \to T$ given by $a \mapsto (a, a^{-1})$. It is a right-inverse of Ind. Applied to the basic representation of $U(1)$ it gives the standard representation of $U(2)$ on $\mathbb{C}^2$, while applied to its nth power it gives the irreducible representation of dimension $n+1$ of $U(2)$, realized by binary forms of degree n.

Given the link L in $\mathbb{R}^3$, we consider the following isomorphisms

$$\text{the Bott isomorphism } \beta\colon K_{U(1)}(L) \to K_{U(1)}(L'), \tag{4.1}$$

$$\text{the homomorphism } K_{U(1)}(N') \to K_{U(1)\times U(1)}(N' \times U(1)), \tag{4.2}$$

given by $x \mapsto x \otimes 1$, where $1 \in K_{U(1)}(L)$ is the image of $\mathbb{Z}^n \to \mathbb{Z}^m$ under the diagonal embedding.

$$\begin{gathered}\text{the automorphism } w^* \text{ of } K \text{ induced by the map}\\ w\colon (x, y, z) \mapsto (z, y^{-1}, x^{-1}) \text{ on } U(1)^3.\end{gathered} \tag{4.3}$$

We now define the **Jones polynomial** $J(L)$ by

$$J(L) = \beta^{-1}(w^*(\beta(1) \otimes 1)). \tag{4.4}$$

If $\sqrt{t} \in K_{U(1)}(S^2)$ denotes the Hopf line bundle with Chern class $+1$, then elements of $K_{U(2)}(S^3) = K_{U(1)}(S^2)$ are just Laurent series in $\sqrt{t}$ with integer coefficients. Thus $J(L)$ is such a Laurent series.

5. The Jones Polynomial

In this section, we will derive the basic properties of the link invariant $J(L)$ that we defined in Sec. 4, thus establishing that it is the same as the original Jones invariant defined in [4]. In particular our invariant is chiral, with the variable t replaced by $1/t$ under reflection.

Formula (4.4), which defines $J(L) \in K_{U(1)}(L')$, shows that we switch (via w^*) from the trivialization based on o to that based on ∞.

All that we have to prove is that $J(L)$ satisfies the skein relation (1.1). For this, we just move the origin o. This isotopy leaves $J(L)$ unaltered but the change of origin yields two different plane projections from which (1.1) follows.

Thus we have given a direct definition of the Jones polynomial of an oriented link L in $\mathbb{R}^3$ which has the following properties:

(1) It is 3-dimensional and chiral.
(2) It is a finite Laurent series in $\sqrt{t}$ with integer coefficients.
(3) It is rigorous and simple.

While there are many different definitions of the Jones polynomial there are none that have all the three properties above. Our definition rests on equivariant K-theory as developed in [1, 2].

This completes the task we set ourselves.

6. Comments

In this final section, we shall make a few comments about our new definition of the Jones polynomial and its ramifications.

The Jones polynomial was one of the first *quantum* invariants. This was highlighted in [3], together with the Donaldson invariants. The challenge of understanding these new invariants was taken up by Witten [5] and has since exploded. Many Field Medals were awarded in this explosion, which still reverberates.

Our definition does not mention the word quantum, stressing the fact that it is purely geometrical, so it will be interesting to see its impact on mathematics and what light it sheds on the relation with quantum theory. Here we enumerate a few particular directions for further investigation:

(6.1) There are obvious algebraic extensions of our definition of the Jones polynomial which involve higher representations of $U(2)$.

(6.2) There are interesting extensions which define Jones-like polynomials of 2 and 3 variables.

(6.3) There are mathematical relations between our definition and many topics which have been labeled *quantum*: quantum groups, quantum Yang–Baxter, ..., suggesting a deeper relation with quantum physics.

(6.4) Witten [5] gave a quantum field theory definition of the Jones polynomial, providing an umbrella that shelters the topics in point (6.3) above.

(6.5) There are several ways to deal with the foundational and computational aspects of point (6.4). Some involve combinatorics, some involve the Hamiltonian and some involve Wick rotation to statistical mechanics and the work of Baxter.

(6.6) How does our definition avoid any reference to physics and quantum theory? The answer lies in K-theory which, from its early days, had the Dirac equation at its heart. This is the *Rosetta Stone* helping the translation between geometry and physics. It was always clear that K-theory, built globally on linear spaces, was a quantum analogue of cohomology, built on local data.

(6.7) The framework from which our definition emerges has a universal structure which applies across the board. Many recent and forthcoming papers exemplify this, see [6–8].

Acknowledgment

This work was supported by the Leverhulme Foundation.

References

[1] M. F. Atiyah and G. B. Segal, Equivariant K-theory and completion, *J. Differential Geom.* **3** (1969) 1–18.
[2] M. F. Atiyah, Power operations in K-theory, *Quat. J. Math. Oxford (2)* **17** (1966) 165–193.
[3] M. F. Atiyah, New invariants of three- and four-dimensional manifolds, in *The Mathematical Heritage of Hermann Weyl (Durham, 1987)*, Proc. Symposia Pure Math, Vol 48 (Amer. Math. Soc., 1987), pp. 285–299.
[4] V. Jones, Hecke algebra representations of braid groups and link polynomials, *Ann. of Math. (2)* **126**(2) (1987) 335–388.
[5] E. Witten, Quantum field theory and the Jones polynomial, *Comm. Math. Phys.* **121**(3) (1989) 351–399.
[6] M. F. Atiyah, The non-existent 6-sphere, to appear in *Int. J. Geom. Methods Mod. Phys.*
[7] M. F. Atiyah, Groups of odd order, submitted to *J. Geom. Phys.*
[8] M. F. Atiyah and J. Berndt, The Kervaire invariant and the magic square, submitted to the *Journal of Topology of the London Mathematical Society*.

Baxter Operator and Baxter Equation for q-Toda and Toda$_2$ Chains

Olivier Babelon*,§, Karol K. Kozlowski†,¶ and Vincent Pasquier‡,||

*Sorbonne Université, UPMC Univ Paris 06, CNRS, UMR 7589, LPTHE, F-75005 Paris, France

†Univ Lyon, ENS de Lyon, Univ Claude Bernard Lyon 1, CNRS, Laboratoire de Physique, F-69342 Lyon, France

‡Université Paris Saclay, CNRS, CEA, IPhT, F-91191 Gif-sur-Yvette, France

§babelon@lpthe.jussieu.fr

¶karol.kozlowski@ens-lyon.fr

||vincent.pasquier@cea.fr

We construct the Baxter operator $\mathtt{Q}(\lambda)$ for the q-Toda chain and the Toda$_2$ chain (the Toda chain in the second Hamiltonian structure). Our construction builds on the relation between the Baxter operator and Bäcklund transformations that were unravelled in [13]. We construct a number of quantum intertwiners ensuring the commutativity of $\mathtt{Q}(\lambda)$ with the transfer matrix of the models and of the $\mathtt{Q}$'s between each other. Most importantly, $\mathtt{Q}(\lambda)$ is modular invariant in the sense of Faddeev. We derive the Baxter equation for the eigenvalues $q(\lambda)$ of $\mathtt{Q}(\lambda)$ and show that these are entire functions of λ. This last property will ultimately lead to the quantization of the spectrum for the considered Toda chains, in a subsequent publication [1].

This work is dedicated to the memory of L. D. Faddeev

1. Introduction

The Toda chain was introduced by Toda [30] on the classical level and constituted, for over 50 years, a toy model for developing various exact solvability techniques on either classical or quantum level. In particular, it is this model that saw the birth of the classical separation of variables in the approach using analytical properties on the spectral curve [6, 11, 25] and of the quantum separation of variables [29] following the earlier ideas of separability developed in the series of papers by Gutzwiller [14, 15]. Due to its natural relation with the representation theory of $\mathfrak{gl}_n$ [21, 27], the model can also be solved by harmonic analysis on the associated Lie group, see [33] for a review. In the modern language, the quantum integrability of the Toda chain is related with the Yang–Baxter equation associated with the representation theory of the Yangian $\mathcal{Y}(\widehat{\mathfrak{sl}}_2)$, as first established in Gaudin's book [12]. This aroused the natural question of constructing the Baxter operator for the Toda chain. In the process of answering this question, a relation with Bäcklund transformation was unveiled in [13]. This triggered a series of deep works on the subject by Sklyanin and Kuznetsov, see e.g. [23].

It is natural to study the q-deformations of the Toda chain, which are related to the representation theory of $U_q(\widehat{\mathfrak{sl}}_2)$. These include the q-Toda chain, first introduced in [28], and the much less known Toda$_2$ model which corresponds to the quantization of the Toda chain in the *second* Hamiltonian structure [2]. One of the many interesting features of these models consists in their modular invariance as advocated by L. D. Faddeev [8, 9]. This was already used in [20] to construct the eigenfunctions of the open q-Toda chain, and it is also crucial in our construction Baxter operator. The Toda$_2$ chain is interesting by its direct relation to a discretization of the Virasoro algebra [2]. The q-Toda chain has recently received a lot of attention in various domains see e.g. [16, 17, 19, 24, 26]. In particular, the work [26] proposed a construction of a $\mathtt{Q}$ operator for the q-Toda. This construction was given in terms of a formal series in the spectral parameter with operator-valued combinatorial coefficients. When compared with our construction, this result seems much more implicit; in particular, the modular invariance of that result is not explicitly manifest.

The paper is organized as follows. Section 2 introduces the models of interest and lists several of their properties of relevance to our study. Section 3 focuses on the construction of various intertwining operators that play an important role in the construction of the Baxter $\mathtt{Q}$-operator. Section 4 is devoted to the construction of the $\mathtt{Q}$-operator as well as to the characterization of its main properties. In particular, we construct there the operator valued Baxter equation and use it to obtain the scalar $t-q$ equation for the eigenvalues $q(\lambda)$ of the operator $\mathtt{Q}(\lambda)$. We then use the explicit form of the operator $\mathtt{Q}(\lambda)$ to prove that $q(\lambda)$ are *entire* functions of λ. This last property will ultimately yield [1] the quantization conditions for the spectrum of the considered Toda chains along the lines developed in [13, 22].

Various technical aspects of the analysis developed in the paper are gathered into several appendices. Appendix A presents the Bäcklund transformation. Appendix B provides a general scheme for constructing intertwiners on the quantum space for the class of Lax matrices of interest to the study. Appendix C summarizes the main properties and definitions of the special functions of interest to the present study.

2. The Models of Interest

2.1. *The Lax and transfer matrices*

The q-Toda and the Toda$_2$ chains are most conveniently constructed by means of the quantum inverse scattering method. The local Lax matrix which encompasses both models takes the form

$$\mathrm{L}_{0V_{x_n}}(\lambda) = \begin{pmatrix} \mathrm{e}^{-\frac{2\pi}{\omega_2}\lambda} - \mathrm{e}^{-\omega_1 \mathtt{X}_n} & q^2 \mathrm{e}^{-\frac{2\pi}{\omega_2}\lambda}[d_2 + d_1 q \mathrm{e}^{-\omega_1 \mathtt{X}_n}]\mathrm{e}^{-\frac{2\pi}{\omega_2}\mathtt{x}_n} \\ -q^{-2}\mathrm{e}^{\frac{2\pi}{\omega_2}\mathtt{x}_n} & -d_2 \end{pmatrix}. \tag{2.1}$$

The Lax operator is realized as a 2×2 matrix on the auxiliary space $V_0 \simeq \mathbb{C}^2$ while its entries act on the local quantum space $V_{x_n} \simeq L^2(\mathbb{R})$. Above, $\mathtt{X}_n$, $\mathtt{x}_n$ are canonically conjugated operators

$$[\mathtt{X}_n, \mathtt{x}_n] = -\mathrm{i}$$

so that the operator part of the entries of the above Lax matrix form a Weyl pair:

$$\mathrm{e}^{-\frac{2\pi}{\omega_2}\mathtt{x}_n}\mathrm{e}^{-\omega_1 \mathtt{X}_n} = q^2 \mathrm{e}^{-\omega_1 \mathtt{X}_n}\mathrm{e}^{-\frac{2\pi}{\omega_2}\mathtt{x}_n}, \quad \text{with} \quad q = \mathrm{e}^{\mathrm{i}\pi\frac{\omega_1}{\omega_2}}. \tag{2.2}$$

We shall henceforth realize $\mathtt{x}_n$ as a multiplication operator on V_{x_n}.

The remaining parameters d_1 and d_2 are free and their specializations provide one with the Lax matrices of

- q-Toda ($d_2 = 0$)

$$\mathrm{L}^{q\text{-Toda}}_{0V_{x_n}}(\lambda) = \begin{pmatrix} \mathrm{e}^{-\frac{2\pi}{\omega_2}\lambda} - \mathrm{e}^{-\omega_1 \mathtt{X}_n} & q^3 \mathrm{e}^{-\frac{2\pi}{\omega_2}\lambda} d_1 \mathrm{e}^{-\omega_1 \mathtt{X}_n}\mathrm{e}^{-\frac{2\pi}{\omega_2}\mathtt{x}_n} \\ -q^{-2}\mathrm{e}^{\frac{2\pi}{\omega_2}\mathtt{x}_n} & 0 \end{pmatrix}, \tag{2.3}$$

- Toda$_2$ ($d_1 = 0$)

$$\mathrm{L}^{\mathrm{Toda}_2}_{0V_{x_n}}(\lambda) = \begin{pmatrix} \mathrm{e}^{-\frac{2\pi}{\omega_2}\lambda} - \mathrm{e}^{-\omega_1 \mathtt{X}_n} & \mathrm{e}^{-\frac{2\pi}{\omega_2}\lambda} q^2 d_2 \mathrm{e}^{-\frac{2\pi}{\omega_2}\mathtt{x}_n} \\ -q^{-2}\mathrm{e}^{\frac{2\pi}{\omega_2}\mathtt{x}_n} & -d_2 \end{pmatrix}. \tag{2.4}$$

The Lax matrix $\mathrm{L}_{0V_{x_n}}(\lambda)$ satisfies the usual Yang–Baxter equation

$$R_{12}(\lambda_1 - \lambda_2)\mathrm{L}_{1V_{x_n}}(\lambda_1)\mathrm{L}_{2V_{x_n}}(\lambda_2) = \mathrm{L}_{2V_{x_n}}(\lambda_2)\mathrm{L}_{1V_{x_n}}(\lambda_1)R_{12}(\lambda_1 - \lambda_2) \tag{2.5}$$

with the 4×4 quantum R-matrix acting on the tensor product of two auxiliary spaces 1 and 2:

$$R(\lambda) = \begin{pmatrix} 1 & 0 & 0 & 0 \\ 0 & \dfrac{q^{-1}\sinh\frac{\pi\lambda}{\omega_2}}{\sinh\frac{\pi}{\omega_2}(\lambda + \mathrm{i}\omega_1)} & \dfrac{(q - q^{-1})\mathrm{e}^{-\frac{\pi\lambda}{\omega_2}}}{2\sinh\frac{\pi}{\omega_2}(\lambda + \mathrm{i}\omega_1)} & 0 \\ 0 & \dfrac{(q - q^{-1})\mathrm{e}^{\frac{\pi\lambda}{\omega_2}}}{2\sinh\frac{\pi}{\omega_2}(\lambda + \mathrm{i}\omega_1)} & \dfrac{q\sinh\frac{\pi\lambda}{\omega_2}}{\sinh\frac{\pi}{\omega_2}(\lambda + \mathrm{i}\omega_1)} & 0 \\ 0 & 0 & 0 & 1 \end{pmatrix}. \tag{2.6}$$

This R-matrix is twisted in respect to the "usual" 6vertex model R-matrix.

A first set of integrals of motion for the model is obtained by the standard recipe of the quantum inverse scattering method. The transfer matrix which generates the commuting quantities is defined as the trace of a monodromy matrix

$$\mathtt{t}(\lambda) = \mathrm{Tr}_0\Big[\mathtt{L}_{0V_{x_N}}(\lambda)\cdots\mathtt{L}_{0V_{x_1}}(\lambda)\Big].$$

$\mathtt{t}$ is an operator-valued polynomial of degree N in $\mathrm{e}^{-\frac{2\pi}{\omega_2}\lambda}$.

$$\mathtt{t}(\lambda) = \sum_{j=0}^{N}(-1)^j \mathtt{H}_j \mathrm{e}^{-\frac{2\pi}{\omega_2}(N-j)\lambda}.$$

Explicitly, one has

$$\begin{aligned}
\mathtt{H}_0 &= \mathrm{id}, \\
\mathtt{H}_1 &= \sum_{n=1}^{N}\left\{\left(1+q^{-1}d_1\mathrm{e}^{-\frac{2\pi}{\omega_2}(\mathtt{x}_n-\mathtt{x}_{n-1})}\right)\mathrm{e}^{-\omega_1\mathtt{X}_n} + d_2\mathrm{e}^{-\frac{2\pi}{\omega_2}(\mathtt{x}_n-\mathtt{x}_{n-1})}\right\}, \\
&\ \ \vdots \\
\mathtt{H}_N &= \prod_{n=1}^{N}\mathrm{e}^{-\omega_1\mathtt{X}_n} + d_2^N .
\end{aligned} \tag{2.7}$$

One can exhibit a slightly stronger conserved quantity than $\mathtt{H}_N$, namely, $\mathtt{P}_{\mathrm{tot}} = \sum_{a=1}^{N}\mathtt{X}_a$. Indeed, one can readily check that

$$\mathrm{e}^{\mathrm{i}\gamma\mathtt{X}_a}\mathtt{L}_{0V_{x_n}}(\lambda)\mathrm{e}^{-\mathrm{i}\gamma\mathtt{X}_a} = \mathrm{e}^{-\frac{\pi\gamma}{\omega_2}\sigma_3}\mathtt{L}_{0V_{x_n}}(\lambda)\mathrm{e}^{\frac{\pi\gamma}{\omega_2}\sigma_3}, \tag{2.8}$$

what ensures that, for any γ,

$$\mathrm{e}^{\mathrm{i}\gamma\mathtt{P}_{\mathrm{tot}}}\mathtt{t}(\lambda)\mathrm{e}^{-\mathrm{i}\gamma\mathtt{P}_{\mathrm{tot}}} = \mathtt{t}(\lambda).$$

Then $\big[\mathtt{P}_{\mathrm{tot}}, \mathtt{t}(\lambda)\big] = 0$ follows upon taking the γ-derivative at $\gamma = 0$ of the previous relation.

2.2. *The dual models*

As was already pointed out by many authors and, in particular, by Faddeev [8, 9], since the entries of the Lax matrix $\mathtt{L}_{0V_{x_n}}(\lambda)$ only involve explicitly the Weyl pair $\big(\mathrm{e}^{-\frac{2\pi}{\omega_2}\mathtt{x}_n}, \mathrm{e}^{-\omega_1\mathtt{X}_n}\big)$, the representation induced on V_{x_n} by the algebra generated by the entries of $\mathtt{L}_{0n}(\lambda)$ is reducible. Indeed, any operator belonging to the algebra generated by the dual Weyl pair

$$\mathrm{e}^{-\frac{2\pi}{\omega_1}\mathtt{x}_n}\mathrm{e}^{-\omega_2\mathtt{X}_n} = \widetilde{q}^{\,2}\mathrm{e}^{-\omega_2\mathtt{X}_n}\mathrm{e}^{-\frac{2\pi}{\omega_1}\mathtt{x}_n}, \qquad \widetilde{q} = \mathrm{e}^{\mathrm{i}\pi\frac{\omega_2}{\omega_1}}$$

commutes with the former one. This reducibility makes ambiguous several steps occurring in the resolution of the joint spectral problem associated with $\{\mathtt{H}_j\}_{j=0}^{N}$. A way of avoiding such ambiguities is to directly consider the representation on V_{x_n} of the modular double, namely, by simultaneously considering the Lax matrix $\mathtt{L}_{0V_{x_n}}(\lambda)$ and its dual $\widetilde{\mathtt{L}}_{0V_{x_n}}(\lambda)$ which is obtained from $\mathtt{L}_{0V_{x_n}}(\lambda)$ by exchanging ω_1 and ω_2 and upon replacing the coupling constants by tilded quantities $(d_1, d_2) \to (\widetilde{d_1}, \widetilde{d_2})$. It is straightforward to see that

$$\left[\left[\mathrm{L}_{0V_{x_n}}(\lambda) \right]_{ab}, \left[\widetilde{\mathrm{L}}_{0V_{x_n}}(\lambda) \right]_{cd} \right] = 0.$$

While the two dual algebras generated by $\mathrm{L}_{0V_{x_n}}$ and $\widetilde{\mathrm{L}}_{0V_{x_n}}$ do not see each other, they allow for an unambiguous fixing of various constants which, in case one would solely consider one of the two sub-algebras, would be only fixed to be quasi-constants. In particular, the fact of constructing the joint spectrum of the transfer matrix $\mathtt{t}$ and its dual transfer matrix $\widetilde{\mathtt{t}}$ imposes that all objects used to built up this spectrum have to be modular invariant in the terminology of Faddeev, i.e. invariant under the transformations $\omega_1 \leftrightarrow \omega_2$ and $(d_1, d_2) \to (\widetilde{d_1}, \widetilde{d_2})$. In particular, the Baxter operator will exhibit such a modular invariance.

The dual transfer matrix provides one with the second set of conserved quantities

$$\widetilde{\mathtt{t}}(\lambda) = \sum_{j=0}^{N} (-1)^j \widetilde{\mathtt{H}}_j \mathrm{e}^{-\frac{2\pi}{\omega_1}(N-j)\lambda}$$

where now

$$\begin{aligned}
\widetilde{\mathtt{H}}_0 &= \mathrm{id}, \\
\widetilde{\mathtt{H}}_1 &= \sum_{n=1}^{N} \left\{ \left(1 + \widetilde{q}^{-1} \widetilde{d_1} \mathrm{e}^{-\frac{2\pi}{\omega_1}(\mathtt{x}_n - \mathtt{x}_{n-1})}\right) \mathrm{e}^{-\omega_2 \mathtt{x}_n} + \widetilde{d_2} \mathrm{e}^{-\frac{2\pi}{\omega_1}(\mathtt{x}_n - \mathtt{x}_{n-1})} \right\}, \\
&\ \vdots \\
\widetilde{\mathtt{H}}_N &= \prod_{n=1}^{N} \mathrm{e}^{-\omega_2 \mathtt{x}_n} + \widetilde{d_2}^N .
\end{aligned} \qquad (2.9)$$

Obviously, the Hamiltonians $\mathtt{H}_j$ commute with the Hamiltonians $\widetilde{\mathtt{H}}_k$. It appears that one can now impose two conditions on the parameters at play so as to have interesting reality relations between the Hamiltonians.

(i) ω_1 and ω_2 are real numbers. Then we can impose that the coupling constants verify

$$d_1 = \overline{d_1}, \quad d_2 = \overline{d_2}, \quad \widetilde{d_1} = \overline{\widetilde{d_1}}, \quad \widetilde{d_2} = \overline{\widetilde{d_2}}.$$

In such a case, one gets that

$$\mathtt{H}_j^{\dagger} = \mathtt{H}_j, \quad \widetilde{\mathtt{H}}_j = \widetilde{\mathtt{H}}_j^{\dagger}.$$

(ii) ω_1 and ω_2 are complex numbers such that $\omega_2 = \overline{\omega}_1$. Upon imposing the coupling constants to satisfy

$$\widetilde{d_1} = \overline{d_1}, \quad \widetilde{d_2} = \overline{d_2}$$

one gets that

$$\mathtt{H}_j^{\dagger} = \widetilde{\mathtt{H}}_j .$$

In both cases we are lead to parametrize

$$d_1 = \mathrm{e}^{-\frac{2\pi}{\omega_2}\kappa_1}, \quad d_2 = \mathrm{e}^{-\frac{2\pi}{\omega_2}\kappa_2}, \quad \widetilde{d}_1 = \mathrm{e}^{-\frac{2\pi}{\omega_1}\kappa_1}, \quad \widetilde{d}_2 = \mathrm{e}^{-\frac{2\pi}{\omega_1}\kappa_2}$$

with κ_1 and κ_2 real and modular invariant. This choice of parametrization will be made from now on. The q-Toda and Toda$_2$ Hamiltonians are obtained from $\mathtt{H}_1$ upon enforcing the respective specializations $d_2 = 0$ for q-Toda and $d_1 = 0$ for Toda$_2$ (see [2] for a discussion of this model):

- q-Toda ($d_2 = 0$)

$$\mathtt{H}_1^{q\text{-Toda}} = \sum_{n=1}^{N} \Big[1 + q^{-1}\mathrm{e}^{-\frac{2\pi}{\omega_2}\kappa_1}\mathrm{e}^{-\frac{2\pi}{\omega_2}(\mathtt{x}_n - \mathtt{x}_{n-1})}\Big]\mathrm{e}^{-\omega_1 \mathtt{X}_n};$$

- Toda$_2$ ($d_1 = 0$)

$$\mathtt{H}_1^{\mathrm{Toda}_2} = \sum_{n=1}^{N} \Big\{\mathrm{e}^{-\omega_1 \mathtt{X}_n} + \mathrm{e}^{-\frac{2\pi}{\omega_2}\kappa_2}\mathrm{e}^{-\frac{2\pi}{\omega_2}(\mathtt{x}_n - \mathtt{x}_{n-1})}\Big\}.$$

3. The Various Intertwiners of Interest

3.1. $\mathbb{L}$*LM* $=$ *ML*$\mathbb{L}$*-operator*

The first step in constructing the $\mathtt{Q}$-operator is to build an appropriate, modular invariant, intertwiner for a product of two $\mathtt{L}$ operators, both sharing the same auxiliary space but acting on different quantum spaces. The prototypes of such intertwiners go back to [10, 31]. More precisely, one introduces two representation spaces V_x and V_u endowed with the respective action of Weyl-pairs $(\mathrm{e}^{-\frac{2\pi}{\omega_2}\mathtt{x}}, \mathrm{e}^{-\omega_1 \mathtt{X}})$, $(\mathrm{e}^{-\frac{2\pi}{\omega_2}\mathtt{u}}, \mathrm{e}^{-\omega_1 \mathtt{U}})$ and their duals. Then, as motivated in Appendix A on Bäcklund transformations, one focuses on building an intertwiner for the two Lax operators $\mathtt{L}_{0V_x}(\lambda)$ and $\mathtt{M}_{0V_u}(\lambda; t)$, and their duals, where $\mathtt{M}_{0V_u}(\lambda; t)$ is obtained from $\mathtt{L}_{0V_u}(\lambda)$ by the substitution

$$\mathtt{U} \hookrightarrow \mathtt{U} + \frac{2\pi}{\omega_1\omega_2}t \qquad d_1 \hookrightarrow q^{-2}\mathrm{e}^{\frac{2\pi}{\omega_2}t} \qquad d_2 \hookrightarrow -q^{-1}, \tag{3.1}$$

so that it takes the explicit form

$$\mathtt{M}_{0V_u}(\lambda; t) = \begin{pmatrix} \mathrm{e}^{-\frac{2\pi}{\omega_2}\lambda} - \mathrm{e}^{-\frac{2\pi}{\omega_2}t}\mathrm{e}^{-\omega_1 \mathtt{U}} & -q\mathrm{e}^{-\frac{2\pi}{\omega_2}\lambda}(1 - \mathrm{e}^{-\omega_1 \mathtt{U}})\mathrm{e}^{-\frac{2\pi}{\omega_2}\mathtt{u}} \\ -q^{-2}\mathrm{e}^{\frac{2\pi}{\omega_2}\mathtt{u}} & q^{-1} \end{pmatrix}. \tag{3.2}$$

This Lax matrix corresponds to a slight deformation of the classical formula Eq. (A.5). The extra factors q are dictated by the requirement that $\mathtt{M}_{0V_u}(\lambda; t)$ satisfies the Baxter equation (3.31) and that the Baxter $\mathtt{Q}$ operator we will construct is modular invariant, and thus also intertwines the dual Lax operators.

The construction of the intertwiner

$$\mathbb{L}_{V_x V_u}(t)\, \mathtt{L}_{0V_x}(\lambda)\, \mathtt{M}_{0V_u}(\lambda; t) = \mathtt{M}_{0V_u}(\lambda; t)\, \mathtt{L}_{0V_x}(\lambda)\, \mathbb{L}_{V_x V_u}(t) \tag{3.3}$$

is obtained by the general recipe discussed in Appendix B and more precisely in Sec. B.1. It takes the form

$$\mathbb{L}_{V_x V_u}(t) = C_{\mathbb{L}}(t) \cdot \mathtt{P}_{xu} \cdot \phi_{14}(\mathtt{x}-\mathtt{u}) \cdot \psi_{24}(\mathtt{U}) \cdot \phi_{23}(\mathtt{x}-\mathtt{u}) \tag{3.4}$$

where the building blocks of $\mathbb{L}_{V_x V_u}(t)$ satisfy to first-order finite difference equations. These can be explicitly solved in terms of the double sine function $\mathcal{S}$ whose main properties are recalled in Appendix C.2:

$$\phi_{14}(x) = \mathcal{S}^{-1}\left(x - t + \mathrm{i}\frac{\Omega}{2}\right) \cdot \mathrm{e}^{-\frac{2\mathrm{i}\pi t}{\omega_1\omega_2}x}, \tag{3.5}$$

$$\phi_{23}(x) = \mathcal{S}\left(x + \kappa_1 - \mathrm{i}\frac{\Omega}{2}\right), \tag{3.6}$$

$$\psi_{24}(p) = \mathrm{e}^{\mathrm{i}p\left(\kappa_2 - \mathrm{i}\frac{\Omega}{2}\right)} \cdot \mathcal{S}\left(\frac{\omega_1\omega_2}{2\pi}p + t + \kappa_1 - \kappa_2 - \mathrm{i}\frac{\Omega}{2}\right) \cdot \mathcal{S}^{-1}\left(\frac{\omega_1\omega_2}{2\pi}p\right). \tag{3.7}$$

Here, we introduced the useful shorthand notation

$$\Omega = \omega_1 + \omega_2. \tag{3.8}$$

Finally, the constant prefactor is expressed in terms of the quantum dilogarithm, cf. Sec. C.3 takes the form

$$C_{\mathbb{L}}(t) = \sqrt{\omega_1\omega_2}\,\frac{\mathrm{e}^{-\frac{2\mathrm{i}\pi\alpha_0}{\omega_1\omega_2}\left(\alpha_0 + \frac{\mathrm{i}\Omega}{2}\right)}}{\varpi\left(2\alpha_0 - \frac{\mathrm{i}\Omega}{2}\right)} \quad \text{with} \quad \alpha_0 = \frac{t + \kappa_1 - \kappa_2}{2} - \frac{\mathrm{i}\Omega}{4}. \tag{3.9}$$

Clearly, the value of the constant $C_{\mathbb{L}}(t)$ does not impact the intertwining property of $\mathbb{L}_{V_x V_u}(t)$, but fixing the constant factor to the above value will appear convenient in the following.

The operator product in (3.4) can be recast in a more compact form:

$$\mathbb{L}_{V_x V_u}(t) = C'_{\mathbb{L}}(t)\,\mathtt{P}_{xu}\mathrm{e}^{\mathrm{i}\alpha\mathtt{U}}\mathrm{e}^{\frac{2\mathrm{i}\pi\beta}{\omega_1\omega_2}(\mathtt{u}-\mathtt{x})}\,\frac{S\big(-aq\mathrm{e}^{-\frac{2\pi}{\omega_2}t}(1-\mathrm{e}^{-\omega_1\mathtt{U}})\mathrm{e}^{\frac{2\pi}{\omega_2}(\mathtt{x}-\mathtt{u})}\big)}{S\big(b(1-\mathrm{e}^{-\omega_1\mathtt{U}})\mathrm{e}^{\frac{2\pi}{\omega_2}(\mathtt{x}-\mathtt{u})}\big)} \tag{3.10}$$

where $C'_{\mathbb{L}}(t)$ is some constant whose explicit expression is irrelevant for our purposes,

$$a = \mathrm{e}^{\frac{2\pi}{\omega_2}(\kappa_2 + \frac{\mathrm{i}}{2}\Omega)}, \quad b = \mathrm{e}^{\frac{2\pi}{\omega_2}(\kappa_1 + \frac{\mathrm{i}}{2}\Omega)}, \quad \alpha = \kappa_2 - \frac{\mathrm{i}}{2}\Omega, \quad \beta = \kappa_2 - \kappa_1 + \frac{\mathrm{i}}{2}\Omega. \tag{3.11}$$

The main building block of the formula is the function S defined in (C.12), which is closely related to the double sine function, see Appendix C.2 for more details.

We refer to Sec. B.2 and, in particular Eq. (B.18), for a proof of this representation.

3.2. *Integral kernel of* $\mathbb{L}_{V_x V_u}(t)$

In this subsection, we realize $\mathbb{L}_{V_x V_u}(t)$ as an integral operator acting on the space of functions on the spectrum of the operators $\mathbf{x}\otimes\mathbf{u}$. More precisely, we show that

$$\mathbb{L}_{V_x V_u}(t) = \mathbb{E}_{V_x V_u}\,\mathbb{L}^{(\mathrm{c})}_{V_x V_u}(t)\,\mathbb{E}^{-1}_{V_x V_u} \quad \text{with} \quad \mathbb{E}_{V_x V_u} = \mathrm{e}^{-\frac{3}{4}\Omega(\mathtt{X}+\mathtt{U})}. \tag{3.12}$$

The operator $\mathbb{L}^{(\mathrm{c})}_{V_x V_u}(t)$ is realized as an integral operator on $L^2(\mathbb{R})$:

$$\big(\mathbb{L}^{(\mathrm{c})}_{V_x V_u}(t)\cdot f\big)(x,u) = \int_{\mathbb{R}}\mathrm{d}y\int_{\mathbb{R}}\mathbb{L}_t(x,u;y,v)f(y,v)\mathrm{d}v \tag{3.13}$$

in which the integral kernel takes the manifestly modular invariant form:

$$\mathbb{L}_t(x,u;y,v) = \delta(y-u)\mathrm{e}^{\frac{2\mathrm{i}\pi}{\omega_1\omega_2}t(x-y)}\frac{\mathcal{S}\Big(x-v+\kappa_2-\frac{3\mathrm{i}}{2}\Omega\Big)\mathcal{S}\Big(y-v+\kappa_1-\frac{\mathrm{i}}{2}\Omega\Big)}{\mathcal{S}\Big(y-x-t+\frac{\mathrm{i}}{2}\Omega\Big)\mathcal{S}(x-v+\kappa_1+t-\mathrm{i}\Omega)}. \tag{3.14}$$

Let us specialize this result to the q-Toda and Toda$_2$ cases

- q–Toda ($d_2=0$)

$$\mathbb{L}_t(x,u;y,v) = \delta(y-u)\frac{\mathrm{e}^{\frac{2\mathrm{i}\pi}{\omega_1\omega_2}t(x-y)}\,\mathcal{S}\Big(y-v+\kappa_1-\frac{\mathrm{i}}{2}\Omega\Big)}{\mathcal{S}\Big(y-x-t+\frac{\mathrm{i}}{2}\Omega\Big)\mathcal{S}(x-v+\kappa_1+t-\mathrm{i}\Omega)}; \tag{3.15}$$

- Toda$_2$ ($d_1=0$)

$$\mathbb{L}_t(x,u;y,v) = \delta(y-u)\mathrm{e}^{\frac{2\mathrm{i}\pi}{\omega_1\omega_2}t(x-y)}\frac{\mathcal{S}\Big(x-v+\kappa_2-\frac{3\mathrm{i}}{2}\Omega\Big)}{\mathcal{S}\Big(y-x-t+\frac{\mathrm{i}}{2}\Omega\Big)}. \tag{3.16}$$

The operator form of $\mathbb{L}_{V_x V_u}(t)$ given in (3.4) can be recast as

$$\mathbb{L}_{V_x V_u}(t) = \mathrm{P}_{xu}\cdot\frac{\mathrm{e}^{-\frac{2\mathrm{i}\pi}{\omega_1\omega_2}t(\mathtt{x}-\mathtt{u})}}{\mathcal{S}\Big(\mathtt{x}-\mathtt{u}-t+\mathrm{i}\frac{\Omega}{2}\Big)}\cdot\mathrm{e}^{\mathrm{i}\mathtt{U}\left(\kappa_2-\mathrm{i}\frac{\Omega}{2}\right)}$$
$$\cdot\frac{\mathcal{S}\Big(\frac{\omega_1\omega_2}{2\pi}\mathtt{U}+t+\kappa_1-\kappa_2-\mathrm{i}\frac{\Omega}{2}\Big)}{\mathcal{S}\Big(\frac{\omega_1\omega_2}{2\pi}\mathtt{U}\Big)}\cdot\mathcal{S}\Big(\mathtt{x}-\mathtt{u}+\kappa_1-\mathrm{i}\frac{\Omega}{2}\Big). \tag{3.17}$$

Starting from the Fourier transform of the function $D_\alpha(x)$ which is recalled in (C.18) and observing that

$$\frac{\mathcal{S}\left(p+t+\kappa_1-\kappa_2-\mathrm{i}\frac{\Omega}{2}\right)}{\mathcal{S}(p)} = D_{\alpha_0}\left(p+\beta_0\right)\mathrm{e}^{\frac{2\mathrm{i}\pi}{\omega_1\omega_2}\alpha_0(p+\beta_0)} \tag{3.18}$$

with $2\alpha_0 = t+\kappa_1-\kappa_2-\mathrm{i}\frac{\Omega}{2}$ and $2\beta_0 = t+\kappa_1-\kappa_2+\mathrm{i}\frac{\Omega}{2}$, one has the integral representation

$$\frac{\mathcal{S}\left(p+t+\kappa_1-\kappa_2-\mathrm{i}\frac{\Omega}{2}\right)}{\mathcal{S}(p)} = \frac{\mathcal{A}(\alpha_0)}{\sqrt{\omega_1\omega_2}}\mathrm{e}^{\frac{2\mathrm{i}\pi}{\omega_1\omega_2}\alpha_0\beta_0}\int_{\mathbb{R}}\mathrm{d}v\frac{\mathcal{S}(v-\alpha_0-\mathrm{i}\Omega)}{\mathcal{S}(v+\alpha_0)}\mathrm{e}^{-\mathrm{i}v\frac{2\pi}{\omega_1\omega_2}p}. \tag{3.19}$$

Therefore, given any sufficiently regular function, one has the realization as an integral representation

$$\left(\mathrm{e}^{\mathrm{i}\mathtt{X}\left(\kappa_2-\mathrm{i}\frac{\Omega}{2}\right)}\frac{\mathcal{S}\left(\frac{\omega_1\omega_2}{2\pi}\mathtt{X}+t+\kappa_1-\kappa_2\right)}{\mathcal{S}\left(\frac{\omega_1\omega_2}{2\pi}\mathtt{X}+\mathrm{i}\frac{\Omega}{2}\right)}\cdot f\right)(x)$$

$$= \frac{\mathcal{A}(\alpha_0)}{\sqrt{\omega_1\omega_2}}\mathrm{e}^{\frac{2\mathrm{i}\pi}{\omega_1\omega_2}\alpha_0\beta_0}\int_{\mathbb{R}}\mathrm{d}v\frac{\mathcal{S}\left(x-v+\kappa_2-\frac{3\mathrm{i}}{4}\Omega\right)}{\mathcal{S}\left(x-v+t+\kappa_1-\frac{\mathrm{i}}{4}\Omega\right)}f\left(v-\frac{3\mathrm{i}}{4}\Omega\right) \tag{3.20}$$

where, after acting with the translation operator on f, we made a linear change of variables:

$$v \mapsto x-v+\frac{1}{2}(t+\kappa_1+\kappa_2). \tag{3.21}$$

Thus, upon acting with the multiplication operators appearing to the right and left of (3.17), one gets for any sufficiently regular function of two variables

$$\left(\mathbb{L}_{V_xV_u}(t)\cdot f\right)(x,u) = \int_{\mathbb{R}}\mathrm{d}v\;\mathcal{L}_t\left(x,u;v-\frac{3\mathrm{i}}{4}\Omega\right)f\left(u,v-\frac{3\mathrm{i}}{4}\Omega\right), \tag{3.22}$$

where

$$\mathcal{L}_t(x,u;v) = \mathrm{e}^{\frac{2\mathrm{i}\pi}{\omega_1\omega_2}t(x-u)}\frac{\mathcal{S}\left(x-v+\kappa_2-\frac{3\mathrm{i}}{2}\Omega\right)\mathcal{S}\left(u-v+\kappa_1-\mathrm{i}\frac{\Omega}{2}\right)}{\mathcal{S}(x-v+t+\kappa_1-\mathrm{i}\Omega)\mathcal{S}\left(u-x-t+\mathrm{i}\frac{\Omega}{2}\right)}. \tag{3.23}$$

It remains to observe that

$$\mathcal{L}_t\left(x,u;v-\frac{3\mathrm{i}}{4}\Omega\right) = \mathcal{L}_t\left(x+\frac{3\mathrm{i}}{4}\Omega,u+\frac{3\mathrm{i}}{4}\Omega;v\right)$$

what allows one to obtain the below integral representation

$$\big(\mathbb{E}^{-1}_{V_x V_u}\,\mathbb{L}_{V_x V_u}(t)\mathbb{E}_{V_x V_u} f\big)(x,u) = \int_{\mathbb{R}} \mathrm{d}v\; \mathcal{L}_t\big(x,u;v\big) f\big(u,v\big). \tag{3.24}$$

Thus, representing the action on the u variable as an integral versus a Dirac mass, the claimed form of the kernel follows.

3.3. *The* $\mathtt{M}-\mathtt{M}$ *intertwiner*

In this subsection, we construct the intertwiner $\mathbb{R}_{V_u V_v}(t,t')$ satisfying

$$\mathbb{R}_{V_u V_v}(t,t')\mathtt{M}_{0V_u}(\lambda;t)\mathtt{M}_{0V_v}(\lambda;t') = \mathtt{M}_{0V_v}(\lambda;t')\mathtt{M}_{0V_u}(\lambda;t)\mathbb{R}_{V_u V_v}(t,t'). \tag{3.25}$$

Here λ is the spectral parameter, t, t' are two different Bäcklund parameters. In fact, $\mathbb{R}_{V_u V_v}(t,t')$ can be directly inferred from the intertwiner $\mathbb{L}_{V_u V_v}(t)$ arising in (3.3) owing to the transformation (3.1) that turns $\mathtt{L}_{0V_u}(\lambda)$ into $\mathtt{M}_{0V_u}(\lambda;t)$. One finds

$$\mathbb{R}_{V_u V_v}(t,t') = \mathtt{P}_{uv}\frac{S\big(-q^3\mathrm{e}^{-\frac{2\pi}{\omega_2}t'}(1-\mathrm{e}^{-\omega_1\mathtt{V}})\mathrm{e}^{-\frac{2\pi}{\omega_2}(\mathtt{v}-\mathtt{u})}\big)}{S\big(-q^3\mathrm{e}^{-\frac{2\pi}{\omega_2}t}(1-\mathrm{e}^{-\omega_1\mathtt{V}})\mathrm{e}^{-\frac{2\pi}{\omega_2}(\mathtt{v}-\mathtt{u})}\big)}. \tag{3.26}$$

Here, for convenience, we have set the constant prefactor present in $\mathbb{L}$ to 1 as the latter does not play a role on the intertwining property of $\mathbb{R}_{V_u V_v}(t,t')$. Also, it is manifest from the above formula that $\mathbb{R}_{V_u V_v}(t,t')$ is an invertible operator for almost all values of t,t'.

The proof goes as follows. Starting from (3.3), one makes the substitutions

$$t\hookrightarrow t' \quad \mathtt{U}\hookrightarrow\mathtt{U} \qquad d_1\hookrightarrow q^{-2}\mathrm{e}^{\frac{2\pi}{\omega_2}t} \qquad d_2\hookrightarrow -q^{-1}, \tag{3.27}$$

what recasts (3.3) in the form

$$\begin{aligned}&\mathrm{e}^{\frac{2i\pi t}{\omega_1\omega_2}\mathtt{u}}\mathbb{R}_{V_u V_v}(t,t')\mathrm{e}^{-\frac{2i\pi t}{\omega_1\omega_2}\mathtt{u}}\,\tilde{\mathtt{M}}_{0V_u}(\lambda;t)\mathtt{M}_{0V_v}(\lambda;t')\\ &\quad= \mathtt{M}_{0V_v}(\lambda;t')\tilde{\mathtt{M}}_{0V_u}(\lambda;t)\mathrm{e}^{\frac{2i\pi t}{\omega_1\omega_2}\mathtt{u}}\mathbb{R}_{V_u V_v}(t,t')\mathrm{e}^{-\frac{2i\pi t}{\omega_1\omega_2}\mathtt{u}}\end{aligned} \tag{3.28}$$

where

$$\tilde{\mathtt{M}}_{0V_u}(\lambda;t) = \begin{pmatrix} \mathrm{e}^{-\frac{2\pi}{\omega_2}\lambda}-\mathrm{e}^{-\omega_1\mathtt{U}} & -q\mathrm{e}^{-\frac{2\pi}{\omega_2}\lambda}(1-\mathrm{e}^{\frac{2\pi}{\omega_2}t}\mathrm{e}^{-\omega_1\mathtt{U}})\mathrm{e}^{-\frac{2\pi}{\omega_2}\mathtt{u}} \\ -q^{-2}\mathrm{e}^{\frac{2\pi}{\omega_2}\mathtt{u}} & q^{-1}\end{pmatrix}. \tag{3.29}$$

One can then move $\mathrm{e}^{-\frac{2i\pi t}{\omega_1\omega_2}\mathtt{u}}$ occurring in the right-hand side of (3.28) to the left and $\mathrm{e}^{\frac{2i\pi t}{\omega_1\omega_2}\mathtt{u}}$ occurring in the left-hand side of this equation to the right. Since

$$\mathrm{e}^{-\omega_1\mathtt{U}}\mathrm{e}^{\frac{2i\pi t}{\omega_1\omega_2}\mathtt{u}} = \mathrm{e}^{-\frac{2\pi}{\omega_2}t}\mathrm{e}^{\frac{2i\pi t}{\omega_1\omega_2}\mathtt{u}}\mathrm{e}^{-\omega_1\mathtt{U}} \quad\text{and}\quad \mathrm{e}^{-\frac{2i\pi t}{\omega_1\omega_2}\mathtt{u}}\mathrm{e}^{-\omega_1\mathtt{U}} = \mathrm{e}^{-\frac{2\pi}{\omega_2}t}\mathrm{e}^{-\omega_1\mathtt{U}}\mathrm{e}^{-\frac{2i\pi t}{\omega_1\omega_2}\mathtt{u}} \tag{3.30}$$

one then recovers, upon simplifying the position operator dependent exponents, Eq. (3.25).

3.4. *The* $\mathbb{L} - \mathbb{L}$ *intertwiner*

Using the commutation relations of Eqs. (3.3), (3.25) we can write the transformation

$$\mathtt{L}_{0V_x}(\lambda)\mathtt{M}_{0V_u}(\lambda;t)\mathtt{M}_{0V_v}(\lambda;t') \to \mathtt{M}_{0V_v}(\lambda;t')\mathtt{M}_{0V_u}(\lambda;t)\mathtt{L}_{0V_x}(\lambda)$$

in two different ways. As usual we expect the compatibility relation

$$\mathbb{R}_{V_uV_v}(t,t')\mathbb{L}_{V_xV_v}(t')\mathbb{L}_{V_xV_u}(t) = \mathbb{L}_{V_xV_u}(t)\mathbb{L}_{V_xV_v}(t')\mathbb{R}_{V_uV_v}(t,t'). \tag{3.31}$$

We now establish this relation directly. Writing

$$\mathbb{R}_{V_uV_v}(t,t') = \mathtt{P}_{uv}\check{\mathbb{R}}_{V_uV_v}(t,t'), \quad \mathbb{L}_{V_xV_u}(t) = \mathtt{P}_{xu}\check{\mathbb{L}}_{V_xV_u}(t),$$

Eq. (3.31) becomes equivalent to

$$\check{\mathbb{R}}_{V_xV_u}(t,t')\check{\mathbb{L}}_{V_uV_v}(t')\check{\mathbb{L}}_{V_xV_u}(t) = \check{\mathbb{L}}_{V_uV_v}(t)\check{\mathbb{L}}_{V_xV_u}(t')\check{\mathbb{R}}_{V_uV_v}(t,t') \tag{3.32}$$

where $\check{\mathbb{L}}_{V_uV_v}(t)$ can be read from Eq. (3.10) and $\check{\mathbb{R}}_{V_uV_v}(t,t')$ from Eq. (3.26). In the following, it will appear convenient to introduce the shorthand notations

$$\begin{aligned}\mathcal{U} &= (1-\mathrm{e}^{-\omega_1\mathtt{U}})\mathrm{e}^{-\frac{2\pi}{\omega_2}(\mathtt{u}-\mathtt{x})},\\ \mathcal{V} &= (1-\mathrm{e}^{-\omega_1\mathtt{V}})\mathrm{e}^{-\frac{2\pi}{\omega_2}(\mathtt{v}-\mathtt{x})},\\ \mathcal{W} &= (1-\mathrm{e}^{-\omega_1\mathtt{V}})\mathrm{e}^{-\frac{2\pi}{\omega_2}(\mathtt{v}-\mathtt{u})}.\end{aligned}$$

Given α, β as in (3.11), one uses

$$\mathcal{W}\mathrm{e}^{\mathrm{i}\alpha\mathtt{U}} = -qd_2\mathrm{e}^{\mathrm{i}\alpha\mathtt{U}}\mathcal{W}, \quad \text{and} \quad \mathrm{e}^{-\omega_1\mathtt{U}}\mathrm{e}^{\frac{2\mathrm{i}\pi}{\omega_1\omega_2}\beta(\mathtt{v}-\mathtt{u})} = -qd_1d_2^{-1}\mathrm{e}^{\frac{2\mathrm{i}\pi}{\omega_1\omega_2}\beta(\mathtt{v}-\mathtt{u})}\mathrm{e}^{-\omega_1\mathtt{U}},$$

etc., to push these factors to the left of each side of the above equation, leading to

$$\check{\mathbb{R}}_{V_xV_u}(t,t')\check{\mathbb{L}}_{V_uV_v}(t')\check{\mathbb{L}}_{V_xV_u}(t) = \mathtt{E}\cdot S^{-1}(q^2ba^{-1}\mathcal{W})\cdot\mathtt{C}_L\cdot S\big(-q^3\mathrm{e}^{-\frac{2\pi}{\omega_2}t'}\,\mathcal{W}\big) \tag{3.33}$$

and

$$\check{\mathbb{L}}_{V_uV_v}(t)\check{\mathbb{L}}_{V_xV_u}(t')\check{\mathbb{R}}_{V_uV_v}(t,t') = \mathtt{E}\cdot S^{-1}(q^2ba^{-1}\mathcal{W})\cdot\mathtt{C}_R\cdot S\big(-q^3\mathrm{e}^{-\frac{2\pi}{\omega_2}t'}\,\mathcal{W}\big). \tag{3.34}$$

Here, a, b are as defined in (3.11), and we agree upon

$$\mathtt{E} = C_{\mathbb{L}}(t)C_{\mathbb{L}}(t')\mathrm{e}^{\mathrm{i}(\mathtt{U}+\mathtt{V})\left(\kappa_2-\mathrm{i}\frac{\Omega}{2}\right)}\mathrm{e}^{\frac{2\mathrm{i}\pi}{\omega_1\omega_2}(\mathtt{x}-\mathtt{v})\left(\kappa_1-\kappa_2-\mathrm{i}\frac{\Omega}{2}\right)}\mathrm{e}^{-\frac{2\mathrm{i}\pi}{\omega_1\omega_2}\left(\kappa_2-\mathrm{i}\frac{\Omega}{2}\right)\left(\kappa_1-\kappa_2-\mathrm{i}\frac{\Omega}{2}\right)}. \tag{3.35}$$

Finally,

$$\begin{aligned}\mathtt{C}_L = S(q^2ba^{-1}\mathcal{W})\cdot\frac{S\big(-aq\mathrm{e}^{-\frac{2\pi}{\omega_2}t'}\,\mathcal{U}\big)}{S\big(-aq\mathrm{e}^{-\frac{2\pi}{\omega_2}t}\,\mathcal{U}\big)}\cdot S^{-1}(q^2ba^{-1}\mathcal{W})\\ \times S\big(-q^3\mathrm{e}^{-\frac{2\pi}{\omega_2}t'}\,\mathcal{W}\big)\cdot\frac{S\big(-aq\mathrm{e}^{-\frac{2\pi}{\omega_2}t}\,\mathcal{U}\big)}{S(b\,\mathcal{U})}\cdot S^{-1}\big(-q^3\mathrm{e}^{-\frac{2\pi}{\omega_2}t'}\,\mathcal{W}\big),\end{aligned}$$

and

$$\mathtt{C}_R = S\big(-q^3 \mathrm{e}^{-\frac{2\pi}{\omega_2}t}\,\mathcal{W}\big)\frac{S\big(-aq\mathrm{e}^{-\frac{2\pi}{\omega_2}t'}\,\mathcal{U}\big)}{S\big(b\,\mathcal{U}\big)} \cdot S^{-1}\big(-q^3 \mathrm{e}^{-\frac{2\pi}{\omega_2}t}\,\mathcal{W}\big).$$

$\mathtt{C}_L$ and $\mathtt{C}_R$ can be simplified further by means of the below rewriting of the adjoint action which is valid for any constant c:

$$\begin{aligned} S(c\mathcal{W})\mathcal{U}\,S^{-1}(c\mathcal{W}) &= [1 - S(c\mathcal{W})\mathrm{e}^{-\omega_1 \mathtt{U}}\,S^{-1}(c\mathcal{W})]\mathrm{e}^{-\frac{2\pi}{\omega_2}(\mathtt{u}-\mathtt{x})} \\ &= [1 - S(c\mathcal{W})S^{-1}(q^2 c\mathcal{W})\mathrm{e}^{-\omega_1 \mathtt{U}}]\mathrm{e}^{-\frac{2\pi}{\omega_2}(\mathtt{u}-\mathtt{x})} \\ &= [1 - (1 - c\mathcal{W})\mathrm{e}^{-\omega_1 \mathtt{U}}]\mathrm{e}^{-\frac{2\pi}{\omega_2}(\mathtt{u}-\mathtt{x})} \\ &= \mathcal{U} + cq^{-2}\mathcal{V}\,\mathrm{e}^{-\omega_1 \mathtt{U}}. \end{aligned}$$

Note that, in the intermediate equations, we have used the finite difference equation (C.14). Thence, upon setting $\mathtt{y} = \mathcal{U}$ and $\mathtt{Y} = \mathcal{V}\,\mathrm{e}^{-\omega_1 U}$, one gets

$$\mathtt{C}_L = \frac{S\big(-aq\mathrm{e}^{-\frac{2\pi}{\omega_2}t'}\,(\mathtt{y} + ba^{-1}\mathtt{Y})\big)}{S\big(-aq\mathrm{e}^{-\frac{2\pi}{\omega_2}t}(\mathtt{y} + ba^{-1}\mathtt{Y})\big)} \cdot \frac{S\big(-aq\mathrm{e}^{-\frac{2\pi}{\omega_2}t}(\mathtt{y} - q\mathrm{e}^{-\frac{2\pi}{\omega_2}t'}\mathtt{Y})\big)}{S\big(b(\mathtt{y} - q\mathrm{e}^{-\frac{2\pi}{\omega_2}t'}\mathtt{Y})\big)} \tag{3.36}$$

and

$$\mathtt{C}_R = \frac{S\big(-aq\mathrm{e}^{-\frac{2\pi}{\omega_2}t'}\,(\mathtt{y} - q\mathrm{e}^{-\frac{2\pi}{\omega_2}t}\mathtt{Y})\big)}{S\big(b(\mathtt{y} - q\mathrm{e}^{-\frac{2\pi}{\omega_2}t}\mathtt{Y})\big)}\,. \tag{3.37}$$

Notice that $\mathcal{U}$ and $\mathcal{V}$ commute so that $(\mathtt{y}, \mathtt{Y})$ form again a Weyl pair. It remains to use the Schützenberger relation (C.15)

$$S(c_1\mathtt{y} + c_2\mathtt{Y}) = S(c_1\mathtt{y})S(c_2\mathtt{Y})$$

to recast these operators as

$$\begin{aligned} \mathtt{C}_L &= S^{-1}\big(-bq\mathrm{e}^{-\frac{2\pi}{\omega_2}t}\mathtt{Y}\big)S^{-1}\big(-aq\mathrm{e}^{-\frac{2\pi}{\omega_2}t}\mathtt{y}\big)S\big(-aq\mathrm{e}^{-\frac{2\pi}{\omega_2}t'}\,\mathtt{y}\big)S\big(-bq\mathrm{e}^{-\frac{2\pi}{\omega_2}t'}\,\mathtt{Y}\big) \\ &\quad \times S^{-1}\big(-bq\mathrm{e}^{-\frac{2\pi}{\omega_2}t'}\,\mathtt{Y}\big)S^{-1}(b\mathtt{y})S\big(-aq\mathrm{e}^{-\frac{2\pi}{\omega_2}t}\mathtt{y}\big)S\big(aq^2\mathrm{e}^{-\frac{2\pi}{\omega_2}(t+t')}\mathtt{Y}\big), \\ \mathtt{C}_R &= S^{-1}\big(-bq\mathrm{e}^{-\frac{2\pi}{\omega_2}t}\mathtt{Y}\big)S^{-1}(b\mathtt{y})S\big(-aq\mathrm{e}^{-\frac{2\pi}{\omega_2}t'}\,\mathtt{y}\big)S\big(aq^2\mathrm{e}^{-\frac{2\pi}{\omega_2}(t+t')}\mathtt{Y}\big). \end{aligned}$$

At this stage, it becomes evident that $\mathtt{C}_R = \mathtt{C}_L$. Exactly the same techniques applies so as to show that $\mathbb{R}_{V_u V_v}(t, t')$ satisfies to the Yang–Baxter equation. An alternative proof of a similar identity using only the pentagonal identity can be found in [18].

4. The $\mathtt{Q}$ Operator and the Baxter Equation

4.1. *Definition and basic properties of the* $\mathtt{Q}$ *operator*

In exact parallel to the transfer matrix

$$\mathtt{t}(\lambda) = \mathrm{Tr}_0\big[\mathrm{L}_{0V_{x_N}}(\lambda)\cdots \mathrm{L}_{0V_{x_1}}(\lambda)\big],$$

the Baxter $\mathtt{Q}$-operator is defined, *modulo* a gauge transformation, as a trace of $\mathbb{L}$ matrices

$$\mathtt{Q}(\lambda) = \mathbb{E}^{-1}_{\otimes V_{x_a}} \cdot \mathrm{Tr}_{V_u}\big[\mathbb{L}_{V_{x_N},V_u}(\lambda)\cdots\mathbb{L}_{V_{x_1},V_u}(\lambda)\big]\cdot\mathbb{E}_{\otimes V_{x_a}} \quad \text{with} \quad \mathbb{E}_{\otimes V_{x_a}} = \prod_{a=1}^{N} \mathrm{e}^{-\frac{3}{4}\Omega \mathtt{X}_a}.$$

One has the properties:

(i) $[\mathtt{t}(\lambda), \mathtt{t}(\lambda')] = 0$,
(ii) $[\mathtt{t}(\lambda), \mathtt{Q}(\lambda')] = 0$,
(iii) $[\mathtt{Q}(\lambda), \mathtt{Q}(\lambda')] = 0$.

Property (i) follows from

$$R_{00'}(\lambda,\lambda')\mathtt{L}_{0V_u}(\lambda)\mathtt{L}_{0'V_u}(\lambda') = \mathtt{L}_{0'V_u}(\lambda')\mathtt{L}_{0V_u}(\lambda)R_{00'}(\lambda,\lambda'),$$

property (ii) from the fact that $\big[\mathtt{t}(\lambda), \sum_{a=1}^{N}\mathtt{X}_a\big] = 0$ and the intertwining relation

$$\mathtt{M}_{0V_u}(\lambda;\lambda')\mathtt{L}_{0V_x}(\lambda)\mathbb{L}_{V_xV_u}(\lambda') = \mathbb{L}_{V_xV_u}(\lambda')\mathtt{L}_{0V_x}(\lambda)\mathtt{M}_{0V_u}(\lambda;\lambda'),$$

while property (iii) follows from

$$\mathbb{R}_{V_uV_v}(\lambda,\lambda')\mathbb{L}_{V_xV_v}(\lambda')\mathbb{L}_{V_xV_u}(\lambda) = \mathbb{L}_{V_xV_u}(\lambda)\mathbb{L}_{V_xV_v}(\lambda')\mathbb{R}_{V_uV_v}(\lambda,\lambda').$$

We now establish that the $\mathtt{Q}(\lambda)$ operator is realized concretely as an integral operator on $L^2(\mathbb{R}^N)$:

$$\big(\mathtt{Q}(\lambda)\cdot f\big)(\boldsymbol{x}) = \int_{\mathbb{R}^N} \mathcal{Q}_\lambda(\boldsymbol{x},\boldsymbol{y})f(\boldsymbol{y})\mathrm{d}\boldsymbol{y} \tag{4.1}$$

with $\boldsymbol{x} = (x_1,\ldots,x_N)$ and $\boldsymbol{y} = (y_1,\ldots,y_N)$.

Indeed, by doing a similarity transformation under the trace, one can recast $\mathtt{Q}(\lambda)$ as

$$\mathtt{Q}(\lambda) = \mathrm{Tr}_{V_u}\big[\mathbb{L}^{(\mathrm{c})}_{V_{x_N},V_u}(\lambda)\cdots\mathbb{L}^{(\mathrm{c})}_{V_{x_1},V_u}(\lambda)\big]$$

in which the operator $\mathbb{L}^{(\mathrm{c})}_{V_{x_N},V_u}(\lambda)$ has been introduced in (3.13). Then, for $f \in L^2(\mathbb{R}^N\times\mathbb{R})$, by using the kernel representation Eq. (3.16) one has

$$\begin{aligned}&\big(\mathbb{L}^{(\mathrm{c})}_{V_{x_N},V_u}(\lambda)\cdots\mathbb{L}^{(\mathrm{c})}_{V_{x_1},V_u}(\lambda)\cdot f\big)(\boldsymbol{x},u)\\ &\quad= \int_{\mathbb{R}^N}\mathrm{d}\boldsymbol{y}\int_{\mathbb{R}^N}\mathrm{d}\boldsymbol{v}\prod_{a=1}^{N}\mathbb{L}_\lambda\big(x_a,v_{a+1};y_a,v_a\big)f(\boldsymbol{y},v_1).\end{aligned} \tag{4.2}$$

Here, we agree upon $v_{N+1} = u$. The latter allows one to identify the integral kernel of the above operator and, upon taking the partial trace and agreeing upon periodic boundary conditions $v_{N+1} \equiv v_1$, one gets that

$$\mathcal{Q}_\lambda(\boldsymbol{x},\boldsymbol{y}) = \int_{\mathbb{R}^N}\mathrm{d}\boldsymbol{v}\prod_{a=1}^{N}\mathbb{L}_\lambda\big(x_a,v_{a+1};y_a,v_a\big) = \prod_{a=1}^{N}\mathcal{L}_\lambda(x_a,y_a;y_{a-1}). \tag{4.3}$$

The integrals have been done thanks to the delta functions and the kernel $\mathcal{L}_\lambda$ has been introduced in (3.23). Here, we agree upon the periodic boundary conditions in the variables y: $y_0 = y_N$. Note that $\mathcal{L}_\lambda(x_a, y_a; y_{a-1})$ captures all the dependence of the kernel on the variable x_j attached to the jth site. For further convenience, it is useful to introduce the functions w_a given by

$$w_a(\boldsymbol{x}, \boldsymbol{y}) = \mathcal{L}_\lambda(x_a, y_a; y_{a-1}). \tag{4.4}$$

Here $\boldsymbol{x}$, $\boldsymbol{y}$ are N-dimensional vectors as given below of (4.1).

4.2. *The Baxter equation*

We now derive the operator form of the Baxter equation:

$$\begin{aligned} &\mathtt{t}(\lambda)\mathtt{Q}(\lambda) \\ &\quad = \frac{\mathrm{e}^{-\omega_1 \mathtt{P}_{\mathrm{tot}}} \cdot \mathtt{Q}(\lambda - \mathrm{i}\omega_1)}{\big(-q^3 \mathrm{e}^{-\frac{2\pi}{\omega_2}\lambda}\big)^N} + \big(-q\mathrm{e}^{-\frac{2\pi}{\omega_2}\lambda}\big)^N \big(d_2 + q^{-1} d_1 \mathrm{e}^{-\frac{2\pi}{\omega_2}\lambda}\big)^N \mathtt{Q}(\lambda + \mathrm{i}\omega_1), \end{aligned} \tag{4.5}$$

where we remind that $\mathtt{P}_{\mathrm{tot}} = \sum_{a=1}^N \mathtt{X}_a$. The dual transfer matrix solves the dual equation

$$\begin{aligned} &\widetilde{\mathtt{t}}(\lambda)\mathtt{Q}(\lambda) \\ &\quad = \frac{\mathrm{e}^{-\omega_2 \mathtt{P}_{\mathrm{tot}}} \cdot \mathtt{Q}(\lambda - \mathrm{i}\omega_2)}{\big(-\widetilde{q}^3 \mathrm{e}^{-\frac{2\pi}{\omega_1}\lambda}\big)^N} + \big(-\widetilde{q}\mathrm{e}^{-\frac{2\pi}{\omega_1}\lambda}\big)^N \big(\widetilde{d}_2 + \widetilde{q}^{-1} \widetilde{d}_1 \mathrm{e}^{-\frac{2\pi}{\omega_1}\lambda}\big)^N \mathtt{Q}(\lambda + \mathrm{i}\omega_2). \end{aligned} \tag{4.6}$$

We only discuss the proof of Eq. (4.5) as the dual case follows, for instance, upon making the duality transformation on the level of Eq. (4.5) and using that $\mathtt{Q}(\lambda)$ is modular invariant.

One has

$$\big(\mathtt{t}(\lambda) \cdot Q_\lambda\big)(\boldsymbol{x}, \boldsymbol{y}) = \mathrm{Tr}_0\big[\big(\mathrm{L}_{0V_{x_N}}(\lambda) \cdot w_N\big) \cdots \big(\mathrm{L}_{0V_{x_1}}(\lambda) \cdot w_1\big)\big](\boldsymbol{x}, \boldsymbol{y})$$

where $\big(\mathrm{L}_{0j}(\lambda) \cdot w_j\big)(\boldsymbol{x}, \boldsymbol{y})$ is the below matrix function

$$\begin{aligned} &\big(\mathrm{L}_{0j}(\lambda) \cdot w_j\big)(\boldsymbol{x}, \boldsymbol{y}) \\ &= \begin{pmatrix} (\mathrm{e}^{-\frac{2\pi}{\omega_2}\lambda} - \mathrm{e}^{-\omega_1 \mathtt{X}_j})\mathcal{L}_\lambda(x_j, y_j; y_{j-1}) & q^2 \mathrm{e}^{-\frac{2\pi}{\omega_2}\lambda}(d_2 + q d_1 \mathrm{e}^{-\omega_1 \mathtt{X}_j})\mathrm{e}^{-\frac{2\pi}{\omega_2}x_j}\mathcal{L}_\lambda(x_j, y_j; y_{j-1}) \\ -q^{-2}\mathrm{e}^{\frac{2\pi}{\omega_2}x_j}\mathcal{L}_\lambda(x_j, y_j; y_{j-1}) & -d_2 \mathcal{L}_\lambda(x_j, y_j; y_{j-1}) \end{pmatrix}. \end{aligned}$$

In the 11-matrix element, we use

$$\begin{aligned} &\big(\mathrm{e}^{-\omega_1 \mathtt{X}_j}\mathcal{L}_\lambda\big)(x_j, y_j; y_{j-1}) \\ &\quad = \mathcal{L}_\lambda(x_j + \mathrm{i}\omega_1, y_j; y_{j-1}) \\ &\quad = q^{-1}\frac{(q\mathrm{e}^{-\frac{2\pi}{\omega_2}\lambda} + \mathrm{e}^{-\frac{2\pi}{\omega_2}(y_j - x_j)})(1 + q d_2 \mathrm{e}^{-\frac{2\pi}{\omega_2}(x_j - y_{j-1})})}{1 - d_1 \mathrm{e}^{-\frac{2\pi}{\omega_2}\lambda}\mathrm{e}^{-\frac{2\pi}{\omega_2}(x_j - y_{j-1})}}\mathcal{L}_\lambda(x_j, y_j; y_{j-1}) \end{aligned} \tag{4.7}$$

and in the 12-matrix element we use

$$\mathrm{e}^{-\omega_1 \mathrm{x}_j} \mathrm{e}^{-\frac{2\pi}{\omega_2} x_j} \mathcal{L}_\lambda(x_j, y_j; y_{j-1}) = q^{-2} \mathrm{e}^{-\frac{2\pi}{\omega_2} x_j} \mathcal{L}_\lambda(x_j + \mathrm{i}\omega_1, y_j; y_{j-1}).$$

The q-factors adapt themselves to ensure the triangularization property which is at the core of the Baxter–Bäcklund approach

$$\big(\mathtt{L}_{0j}(\lambda) \cdot w_j\big)(\boldsymbol{x}, \boldsymbol{y}) = \begin{pmatrix} 1 & 0 \\ q^{-1} \mathrm{e}^{\frac{2\pi}{\omega_2} y_j} & 1 \end{pmatrix} \begin{pmatrix} A_j & B_j \\ 0 & D_j \end{pmatrix} \begin{pmatrix} 1 & 0 \\ -q^{-1} \mathrm{e}^{\frac{2\pi}{\omega_2} y_{j-1}} & 1 \end{pmatrix}$$

where (compare with Eq. (A.7))

$$A_j = -q^{-1} \mathrm{e}^{-\frac{2\pi}{\omega_2}(y_j - x_j)} \big(1 + q d_2 \mathrm{e}^{-\frac{2\pi}{\omega_2}(x_j - y_{j-1})}\big) \mathcal{L}_\lambda(x_j, y_j; y_{j-1}),$$

$$D_j = -(d_2 + q^{-1} d_1 \mathrm{e}^{-\frac{2\pi}{\omega_2}\lambda}) \mathrm{e}^{-\frac{2\pi}{\omega_2}(x_j - y_j)} \frac{\big(q \mathrm{e}^{-\frac{2\pi}{\omega_2}\lambda} + \mathrm{e}^{-\frac{2\pi}{\omega_2}(y_j - x_j)}\big)}{1 - d_1 \mathrm{e}^{-\frac{2\pi}{\omega_2}\lambda} \mathrm{e}^{-\frac{2\pi}{\omega_2}(x_j - y_{j-1})}} \mathcal{L}_\lambda(x_j, y_j; y_{j-1}).$$

Hence,

$$\big(\mathtt{t}(\lambda) \cdot Q_\lambda\big)(\boldsymbol{x}, \boldsymbol{y}) = \prod_{j=1}^{N} A_j + \prod_{j=1}^{N} D_j. \tag{4.8}$$

It is easy to see using the explicit formula, Eq. (3.23), that

$$\begin{aligned} A_j &= -q^{-3} \mathrm{e}^{\frac{2\pi}{\omega_2}\lambda} \mathcal{L}_{\lambda - \mathrm{i}\omega_1}(x_j + \mathrm{i}\omega_1, y_j; y_{j-1}), \\ D_j &= -q \mathrm{e}^{-\frac{2\pi}{\omega_2}\lambda} \big(d_2 + q^{-1} d_1 \mathrm{e}^{-\frac{2\pi}{\omega_2}\lambda}\big) \mathcal{L}_{\lambda + \mathrm{i}\omega_1}(x_j, y_j; y_{j-1}). \end{aligned}$$

Inserting this back into Eq. (4.8), upon using $\boldsymbol{e} = (1, \ldots, 1) \in \mathbb{R}^N$, we obtain that

$$\begin{aligned} \big(\mathtt{t}(\lambda) \cdot Q_\lambda\big)(\boldsymbol{x}, \boldsymbol{y}) &= \big(-q^3 \mathrm{e}^{-\frac{2\pi}{\omega_2}\lambda}\big)^{-N} Q_{\lambda - \mathrm{i}\omega_1}(\boldsymbol{x} + \mathrm{i}\omega_1 \boldsymbol{e}, \boldsymbol{y}) \\ &\quad + \big(-q \mathrm{e}^{-\frac{2\pi}{\omega_2}\lambda}\big)^N \big(d_2 + q^{-1} d_1 \mathrm{e}^{-\frac{2\pi}{\omega_2}\lambda}\big)^N Q_{\lambda + \mathrm{i}\omega_1}(\boldsymbol{x}, \boldsymbol{y}). \end{aligned} \tag{4.9}$$

This yields the operator $\mathtt{t} - \mathtt{Q}$ equation given in (4.5). This form of the equation is somewhat impractical in that it involves the operator $\mathtt{P}_{\mathrm{tot}}$ which has a purely continuous spectrum. Due to translation invariance, the same holds for $\mathtt{t}(\lambda)$ and $\mathtt{Q}(\lambda)$. In order to factor out the continuous part of the spectrum and work with a spectral problem associated with the pointwise part of the spectrum, one considers the unitary map $\mathcal{F}$: $L^2(\mathbb{R}^{N-1} \times \mathbb{R}) \mapsto L^2(\mathbb{R}^N)$ such that

$$\mathcal{F}[\varphi](\boldsymbol{x}) = \int_{\mathbb{R}} \frac{\mathrm{d}p}{2\pi} \mathrm{e}^{\mathrm{i}p x_N} \varphi(x_{1N}, \ldots, x_{N-1N}; p) \tag{4.10}$$

for functions belonging to a suitable dense subset in $L^2(\mathbb{R}^{N-1} \times \mathbb{R})$ and where we agree upon $x_{ab} = x_a - x_b$. Then, on an appropriate dense subspace, one has

$$\mathrm{e}^{-\omega_1 \mathrm{P}_{\mathrm{tot}}} \cdot \mathcal{F}[\varphi](\boldsymbol{x}) = \int_{\mathbb{R}} \frac{\mathrm{d}p}{2\pi} \mathrm{e}^{\mathrm{i}px_N} \mathrm{e}^{-\omega_1 p}\, \varphi(x_{1N}, \ldots, x_{N-1N}; p). \tag{4.11}$$

Likewise, it holds

$$\mathtt{Q}(\lambda) \cdot \mathcal{F}[\varphi](\boldsymbol{x}) = \int_{\mathbb{R}} \frac{\mathrm{d}p}{2\pi} \int_{\mathbb{R}^N} \mathrm{d}\boldsymbol{y} Q_\lambda(\boldsymbol{x}, \boldsymbol{y}) \mathrm{e}^{\mathrm{i}py_N} \varphi(y_{1N}, \ldots, y_{N-1N}; p). \tag{4.12}$$

Changing the variables to $\boldsymbol{y} \mapsto \boldsymbol{z} + (\boldsymbol{0}, y)$ where $\boldsymbol{z} = (y_{1N}, \ldots, y_{N-1N}, 0)$ $y = y_N$, and setting $\boldsymbol{x} = \boldsymbol{u} + x_N \boldsymbol{e}$ where $\boldsymbol{u} = (x_{1N}, \ldots, x_{N-1N}, 0)$ and $\boldsymbol{e} = (1, \ldots, 1) \in \mathbb{R}^N$, one gets

$$\mathtt{Q}(\lambda) \cdot \mathcal{F}[\varphi](\boldsymbol{x}) = \int_{\mathbb{R}} \frac{\mathrm{d}p}{2\pi} \int_{\mathbb{R}^{N-1}} \mathrm{d}\boldsymbol{z} \int_{\mathbb{R}} \mathrm{d}y\, Q_\lambda(\boldsymbol{u} + x_N \boldsymbol{e}, \boldsymbol{z} + y\boldsymbol{e}) \mathrm{e}^{\mathrm{i}py}\, \varphi(z_1, \ldots, z_{N-1}; p). \tag{4.13}$$

Upon using the translation invariance of $\mathtt{Q}$'s kernel $Q_\lambda(\boldsymbol{x} + r\boldsymbol{e}, \boldsymbol{y} + r\boldsymbol{e}) = Q_\lambda(\boldsymbol{x}, \boldsymbol{y})$ and changing variables, one gets

$$\mathtt{Q}(\lambda) \cdot \mathcal{F}[\varphi](\boldsymbol{x}) = \int_{\mathbb{R}} \frac{\mathrm{d}p}{2\pi} \mathrm{e}^{\mathrm{i}px_N} \int_{\mathbb{R}^{N-1}} \mathrm{d}\boldsymbol{z}\, \overline{Q}_\lambda(\boldsymbol{u}, \boldsymbol{z}; p)\, \varphi(z_1, \ldots, z_{N-1}; p) \tag{4.14}$$

where

$$\overline{Q}_\lambda(\boldsymbol{u}, \boldsymbol{z}; p) = \int_{\mathbb{R}} \mathrm{d}y\, Q_\lambda(\boldsymbol{u}, \boldsymbol{z} + y\boldsymbol{e}) \mathrm{e}^{\mathrm{i}py}. \tag{4.15}$$

Likewise, due to translation invariance, $\mathtt{t}$ passes through the action of $\mathcal{F}$, namely

$$\mathtt{t}(\lambda) \cdot \mathcal{F}[\varphi](\boldsymbol{x}) = \int_{\mathbb{R}} \frac{\mathrm{d}p}{2\pi} \mathrm{e}^{\mathrm{i}px_N} \big(\overline{\mathtt{t}}(\lambda; p) \cdot \varphi\big)(\boldsymbol{u}; p), \tag{4.16}$$

with $\boldsymbol{u}$ defined as above. Above, $\overline{\mathtt{t}}(\lambda; p)$ is an operator on functions depending on the reduced set of variables and is a multiplication operator in p. Thence, by projecting out, one gets a reduced $\mathtt{t} - \mathtt{Q}$ equation on a sector with a fixed continuous eigenvalue p_0 of $\mathrm{P}_{\mathrm{tot}}$:

$$\overline{\mathtt{t}}(\lambda; p_0) \overline{\mathtt{Q}}(\lambda; p_0) = \frac{\mathrm{e}^{-\omega_1 p_0} \cdot \overline{\mathtt{Q}}(\lambda - \mathrm{i}\omega_1; p_0)}{\big(-q^3 \mathrm{e}^{-\frac{2\pi}{\omega_2}\lambda}\big)^N} + \big(-q\mathrm{e}^{-\frac{2\pi}{\omega_2}\lambda}\big)^N \big(d_2 + q^{-1} d_1 \mathrm{e}^{-\frac{2\pi}{\omega_2}\lambda}\big)^N \overline{\mathtt{Q}}(\lambda + \mathrm{i}\omega_1; p_0). \tag{4.17}$$

Clearly, the dual reduced equation holds as well. It seems already reasonable to assume that the reduced operators $\overline{\mathtt{t}}(\lambda; p_0)$ and $\overline{\mathtt{Q}}(\lambda; p_0)$ have, for fixed λ, a point spectrum. We will make this assumption in the following and leave its proof to some subsequent work.

4.3. *Analytic properties of the solution of Baxter equation*

We now argue that for both limiting cases, the q-Toda and the Toda$_2$ chains, any Eigenvalue $\mathfrak{q}(\lambda)$ of the reduced operator $\overline{\mathsf{Q}}(\lambda;p)$ is an entire function of the spectral parameter. However, for generic values of the coupling constants d_1 and d_2, q is a meromorphic function of the spectral parameter and has Nth order poles on the lattice

$$\kappa_2 - \kappa_1 + \mathrm{i}\frac{\Omega}{2} + \mathrm{i}\mathbb{N}\omega_1 + \mathrm{i}\mathbb{N}\omega_2.$$

Thus, the two limiting cases seem to be very special in respect to their analytic structure. The pattern of poles can be argued by following the reasoning developed by Bytsko–Teschner [4]. Let $\boldsymbol{z} \mapsto \Psi_{\mathfrak{q}}(\boldsymbol{z};p)$ be an Eigenfunction of $\overline{\mathsf{Q}}(\lambda;p)$ associated with the Eigenvalue $\mathfrak{q}(\lambda)$ and let φ be a test function. The pointwiseness of the spectrum on the reduced space ensures that $\Psi_{\mathfrak{q}}(\boldsymbol{z};p)$ decays fast enough at infinity. Although a precise estimate of this decay would demand additional work, for the purpose of the handlings below, we simply assume that it is fast enough for our needs and leave the study of this question for later investigation. Then, it holds,

$$\mathfrak{q}(\lambda)\cdot(\varphi,\Psi_{\mathfrak{q}}) = \int_{\mathbb{R}^{N-1}} \mathrm{d}\boldsymbol{u} \int_{\mathbb{R}^{N-1}} \mathrm{d}\boldsymbol{z} \int_{\mathbb{R}} \mathrm{d}v \mathrm{e}^{\mathrm{i}pv} Q_\lambda(\boldsymbol{u}_N, \boldsymbol{z}_N + v\boldsymbol{e})\varphi^*(\boldsymbol{u})\Psi_{\mathfrak{q}}(\boldsymbol{z};p). \quad (4.18)$$

Here $(\cdot,\cdot)$ stands for the canonical $L^2\big(\mathbb{R}^{N-1}\big)$ scalar product, $\boldsymbol{z}_N$ and $\boldsymbol{u}_N$ stand for the canonical embeddings of $\boldsymbol{z}, \boldsymbol{u} \in \mathbb{R}^{N-1}$ into $\mathbb{R}^N$: $\boldsymbol{z}_N = (\boldsymbol{z}, 0)$, $\boldsymbol{u}_N = (\boldsymbol{u}, 0)$. Also, we remind that $\boldsymbol{e} = (1, \ldots, 1) \in \mathbb{R}^N$. Upon a change of variables, we get

$$\mathfrak{q}(\lambda)\cdot(\varphi,\Psi_{\mathfrak{q}}) = \int_{\mathbb{R}^{N-1}} \mathrm{d}\boldsymbol{u} \int_{\mathbb{R}^{N}} \mathrm{d}\boldsymbol{w} \mathrm{e}^{\mathrm{i}pw_N} Q_\lambda(\boldsymbol{u}_N, \boldsymbol{w})\varphi^*(\boldsymbol{u}_{w_N})\Psi_{\mathfrak{q}}(\boldsymbol{z};p) \quad (4.19)$$

where $\boldsymbol{u}_{w_N} = \big(u_1 + w_N, \ldots, u_{N-1} + w_N\big)$.

Since $\Psi_{\mathfrak{q}}(\boldsymbol{z};p)$ is independent of λ and owing to the good decay properties at ∞ of the test function φ and the Eigenfunction $\Psi_{\mathfrak{q}}$, the sole mechanism that can give rise to a pole of $\mathfrak{q}(\lambda)$ is when the integration contour gets pinched between two poles of the integral kernel of the Q-operator, one coming from the upper and the other from the lower half plane. Agreeing below on the convention $u_N = 0$, the integral kernel $Q_\lambda(\boldsymbol{u}_N, \boldsymbol{w})$, as follows from inspection of Eqs. (4.3), (3.23), has poles at

$$u_a - w_{a-1} + \kappa_2 - \frac{3\mathrm{i}\Omega}{2} = -\mathrm{i}m\omega_1 - \mathrm{i}n\omega_2, \quad m,n \geq 1,$$

$$w_a - w_{a-1} + \kappa_1 - \frac{\mathrm{i}\Omega}{2} = -\mathrm{i}m\omega_1 - \mathrm{i}n\omega_2, \quad m,n \geq 1,$$

$$u_a - w_{a-1} + \kappa_1 + \lambda - \mathrm{i}\Omega = \mathrm{i}m'\omega_1 + \mathrm{i}n'\omega_2, \quad m',n' \geq 0,$$

$$w_a - u_a - \lambda + \frac{\mathrm{i}\Omega}{2} = \mathrm{i}m'\omega_1 + \mathrm{i}n'\omega_2, \quad m',n' \geq 0.$$

The first two sets of poles are located solely in the upper half-plane, while the third one is only located in the lower one, just as most of the poles belonging to the last set. As can be inferred from doing contour deformation or through computing explicitly the boundary values, the pinching in λ which generates effective poles will only occur at those values of λ which are independent of $\boldsymbol{u}, \boldsymbol{w}$. Thus, this can happen only when the closest to the real axis pole of the first set pinches with some pole present in the third set of pole, namely when

$$\lambda = \kappa_2 - \kappa_1 + \frac{\mathrm{i}}{2}\Omega + \mathrm{i}m''\omega_1 + \mathrm{i}n''\omega_2, \quad m'', n'' \geq 0.$$

Since there are N factors in the kernel giving rise to such poles, all-in-all, this gives rise to an Nth order pole.

Note that, in the q-Toda or Toda$_2$ cases one of the two building blocks of the kernel which are both responsible for the generation of these poles is absent (either $\mathcal{S}(x_j - y_{j-1} + \kappa_2 - \frac{3\mathrm{i}}{2}\Omega)$ or $\mathcal{S}(x_j - y_{j-1} + \kappa_1 + \lambda - \mathrm{i}\Omega)$) so that no pinching can arise. This ensures that $\mathfrak{q}(\lambda)$ is an *entire* function of λ in these two cases.

4.4. *Quantization of the spectrum by means of a scalar* $t - \mathtt{Q}$ *equation*

The hypothesis of a pointwise spectrum for the reduced operators $\overline{\mathtt{t}}(\lambda; p_0)$ and $\overline{\mathtt{Q}}(\lambda; p_0)$ and their commutativity allows one to project the reduced $\mathtt{t} - \mathtt{Q}$ equation (4.17) onto a given joint Eigenvector of $\overline{\mathtt{t}}(\lambda; p_0)$ and $\overline{\mathtt{Q}}(\lambda; p_0)$ associated with the respective eigenvalues $t(\lambda)$ and $\mathfrak{q}(\lambda)$. Here, for the sake of compactness of notations, we drop the p_0 dependence of these eigenvalues since p_0 appears explicitly in the equation. This scalar $t - q$ equation reads:

$$\begin{aligned} t(\lambda)\mathfrak{q}(\lambda) = {} & (-q^3)^{-N} \mathrm{e}^{\frac{2\pi}{\omega_2}N\lambda} \mathrm{e}^{-\omega_1 p_0} \mathfrak{q}(\lambda - \mathrm{i}\omega_1) \\ & + (-q)^N \mathrm{e}^{-\frac{2\pi}{\omega_2}N\lambda} (d_2 + q^{-1} d_1 \mathrm{e}^{-\frac{2\pi}{\omega_2}\lambda})^N \mathfrak{q}(\lambda + \mathrm{i}\omega_1). \end{aligned} \tag{4.20}$$

Note that the *same* $\mathfrak{q}$ satisfies as well the dual $t - q$ equation. We now specialize this equation to the q-Toda and Toda$_2$ cases. In both cases, we recast the equation in a canonical form what will allow us to simplify the analysis to come. Finally, note that the dual results holds upon making a modular transformation.

- The q-Toda case ($d_2 = 0$)
  The Baxter equation then takes the form

$$t(\lambda)\mathfrak{q}(\lambda) = (-1)^N q^{-3N} \mathrm{e}^{\frac{2\pi}{\omega_2}N\lambda} \mathrm{e}^{-\omega_1 p_0} \mathfrak{q}(\lambda - \mathrm{i}\omega_1) + (-1)^N d_1^N \mathrm{e}^{-\frac{4\pi}{\omega_2}N\lambda} \mathfrak{q}(\lambda + \mathrm{i}\omega_1).$$

  Recall that in the case of interest, $t(\lambda)$ is of the form

$$t(\lambda) = \left[\mathrm{e}^{-\frac{2\pi}{\omega_2}N\lambda} + \cdots + (-1)^N \mathrm{e}^{-\omega_1 p_0}\right].$$

Upon multiplying by $\mathrm{e}^{\frac{\omega_1}{2}p_0}\mathrm{e}^{\frac{N\pi}{\omega_2}\lambda}$, one gets a more convenient factorization

$$\begin{aligned}\mathrm{e}^{\frac{\omega_1}{2}p_0}\mathrm{e}^{\frac{N\pi}{\omega_2}\lambda}\,t(\lambda) &= \left[\mathrm{e}^{-\frac{N\pi}{\omega_2}\lambda}\mathrm{e}^{\frac{\omega_1}{2}p_0}+\cdots+(-1)^N\mathrm{e}^{\frac{N\pi}{\omega_2}\lambda}\mathrm{e}^{-\frac{\omega_1}{2}p_0}\right]\\ &= (-1)^N\prod_{k=1}^{N}\left\{2\sinh\frac{\pi}{\omega_2}(\lambda-\tau_k)\right\}\end{aligned}$$

where the τ_k are subjected to the constraint

$$\prod_{k=1}^{N}\mathrm{e}^{-\frac{\pi}{\omega_2}\tau_k}=\mathrm{e}^{-\frac{1}{2}\omega_1 p_0}, \tag{4.21}$$

as can be inferred from (2.7). This handling transforms the Baxter equation into

$$\begin{aligned}\prod_{k=1}^{N}\left\{2\sinh\frac{\pi}{\omega_2}(\lambda-\tau_k)\right\}\cdot\mathfrak{q}(\lambda) &= (q^3)^{-N}\mathrm{e}^{\frac{3\pi}{\omega_2}N\lambda}\mathrm{e}^{-\frac{\omega_1}{2}p_0}\mathfrak{q}(\lambda-\mathrm{i}\omega_1)\\ &\quad+d_1^N\mathrm{e}^{-\frac{3\pi}{\omega_2}N\lambda}\mathrm{e}^{\frac{\omega_1}{2}p_0}\mathfrak{q}(\lambda+\mathrm{i}\omega_1).\end{aligned} \tag{4.22}$$

In order to put this equation in a canonical form, it is convenient to make a change of unknown function

$$\mathfrak{q}(\lambda)=\mathrm{e}^{-\frac{3\mathrm{i}\pi N}{2\omega_1\omega_2}\lambda^2+\left(-\frac{3\pi N}{2\omega_1\omega_2}\Omega+\frac{\mathrm{i}}{2}p_0-\frac{\mathrm{i}\pi N}{\omega_1\omega_2}\kappa_1\right)\lambda}q(\lambda).$$

This recasts the Baxter equation Eq. (4.22) as

$$\prod_{k=1}^{N}\left\{2\sinh\frac{\pi}{\omega_2}(\lambda-\tau_k)\right\}q(\lambda)=\mathrm{e}^{-\frac{\pi\kappa_1}{\omega_2}N}\left((-\mathrm{i})^N q(\lambda-\mathrm{i}\omega_1)+\mathrm{i}^N q(\lambda+\mathrm{i}\omega_1)\right). \tag{4.23}$$

- The Toda$_2$ case ($d_1 - 0$)
  The scalar Baxter equation takes the form

$$\begin{aligned}&t(\lambda)\mathfrak{q}(\lambda)\\ &\quad=(-q^3)^{-N}\mathrm{e}^{\frac{2N\pi}{\omega_2}\lambda}\mathrm{e}^{-\omega_1 p_0}\mathfrak{q}(\lambda-\mathrm{i}\omega_1)+(-q)^N\mathrm{e}^{-\frac{2N\pi}{\omega_2}\lambda}d_2^N\mathfrak{q}(\lambda+\mathrm{i}\omega_1).\end{aligned} \tag{4.24}$$

Now $t(\lambda)$ is of the form

$$t(\lambda)=\left[\mathrm{e}^{-\frac{2\pi N}{\omega_2}\lambda}+\cdots+(-1)^N(\mathrm{e}^{-\omega_1 p_0}+d_2^N)\right]=\prod_{k=1}^{N}\left\{\mathrm{e}^{-\frac{2\pi}{\omega_2}\lambda}-\mathrm{e}^{-\frac{2\pi}{\omega_2}\tau_k}\right\},$$

and the roots τ_k are now subjected to the constraint

$$\prod_{k=1}^{N}\mathrm{e}^{-\frac{2\pi}{\omega_2}\tau_k}=\mathrm{e}^{-\omega_1 p_0}+d_2^N, \tag{4.25}$$

as can be inferred from (2.7). It is again convenient to make a change of unknown function

$$\mathsf{q}(\lambda) = \mathrm{e}^{-\frac{\mathrm{i}\pi N}{\omega_1\omega_2}\lambda^2 + \left(-\frac{2\pi N}{\omega_1\omega_2}\Omega - \frac{\mathrm{i}\pi N}{\omega_1\omega_2}\kappa_2\right)\lambda} q(\lambda). \tag{4.26}$$

This puts the Baxter equation (4.24) in the form

$$\prod_{k=1}^{N}\{\mathrm{e}^{-\frac{2\pi}{\omega_2}\lambda} - \mathrm{e}^{-\frac{2\pi}{\omega_2}\tau_k}\}q(\lambda)$$
$$= (-1)^N \mathrm{e}^{-\frac{\pi}{\omega_2}N\kappa_2}\mathrm{e}^{-\frac{\omega_1}{2}p_0}\left(\mathrm{e}^{-\omega_1 p_0}q(\lambda - \mathrm{i}\omega_1) + q(\lambda + \mathrm{i}\omega_1)\right). \tag{4.27}$$

The Baxter equations for both models can thus be put in the canonical form

$$t_{\boldsymbol{\tau}}(\lambda)q(\lambda) = g^{N\omega_1}\left(\sigma\,\varkappa^{\omega_1}q(\lambda - \mathrm{i}\omega_1) + \sigma^{-1}q(\lambda + \mathrm{i}\omega_1)\right), \tag{4.28}$$

where

- q-Toda

$$t_{\boldsymbol{\tau}}(\lambda) = \prod_{k=1}^{N}\left\{2\sinh\frac{\pi}{\omega_2}(\lambda - \tau_k)\right\}, \quad \sigma = (-\mathrm{i})^N, \quad g = \mathrm{e}^{-\frac{\pi\kappa_1}{\omega_1\omega_2}}, \quad \varkappa = 1; \tag{4.29}$$

- Toda$_2$

$$t_{\boldsymbol{\tau}}(\lambda) = \prod_{k=1}^{N}\{\mathrm{e}^{-\frac{2\pi}{\omega_2}\lambda} - \mathrm{e}^{-\frac{2\pi}{\omega_2}\tau_k}\}, \quad \sigma = (-1)^N, \quad g = \mathrm{e}^{-\frac{\pi\kappa_2}{\omega_1\omega_2}}, \quad \varkappa = \mathrm{e}^{-p_0}. \tag{4.30}$$

The main difference between the q-Toda and Toda$_2$ chains is that, in the former model, $\varkappa$ depends explicitly on the zero mode p_0 and the transfer matrix eigenvalue polynomial only grows in one direction $\Re(\lambda/\omega_2) \to -\infty$. For further applications, it will be important to study the regularity properties in p_0 of the solution q to the $t - q$ equations governing the spectrum of the Toda$_2$ chain. We leave this to a subsequent publication.

5. Conclusion

In this paper, we have constructed the Baxter operator $\mathtt{Q}(\lambda)$ for the q-Toda and Toda$_2$ chains, two q-deformations of the Toda chain. We used, as a starting point, the relation, in the classical limit, between Baxter operator and Bäcklund transformations. However, in the quantum case, we used Faddeev's modular invariance as a guideline to properly define the fully quantum operator $\mathtt{Q}(\lambda)$. We then derived Baxter $t-q$ equation and showed, using our operator $\mathtt{Q}(\lambda)$, that its solutions should be requested to be *entire* functions of λ. This last property paves the way to the quantization conditions yielding the spectrum of the q-Toda and Toda$_2$ chains. This matter will be investigated in a forthcoming publication [1].

Acknowledgment

K.K.K. acknowledges support from CNRS and ENS de Lyon. The authors are indebted to R. Kashaev, J. Teschner and G. Niccoli for stimulating discussions at various stages of the project.

Appendix A. Bäcklund Transformation

By taking the $q \to 1$ limit of Eq. (2.1), one gets the classical Lax matrix

$$L(\lambda \,|\, \hat{x}_n, \hat{X}_n) = \begin{pmatrix} \lambda - \hat{X}_n & \lambda[d_2 + d_1 \hat{X}_n]\hat{x}_n \\ -\hat{x}_n^{-1} & -d_2 \end{pmatrix} \tag{A.1}$$

which is expressed in terms of exponents

$$\hat{X}_n = \mathrm{e}^{X_n}, \quad \hat{x}_n = \mathrm{e}^{x_n}$$

of Darboux canonical coordinates $\{x_n, X_m\} = \delta_{nm}$, in which $\{*, *\}$ is a Poisson bracket. Note that the notation is consistent since all the dependence on n of the classical Lax matrix is contained in the Darboux coordinates x_n, X_n or, rather, its exponentiated counterparts $\hat{x}_n, \hat{X}_n$. To construct Baxter's $\mathtt{Q}$ operator, we will use its relation to Bäcklund transformations. The main observation of [13] is that Bäcklund transformations are related to the triangulation of the matrix $L(\lambda \,|\, \hat{x}_n, \hat{X}_n)$ by a gauge transformation, just as Baxter constructed his $\mathtt{Q}$ operator.

Bäcklund transformations are canonical transformations $(x_n, X_n) \to (y_n, Y_n)$, $n = 1, \ldots, N$, that preserve the form of the Hamiltonians. This last property is achieved if the transformation acts on the Lax matrix by a gauge transformation i.e. there exist matrices $M(\lambda; t \,|\, \hat{u}_n, \hat{U}_n)$, depending on the dynamical variables $\hat{u}_n, \hat{U}_n$ and on a parameter t, such that

$$L(\lambda \,|\, \hat{x}_n, \hat{X}_n) M(\lambda; t \,|\, \hat{u}_n, \hat{U}_n) = M(\lambda; t \,|\, \hat{u}_{n+1}, \hat{U}_{n+1}) L(\lambda \,|\, \hat{y}_n, \hat{Y}_n). \tag{A.2}$$

A nice way to derive the Bäcklund transformation and the matrix $M(\lambda; t \,|\, \hat{u}_n, \hat{U}_n)$ was devised by Kuznetsov and Sklyanin [23]. It is explained below, but let us first state the result in the form

$$\hat{X}_n = \frac{(1 + d_1 \hat{x}_{n+1} \hat{x}_n^{-1})}{(1 + d_1 \hat{x}_n \hat{x}_{n-1}^{-1})} \frac{(\hat{t} + \hat{x}_n \hat{y}_n^{-1})(1 + d_2 \hat{y}_{n+1} \hat{x}_n^{-1})}{(1 - t d_1 \hat{y}_{n+1} \hat{x}_n^{-1})}, \tag{A.3}$$

$$\hat{Y}_n = \frac{(t + \hat{x}_n \hat{y}_n^{-1})(1 + d_2 \hat{y}_n \hat{x}_{n-1}^{-1})}{(1 - t d_1 \hat{y}_n \hat{x}_{n-1}^{-1})}. \tag{A.4}$$

Here $\hat{x}_n = \mathrm{e}^{x_n}$, $\hat{X}_n = \mathrm{e}^{X_n}$, etc., and t is the parameter of the Bäcklund transformation. This is a slight generalization of the results of [3] or a limit of the results of [23].

The matrix $M(\lambda; t \,|\, \hat{u}_n, \hat{U}_n)$ reads

$$M(\lambda; t \,|\, \hat{u}_n, \hat{U}_n) = \begin{pmatrix} \lambda - t\hat{U}_n & -\lambda(1 - \hat{U}_n)\hat{u}_n \\ -\hat{u}_n^{-1} & 1 \end{pmatrix}, \tag{A.5}$$

where the dynamical variables are expressed as

$$\hat{u}_{n+1} = \hat{x}_n, \qquad \hat{U}_n = \frac{(1 + d_1\hat{x}_n\hat{x}_{n-1}^{-1})(1 + d_2\,\hat{y}_n\hat{x}_{n-1}^{-1})}{1 - d_1 t\,\hat{y}_n\hat{x}_{n-1}^{-1}}.$$

One can check by direct calculation that indeed Eq. (A.2) holds.

The relation of (A.2) with the triangulation of the Lax matrix is as follows. Since

$$\det\left[M(\lambda; t\,|\,\hat{u}_n, \hat{U}_n)\right] = (\lambda - t)\hat{U}_n,$$

the matrix $M(t; t\,|\,\hat{u}_n, \hat{U}_n)$ is of rank one. Then, the kernel is

$$M(t; t\,|\,\hat{u}_n, \hat{U}_n)\begin{pmatrix} 1 \\ \hat{x}_{n-1}^{-1} \end{pmatrix} = 0.$$

Since the kernel is one-dimensional, Eq. (A.2) implies that

$$L(t\,|\,\hat{y}_n, \hat{Y}_n)\begin{pmatrix} 1 \\ \hat{x}_{n-1}^{-1} \end{pmatrix} \propto \begin{pmatrix} 1 \\ \hat{x}_n^{-1} \end{pmatrix}.$$

As a consequence, we have the triangulation property by a gauge transformation

$$\begin{pmatrix} 1 & 0 \\ -\hat{x}_n^{-1} & 1 \end{pmatrix} L_n(t\,|\,\hat{y}_n, \hat{Y}_n)\begin{pmatrix} 1 & 0 \\ \hat{x}_{n-1}^{-1} & 1 \end{pmatrix} = \begin{pmatrix} A_n & B_n \\ 0 & D_n \end{pmatrix}.$$

Straightforward algebra then yields

$$A_n = -\hat{x}_n\hat{y}_n^{-1}(1 + d_2\hat{y}_n\hat{x}_{n-1}^{-1}), \tag{A.6}$$

$$D_n = -(d_2 + d_1 t)\hat{y}_n\hat{x}_n^{-1}\frac{t + \hat{x}_n\hat{y}_n^{-1}}{1 - d_1 t\hat{y}_n\hat{x}_{n-1}^{-1}}. \tag{A.7}$$

We now recall the Kuztnetsov–Sklyanin construction leading to the non-intuitive formulae Eqs. (A.19), (A.20) for the Bäcklund transformation of our model.

One starts with the matrices $L(\lambda\,|\,\hat{x}_n, \hat{X}_n)$, $M(\lambda, t\,|\,\hat{u}_n, \hat{U}_n)$, $L(\lambda\,|\,\hat{y}_n, \hat{Y}_n)$ and $M(\lambda, t\,|\,\hat{v}_n, \hat{V}_n)$ all satisfying the Sklyanin bracket (symplectic orbits of dimension 2):

$$\{N_1(\lambda_1), N_2(\lambda_2)\} = [r_{12}(\lambda_1, \lambda_2), N_1(\lambda_1)N_2(\lambda_2)]. \tag{A.8}$$

$r_{12}(\lambda_1, \lambda_2)$ appearing above is the classical limit of the quantum R-matrix Eq. (2.6):

$$r_{12}(\lambda_1, \lambda_2) = \frac{1}{\lambda_2 - \lambda_1}\begin{pmatrix} 0 & 0 & 0 & 0 \\ 0 & \lambda_2 & -\lambda_1 & 0 \\ 0 & -\lambda_2 & \lambda_1 & 0 \\ 0 & 0 & 0 & 0 \end{pmatrix}. \tag{A.9}$$

Then the matrices $L(\lambda\,|\,\hat{x}_n, \hat{X}_n)M(\lambda, t\,|\,\hat{u}_n, \hat{U}_n)$ and $M(\lambda, t\,|\,\hat{v}_n, \hat{V}_n)L(\lambda\,|\,\hat{y}_n, \hat{Y}_n)$ also satisfy the bracket Eq. (A.8). The equation

$$L(\lambda\,|\,\hat{x}_n, \hat{X}_n)M(\lambda; \hat{u}_n, \hat{U}_n) = M(\lambda; \hat{v}_n, \hat{V}_n)L(\lambda\,|\,\hat{y}_n, \hat{Y}_n) \tag{A.10}$$

defines a symplectic transformation $(\hat{x}_n, \hat{X}_n, \hat{u}_n, \hat{U}_n) \to (\hat{y}_n, \hat{Y}_n, \hat{v}_n, \hat{V}_n)$ on a symplectic leaf of Sklyanin bracket. Imposing the constraints

$$\hat{v}_n = \hat{u}_{n+1}, \quad \hat{V}_n = \hat{U}_{n+1}$$

yields a solution of Eq. (A.2). The success of this approach relies on the proper choice of the elementary solutions of Eq. (A.8) which we now describe.

The most elementary L-matrices solutions of Eq. (A.8) depending on a single Weyl pair (viz. $\{\hat{x}, \hat{y}\} = \hat{y}\hat{x}$) are

$$L(\lambda, P) = \begin{pmatrix} \alpha & \beta\hat{x} \\ \gamma\lambda^{-1}\hat{x}^{-1}\hat{y} & \delta\hat{y} \end{pmatrix}, \quad P = \begin{pmatrix} \alpha & \beta \\ \gamma\lambda^{-1} & \delta \end{pmatrix}. \tag{A.11}$$

Here λ is the spectral parameter and the coefficients of P are otherwise arbitrary constants. From this elementary L-matrix, we can construct a more general family of L-matrices still depending on a single Weyl pair. Start with two independent Weyl pairs

$$\{\hat{x}_i, \hat{y}_j\} = \hat{y}_j\hat{x}_i\delta_{ij}, \quad i = 1, 2 \tag{A.12}$$

one has that the product of two elementary L-matrices multiplied by λ

$$L(\lambda, P_1, P_2) = \lambda \begin{pmatrix} \alpha_1 & \beta_1\hat{x}_1 \\ \gamma_1\lambda^{-1}\hat{x}_1^{-1}\hat{y}_1 & \delta_1\hat{y}_1 \end{pmatrix} \begin{pmatrix} \alpha_2 & \gamma_2\hat{x}_2 \\ \beta_2\lambda^{-1}\hat{x}_2^{-1}\hat{y}_2 & \delta_2\hat{y}_2 \end{pmatrix} \tag{A.13}$$

also satisfies the bracket Eq. (A.8). Now, this L-matrix is invariant under the group action

$$\hat{x}_1 \to s\hat{x}_1, \quad \hat{y}_1 \to s\hat{y}_1, \quad \hat{y}_2 \to s^{-1}\hat{y}_2, \quad \hat{x}_2 \to \hat{x}_2$$

which preserves the quadratic bracket Eq. (A.12). The generator of the group action is

$$h = \log \hat{h} \quad \text{with} \quad \hat{h} = \hat{x}_1\hat{y}_1^{-1}\hat{x}_2^{-1}.$$

Invariant functions are generated by $\hat{h}$ itself (which is set to 1), $\hat{x} = \hat{x}_2$ and $\hat{X} = \hat{y}_1\hat{y}_2$. The L-matrix being invariant, it can be expressed in terms of the three invariant functions $\hat{h}, \hat{x}, \hat{X}$ and hence lives on the reduced phase space where it still satisfies Eq. (A.8). Hence the Poisson bracket of the L-matrix depends only on the Poisson brackets of these three quantities. $\hat{h}$ being the generator of the group, it Poisson commutes with the invariant functions $\hat{x}$ and $\hat{X}$, and of course with itself. So it plays no role and can be set to a constant. Here, we choose $\hat{h} = 1$, meaning that $\hat{y}_1\hat{x}_2 = \hat{x}_1$.

Only the Poisson bracket of $\hat{x}$ and $\hat{X}$ matters and turns out to be

$$\{\hat{x}, \hat{X}\} = \hat{x}\hat{X}.$$

After the reduction, we can finally write

$$L(\lambda, P_1, P_2) = \begin{pmatrix} 1 & \\ & \hat{x}^{-1} \end{pmatrix} P_1 \begin{pmatrix} 1 & \\ & \hat{X} \end{pmatrix} P_2^t \begin{pmatrix} 1 & \\ & \hat{x} \end{pmatrix},$$

$$P_1 = \begin{pmatrix} \alpha_1\lambda & \beta_1 \\ \gamma_1 & \delta_1 \end{pmatrix}, \quad P_2 = \begin{pmatrix} \alpha_2 & \beta_2 \\ \gamma_2 & \delta_2\lambda \end{pmatrix}.$$

The transposition on P_2 was introduced for later convenience. The Lax matrix Eq. (A.1) can indeed be written in this factorized form:

$$L(\lambda\,|\,\hat{x},\hat{X}) = \begin{pmatrix}1 & \\ & \hat{x}^{-1}\end{pmatrix}\begin{pmatrix}\lambda & 1\\ -1 & 0\end{pmatrix}\begin{pmatrix}1 & \\ & \hat{X}\end{pmatrix}\begin{pmatrix}1 & d_2\\ -1 & d_1\lambda\end{pmatrix}\begin{pmatrix}1 & \\ & \hat{x}\end{pmatrix}.$$

We look for $M(\lambda,t\,|\,\hat{u},\hat{U})$ in this general set of factorized matrices

$$M(\lambda,t\,|\,\hat{u},\hat{U}) = \begin{pmatrix}1 & 0\\ 0 & \hat{u}^{-1}\end{pmatrix}\begin{pmatrix}\alpha_1\lambda & \beta_1\\ \gamma_1 & \delta_1\end{pmatrix}\begin{pmatrix}1 & 0\\ 0 & \hat{U}\end{pmatrix}\begin{pmatrix}\alpha_2 & \gamma_2\\ \beta_2 & \delta_2\lambda\end{pmatrix}\begin{pmatrix}1 & 0\\ 0 & \hat{u}\end{pmatrix}$$

and just replacing $(\hat{u},\hat{U}) \to (\hat{v},\hat{V})$ for $M(\lambda,t\,|\,\hat{v},\hat{V})$. Then, Eq. (A.10) can be solved for $(\hat{y}_n,\hat{Y}_n,\hat{v}_n,\hat{V}_n)$ as functions of $(\hat{x}_n,\hat{X}_n,\hat{u}_n,\hat{U}_n)$. These provide $4N$ equations relating the $8N$ variables

$$(\hat{x}_n,\hat{X}_n,\hat{u}_n,\hat{U}_n,\hat{y}_n,\hat{Y}_n,\hat{v}_n,\hat{V}_n). \tag{A.14}$$

In particular, we find

$$\hat{v}_n = -\hat{x}_n\frac{\beta_1\gamma_1\hat{u}_n + \gamma_1 d_2\delta_1\hat{x}_n - \alpha_1\delta_1\hat{u}_n\hat{X}_n + \gamma_1 d_1\delta_1\hat{x}_n\hat{X}_n}{\alpha_1(d_2\delta_1\hat{x}_n + \beta_1\hat{u}_n)}. \tag{A.15}$$

Next, we have to impose the constraints

$$\hat{v}_n = \hat{u}_{n+1}, \qquad \hat{V}_n = \hat{U}_{n+1}. \tag{A.16}$$

This adds $2N$ equations to our $4N$ equations. Hence we can solve everything in terms of the $2N$ variables, say $(\hat{x}_n,\hat{y}_n)$. The first condition in Eq. (A.16) combined with Eq. (A.15) in general leads to highly non local and untractable formulae. However, we remark that, if

$$\delta_1 = 0, \tag{A.17}$$

then, the equation simplifies drastically

$$\hat{u}_{n+1} = -\frac{\gamma_1}{\alpha_1}\hat{x}_n.$$

The latter entails that the rest of the equations also simplifies and we get

$$\hat{U}_n = -\frac{\alpha_1\gamma_2}{\beta_1\delta_2}\frac{(1+d_1\hat{x}_n\hat{x}_{n-1}^{-1})(1+d_2 s\,\hat{y}_n\hat{x}_{n-1}^{-1})}{(1-d_1 ts\,\hat{y}_n\hat{x}_{n-1}^{-1})}, \tag{A.18}$$

$$\hat{X}_n = \frac{(1+d_1\hat{x}_{n+1}\hat{x}_n^{-1})}{(1+d_1\hat{x}_n\hat{x}_{n-1}^{-1})}\frac{(t+s^{-1}\hat{x}_n\hat{y}_n^{-1})(1+d_2 s\,\hat{y}_{n+1}\hat{x}_n^{-1})}{(1-td_1 s\,\hat{y}_{n+1}\hat{x}_n^{-1})}, \tag{A.19}$$

$$\hat{Y}_n = \frac{(t+s^{-1}\hat{x}_n\hat{y}_n^{-1})(1+d_2 s\,\hat{y}_n\hat{x}_{n-1}^{-1})}{(1-td_1 s\,\hat{y}_n\hat{x}_{n-1}^{-1})}, \tag{A.20}$$

where

$$t = \frac{\gamma_2\beta_2}{\alpha_2\delta_2}, \qquad s = \frac{\alpha_1\alpha_2}{\gamma_1\gamma_2}.$$

The parameter s can be eliminated by a rescaling of the variables $\hat{y}_n$ which is indeed a symmetry of the theory, so we can set it equal to one. The Bäcklund transformation obtained in this way is derived from the generating function

$$\log \hat{X}_n = \hat{x}_n \frac{\partial}{\partial \hat{x}_n} F, \quad \log \hat{Y}_n = -\hat{y}_n \frac{\partial}{\partial \hat{y}_n} F$$

where F reads

$$F = \sum_n \int^{\frac{\hat{x}_n}{\hat{y}_n}} \frac{dx}{x} \log(t+x) - \int^{\frac{\hat{y}_{n+1}}{\hat{x}_n}} dx \log(1+d_2 x)$$

$$+ \int^{\frac{\hat{y}_{n+1}}{\hat{x}_n}} \frac{dx}{x} \log(1-d_1 t x) - \int^{\frac{\hat{x}_{n+1}}{\hat{x}_n}} \frac{dx}{x} \log(1+d_1 x).$$

The matrices M in (A.2) can be taken as

$$M(\lambda, t \,|\, \hat{u}_n, \hat{U}_n) = \begin{pmatrix} 1 & 0 \\ 0 & \hat{u}_n^{-1} \end{pmatrix} \begin{pmatrix} \lambda & 1 \\ -1 & 0 \end{pmatrix} \begin{pmatrix} 1 & 0 \\ 0 & \hat{U}_n \end{pmatrix} \begin{pmatrix} 1 & -1 \\ -t & \lambda \end{pmatrix} \begin{pmatrix} 1 & 0 \\ 0 & \hat{u}_n \end{pmatrix}$$

$$= \begin{pmatrix} \lambda - t\hat{U}_n & -\lambda(1-\hat{U}_n)\hat{u}_n \\ -\hat{u}_n^{-1} & 1 \end{pmatrix}. \tag{A.21}$$

At the quantum level, the Bäcklund canonical transformation is replaced by a similarity transformation

$$(y, Y, v, V) = \mathbb{L}^{-1}_{V_x V_u}(t)(x, X, u, U)\mathbb{L}_{V_x V_u}(t),$$

where now $(\mathrm{e}^{-\frac{2\pi}{\omega_2}x}, \mathrm{e}^{-\omega_1 X})$ and $(\mathrm{e}^{-\frac{2\pi}{\omega_2}u}, \mathrm{e}^{-\omega_1 U})$ are Weyl pairs:

$$\mathrm{e}^{-\frac{2\pi}{\omega_2}x}\mathrm{e}^{-\omega_1 X} = q^2 \mathrm{e}^{-\omega_1 X}\mathrm{e}^{-\frac{2\pi}{\omega_2}x}, \quad \mathrm{e}^{-\frac{2\pi}{\omega_2}u}\mathrm{e}^{-\omega_1 U} = q^2 \mathrm{e}^{-\omega_1 U}\mathrm{e}^{-\frac{2\pi}{\omega_2}u}.$$

Hence Eq. (A.2) becomes

$$\mathbb{L}_{V_x V_u}(t)\mathtt{L}_{0V_x}(\lambda)\mathtt{M}_{0V_u}(\lambda; t) = \mathtt{M}_{0V_u}(\lambda; t)\mathtt{L}_{0V_x}(\lambda)\mathbb{L}_{V_x V_u}(t) \tag{A.22}$$

where the operator $\mathbb{L}_{V_x V_u}(t)$ is independent of λ but depends on the Bäcklund parameter t, and $\mathtt{L}_{0V_x}(\lambda)$, respectively $\mathtt{M}_{0V_u}(\lambda; t)$, is the quantum deformation of the classical object $L(\lambda \,|\, \hat{x}, \hat{X})$, respectively $M(\lambda, t \,|\, \hat{u}, \hat{U})$.

Appendix B. Quantum Intertwiners

The purpose of this Appendix is to construct the intertwiner $\mathbb{L}_{V_x V_u}(t)$ arising in (3.3). Starting from the factorization insight issuing from the implementation of the Bäcklund transformations, we construct $\mathbb{L}_{V_x V_u}(t)$ by means of the representation theory of the symmetric group $\mathfrak{S}_4$ following the strategy devised by Derkachov in [5] where he considered the cases of the $\mathfrak{sl}(2,\mathbb{C})$ and $\mathfrak{sl}(3,\mathbb{C})$ quantum XXX magnet.

B.1. *The double sine function representation*

Consider an auxiliary Lax matrix depending on two matrices

$$P_i = \begin{pmatrix} \alpha_i & \beta_i \\ \gamma_i & \delta_i \end{pmatrix}$$

and given by

$$\mathtt{L}(P_1, P_2) = \begin{pmatrix} 1 & 0 \\ 0 & e^{\frac{2\pi}{\omega_2}\mathtt{x}} \end{pmatrix} P_1 \begin{pmatrix} 1 & 0 \\ 0 & e^{-\omega_1 \mathtt{X}} \end{pmatrix} P_2^t \begin{pmatrix} 1 & 0 \\ 0 & e^{-\frac{2\pi}{\omega_2}\mathtt{x}} \end{pmatrix}.$$

Here $(e^{-\frac{2\pi}{\omega_2}\mathtt{x}}, e^{-\omega_1\mathtt{X}})$ is a Weyl pair

$$e^{-\frac{2\pi}{\omega_2}\mathtt{x}} e^{-\omega_1\mathtt{X}} = q^2 e^{-\omega_1\mathtt{X}} e^{-\frac{2\pi}{\omega_2}\mathtt{x}}.$$

The matrices $\mathtt{L}_{0,V_x}(\lambda)$ and $\mathtt{M}_{0,V_u}(\lambda;t)$ appearing in Sec. 3.1 are of this form:

$$\mathtt{L}_{0,V_x}(\lambda) = \begin{pmatrix} 1 & 0 \\ 0 & e^{\frac{2\pi}{\omega_2}\mathtt{x}} \end{pmatrix} \begin{pmatrix} e^{-\frac{2\pi}{\omega_2}\lambda} & 1 \\ -q^{-2} & 0 \end{pmatrix} \begin{pmatrix} 1 & 0 \\ 0 & e^{-\omega_1\mathtt{X}} \end{pmatrix} \begin{pmatrix} 1 & q^2 d_2 \\ -1 & q^3 e^{-\frac{2\pi}{\omega_2}\lambda} d_1 \end{pmatrix} \begin{pmatrix} 1 & 0 \\ 0 & e^{-\frac{2\pi}{\omega_2}\mathtt{x}} \end{pmatrix},$$

$$\mathtt{M}_{0,V_u}(\lambda;t) = \begin{pmatrix} 1 & 0 \\ 0 & e^{\frac{2\pi}{\omega_2}\mathtt{u}} \end{pmatrix} \begin{pmatrix} e^{-\frac{2\pi}{\omega_2}\lambda} & 1 \\ -q^{-2} & 0 \end{pmatrix} \begin{pmatrix} 1 & 0 \\ 0 & e^{-\omega_1\mathtt{U}} \end{pmatrix} \begin{pmatrix} 1 & -q \\ -e^{-\frac{2\pi}{\omega_2}t} & q e^{-\frac{2\pi}{\omega_2}\lambda} \end{pmatrix} \begin{pmatrix} 1 & 0 \\ 0 & e^{-\frac{2\pi}{\omega_2}\mathtt{u}} \end{pmatrix}.$$

In order to build $\mathbb{L}_{V_x V_u}(t)$, it is thus enough to look for an intertwiner $\mathbb{L}(P_1, P_2, P_3, P_4)$ such that

$$\mathbb{L}(P_1,P_2,P_3,P_4)\mathtt{L}(P_1,P_2)\mathtt{L}'(P_3,P_4) = \mathtt{L}'(P_3,P_4)\mathtt{L}(P_1,P_2)\mathbb{L}(P_1,P_2,P_3,P_4). \tag{B.1}$$

Above, the $'$ indicates that the corresponding Lax matrix is defined in terms of a Weyl pair built up from independent operators $\mathtt{x}'$ and $\mathtt{X}'$:

$$(e^{-\frac{2\pi}{\omega_2}\mathtt{x}'}, e^{-\omega_1\mathtt{X}'}) \qquad \text{so that} \qquad e^{-\frac{2\pi}{\omega_2}\mathtt{x}'} e^{-\omega_1\mathtt{X}'} = q^2 e^{-\omega_1\mathtt{X}'} e^{-\frac{2\pi}{\omega_2}\mathtt{x}'}.$$

It is convenient to look for a solution to (B.1) in the form

$$\mathbb{L}(P_1,P_2,P_3,P_4) = \mathtt{P}_{xx'} \cdot \check{\mathbb{L}}(P_1,P_2,P_3,P_4) \tag{B.2}$$

where $\mathtt{P}_{xx'}$ is the permutation operator

$$\mathtt{P}_{xx'}\mathtt{x} = \mathtt{x}'\mathtt{P}_{xx'}, \quad \mathtt{P}_{xx'}\mathtt{X} = \mathtt{X}'\mathtt{P}_{xx'}, \quad \mathtt{P}_{xx'}^2 = \mathrm{id}.$$

This operator intertwines between the two irreducible representations of the Weyl algebra and its dual.

Then, $\check{\mathbb{L}}(P_1, P_2, P_3, P_4)$ solves the equation

$$\check{\mathbb{L}}(P_1,P_2,P_3,P_4)\mathtt{L}(P_1,P_2)\mathtt{L}'(P_3,P_4) = \mathtt{L}(P_3,P_4)\mathtt{L}'(P_1,P_2)\check{\mathbb{L}}(P_1,P_2,P_3,P_4). \quad (B.3)$$

The main idea for constructing $\check{\mathbb{L}}(P_1, P_2, P_3, P_4)$ is to first find quantum intertwiners of more elementary objects, namely the solutions $\Psi_{P_1P_2}$ to

$$\Psi_{P_1P_2}\mathtt{L}(P_1,P_2) = \mathtt{L}(P_2,P_1)\Psi_{P_1P_2} \quad (B.4)$$

and $\Phi_{P_2P_3}$ to

$$\Phi_{P_2P_3}\mathtt{L}(P_1,P_2)\mathtt{L}'(P_3,P_4) = \mathtt{L}(P_1,P_3)\mathtt{L}'(P_2,P_4)\Phi_{P_2P_3}. \quad (B.5)$$

It seems natural to look for the solution $\Psi_{P_1P_2}$ to (B.4) as a sole function of $\mathrm{e}^{-\omega_1\mathtt{X}}$, viz. $\Psi_{P_1P_2} = \check{\psi}_{P_1P_2}(\mathtt{X})$. Then, (B.4) reduces to

$$\begin{pmatrix} \check{\psi}_{P_1P_2}(\mathtt{X}) & 0 \\ 0 & \check{\psi}_{P_1P_2}\left(\mathtt{X} - \frac{2\mathrm{i}\pi}{\omega_2}\right) \end{pmatrix} P_1 \begin{pmatrix} 1 & 0 \\ 0 & \mathrm{e}^{-\omega_1\mathtt{X}} \end{pmatrix} P_2^t$$
$$= P_2 \begin{pmatrix} 1 & 0 \\ 0 & \mathrm{e}^{-\omega_1\mathtt{X}} \end{pmatrix} P_1^t \begin{pmatrix} \check{\psi}_{P_1P_2}(\mathtt{X}) & 0 \\ 0 & \check{\psi}_{P_1P_2}\left(\mathtt{X} - \frac{2\mathrm{i}\pi}{\omega_2}\right) \end{pmatrix} \quad (B.6)$$

i.e. the above matrix should be symmetric. Writing this condition yields the finite difference equation

$$\frac{\check{\psi}_{P_1P_2}\left(\mathtt{X} - \frac{2\mathrm{i}\pi}{\omega_2}\right)}{\check{\psi}_{P_1P_2}(\mathtt{X})} = \frac{\alpha_1\gamma_2 + \beta_1\delta_2\mathrm{e}^{-\omega_1\mathtt{X}}}{\gamma_1\alpha_2 + \delta_1\beta_2\mathrm{e}^{-\omega_1\mathtt{X}}} \quad (B.7)$$

along with it modular dual. The finite difference equation can be readily solved in terms of the double sine function $\mathcal{S}$, or equivalently, upon a slight change of parametrization, in terms of the functions S which are both discussed in Appendix C.2. This will be done later on.

Similarly, when solving Eq. (B.5) it seems natural to assume that $\Phi_{P_2P_3} = \check{\phi}_{P_2P_3}(\mathtt{x} - \mathtt{x}')$. The equation becomes

$$\begin{pmatrix} \check{\phi}_{P_2P_3}(\mathtt{x} - \mathtt{x}') & \\ & \check{\phi}_{P_2P_3}(\mathtt{x} - \mathtt{x}' - \mathrm{i}\omega_1) \end{pmatrix} P_2^t \begin{pmatrix} 1 & 0 \\ 0 & \mathrm{e}^{-\frac{2\pi}{\omega_2}(\mathtt{x}-\mathtt{x}')} \end{pmatrix} P_3$$
$$= P_3^t \begin{pmatrix} 1 & 0 \\ 0 & \mathrm{e}^{-\frac{2\pi}{\omega_2}(\mathtt{x}-\mathtt{x}')} \end{pmatrix} P_2 \begin{pmatrix} \check{\phi}_{P_2P_3}(\mathtt{x} - \mathtt{x}') & \\ & \check{\phi}_{P_2P_3}(\mathtt{x} - \mathtt{x}' - \mathrm{i}\omega_1) \end{pmatrix}$$

saying that the matrix is symmetric. Writing this condition explicitly yields

$$\frac{\check{\phi}_{P_2P_3}(\mathtt{x} - \mathtt{x}' - \mathrm{i}\omega_1)}{\check{\phi}_{P_2P_3}(\mathtt{x} - \mathtt{x}')} = \frac{\alpha_2\beta_3 + \gamma_2\delta_3\mathrm{e}^{-\frac{2\pi}{\omega_2}(\mathtt{x}-\mathtt{x}')}}{\beta_2\alpha_3 + \delta_2\gamma_3\mathrm{e}^{-\frac{2\pi}{\omega_2}(\mathtt{x}-\mathtt{x}')}}. \quad (B.8)$$

The two elementary operators $\check{\psi}$ and $\check{\phi}$ allow one to build the more complicated intertwiner $\check{\mathbb{L}}(P_1, P_2, P_3, P_4)$ as

$$\check{\mathbb{L}}(P_1, P_2, P_3, P_4) = \check{C}\Phi_{14}\Psi'_{24}\Psi_{13}\Phi_{23}. \tag{B.9}$$

Here $\check{C}$ is a constant yet to be fixed and we used the shorthand notation

$$\Phi_{ij} = \check{\phi}_{P_iP_j}(\mathtt{x} - \mathtt{x}'), \qquad \Psi_{ij} = \check{\psi}_{P_iP_j}(\mathtt{X}) \quad \text{and} \quad \Psi'_{ij} = \check{\psi}_{P_iP_j}(\mathtt{X}').$$

The proof of this identity goes through a successive exchange of the matrices P_i:

$$\begin{aligned}
\Phi_{23}\mathrm{L}(P_1, P_2)\mathrm{L}'(P_3, P_4) &= \mathrm{L}(P_1, P_3)\mathrm{L}'(P_2, P_4)\Phi_{23} \\
\Psi_{13}\Phi_{23}\mathrm{L}(P_1, P_2)\mathrm{L}'(P_3, P_4) &= \mathrm{L}(P_3, P_1)\mathrm{L}'(P_2, P_4)\Psi_{13}\Phi_{23} \\
\Psi'_{24}\Psi_{13}\Phi_{23}\mathrm{L}(P_1, P_2)\mathrm{L}'(P_3, P_4) &= \mathrm{L}(P_3, P_1)\mathrm{L}'(P_4, P_2)\Psi'_{24}\Psi_{13}\Phi_{23} \\
\Phi_{14}\Psi'_{24}\Psi_{13}\Phi_{23}\mathrm{L}(P_1, P_2)\mathrm{L}'(P_3, P_4) &= \mathrm{L}(P_3, P_4)\mathrm{L}'(P_1, P_2)\Phi_{14}\Psi'_{24}\Psi_{13}\Phi_{23}.
\end{aligned}$$

Having in mind the application of this analysis to the construction of the intertwiner $\mathbb{L}_{V_xV_u}(t)$ in (3.3), we can simplify the expressions of the building blocks of $\check{\mathbb{L}}(P_1, P_2, P_3, P_4)$ by specializing the parameters as

$$\begin{aligned}
&\alpha_1 = \alpha_3 = \mathrm{e}^{-\frac{2\pi}{\omega_2}\lambda}, \quad \beta_1 = \beta_3 = 1, \\
&\alpha_2 = \alpha_4 = 1, \quad \beta_2 = -1, \quad \beta_4 = -\mathrm{e}^{-\frac{2\pi}{\omega_2}t},
\end{aligned}$$

as well as

$$\begin{aligned}
&\gamma_1 = \gamma_3 = -q^{-2}, \quad \gamma_2 = q^2 d_2, \quad \gamma_4 = -q, \\
&\delta_1 = \delta_3 = 0, \quad \delta_2 = q^3 d_1 \mathrm{e}^{-\frac{2\pi}{\omega_2}\lambda}, \quad \delta_4 = q\mathrm{e}^{-\frac{2\pi}{\omega_2}\lambda}.
\end{aligned}$$

This puts the finite difference equations in the form

$$\frac{\check{\phi}_{P_1P_4}(x - \mathrm{i}\omega_1)}{\check{\phi}_{P_1P_4}(x)} = \mathrm{e}^{-\frac{2\pi}{\omega_2}(\lambda + t + \mathrm{i}\frac{\omega_2}{2})}\left(1 - \mathrm{e}^{\frac{2\pi}{\omega_2}(t - x - \mathrm{i}\frac{\Omega}{2})}\right) \tag{B.10}$$

$$\frac{\check{\phi}_{P_2P_3}(x - \mathrm{i}\omega_1)}{\check{\phi}_{P_2P_3}(x)} = \frac{\mathrm{e}^{\frac{2\pi}{\omega_2}(\lambda + \mathrm{i}\frac{\omega_2}{2})}}{1 - \mathrm{e}^{-\frac{2\pi}{\omega_2}(x + \kappa_1 - \mathrm{i}\frac{\Omega}{2})}} \tag{B.11}$$

$$\frac{\check{\psi}_{P_1P_3}\left(X - \frac{2\mathrm{i}\pi}{\omega_2}\right)}{\check{\psi}_{P_1P_3}(X)} = 1 \tag{B.12}$$

$$\frac{\check{\psi}_{P_2P_4}\left(X - \frac{2\mathrm{i}\pi}{\omega_2}\right)}{\check{\psi}_{P_2P_4}(X)} = \mathrm{e}^{\frac{2\pi}{\omega_2}(\kappa_2 - \mathrm{i}\frac{\Omega}{2})}\frac{1 - \mathrm{e}^{-\frac{2\pi}{\omega_2}(\frac{\omega_1\omega_2}{2\pi}X + \lambda + \mathrm{i}\frac{\omega_2}{2})}}{1 - \mathrm{e}^{-\frac{2\pi}{\omega_2}(\frac{\omega_1\omega_2}{2\pi}X + \lambda + t + \kappa_1 - \kappa_2 - \mathrm{i}\frac{\omega_1}{2})}} \tag{B.13}$$

along with the dual equations. Since the construction is independent of constants, one can always choose $\check{\psi}_{13}(z) = 1$. The other functions are uniquely fixed, up to a constant, to be

$$\check{\psi}_{P_2P_4}(X) = \mathrm{e}^{\mathrm{i}X(\kappa_2 - \mathrm{i}\frac{\Omega}{2})} \frac{\mathcal{S}\big(\frac{\omega_1\omega_2}{2\pi}X + t + \kappa_1 - \kappa_2 + \lambda - \mathrm{i}\frac{\omega_1}{2}\big)}{\mathcal{S}\big(\frac{\omega_1\omega_2}{2\pi}X + \lambda + \mathrm{i}\frac{\omega_2}{2}\big)},$$

$$\check{\phi}_{P_1P_4}(x) = \frac{\mathrm{e}^{-\frac{2\mathrm{i}\pi x}{\omega_1\omega_2}(\lambda + t + \mathrm{i}\frac{\omega_2}{2})}}{\mathcal{S}\big(x - t + \mathrm{i}\frac{\Omega}{2}\big)},$$

$$\check{\phi}_{P_2P_3}(x) = \mathrm{e}^{\frac{2\mathrm{i}\pi x}{\omega_1\omega_2}(\lambda + \mathrm{i}\frac{\omega_2}{2})} \mathcal{S}\left(x + \kappa_1 - \mathrm{i}\frac{\Omega}{2}\right).$$

Starting from these representations, taking explicitly the operator products and moving the λ and operator dependent parts so as to cancel them out, one eventually obtains that

$$\tilde{\mathbb{L}}_{V_xV_u}(t) = C(\lambda) \cdot \mathtt{P}_{xu} \cdot \phi_{14}(\mathtt{x} - \mathtt{u}) \cdot \psi_{24}(\mathtt{U}) \cdot \phi_{23}(\mathtt{x} - \mathtt{u}) \tag{B.14}$$

in which the building blocks are as defined in Eqs. (3.5)–(3.7) intertwines $\mathtt{LM}$. Here, $C(\lambda)$ is some λ dependent constant that can be made explicit but is irrelevant to the intertwining property. Hence, upon changing this constant prefactor to the desired value, one indeed gets that (3.4) does enjoy the sought intertwiner property.

B.2. *A compact representation*

It is however useful to provide a second representation for the intertwiner, this time in terms of the S-function introduced in (C.12):

$$\mathtt{P}_{xu}\mathbb{L}_{V_xV_u}(t) = C_L(t)\mathrm{e}^{-\frac{2\mathrm{i}\pi t}{\omega_1\omega_2}(\mathtt{x}-\mathtt{u})}\mathrm{e}^{\mathrm{i}\mathtt{U}(\kappa_2 - \mathrm{i}\frac{\Omega}{2})} S^{-1}\big(q^{-2}\mathfrak{a}\mathrm{e}^{-\frac{2\pi}{\omega_2}(\mathtt{x}-\mathtt{u})}\big)$$
$$\times \frac{S\big(\mathfrak{b}\mathfrak{a}^{-1}\mathrm{e}^{-\omega_1\mathtt{U}}\big)}{S\big(\mathrm{e}^{-\omega_1\mathtt{U}}\big)} S\big(q^{-2}\mathfrak{b}\mathrm{e}^{-\frac{2\pi}{\omega_2}(\mathtt{x}-\mathtt{u})}\big).$$

Here, we have set

$$\mathfrak{a} = d_2 q^2 \mathrm{e}^{\frac{2\pi}{\omega_2}t} \quad \text{and} \quad \mathfrak{b} = -q^3 d_1. \tag{B.15}$$

We can simplify the above formula by using the Schützenberger relation given in (C.15). In order to do so, we have to use the relation

$$S(z) = G(z)S^{-1}(q^2z^{-1})$$

where $G(z)$ is such that

$$\frac{G(q^2z)}{G(z)} = -z^{-1}. \tag{B.16}$$

Upon using

$$G^{-1}\big(q^{-2}\mathfrak{b}\mathrm{e}^{\frac{2\pi}{\omega_2}(\mathtt{x}'-\mathtt{x})}\big)\mathrm{e}^{-\omega_1\mathtt{X}'}G\big(q^{-2}\mathfrak{b}\mathrm{e}^{\frac{2\pi}{\omega_2}(\mathtt{x}'-\mathtt{x})}\big) = -q^4\mathfrak{b}^{-1}\mathrm{e}^{-\omega_1\mathtt{X}'}\mathrm{e}^{-\frac{2\pi}{\omega_2}(\mathtt{x}'-\mathtt{x})}$$

we obtain

$$
\begin{aligned}
\mathrm{P}_{xu}\mathbb{L}_{V_xV_u}(t) = {} & C_{\mathbb{L}}(t)\mathrm{e}^{-\frac{2\mathrm{i}\pi t}{\omega_1\omega_2}(\mathtt{x}-\mathtt{u})}\mathrm{e}^{\mathrm{i}\mathtt{U}(\kappa_2-\mathrm{i}\frac{\Omega}{2})}G^{-1}(q^{-2}\mathfrak{a}\mathrm{e}^{\frac{2\pi}{\omega_2}(\mathtt{u}-\mathtt{x})})G(q^{-2}\mathfrak{b}\mathrm{e}^{\frac{2\pi}{\omega_2}(\mathtt{u}-\mathtt{x})}) \\
& \times S\big(q^4\mathfrak{a}^{-1}\mathrm{e}^{\frac{2\pi}{\omega_2}(\mathtt{x}-\mathtt{u})}\big)S\big(-q^4\mathfrak{a}^{-1}\mathrm{e}^{-\omega_1\mathtt{U}}\mathrm{e}^{\frac{2\pi}{\omega_2}(\mathtt{x}-\mathtt{u})}\big) \\
& \times \big[S\big(q^4\mathfrak{b}^{-1}\mathrm{e}^{\frac{2\pi}{\omega_2}(\mathtt{x}-\mathtt{u})}\big)S\big(-q^4\mathfrak{b}^{-1}\mathrm{e}^{-\omega_1\mathtt{U}}\mathrm{e}^{\frac{2\pi}{\omega_2}(\mathtt{x}-\mathtt{u})}\big)\big]^{-1}.
\end{aligned}
\tag{B.17}
$$

We can now use Schützenberger relation and the composition

$$
G^{-1}\big(q^{-2}\mathfrak{a}\mathrm{e}^{\frac{2\pi}{\omega_2}(\mathtt{u}-\mathtt{x})}\big)G\big(q^{-2}\mathfrak{b}\mathrm{e}^{\frac{2\pi}{\omega_2}(\mathtt{u}-\mathtt{x})}\big) = C'\mathrm{e}^{-\frac{2\mathrm{i}\pi}{\omega_1\omega_2}(\mathtt{u}-\mathtt{x})(t+\kappa_1-\kappa_2-\mathrm{i}\frac{\Omega}{2})}
$$

for some constant C' to obtain

$$
\begin{aligned}
& \mathrm{P}_{xu}\mathbb{L}_{V_xV_u}(t) \\
& = C_{\mathbb{L}}(t)C'\mathrm{e}^{-\frac{2\mathrm{i}\pi}{\omega_1\omega_2}(\mathtt{x}-\mathtt{u})(\kappa_2-\kappa_1+\mathrm{i}\frac{\Omega}{2})}\mathrm{e}^{\mathrm{i}\mathtt{U}(\kappa_2-\mathrm{i}\frac{\Omega}{2})}\frac{S\big(q^4\mathfrak{a}^{-1}(1-\mathrm{e}^{-\omega_1\mathtt{U}})\mathrm{e}^{\frac{2\pi}{\omega_2}(\mathtt{x}-\mathtt{u})}\big)}{S\big(q^4\mathfrak{b}^{-1}(1-\mathrm{e}^{-\omega_1\mathtt{U}})\mathrm{e}^{\frac{2\pi}{\omega_2}(\mathtt{x}-\mathtt{u})}\big)}.
\end{aligned}
\tag{B.18}
$$

Appendix C. Special Functions

C.1. *q products*

Given $|p|<1$ one denotes

$$
(z;p) = \prod_{k\geq 0}(1-zp^k). \tag{C.1}
$$

This allows one to define the θ function and its dual as

$$
\theta(\lambda) = \big(\mathrm{e}^{-\frac{2\pi}{\omega_2}\lambda};q^2\big)\cdot\big(q^2\mathrm{e}^{\frac{2\pi}{\omega_2}\lambda};q^2\big) \quad\text{and}\quad \tilde{\theta}(\lambda) = \big(\mathrm{e}^{\frac{2\pi}{\omega_1}\lambda};\tilde{q}^{-2}\big)\cdot\big(\tilde{q}^{-2}\mathrm{e}^{-\frac{2\pi}{\omega_1}\lambda};\tilde{q}^{-2}\big).
\tag{C.2}
$$

Note that, up to a constant and an exponential prefactor, $\theta(\lambda)$ coincides with the usual theta function $\theta_1(\lambda\,|\,\tau)$. The modular transformation formula for $\theta_3(\lambda\,|\,\tau)$ translates into

$$
\theta(\lambda) = \tilde{\theta}(\lambda)\mathrm{e}^{\mathrm{i}B(\lambda)} \tag{C.3}
$$

in which

$$
B(z) = \frac{\pi}{\omega_1\omega_2}z^2 + \mathrm{i}\frac{\pi\Omega}{\omega_1\omega_2}z - \frac{\pi}{6\omega_1\omega_2}\big(\omega_1^2+3\omega_1\omega_2+\omega_2^2\big) \tag{C.4}
$$

and where we introduced the useful quantity

$$
\Omega = \omega_1+\omega_2.
$$

C.2. *The double sine function*

The double sine function $\mathcal{S}$ is defined by the integral representation

$$\ln \mathcal{S}(z) = \int_{\mathbb{R}+\mathrm{i}0+} \frac{\mathrm{d}t}{t} \frac{\mathrm{e}^{\mathrm{i}zt}}{\left(\mathrm{e}^{\omega_1 t}-1\right)\left(\mathrm{e}^{\omega_2 t}-1\right)}. \tag{C.5}$$

$\mathcal{S}$ can be represented as a convergent infinite product in the case where $\Im\left(\frac{\omega_1}{\omega_2}\right) > 0$, i.e. $|q| < 1$ and $|\tilde{q}| > 1$:

$$\mathcal{S}(\lambda) = \frac{\left(\mathrm{e}^{-\frac{2\pi}{\omega_2}\lambda}; q^2\right)}{\left(\widetilde{q}^{-2}\mathrm{e}^{-\frac{2\pi}{\omega_1}\lambda}; \widetilde{q}^{-2}\right)} = \mathrm{e}^{\mathrm{i}B(\lambda)} \cdot \frac{\left(\mathrm{e}^{\frac{2\pi}{\omega_1}\lambda}; \widetilde{q}^{-2}\right)}{\left(q^2\mathrm{e}^{\frac{2\pi}{\omega_2}\lambda}; q^2\right)}. \tag{C.6}$$

The equivalence of these two representation is a consequence of the modular transformation relation for theta functions (C.3).

The double sine function satisfies the quasi-periodicity relations

$$\frac{\mathcal{S}(z-\mathrm{i}\omega_1)}{\mathcal{S}(z)} = \frac{1}{1-\mathrm{e}^{-\frac{2\pi}{\omega_2}z}}, \quad \frac{\mathcal{S}(z-\mathrm{i}\omega_2)}{\mathcal{S}(z)} = \frac{1}{1-\mathrm{e}^{-\frac{2\pi}{\omega_1}z}} \tag{C.7}$$

and enjoys a reflection property

$$\mathcal{S}(\lambda)\mathcal{S}(-\lambda-\mathrm{i}\Omega) = \mathrm{e}^{\mathrm{i}B(\lambda)}. \tag{C.8}$$

The zeroes and poles of $\mathcal{S}(z)$ are all simple and located on the lattices

$$\mathrm{i}m\omega_1 + \mathrm{i}n\omega_2, \quad m,n, \geq 0 \quad \text{zeroes} \tag{C.9}$$

$$\mathrm{i}m\omega_1 + \mathrm{i}n\omega_2, \quad m,n, \leq -1 \quad \text{poles.} \tag{C.10}$$

Finally, $\mathcal{S}$ has the $\lambda \to \infty$ asymptotics

$$\mathcal{S}(\lambda) \sim \begin{cases} 1 & \arg(\omega_1) - \frac{\pi}{2} < \arg(\lambda) < \arg(\omega_2) + \frac{\pi}{2} \\ \mathrm{e}^{\mathrm{i}B(\lambda)} & \arg(\omega_1) - 3\frac{\pi}{2} < \arg(\lambda) < \arg(\omega_2) - \frac{\pi}{2} \\ \mathrm{e}^{\mathrm{i}B(\lambda)}\left(q^2\mathrm{e}^{\frac{2\pi}{\omega_2}\lambda}; q^2\right)^{-1} & \arg(\omega_2) - \frac{\pi}{2} < \arg(\lambda) < \arg(\omega_1) - \frac{\pi}{2} \\ \left(q^2\mathrm{e}^{-\frac{2\pi}{\omega_2}\lambda}; q^2\right) & \arg(\omega_2) + \frac{\pi}{2} < \arg(\lambda) < \arg(\omega_1) + \frac{\pi}{2} \end{cases}. \tag{C.11}$$

It is sometimes more convenient to work with a closely related function which is denoted by $S(x)$ and is defined as

$$S(\mathrm{e}^{-\frac{2\pi}{\omega_2}z}) = \mathcal{S}(z) \tag{C.12}$$

so that, for $|q| < 1$,

$$S(x) = \frac{(x; q^2)}{\left(\tilde{q}^{-2}x^{\frac{\omega_2}{\omega_1}}; \tilde{q}^{-2}\right)}. \tag{C.13}$$

Since $q^{\frac{2\omega_2}{\omega_1}} = 1$, S satisfies the functional equation

$$\frac{S(q^2x)}{S(x)} = \frac{1}{1-x}. \tag{C.14}$$

Volkov [32] has argued that it satisfies Schützenberger relation

$$S(\mathtt{x}+\mathtt{X}) = S(\mathtt{x})S(\mathtt{X}), \quad \text{provided that} \quad \mathtt{x}\mathtt{X} = q^2\mathtt{X}\mathtt{x}. \tag{C.15}$$

This identity has been rigorously established by Woronowicz in [34]. Then, one should understand $\mathtt{x}+\mathtt{X}$ appearing in the left-hand side and $\mathtt{x}$, $\mathtt{X}$ appearing in the right-hand side as the self-adjoint extension of the respective operators.

C.3. *The quantum dilogarithm*

The quantum dilogarithm [7] ϖ is a meromorphic function that is directly related to the double sine function $\mathcal{S}$:

$$\varpi\left(\lambda + \mathrm{i}\frac{\Omega}{2}\right) = \mathrm{e}^{-\mathrm{i}\frac{B(\lambda)}{2}}\mathcal{S}(\lambda). \tag{C.16}$$

The below ratio of quantum dilogarithms enjoys nice Fourier transformation properties

$$D_\alpha(x) = \frac{\varpi(x+\alpha)}{\varpi(x-\alpha)}, \tag{C.17}$$

namely, it holds

$$D_\alpha(p) = \frac{\mathcal{A}(\alpha)}{\sqrt{\omega_1\omega_2}}\int_{\mathbb{R}} D_{\alpha^\star}(v)\mathrm{e}^{-\frac{2\mathrm{i}\pi}{\omega_1\omega_2}vp}\cdot \mathrm{d}v \quad \text{with} \quad \begin{cases} \alpha^\star = -\alpha - \mathrm{i}\dfrac{\Omega}{2} \\ \mathcal{A}(\alpha) = \varpi(\alpha - \alpha^\star) \end{cases}. \tag{C.18}$$

References

[1] O. Babelon, K. K. Kozlowski and V. Pasquier, Solution of Baxter equation for the q-Toda and Toda$_2$ chains by NLIE, in preparation.
[2] O. Babelon, K. K. Kozlowski and V. Pasquier, The Toda$_2$ chain, in preparation.
[3] M. Bruschi and O. Ragnisco, Recursion operator and Bäcklund transformation for Ruijsenaars–Toda lattice, *Phys. Lett. A* **129** (1988) 21–25.
[4] A. Bytsko and J. Teschner, The integrable structure of nonrational conformal field theory, *Adv. Theor. Math. Phys.* **17** (2013) 701–740, arXiv:0902.4825.
[5] S. E. Derkachov, Factorization of the R-matrix. I, *J. Math. Sci.* **143** (2007) 2773–2790.
[6] B. Dubrovin, I. Krichever and S. Novikov, *Integrable Systems. I*, Encyclopedia of Mathematical Sciences, Dynamical Systems IV (Springer, 1990).
[7] L. D. Faddeev and R. Kashaev, Quantum dilogarithm, *Mod. Phys. Lett. A* **9** (1994) 427–434.
[8] L. D. Faddeev, Discete Heisenberg–Weyl group and modular group, *Lett. Math. Phys.* **34** (1995) 249–254.
[9] L. D. Faddeev, Modular double of a quantum group, *Math. Phys. Stud.* **21** (2000) 149–156.

[10] V. Fateev and A. B. Zamolodchikov, Self-dual solutions of the star-triangle relations in Z_N-models, *Phys. Lett.* **92** (1982) 37–39.

[11] H. Flaschka, The Toda lattice I: Existence of integrals, *Phys. Rev. B* **9** (1974) 1924–1925; The Toda lattice II: Inverse scattering solution, *Prog. Theor. Phys.* **51** (1974) 703–716.

[12] M. Gaudin, *La fonction d'onde de Bethe*, Collection CEA, Série Scientifique (Masson, 1983).

[13] M. Gaudin and V. Pasquier, The periodic Toda chain and a matrix generalization of the Bessel function recursion relations, *J. Phys. A* **25** (1992) 5243–5252.

[14] M. C. Gutzwiller, The quantum mechanical Toda lattice, *Ann. Phys.* **124** (1980) 347–387.

[15] M. C. Gutzwiller, The quantum mechanical Toda lattice II, *Ann. Phys.* **133** (1981) 304–331.

[16] M. Hallnäs and S. Ruijsenaars, Kernel functions and Bäcklund transformations for relativistic Calogero–Moser and Toda systems, *J. Math. Phys.* **53** (2012) 123512.

[17] Y. Hatsuda and M. Marino, Exact quantization conditions for the relativistic Toda lattice, arXiv:1511.02860 [hep-th].

[18] R. Kashaev, The Yang–Baxter relation and gauge invariance, arXiv:1510.03043.

[19] R. Kashaev and S. M. Sergeev, Spectral equations for the modular oscillator, arXiv:1703.06016.

[20] S. Kharchev, D. Lebedev and M. Semenov-Tian-Shansky, Unitary representations of $U_q(\mathfrak{sl}(2,\mathbb{R}))$, the modular double and the multiparticle q-deformed Toda chains, *Comm. Math. Phys.* **225** (2002) 573–609.

[21] B. Kostant, The solution to the generalized Toda lattice and representation theory, *Adv. Math.* **34** (1979) 195–338.

[22] K. K. Kozlowski and J. Teschner, TBA for the Toda chain, in *Infinite Analysis 09: New Trends in Quantum Integrable Systems, Festschrift Volume for Tetsuji Miwa* (World Scientific, 2010), pp. 195–219.

[23] V. B. Kuznetsov and E. K. Sklyanin, Few remarks on Bäcklund transformations for many-body systems, *J. Phys. A* **31** (1998) 2241–2251.

[24] A. K. Kashani-Poor, Quantization condition from exact WKB for difference equations, *J. High Energy Phys.* **2016** (2016) 180.

[25] P. van Moerbeke, The spectrum of Jacobi matrices, *Invent. Math.* **37** (1976) 45–81; P. van Moerbeke and D. Mumford, The spectrum of difference operators and algebraic curves, *Acta. Math.* **143** (1979) 93–154.

[26] G. Pronko and S. M. Sergeev, Quantum relativistic Toda chain, *J. Appl. Math.* **1** (2001) 47–68.

[27] A. Reyman and M. Semenov-Tian-Shansky, Group–theoretical methods in the theory of finite dimensional integrable systems, in *Dynamical Systems VIII*, Encyclopaedia of Mathematical Sciences, Vol. 16 (Springer Verlag, 1994), pp. 116–225.

[28] S. Ruijsenaars, The relativistic Toda systems, *Comm. Math. Phys.* **133** (1990) 217–247.

[29] E. K. Sklyanin, The quantum Toda chain, in *Nonlinear Equations in Classical and Quantum Field Theory* (*Meudon/Paris, 1983/1984*), Lecture Notes in Physics, Vol. 226 (Springer, 1985), pp. 196–233.

[30] W. Toda, Wave propagation in anharmonic lattices, *J. Phys. Soc. Jpn.* **23** (1967) 501–506.

[31] A. Volkov, Quantum Volterra model, *Phys. Lett. A* **167** 345–355 (1992).

[32] A. Volkov, Noncommutative hypergeometry, *Comm. Math. Phys.* **258** (2005) 257–273.

[33] N. R. Wallach, *Real Reductive Groups II* (Academic Press, Inc., 1992); Pure and Applied Mathematics, **132-II**.
[34] S. L. Woronowicz, Quantum exponential function, *Rev. Math. Phys.* **12**(6) (2000) 873–920.

Poisson Sigma Model and Semiclassical Quantization of Integrable Systems

Alberto S. Cattaneo[*,§], Pavel Mnev[†,¶] and Nicolai Reshetikhin[‡,||]

[*]*Institut für Mathematik, Winterthurerstrasse 190, 8057 Zurich, Switzerland*

[†]*University of Notre Dame, Notre Dame, Indiana 46556, USA*
and
St. Petersburg Department of V. A. Steklov Institute of Mathematics of the Russian Academy of Sciences, Fontanka 27, St. Petersburg, 191023 Russia

[‡]*Department of Mathematics, University of California, Berkeley, CA 94720, USA*
and
Physics Department, St. Petersburg University, Russia
and
KdV Institute for Mathematics, University of Amsterdam, Science Park 904, 1098 XH Amsterdam, The Netherlands

[§]*cattaneo@math.uzh.ch*

[¶]*pmnev@nd.edu*

[||]*reshetik@math.berkeley.edu*

In this paper, we outline the construction of semiclassical eigenfunctions of integrable models in terms of the semiclassical path integral for the Poisson sigma model with the target space being the phase space of the integrable system. The semiclassical path integral is defined as a formal power series with coefficients being Feynman diagrams. We also argue that in a similar way one can obtain irreducible semiclassical representations of Kontsevich's star product.

Dedicated to the memory of L. D. Faddeev

Introduction

This paper is motivated by two problems. The first one is how to describe the full semiclassical expansion of eigenfunctions of a complete set of quantum Hamiltonians for a quantum integrable system. More generally, the problem is how to describe full semiclassical asymptotics of the scalar products of two integrable systems on the same space of geometric quantization, or corresponding cyclic amplitudes. The second problem is how to construct an irreducible semiclassic representation of Kontsevich's star product on the algebra of functions on a Poisson manifold.

It turns out that both questions can be answered in terms of the semiclassical quantization of the Poisson sigma model with special boundary conditions. This connects the structure of the semiclassical eigenfunctions of any given integrable system to the quantization of a very special topological gauge theory.

This paper is a research announcement. An extended version of this paper with proofs and mathematically sound construction of Feynman diagrams will be published separately.

We would like to dedicate this paper to the memory of L. D. Faddeev. He made fundamental contributions to both, quantization of gauge theories and quantization of integrable systems [9, 8]. As we demonstrate here, they are related in a very direct way. For other relations between gauge theories and integrable systems, see [7, 12, 10, 13].

1. Semiclassical Asymptotic in Integrable Systems and Topological Quantum Mechanics

1.1. *Semiclassical asymptotic in integrable systems*

Let (M,ω) be a symplectic manifold. Fix the geometric quantization data which consist of the following.

- A line bundle L (a prequantization line bundle) with Hermitian structure on fibers and a Hermitian connection α on L such that the symplectic form ω is the curvature of α, i.e. $d\alpha = \omega$ and the Hermitian product is covariantly constant.
- A real polarization $P \subset TM$ which is an integrable tangent distribution on M such that each generic leaf is a Lagrangian submanifold in M. We assume that the space of leaves $B = M/P$ is almost everywhere smooth, i.e. that the polarization is a Lagrangian fibration $\pi\colon M \to B$ (generic fibers are Lagrangian).

The space of geometric quantization $H_P^{(1/2)}$ is the space of half-densities on M which are covariantly constant (with respect to the connection α) along P. Locally, it can be identified with functions on M/P.

Let $C_h(M)$ be quantized algebra of functions on M. Assume it acts on the space $H_P^{(1/2)}$.

A classical integrable system on M is a Lagrangian fibration[a] $\pi\colon M \to B$ which defines a Poisson commuting subalgebra $C(M,B) \subset C(M)$ in the algebra of functions on M, $C(M,B) = \pi^*(C(B))$. Assume that the subalgebra $C(M,B)$ is quantized, i.e. deformed into a maximal commutative subalgebra $C_h(M,B) \subset C_h(M)$, where $C_h(M)$ is a deformation quantization of $C(M)$ given by a star product (see [11] and references therein). For $2n$-dimensional M, such subalgebra has rank n. For the purpose of this paper, one can think that $M = T^*Q$ for some smooth n-dimensional manifold Q, the algebra $C_h(M)$ is the algebra of differential

[a]The generic fibers are Lagrangian submanifolds.

operators where derivatives are multiplied by $-ih$ and the algebra $C(M)$ is the algebra of smooth functions on Q which are polynomials in the cotangent directions. If h is a formal variable, we assume that vector spaces $C_h(M)$ and $C(M)[[h]]$ are isomorphic and the isomorphism is given by placing derivatives to the right from smooth coefficient functions. The commutative subalgebra $C_h(M,B)$ is the algebra of commuting differential operators with the principal symbol from $\pi^*(C(B))$.

Now, assume that we have a real polarization P with the corresponding Lagrangian fibration π: $M \to B = M/P$. Assume we have two integrable systems corresponding to the Lagrangian fibrations $\pi_{1,2}$: $M \to B_{1,2}$.[b] Assume that fibers of both projections are generically transverse and generically transverse with leaves of P. Assume that the algebra $C_h(M)$ acts on the space $H_P^{(1/2)}$. When $M = T^*Q$, P is the real polarization given by the Lagrangian fibration $T^*Q \to Q$ and the space $H_P^{(1/2)}$ can be naturally identified with functions on Q.

We will say a vector $\psi_\chi \in H_P^{(1/2)}$ is an eigenvector of $C_h(M,B_2)$ corresponding to the character χ: $C_h(M,B_2) \to \mathbb{R}$ if $a\psi_\chi = \chi(a)\psi_\chi$ for any $a \in C_h(M,B_2)$. Semiclassically, as $h \to 0$ the set of characters can be identified with B_2. Similarly, we have eigen-half-densities for $C_h(M,B_1)$. Semiclassically, characters of $C_h(M,B_1)$ can be identified with points of B_1. Denote by $\psi_{b_1}^{(1)}$ and $\psi_{b_2}^{(2)}$ semiclassical eigen-half-densities of commutative subalgebras $C_h(M,B_1)$ and $C_h(M,B_2)$ acting in $H_P^{(1/2)}$.

Conjecture 1 ([14]). *Semiclassical asymptotics of the scalar product of eigen-half-densities of $C_h(M,B_2)$ and $C_h(M,B_1)$ in the space $H_P^{(1/2)}$ has the following structure:*

$$(\psi_{b_2}^{(2)}, \psi_{b_1}^{(1)}) = \frac{C}{(2\pi h)^{n/2}} \sum_{c \in \mathcal{L}_{b_1}^{(1)} \cap \mathcal{L}_{b_2}^{(2)}} e^{\frac{i}{h} S_{\gamma_1,\gamma_2}(c,b_1,b_2) + \frac{i\pi}{2}\mu_{\gamma_1,\gamma_2}(c)}$$
$$\times \left| det\left(\frac{\partial^2 S_{\gamma_1,\gamma_2}(c,b_1,b_2)}{\partial b_1^i \partial b_2^j} \right) \right|^{1/2} (1 + O(h))\sqrt{|db_1 db_2|}. \quad (1)$$

Here as we have made a choice of reference points $x_0^{(i)} \in \mathcal{L}_{b_i}^{(i)}$, $S_{\gamma_1,\gamma_2}(c,b_1,b_2) = S_{\gamma_1}^{(1)} - S_{\gamma_2}^{(2)}$ where $S_{\gamma_i}^{(i)} = \int_{\gamma_i \subset \mathcal{L}^{(i)}} \alpha$ and α is the prequantization connection, γ_1 and γ_2 are paths in $\mathcal{L}_{b_1}^{(1)}$ and $\mathcal{L}_{b_2}^{(2)}$, respectively, connecting corresponding reference points[c]

[b]In other words, we have three mutually transverse polarizations.

[c]Here we assume that paths are contained in a neighborhood where the bundle is trivialized and α is a 1-form on M. Otherwise, we should lift paths to the prequantization bundle. The result gives the parallel transport in the prequantization line bundle which is consistent with the fact that globally, $\psi_{b_a}^{(a)}$ are not functions in q and b_a, but sections of the prequantization line bundle. Globally, the exponent in formula (1) is also not a function in b_1 and b_2, but a section of a line bundle.

Changing $\alpha \mapsto \alpha + df$ is an automorphism of the prequantization line bundle. It corresponds to the shift $-ih\frac{\partial}{\partial q} \mapsto ih\frac{\partial}{\partial q} + \frac{\partial f}{\partial q}$ in the deformation quantization of $C(T^*Q)$ by differential operators. In (1), it results in the phase change $(\psi_{b_2}^{(2)}, \psi_{b_1}^{(1)}) \mapsto (\psi_{b_2}^{(2)}, \psi_{b_1}^{(1)}) \exp(\frac{i}{h}(f(x_1) - f(x_2)))$.

and c. Lagrangian submanifolds $\mathcal{L}^{(1)}_{b_1}$ and $\mathcal{L}^{(2)}_{b_2}$ are fibers over $b_1 \in B_1$ and $b_2 \in B_2$, respectively. We assume that $\mathcal{L}^{(1)}_{b_1}$ and $\mathcal{L}^{(2)}_{b_2}$ are transverse. The number $\mu_{\gamma_1,\gamma_2}(c)$ is the weighted number of points along γ_2 between x_2 and c where the tangent space to $\mathcal{L}^{(2)}_{b_2}$ intersect a fiber of π_1 over a line. The point counts with plus if it is crossed in the positive direction and with the minus if it is crossed in the negative direction, i.e. μ is the Maslov index. When Bohr–Sommerfeld quantization conditions for b_1 and b_2 hold, the exponent does not depend on the choice of γ_1 and γ_2. The Hessian and the higher order terms do not depend on the choice of reference points.

Note that to write formula (1), we made a choice of a reference point on every fiber of π_1 and of π_2. Consistent choice of such reference points can be inferred, e.g., from fixing a Lagrangian submanifold Λ intersecting fibers of π_1 and π_2 generically transversally. A change of Λ changes the formula for the scalar product by a constant factor (each term in the sum changes by the same factor). Thus, the eigenfunctions depend on Λ only projectively. The choice of Λ is also convenient for making expression (1) invariant with respect to automorphisms of the prequantization line bundle. Let us multiply the scalar product by the factor $\exp(\frac{i}{h}\int_{\gamma\in\Lambda}\alpha)$ where the path γ connects intersection points $\Lambda\cap\mathcal{L}_1$ and $\Lambda\cap\mathcal{L}_2$ in Λ. From the spectral point of view, this is just the multiplication of eigenfunctions by constants. However, it makes semiclassical scalar products invariant with respect to automorphisms of the prequantization line bundle.

A natural object which does not depend on the choice of Λ is the transition probability, i.e. the square of the absolute value of the scalar product:

$$\begin{aligned}|(\psi^{(2)}_{b_2},\psi^{(1)}_{b_1})|^2 = \frac{1}{2\pi h^n} &\sum_{a,c\in\mathcal{L}^{(1)}_{b_1}\cap\mathcal{L}^{(2)}_{b_2}} \left| det\left(\frac{\partial^2 S_\gamma(c,b_1,b_2)}{\partial b_1^i \partial b_2^j}\right) det\left(\frac{\partial^2 S_\gamma(a,b_1,b_2)}{\partial b_1^i \partial b_2^j}\right)\right|^{1/2} \\ &\times \exp\left(\frac{i}{h}S_{\gamma_1,\gamma_2}(b_1,b_2) + \frac{i\pi}{2}(\mu_{\gamma_1,\gamma_2}(c) - \mu_{\gamma_1,\gamma_2}(a))\right) \\ &\times (1+O(h))|db_1 db_2|. \end{aligned} \tag{2}$$

Here γ_1 is a path in $\mathcal{L}^{(1)}_{b_1}$ connecting points a and c and γ_2 is a path in $\mathcal{L}^{(2)}_{b_2}$ connecting these points. The exponent $S_{\gamma_1,\gamma_2}(b_1,b_2) = \int_{D_{a,c}}\omega$, where $D_{a,c}$ is a disc bounded by γ_1 and γ_2, does not depend on the choice of paths if Bohr–Sommerfeld quantization conditions hold. The Hessian also does not depend on the choice of γ's.

More generally, if $\pi_a\colon M \to B_a$ are Lagrangian fibrations, the cyclic product

$$(\psi^{(1)}_{b_1},\psi^{(k)}_{b_k})\cdots(\psi^{(3)}_{b_3},\psi^{(2)}_{b_2})(\psi^{(2)}_{b_2},\psi^{(1)}_{b_1})$$

does not depend on the choice of $U(1)$-ambiguities of normalized eigenfunctions. We will call such products *cyclic amplitudes*. For the semiclassical asymptotic of a

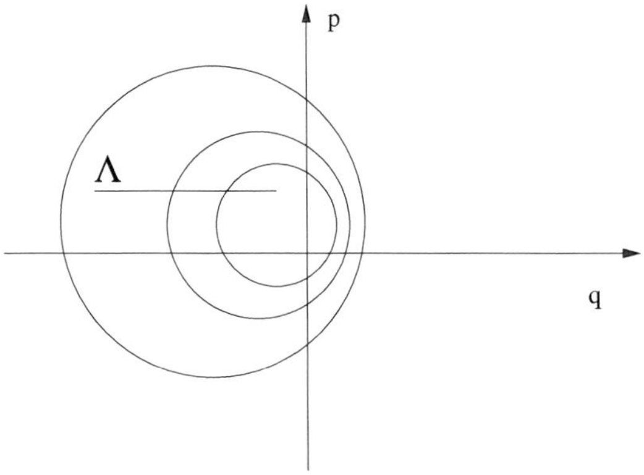

Fig. 1. Choice of reference points corresponding to an integrable system on $\mathbb{R}^2 = T^*\mathbb{R}$ with $\alpha = pdq$ and the standard Lagrangian fibration $T^*\mathbb{R} \to \mathbb{R}$. In this case, $\Lambda = \{p = 0\}$. Locally, Λ is a section of the Lagrangian fibration given by level curves of the Hamiltonian.

cyclic amplitude, we have:

$$(\psi^{(1)}_{b_1}, \psi^{(k)}_{b_k}) \cdots (\psi^{(3)}_{b_3}, \psi^{(2)}_{b_2})(\psi^{(2)}_{b_2}, \psi^{(1)}_{b_1})$$

$$= \frac{1}{2\pi h^{kn/2}} \sum_{c_1 \in \mathcal{L}^{(1)}_{b_1} \cap \mathcal{L}^{(2)}_{b_2}, \ldots, c_k \in \mathcal{L}^{(k)}_{b_k} \cap \mathcal{L}^{(1)}_{b_1}} \left| \prod_{a=1}^{k} det \left(\frac{\partial^2 S}{\partial b^i_a \partial b^j_{a+1}} \right) \right|^{1/2}$$

$$\times \exp\left(\frac{i}{h} S(c_1, \ldots, c_k) + \frac{i\pi}{2}\mu \right) (1 + O(h)) \prod_{a=1}^{k} |db_a|. \tag{3}$$

Here $S(c_1, \ldots, c_k) = \sum_{a=1}^{k} \int_{\gamma_{a,a+1}} \alpha|_{\mathcal{L}^{(a)}_{b_a}}$ where $\gamma_{a,a+1}$ is a path in $\mathcal{L}^{(a)}_{b_a}$ connecting c_a and c_{a+1} and μ is the corresponding Maslov index. The exponent does not depend on the choice of γ's if Bohr–Sommerfeld conditions hold. By the Stokes theorem, it is clear that $S(c_1, \ldots, c_k) = \int_D \omega$ where D is a disk bounded by paths $\gamma_{a,a+1}$.

1.2. *Topological quantum mechanics*

1.2.1. *Scalar product as a path integral for topological quantum mechanics*

We want to write formula (1) as the semiclassical expansion of a path integral. The natural candidate is the topological quantum mechanics

$$(\psi^{(2)}_{b_2}, \psi^{(1)}_{b_1}) = C \int_{\gamma(0) \in \mathcal{L}_1, \gamma(1) \in \mathcal{L}_2} e^{\frac{i}{h}\left(\int_\gamma \alpha + f^{(1)}_{b_1}(\gamma(0)) - f^{(2)}_{b_2}(\gamma(1)) + \int_{\gamma_0 \subset \Lambda} \alpha\right)} D\gamma. \tag{4}$$

Here γ_0 connects reference points x_1 and x_2[d] and $\mathcal{L}_a = \mathcal{L}_{b_a}^{(a)}$ where $a = 1, 2$ and $f_{b_a}^a$ are boundary corrections with defining property

$$df_{b_a}^{(a)} = \iota_a^*(\alpha) \tag{5}$$

where ι_a: $\mathcal{L}_a \to M$ are embeddings of Lagrangian submanifolds to M. These functions can be chosen as follows. For $x_a \in \mathcal{L}_a$, fix a path γ_a on $\mathcal{L}_a$ connecting x with a reference point $x_a \in \mathcal{L}_a$. Define

$$f_{b_a}^{(a)}(x) = \int_{\gamma_a} \alpha.$$

This function satisfies (5) and the exponent in (4) does not depend on γ_a if Bohr–Sommerfeld quantization conditions hold. An arbitrary constant C changes by the factor $\exp(\frac{i}{h}(\int_{x_1}^{y_1} \alpha - \int_{x_2}^{y_2} \alpha))$ when we change reference points from x to y.

This integral should be understood semiclassically, as a formal power series supported at critical points of the exponent. Changing the reference points x_a changes the integral by a constant factor. A consistent choice of reference points on fibers of Lagrangian fibrations π_a: $M \to B_a$ is given a Lagrangian submanifold Λ intersecting transversally each fiber.

It is easy to see that critical points are exactly the intersection points $\mathcal{L}_1 \cap \mathcal{L}_2$. Indeed if we choose local coordinates x^i so that $\alpha = \sum_i \alpha_i(x) dx^i$ and choose a parametrization of γ, we will have the following formula for the variation of the integral in the exponent:

$$\delta \int_\gamma \alpha = \int_0^1 \sum_{ij} \left(\frac{\alpha_i}{\partial x^j} - \frac{\alpha_j}{\partial x^i} \right) \delta x^j(t) \frac{dx^i}{dt}(t) dt + \sum_i (\alpha_i(x(1)) \delta x^i(1) - \alpha_i(x(0)) \delta x^i(0)).$$

This implies that $\frac{dx^i}{dt} = 0$, i.e. critical points are constant trajectories and therefore, taking into account boundary conditions for γ, we conclude that the critical points of the exponent in (4) are the intersection points $\mathcal{L}_1 \cap \mathcal{L}_2$. However, to develop the perturbation theory for such path integrals it is convenient to reformulate them in terms of two-dimensional topological field theory. So, the next step is to reformulate the integral (4) as the partition function for the Poisson sigma model with special boundary conditions. We will do this in Sec. 2.

1.2.2. *Transition probabilities and cyclic amplitudes*

Transition probabilities can also be written in terms of topological quantum mechanics. They do not depend on the normalization of eigenfunctions. Semiclassically, this leads to an unambiguous semiclassical formula (2) for these probabilities which does not require a choice of references points. It is natural to expect the following

[d] The last summand in the exponent is making it to be invariant with respect to automorphisms of the prequantization line bundle. This means that the integral and expression (1) with this factor depend only on the curvature $\omega = d\alpha$ of α.

path integral for (2):

$$|(\psi_{b_2}^{(2)}, \psi_{b_1}^{(1)})|^2 = \int_{\gamma_1(0),\gamma_2(0)\in\mathcal{L}_1;\gamma_1(1),\gamma_2(1)\in\mathcal{L}_2} \times \exp\left(\frac{i}{h}\left(\int_{\gamma_1}\alpha - \int_{\gamma_2}\alpha + \int_{\sigma_1}\alpha - \int_{\sigma_2}\alpha\right)\right) D\gamma D\sigma. \quad (6)$$

Here path σ_1 connecting $\gamma_1(0)$ and $\gamma_2(0)$ in $\mathcal{L}_1$ and σ_2 connecting $\gamma_1(1)$ and $\gamma_2(1)$ in $\mathcal{L}_2$. As for the amplitudes, this path integral should be understood semiclassically. We argue that it reproduces formula (2).

Similarly we expect that the cyclic amplitudes can be written as the path integral

$$(\psi_{b_1}^{(1)}, \psi_{b_k}^{(k)}) \cdots (\psi_{b_3}^{(3)}, \psi_{b_2}^{(2)})(\psi_{b_2}^{(2)}, \psi_{b_1}^{(1)}) = \int \exp\left(\frac{i}{h}\sum_{a=1}^{k}\left(\int_{\gamma_a}\alpha + \int_{\sigma_a}\alpha\right)\right)\prod_{a+1}^{k} D\gamma_a D\sigma_a. \quad (7)$$

Here $\gamma_a\colon I = [0,1] \to M$, such that $\gamma_a(0) \in \mathcal{L}_a$, $\gamma_a(1) \in \mathcal{L}_{a+1}$ and σ_a is a path in $\mathcal{L}_a$ such that $\sigma_a(0) = \gamma_{a-1}(1), \sigma_a(1) = \gamma_a(0)$.

1.2.3. *Two-dimensional version of the path integral for the topological quantum mechanics*

Here we reformulate the topological classical mechanics as a two-dimensional topological field theory. As before, M is a symplectic manifold with a prequantization line bundle with connection α such that $\omega = d\alpha$. Consider the classical field theory on a disk D with fields $X\colon D \to M$ and with the action functional[e]

$$S[X] = \frac{1}{2}\int_D X^*(\omega) = \frac{1}{2}\int_D \omega_{ij}(X)dX^i \wedge dX^j.$$

Using Stokes theorem, we can write it as:

$$\int_D X^*(d\alpha) = \int_{\partial D} X^*(\alpha),$$

where α is the prequantization connection. Now let us organize the boundary conditions in such a way that the integral over the boundary ∂D would give the exponent in (4). We assume that Λ is a Lagrangian submanifold defining the reference points. It should be transversal to $\mathcal{L}_{b_1}$ and $\mathcal{L}_{b_2}$ for generic (b_1, b_2).

Assume the boundary is partitioned in four intervals $\partial D = I \sqcup I_1 \sqcup I_2 \sqcup I_{12}$ with no restriction on X on I and

$$X|_{I_1} \in \mathcal{L}_1, \quad X|_{I_2} \in \mathcal{L}_2, \quad X|_{I_{12}} \in \Lambda.$$

This is illustrated in Fig. 2.

[e]This action is also the topological part of the A-model [15].

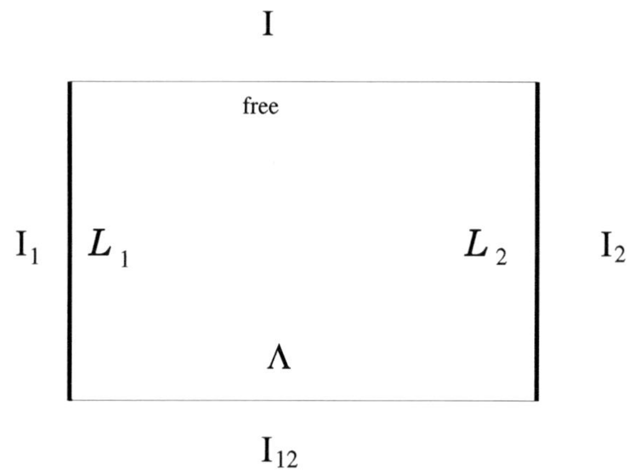

Fig. 2. Partitioning the boundary of a disk and the boundary conditions for the field X corresponding to amplitudes in topological quantum mechanics.

Then the integral over the boundary can be written as

$$\int_{\partial D}\alpha=\int_I\alpha+\int_{I_1}\alpha+\int_{I_2}\alpha+\int_{I_{12}}\alpha=\int_I\alpha+f^{(1)}_{b_1}(\gamma(0))-f^{(2)}_{b_2}(\gamma(1))+\int_{I_{12}}\alpha$$

where f_1 and f_2 are generating functions for the connection α restricted to the Lagrangian submanifolds $\mathcal{L}_1$ and $\mathcal{L}_2$, respectively, as in (4), with reference points $x_i=\mathcal{L}_i\cap\Lambda$. The last term corresponds to an arbitrary constant C in (5). Denote by $\gamma=\{X(s)\in M|s\in I\}$ the parametrized path in M corresponding to $I\subset\partial D$, then we can write

$$\int_I X^*(\alpha)=\int_\gamma\alpha,\quad f^{(i)}_{b_i}(x)=\int_{\gamma_{x,x_i}}\alpha.$$

Here $x\in\mathcal{L}_i$, $x_i=\mathcal{L}_i\cap\Lambda$.

Critical points of $S[X]$ are mappings which are constant on I and satisfy the boundary conditions on the other parts of ∂D.

The path integral representing the scalar product (4) can be written then up to (an infinite) constant as

$$(\psi^{(2)}_{b_2},\psi^{(1)}_{b_1})=\int_{X:D\to M}e^{\frac{i}{\hbar}\int_D X^*(\omega)}DX \tag{8}$$

with the boundary conditions described above.

This path integral should be treated as a semiclassical path integral given by oscillatory exponential contributions and power series with coefficients being Feynman diagrams. It is a second-order topological quantum field theory. To deal with the boundary conditions using the Hamiltonian framework, we will reformulate it as the first-order Poisson sigma model.

Note that we expect to reproduce formula (1) where we initially fixed the reference points $x_a\in\mathcal{L}_{b_a}\cap\Lambda$. It means that we should fix $X(u_a)=x_a$, where

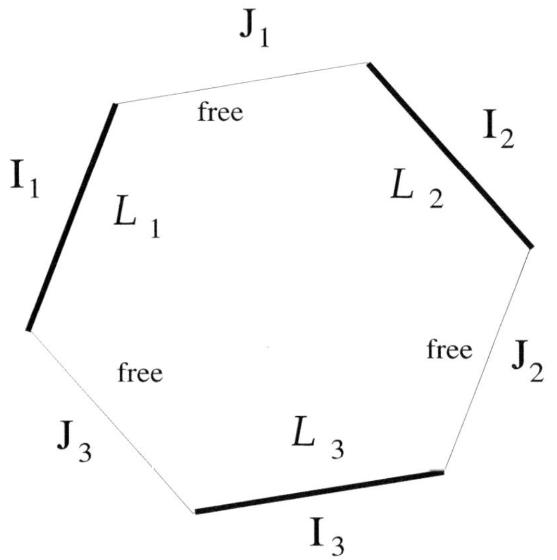

Fig. 3. A partitioning of the boundary of a disk and the boundary conditions for the X-field corresponding to cyclic amplitudes, $k = 3$.

$u_a = I_a \cap I_{12}$, in formula (8). The path integral then is defined in terms of Feynman diagram contributions for each critical point of the action with fixed x_1 and x_2. Varying x_1 and x_2 results in an overall $U(1)$ factor and does not change the Feynman diagram contributions.

Similarly, the path integral for the topological quantum mechanics describing the cyclic amplitudes can be written as

$$(\psi_{b_1}^{(1)}, \psi_{b_k}^{(k)}) \cdots (\psi_{b_3}^{(3)}, \psi_{b_2}^{(2)})(\psi_{b_2}^{(2)}, \psi_{b_1}^{(1)}) = \int_{X:D \to M} e^{\frac{i}{\hbar} \int_D \omega} DX \tag{9}$$

where the boundary of the disc is partitioned to intervals $I_1, \ldots, I_k$ and $J_1, \ldots, J_k$, see Fig. 3 with boundary conditions $X|_{I_a} \in \mathcal{L}_a$ and X is free on intervals J_a.

2. Topological Quantum Mechanics as the Poisson Sigma Model

2.1. *Classical Poisson sigma model corresponding to topological quantum mechanics*

2.1.1. *Poisson sigma model and boundary conditions*

Recall that in the Poisson sigma model, the fields are pairs (X, η) representing components of the mapping of bundles (X, η): $TD \to T^*M$. Because the main difficulty of the problem is not in the geometric structure of M, we assume that $M = \mathbb{R}^N$ for some even N, equipped with possibly nonconstant symplectic structure. Then X: $D \to \mathbb{R}^N$ and $\eta \in \Omega^1(D) \otimes \mathbb{R}^N$. Here we have identified $\mathbb{R}^N \simeq (\mathbb{R}^N)^*$. We will denote the components of these fields $X^i(u) \in \Omega^0(D)$ and $\eta_j(u) \in \Omega^1(D)$.

The action is:

$$S[X,\eta] = \int_D \left(\eta_i \wedge dX^i + \frac{1}{2}\omega^{ij}(X)\eta_i \wedge \eta_j \right).$$

Here $\omega^{ij}(X)$ is the Poisson tensor corresponding to ω_{ij}, i.e. the inverse matrix to $\omega_{ij}(X)$. We will follow the agreement of summing up over repeated contravariant and covariant indices.

As we will see below, the following boundary conditions for the field (X,η) correspond to the amplitudes (4) in topological quantum mechanics.

- For $s \in I$, $(X,\eta) \in M \subset T^*M$ where M is regarded as a zero-section Lagrangian submanifold of T^*M, i.e. no condition on X and $\eta|_I = 0$.
- For $s \in I_1$, $X(s) \in \mathcal{L}_1$ and $\eta(s) \in N^*_{X(s)}\mathcal{L}_1 \otimes T^*_s D$ where $N^*\mathcal{L}_1$ is the conormal bundle to $\mathcal{L}_1$. Its fiber over $X(s)$ is the Lagrangian subspace in $T^*_{X(s)}M$, defined as $\{\eta(s) \in T^*_{X(s)}M | \langle a, \eta(s)\rangle|_{I_1} = 0\}$ for all $a \in T_{X(s)}\mathcal{L}_1$. By restriction of a form to the boundary, we always mean the pull-back. Here $\langle -,-\rangle$ is the pairing of tangent and cotangent vectors.
- Similarly, for $s \in I_2$, $X(s) \in \mathcal{L}_2$ and $\eta(s) \in N^*_{X(s)}\mathcal{L}_2 \otimes T^*_s D$.
- For $s \in I_{12}$, $X(s) \in \Lambda$ and $\eta(s) \in N^*_{X(s)}\Lambda \otimes T^*_s D$.

2.1.2. *The relation to topological quantum mechanics*

The topological sigma model described above is related to the second-order topological field theory introduced earlier as follows. Consider the critical point of $S[X,\eta]$ in η. The functional is quadratic in η, therefore such a critical point is unique and it is given by

$$\overline{\eta}_i = -\omega_{ij} dX^j. \tag{10}$$

Let V be an isotropic submanifold M. It is clear that if the boundary condition for X is chosen in such a way that $X(s) \in V$ when $s \in I \subset \partial D$ then $\overline{\eta}(s) \in N^*V$ when $s \in I$. Thus, this critical point agrees with boundary conditions above. Computing the action of the Poisson sigma model at the critical point for η, we recover the second-order action for the X-field:

$$S[X,\overline{\eta}] = \frac{1}{2}\int_D \omega_{ij}(X) dX^i \wedge dX^j.$$

Now let us describe the critical points of the action of Poisson sigma model in both X and η. The Euler–Lagrange equations are

$$dX^i + \omega^{ij}(X)\eta_j = 0, \quad d\eta_i + \frac{1}{2}\partial_i \omega^{kl}\eta_k \wedge \eta_l = 0.$$

It is easy to see that the first equation implies the second one. Note that here we used the invertibility of the Poisson tensor, i.e. the fact that the Poisson sigma model has a symplectic manifold as the target space.

If Euler–Lagrange equations hold, the variation of the action is given only by boundary terms:

$$\delta S[X,\eta]_{\text{boundary}} = \int_{\partial D} \eta_i \delta X^i.$$

This integral vanishes because of the boundary conditions. Indeed, $\delta X|_{I_a} \in T_X\mathcal{L}_a$, $\eta|_{I_a} \in N^*_X\mathcal{L}_a$, which implies $\eta_i\delta X|_{I_a} = 0$, similarly for I_{12} with $\mathcal{L}_a$ being replaced by Λ, and on I the boundary term vanishes because $\eta = 0$.

Thus, the critical points are given by (10) with an arbitrary smooth function X: $D \to M$ satisfying the boundary conditions described above. Boundary conditions on η, together with the Euler–Lagrange equations, imply

$$X(s) = c, \quad s \in I, \quad c \in \mathcal{L}_1 \cap \mathcal{L}_2.$$

Thus, a critical point of $S[X,\eta]$ is a pair $(\overline{X},\overline{\eta})$ where $\overline{\eta}$ is given by (10) and $\overline{X}$ is any smooth function on D satisfying the boundary conditions and $X|_I = c$, $c \in \mathcal{L}_1\cap\mathcal{L}_2$.

2.1.3. *Gauge invariance*

The Poisson sigma model is gauge invariant with respect to the following infinitesimal gauge transformations, i.e. vector fields on the space of fields

$$\delta_\beta X^i = \omega^{ij}\beta_j, \quad \delta_\beta\eta_i = -d\beta_i - \partial_i\omega^{jk}\eta_j\beta_k.$$

Here $\beta \in \Omega^0(D)\otimes\mathbb{R}^N$ (in general $\beta(u) \in T^*_{X(u)}M$). The Lie subalgebra with $\beta|_I = 0$, $\beta(s) \in N^*_{X(s)}\mathcal{L}_1$ when $s \in I_1$, $\beta(s) \in N^*_{X(s)}\mathcal{L}_2$ when $s \in I_2$ and $\beta(s) \in N^*_{X(s)}\Lambda$ when $s \in I_{12}$ preserves the boundary conditions corresponding to the topological quantum mechanics.

The gauge symmetry of the Poisson sigma model is directly related to the reparametrization invariance of topological quantum mechanics.

2.2. *Amplitudes*

Thus, up to constants, we expect the following formula for scalar products (amplitudes) in terms of the Poisson sigma model:

$$(\psi^{(2)}_{b_2}, \psi^{(1)}_{b_1}) = \int_{(X,\eta):D\to T^*M} e^{\frac{i}{\hbar}S(X,\eta)} DXD\eta.$$

Here we assume the boundary conditions described in Sec. 2.1.1 and the path integral should be understood semiclassically, as a formal power series of Feynman diagrams. Because the action is gauge invariant and therefore highly degenerate, to make sense of such a path integral, we should do the gauge fixing. We will do it later using Lorenz gauge fixing and an extra gauge fixing on the boundary.

As it was discussed before, changing the auxiliary Lagrangian submanifold Λ will only produce a constant $U(1)$-factor.

2.3. *Cyclic amplitudes*

We can do the same to the path integral describing cyclic amplitudes as we did for amplitudes. This gives the following representation of cyclic amplitudes in terms of the Poisson sigma model:

$$(\psi_{b_1}^{(1)}, \psi_{b_k}^{(k)}) \cdots (\psi_{b_3}^{(3)}, \psi_{b_2}^{(2)})(\psi_{b_2}^{(2)}, \psi_{b_1}^{(1)}) = \int_{(X,\eta):D\to T^*M} e^{\frac{i}{\hbar}S(X,\eta)} DX D\eta. \tag{11}$$

Here we impose boundary conditions

- $\eta = 0$ and X is not fixed on J_a,
- $X(s) \in \mathcal{L}_a$ and $\eta(s) \in N^*_{X(s)}\mathcal{L}_a \otimes T^*_s I_a$, when $s \in I_a$.

The natural question about formula (11) is why does the path integral factorize into the product of path integrals describing the amplitudes? Roughly speaking, this happens because the field theory is topological. This implies that we can deform the disc in Fig. 3 into a semi-infinite strip with the side corresponding to the auxiliary Lagrangian Λ (which we have to choose) at infinity. In terms of the gauge fixing, this means choosing a very special metric on D. In this limit, the critical points are simply $X(u) = c \in \mathcal{L}_1 \cap \mathcal{L}_2$ and $\eta = 0$. Feynman diagrams vanish if any of their edges extend to infinity. Thus, all Feynman diagrams are supported at the end of the semi-infinite strip near the boundary segment I.

For the same reasons, a disc with alternating boundary conditions (corresponding to cyclic amplitudes) can be deformed into a "star" with semi-infinite "tentacles", as it is shown in Fig. 4. In this limit, critical points can be chosen as constant maps $X(u) = c \in \mathcal{L}_a \cap \mathcal{L}_{a+1}$ near the end of the star's tentacles. Also, in this limit only the Feynman diagrams localized near the ends of the star will contribute and these contributions are exactly the amplitudes of the topological

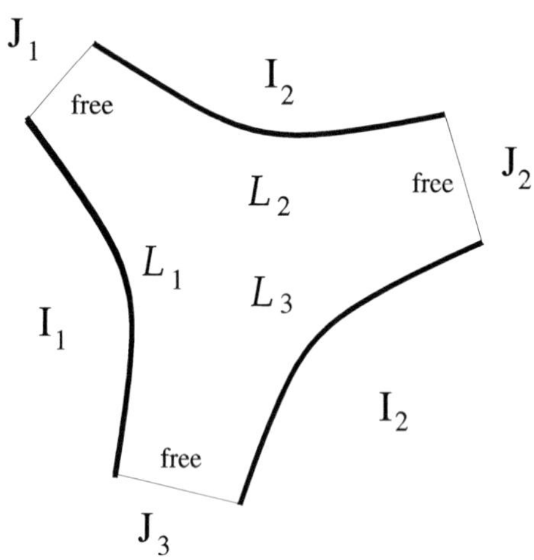

Fig. 4. Deformation of the disc into a star with infinitely long tentacles for $k = 3$.

quantum mechanics discussed above. Thus, in this limit the path integral (11) factorizes as expected. But since the field theory is topological, this factorization is exact.

This is the heuristics of the factorization. The factorization in terms of Feynman diagrams involves direct manipulations with the diagrams (as in case of the associativity of the star product in [11]) and will be done in a separate publication.

3. Classical BV Extended Poisson Sigma Model and Gauge Fixing

3.1. *Classical BV-BFV Poisson sigma model*

Recall that the BV extended Poisson sigma model is an AKSZ model where fields on D which can be naturally organized in supermultiplets [2]:

$$\widetilde{X}^i = X^i + \eta^{\dagger,i} + \beta^{\dagger,i}, \quad \widetilde{\eta}_i = \beta_i + \eta_i + X_i^{\dagger}.$$

The space of fields $\mathcal{F}_D$ on the bulk D is $\mathbb{Z}$-graded by the ghost number: $gh(X) = gh(\eta) = 0$, $gh(\eta^\dagger) = gh(X^\dagger) = -1$, $gh(\beta) = 1$, $gh(\beta^\dagger) = -2$. The fields are differential forms $X, \beta \in \Omega^0(D, \mathbb{R}^N)$, $\eta, \eta^\dagger \in \Omega^1(D, \mathbb{R}^N)$ and $\beta^\dagger, X^\dagger \in \Omega^2(D, \mathbb{R}^N)$. The space of fields $\mathcal{F}_{\partial D}$ on the boundary ∂D consists of pull backs of $X, \eta, \eta^\dagger, \beta$.

The space of fields on D can be naturally regarded as a shifted cotangent bundle to the space of fields X, η and β.

Both $\mathcal{F}_D$ and $\mathcal{F}_{\partial D}$ have natural symplectic structures:

$$\omega_D = \int_D \delta X_i^\dagger \wedge \delta X^i + \int_D \delta \eta^{\dagger,i} \wedge \delta \eta_i + \int_D \delta \beta^{\dagger,i} \wedge \delta \beta_i,$$

$$\omega_{\partial D} = \int_{\partial D} \delta \eta_i \wedge \delta X^i + \int_D \delta \beta_i \wedge \delta \eta^{\dagger,i}.$$

The action functional of the BV extended Poisson sigma model is

$$S_D = \int_D \left(\widetilde{\eta}_i \wedge d\widetilde{X}^i + \frac{1}{2}\omega^{ij}(\widetilde{X})\widetilde{\eta}_i \wedge \widetilde{\eta}_j \right). \tag{12}$$

The boundary action

$$\begin{aligned} S_{\partial D} &= \int_{\partial D} \left(\tilde{\eta}_i \wedge d\tilde{X}^i + \frac{1}{2}\omega^{ij}(\tilde{X})\tilde{\eta}_i \wedge \tilde{\eta}_j \right) \\ &= \int_{\partial D} \beta_i dX^i + \omega^{ij}(X)\beta_i \eta_j + \frac{1}{2}\partial_k \omega^{ij}(X)\eta^{\dagger,k}\beta_i\beta_j \end{aligned}$$

together with cohomological vector fields

$$Q_D = \int_D (d\widetilde{X}^i + \omega^{ij}(\widetilde{X}) \wedge \widetilde{\eta}_j) \wedge \frac{\delta}{\delta \widetilde{X}^i} + \int_D \left(d\widetilde{\eta}_i + \frac{1}{2}\partial_i \omega^{kl}(\widetilde{X}) \wedge \widetilde{\eta}_k \wedge \widetilde{\eta}_l \right) \wedge \frac{\delta}{\delta \widetilde{\eta}_i}$$

and

$$Q_{\partial D}F = \{S_{\partial D}, F\}_{\partial D}$$

defines the BV-BFV structure of the models [5].

Now let us discuss boundary conditions for the BV extended model.

3.2. *Boundary conditions corresponding to the topological quantum mechanics*

Boundary conditions for BV extended field theories are Q_∂-invariant Lagrangian submanifolds in the space of boundary BFV fields [5].

Boundary conditions corresponding to amplitudes in topological quantum mechanics are:

- on I, no conditions on $\tilde{X}$ and $\tilde{\eta} = 0$, i.e. fields are in the zero section Lagrangian,
- when $s \in I_1$, $X(s) \in \mathcal{L}_1$, $\eta^\dagger(s)|_{I_1} \in T_{X(s)}[-1]\mathcal{L}_1 \otimes T_s^* I_1$, $\eta(s)|_{I_1} \in N^*_{X(s)}\mathcal{L}_1 \otimes T_s^* I_1$, and $\beta(s)|_{I_1} \in N^*_{X(s)}[1]\mathcal{L}_1 \otimes T_s^* I_1$,
- when $s \in I_2$, $X(s) \in \mathcal{L}_2$, $\eta^\dagger(s)|_{I_2} \in T_{X(s)}[-1]\mathcal{L}_2 \otimes T_s^* I_2$, $\eta(s)|_{I_2} \in N^*_{X(s)}\mathcal{L}_2 \otimes T_s^* I_2$, and $\beta(s)|_{I_2} \in N^*_{X(s)}[1]\mathcal{L}_2 \otimes T_s^* I_2$,
- and similarly, when $s \in I_{12}$, $X(s) \in \Lambda$, $\eta^\dagger(s)|_{I_{12}} \in T_{X(s)}[-1]\Lambda \otimes T_s^* I_{12}$, $\eta(s)|_{I_{12}} \in N^*_{X(s)}\Lambda \otimes T_s^* I_{12}$, and $\beta(s)|_{I_{12}} \in N^*_{X(s)}[1]\Lambda \otimes T_s^* I_{12}$.

Here, as before $\mathcal{L}_a = \mathcal{L}^{(a)}_{b_a}$. Note that the boundary conditions for β and $\eta^\dagger$ do not follow from the boundary conditions for X and η. Our choice of boundary conditions is such that the boundary Lagrangian submanifold is tangent to the vector field $Q_{\partial D}$. In other words, gauge transformations preserve this boundary condition.

Boundary conditions corresponding to cyclic amplitudes on intervals J_a are the same as the above on I, and on intervals I_a, are the same as the above for I_1, I_2 or I_{12} for corresponding Lagrangian submanifolds. For more details, see [3, 4].

3.3. *Gauge fixing*

Recall that in the BV framework gauge fixing means a choice of a Lagrangian submanifold in the odd-symplectic space of bulk fields. In case when the gauge symmetry is the result of a gauge group action, gauge fixing is a choice of a section of the projection from the space of fields to gauge classes of fields. This is equivalent to choosing a Lagrangian submanifold in a BV extended field theory. In the case of Poisson sigma model, where the gauge symmetry is not given by a group action, choosing such a Lagrangian submanifold is the analog of choosing a gauge fixing section.

We choose the Lorenz gauge fixing which means we fix the following Lagrangian submanifold in $\mathcal{F}_D$:

$$L_{gf} = \{X_i^\dagger = \beta^{\dagger,i} = 0, \quad \eta^{\dagger,i} = *d\gamma^i, \quad d^*\eta_i = 0\}.$$

In particular, this means $d^*\widetilde{\eta}_i = 0$ and $d^*\widetilde{X}^i = 0$. Because D is simply-connected, the condition $d * \eta_i = 0$ implies that $\eta_i = *df_i$ for some function f.

Let us check that this subspace with the boundary conditions corresponding to the topological quantum mechanics is indeed Lagrangian. Consider the restriction of the symplectic form

$$\omega_D = \int_D \delta X_i^\dagger \wedge \delta X^i + \int_D \delta\eta^{\dagger,i} \wedge \delta\eta_i + \int_D \delta\beta^{\dagger,i} \wedge \delta\beta_i$$

to L_{gf}. The first term is zero because $X^\dagger = 0$ on L_{gf}, the third term is zero because on this submanifold, $\beta^\dagger = 0$. The second term restricted to L_{gf} can be written as:

$$\int_D \delta\eta^{\dagger,i} \wedge \delta\eta_i = \int_D *d\delta\gamma^i \wedge \delta\eta_i = -\int_D \delta\gamma^i d * \delta\eta_i + \int_{\partial D} \delta\gamma^i * \delta\eta_i.$$

The first integral is zero on L_{gf}. In order for the boundary term in this integral to vanish, one needs to put an extra boundary condition $\gamma(s) \in T_{X(s)}L_a$ for $s \in I_a$ and $\gamma = 0$ on I (where $\eta = 0$). In particular, this condition implies that the normal component of η^+ is in $T_{X(s)}L_a$ for $s \in I_a$ and vanishes for $s \in I$.[f]

The gauge fixing submanifold L_{gf} is a Lagrangian submanifold of conormal type if we will impose the "strictification" of boundary conditions discussed above. One can define an equivalent gauge fixing using the "gauge fixing fermion". For this, one needs to introduce a quadruplet of auxiliary fields γ, $\overline{\nu}$ (Lagrange multiplier), $\gamma^\dagger$ and $\nu^\dagger$. Such extra fields are typical for the BV extended version of Faddeev–Popov gauge fixing with Lagrange multipliers. For details, see [2].

3.4. *Solutions to the Euler–Lagrange equations*

The gauge fixed action is the restriction of the action to L_{gf} which was computed in [2]:

$$S_{gf}[\tilde{X}, \tilde{\eta}] = \int_D \Big(\eta_i \wedge dX^i + \frac{1}{2}\omega^{ij}(X)\eta_i \wedge \eta_j - \eta^{\dagger,i}$$
$$\wedge \left(d\beta_i + \partial_i\omega^{kl}(X)\eta_k\beta_l\right) - \frac{1}{4}\eta^{\dagger,i} \wedge \eta^{\dagger,j}\partial_i\partial_j\omega^{kl}\beta_k\beta_l \Big).$$

[f] In order for the space of fields, subject to the boundary conditions corresponding to topological quantum mechanics to be closed under d and d^*, one needs to refine (or "strictify") the boundary conditions.

On I, one should impose the ultra-Dirichlet boundary condition for $\tilde{\eta}$ and ultra-Neumann for $\tilde{X}$, cf. [6, Appendix A.1] (see also Appendix A.3.2 in the same reference for the logic of strictification of the boundary conditions). To refine boundary conditions for $s \in I_a$, one should split T^*M into the direct sum of $N^*\mathcal{L}_a$ and a complementary vector space. This can be achieved by choosing an almost complex structure J on TM. The refined boundary conditions can be chosen then as even normal derivatives of $\tilde{X}|_{I_a}$ and odd normal derivatives of $(*\tilde{X})|_{I_a}$ are in $T\mathcal{L}_a$; odd normal derivatives of $\tilde{X}|_{I_a}$ and even normal derivatives of $(*\tilde{X})|_{I_a}$ are in $J(T\mathcal{L}_a)$; even normal derivatives of $\tilde{\eta}|_{I_a}$ and odd normal derivatives of $(*\tilde{\eta})|_{I_a}$ are in $N^*\mathcal{L}_a$; odd normal derivatives of $\tilde{\eta}|_{I_a}$ and even normal derivatives of $(*\tilde{\eta})|_{I_a}$ are in $J^*(N^*\mathcal{L}_a)$.

To be more precise, we need these boundary conditions for the fluctuations around classical solutions, but we will defer this discussion to the extended version of this note.

In this action, $\eta_i = *df_i$ and $\eta^{\dagger,i} = *d\gamma^i$. For the variation, we have:

$$\delta S_{gf}[\tilde{X},\tilde{\eta}] = \text{ bulk terms} + \int_D (\eta_i \delta X^i + \eta^{\dagger,i}\delta\beta). \tag{13}$$

As usual, it consists of two terms: the bulk term and the boundary term. The bulk terms vanishes if and only if the Euler–Lagrange equations hold:

$$d^*(dX^i + \omega^{ij}(X)\eta_j - \eta^{\dagger,l}\wedge \partial_l\omega^{ij}(X)\beta_j) = 0,$$

$$d\eta_i + \frac{1}{2}\partial_i\omega^{kl}\eta_k\wedge\eta_l - \eta^{\dagger,l}\wedge\partial_l\partial_i\omega^{jk}(X)\eta_j\beta_k - \frac{1}{4}\eta^{\dagger,k}\wedge\eta^{\dagger,l}\partial_k\partial_l\partial_i\omega^{st}\beta_s\beta_t = 0,$$

$$d^*(d\beta_i + \partial_i\omega^{jk}(X)\eta_j\beta_k + \frac{1}{2}\eta^{\dagger,j}\partial_i\partial_j\omega^{km}(X)\beta_k\beta_m) = 0,$$

$$d\eta^{\dagger,i} + \eta^{\dagger,j}\wedge\partial_j\omega^{ki}(X)\eta_k + \frac{1}{2}\eta^{\dagger,l}\wedge\eta^{\dagger,j}\partial_l\partial_j\omega^{ki}(X)\beta_k = 0.$$

The boundary terms in (13) vanish because of the boundary conditions described earlier. These equations with boundary conditions described earlier still have infinitely many solutions because the gauge symmetry on the boundary is not yet fixed, see the discussion below.

3.5. *The case of constant symplectic structure on* $\mathbb{R}^{2n}$

From now on, we will focus on the simplest but already non-trivial case when $M = \mathbb{R}^{2n}$ with constant symplectic structure. In this case, the Euler–Lagrange equations become:

$$d^*(dX^i + \omega^{ij}\eta_j) = 0, \quad d\eta_j = 0, \quad d^*d\beta_i = 0, \quad d\eta^{\dagger,i} = 0.$$

We also should take into account the gauge fixing conditions $\eta^{\dagger,i} = *d\gamma^i$, $d^*\eta = 0$ and the boundary conditions. This gives

(1)
$$d^*dX^i = 0, \quad X|_{I_\alpha} \in \mathcal{L}_\alpha, \quad X_{I_{12}} \in \Lambda, \quad dX^i|_I = 0.$$
Here and below $\alpha = 1, 2$.

(2) For η, we have
$$d\eta = 0$$
with boundary conditions $\eta|_I = 0$, $\eta(s)|_{I_a} \in N^*_{X(s)}\mathcal{L}_a \otimes T^*_s I_a$ when $s \in I_\alpha$ and $\eta(s)|_{I_{12}} \in N^*_{X(s)}\Lambda \otimes T^*_s I_{12}$ when $s \in I_{12}$.

(3) For β_i we have the linear boundary value problem
$$d^*d\beta_i = 0, \quad \beta_i|_I = 0,$$
with $\beta(s)|_{I_a} \in N^*_{X(s)}\mathcal{L}_a \otimes T^*_s I_a$ when $s \in I_a$, $\beta(s)|_{I_{12}} \in N^*_{X(s)}\Lambda \otimes T^*_s I_{12}$ when $s \in I_{12}$, and $\beta|_I = 0$.

(4) For the fields γ^i, we obtain linear problem:

$$d^* d\gamma^i = 0$$

with boundary conditions $*d\gamma(s)|_{I_a} \in T_{X(s)}\mathcal{L}_a \otimes T^*_s I_a$ when $s \in I_a$, $*d\gamma(s)|_{I_{12}} \in T_{X(s)}\Lambda \otimes T^*_s I_{12}$ when $s \in I_{12}$, and free boundary conditions on I.

Note that Euler–Lagrange equations are linear, but boundary conditions are not. It is clear that these boundary conditions are incomplete: we still have infinitely many solutions to the Euler–Lagrange equations with these boundary conditions. This is due to the remaining gauge symmetry on the boundary. To observe the remaining gauge symmetry, let us focus on fields with $gh = 0$, i.e. on X and η. When the symplectic structure is constant, infinitesimal gauge transformations are

$$\delta X^i = \omega^{ij}\beta_j, \quad \delta\eta_i = -d\beta_i.$$

Such a vector field is parallel (tangent) to the Lorenz gauge fixing section $d * \eta = 0$ only if $d^* d\beta_i = 0$, i.e. when β is harmonic. Boundary values of β are constrained by conditions on X. Since $X(s) \in \mathcal{L}_a$ when $s \in I_a$, we have $\delta X(s) \in T_{X(s)}\mathcal{L}_a$ and therefore $\beta \in N^*_{X(s)}\mathcal{L}_a$. This agrees with the boundary conditions on η. However, there are infinitely many such harmonic functions. They form the remaining boundary gauge symmetry. We call it boundary gauge symmetry because a harmonic function is determined by its boundary value.

It is easy to see that the refined boundary conditions discussed in the previous section fix this boundary gauge symmetry.

4. Quantization of the BV Extended Poisson Sigma Model

Here we will focus on the Poisson sigma model with boundary conditions corresponding to amplitudes and cyclic amplitudes in the topological quantum mechanics.

According to the general philosophy of BV quantization, we should choose a gauge fixing Lagrangian submanifold in the space of bulk BV fields in such a way that the BV extended action has isolated critical points on this Lagrangian. Then the perturbative partition function is defined by the sum of corresponding Feynman diagrams. Thus, our main object, should be

$$\int_{L_{gf}} \exp\left(\frac{i}{h} S_{gf}[\widetilde{X}, \widetilde{\eta}]\right) D\widetilde{X} D\widetilde{\eta}$$

with boundary conditions described above.

However, in our case Lorenz gauge fixing does not provide isolated critical points because of the remaining gauge transformations which are parallel to the Lagrangians $\mathcal{L}_a$ (unless we consider refined boundary conditions). In addition to this, we have nonlinear boundary conditions. One way to tackle both problems is to treat the boundary conditions as boundary constraints using Lagrange multipliers.

The main idea in the rest of this section is to introduce Lagrange multipliers λ for the boundary constraints on X, keep X free on the boundary and impose boundary conditions $\eta = 0$.[g] The usual mechanism with Lagrange multipliers will constrain field X to $\mathcal{L}_a$ on I_a. Lagrange multipliers λ parametrize the "true" boundary values of η. Thus, Lagrange multipliers will effectively change the boundary condition $\eta = 0$ and free X on each I_a to the correct boundary condition corresponding to the Lagrangian submanifold $\mathcal{L}_a$. We conjecture that this gives the correct Feynman diagram expansion in the path integral. Here we consider the case of the amplitude in the asymptotic metric, so $a = 1, 2$.

Assume that the Lagrangian fibers $\mathcal{L}^{(a)}_{b_a}$ are the level surfaces of the Poisson commuting functions $H^{(a)}_\alpha \in C(M)$, where $\alpha = 1, \ldots, dim(M)/2$. This is a reasonable assumption in the context of integrable systems where these functions are commuting Hamiltonians.

In the BV framework, we want to introduce an extra term $\widetilde{S}_\partial$ in the action with Lagrange multipliers and their BV counterparts in such a way that the total action satisfies the classical master equation. In order to do this, we introduce boundary fields $\lambda^a_\alpha \in \Omega^1(I_a)$, $\lambda^{\alpha,\dagger}_a \in \Omega^0(I_a)[-1]$, $c^a_\alpha \in \Omega^0(I_a)[1]$, $c^{\alpha,\dagger}_a \in \Omega^1(I_a)[-2]$, $\overline{c}^\alpha_a \in \Omega^0(I_a)[-1]$, $\overline{c}^{a,\dagger}_\alpha \in \Omega^1(I_a)$, $\mu^\alpha_a \in \Omega^0(I_a)$, $\mu^{a,\dagger}_\alpha \in \Omega^1(I_a)[-1]$. Denote this space of fields $\widetilde{F}_{\partial D}$. Naturally,

$$\widetilde{F}_{\partial D} = \oplus_a \widetilde{F}_{I_a}.$$

Each space $\widetilde{F}_{I_a}$ has a natural symplectic structure of degree -1:

$$\widetilde{\omega}_{I_a} = \int_{I_a} \sum_\alpha (\delta\lambda^{\alpha,\dagger}_a \wedge \delta\lambda^a_\alpha + \delta c^{\alpha,\dagger}_a \wedge \delta c^a_\alpha + \delta\overline{c}^{a,\dagger}_\alpha \wedge \delta\overline{c}^\alpha_a + \delta\mu^{a,\dagger}_\alpha \wedge \delta\mu^\alpha_a).$$

Note that the symplectic space $(\widetilde{F}_{\partial D}, \widetilde{\omega}_{\partial D})$ is not the space of BFV fields, but the space of BV extended Lagrange multipliers.

Denote by $F^{(0)}_D$ the space of bulk fields with boundary conditions $\eta|_{\partial D} = 0$ and $X|_{\partial D}$ is free and define the total space of fields as $\mathcal{F}_D = F^{(0)}_D \oplus \widetilde{F}_{\partial D}$. This space has a natural symplectic form of degree -1 which is the direct sum of symplectic forms on $F^{(0)}_D$ and $\widetilde{F}_{\partial D}$.

Define the boundary correction to the action as

$$\widetilde{S}_\partial = \sum_a \sum_\alpha \int_{I_a} (\lambda^a_\alpha (H^{(\alpha)}_a - b^\alpha_a) + \lambda^{\alpha,\dagger}_a dc^a_\alpha + c^a_\alpha \partial_i H^{(\alpha)}_a \eta^{i,\dagger} + \mu^\alpha_a \overline{c}^{a,\dagger}_\alpha).$$

Again, this is not the BFV action, this is a BV extended term with Lagrange multiplier enforcing the boundary conditions on X.

It is easy to check that the modified action $S + \widetilde{S}_\partial$ satisfies the classical master equation.

$$\{S + \widetilde{S}_\partial, S + \widetilde{S}_\partial\}_D = 0$$

[g] To be more precise, we should do this only to fluctuations, but we will discuss this, as many other missing details in this note, in its extended version.

where the Poisson bracket is defined by the symplectic structure on the total space of BV fields $\mathcal{F}_D$.

We will choose Lorenz gauge fixing in $F_D^{(0)}$, as in [2]. The additional gauge fixing for BV Lagrange multipliers can be chosen as a version of the Lorenz gauge fixing. Corresponding Lagrange submanifold in $\widetilde{\mathcal{F}}_\partial$ is the graph of the function

$$\psi = \sum_a \int_{I_a} \sum_\alpha \overline{c}_a^\alpha d^* \lambda_\alpha^a.$$

The Lagrangian submanifold is defined as $a^\dagger = \frac{\delta\psi}{\delta a}$ where a are boundary fields which translates to:

$$\lambda_a^{\alpha,\dagger} = -d^* \overline{c}_a^\alpha, \quad c_a^{\alpha,\dagger} = 0,$$

$$\overline{c}_\alpha^{a,\dagger} = d^* \lambda_\alpha^a, \quad \mu_a^{\alpha,\dagger} = 0.$$

For the restriction of the boundary action to this Lagrangian submanifold, we have

$$\widetilde{S}_{\partial,gf} = \sum_a \sum_\alpha \int_{I_a} (\lambda_\alpha^a (H_a^{(\alpha)} - b_a^\alpha) - d^* \overline{c}_a^\alpha dc_\alpha^a + c_\alpha^a \partial_i H_a^{(\alpha)} \eta^{i,\dagger} + \mu_a^\alpha d^* \lambda_\alpha^a).$$

Note that if we ignore all boundary fields except λ, the Euler–Lagrange equations for $S + \widetilde{S}_{\partial D}$ induce the Euler–Lagrange equations discussed before in the bulk. On the boundary, on $I_a \in \partial D$, we will have $H_a^{(\alpha)}(X) = b_a^\alpha$ which means $X \in \mathcal{L}_a$ and we will also have $\eta_i = \sum_\alpha \lambda_\alpha \partial_i H_a^{(\alpha)}$ which means $\eta(s) \in N^*_{X(s)} \mathcal{L}_a$.

4.1. *Gluing*

Two Lagrangian fibrations $\pi_a \colon M \to B_a$ on M define two real polarizations P_1, P_2. Denote by $U_{P_2,P_1} \colon H_{P_1}^{1/2} \to H_{P_2}^{1/2}$ the integral operator with the integral kernel $(\psi_{b_2}^{(2)}, \psi_{b_1}^{(1)})$. Here, by an integral operator we mean a formal integral operator where the integration is given by the formal power series in Feynman diagrams. This is a semiclassical version of the Blattner–Kostant–Sterberg kernels in geometric quantization.

Path integral arguments naturally lead to the composition rule

$$U_{P_3,P} * U_{P,P_1} = U_{P_3,P_1} \exp\left(\frac{i\pi}{4} \mu(P_1, P, P_3)\right)$$

where $\mu(P_1, P, P_3)$ is the Maslov index. Here the composition is the formal stationary phase integral over B_2 given by contributions from critical points. We conjecture that this identity holds as an identity of formal powers series with coefficients being Feynman diagrams for the Poisson sigma model with appropriate boundary conditions. In terms of the path integral, it is a two-step argument. First because our field theory is topological, we can deform the gauge fixing choice of the metric to a metric where I_2 shrinks to a point. Then integrating over B_2 allows all possible values of X on the interval I_2 and thus, removes the constraint.

Pictorially, this identity corresponds to gluing identity from Fig. 5.

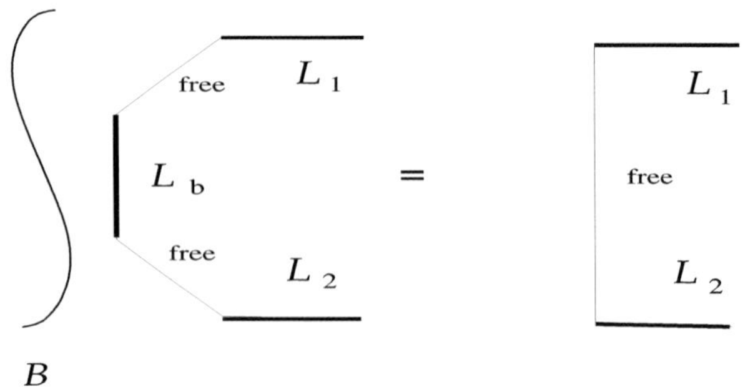

Fig. 5. Composing two BKS kernels.

5. Geometric Representation of Kontsevich's Star Products for Poisson Manifolds

Here we will denote by $\mathcal{P}$ a Poisson manifold, i.e. a smooth manifold with a bivector field $p \in \Gamma(\wedge^2 T^*\mathcal{P})$ such that the bracket

$$\{f, g\} = p(df \wedge dg)$$

defines a Lie algebra structure on the space $C(\mathcal{P})$ of smooth functions on $\mathcal{P}$.

Recall that a star product of two smooth functions on $\mathcal{P}$ is an associative product on $C(\mathcal{P})[[h]]$ defined as

$$f * g = fg + \frac{ih}{2}\{f, g\} + \sum_{n\geq 2} m_n(f, g)(ih)^n$$

where coefficients m_n are bidifferential operators. The associativity of the star product is equivalent to certain bilinear identities for coefficients m_n. Kontsevich in [11] gave an explicit construction of such star product where the bidifferential operators m_n are described explicitly in terms of integrals on configuration spaces.

5.1. *Kontsevich's star product as Poisson sigma model*

In this section, as above, we assume that D is an oriented disc. We think of D as a disc in $\mathbb{R}^2$ and assume the counterclockwise orientation.

The Poisson σ-model with boundary conditions $\eta = 0$ and X free [2] describes Kontsevich's star product [11] on the space of smooth functions on a Poisson manifold $\mathcal{P}$. Recall this construction. The star product of two smooth functions f and g can be regarded as the path integral

$$f * g = \int_{X(\bullet)=x} \exp\left(\frac{i}{h} S[X, \eta]\right) f(X(0))g(X(1))DXD\eta.$$

Here $\bullet$, 0, 1 are three points on the boundary of the disc (see Fig. 6). The integral is understood as a formal power series of Feynman integral contributions from fluctuations around the critical point $X(u) = x$, $\eta(u) = 0$. The description of Feynman diagrams in the Lorenz gauge in BV framework was derived in [2].

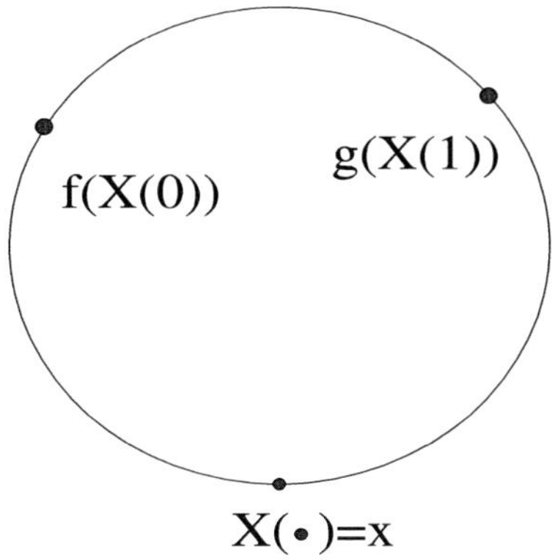

Fig. 6. The disc with three points $\bullet, 0, 1$ on the boundary and insertions of functions f and g evaluated at $X(0)$ and $X(1)$ in the integral.

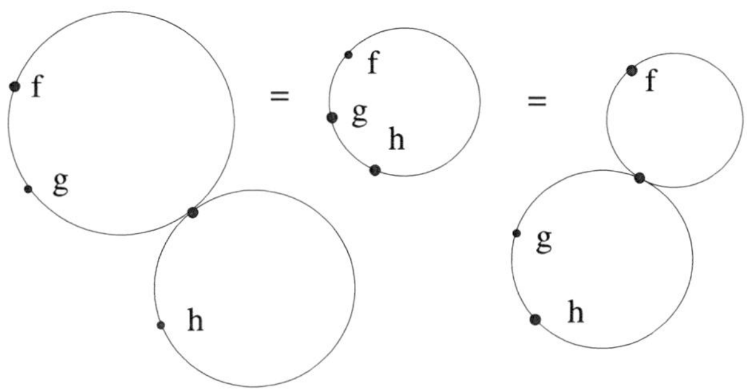

Fig. 7. The path integral interpretation of the associativity.

The associativity of this star product can be proven directly as in [11]. It has a simple heuristic path integral interpretation. The associativity of the star product can be interpreted as the computation of the path integral in two different ways as it is shown in Fig. 7 where all path integrals should be understood perturbatively, i.e. as formal series in h with coefficients given by Feynman diagrams. Note that, of course, this path integral interpretation should be regarded only as a guideline for how to prove the identity between the Feynman diagrams, but not as a proof itself. The proof remains to be the direct proof given in [11].

Making formal deformations of ω in Kontsevich's star product one can obtain all non-equivalent $*$-products. Changing coordinates on the target space by formal diffeomorphisms will obtain all equivalent star products.

5.2. *Poisson sigma model with Lagrangian boundary conditions*

Let $S \subset \mathcal{P}$ be a symplectic leaf. Assume we have three Lagrangian fibrations on S, π: $S \to B$ and $\pi_{1,2}$: $S \to B_{1,2}$.

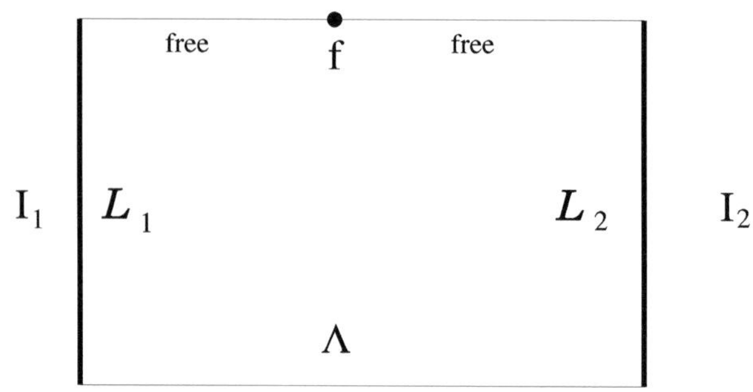

Fig. 8. The boundary conditions for the path integral representing matrix elements $(\psi_{b_2}^{(2)}, \pi(f)\psi_{b_1}^{(1)})$.

Thus, as in the first section, we can define the vector space of 1/2-densities $H_P^{1/2}$ and two integrable systems corresponding to two other Lagrangian fibrations. We want to define the action of the associative algebra $(C(\mathcal{P})[[h]], *)$ on the vector space $H_P^{1/2}$. That is, to each $f \in C(\mathcal{P})[[h]]$, we want to assign $\pi(f)\colon H_P^{1/2} \to H_P^{1/2}$, in such a way that $\pi(f * g) = \pi(f)\pi(g)$.

Define the matrix elements of $\pi(f)$ between semiclassical eigen-half-densities as the path integral

$$(\psi_{b_2}^{(2)}, \pi(f)\psi_{b_1}^{(1)}) = \int \exp\left(\frac{i}{h}S[X,\eta]\right) f(X(\bullet))DXD\eta. \tag{14}$$

Here the integral is taken over fields of the Poisson sigma model with the target space $\mathcal{P}$ with boundary conditions $X|_{I_a} \in \mathcal{L}_a \subset S \subset \mathcal{P}$, $\eta|_{I_a} \in N_X^*\mathcal{L}_a$, $a = 1,2$ where $\mathcal{L}_a = \mathcal{L}_{b_a}^{(1)}$ are Lagrangian fibers of $\pi_{1,2}$. On the lower part of the boundary we have the same boundary condition with $\mathcal{L}_a$ being replaced by Λ and on the upper part of the boundary conditions are $\eta = 0$ and X is any. These boundary conditions and the insertion of the function f are illustrated in Fig. 8.

Note that matrix elements (14) do not depend on the choice of α. In particular, for $M = T^*Q$, a combination of a formal diffeomorphism and a formal deformation of ω brings this star product to the one given by differential operators (the one we used in the first section).

Boundary conditions corresponding to the dual map $\pi(f)^*$ are shown in Fig. 9. It is natural to consider (14) as the kernel of an integral operator $\pi_{21}(f)\colon H_{P_1}^{1/2} \to H_{P_2}^{1/2}$. In particular, when $P_1 = P_2 = P$, $\pi(F)$ is an endomorphism of $H_P^{1/2}$. But in this, the semiclassical limit becomes quite singular. It is clear that the following identities hold:

$$U_{P_3P_2}\pi_{21}(f) = \pi_{31}(f) = \pi_{32}(f)U_{P_2P_1}$$

where $U_{P_2P_1}$ is the Blattner–Kostant–Sternberg kernel given by the scalar product which was the subject of previous sections. For the semiclassical asymptotic of (14),

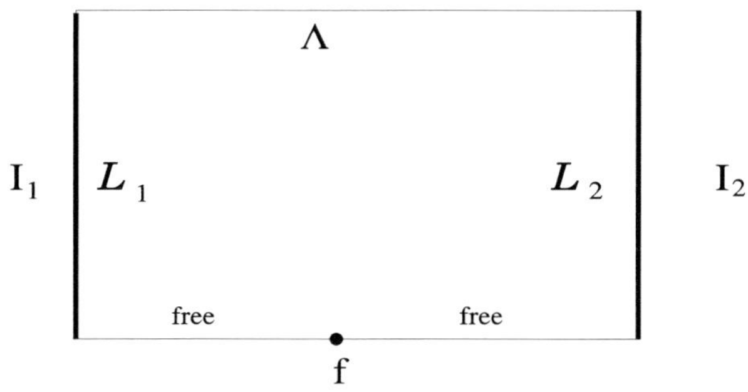

Fig. 9. Boundary conditions for the path integral representing $(\psi^{(1)}_{b_1}, \pi(f)^*\psi^{(2)}_{b_2}) = \overline{(\psi^{(2)}_{b_2}, \pi(f)\psi^{(1)}_{b_1})}$.

we have

$$\pi_{21}(f)(b_2, b_1) = \frac{C}{(2\pi h)^{n/2}} \sum_{c \in \mathcal{L}^{(1)}_{b_1} \cap \mathcal{L}^{(2)}_{b_2}} e^{\frac{i}{h} S_{\gamma_1,\gamma_2}(c,b_1,b_2) + \frac{i\pi}{2}\mu_{\gamma_1,\gamma_2}(c)}$$

$$\times \left| det \left(\frac{\partial^2 S_{\gamma_1,\gamma_2}(c, b_1, b_2)}{\partial b^i_1 \partial b^j_2} \right) \right|^{1/2} f(c)(1 + O(h))\sqrt{|db_1 db_2|} \quad (15)$$

where everything has the same meaning as in (1).

The path integral arguments used in the composition law of amplitudes also imply the following important property of matrix elements (14). If f is constant of fibers $\mathcal{L}^{(1)}_{b_1}$ and g is constant on fibers $\mathcal{L}^{(2)}_{b_2}$ then

$$\pi_{21}(f)(b_2, b_1) = U_{P_2P_1}(b_2, b_1)f(b_1), \quad \pi_{21}(g)(b_2, b_1) = g(b_2)U_{P_2P_1}(b_2, b_1).$$

Now we should check the homomorphism property:

$$\pi_{21}(f * g)(b_2, b_1) = \int_B \pi_{20}(f)_{b_2,b}\pi_{01}(g)(b, b_1) \quad (16)$$

where index 0 corresponds to the Lagrangian fibration P and $*$ is the Kontsevich's star product.

This identity has the following path integral interpretation. The partition function corresponding to Fig. 10 is equal to the partition function corresponding to Figs. 11 and 12. The arguments are the same as in case of the gluing properties of amplitudes.

According to the general philosophy of quantization of Poisson manifolds, symplectic leaves correspond to irreducible representations.[h] Thus, we expect that representations constructed in this way are irreducible representations of $C_h(\mathcal{P})$.

[h]When $\mathcal{P}$ is the dual space to a Lie algebra, this correspondence is at the heart of the orbit method, which assigns irreducible representations of a Lie group to co-adjoint orbits. This is known as the Kirillov–Kostant method.

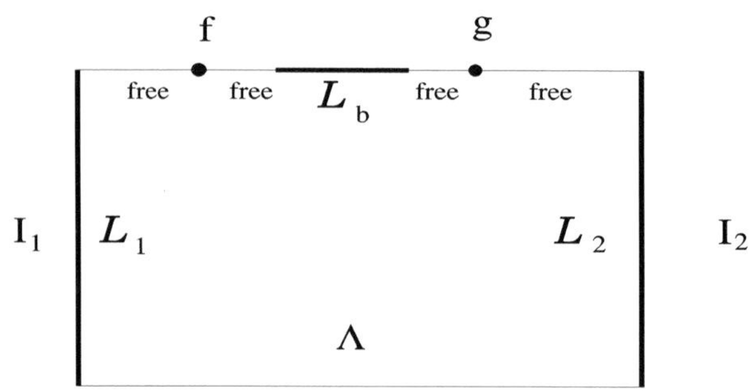

Fig. 10. Boundary conditions for the path integral interpretation of the integrand in (16).

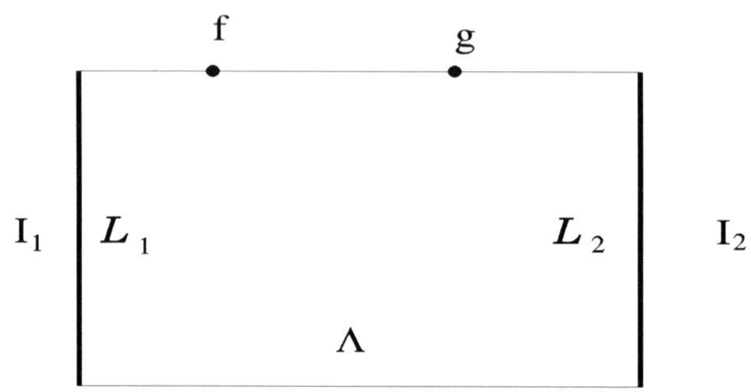

Fig. 11. Boundary conditions for the path integral interpretation of the result of integration in the right-hand side of (16).

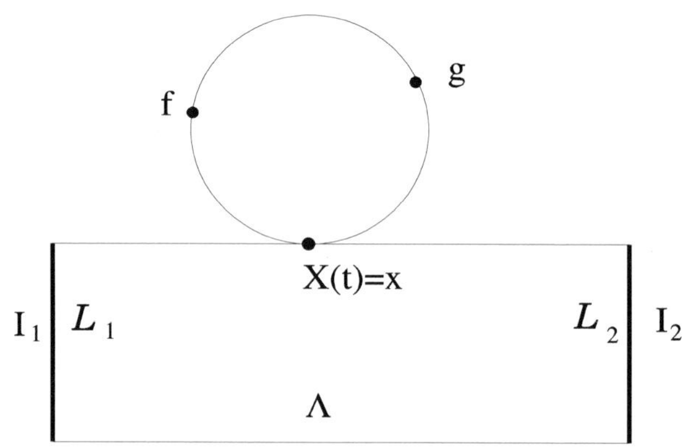

Fig. 12. Boundary conditions for the path integral interpretation of the left-hand side of (16).

6. Conclusion

Throughout this paper, we use real polarizations which are given by Lagrangian fibrations. Such polarizations have clear semiclassical meaning in terms of real symplectic geometry and analytical mechanics.

However, such polarizations are typically quite singular. Complex polarizations have richer mathematical structures, are less singular, allow tools from complex geometry, and in many important cases, from algebraic geometry. This is why complex polarizations are more in use in mathematical literature on geometric quantization. In physics literature, the corresponding representations of quantum observables are known as the holomorphic realization of quantum mechanics. Topological quantum mechanics in holomorphic representation is closely related the A-model in string theory [15] (see also [1]).

Note that the path integral formula for (1) suggests that the torsion for the Poisson sigma model on a disc with boundary conditions corresponding to TQM is given by the semiclassical Hessian in (1).

This paper is only an outline. The detailed description of Feynman diagrams with proofs of basic statements will be done in a separate publication.

Acknowledgments

A. S. C. acknowledges partial support of SNF Grant No. 200020-172498/1. This research was (partly) supported by the NCCR SwissMAP, funded by the Swiss National Science Foundation, and by the COST Action MP1405 QSPACE, supported by COST (European Cooperation in Science and Technology). P. M. acknowledges partial support of RFBR Grant No. 17-01-00283a. The work of N. R. was supported by the NSF grant DMS-1601947.

References

[1] F. Bonechi, A. S. Cattaneo and R. Iraso, A-model, comparing Poisson sigma model with A-model, *JHEP* **10** (2016) 133; arXiv:1607.03411.

[2] A. S. Cattaneo and G. Felder, A path integral approach to the Kontsevich quantization formula, *Comm. Math. Phys.* **212** (2000) 591–611.

[3] A. S. Cattaneo and G. Felder, Coisotropic submanifolds in Poisson geometry and branes in the Poisson sigma model, *Lett. Math. Phys.* **69** (2004) 157–175.

[4] A. S. Cattaneo and G. Felder, Relative formality theorem and quantization of coisotropic submanifolds, *Adv. Math.* **208**(2) (2007) 521–548.

[5] A. S. Cattaneo, P. Mnev and N. Reshetikhin, Classical BV theories on manifolds with boundary, *Comm. Math. Phys.* **332**(2) (2014) 535–603; arXiv:1201.0290.

[6] A. S. Cattaneo, P. Mnev and N. Reshetikhin, Perturbative quantum gauge theories on manifolds with boundary, *Comm. Math. Phys.* **357**(2) (2018) 631–730; arXiv:1507.01221.

[7] K. Costello, E. Witten and M. Yamazaki, Gauge theory and integrability, I: arXiv:1709.09993; II: arXiv:1802.01579.

[8] L. D. Faddeev and L. A. Takhtajan, *Hamiltonian Methods in Soliton Theory* (Springer, 1987).

[9] L. D. Faddeev and V. N. Popov, Feynman diagrams for the Yang–Mills field, *Phys. Lett. B* **25**(1) (1965) 29.

[10] V. Kazakov, Quantum spectral curve of γ-twisted $N = 4$ SYM theory and fishnet CFT, arXiv:1802.02160.

[11] M. Konstevich, Deformation quantization of Poisson manifolds, *Lett. Math. Phys.* **66** (2003) 157–216; arXiv:q-alg/9709040.
[12] L. N. Lipatov, Integrability of scattering amplitudes in $N = 4$, SUSY, *J. Phys. A* **42** (2009) 304020; L. N. Lipatov, High energy asymptotics of multi–colour QCD and exactly solvable lattice models, arXiv:hep-th/9311037.
[13] N. A. Nekrasov and S. L. Shatashvili, Supersymmetric vacua and Bethe ansatz, *Nucl. Phys. Proc. Suppl.* **192/193** (2009) 91–112; arXiv:0901.4744.
[14] N. Reshetikhin, Semiclassical geometry of integrable systems, arXiv:1802.00416.
[15] E. Witten, Topological sigma model, *Comm. Math. Phys.* **118** (1988) 411–449.

Orthogonal Polynomials in Mathematical Physics

Chuan-Tsung Chan[*,||], A. Mironov[†,‡,§,**], A. Morozov[‡,§,††] and A. Sleptsov[‡,§,¶,‡‡]

[*] *Department of Applied Physics, Tunghai University, Taichung, 40704, Taiwan*
[†] *Lebedev Physics Institute, Moscow 119991, Russia*
[‡] *ITEP, Moscow 117218, Russia*
[§] *Institute for Information Transmission Problems, Moscow 127994, Russia*
[¶] *Laboratory of Quantum Topology, Chelyabinsk State University, Chelyabinsk 454001, Russia*
[||] *ctchan@go.thu.edu.tw*
[**] *mironov@lpi.ru; mironov@itep.ru*
[††] *morozov@itep.ru*
[‡‡] *sleptsov@itep.ru*

This is a review of (q-)hypergeometric orthogonal polynomials and their relation to representation theory of quantum groups, to matrix models, to integrable theory, and to knot theory. We discuss both continuous and discrete orthogonal polynomials, and consider their various generalizations. The review also includes the orthogonal polynomials into a generic framework of (q-)hypergeometric functions and their integral representations. In particular, this gives rise to relations with conformal blocks of the Virasoro algebra.

To the memory of Ludwig Dmitrievich Faddeev

Contents

1. Introduction

Orthogonal polynomials [1] are a classical chapter in natural science, and still a somewhat mystical one. They were originally introduced in the 19th century in a place, which may look strange from today's perspectives: in the theory of continued fraction, which is now far away from the mainstreams in mathematical physics. Then some of them appeared in study of the hypergeometric series and entered the textbooks, as the simplest solutions of the most relevant equations, both in classical and quantum physics at ordinary energies. In the 20th century, they became for some time the main tool in the study of matrix models, the basic example and source of inspiration for modern quantum field and string theories. Quite recently, the hypergeometric polynomials and their mathematically evident q-deformations re-emerged in the advanced theory of conformal blocks [2] and knot invariants [3], they also re-appeared in matrix model in a new avatar [4]. This is one of the immediate reasons for this paper. It is basically a review of the well and not-so-well known facts from the theory of orthogonal polynomials, with the emphasize on particular aspects needed for a progress in Painléve and Racah theories, and in matrix models.

We are interested in the interplay of the following subjects:

- Integral representations (e.g., hypergeometric functions, matrix models).
- Ward identities (Virasoro/string equations) from invariance under the change of integration variables.
- Gauss relations (for hypergeometric functions).
- 3-term relations.
- Orthogonal polynomials and matrices.
- Differential equations (of hypergeometric type, spectral curves).
- Free-field formalism in representation and quantum field theories (including hypergeometric screening integrals).

It is the last item in the list which makes functions of the hypergeometric type very much distinguished in quantum field theory, while what mattered in classical physics was rather the second-order differential equations which they often satisfy. This promotion from classical to quantum level dictates also a change of view on orthogonal polynomials arising from the hypergeometric series

$$\sum_k \frac{\prod_i (a_i)_k}{\prod_j (b_j)_k} \cdot \frac{z^k}{k!}. \tag{1.1}$$

The reduction to polynomials happens whenever any of the parameters a is negative integer, $a = -n$ since then $(a)_k = \frac{\Gamma(a+k)}{\Gamma(a)}$ vanishes for all $k > n$. The resultant from this truncation is a polynomial both in all other parameters a and in the argument z. The difference is only that the dependence on a's is "quantized", while on z is classical, in fact, z can be considered as an infinitely-large parameter so that asymptotically $(z)_k \sim z^k$. Accordingly, a class of polynomials as functions of z is nowadays called "very classical", while those considered as functions of a, just "classical" (with "quantum" in this context referring to q-deformations).

Thus, natural for physical applications are the hypergeometric functions, and natural for hypergeometric functions is a reduction to polynomials. However, what is much less natural for these polynomials is orthogonality. Orthogonal polynomials are distinguished by the peculiar property known as the 3-term relation, while natural for the hypergeometric polynomials is to satisfy a p-term relation, with p depending on the number of a and b parameters, i.e. on the type (r, s) of the hypergeometric function ${}_rF_s$. As we explain in the paper, for the most interesting case of $r = s+1$ (other values of $r-s$ can be achieved by taking certain limits of these reference functions) the parameter p is equal to 3 for $s \leq 3$, while in general it is rather equal to $2s-3$, i.e. exceeds 3 for large s. Moreover, if one insists on polynomiality of the z-dependence, i.e. on the "very classical" polynomials, the $p=3$ bound actually shifts further down to $s=2$, i.e. to the ordinary hypergeometric functions. It is *this* property that explains at the quantum level the somewhat mysterious limitations on s allowing appearance of orthogonal polynomials in the celebrated Askey scheme [5, 6]. This, in turn, raises a very interesting question of the role of polynomials with p-term relations: clearly it should exist and be quite important in quantum field theory.

A natural point to look at from this perspective would be matrix-model applications of orthogonal polynomials, where the natural conjecture could be that the orthogonality ($p=3$) is related to integrability, the basic property of matrix model partition functions. Perhaps, increasing p could be still another natural deformation of integrability.

Another side of the story at $p=3$ concerns the orthogonality measure of the polynomials. Extremely interesting is the case when the weight function reduces to a *finite* sum of delta-functions, and the associated finite set of orthogonal polynomials actually provide orthogonal matrices. Moreover, when they are hypergeometric, these orthogonal matrices are nothing but the simplest Racah matrices ($6j$-symbols)! Thus, the interplay between the hypergeometric functions, $p=3$, and the discreteness of the weight function becomes crucial for the Racah level of representation theory, and, hence, for the corresponding applications like modular properties of conformal blocks and knot invariants.

This list of ideas explains the logic of the present paper and its composition. We begin in Sec. 2 from recalling the basic properties of orthogonal polynomials *per se*, especially the continuous-fraction relation between the 3-term relation and the measure, which was the birthplace of the theory of orthogonal polynomials in the 19th century. Then in Sec. 3, we describe the standard classical viewpoint (Askey scheme) on the *hypergeometric* orthogonal polynomials, which puts the main emphasize on the second-order differential (or difference, if the q-deformation is allowed) equation for them. In Sec. 4, we explain the alternative "quantum" viewpoint, where the relations are generically p-term, and only restriction to $s=3$ or 2 makes $p=3$ and polynomial orthogonal. In Sec. 5, we claim that the explanation of p-term relations can be achieved through integral representations of hypergeometric series and is associated with division of the integration contour in fragments; this, however, is not very well elaborated in the present paper and deserves further work. In Sec. 6, we discuss the discrete orthogonal polynomials, the typical example being the Racah polynomials, and discuss the relation of these latter with the Askey–Wilson polynomials. In Sec. 7, we describe various extensions of the classical orthogonal polynomials: to multivariable polynomials, which is essential for constructing the Racah matrices for rectangular representations of $SL_q(N)$, and to Askey–Wilson functions, which solve the same second order differential (difference) equation as the Askey–Wilson polynomials do. We consider the q-deformations of the Askey–Wilson functions both in the case of $0 < q < 1$ and in the case of $|q| = 1$. These functions are important for constructing the $6j$-symbols for various infinite-dimensional representations, which we discuss in Sec. 9 in applications to knot theory. Finally, Secs. 8 and 9 remind and develop the standard applications of orthogonal polynomials respectively to matrix models and to Racah matrices and knots, in terms which seem most adequate for further thinking about the role of $p = 3$.

It is this last application to knots, which brings the present paper closest to the interests and achievements of Ludwig Faddeev. Still for us the main inspiration from Ludwig was the need to search for deep cross-disciplinary links between different branches of theoretical physics. Hopefully this paper provides a good example of this kind, despite it being a very long way from the feelings that we express to show clear and rigorous results. Making subtle things clear was and remains the trademark of Ludwig Faddeev's scientific school. We can only dream about approaching these high standards, but we try at least to clearly describe the *setting* and the research domain which looks both traditional and extremely intriguing. We hope that Ludwig would like and appreciate the emerging challenges.

2. Orthogonal Polynomials: A Review of Standard Properties

Standard properties of orthogonal polynomials can be found in the books [1, 7], and those in the q-case in [6, 8] (see also [9] and references therein).

2.1. *Orthogonal polynomials*

Given a non-decreasing measure function, $\mu(x)$, the orthogonal polynomials can be constructed through the Gram–Schmidt process,

$$\int P_n(x)P_m(x)d\mu(x) = \|P_n\|^2\delta_{m,n}, \quad n = 0, 1, 2, \ldots, \tag{2.1}$$

where $\|P_n\|$ is the norm of the nth orthogonal polynomial. This set of functions can be viewed as a complete basis of the Hilbert space and there are several equivalent representations:

- Expression through the moments, $m_n := \int z^n\, d\mu(z)$,

$$P_n(x) = \det_{(n+1)\times(n+1)} \begin{pmatrix} x^n & \dots & m_{n+2} & m_{n+1} & m_n \\ x^{n-1} & \dots & m_{n+1} & m_n & m_{n-1} \\ \vdots & & \vdots & \vdots & \vdots \\ x & \dots & m_3 & m_2 & m_1 \\ 1 & \dots & m_2 & m_1 & m_0 \end{pmatrix}. \tag{2.2}$$

- Expression of polynomials through n-fold integrals

$$P_n(x) = \frac{1}{n!} \int \prod_{i<j} (z_i - z_j)^2 \prod_{i=1}^{n} (x - z_i) d\mu(z_i). \tag{2.3}$$

- The 3-term recursive relation among monic polynomials (i.e. those with unit leading coefficient)

$$x\mathcal{P}_n(x) = \mathcal{P}_{n+1}(x) + K_n \mathcal{P}_n(x) + R_n \mathcal{P}_{n-1}(x), \quad R_n > 0. \tag{2.4}$$

  Hereafter, we denote the monic polynomials by calligraphic letters.

In fact, if the system of polynomials $P_n(x)$ satisfy the 3-term relations, there is always a measure, for which they are orthogonal and

$$\sum_{n=0}^{\infty} \frac{P_n(x)P_n(y)}{\|P_n\|^2} = \delta(x-y) \cdot \frac{dx}{d\mu(x)}. \tag{2.5}$$

In addition, there exists a continued fraction expression for the generating function of the moments, namely, the resolvent,

$$d\rho(x) := \left\langle \frac{dx}{x-z} \right\rangle = \sum_{k=0}^{\infty} m_k \frac{dx}{x^{k+1}} = \cfrac{dx}{x - K_0 - \cfrac{R_1}{x - K_1 - \cfrac{R_2}{x - K_2 - \cdots}}}. \tag{2.6}$$

Hereafter, we denote through $\langle \cdots \rangle$ the averaging with the measure $\mu(z)$:

$$\langle \mathcal{O} \rangle = \int \mathcal{O}\, d\mu(z). \tag{2.7}$$

2.2. *Measure from the 3-term relations*

2.2.1. *Uniqueness of measure*

Explicit shape of the measure $d\mu(z)$ depends on the integration contour:

$$\langle f \rangle = \int_C f(z) d\mu_C(z). \tag{2.8}$$

All measures giving rise to the same average are considered equivalent. Monic orthogonal polynomials are described by a unique orthogonality measure [10, p. 59] if

$$\sum_n \frac{1}{\sqrt{R_n}} = \infty. \tag{2.9}$$

Furthermore, if the orthogonality measure has a finite support, it is also unique [7, Chap. 2, Theorem 5.6]. An example of a non-unique measure can be found in [9, Sec. 3.2].

2.2.2. *Resolvent measure and continued fraction*

If moments of the z-variable are known, then the resolvent given by the Stieltjes kind transform,

$$d\rho(x) = dx \int_C \frac{d\mu_C(z)}{x-z} = \left\langle \frac{dx}{x-z} \right\rangle = \sum_{k=0}^{\infty} \frac{\langle z^k \rangle}{x^{k+1}} dx, \tag{2.10}$$

can be treated as a measure with the contour surrounding infinity:

$$\langle z^k \rangle = -\frac{1}{2\pi i} \oint_\infty x^k d\rho(x). \tag{2.11}$$

The inverse transformation from $d\rho(x)$ to $d\mu_C(x)$, including the case of $C = R$ the real line, is more tricky.

If we have a set of monic polynomials, satisfying 3-term relation,

$$z\mathcal{P}_n(z) = \mathcal{P}_{n+1}(z) + K_n\mathcal{P}_n(z) + R_n\mathcal{P}_{n-1}(z), \tag{2.12}$$

then

$$\begin{aligned}
\mathcal{P}_0 &= 1 \\
\mathcal{P}_1 &= z - K_0 \\
\mathcal{P}_2 &= z^2 - (K_0 + K_1)z + (K_0K_1 - R_1) \\
\mathcal{P}_3 &= z^3 - (K_0 + K_1 + K_2)z^2 + (K_0K_1 + K_2K_2 + K_1K_2 - R_1 - R_2)z \\
&\quad - (K_0K_1K_2 - K_2R_1 - K_0R_2) \\
&\dots
\end{aligned} \tag{2.13}$$

and

$$\begin{aligned}
\langle z \rangle &= K_0, \\
\langle z^2 \rangle &= K_0^2 + R_1, \\
\langle z^3 \rangle &= K_0^3 + (2K_0 + K_1)R_1, \\
&\dots
\end{aligned} \tag{2.14}$$

It is easy to check that these moments guarantee orthogonality of all the polynomials at once,

$$\langle \mathcal{P}_n(z)\mathcal{P}_m(z) \rangle = \delta_{n,m} \cdot \|\mathcal{P}_0\|^2 \cdot \prod_{i=1}^{n} R_i. \tag{2.15}$$

These formulas give rise to the resolvent in the peculiar form of the continued fraction:

$$d\rho(x) = \left\langle \frac{dx}{x-z} \right\rangle = \cfrac{dx}{x - K_0 - \cfrac{R_1}{x - K_1 - \cfrac{R_2}{x - K_2 - \cfrac{R_3}{x - K_3 - \cdots}}}}$$
$$= \frac{dx}{x-K_0} + \frac{R_1 dx}{(x-K_0)^2(x-K_1)} + \left[\frac{R_1^2 dx}{(x-K_0)^3(x-K_1)^2} + \frac{R_1 R_2 dx}{(x-K_0)^2(x-K_1)^2(x-K_2)} \right] + \cdots. \tag{2.16}$$

Example. If all $K_n = 0$ and $R_n = R$ are independent of n, then continuous fraction in (2.16) satisfies $\rho' = 1/(x - R\rho')$ and

$$d\rho(x) = \frac{x - \sqrt{x^2 - 4R}}{2R} dx \tag{2.17}$$

which is the genus-zero resolvent of the Gaussian matrix model. In fact, this resolvent describes the measure

$$d\mu(z) = \sqrt{1 - z^2/(4R)} dz \tag{2.18}$$

at the segment $[-2\sqrt{R}, 2\sqrt{R}]$ and the system of polynomials orthogonal with respect to this measure is the Chebyshev polynomials of the second type

$$\mathcal{P}_n(z) = R^{n/2} \sum_{0}^{[\frac{n}{2}]} \frac{(-1)^m (n-m)!}{m!(n-2m)!} \left(\frac{z}{2\sqrt{R}} \right)^{n-2m} = P_n^{(\frac{1}{2},\frac{1}{2})} \left(\frac{z}{2\sqrt{R}} \right) \tag{2.19}$$

where we denote $P_n^{(\alpha,\beta)}(z)$ the Jacobi polynomials (see Appendix B).

For orthogonal polynomials, higher resolvents are expressed through the first one and literally have nothing to do with the matrix model multi-resolvents, their genus expansion and AMM/EO topological recursion.

2.2.3. *Summing continued fractions with the help of orthogonal polynomials*

However, in general, summation of continued fractions is a far less trivial problem. In fact, it was the one that originated the interest to orthogonal polynomials. Let us see how this works in the case of the polynomials given on the real axis. Let us consider the general 3-term recurrence relation

$$P_{n+1}(z) = (A_n z + B_n) P_n(z) - C_n P_{n-1}(z), \quad n \geq 0. \tag{2.20}$$

A system of orthogonal polynomials $\{P_n(z)\}$ satisfies this relations with the initial conditions

$$P_{-1}(z) = 0, \quad P_0(z) = 1. \tag{2.21}$$

We call these polynomials *primary*.

However one can generate another set of polynomials $\{P_n^*(z)\}$ satisfying the same recurrence relation (2.20) with different initial conditions

$$P_{n+1}^*(z) = (A_n z + B_n)P_n^*(z) - C_n P_{n-1}^*(z), \quad n > 0,$$
$$P_0^*(z) = 0, \quad P_1^*(z) = A_0.$$

They are also orthogonal polynomials with respect to the same measure $d\mu(z)$. We call them *secondary* polynomials. The secondary polynomials are related to the primary ones with the help of Stieltjes transformation

$$P_n^*(z) = \int_{\mathbb{R}} \frac{P_n(t) - P_n(z)}{t - z} d\mu(t). \tag{2.22}$$

In this non-monic case, the resolvent is given by the continued fraction

$$d\rho(x) = \cfrac{A_0}{A_0 x + B_0 - \cfrac{C_1}{A_1 x + B_1 - \cfrac{C_2}{A_2 x + B_2 - \cdots}}} dx\,. \tag{2.23}$$

If this fraction converges, then

$$d\rho(z) = \lim_{n\to\infty} \frac{P_n^*(z)}{P_n(z)} dz \tag{2.24}$$

and the measure $d\mu(z)$ is related with the continued fraction (resolvent) $d\rho(z)$ by the Stieltjes transformation (2.10)

$$d\rho(x) = dx \int_{\mathbb{R}} \frac{d\mu_{\mathbb{R}}(z)}{x - z}. \tag{2.25}$$

In order to find the measure $d\mu(z)$, one should use the inverse Stieltjes transformation due to the Stieltjes–Perron formula:

$$\boxed{d\mu(z) = \lim_{\varepsilon\to 0^+} \frac{d\rho(z - i\varepsilon) - d\rho(z + i\varepsilon)}{2\pi i}}\,. \tag{2.26}$$

One can see that, indeed, the measure (2.18) can be obtained from the resolvent (2.17) this way.

Example: Hermite Polynomials. Let us consider the Hermite polynomials and their recurrence relation:

$$H_{n+1}(z) = 2zH_n(z) - 2nH_{n-1}(z),$$
$$A_n = 2,\ B_n = 0,\ C_n = 2n.$$

According to (2.23), one can write the corresponding continued fraction:

$$\frac{d\rho(z)}{dz} = \cfrac{2}{2z - \cfrac{2}{2z - \cfrac{4}{2z - \cfrac{6}{2z - \cfrac{8}{2z - \cdots}}}}}$$

$$\overset{z=1/\xi}{=} \xi + \frac{1}{2}\xi^3 + \frac{3}{4}\xi^5 + \frac{15}{8}\xi^7 + \frac{105}{16}\xi^9 + \frac{945}{32}\xi^{11} + O(\xi^{12}). \tag{2.27}$$

On the other hand, the first few Hermite polynomials and the corresponding secondary polynomials are explicitly given as:

$$\begin{aligned} &P_0(z) = 1, && P_0^*(z) = 0, \\ &P_1(z) = 2z, && P_1^*(z) = 2, \\ &P_2(z) = 4z^2 - 2, && P_2^*(z) = 4z, \\ &P_3(z) = 8z^3 - 12z, && P_3^*(z) = 8z^2 - 8, \\ &P_4(z) = 16z^4 - 48z^2 + 12, && P_4^*(z) = 16z^3 - 40z, \\ &P_5(z) = 32z^5 - 160z^3 + 120z. && P_5^*(z) = 32z^4 - 144z^2 + 64. \end{aligned} \tag{2.28}$$

By computing the first few ratios between the secondary and primary polynomials,

$$\frac{P_1^*(z)}{P_1(z)} = \frac{2}{2z} \overset{z=1/\xi}{=} \xi$$

$$\frac{P_2^*(z)}{P_2(z)} = \frac{4z}{4z^2-2} = \cfrac{2}{2z - \cfrac{2}{2z}} \overset{z=1/\xi}{=} \xi + \frac{\mathbf{1}}{\mathbf{2}}\xi^{\mathbf{3}} + \frac{1}{4}\xi^5 + O(\xi^7)$$

$$\frac{P_3^*(z)}{P_3(z)} = \frac{8z^2-8}{8z^3-12z} = \cfrac{2}{2z - \cfrac{2}{2z - \cfrac{4}{2z}}} \overset{z=1/\xi}{=} \xi + \frac{\mathbf{1}}{\mathbf{2}}\xi^{\mathbf{3}} + \frac{\mathbf{3}}{\mathbf{4}}\xi^{\mathbf{5}} + \frac{9}{8}\xi^7 + \frac{27}{16}\xi^9 + O(\xi^{11})$$

$$\frac{P_4^*(z)}{P_4(z)} = \frac{16z^3-40z}{16z^4-48z^2+12} = \cfrac{2}{2z - \cfrac{2}{2z - \cfrac{4}{2z - \cfrac{6}{2z}}}} \overset{z=1/\xi}{=} \xi + \frac{\mathbf{1}}{\mathbf{2}}\xi^{\mathbf{3}} + \frac{\mathbf{3}}{\mathbf{4}}\xi^{\mathbf{5}} + \frac{\mathbf{15}}{\mathbf{8}}\xi^{\mathbf{7}} + \frac{81}{16}\xi^9 + O(\xi^{11})$$

one can see that the first n terms of $\frac{P_n^*(z)}{P_n(z)}$ coincide with the first n terms of the series (2.27), which illustrates (2.24). It is also clear that, in the limit of $n \to \infty$, this ratio approaches to the following series,

$$\frac{d\rho(z)}{dz} = \frac{P_n^*(z)}{P_n(z)} \to \frac{1}{\pi} \sum_k \Gamma\left(k + \frac{1}{2}\right) z^{-2k-1}. \tag{2.29}$$

The Borel sum of this series is

$$\frac{1}{\pi} \int_0^\infty dt\, e^{-t} \sum_k \Gamma\left(k + \frac{1}{2}\right) z^{-2k-1} \frac{t^k}{k!} = \frac{1}{\pi} \int_0^\infty \frac{dt\, e^{-t}}{\sqrt{z^2 - t}}. \tag{2.30}$$

This integral converges at pure imaginary $z = i\zeta$ and is equal (upon substitution $t \to y^2 + 2y\zeta$) to

$$\frac{d\rho(z)}{dz} = \frac{1}{i\pi} e^{\zeta^2} \operatorname{erf}(-\zeta) = \frac{1}{i\pi} e^{-z^2} \operatorname{erf}(iz) = \frac{1}{i\pi} \int_{-\infty}^\infty \frac{e^{-x^2}}{z - x} dx \tag{2.31}$$

i.e. from (2.25), the measure is the Gaussian one, $d\mu_{\mathbb{R}}(x) = e^{-x^2}$.

2.3. *Orthogonal polynomials of discrete variable*

If the measure $\mu(x)$ in (2.1) is a non-decreasing step function with jumps $\mu_j > 0$ at $x = x_j$, $j = 1, 2, \ldots, N$,

$$d\mu(x) = \sum_{j=1}^{N} \mu_j \delta(x - x_j) dx. \tag{2.32}$$

then (2.1) acquires the form

$$\sum_{j=1}^{N} P_n(x_j) P_m(x_j) \mu_j = \|P_n\|^2 \delta_{m,n}. \tag{2.33}$$

The counterpart of (2.5) in this case is

$$\sum_n \frac{P_n(x_i) P_n(x_j)}{\|P_n\|^2} = \frac{\delta_{i,j}}{\mu_j}. \tag{2.34}$$

The usual boundary conditions

$$P_{-1}(x_j) = 0, \quad P_0(x_j) = 1, \quad \forall j \tag{2.35}$$

is often added with

$$P_N(x_j) = 0, \quad \forall j, \tag{2.36}$$

at some N. An example of a system of such polynomials can be found in Sec. 6.1.

In the case of polynomials orthogonal with discrete measure (2.33),

$$U_{ij} = P_i(x_j) \cdot \frac{\sqrt{\mu_j}}{\|P_i\|}, \tag{2.37}$$

is an $N \times N$ orthogonal (unitary) matrix.

3. Classical Orthogonal Polynomials

3.1. *Generalities*

Considering orthogonal polynomials with the measure of generic type is a too unspecified problem, which, hence, allows one to make only very general statements about the polynomials, which we listed in the previous section. There are, however, various possibilities of specifying the family of orthogonal polynomials. In this section, we briefly describe the family that is called the classical orthogonal polynomials. Following T. Koornwinder, we differentiate between the *classical* and *very classical* polynomials: the latter ones are those considered in the standard textbooks (see Appendix B), while the classical polynomials are those described by the Askey scheme (see Sec. 3.3). In fact, all classical polynomials can be associated either with the hypergeometric functions or with their q-deformation. Hence, sometimes we call polynomials of such a type hypergeometric.

Being hypergeometric, all classical polynomials have integral representations which belong to two large (intersecting) classes:

- The very classical polynomials are all specializations of the Jacobi polynomials, which are standard hypergeometric functions ${}_2F_1(x)$ and, hence, can be realized as one-dimensional integrals,

$$P_n(x) = \int (x-z)^n d\nu_n(z), \tag{3.1}$$

with some measure $d\nu_n(z)$, which depends on n in a relatively simple way.
- Other classical polynomials are of the Askey–Wilson (AW) type (hence, the name AW type polynomials), where the dependence on x and n is in distinct factors (and can be recast to the powers of linear functions in z as we explain below):

$$P_n(x) = \int z^n \psi(z,x) d\eta_n(z). \tag{3.2}$$

These two families of polynomials are related with the idea that, in hypergeometric case, the polynomial is of the form like

$$\sum_{k=0}^{n} \frac{n!}{k!(n-k)!} \frac{\Gamma(u+k)}{\Gamma(u)} \cdot s_k z^k, \tag{3.3}$$

i.e. is a polynomial both in z (the very classical polynomial) and in u (the remaining classical polynomials), but the methods to make them orthogonal differ in the two classes.

3.2. *Classical orthogonal polynomials as hypergeometric polynomials*

The hypergeometric functions of one variable are defined as (see Appendix A)

$${}_rF_s\left(\begin{array}{c} a_1,\dots,a_r \\ b_1,\dots,b_s \end{array}\middle| z\right) := \sum_{k=0}^{\infty} \frac{(a_1)_k \dots (a_r)_k}{(b_1)_k \dots (b_s)_k} \cdot \frac{z^k}{k!}. \tag{3.4}$$

These functions become polynomials when any of $a_i = -n$: in these cases, they are polynomials both in z (the very classical case) and in any other variable a_j (the AW type case).

3.2.1. *The very classical case*

The hypergeometric functions satisfy the following equations

$$\left(z\frac{\partial}{\partial z}+a_i\right) {}_rF_s\left(\begin{array}{c} a_1,\dots,a_i,\dots,a_r \\ b_1,\dots,b_s \end{array}\middle| z\right) = a_i \; {}_rF_s\left(\begin{array}{c} a_1,\dots,a_i+1,\dots,a_r \\ b_1,\dots,b_s \end{array}\middle| z\right),$$

$$\left(z\frac{\partial}{\partial z}+b_j-1\right) {}_rF_s\left(\begin{array}{c} a_1,\dots,a_r \\ b_1,\dots,b_j,\dots,b_s \end{array}\middle| z\right) = (b_j-1) \; {}_rF_s\left(\begin{array}{c} a_1,\dots,a_r \\ b_1,\dots,b_j-1,\dots,b_s \end{array}\middle| z\right), \tag{3.5}$$

and

$$\frac{\partial}{\partial z}\,{}_rF_s\left(\begin{matrix}a_1,\ldots,a_r\\ b_1,\ldots,b_s\end{matrix}\middle|\,z\right)=\frac{\prod a_i}{\prod b_j}\,{}_rF_s\left(\begin{matrix}a_1+1,\ldots,a_r+1\\ b_1+1,\ldots,b_s+1\end{matrix}\middle|\,z\right). \tag{3.6}$$

From the point of view of matrix models these are just the lowest of the Virasoro constraints, see [2].

From these equations, we deduce the following consequences:

- the hypergeometric differential equations of order $\max(r, s+1)$, which follows from multiplying (3.5) for a_i shifts and comparing with the product of (3.6) and all (3.5) for b_j;
- linear algebraic relations (Gauss' contiguous relations) between F with shifted parameters a and b.

The Gauss relations imply 3-term relations for the case $(r,s)=(2,1)$ which can be recast into the form (2.4) provided $a_1=-n$ and $a_2=n+\alpha$, where α is an arbitrary constant. This means that the polynomials ${}_2F_1\left(\begin{matrix}-n,n+\alpha\\ \beta\end{matrix}\middle|\,z\right)$ satisfy 3-term relations and, hence, they are orthogonal polynomials. These are the most general very classical orthogonal polynomials, the Jacobi polynomials (see Appendix B), which we denote as $P_n^{(a,b)}(z)$:

$$P_n^{(a,b)}(x):=\frac{(a+1)_n}{n!}\,{}_2F_1\left(\begin{matrix}-n, & a+b+n+1\\ & a+1\end{matrix}\middle|\,\frac{1}{2}(1-x)\right). \tag{3.7}$$

3.2.2. *Askey–Wilson polynomials*

Of all the AW type polynomials, the most general are the AW polynomials themselves [11], all other are obtained upon suitable specializations. Hence, we describe here the AW polynomials and their properties following [8, Sec. 7.5]. In particular, for the sake of generality, we discuss the q-deformation of the polynomials, which is not too much different from the AW polynomials at $q=1$. All the definitions of the basic hypergeometric series, ${}_r\phi_s$, and other q-functions can be found in Appendix A.

The AW polynomials are defined through the balanced ${}_4\phi_3$ basic hypergeometric series,

$$\begin{aligned}&P_n^{AW}(x=\cos\theta|a,b,c,d)\\ &\quad:=a^{-n}(ab,ac,ad;q)_n\cdot{}_4\phi_3\left(\begin{matrix}q^{-n},abcdq^{n-1},ae^{i\theta},ae^{-i\theta}\\ ab,ac,ad\end{matrix}\middle|\,z=q\right).\end{aligned} \tag{3.8}$$

They are symmetric in a,b,c,d, which can be proved using the Sears transformation formula between two terminating balanced ${}_4\phi_3$ series [8]

$${}_4\phi_3\left(\begin{matrix}q^{-n},a,b,c\\ d,e,f\end{matrix}\middle|\,z=q\right)=a^n\frac{(e/a,f/a;q)_n}{(e,f;q)_n}\,{}_4\phi_3\left(\begin{matrix}q^{-n},a,d/b,d/c\\ d,q^{1-n}a/e,q^{1-n}a/f\end{matrix}\middle|\,z=q\right). \tag{3.9}$$

The polynomials satisfy the duality relation

$$P_n^{AW}(aq^m|a,b,c,d)=P_m^{AW}(\hat{a}q^n|\hat{a},\hat{b},\hat{c},\hat{d}), \quad \hat{a}:=\sqrt{\frac{abcd}{q}},\ \hat{b}:=\frac{ab}{\hat{a}},\ \hat{c}:=\frac{ac}{\hat{a}},\ \hat{d}:=\frac{ad}{\hat{a}}. \tag{3.10}$$

The 3-term relation is of the form

$$2xP_n^{AW}(x) = A_nP_{n+1}^{AW}(x) + B_nP_n^{AW}(x) + C_nP_{n-1}^{AW}(x), \tag{3.11}$$

with $P_{-1}^{AW}(x) = 0$, $P_0^{AW}(x) = 1$, and

$$\begin{aligned} A_n &= \frac{1-abcdq^{n-1}}{(1-abcdq^{2n-1})(1-abcdq^{2n})}, \\ C_n &= \frac{(1-q^n)(1-abq^{n-1})(1-acq^{n-1})(1-adq^{n-1})}{(1-abcdq^{2n-2})(1-abcdq^{2n-1})} \\ &\quad \cdot (1-bcq^{n-1})(1-bdq^{n-1})(1-cdq^{n-1}), \\ B_n &= a+\frac{1}{a}-A_n\frac{(1-abq^n)(1-acq^n)(1-adq^n)}{a} \\ &\quad - C_n\frac{a}{(1-abq^{n-1})(1-acq^{n-1})(1-adq^{n-1})}. \end{aligned} \tag{3.12}$$

The orthogonality condition at $|q| < 1$ and with none of the pairwise products of the parameters a,b, c, d equal to q^{-k}, $k \in \mathbb{Z}_{\geq 0}$ is

$$\begin{aligned} &\frac{1}{2\pi i}\int_{\mathcal{C}} P_m^{AW}\left(\frac{z+z^{-1}}{2}\right)P_n^{AW}\left(\frac{z+z^{-1}}{2}\right) \\ &\quad\times \frac{(z^2,z^{-2};q)_\infty}{(az,az^{-1},bz,bz^{-1},cz,cz^{-1},dz,dz^{-1};q)_\infty}\frac{dz}{z} = 2\frac{\delta_{m,n}}{h_n}\int_{-1}^{1} w(x)dx, \end{aligned} \tag{3.13}$$

where $\mathcal{C}$ is the unit circle traversed in positive direction with suitable deformations to separate the sequences of poles converging to zero from the sequences of poles diverging to ∞ and

$$w(x|a,b,c,d) = \frac{h(x|1,-1,\sqrt{q},-\sqrt{q})}{h(x|a,b,c,d)\sqrt{1-x^2}}, \tag{3.14}$$

$$h(x|a_1,\ldots,a_m) := \prod_{i=1}^{m}(a_ie^{i\theta},a_ie^{-i\theta};q)_\infty = \prod_{i=1}^{m}\prod_{n=0}^{\infty}(1-2a_ixq^n+a_i^2q^{2n}), \tag{3.15}$$

$$\int_{-1}^{1} w(x)dx = \frac{2\pi(abcd;q)_\infty}{(q,ab,ac,ad,bc,bd,cd;q)_\infty}, \tag{3.16}$$

$$h_n = \frac{(1-abcdq^{2n-1})}{(1-abcdq^{-1})}\frac{(abcdq^{-1};q)_n}{(q,ab,ac,ad,bc,bd,cd;q)_n}. \tag{3.17}$$

If two, one or zero pairs of complex conjugate among the parameters a, b, c, d with remaining parameters real and $|a|$, $|b|$, $|c|$, $|d|$ are all less than 1, the contour is just the unit circle and the orthogonality relation takes the form

$$\int_{-1}^{1} P_m^{AW}(x)P_n^{AW}(x)w(x)dx = \frac{\delta_{m,n}}{h_n}\int_{-1}^{1} w(x)dx. \tag{3.18}$$

Assume that one of the parameters, say, $|a| > 1$. Then, one can again deform the contour to the unit circle and add residues at the poles at all aq^k lying outside the circle, the result being

$$\int_{-1}^{1} P_m^{AW}(x)P_n^{AW}(x)w(x)dx + 2\pi\sum_k P_m^{AW}(x_k)P_n^{AW}(x_k)w_k = \frac{\delta_{m,n}}{h_n}\int_{-1}^{1} w(x)dx \tag{3.19}$$

where

$$w_k = \frac{1-a^2q^{2k}}{1-a^2}\frac{(a^{-2};q)_\infty}{(q,ab,ac,ad,b/a,c/a,d/a;q)_\infty}\frac{(a^2,ab,ac,ad;q)_k}{(q,aq/b,aq/c,aq/d;q)_k}\left(\frac{q}{abcd}\right)^k \tag{3.20}$$

and the sum runs over the points $x_k = (aq^k + q^{-k}/a)/2$ with aq^k lying outside the unit circle.

3.3. *Askey scheme*

As we already noticed, the Askey–Wilson polynomials are the maximal ones within the so-called (q)-Askey scheme [5, 6], which describes the classical orthogonal polynomials. These polynomials all have a representation through the Rodriguez formula, and their measure is determined from the Pearson operator equation [6].

Let us start with not orthogonal polynomials with $q = 1$. Consider polynomials that solve an eigenvalue problem of the form (all classical polynomials are of this kind),

$$A(x)P_n(Q(x+\gamma)) + B(x)P_n(Q(x)) + C(x)P_n(Q(x-\gamma)) = \lambda_n P_n(Q(x)), \tag{3.21}$$

where $Q(x)$ is a second degree polynomial and γ is a parameter. Then, it turns out that there are only 13 families of orthogonal polynomials that solve the eigenvalue problem (3.21). They depend on at most four parameters and all are obtained by the maximal member of the family, the Askey polynomial. Hence, all of them are expressed through hypergeometric functions, at most, ${}_4F_3$. An important property of this kind of polynomials is that all they can be expressed from the measure through the Rodriguez formula.

The same scheme is equally well applicable in the q-deformed case (note that, apart from limit $q \to 1$, there is an interesting limit $q = -1$, where more "classical" polynomials emerge [12]). In this case, we consider the eigenvalue problem

$$\phi(x)\hat{H}_q\hat{H}_{1/q}(x) + \psi(x)\hat{H}_q P_n(x) = \lambda_n P_n(qx+\omega) \tag{3.22}$$

where $\hat{H}_q$ is the Hanh operator:

$$\hat{H}_q f(x) := \frac{f(qx+\omega)-f(x)}{qx+\omega-x} = \frac{\hat{D}f(x)-f(x)}{\hat{D}x-x} \tag{3.23}$$

and $\hat{D}f(x) := f(qx+\omega)$ is the dilatation operator. One can prove that $\phi(x)$ can be at most quadratic polynomial and $\psi(x)$, at most linear one:

$$\phi(x) = \phi_2 x^2 + 2\phi_1 x + \phi_0, \qquad \psi(x) = 2\psi_1 x + \psi_0. \tag{3.24}$$

Then the eigenvalues are,

$$\lambda_n = [n]_q \left(\frac{\phi_2}{q^n}[n-1]_q + 2\psi_1 \right), \quad \text{where } [n]_q := \frac{1-q^n}{1-q}. \tag{3.25}$$

In this case, the Rodriguez formula is still satisfied. For the monic orthogonal polynomial is looks as

$$\mathcal{P}_n(x) = \frac{K_n}{\mu'(x)} \hat{H}\hat{D}^{-n}\mu'(x) \prod_{i=1}^{n} \left(\hat{D}^{-i-1}\phi(x) \right) \tag{3.26}$$

where $d\mu(x) = \mu'(x)dx$ is the orthogonality measure, and

$$K_n = q^n \prod_{i=1}^{n} \frac{[i]}{\lambda_n - \lambda_{n-i}}. \tag{3.27}$$

In the limit to the classical orthogonal polynomials $q \to 1$, $\omega \to 0$, one obtains the standard Rodriguez formula

$$\mathcal{P}_n(x) = \frac{K_n}{\mu'(x)} \frac{d^n}{dx^n} \left(\mu'(x)\phi(x)^n \right). \tag{3.28}$$

Note that the weight function $\mu'(x)$ satisfies the equation (which follows from the Pearson operator equation [6])

$$\hat{H}\left(\left(D^{-k}\mu'(x) \right) \left(\hat{D}^{-k-1}\phi(x) \right) \right) = q^{-k} \left(\hat{D}^{-k+1}\mu'(x) \right) \left(\hat{D}^{-k}\psi(x) \right). \tag{3.29}$$

Example: The Rodriguez Formula for the Askey–Wilson Polynomials. In the case of the Askey–Wilson polynomials (3.8), the difference equation looks as

$$\begin{aligned} &(q^{-n}-1)(1-abcdq^{n-1})P_n^{AW}(z) \\ &\quad = A(z)P_n^{AW}(qz) - [A(z)+A(z^{-1})]P_n^{AW}(z) + A(z^{-1})P_n^{AW}(q^{-1}z), \end{aligned} \tag{3.30}$$

where $z = e^{i\theta}$ and

$$A(z) := \frac{(1-az)(1-bz)(1-cz)(1-dz)}{(1-z^2)(1-qz^2)}. \tag{3.31}$$

On the other hand, the Rodriguez formula is

$$\mu'(x;a,b,c,d;q)P_n^{AW}(x) = \left(\frac{q-1}{2} \right)^n q^{n(n-1)/4} \hat{D}_q^n \mu'(x; a\sqrt{q^n}, b\sqrt{q^n}, c\sqrt{q^n}, d\sqrt{q^n}; q), \tag{3.32}$$

$\hat{D}_q = \hat{H}_q|_{\omega=0}$ and $w = \mu'$ is the orthogonality weight (3.14).

4. Hypergeometric Orthogonal Polynomials

4.1. *Hypergeometric polynomials*

In this section, we are going to demonstrate that there is no immediate generalization of the orthogonal polynomials to the hypergeometric functions higher than ${}_4\phi_3$. That is, for an ansatz generalizing the AW polynomial, we check the 3-term relation and find that it is fulfill at most for the Askey–Wilson polynomials. Specifically, we consider two different ansatzes: polynomials in the hypergeometric variable z and in parameters of hypergeometric functions at constant z.

• Polynomials of the first type are of the following q-hypergeometric form

$$ {}^{I}Q_n^{(r,s)}(B,\{b_i\},\{c_i\};x) := {}_r\phi_s\left(\begin{matrix} q^{-n}, & Bq^n, b_1, \ldots, b_{r-2} \\ & c_1, \ldots, c_s \end{matrix}\middle| z = x\right), \tag{4.1} $$

where $\{b_i\}_{i=1}^{r-2}$ and $\{c_i\}_{i=1}^{s}$ are arbitrary parameters, which may depend on n and x, while the parameter B is independent of n and x.

• Polynomials of the second type are ($x := \cos\theta$),

$$ \begin{aligned} {}^{II}Q_n^{(r,s)}(a,B,\{b_i\},\{c_i\};x) &:= {}_r\phi_s\left(\begin{matrix} q^{-n}, ae^{i\theta}, ae^{-i\theta}, & Bq^n, b_1, \ldots, b_{r-4} \\ & c_1, \ldots, c_s \end{matrix}\middle| z = q\right) \\ &= \sum_{k=0}^{n} \frac{(q^{-n};q)_k h(x|a)_k \cdot (Bq^n;q)_k \prod_{i=1}^{r-4}(b_i;q)_k}{(q;q)_k \prod_{i=1}^{s}(c_i;q)_k} \\ &\quad \cdot q^k\left[(-)^k q^{k(k-1)/2}\right]^{1+s-r}, \end{aligned} \tag{4.2} $$

with

$$ h(x|a)_k := (ae^{i\theta};q)_k(ae^{-i\theta};q)_k = \prod_{j=0}^{k-1}(1 - 2axq^j + a^2q^{2j}), \tag{4.3} $$

where parameters a and B are independent on n and x, while $\{b_i\}_{i=1}^{r-4}$ and $\{c_i\}_{i=1}^{s}$ can be arbitrary.

Since one of the upper parameters in the q-hypergeometric functions (4.1) and (4.2) is q^{-n} with $n \in \mathbb{Z}_+$, these are, indeed, polynomials.

4.2. *Criteria for the 3-term relation*

Let us consider a polynomial $Q_n(x)$ of degree n in variable x. We denote a coefficient of the polynomial in front of x^m by $[Q_n]_m$. If polynomials $Q_n(x)$ satisfy the 3-term relation, then all 4×4 determinants vanish for any m such that $3 \le m \le n+1$:

$$ \det_{4\times 4}\begin{pmatrix} [x\cdot Q_n]_m & [Q_{n+1}]_m & [Q_n]_m & [Q_{n-1}]_m \\ [x\cdot Q_n]_{m-1} & [Q_{n+1}]_{m-1} & [Q_n]_{m-1} & [Q_{n-1}]_{m-1} \\ [x\cdot Q_n]_{m-2} & [Q_{n+1}]_{m-2} & [Q_n]_{m-2} & [Q_{n-1}]_{m-2} \\ [x\cdot Q_n]_{m-3} & [Q_{n+1}]_{m-3} & [Q_n]_{m-3} & [Q_{n-1}]_{m-3} \end{pmatrix} = 0. \tag{4.4} $$

This vanishing is not sufficient, but a necessary condition for the 3-term relation. For polynomials (4.2), we denote these determinants $D_{n,m}^{(r,s)}$. They actually factorize on a codimension-one variety in the space of parameters $b_1, \ldots, b_{r-4}$ and $c_1, \ldots, c_s$, and vanish altogether at a codimension-two variety.

Examples.

- **For** $(r,s)=(4,3)$,

$$D_{2,3}^{(4,3)} \sim (Ba^2q - c_1c_2c_3) = 0 \quad \text{at } B = \frac{c_1c_2c_3}{a^2q}, \tag{4.5}$$

  and the same remains true for all other $D_{n,m}^{(4,3)}$. These Askey–Wilson polynomials are indeed orthogonal.

- **For** $(r,s)=(5,4)$,

$$\begin{aligned} D_{2,3}^{(5,4)} &\sim \prod_{i=2}^{5}(Bq^i - 1)\prod_{j=0}^{1}(b_1q^j - 1), \\ D_{7,8}^{(5,4)} &\sim \prod_{i=6}^{15}(Bq^i - 1)\prod_{j=0}^{6}(b_1q^j - 1), \end{aligned} \tag{4.6}$$

  but b_1 cannot be equal to q^{-k} with $k < n$, because it reduces the degree of the polynomial $Q_n(x)$. Thus, there are no counterparts of the Askey–Wilson polynomials in this case.

4.3. *Conjecture in generic case*

In general, we expect the following pattern for the q-hypergeometric polynomials (4.1) and (4.2):

Type of $Q(x)$, parameters	Name in q-Askey scheme	(r,s) for which the 3-term relation is satisfied
type I	little q-Jacobi	(2, 1)
type I, $B = \dfrac{\tilde{B}}{xq^{n-1}}$	big q-Laguerre	(2, 1)
type I, $b_1 = x$	big q-Jacobi	(3, 2)
type I, $b_1 = \dfrac{1}{x}\dfrac{c_1c_2}{Bq}$	equiv. to big q-Jacobi	(3, 2), see Remark
type II	Askey–Wilson	(4, 3)
	...	...

Remark. These polynomials are equivalent to the big q-Jacobi ones with the help of a particular case of Sear's identity (see [8, Eq. (3.2.2)]). We write down the corresponding 3-term recurrence relation, because we did not find it in the literature:

$$A_n Q_{n+1} - (B_n x - C_n) Q_n - D_n Q_{n-1} = 0,$$

$$\begin{aligned} A_n &= b q^{2n+2}(q^{4n-2}b-1)(q^{2n}b-1)(q^{2n}c_1-1)(q^{2n}c_2-1),\\ B_n &= b(q^{4n-2}b-1)(q^{4n}b-1)(q^{4n+2}b-1),\\ C_n &= q^{2n}(q^{4n}b-1)\cdot(q^{4n+2}b^2+q^{4n}b(bc_1+bc_2+c_1c_2)-q^{2n+2}b(b+c_1+c_2)\\ &\quad -q^{2n}b(c_1c_2+b+c_1+c_2)-q^{2n-2}bc_1c_2+q^2b+bc_1+bc_2+c_1c_2),\\ D_n &= q^n\{q\}[n](q^{4n+2}b-1)(q^{2n}b-c_1)(q^{2n}b-c_2). \end{aligned} \tag{4.7}$$

Thus, with the help of determinant criteria, one can check what type of recurrence relations emerge for polynomials. For the polynomials of first type, we have the following recurrence relations:

(r, s)	Number of terms in recurrence relation
(2, 1)	3
(3, 2)	5
(4, 3)	5
(5, 4)	7
(6, 5)	7
(7, 6)	>7

In order to form a similar table for the polynomials of the second type, one also needs to check if the stability condition for the q-hypergeometric function is satisfied. The stability condition for the q-hypergeometric function (A.7) reads

$$z\prod_{i=1}^{r} a_i = \prod_{j=1}^{s} b_j. \tag{4.8}$$

Then, we get the following table for the polynomials of the second type:

(r, s)	Number of terms in recurrence relation	Imposed stability condition
(3, 2)	3	*no*
(4, 3)	3	*yes*
(5, 4)	5	*no*
(6, 5)	5	*yes*
(7, 6)	7	*no*
(8, 7)	7	*yes*

It is clear that if one obtains an s-term relation for polynomials ${}_{s+1}\phi_s(\dots|z=q)$ under the stability condition, one can put $a_r = b_s = 0$ and obtain polynomials ${}_s\phi_{s-1}(\dots|z=q)$ without the stability condition. This is precisely what we observe in the table above. Condition (4.8) is the one which allows us to substitute all $(a)_n$ and $(b)_n$ in the definition of hypergeometric series by q-numbers $\frac{[a+n-1]!}{[a-1]!}$ and $\frac{[b+n-1]!}{[b-1]!}$ without any extra powers of q.

Thus, we see that,

> among the hypergeometric polynomials, the "biggest" available example of the 3-term relations and, hence, of the orthogonal polynomials is the Askey–Wilson polynomials.

4.4. *Orthogonal polynomials made from higher q-hypergeometric functions*

Thus, there are no higher q-hypergeometric functions that are orthogonal polynomials. However, their linear combinations can easily form a system of orthogonal polynomials. We conclude this section with a simplest example of this kind, the Ismail–Rahman polynomials [13]. These are the systems of polynomials $\mathfrak{z}_n^{(\alpha,\beta)}(x,\vec{t};q)$ of degree n in $x = 1/2(z+z^{-1})$ parametrized by parameters (α,t_1,t_2,t_3,t_4) and $\beta = 0,1$:

$$\mathfrak{z}_n^{(\alpha,\beta)}(x,\vec{t};q) := \frac{zt_1^{1-2\alpha}}{1-t_1t_2t_3t_4q^{2\alpha-2}} \frac{(q^\alpha,t_1z;q)_\infty}{(t_2t_3t_4/z,t_1t_2t_3t_4q^{\alpha-1};q)_\infty} \frac{\prod_{2\le j<k\le 4}(t_jt_k,t_jt_kq^{\alpha-1};q)_\infty}{\prod_{k=2}^4(t_1t_kq^\alpha,t_kz;q)_\infty}$$
$$\times\Big[\Big(s_{\alpha-1}(z)-\beta s_\alpha(z)\Big)r_{n+\alpha}(z)-\Big(r_{\alpha-1}(z)-\beta r_\alpha(z)\Big)s_{n+\alpha}(z)\Big], \tag{4.9}$$

where

$$r_\alpha(z) := \frac{(t_2t_3t_4q^\alpha/z;q)_\infty}{(t_1zq^\alpha;q)_\infty} \frac{\prod_{k=2}^4(t_1t_kq^\alpha;q)_\infty}{\prod_{2\le j<k\le 4}(t_jt_kq^\alpha;q)_\infty}$$
$$\times\left(\frac{t_1}{z}\right)^\alpha {}_8W_7\Big(t_2t_3t_4/(qz);t_2/z,t_3/z,t_4/z,t_1t_2t_3t_4q^{\alpha-1},q^{-\alpha};q,qz/t_1\Big),$$

$$s_\alpha(z) := \frac{(t_1t_2t_3t_4q^{2\alpha},t_2t_3t_4zq^\alpha;q)_\infty}{(t_2t_3t_4zq^{2\alpha+1},q^{\alpha+1};q)_\infty} \frac{\prod_{k=2}^4(t_kq^{\alpha+1};q)_\infty}{\prod_{2\le j<k\le 4}(t_jt_kq^\alpha;q)_\infty}$$
$$\times (t_1z)^\alpha {}_8W_7(t_2t_3t_4zq^{2\alpha};t_2t_3q^\alpha,t_2t_4q^\alpha,t_3t_4q^\alpha,q^{\alpha+1},zq/t_1;q,t_1z). \tag{4.10}$$

5. Integral Representations of Orthogonal Polynomials, Ward Identities and Conformal Blocks

The hypergeometric series possess integral representations with multiple integrations, which also suggest an interpretation in terms of conformal blocks. At subvarieties of high codimension in the space of parameters, the hypergeometric series ${}_rF_s$ reduces to those at much smaller values of r and s and thus can contain less integrations. In the extreme case of the very-well-poised subvarieties (see Sec. 5.2.1 and Appendix A.2) just a single integration remains, i.e. the corresponding conformal block contains just a single screening-charge insertion. In this case, a natural splitting of the integration contour into three segments provides non-trivial identities, which can be interpreted as 3-term relations. This gives rise to non-trivial families of orthogonal polynomials including the Askey–Wilson/Racah as the most general known example.

In this section, we also discuss the modular transformations of the hypergeometric functions, which directly follow from their integral representations. On one hand, this is important for the issue of conformal blocks [14–16], on the other hand, these encodes essential properties of the corresponding hypergeometric and polynomial determinant solutions to the Painléve [17] and discrete Painléve equations [2, 18, 19] (included into the general framework of CFT/Painléve correspondence [20, 21]).

5.1. *Dotsenko–Fateev representations of hypergeometric functions*

Using the integral representation of the Euler Beta-function,

$$B(x,y) := \int_0^1 dt\; t^{x-1}(1-t)^{y-1} = \frac{\Gamma(x)\Gamma(y)}{\Gamma(x+y)}, \tag{5.1}$$

and the definition of the hypergeometric series (A.1), one can directly derive the recursion relation

$$\begin{aligned}&{}_rF_s\left(\begin{matrix}a_1,\ldots,a_r\\ b_1,\ldots,b_s\end{matrix}\middle|\, z\right)\\ &= \frac{\Gamma(b_s)}{\Gamma(a_r)\Gamma(b_s-a_r)}\int_0^1 dt\, t^{a_r-1}(1-t)^{b_s-a_r-1}\;{}_{r-1}F_{s-1}\left(\begin{matrix}a_1,\ldots,a_{r-1}\\ b_1,\ldots,b_{s-1}\end{matrix}\middle|\, tz\right)\end{aligned} \tag{5.2}$$

which implies that there is an integral representation with just $\min(r,s)$ integrals, which, however, does not explicitly respect the $S_r \otimes S_s$ symmetry between the parameters a_i and b_i of the hypergeometric function. In particular,

$$\begin{aligned}{}_{s+1}F_s\left(\begin{matrix}a_0,a_1,\ldots,a_s\\ b_1,\ldots,b_s\end{matrix}\middle|\, z\right) &= \left(\prod_{i=1}^{s}\frac{\Gamma(b_i)}{\Gamma(a_i)\Gamma(b_i-a_i)}\int_0^1 t_i^{a_i-1}(1-t_i)^{b_i-a_i-1}dt_i\right)\\ &\quad\cdot\overbrace{(1-t_1\cdots t_s z)^{-a_0}}^{\sum_k \frac{(a_0)_k}{k!}\cdot(t_1\ldots t_s z)^k},\end{aligned} \tag{5.3}$$

or, after the change of variables $t_i \to t_i/t_{i+1}$,

$$\begin{aligned}{}_{s+1}F_s\left(\begin{matrix}a_0,a_1,\ldots,a_s\\ b_1,\ldots,b_s\end{matrix}\middle|\, z\right) &= \left(\prod_{j=1}^{s}\frac{\Gamma(b_j)}{\Gamma(a_j)\Gamma(b_j-a_j)}\int_0^{t_{j+1}} dt_j t_j^{a_j-a_{j-1}-1}\right.\\ &\quad\left.\times\left(1-\frac{t_j}{t_{j+1}}\right)^{b_j-a_j-1}\right)\cdot t_1^{a_0}(1-t_1z)^{-a_0},\end{aligned} \tag{5.4}$$

where we put $t_{s+1}=1$.

These representations can be associated with the free field representation of the conformal blocks [23–25] (see [26] for the q-deformation), see Sec. 5.1.2.

5.1.1. *Example: Ward and bilinear identities*

As soon as hypergeometric function have integral representations, various relations between these functions can be interpreted as Ward identities, as explained in [2]. The simplest examples of such relations is

$$
\begin{aligned}
{}_2F_1\left(\begin{matrix} a,b \\ c \end{matrix}\middle| x\right) &= \frac{\Gamma(c)}{\Gamma(b)\Gamma(c-b)}\int_0^1 (1-xt)^{-a}t^{b-1}(1-t)^{c-b-1}dt \\
&= (1-x)^{c-a-b}\cdot {}_2F_1\left(\begin{matrix} c-a,c-b \\ c \end{matrix}\middle| x\right), \qquad (5.5)
\end{aligned}
$$

which can be obtained by the change of integration variable $t \longrightarrow \frac{1-t}{1-xt}$.

Another example is provided by the replace $t \longrightarrow 1-t$ and leads to the Euler transformation formula

$$
\begin{aligned}
{}_2F_1\left(\begin{matrix} a,b \\ c \end{matrix}\middle| x\right) &= \frac{\Gamma(c)}{\Gamma(b)\Gamma(c-b)}\int_0^1 (1-xt)^{-a}t^{b-1}(1-t)^{c-b-1}dt \\
&= (1-x)^{-a}\,{}_2F_1\left(\begin{matrix} a,c-b \\ c \end{matrix}\middle| \frac{x}{x-1}\right). \qquad (5.6)
\end{aligned}
$$

Since

$$
(1-x)^{-\alpha} = \sum_{n=0}^{\infty}\frac{(\alpha)_n}{n!}x^n \qquad (5.7)
$$

is also a hypergeometric series, one can also interpret (5.5) as a kind of bilinear identity [27]. In the lowest orders, it is trivial:

$$
1+\frac{ab}{c}x+\cdots = \Big(1+(a+b-c)x+\cdots\Big)\left(1+\frac{(c-a)(c-b)}{c}x+\cdots\right), \qquad (5.8)
$$

i.e. $ab=(c-a)(c-b)+c(a+b-c)$ etc, but in general it looks more interesting:

$$
\begin{aligned}
\frac{(a)_n(b)_n}{n!\,(c)_n} &= \sum_{j=0}^{n}\frac{(c-a)_j(c-b)_j}{(c)_j}\cdot\frac{(a+b-c)_{n-j}}{j!(n-j)!} \\
&= \frac{(c-a-b)_n}{n!}\sum_{j=0}^{n}\frac{(-n)_j(c-a)_j(c-b)_j}{(c)_j(c-a-b-n+1)_j\,j!}, \qquad (5.9)
\end{aligned}
$$

i.e.

$$
{}_3F_2\left(\begin{matrix} -n,c-a,c-b \\ c,c-a-b-n+1 \end{matrix}\middle| x=1\right) = \frac{(a)_n(b)_n}{(c)_n(a+b-c)_n}, \qquad (5.10)
$$

which is nothing but Pfaff–Saalschütz's identity, [8, Sec. 1.7].

5.1.2. *Hypergeometric versus modular transformations*

Now let us discuss the relation with conformal blocks. Consider the conformal block at $c = 1$, $\alpha_4 = -\alpha_1 - \alpha_2 - \alpha_3 - 1$ [17]:

$$B(\alpha_i, \alpha; x)\big|_{\alpha=\alpha_1+\alpha_2+1} = x^{2\alpha_1\alpha_2+2\alpha_1+2\alpha_2+1}(1-x)^{2\alpha_2\alpha_3}\,{}_2F_1\left(\begin{matrix}-2\alpha_3, 2\alpha_1+1\\ 2\alpha_1+2\alpha_2+2\end{matrix}\middle| x\right),$$

$$B(\alpha_i, \alpha; x)\big|_{\alpha=\alpha_1+\alpha_2} = x^{2\alpha_1\alpha_2}(1-x)^{2\alpha_2\alpha_3}\,{}_2F_1\left(\begin{matrix}-2\alpha_1-2\alpha_2-2\alpha_3-1, -2\alpha_2\\ -2\alpha_1-2\alpha_2\end{matrix}\middle| x\right), \tag{5.11}$$

where α_i, $i = 1, 2, 3, 4$ parametrize the external conformal dimensions, $\Delta_i = \alpha_i^2$, while α, the intermediate one, $\Delta = \alpha^2$.

Then, (5.5) can be rewritten as

$$B(\alpha_1, \alpha_2, \alpha_3, \alpha_4, \alpha; x) = (1-x)^{2(\alpha_1+\alpha_2)(\alpha_1+\alpha_3)-2\alpha_4}B(\alpha_2, \alpha_1, \alpha_4, \alpha_3, \alpha; x). \tag{5.12}$$

It is nothing but a result of modular transformation of the conformal block [28]. Indeed, the generic four-point function is [29, 30]

$$\langle V_{\Delta_1}(z_1)V_{\Delta_2}(z_2)V_{\Delta_3}(z_3)V_{\Delta_4}(z_4)\rangle = \prod_{i>j}^{4}(z_i - z_j)^{d_{ij}}(\bar z_i - \bar z_j)^{\bar d_{ij}}G(x, \bar x), \tag{5.13}$$

where d_{ij} are determined by the equations

$$\sum_{j\neq i} d_{ij} = 2\Delta_i \tag{5.14}$$

$x = \frac{(z_1-z_2)(z_3-z_4)}{(z_1-z_3)(z_2-z_4)}$ and similarly for the complex conjugated part. $G(x, \bar x)$ is a bilinear combination of the conformal blocks $B(\{\Delta_i\}, \Delta, , c; x)$ and $B(\{\bar\Delta_i\}, \bar\Delta, c; \bar x)$. Due to the projective invariance, one can choose three of these four points arbitrarily. If choosing $z_4 = \infty$, $z_3 = 1$ and $z_1 = 0$, one obtains $z_2 = x$, and the 4-point correlator (5.13) becomes

$$(1-x)^{d_{32}}\bar x^{\bar d_{21}}(1-\bar x)^{\bar d_{32}}x^{d_{21}}G(x, \bar x). \tag{5.15}$$

The simultaneous interchange of external lines $\vec\alpha_1 \leftrightarrow \vec\alpha_2$ and $\vec\alpha_3 \leftrightarrow \vec\alpha_4$ in formula (5.13) does not change x and $\bar x$ and leads to the factor of $(1-x)^{d_{14}}$ instead of $(1-x)^{d_{23}}$, which is exactly the factor $(1-x)^{2(\alpha_1+\alpha_2)(\alpha_1+\alpha_3)-2\alpha_4}$ in (5.12):

$$d_{23} - d_{14} = \Delta_2 + \Delta_2 - \Delta_1 - \Delta_4 = 2(\alpha_1+\alpha_2)(\alpha_1+\alpha_3) - 2\alpha_4. \tag{5.16}$$

Similarly, the interchange of only external lines $\vec\alpha_1 \leftrightarrow \vec\alpha_2$ in formula (5.13) changes $x \to x/(x-1)$ and analogously $\bar x$ and is described by the Euler transformation formula (5.6).

Note that another modular transformation, which replaces $x \to 1-x$, is far less trivial [14, 31, 32] and corresponds in terms of hypergeometric functions to the identity

$$
{}_2F_1\left(\begin{matrix} a,b \\ c \end{matrix}\middle| x\right) = \frac{\Gamma(c)\Gamma(c-a-b)}{\Gamma(c-a)\Gamma(c-b)}\,{}_2F_1\left(\begin{matrix} a,b \\ a+b-c+1 \end{matrix}\middle| 1-x\right)
$$

$$
+ (1-x)^{c-a-b}\frac{\Gamma(c)\Gamma(a+b-c)}{\Gamma(a)\Gamma(b)}\,{}_2F_1\left(\begin{matrix} c-a,c-b \\ c-a-b+1 \end{matrix}\middle| 1-x\right). \tag{5.17}
$$

In this case, this transformation maps the conformal block to another one with different conformal dimensions. One can apply the same transformation again and obtain a sequence of conformal blocks with different conformal dimensions. This sequence is infinite unless the conformal dimensions are specifically constrained. If they are properly constrained, the sequence becomes finite, the modular transformation becomes a linear transformation in the finite-dimensional space of the corresponding conformal blocks, and this is exactly the case of minimal models [23, 29].

5.1.3. *Integral representation of the very classical polynomials*

Jacobi Polynomials. One could try to use a particular example of formula (5.3) at $s=1$ (i.e. the Gaussian hypergeometric function)

$$
{}_2F_1\left(\begin{matrix} a,b \\ c \end{matrix}\middle| z\right) = \frac{\Gamma(c)}{\Gamma(b)\Gamma(c-b)}\int_0^1 dt\, t^{b-1}(1-t)^{c-b-1}(1-tz)^{-a}, \tag{5.18}
$$

and the definition of the Jacobi polynomials (B.1) in order to obtain an integral representation of these latter

$$
P_n^{(a,b)}(z) = \frac{(a+n)!}{2^n n!(a+b+n)!\Gamma(-b-n)}\int_0^1 dt\, t^{a+b+n}(1-t)^{-b-n-1}(zt-t+2)^n. \tag{5.19}
$$

Unfortunately, this integral is never defined at all natural n at once. However, we can proceed in a more tricky way [33]. First of all, one can obtain Bateman's integral

$$
{}_2F_1\left(\begin{matrix} a,b \\ c \end{matrix}\middle| z\right) = \frac{1}{B(c',c-c')}\int_0^1 dt\, t^{c'-1}(1-t)^{c-c'-1}\,{}_2F_1\left(\begin{matrix} a,b \\ c' \end{matrix}\middle| zt\right), \tag{5.20}
$$

just from the definition of the hypergeometric function as a series (A.1) and evaluating the integrals with the help of (5.1). Now, combining this formula with (B.1) and with the Euler transformation (5.6), one immediately arrives at

$$
P_n^{(a,b)}(z) = \frac{2^{a-b}}{B(a-b,b+n)}\frac{(1+z)^{b+n+1}}{(1-z)^a}
$$

$$
\times \int_z^1 dt(1-t)^b(1+t)^{-a-n-1}(t-z)^{a-b-1}P_n^{(b,b)}(t). \tag{5.21}
$$

Making the substitution $\xi^2 = \frac{(1-t)(1+z)}{(1+t)(1-z)}$, one obtains the integral formula that expresses the Jacobi polynomials $P_n^{(a,b)}(z)$ through $P_n^{(b,b)}(z)$, which are equal to **the Gegenbauer (or ultraspherical) polynomials** up to a normalization:

$$\begin{aligned} P_n^{(a,b)}(z) &= \frac{(-1)^n 2^{1-n}}{B(a-b,b+n+1)}(1+z)^2 \\ &\quad \times \int_0^1 d\xi\; \xi^{2b+1}\Big((z-1)\xi^2 - z - 1\Big)^n (1-\xi^2)^{a-b-1} P_n^{(b,b)} \\ &\quad \times \left(\frac{(1-z^2)(\xi^2-1)}{\xi^2(z-1)-z-1}\right). \end{aligned} \tag{5.22}$$

$P_n^{(b,b)}(z)$ have the integral representation[a]

$$P_n^{(b,b)}(z) = \frac{\Gamma(b+n+1)}{n!\Gamma\Big(b+\frac{1}{2}\Big)\Gamma\Big(\frac{1}{2}\Big)} \int_0^\pi d\theta\; \sin^{2b}\theta\Big[z + i\sqrt{1-z^2}\cos\theta\Big]^n. \tag{5.23}$$

Hence, one finally obtains an integral representation of the Jacobi polynomials [34]

$$\begin{aligned} P_n^{(a,b)}(z) &= \frac{2^{1-n}\Gamma(a+n+1)}{n!\Gamma(a-b)\Gamma\Big(b+\frac{1}{2}\Big)\Gamma\Big(\frac{1}{2}\Big)} \\ &\quad \times \int_0^1 d\xi \int_0^\pi d\theta (1-\xi^2)^{a-b-1}\xi^{2b+1}\sin^{2b}\theta \\ &\quad \times \Big[1+z-(1-z)\xi^2 + 2i\sqrt{1-z^2}\;\xi\cos\theta\Big]^n. \end{aligned} \tag{5.24}$$

This integral exists at $a > b > -1/2$. At a wider range of parameters $a, b > -1/2$, there is another, more symmetric double integral representation for the Jacobi

[a]This can be checked by expanding (5.23) into the Newton binomial and using the integrals

$$\int_0^\pi d\theta \sin^{2b}\theta\cos^k\theta = \begin{cases} 0 & \text{for } k = 2n+1, \\ 4^b B\Big(b+\dfrac{1}{2}, b+\dfrac{1}{2}\Big)(b+1)_n \dfrac{\Gamma\Big(n+\frac{1}{2}\Big)}{\sqrt{\pi}} & \text{for } k = 2n. \end{cases}$$

This leaves us with a sum of terms $z^{2n-2k}(1-z^2)^k$ which, upon with the substitution $(1-z)/2 = y$, i.e. $z = 1-2y$, $1-z^2 = 4y(1-y)$ can be expanded into degrees of y. Now one can check that the coefficient in front of each y_p coincides with the corresponding coefficient of $P_n^{(b,b)}(z)$ as a hypergeometric series (A.1) of $y = (1-z)/2$.

polynomials,[b] [35]

$$P_n^{(a,b)}(1-2y^2) = \frac{(-4)^n}{\pi(2n)!}\frac{\Gamma(a+n+1)\Gamma(b+n+1)}{\Gamma\left(a+\frac{1}{2}\right)\Gamma\left(b+\frac{1}{2}\right)}\int_{-1}^{1}d\xi_1\int_{-1}^{1}d\xi_2$$

$$\times\left[y\xi_1 \pm i\sqrt{1-y^2}\xi_2\right]^{2n}(1-\xi_1^2)^{a-1/2}(1-\xi_2^2)^{b-1/2}. \tag{5.25}$$

In the following, we shall derive integral representations of the Laguerre and Hermite polynomials, which is of relevance when we discuss the use of orthogonal polynomials in the study of the matrix model in Sec. 8.

Laguerre Polynomials. Another particular application of formula (5.2) is to the case of the degenerate hypergeometric function ${}_1F_1$:

$$ {}_1F_1\left(\begin{matrix}a\\b\end{matrix}\middle| z\right) = \frac{\Gamma(b)}{\Gamma(a)\Gamma(b-a)}\int_0^1 dt\, t^{a-1}(1-t)^{b-a-1}\,{}_0F_0(tz)$$

$$= \frac{\Gamma(b)}{\Gamma(a)\Gamma(b-a)}\int_0^1 dt\, e^{tz}t^{a-1}(1-t)^{b-a-1}. \tag{5.26}$$

One may try to use the definition of the Laguerre polynomials as degenerate hypergeometric functions (B.8) in order to obtain their integral representation

$$L_n^{(a)}(z) = \frac{1}{\Gamma(-n)\Gamma(1-n)}\int_0^1 dt\, e^{tz}t^{-n-1}(1-t)^{n+a}. \tag{5.27}$$

However, this formula requires a regularization because of the zeroes at natural n due to $\Gamma(-n)$ in the denominator and divergency of the integral at zero. Hence, we use another formula,

$$ {}_1F_1\left(\begin{matrix}a\\b\end{matrix}\middle| z\right) = \frac{\Gamma(b)}{\Gamma(b-a)}e^z z^{(1-b)/2}\int_0^\infty dt\, e^{-t}t^{(b-1)/2-a}J_{b-1}(2\sqrt{tz}), \tag{5.28}$$

where $J_b(z)$ is the Bessel function of order b. This formula is proved by expanding the Bessel function into the series

$$J_b(z) = \left(\frac{z}{2}\right)^b\sum_{k=0}^{\infty}\frac{(-1)^k}{k!\Gamma(k+b+1)}\left(\frac{z}{2}\right)^{2k}, \tag{5.29}$$

[b]The simplest way to prove this formula is to expand $(\ldots)^{2n}$ in the integrand into the Newton binomial, use (5.1) and then use the Newton binomial expansion for $(1-y^2)^k$ again. Then, using

$$\sum_{l\le k\le n}\frac{k!\Gamma\left(k+\frac{1}{2}\right)\Gamma\left(n-k+\frac{1}{2}\right)}{(k-l)!(2k)!(2n-2k)!\Gamma(k+a+1)\Gamma(n-k+b+1)}$$

$$= \frac{4^{-n}\pi\,\Gamma(a+b+2n-l+1)}{\Gamma(a+n+1)\Gamma(b+n+1-l)\Gamma(a+b+n+1)(n-l)!},$$

one easily checks that the coefficients of expansion coincide with the coefficients of series (B.1) in $(1-x)/2 = y^2$.

and calculating the integrals with the help of the first formula in (A.6). The series obtained coincides with the degenerate hypergeometric series at the l.h.s. Using now the definition of the Laguerre polynomials (B.8), one finally obtains from (5.28)

$$L_n^{(a)}(z) = \frac{1}{n!} e^z z^{-a/2} \int_0^\infty dt\, e^{-t} t^{n+a/2} J_a(2\sqrt{tz}). \tag{5.30}$$

Hermite Polynomials. In this case, we can use the definition (B.15) and note that

$$H_n(z) = \begin{cases} (-4)^m m!\, L_m^{(-\frac{1}{2})}(z^2), & n = 2m, \\ 2(-4)^m m!\, z L_m^{(\frac{1}{2})}(z^2), & n = 2m+1. \end{cases} \tag{5.31}$$

Then, the integral representation is read off from (5.30):

$$H_n(z) = \frac{2^{n+1}}{\sqrt{\pi}} e^{z^2} \int_0^\infty dt\, e^{-t^2} t^n \cos\left(2zt - \frac{n\pi}{2}\right) = \frac{(2i)^n}{\sqrt{\pi}} \int_{-\infty}^\infty dt\, t^n e^{(z+it)^2}, \tag{5.32}$$

where we have taken into account that

$$J_{\pm\frac{1}{2}}(z) = \sqrt{\frac{2}{\pi z}} \sin\left(z + \frac{\pi(1 \pm 1)}{4}\right), \tag{5.33}$$

and changed the integration variable $t \to t^2$.

5.2. *A single-integral representation for the AW type polynomials*

This is another integral representation for the AW type polynomials, which requires just one integration. Defining the Jackson integral

$$\int_a^b f(t) d_q t = (1-q) \sum_{n=0}^\infty q^n \Big[b f(bq^n) - a f(aq^n) \Big], \tag{5.34}$$

the q-hypergeometric function at $z = q$ is

$$\begin{aligned} {}_r\phi_s \left(\begin{matrix} a_1, \ldots, a_r \\ b_1, \ldots, b_s \end{matrix} \middle| z \right) &= \sum_{n=0}^\infty \frac{(a_1;q)_n \cdots (a_r;q)_n}{(b_1;q)_n \cdots (b_s;q)_n} \frac{z^n}{(q;q)_n} \\ &\xrightarrow{z=q} \frac{1}{1-q} \cdot \frac{(a_1, \ldots, a_r; q)_\infty}{(b_1, \ldots, b_s, q; q)_\infty} \\ &\quad \times \int_0^1 d_q t \frac{(b_1 t, \ldots, b_s t, qt; q)_\infty}{(a_1 t, \ldots, a_r t; q)_\infty}, \end{aligned} \tag{5.35}$$

i.e. the q-hypergeometric series at fixed $z = q$ is always a single (Jackson) integral. This gives us an integral representation for any orthogonal polynomials of the second type.

5.2.1. *Example: 3-term relation and integral representation for the Askey–Wilson polynomials*

Consider the very-well-poised hypergeometric series ${}_8W_7$ (A.11) at a special point, where it has an integral representation with just one integration:

$$
\begin{aligned}
&{}_8W_7\left(a; b, c, d, e, f \,\middle|\, q, \frac{a^2q^2}{bcdef}\right) \\
&\quad = \frac{aq - def}{adefq(1-q)} \frac{(aq, d, e, f, aq/bc, aq/de, aq/df, aq/ef; q)_\infty}{(q, aq/b, aq/c, aq/d, aq/e, aq/f, def/aq, aq/def; q)_\infty} \\
&\qquad \times \int_{aq}^{def} \frac{(t/a, qt/def, aqt/bdef, aqt/cdef; q)_\infty}{(t/de, t/df, t/ef, aqt/bcdef; q)_\infty} d_qt.
\end{aligned} \tag{5.36}
$$

Dividing the integration segment into two pieces:

$$
\int_{aq}^{def} = \int_{aq}^{bdef/a} + \int_{bdef/a}^{def}, \tag{5.37}
$$

one obtains the Bailey 3-term identity:

$$
\begin{aligned}
&\frac{1}{a} \frac{(aq/d, aq/e, aq/f, q/ad, q/ae, q/af; q)_\infty}{(qa^2, ab, ac, b/a, c/a; q)_\infty} {}_8W_7(a^2; ab, ac, ad, ae, af | q, q^2/abcdef) \\
&\quad + (a \leftrightarrow b) + (a \leftrightarrow c) = 0.
\end{aligned} \tag{5.38}
$$

It remains to note that

$$
\begin{aligned}
&{}_8W_7(a, b, c, d, e, q^{-n} | q, a^2q^{2+n}/bcde) \\
&\quad = \frac{(aq, aq/de; q)_n}{(aq/d, aq/e)_n} {}_4\phi_3\left(\begin{matrix} q^{-n}, d, e, aq/bc \\ aq/b, aq/c, deq^{-n}/a \end{matrix} \,\middle|\, q, q\right).
\end{aligned} \tag{5.39}
$$

Hence, the Askey–Wilson polynomials,

$$
\begin{aligned}
&P_n(x = \cos\theta; a, b, c, d; q) \\
&\quad = (ab, ac, ad; q)_n a^{-n} {}_4\phi_3\left(\begin{matrix} q^{-n}, abcdq^{n-1}, ae^{i\theta}, ae^{-i\theta} \\ ab, ac, ad \end{matrix} \,\middle|\, q, q\right),
\end{aligned} \tag{5.40}
$$

satisfy the 3-term recurrence relation that follows from (5.38) and, hence, they are orthogonal polynomials.

There are three technical points behind this line: (i) integral representation (5.36), the proof is in [8, Sec. 2.10]; (ii) the relation between ${}_8W_7$ and ${}_4\phi_3$, the proof is in [8, Sec. 2.5] and (iii) the orthogonality measure, the manifest expression being available from [8, Sec. 7.5].

From these formulas, we also immediately obtain the integral representation for the Askey–Wilson polynomials:

$$P_n^{AW}(x = \cos\theta|a,b,c,d) = \frac{2id}{q(1-q)} \frac{(ab,ac,bc;q)_n}{(ab,ac,bc,q;q)_\infty h(x|d)w(x|a,b,c,d)}$$
$$\times \int_{\frac{qe^{i\theta}}{d}}^{\frac{qe^{-i\theta}}{d}} d_q u \, \frac{(due^{i\theta}, due^{-i\theta}, abcdu/q;q)_\infty}{(adu/q, bdu/q, cdu/q;q)_\infty}$$
$$\times \frac{(q/u;q)_n}{(abcdu/q;q)_n}\left(\frac{du}{q}\right)^n. \tag{5.41}$$

5.2.2. *Mellin–Barnes representation*

Equation (5.35) with arbitrary z is a counterpart of the Mellin–Barnes representation for the classical ($q = 1$) hypergeometric series:

$$\frac{\prod_{k=1}^r \Gamma(a_k)}{\prod_{k=1}^s \Gamma(b_k)} \, {}_rF_s(a_1, ..., a_r;\ b_1, \ldots, b_s \,|\, z)$$
$$= \frac{1}{2\pi i}\int_{-i\infty}^{i\infty} \frac{\prod_{k=1}^r \Gamma(a_k + t)}{\prod_{k=1}^s \Gamma(b_k + t)} \Gamma(-t)(-z)^t dt. \tag{5.42}$$

It is extended to the q-hypergeometric series in the form of an ordinary contour integral:

$${}_{s+1}\phi_s\left(\begin{matrix} a_1, \ldots, a_{s+1} \\ b_1, \ldots, b_s \end{matrix}\,\middle|\, z\right) = \frac{(a_1, a_2, \ldots, a_{s+1}; q)_\infty}{(b_1, \ldots, b_s, q\,; q)_\infty}\left(-\frac{1}{2\pi i}\right)\int_{-i\infty}^{i\infty}$$
$$\times \frac{(q^{t+1}, b_1q^t, \ldots, b_sq^t; q)_\infty}{(a_1q^t, a_2q^t, \ldots, a_{s+1}q^t; q)_\infty} \frac{\pi(-z)^t dt}{\sin \pi t}. \tag{5.43}$$

6. Racah as Hypergeometric Discrete Orthogonal Polynomials

6.1. *Racah polynomials as orthogonal polynomials of discrete variable*

We describe here the q-Racah polynomials [8] that are maximal discrete hypergeometric polynomials, similarly to the Askey–Wilson polynomials being maximal continuous ones. Consider the Askey–Wilson polynomials (3.8) without the normalizing factor and impose the requirement $ac = q^{-N}$ with some natural N. Then, one can define the q-Racah polynomials (with redefined parameters as compared with (3.8)):

$$\mathfrak{R}_n(j;a,b,c,N;q) = {}_4\phi_3\left(\begin{matrix} q^{-n}, abq^{n+1}, q^{-j}, cq^{j-N} \\ aq, q^{-N}, bcq \end{matrix}\,\middle|\, q, q\right). \tag{6.1}$$

These are polynomials of the discrete variable $x_j = q^{-j} + cq^{j-N}$ given on a finite set $j = 0, 1, \ldots, N$. In fact, they are also polynomials of degree j in $y_n = q^{-n} +$

abq^{n+1}, which is a counterpart of the duality relation (3.10) for the Askey–Wilson polynomials.

The orthogonal relations (2.33),

$$\sum_{j=1}^{N} \mathfrak{R}_n(x_j)\mathfrak{R}_m(x_j)\mu_j = \|\mathfrak{R}_n\|^2\delta_{m,n} \tag{6.2}$$

is now satisfied with the weight, $\mu_j = \mu(x_j)$, defined as

$$\mu_j = \frac{1-cq^{2j-N}}{1-cq^{-N}} \frac{(cq^{-N}, aq, bcq, q^{-N}; q)_j}{(q, cq, cq^{-N}/a, q^{-N}/b; q)_j} (abq)^{-j} \tag{6.3}$$

and the norms of the polynomials are

$$\|\mathfrak{R}_n\|^2 = \frac{1-abq}{1-abq^{2n+1}} \frac{(abq^2, 1/c; q)_N}{(bq, aq/c; q)_N} \frac{(q, bq, aq/c, abq^{N+2}; q)_n}{(aq, abq, bcq, q^{-N}; q)_n} (cq^{-N})^n. \tag{6.4}$$

The parameters a, b, c are such that (6.3) and (6.4) are positive.

Note that (2.34) implies that considered as polynomials in $y_n = q^{-n} + abq^{n+1}$, the q-Racah polynomials are also orthogonal with the weight $1/\|\mathfrak{R}_n\|^2$ and the norm square $1/\mu_j$:

$$\sum_n \frac{\mathfrak{R}_n(x_i)\mathfrak{R}_n(x_j)}{\|\mathfrak{R}_n\|^2} = \sum_n \frac{\tilde{\mathfrak{R}}_i(y_n)\tilde{\mathfrak{R}}_j(y_n)}{\|\tilde{\mathfrak{R}}_n\|^2} = \frac{\delta_{i,j}}{\mu_j}. \tag{6.5}$$

Consider now the matrix U with the matrix elements

$$U_{ij} \equiv \frac{\sqrt{\mu_j}}{\|\mathfrak{R}_i\|}\mathfrak{R}_i(x_j) \tag{6.6}$$

Then, from the orthogonality relation

$$\sum_{j=0}^{N} \mathfrak{R}_i(x_j)\mathfrak{R}_k(x_j)\mu_j = \|\mathfrak{R}_i\|^2\delta_{i,k} \tag{6.7}$$

it follows that U is orthogonal. The dual orthogonality relation (2.34) is the same orthogonality relation for the transposed matrix U^t.

6.2. *Relation between* $\mathfrak{R}_n$ *and* P_n^{AW}

From

$$\begin{aligned} \mathfrak{R}_n(\tilde{x}_j) &= {}_4\phi_3\left(\begin{matrix} q^{-n}, \tilde{a}\tilde{b}q^{n+1}, q^{-j}, \tilde{c}q^{j-N} \\ \tilde{a}q, q^{-N}, \tilde{b}\tilde{c}q \end{matrix} \middle| z=q\right) \\ &= {}_4\phi_3\left(\begin{matrix} q^{-n}, abcdq^{n-1}, ae^{i\theta}, ae^{-i\theta} \\ ab, ac, ad \end{matrix} \middle| z=q\right) \\ &= \frac{a^n}{(ab, ac, ad; q)_n} P_n^{AW}(x|a,b,c,d) \end{aligned} \tag{6.8}$$

and $\tilde{x}_j = q^{-j} + \tilde{c}q^{j-N}$, $x = \cos\theta$ we have:

$$a^2 = \tilde{c}q^{-N}$$
$$b = \frac{\tilde{a}q}{a}, \quad c = \frac{q^{-N}}{a}, \quad d = \frac{\tilde{b}\tilde{c}q}{a} \tag{6.9}$$
$$\Longrightarrow abcd = \frac{\tilde{a}\tilde{b}\tilde{c}q^{2-N}}{a^2} = \tilde{a}\tilde{b}q^2$$

and

$$\tilde{x}_j = a(aq^j + a^{-1}q^{-j}) = 2a\cos\theta = 2ax \tag{6.10}$$

Change of variables does not affect (just redefines) the 3-term relation. As for the measures in these two cases, let us note that the orthogonality relation (3.18) for the Askey–Wilson polynomials is singular at $ac = q^{-N}$ for the polynomials with $n \leq N$ being proportional to $\frac{(q^{-N};q)_n}{(q^{-N};q)_\infty}$. However, this is exactly the case of the Racah polynomials. In fact, as it was already noted in Sec. 5.40, in the case of $|a| > 1$, the orthogonality relation (3.19) contains an additional sum as compared with (3.18) coming from pole contributions outside the unit circle

$$\int_{-1}^{1} P_m^{AW}(x)P_n^{AW}(x)w(x)dx + 2\pi\sum_k P_m^{AW}(x_k)P_n^{AW}(x_k)w_k = \frac{\delta_{m,n}}{h_n}\int_{-1}^{1} w(x)dx \tag{6.11}$$

and the sum runs over the points $x_k = (aq^k + q^{-k}/a)/2$. Multiplying the both sides of this identity with the factor $\frac{(q^{-N};q)_\infty}{(q^{-N};q)_n}$, one makes the orthogonality relation non-singular, but at $n \leq N$ the first term at the left-hand side vanishes, and one remains with the orthogonality relation for the Racah polynomials with measure (6.3) (one has to take into account changing the normalization of the Racah polynomials as compared with the Askey–Wilson ones), the points $x_k = (aq^k + q^{-k}/a)/2$ becoming exactly the proper Racah polynomial variables upon identification (6.10).

7. Generalizations of Orthogonal Polynomials

In this section, we consider various generalizations of the hypergeometric polynomials: to the multivariable polynomials and to functions instead of polynomials. In fact, there are two distinct multivariable generalizations, we consider also discrete and continuous orthogonal multivariable polynomials.

7.1. *Multivariable extensions of Askey–Wilson and Racah polynomials*

7.1.1. *BC Koornwinder–Macdonald polynomials*

The Askey–Wilson (q-Racah) polynomials can be described as simplest (one-variable, or one-row) symmetric polynomials for the systems of roots of the BC_n type [36]. This has been first realized by Koornwinder [37] (hence, the name

Koornwinder–Macdonald polynomials). These symmetric polynomials are constructed from the monomials

$$m_\lambda = \sum_{\mu \in G(\lambda)} z_1^{\mu_1} z_2^{\mu_2} \cdots z_n^{\mu_s}, \tag{7.1}$$

where the sum goes over the orbit $G(\lambda)$ of the partition $\lambda = \{\lambda_1 \geq \lambda_2 \geq \cdots \geq 0\}$ under the action of the group $G = S_s \times \mathbb{Z}_2$ which permutes λ_i and changes their signs. This is nothing but the BC_s-type Weyl group. Now defining the second-order difference operator [37]

$$\hat{\mathfrak{D}} = \sum_j \left(P(z; z_j)(\hat{T}_j - 1) + P(z; z_j^{-1})(T_j^{-1} - 1) \right), \tag{7.2}$$

where

$$P(z; z_j) = \frac{\prod_{a=0}^{3}(1 - t_a z_j)}{(1 - z_j^2)(1 - q z_j^2)} \prod_{k \neq j} \frac{(1 - t z_j z_k)(1 - t z_j z_k^{-1})}{(1 - z_j z_k)(1 - z_j z_k^{-1})},$$

$$\hat{T}_j f(z_1, z_2, \ldots) = f(z_1, z_2, \ldots, q z_j, \ldots) \tag{7.3}$$

one can construct the set of BC-Macdonald (or Koornwinder–Macdonald) polynomials

$$P_\lambda^{KM}(z|t_0, t_1, t_2, t_3, t) = \left(\prod_{\mu \leq \lambda} \frac{\hat{\mathfrak{D}} - E_\mu}{E_\lambda - E_\mu} \right) m_\lambda, \tag{7.4}$$

where

$$E_\lambda = \sum_{j=1}^{s} q^{-1} t_0 t_1 t_2 t_3 t^{2n-j-1}(q^{\lambda_j} - 1) + t^{j-1}(q^{-\lambda_j} - 1) \tag{7.5}$$

and $\mu \leq \lambda$ is understood as $\sum_{j=1}^{k}(\lambda_j - \mu_j) \geq 0$ for all $k = 1, \ldots, s$. These polynomials are the eigenfunctions of the difference operator $\hat{\mathfrak{D}}$ (7.2) with the eigenvalues E_λ:

$$\hat{\mathfrak{D}} P_\lambda^{KM} = E_\lambda P_\lambda^{KM}. \tag{7.6}$$

Considering the case of one variable z, i.e. $s = 1$, which is equivalent to one-line Young diagram λ corresponding to symmetric representations, one arrives at the Askey–Wilson polynomials (3.8) with $z = e^{i\theta}$ and $a = t_0$, $b = t_1$, $c = t_2$, $d = t_3$:

$$P_{[n]}^{KM}(e^{i\theta}|t_0, t_1, t_2, t_3, t) = P_n^{AW}(\cos\theta|a, b, c, d). \tag{7.7}$$

Note that, in this case, the t-dependence automatically disappears from the polynomial.

7.1.2. *Multivariable Askey–Wilson and Racah polynomials*

Multivariable Askey–Wilson Polynomials. There is another multivariable generalization of the Askey–Wilson polynomials [38–40]:

$$P_{\mathbf{n}}^{AW}(\mathbf{x}|\alpha_i) = \prod_{i=1}^{s} P_{n_i}^{AW}\left(x_i = \cos\theta_j \middle| \alpha_i q^{N_{i-1}}, \frac{\alpha_i}{\alpha_0^2}, \frac{\alpha_{j+1}}{\alpha_j}e^{i\theta_{j+1}}, \frac{\alpha_{j+1}}{\alpha_j}e^{-i\theta_{j+1}}\right) \quad (7.8)$$

where $\mathbf{n} = (n_1, \ldots, n_s) \in (\mathbb{Z}_{\geq 0})^s$, $\mathbf{x} = (x_1, \ldots, x_s)$, $N_k := \sum_{i=0}^{k} n_k$ with the conventions $N_0 := 0$, $e^{i\theta_0} = \alpha_0$ and $e^{i\theta_{s+1}} = \alpha_{s+2}$.

A counterpart of the conditions $|a| < 1$,$|b| < 1$, $|c| < 1$, $|d| < 1$ for the Askey–Wilson polynomial (3.8) are the conditions

$$\begin{gathered} 0 < |\alpha_{s+1}| < |\alpha_s| < \cdots < |\alpha_1| < \min(1, |\alpha_0|^2), \\ \frac{|\alpha_{s+1}|}{|\alpha_s|} < |\alpha_{s+2}| < \frac{|\alpha_s|}{|\alpha_{s+1}|}. \end{gathered} \quad (7.9)$$

The orthogonality conditions in this case are

$$\int_0^1 \frac{dx_1}{2\pi\sqrt{1-x_1^2}} \cdots \int_0^1 \frac{dx_s}{2\pi\sqrt{1-x_s^2}} W(\mathbf{x}) P_{\mathbf{n}}^{AW}(\mathbf{x}) P_{\mathbf{m}}^{AW}(\mathbf{x}) = H_{\mathbf{n}}\delta_{\mathbf{m},\mathbf{n}} \quad (7.10)$$

with the weight and norm squared given as:

$$\begin{aligned} W(\mathbf{x}) &= \frac{\prod_{i=1}^{s}(e^{2i\theta_i}, e^{-2i\theta_i}; q)_\infty}{\prod_{j=0}^{s}\prod_{\epsilon_{1,2}=\pm 1}(\alpha_{k+1}\alpha_k^{-1}e^{i\epsilon_1\theta_{k+1}+i\epsilon_2\theta_k}; q)_\infty}, \\ H_{\mathbf{n}} &= \prod_{i-1}^{s} \frac{\left(\frac{\alpha_{k+1}^2}{\alpha_0^2}q^{N_{k-1}+N_k-1}; q\right)_{n_k}\left(\frac{\alpha_{k+1}^2}{\alpha_0^2}q^{2N_k}; q\right)_\infty}{\left(q^{n_{k+1}}, \frac{\alpha_k^2}{\alpha_0^2}q^{N_{k-1}+N_k}; \frac{\alpha_{k+1}^2}{\alpha_k^2}q^{n_k}; q\right)_\infty} \\ &\quad \times \prod_{\epsilon=\pm 1}\left(\alpha_{s+1}\alpha_{s+2}^{\epsilon}q^{N_s}, \alpha_{s+1}\alpha_{s+2}^{\epsilon}\alpha_0^{-2}q^{N_s}; q\right)_\infty^{-1}. \end{aligned} \quad (7.11)$$

Multivariable Racah Polynomials. Similarly, at a special value of parameter, one can obtain discrete orthogonal polynomials, which are multivariable Racah polynomials [38–40]. Indeed, consider

$$\frac{\alpha_{s+2}}{\alpha_{s+1}} = q^N \quad (7.12)$$

with a non-negative integer N and put

$$q^{J_k} = e^{i\theta_k}\alpha_k^{-1} \quad (7.13)$$

with some integer J_k's. Then, the polynomials

$$\mathfrak{R}_{\mathbf{n}}(\mathbf{x}, | N, \alpha_i) := \prod_{k=1}^{s} \tilde{\mathfrak{R}}_{n_k}\left(J_k - N_{k-1} \,\middle|\, \frac{\alpha_k^2}{\alpha_0^2} q^{2N_{k_1}-1}, \frac{\alpha_{k+1}^2}{q\alpha_k^2}, \alpha_k^2 q^{J_{k+1}+N_{k-1}}, J_{k+1} - N_{k-1}\right),$$
$$\tilde{\mathfrak{R}}_n(j|a,b,c,N) := (aq, bcq, q^{-N}; q)_n \, q^{Nn/2} c^{-n/2} \, \mathfrak{R}_n(j; a, b, c, N; q), \tag{7.14}$$

are orthogonal on $0 := J_0 \le J_1 \le \cdots \le J_s \le J_{s+1} := N$ with the discrete weight

$$\mu_{\mathbf{J}} = \prod_{k=0}^{s} \frac{\left(\frac{\alpha_{k+1}^2}{\alpha_k^2}; q\right)_{J_{k+1}-J_k}}{(q;q)_{J_{k+1}-J_k}} \frac{\left(\alpha_{k+1}^2; q\right)_{J_{k+1}+J_k}}{(q\alpha_k^2; q)_{J_{k+1}+J_k}} \left(1 - \alpha_k^2 q^{2J_k}\right) \left(\frac{\alpha_{k-1}}{\alpha_k}\right)^{2J_k} \tag{7.15}$$

and the norm squared

$$\|\mathfrak{R}_{\mathbf{n}}\|^2 = \frac{\left(q\alpha_s^2, \frac{\alpha_{s+1}^2}{\alpha_0^2}; q\right)_N}{(\alpha_{s+1}^2, q\alpha_0^2; q)_N} \left(\frac{\alpha_{s+1}^2}{\alpha_0^2} q^N, \alpha_{s+1}^2 q^N, \frac{q^{-N}}{\alpha_0^2}, q^{-N}; q\right)_{N_s} \left(\frac{\alpha_0^2}{\alpha_s^2}\right)^N$$
$$\times \prod_{k=1}^{s} \frac{(q, \alpha_{k+1}^2; q)_{n_k} \left(\frac{\alpha_s^2}{\alpha_0^2}; q\right)_{N_k+N_{k-1}}}{\left(\frac{\alpha_{k+1}^2}{q\alpha_0^2}; q\right)_{N_k+N_{k-1}}} \frac{q\alpha_0^2 - \alpha_{k+1}^2}{q\alpha_0^2 - q^{2N_k}\alpha_{k+1}^2}. \tag{7.16}$$

7.2. *Askey–Wilson functions*

The Askey–Wilson function is defined by [41–44]

$$\Psi_\lambda(z|a,b,c,d) := \widehat{h}\left(z \,\middle|\, \frac{qa\lambda}{\tilde{d}}\right) \frac{(q/(\tilde{d}\lambda), ab, ac, qa/d; q)_\infty}{(\tilde{a}\tilde{b}\tilde{c}\lambda; q)_\infty}$$
$$\cdot {}_8W_7\left(\frac{\tilde{a}\tilde{b}\tilde{c}\lambda}{q}; az, \frac{a}{z}, \tilde{a}\lambda, \tilde{b}\lambda, \tilde{c}\lambda \,\middle|\, q, \frac{q}{\tilde{d}\lambda}\right), \tag{7.17}$$

where

$$\widehat{h}(z|a) := h(z + z^{-1}|a) \tag{7.18}$$

and $h(z|a)$ is defined in (3.15). Here we follow the definitions of $\tilde{a}, \tilde{b}, \tilde{c}, \tilde{d}$ as in Eq. (6.9). This function is symmetric in a, b, c and q/d. Using Baily's transformation [8], one can express the Askey–Wilson function through the sum of two balanced ${}_4\phi_3$ series,

$$\Psi_\lambda(z|a,b,c,d) = \widehat{h}\left(z \,\middle|\, \frac{q}{d}\right) \widehat{h}\left(\lambda \,\middle|\, \frac{q}{\tilde{d}}\right) \frac{(ab, ac; q)_\infty}{(q/(ad); q)_\infty} \, {}_4\phi_3\left(\begin{matrix} az, a/z, \tilde{a}\lambda, \tilde{a}/\lambda \\ ab, ac, ad \end{matrix} \,\middle|\, z = q\right)$$
$$+ \widehat{h}(z|a)\widehat{h}(\lambda|\tilde{a}) \frac{(qb/d, qc/d; q)_\infty}{(ad/q; q)_\infty} \, {}_4\phi_3$$
$$\times \left(\begin{matrix} qz/d, q/(dz), q\lambda/\tilde{d}, q/(\lambda\tilde{d}) \\ qb/d, qc/d, q^2/(ad) \end{matrix} \,\middle|\, z = q\right). \tag{7.19}$$

We will sometimes also consider the Askey–Wilson function which differs by a factor

$$\tilde{\Psi}_\lambda(z|a,b,c,d) = \frac{1}{\widehat{h}\left(z\,\middle|\,\frac{q}{d}\right)}\Psi_\lambda(z|a,b,c,d) \tag{7.20}$$

which satisfies the same Askey–Wilson difference equation (3.30) and, at $\lambda = \tilde{a}q^m$ with positive integer m, becomes the balanced terminating series ${}_4\phi_3$ and coincides with the Askey–Wilson polynomial up to a constant:

$$\tilde{\Psi}_{\tilde{a}q^m}(z|a,b,c,d) = (-d)^m q^{m(m-1)/2}(abq^m, acq^m, bcq^m;q)_\infty P_m^{AW}\left(z+\frac{1}{z}\,\middle|\,a,b,c,d\right). \tag{7.21}$$

It also has a single integral representation that can be read off from (5.36) and is a counterpart of the Mellin–Barnes representation (Sec. 5.2.2)

$$\tilde{\Psi}_\lambda(z|a,b,c,d) = \frac{(a/z,b/z;q)_\infty}{(q/(cz),q/(dz);q)_\infty}\frac{(q,ab\lambda/\tilde{a},ac\lambda/\tilde{a},bc\lambda/\tilde{a})_\infty}{(q\lambda/\tilde{a})_\infty}$$
$$\times\int_{S^1}\frac{\theta\left(\frac{c}{y}\sqrt{\frac{ab}{q\lambda}};q\right)\left(\sqrt{\frac{\lambda}{dc}}\frac{q}{zy},\sqrt{qab\lambda}zy,\sqrt{\frac{c}{d\lambda}}qy;q\right)_\infty}{\left(\sqrt{\frac{qb\lambda}{a}}y,\sqrt{\frac{qa\lambda}{b}}y,\sqrt{\frac{c\lambda}{d}}\frac{1}{y},\sqrt{\frac{ab}{q\lambda}}\frac{1}{zy},\frac{qyz}{\sqrt{dc\lambda}};q\right)_\infty}\frac{dy}{y}, \tag{7.22}$$

where the integral runs over the unit circle S^1 and the renormalized Jacobi θ-function is

$$\theta(x;q) := (x,q/x;q)_\infty. \tag{7.23}$$

Note that the Askey–Wilson function (7.17) satisfies the duality relation similar to (3.10)

$$\Psi_\lambda(z|a,b,c,d) = \Psi_z(\lambda|\tilde{a},\tilde{b},\tilde{c},\tilde{d}) \tag{7.24}$$

and another duality relation

$$\Psi_\lambda(z|a\xi,by,b/y,q\xi/a) = \Psi_\xi(y|a\lambda,bz,b/z,q\lambda/a). \tag{7.25}$$

Introduce a discrete set Γ parametrized by an arbitrary parameter $\nu>0$: $\{\nu q^{-k}>1,\ k\in\mathbb{Z}\}$ and its dual $\tilde{\Gamma}$ parametrized by the dual parameter $\tilde{\nu}=\tilde{d}/(a\nu)$. Then, an essential property of the Askey–Wilson functions is that they are orthogonal for $\lambda\in\tilde{\Gamma}$,

$$\begin{aligned}\langle\Psi_\lambda,\Psi_{\lambda'}\rangle &= \frac{1}{2i}\int_{S^1}\Psi_\lambda(z)\overline{\Psi_{\lambda'}(z)}\frac{h(z^2|1)}{\widehat{h}(z|a,b,c,q/d)\theta(\nu z;q)\theta(\nu/z;q)}\frac{dz}{z}\\ &\quad+2\pi\sum_{x\in\Gamma}\Psi_\lambda(x)\overline{\Psi_{\lambda'}(x)}W_k\\ &= \|\Psi_\lambda\|^2\delta_{\lambda,\lambda'},\quad \lambda,\lambda'\in\tilde{\Gamma}\end{aligned} \tag{7.26}$$

where

$$W_k := \frac{1}{\widehat{h}(\nu|a,b,c,q/d)(q,q;q)_\infty} \frac{(1-q^{2k}\nu^{-2})(a/\nu,b/\nu,c/\nu,q/(d\nu);q)_k}{(q/(a\nu),q/(b\nu),q/(c\nu),d/\nu;q)_k} q^{k(k+1)} \times \left(\frac{abc\nu^2 q}{d}\right)^{-2k}. \tag{7.27}$$

This gives the Askey–Wilson transform from functions with the scalar product (7.26) to the scalar product with the dual parameters $(a,b,c,d,\nu) \longrightarrow (\tilde{a},\tilde{b},\tilde{c},\tilde{d},\tilde{\nu})$:

$$\langle \Psi_\lambda(x), f(x)\rangle \to \tilde{f}(\lambda). \tag{7.28}$$

This transform is unitary with inverse given by the Askey–Wilson functions with the dual parameters.

7.3. *Askey–Wilson functions at* $|q|=1$

Extending the notion of the Askey–Wilson function to the case of $|q| = 1$ (for extension of the Askey–Wilson polynomials to the roots of unity, see [45]), i.e. to $q = e^{2\pi i\tau}$ with a real parameter τ is not immediate, since the main building block, the Pocchammer symbol $(a;q)_\infty$ requires some regularization. The most convenient to regularize is, however, a differently normalized quantity[c] in logarithmic variables $x \to q^x$:

$$\begin{gathered} \frac{q^{-\frac{(2x-1)^2}{16}}}{(-q^x;q)_\infty} \to \Gamma_\tau(x) := e^{i\gamma_\tau(x)}, \\ \gamma_\tau(x) := \frac{1}{2i}\int_0^\infty \frac{dy}{y}\left(\frac{2x-1}{y} - \frac{\sinh\big[(2x-1)\tau y\big]}{\sinh y\cdot\sinh\tau y}\right). \end{gathered} \tag{7.30}$$

This function satisfies

$$\Gamma_\tau(x+1) = 2\cos\pi x\tau\cdot\Gamma_\tau(x) \tag{7.31}$$

and has the modular properties

$$\Gamma_\tau(x) = \Gamma_{\tau^{-1}}(-\tau x+\tau/2+1/2), \qquad \Gamma_\tau(-x+1) = \Gamma_\tau^{-1}(x). \tag{7.32}$$

Now one can define [43] the Askey–Wilson function at $|q| = 1$ by the integral like (7.22)

$$\Psi_\lambda^\tau(x|a,b,c,d) := \frac{2\,\Gamma_\tau\left(\frac{1}{2\tau}-\gamma-\frac{x}{2}+1\right)\Gamma_\tau\left(\frac{1}{2\tau}-\delta-\frac{x}{2}+1\right)}{\Gamma_\tau\left(\frac{1}{2\tau}+\alpha-\frac{x}{2}\right)\Gamma_\tau\left(\frac{1}{2\tau}+\beta-\frac{x}{2}\right)} \int_C dz\, F_1(z)F_2(z), \tag{7.33}$$

[c]The q-Γ-function is usually defined as [8]

$$\Gamma_q(x) := (1-q)^{1-x}\frac{(q;q)_\infty}{(q^x;q)_\infty}. \tag{7.29}$$

Hence, the notation.

where

$$F_1(z) := \frac{\Gamma_\tau\Big(\frac{1}{2\tau} - l + \frac{x}{2} + z - \delta - \gamma + 1\Big)\Gamma_\tau\Big(\frac{1}{2\tau} - l - \frac{x}{2} - z + \alpha + \beta - 1\Big)}{\Gamma_\tau\Big(-\frac{1}{2\tau} + l - \frac{x}{2} - z - \delta - \gamma + 1\Big)\Gamma_\tau\Big(-\frac{1}{2\tau} + l + \frac{x}{2} + z + \alpha + \beta\Big)},$$

$$F_2(z) := \frac{\Gamma_\tau\Big(-\frac{1}{2\tau} + l - z + \gamma - \delta\Big)\Gamma_\tau\Big(-\frac{1}{2\tau} + l + z + \alpha - \beta + \frac{1}{2}\Big)}{\Gamma_\tau\Big(-\frac{1}{2\tau} - l + z + \gamma - \delta + 1\Big)\Gamma_\tau\Big(-\frac{1}{2\tau} - l - z + \alpha - \beta + \frac{1}{2}\Big)} \tag{7.34}$$

and

$$q = e^{2\pi i\tau}, \quad \lambda = q^{2l}, \quad (a,b,c,d) = (q^{2\alpha}, q^{2\beta}, q^{2\gamma}, q^{2\delta}). \tag{7.35}$$

The contour $\mathcal{C}$ here is a deformation of the imaginary axis such that $F_1(z)$ and $F_2(z)$ are analytic on the stirp around $\mathcal{C}$ of width not less than 4. This function solves the Askey–Wilson difference equation (3.30), and is a counterpart of the Mellin–Barnes representation (Sec. 5.2.2) at $|q| = 1$.

8. Orthogonal Polynomials, Matrix Models and Integrable Systems

8.1. *Orthogonal polynomials and matrix models*

Orthogonal polynomials are a popular tool for investigation of the eigenvalue matrix models [46–49]. Partition functions of these models are

$$Z_N[d\mu;\mathcal{C}] = \langle\!\langle 1\rangle\!\rangle_N := \frac{1}{N!}\int_{\mathcal{C}} \Delta^2(z)\prod_{i=1}^N d\mu(z_i) = \det_{1\le i,j\le N}\int z^{i+j-2}d\mu(z), \tag{8.1}$$

where $\Delta(z) := \det_{i,j} z_i^{j-1} = \prod_{i<j}(z_i - z_j)$ is the Vandermonde determinant. Hereafter, for the sake of brevity, we omit the integration contour $\mathcal{C}$ from the arguments of the partition function in most cases. By suitable rearrangement of the Vandermonde determinant, we can show that these partition functions are consecutive product of the norms of monic polynomials orthogonal with respect to the measure $d\mu(x)$,

$$\int \mathcal{P}_n(x)\mathcal{P}_m(x)d\mu(x) = \|\mathcal{P}_n\|^2 \cdot \delta_{m,n}, \tag{8.2}$$

$$Z_N[d\mu] = \prod_{i=0}^{N-1} \|\mathcal{P}_i\|^2. \tag{8.3}$$

In addition, the monic orthogonal polynomials can be understood as expectation value of the resolvent determinant with respect to the matrix integral (see (2.3)),

$$\mathcal{P}_n(x) = \frac{\langle\!\langle \prod_{i=1}^n (x - z_i)\rangle\!\rangle_n}{\langle\!\langle 1\rangle\!\rangle_n} = x^n + \cdots. \tag{8.4}$$

Since multiplication by x is a Hermitian operation in the Hilbert space expanded in terms of orthogonal polynomials:

$$\int \mathcal{P}_n(x) \cdot x\mathcal{P}_m(x) d\mu(x) = 0 \quad \text{for } m < n-1, \tag{8.5}$$

these polynomials satisfy the 3-term recursion relation,

$$x\mathcal{P}_n(x) = \mathcal{P}_{n+1}(x) + K_n \mathcal{P}_n(x) + R_n \mathcal{P}_{n-1}(x). \tag{8.6}$$

From

$$\int \mathcal{P}_n(x) \cdot x\mathcal{P}_{n-1}(x) d\mu(x) = \|\mathcal{P}_n\|^2 = R_n \cdot \|\mathcal{P}_{n-1}\|^2, \tag{8.7}$$

we deduce that

$$\|\mathcal{P}_n\|^2 = \|\mathcal{P}_0\|^2 \prod_{k=1}^{n} R_k, \quad \text{and} \quad Z_N[d\mu] = \|\mathcal{P}_0\|^{2N} \prod_{n=1}^{N-1} \prod_{k=1}^{n} R_k. \tag{8.8}$$

To make a q-deformation of the eigenvalue matrix model, we only need to change the integral into the Jackson sum

$$\int_0^1 f(z)dz \to \int f(z) d_q(z) = \sum_{k=0}^{\infty} (1-q) q^k f(q^k), \tag{8.9}$$

and the Vandermonde determinant is q-deformed when matrix models are promoted to the β-ensembles, with $\Delta^{2\beta}(z) \to \Delta_\beta^{(2)}(z;q)$ equal to

$$\Delta_\beta^{(2)}(z;q) := \prod_{i<j} \left(z_i/z_j;q\right)_\beta \left(z_j/z_i;q\right)_\beta \tag{8.10}$$

for positive integer $\beta \in N$, and to

$$\Delta_\beta^{(2)}(z;q) := \prod_{i<j} \frac{\left(z_i/z_j;q\right)_\infty}{\left(q^\beta z_i/z_j;q\right)_\infty} \frac{\left(z_j/z_i;q\right)_\infty}{\left(q^\beta z_j/z_i;q\right)_\infty} \tag{8.11}$$

otherwise.

8.2. *Orthogonal polynomials and integrable systems*

The relation between matrix models and integrable systems allows one to make a bridge between orthogonal polynomials and integrable systems. Essentially, the system of orthogonal polynomials is nothing but the Baker–Akhiezer function (up to an exponential factor) of an integrable forced hierarchy of the Toda type (i.e. with a discrete type variable) [47, 48, 50]. Let us explain how it works in the simplest case of the model (8.1).

First of all, one can notice [47] that the matrix model partition function (8.1) is the τ-function of the forced [48, 50, 51] Toda chain hierarchy. Indeed, this τ-function has the generic Töpliz form [50]

$$\tau_n = \det_{1\le i,j\le n} \partial^{i+j-2} C(t), \tag{8.12}$$

where $C(t)$ is the moment matrix that depends on infinitely many time variables t_k, and ∂ denotes the derivative with respect to the first time, the dependence on higher times being determined from the compatibility condition

$$\frac{\partial C}{\partial t_k} = \partial^k C \tag{8.13}$$

and the initial conditions are

$$\tau_0 = 1, \qquad \tau_{-1} = 0. \tag{8.14}$$

This τ-function satisfies an infinite set of Hirota bilinear identities: the differential-difference bilinear Toda equation

$$\tau_n \partial^2 \tau_n - (\partial \tau_n)^2 = \tau_{n+1}\tau_{n-1} \tag{8.15}$$

and an infinite hierarchy of differential bilinear equations of the KP type with respect to the time variables t_k [50].

The general solution to (8.13) in the Fourier representation is

$$C(t) = \int_C d\mu_0(z) \exp\left(\sum_k t_k z^k\right) \tag{8.16}$$

which can be easily related to the general point of the infinite-dimensional Grassmannian describing the space of solutions to this infinite hierarchy [50, Appendix B]. On the other hand, comparing (8.1) with (8.12) and (8.16), one immediately arrives at an identification of the orthogonality measure $d\mu(z)$ with the integrand in (8.16):

$$d\mu(z) = d\mu_0(z) \exp\left(\sum_k t_k z^k\right) \tag{8.17}$$

and the Toda chain τ-function (8.12) with the matrix model partition function (8.1):

$$\tau_n = Z_n \left[d\mu(z) = d\mu_0(z) \exp\left(\sum_k t_k z^k\right)\right] \tag{8.18}$$

i.e. the integrable flows are just a proper parametrization of the orthogonality measures so that the concrete solution of the integrable hierarchy (i.e. the concrete point of the Grassmannian) is fixed by the choice of $d\mu_0(z)$. In matrix model theory, this choice is usually done via a set of Virasoro constraints (in fact, the lowest constraint plus integrable equations are sufficient to fix the whole algebra of constraints; this lowest constraint is called the string equation).

There is, however, much more direct connection between the orthogonal polynomials and the integrable systems, not just through the measure, or through the τ-function being a product of the norms: the orthogonal polynomials are nothing but (up to a factor) the Baker–Akhiezer functions of the integrable system [47]. Indeed, the Lax representation for the Toda system is

$$\frac{\partial \mathbf{L}}{\partial t_k} = \frac{1}{2}[\mathcal{R}\mathbf{L}^k, \mathbf{L}] \tag{8.19}$$

where the Lax matrix is

$$\mathbf{L}_{mn} := \delta_{m,n-1} + K_n\delta_{m,n} + R_{n+1}\delta_{m,n+1} \tag{8.20}$$

and the $\mathcal{R}$-matrix acts on the matrix A_{ij} in accordance with

$$\mathcal{R}A_{ij} := \begin{cases} A_{ij} & \text{if } i > j, \\ -A_{ij} & \text{if } i \leq j. \end{cases} \tag{8.21}$$

Then, the 3-term relation (8.6) can be treated as an eigenvalue problem

$$\mathbf{L}\Psi(z;t) = z\Psi(z;t), \tag{8.22}$$

where z plays the role of the spectral parameter. The eigenfunction of the Lax operator is called the Baker–Akhiezer function, and it is just the orthogonal polynomial in our case. The only subtlety is that in (8.22) the normalization of the eigenfunction is not fixed, and the normalization of the Baker–Akhiezer function has an additional factor of $\exp\left(\frac{1}{2}\sum_k t_k z^k\right)$ as compared with the orthogonal polynomial:

$$\Psi_n(z;t) = \exp\left(\frac{1}{2}\sum_k t_k z^k\right)\mathcal{P}_n(z;t). \tag{8.23}$$

8.3. *Orthogonal polynomials as a bridge to matrix models of Kontsevich type*

Representations of matrix model partition functions of type (8.1) in terms of orthogonal polynomials allows one to establish their connections with matrix models of the Kontsevich type. Let us consider again $\tau_n = Z_n\left[d\mu(z) = d\mu_0(z)\exp\left(\sum_k t_k z^k\right)\right]$ with the sum including the term $t_0 \log z$ and make the change of time variables t_k to the Miwa variables λ_a, $a = 1, \ldots, N$ and N is large enough [52, 53]:

$$t_k = \frac{1}{k}\sum_a \lambda_a^{-k}, \qquad t_0 = \sum_a \log \lambda_a. \tag{8.24}$$

Now, using the integral representation in formula (8.1), one can easily derive [4, 50] that

$$Z_n\left[d\mu(z) = d\mu_0(z)\exp\left(\sum_k t_k z^k\right)\right] = \tau_n[d\mu(z)] = \tau_n[d\mu_0(z)] \times \frac{\det_{a,b}\mathcal{P}_{n+a-1}(\lambda_b)}{\Delta(\lambda)}, \tag{8.25}$$

where $\mathcal{P}_n$ here are the monic orthogonal polynomials with respect to the measure $d\mu_0(z)$.

This is the key formula that we need. Now our strategy will be to reduce a matrix model to the eigenvalue model of type (8.1), and a matrix model of the Kontsevich type to the right-hand side of (8.25). In fact, one can consider more general models than (8.1), this question will be addressed elsewhere.

8.3.1. *Hermitian versus Kontsevich models*

Let us consider now as our reference example the partition function of the Gaussian Hermitian one-matrix model [49, 54]

$$\mathbf{Z}_n^{\mathbb{H}}(t) := \frac{1}{\mathrm{Vol}(U_n)} \int D_n M \exp\left(-\frac{1}{2}\mathrm{tr}\, M^2 + \sum_{k=0} t_k \,\mathrm{tr}\, M^k\right)$$
$$= Z_n\left[\exp\left(-\frac{1}{2}z^2 + \sum_k t_k z^k\right); \mathbb{R}\right] \tag{8.26}$$

i.e. the Toda chain τ-function (8.12) with $d\mu_0(z) = \exp(-z^2/2)$ and the integration contour $\mathcal{C} = \mathbb{R}$. Here $\mathrm{Vol}(U_n)$ is the Haar measure of the unitary group $U(n)$ and $D_n M$ is the Cartesian measure on the $n \times n$ Hermitian matrices. This formula implies that one has to consider the Hermite polynomials in (8.25) in order to deal with this model. Note that the monic Hermite polynomials have the integral representation (see (5.32))

$$\mathcal{H}_n(x) := He_n(x) = 2^{-n/2} H_n\left(\frac{x}{\sqrt{2}}\right) = \frac{i^n}{\sqrt{2\pi}} \int_{-\infty}^{+\infty} dy y^n e^{x^2/2 - ixy - y^2/2}. \tag{8.27}$$

Note also that

$$Z_n\left[\exp\left(-\frac{1}{2}z^2\right); \mathbb{R}_+\right] = (2\pi)^{n/2} \prod_{k=0}^{n-1} k! := \mathbf{Z}_n^{\mathbb{H}}(0). \tag{8.28}$$

The Kontsevich type matrix model is given by an integral over the $N \times N$ Hermitian matrix X and depends on an external $N \times N$ matrix Λ

$$\mathbf{Z}_N^{K\mathbb{H}}(n, \Lambda) :- (2\pi)^{(n-N)/2} i^{nN} \left(\prod_{k=0}^{n-1} k!\right)$$
$$\times \frac{c^{\frac{\mathrm{Tr}\,\Lambda^2}{2}}}{\mathrm{Vol}(U_N)} \int D_N X \exp\left(-\frac{1}{2}\mathrm{Tr}\, X^2 + n\, \mathrm{Tr}\, \log X - i\, \mathrm{Tr}\, X\Lambda\right). \tag{8.29}$$

Using the Itzykson–Zuber formula [55],

$$\int D_N U \exp(\mathrm{Tr}\, XUYU^\dagger) = \mathrm{Vol}(U_N) \frac{\det_{a,b} e^{x_a y_b}}{\Delta(x)\Delta(y)}, \tag{8.30}$$

where $D_N U$ is the integral over the unitary $N \times N$ matrix U, and x_a, y_a are eigenvalues of the matrices X, Y, we find that the partition function depends only on the eigenvalues λ_a of the matrix Λ. The explicit result is given as

$$\mathbf{Z}_N^{K\mathbb{H}}(n, \Lambda) = \mathbf{Z}_n^{\mathbb{H}}(0) \times \frac{\det_{a,b} \Phi_a^{\mathbb{H}}(\lambda_b)}{\Delta(\lambda)}, \tag{8.31}$$

where

$$\Phi_a^{\mathbb{H}}(z) := \frac{i^n}{\sqrt{2\pi}} e^{\frac{z^2}{2}} \int_{-\infty}^{+\infty} dx x^{a-1} e^{-x^2/2 + n \log x - ixz} = \mathcal{H}_{n+a-1}(z), \tag{8.32}$$

see (8.27). Now, using (8.25)–(8.26) and (8.31)–(8.32), we obtain that

$$\boxed{\mathbf{Z}_n^{\mathbb{H}}(t) = \mathbf{Z}_N^{K\mathbb{H}}(n, \Lambda)}\,, \tag{8.33}$$

with

$$t_k = \frac{1}{k}\operatorname{Tr}\Lambda^{-k}, \qquad t_0 = \operatorname{Tr}\log\Lambda. \tag{8.34}$$

8.3.2. *Complex Gaussian versus rectangular Kontsevich models*

Another example of a similar correspondence is the complex matrix model case. Let us consider instead of (8.26) the complex Gaussian one-matrix model [54, 56, 57]

$$\begin{aligned}\mathbf{Z}_n^{\mathbb{C}}(\nu, t) := {} & \frac{1}{(\operatorname{Vol}(U_n))^2}\int D_n\mathcal{M}D_n\mathcal{M}^\dagger \\ & \times \exp\left(-\operatorname{tr}(\mathcal{M}\mathcal{M}^\dagger) + \nu\operatorname{tr}\log(\mathcal{M}\mathcal{M}^\dagger) + \sum_{k=0} t_k\operatorname{tr}(\mathcal{M}\mathcal{M}^\dagger)^k\right) \\ = {} & Z_n\left[z^\nu\exp\left(-z + \sum_k t_k z^k\right); \mathbb{R}_+\right],\end{aligned} \tag{8.35}$$

where the integral runs over the complex matrix $\mathcal{M}$ with the Cartesian integration measure. This formula implies that one has to consider this time the Laguerre polynomials in (8.25). The monic Laguerre polynomials have the integral representation (see (5.30))

$$\mathcal{L}_n^{(\nu)}(x) = (-1)^n e^x x^{-\nu/2}\int_0^\infty dz e^{-z} z^{\nu/2+n} J_\nu\big(2\sqrt{xz}\big). \tag{8.36}$$

Note that

$$Z_n\big[z^\nu\exp(-z); \mathbb{R}_+\big] = \prod_k^{n-1}(k+\nu)!\;k! := \mathbf{Z}_n^{\mathbb{C}}(\nu, 0). \tag{8.37}$$

Now we compare (8.35) with the corresponding rectangular complex Kontsevich model [4]

$$\begin{aligned}\mathbf{Z}_{N_1,N_2}^{K\mathbb{C}}(n, \Xi\Xi^\dagger) = {} & \frac{\mathfrak{N}_{N_1,N_2}\, e^{-\operatorname{Tr}_{N_1}(\Xi\Xi^\dagger)}}{\operatorname{Vol}(U_{N_1})\operatorname{Vol}(U_{N_2})} \\ & \times\int D\mathcal{X}D\mathcal{X}^\dagger e^{(-\operatorname{Tr}_{N_1}(\mathcal{X}\mathcal{X}^\dagger)+\operatorname{Tr}_{N_1}(\Xi\mathcal{X}^\dagger)+\operatorname{Tr}_{N_1}(\mathcal{X}\Xi^\dagger)+(n+\nu)\operatorname{Tr}_{N_1}\log(\Xi\Xi^\dagger))} \\ \mathfrak{N}_{N_1,N_2} = {} & 2^{N_1(N_1+1)-\nu(\nu-1)/2}\left(\prod_l^{n-1}(l+\nu)!\;l!\right)\left(\prod_{k=0}^{\nu-1}k!\right), \qquad \nu := N_1 - N_2,\end{aligned} \tag{8.38}$$

where the integral runs over the complex rectangular $N_1 \times N_2$ matrix $\mathcal{X}$ and its $N_2 \times N_1$ Hermitian conjugate $\mathcal{X}^\dagger$. Hence, the external matrix Ξ is this case an

$N_1 \times N_2$ matrix. We could also choose all the traces in (8.38) to be taken over the matrices of size $N_2 \times N_2$ by inverting the order of all terms in the exponentials in (8.38). Without loss of generality, we put $N_1 \geq N_2$. The matrix Ξ can be rotated by unitary matrices to a matrix with N_2 non-zero entries ξ_a on a diagonal, and $\mathrm{Tr}_{N_1} f(\Xi\Xi^\dagger) = \sum_{a=1}^{N_2} f(\xi_a \xi_a^\dagger)$.

This integral can be also dealt with using the generalization of the Itzykson–Zuber formula [58]

$$\int DU_1 \int DU_2 \exp\left(\frac{1}{2}\mathrm{Tr}\left[U_1 \mathcal{X} U_2 \mathcal{Y} + \mathcal{Y}^\dagger U_2^\dagger \mathcal{X}^\dagger U_1^\dagger\right]\right)$$
$$= \mathrm{Vol}(U_{N_1})\mathrm{Vol}(U_{N_2}) \frac{2^{N_1(N_2-1)-\nu(\nu+1)/2}}{\prod_{k=0}^{\nu-1} k! \prod_a^{N_2} (x_a y_a)^\nu} \frac{\det_{a,b} J_\nu(x_a y_b)}{\Delta(x^2)\Delta(y^2)}, \tag{8.39}$$

where U_1 and U_2 are the unitary matrices of sizes $N_1 \times N_1$ and $N_2 \times N_2$ respectively and x_a, y_a are the eigenvalues of the matrices $\mathcal{X}$ and $\mathcal{Y}$ respectively. Hence, one immediately obtains

$$\mathbf{Z}^{K\mathbb{C}}_{N_1,N_2}(n, \Xi, \Xi^\dagger) = \mathbf{Z}^{\mathbb{C}}_n(\nu, 0) \times \frac{\det_{a,b\leq N_2} \Phi^{\mathbb{C}}_a(\xi_b \xi_b^\dagger)}{\Delta(\xi\xi^\dagger)}, \tag{8.40}$$

where

$$\Phi^{\mathbb{C}}_a(z) = e^z z^{-\nu/2} \int_0^{+\infty} dx x^{a-1-\nu/2} e^{-x+(n+\nu)\log x} J_\nu(2\sqrt{xz}) = (-1)^{n+a-1} \mathcal{L}^{(\nu)}_{n+a-1}(z), \tag{8.41}$$

see (8.36). Now, using (8.25), (8.35) and (8.40)–(8.41), we obtain that

$$\boxed{\mathbf{Z}^{\mathbb{C}}_n(\nu, t) = \mathbf{Z}^{K\mathbb{C}}_{N_1,N_2}(n, \Xi\Xi^\dagger)}, \tag{8.42}$$

with

$$t_k = \frac{1}{k}\mathrm{Tr}_{N_1}(\Xi\Xi^\dagger)^{-k}, \qquad t_0 = \mathrm{Tr}_{N_1}\log(\Xi\Xi^\dagger). \tag{8.43}$$

9. Orthogonal Polynomials, 6*j*-Symbols and Knots

9.1. *Coloring knots with orthogonal polynomials*

The old classical problem of effective constructing knot polynomials [59–61] has got a new momentum during last decades and now is one of the most challenging problems in modern mathematical physics.

One of the most effective ways to construct knot polynomials is within the Reshetikhin–Turaev (RT) approach that was suggested in [62, 63] and further developed in [64, 65]. The idea is to consider a plane projection of a given knot, to put an R-matrix or its inverse at each crossing depending on the type of the crossing and take a specifically defined trace of the product of the R-matrices (in fact, this is the usual trace with insertions of specific operators). This trace gives a knot polynomial (the HOMFLY-PT polynomial for the $SL_q(N)$ R-matrices). The topological invariance of this expression is guaranteed by invariance with respect to three Reidemeister moves [66]. The second Reidemeister move is satisfied, since the opposite crossing is described by the inverse R-matrix. The third move is nothing but the Yang–Baxter equation for the R-matrices. The most non-trivial is the first move, it is guaranteed by a properly defined trace.

In fact, technically there are two different descriptions: every knot has a representation as a braid so that there has been developed a procedure of calculating the knot polynomial for its representation as a braid [65]. There is also a large class of arborescent knots [60, 67], which can be calculated in an easier way [3, 68]. Moreover, there is also a mixture of these two approaches [69] that allows one to evaluate knot polynomials in the most effective way known so far.

The general feature of all these approaches is manifest by constructing the R-matrix acting on the tensor product $V_1 \otimes V_2$ in the basis of irreducible representations emerging in the decomposition of this product: $V_1 \otimes V_2 \to \oplus_i W_i$. In fact, since some of the representation can appear a few times, one works in the space of intertwining operators: $V_1 \otimes V_2 = \oplus_W \mathcal{M}_W^{V_1 V_2} W$. In this basis, some of the R-matrices can be chosen to be diagonal, then, the remaining ones are given by conjugation of these diagonal matrices with the Racah matrices, or the Wigner $6j$-symbols, i.e. matrices intertwining the maps

$$(V_1 \otimes V_2) \otimes V_3 \to V_4 \qquad \text{and} \qquad V_1 \otimes (V_2 \otimes V_3) \to V_4. \tag{9.1}$$

The Racah matrices [70] are always orthogonal (or unitary in the complex case) and, hence, satisfy some orthogonality relations. This means that one can color the knot with orthogonal polynomials. However, one has to check each time that the Reidemeister moves are satisfied, which is quite a non-trivial constraint. For the Racah matrices, they are guaranteed. Hence, we consider in this section various examples of the Racah matrices that give rise to various orthogonal systems and to knot polynomials. Note that, depending on the concrete case, the Racah matrices can be discrete orthogonal polynomials (hence, the discrete index), or instead orthogonal functions (hence, continuous index and distributions). The reference example of the first case is given by the Racah polynomials, while the second example consists of two cases: the Askey–Wilson functions, associated with unitary infinite-dimensional representations at $0 < q < 1$, and the Racah "matrices" constructed from the Barnes functions, corresponding to the case of $|q| = 1$.

3-Strand Braid. To complete this subsection, we briefly describe the Racah matrices that emerge in description of knot polynomials. First consider as an example the knot presented by a three-strand braid. Then, its HOMFLY polynomial is described by the formula [65]

$$\begin{aligned} H_V(q,A) &= \mathrm{Tr}_q\big(R_1^a R_2^b \cdots\big) \\ &= \sum_{Q \vdash 3|V|} \big(R_1^{(Q)}\big)^a_{i_1 i_2} \big(R_2^{(Q)}\big)^b_{i_2 i_3} \cdots \big(R_*^{(Q)}\big)^*_{i_n i_1} \dim_q Q, \quad A := q^N, \end{aligned} \tag{9.2}$$

where $Q \in V^{\otimes 3}$. The R-matrices are equal to ($W \in V^{\otimes 2}$)

$$\left(R_1^{(Q)}\right)_{ij} = \pm q^{\kappa_{W_i}} \delta_{ij}, \qquad \left(R_2^{(Q)}\right)_{ij} = U(V,V,V;Q)_{ik}\left(R_1^{(Q)}\right)_{kj} U(V,V,V;Q)_{ij}. \tag{9.3}$$

Here $U(V,V,V;Q)_{ik}$ are the Racah matrices for $V^{\otimes 3} \to Q$, κ_W is the value of the second Casimir operator in representation $W \in V^{\otimes 2}$, the sign of the eigenvalue of the R-matrix depends on whether W lies in the symmetric or antisymmetric square of V, and $\dim_q Q$ is the quantum dimension of representation Q.

Arborescent Knots. Arborescent knots forms a distinguished class of knots introduced by Conway in 1967 [60, 67]. His purpose was to give a new classification of knots, which could be more systematic and logic than already existed. *Arborescent knot* is a closure of arborescent tangle, which can be obtained from a trivial tangles with the help of certain operations. Any arborescent knot can be represented as a weighted tree with any valent vertices [68]. Edges of the tree are 4-strand braids with two strands having an opposite orientation to the other two. In the vertices of the tree, each edge connects with adjacent edges with 2 strands, therefore, every edge connects with two other edges.

Thus, to deal with the arborescent knots, one needs 4-strand braids, where two strands have an opposite orientation to the other two. From the point of view of representation theory, this means that there is a representation V associated with one orientation and $\bar{V}$ (a conjugate representation) with the opposite one. Closure of these braids into an arborescent knot corresponds to a projection on a singlet representation $\varnothing$, which always lives in $V^{\otimes 2} \otimes \bar{V}^{\otimes 2}$.

Now we have three R-matrices corresponding to the generators of the braid group: R_1, R_2, and R_3. Usually one only has to require that R_1 and R_3 are commuting: $[R_1, R_3] = 0$, i.e. that they have a common system of eigenfunctions. However, the condition of projecting onto the singlet implies that, in the case of multiplicity-free representations (in particular, symmetric ones), they just coincide: $R_1 = R_3$ [68]. At the same time, for R_1 and R_2 (9.3) is still correct, only now W can lie in $V \otimes \bar{V}$ as well.

Thus, considering the case of arborescent knots or links with components colored by same representations, whatever be the representation V, one has only two different R-matrices R_1: T acting in $V \otimes V$, and $\bar{T}$ acting in $V \otimes \bar{V}$. Accordingly, there are also two Racah matrices, because all other Racah matrices arising in the arborescent calculus can be reduced to these two cases:

$$\begin{aligned} \bar{S} &:= U(V,\bar{V},V;V) = \left\{ U_{ij} \begin{bmatrix} V & \bar{V} \\ V & \bar{V} \end{bmatrix} \right\}, \\ S &:= U(V,V,\bar{V};V) = \left\{ U_{ij} \begin{bmatrix} V & V \\ \bar{V} & \bar{V} \end{bmatrix} \right\}. \end{aligned} \tag{9.4}$$

Given a Young diagram corresponding to V, the Young diagram corresponding to $\bar{V}$ is a completion to V of the rectangular diagram of height N and width λ_1, where $V = \{\lambda_1 \geq \lambda_2 \geq \cdots\}$. These two Racah matrices are unitary and are related via

$$\bar{S} = \bar{T}^{-1} S T^{-1} S^{\dagger} \bar{T}^{-1}. \tag{9.5}$$

The elements of (quantum) Racah matrix are related with the (quantum) $6j$-symbols by a simple normalization:

$$U_{ij} = \epsilon_{\{V\}} \sqrt{\dim_q V_i \dim_q V_j} \cdot \left\{ \begin{matrix} V_1 & V_2 & V_i \\ V_3 & V_4 & V_j \end{matrix} \right\}, \tag{9.6}$$

where $\dim_q V$ is a quantum dimension of representation V and $\epsilon_{\{V\}}$ is a sign.

9.2. *Racah polynomials and symmetric representations*

Thus, as we already stressed there are many orthogonal systems but it is not simple to distinguish those that provide the Yang–Baxter equations, which we needs for the purposes of knot theory. In order to find such orthogonal systems, one needs to use the group theory: the orthogonal systems coming as $6j$-symbols, or Racah matrices do provide orthogonal systems with the required properties. First, we consider here the finite-dimensional representations of $SL_q(2)$ when the orthogonal system is just a system of discrete orthogonal polynomials, the Racah polynomials considered in Sec. 6.

9.2.1. *Racah for $SL_q(2)$*

In the case of $SL_q(2)$ representations, all $6j$-symbols, or Racah matrix were computed by Kirillov and Reshetikhin in [71] (see also [31, 72]). The answers can be given in terms of the q-hypergeometric function ${}_4\phi_3$, that is, in terms of the q-Racah polynomials (6.1), hence, the name of these latter. Let us introduce a new notation for the q-Racah polynomial, which will be more convenient throughout this section:

$$\widehat{\mathfrak{R}}_n(x; a, b, c, d) := {}_4\phi_3 \left(\begin{matrix} q^{-2n}, q^{2a+2b+2n+2}, q^{-2x}, q^{2x+2c+2d+2} \\ q^{2a+2}, q^{2c+2}, q^{2b+2d+2} \end{matrix} \middle| q^2, q^2 \right), \tag{9.7}$$

$$\left\{ \begin{matrix} r_1 & r_2 & i \\ r_3 & r_4 & j \end{matrix} \right\} \sim \widehat{\mathfrak{R}}_{\frac{r_1+r_3}{2}-j} \left(\frac{r_1 + r_3}{2} - i \; ; \; -r_3-1, -r_2-1, -\frac{1}{2}(r_1+r_2+r_3+r_4+2), \frac{1}{2}(r_2-r_1+r_4-r_3) \right).$$

The matrix T is a diagonal matrix, see (9.3):

$$T = \mathrm{diag} \left(\epsilon_{R_m} q^{c_2(X_m)} \right), \quad [r_1] \otimes [r_2] = \oplus_m [r_1 + r_2 - 2m] =: \oplus_m X_m, \tag{9.8}$$

where $c_2(R)$ is an eigenvalue of the second Casimir operator and ϵ_R is a sign.

The matrix S coincides with $\bar{S}$ in the case of $SL_q(2)$, since $r = \bar{r}$ in this case, and so does T coinciding with $\bar{T}$. The Racah matrix is equal to

$$S_{ij} = (-1)^{r+i+j} \frac{\sqrt{(2i+1)(2j+1)}\,[r]!^2[2r+1]!}{[r-i]![r-j]![r+i+1]![r+j+1]!} \cdot \widehat{\mathfrak{R}}_{r-j}\,(r-i; -r-1, -r-1, -2r-2, 0)\,. \tag{9.9}$$

9.2.2. *Racah for symmetric representation of $SL_q(N)$*

For symmetric representations of $SL_q(N)$, the diagonal matrices T and $\bar{T}$, which become different, take the following form

$$T = \mathrm{diag}\left((-1)^{m+1}\frac{q^{-r^2+m^2+m}}{A^r}\right), \quad m = 0, \dots, r, \qquad \bar{T} = \mathrm{diag}\big((-q^{m-1}A)^m\big), \quad m = 0, \dots, r \tag{9.10}$$

where $A := q^N$.

• Let us consider the matrix $\bar{S}$ in the case of $V_1 = V_3 = [r]$, $V_2 = V_4 = [\bar{r}]$ so that V_{12} and V_{23} are representations of the type $\mathcal{V}_n = [2n, n^{N-2}]$ of $SL_q(N)$, which emerge in the decomposition:

$$[r] \otimes [\bar{r}] = \oplus_{n=0}^{r} \mathcal{V}_n. \tag{9.11}$$

We denote $V_{12} = \mathcal{V}_i \to i$ and $V_{23} = \mathcal{V}_j \to j$, and the Racah matrix is symmetric in i and j. In this case, the explicit formula was given in [73] and [74] in terms of partial sums. In the paper [3], it was shown that the corresponding $6j$-symbols can be written as

$$\begin{Bmatrix} r & \bar{r} & i \\ r & \bar{r} & j \end{Bmatrix} = \frac{[i]!^2[j]!^2[2r+N-1]![N-1]![N-2]!}{[r+i+N-1]![r+j+N-1]![r-i]![r-j]![i+j-r]![i+j-r+N-2]!} \cdot {}_4\Phi_3\left[\begin{matrix} i-r, i-r, j-r, j-r \\ 1-2r-N, i+j-r+1, i+j-r+N-1 \end{matrix}\middle| q, q^2\right],$$

where we have used a q-hypergeometric function of exponential parameters defined as follows

$${}_r\Phi_s\left(\begin{matrix} a_1, \dots, a_r \\ b_1, \dots, b_s \end{matrix}\middle| q, z\right) = {}_r\phi_s\left(\begin{matrix} q^{2a_1}, \dots, q^{2a_r} \\ q^{2b_1}, \dots, q^{2b_s} \end{matrix}\middle| q^2, z\right). \tag{9.12}$$

In order to obtain alternative expressions for the $6j$-symbols, one can use Sears' transformations for terminating balanced ${}_4\Phi_3\left[\dots; q, q^2\right]$ (see [8, 75]):

$${}_4\Phi_3\left[\begin{matrix} x, y, z, n \\ u, v, w \end{matrix}\middle| q, q^2\right] = \frac{[v-z-n-1]![u-z-n-1]![v-1]![u-1]!}{[v-z-1]![v-n-1]![u-z-1]![u-n-1]!} \cdot {}_4\Phi_3\left[\begin{matrix} w-x, w-y, z, n \\ 1-u+z+n, 1-v+z+n, w \end{matrix}\middle| q, q^2\right] \tag{9.13}$$

where the numbers x, y, z, u, v, w are integer, and the balanced series condition

$$x + y + z + n + 1 = u + v + w \tag{9.14}$$

is fulfilled. Also one can use the invariance of ${}_4\Phi_3\left[\dots; q^2, q^2\right]$ under permutations of x, y, z, n or u, v, w. Then, one gets the following expression for the $6j$-symbols:

$$\begin{Bmatrix} r & \bar{r} & i \\ r & \bar{r} & j \end{Bmatrix} = \frac{[i]![j]![N-1]![N-2]!}{[i+N-2]![j+N-2]!} \frac{[r]![r+N-2]![2r+N-1]!}{[r-i]![r-j]![r+i+N-1]![r+j+N-1]!}$$

$$\cdot\, {}_4\Phi_3 \left[\begin{matrix} j-r, -r-j-N+1, i-r, -r-i-N+1 \\ -r, -r-N+2, -2r-N+1 \end{matrix} \middle| q, q^2 \right]. \tag{9.15}$$

By expanding ${}_4\Phi_3\left[\dots; q^2, q^2\right]$ in terms of q-factorials, one gets a generalization of the most commonly encountered formulas in the literature about $SL(2)$ $6j$-symbols. Such expansions are related by Sears' transformations (9.13) of the balanced hypergeometric series. In terms of the q-Racah polynomials, formula (9.15) takes the form

$$\begin{Bmatrix} r & \bar{r} & i \\ r & \bar{r} & j \end{Bmatrix} = \frac{[i]![j]![N-1]![N-2]!}{[i+N-2]![j+N-2]!} \frac{[r]![r+N-2]![2r+N-1]!}{[r-i]![r-j]![r+i+N-1]![r+j+N-1]!}$$

$$\cdot\, \widehat{\mathfrak{R}}_{r-j}(r-i; -r-1, 1-r-N, -2r-N, 0)\,. \tag{9.16}$$

It was also found a three-term relation for these $6j$-symbols

$$a_1 \begin{Bmatrix} r & \bar{r} & i \\ r & \bar{r} & j-1 \end{Bmatrix} + a_2 \begin{Bmatrix} r & \bar{r} & i \\ r & \bar{r} & j \end{Bmatrix} + a_3 \begin{Bmatrix} r & \bar{r} & i \\ r & \bar{r} & j+1 \end{Bmatrix} = 0 \tag{9.17}$$

$$a_1 = [j]^2\, [j-r-1]\, [r+j+N-1]\, [N+2j]$$

$$a_3 = [j-r]\, [j+N-1]^2\, [N+r+j]\, [N+2j-2]$$

$$a_1 + a_2 + a_3 = -[i]\, [i+N-1]\, [N+2j-2]\, [N+2j-1]\, [N+2j].$$

• Now let us consider the matrix S, when $V_1 = V_2 = [r]$, $V_3 = V_4 = [\bar{r}]$ so that V_{23} is still of the type $\mathcal{V}_n = [2n, n^{N-2}]$ since belongs to the decomposition of $[r] \otimes [\bar{r}]$, while V_{12} belongs to

$$[r] \otimes [r] = \oplus_{n=0}^{r} [r+n, r-n]. \tag{9.18}$$

We denote $V_{12} = [r+i, r-i] \to i$ and $V_{23} = \mathcal{V}_j \to j$ and the Racah matrix is no longer symmetric in i and j. Below we assume that $i < j$, otherwise one needs to change $i \leftrightarrow j$ in all answers. In this case, the manifest formulas also can be found in [73, 74]. With the help of hypergeometric notations and Sears' transformation (9.13), one can convert this into

$$\begin{Bmatrix} r & r & i \\ \bar{r} & \bar{r} & j \end{Bmatrix} = \frac{[r]!^2[N-1]![N-2]![2r+N-1]!}{[r-i]![r-j+N-2]![r+i+N-1]![r+j+N-1]!}$$

$$\cdot\, {}_4\Phi_3 \left[\begin{matrix} j-r, -r-j-1, i-r, -r-i-N+1 \\ -r, -r, -2r-N+1 \end{matrix} \middle| q, q^2 \right], \tag{9.19}$$

which, in terms of the q-Racah polynomials (9.7), takes the form

$$\begin{Bmatrix} r & r & i \\ \bar{r} & \bar{r} & j \end{Bmatrix} = \frac{[r]!^2[N-1]![N-2]![2r+N-1]!}{[r-i]![r-j+N-2]![r+i+N-1]![r+j+N-1]!} \cdot \widehat{\mathfrak{R}}_{r-j}(r-i\,;\,-r-1,-r-1,-2r-N,0)\,. \tag{9.20}$$

The corresponding 3-term recurrence relation is

$$b_1\begin{Bmatrix} r & r & i \\ \bar{r} & \bar{r} & j-1 \end{Bmatrix} - b_2\begin{Bmatrix} r & r & i \\ \bar{r} & \bar{r} & j \end{Bmatrix} + b_3\begin{Bmatrix} r & r & i \\ \bar{r} & \bar{r} & j+1 \end{Bmatrix} = 0 \tag{9.21}$$

with

$$b_1 = [j]^2\,[j+r+1]\,[r-j+N-1]\,[2j+2]$$
$$b_3 = [r-j]\,[j+1]^2\,[N+r+j]\,[2j]$$
$$b_1+b_2+b_3 = [i]\,[i+N-1]\,[2j]\,[2j+1]\,[2j+2].$$

9.3. *Askey–Wilson functions, infinite-dimensional representations and Hikami invariants*

As we already mentioned, an advantage of obtaining orthogonal systems for constructing knot polynomials via the Racah matrices is that it guarantees the third Reidemeister move, i.e. the Yang–Baxter equation. In this way, one can construct both orthogonal polynomials and orthogonal functions. The former case is described by the Racah matrices of finite-dimensional representations and was discussed in the previous subsection. The other case corresponds to the Racah matrices of infinite-dimensional representations (of non-compact groups), and, in that case, one has to check convergency of the proper infinite sums or integrals. One meets this kind of problems when constructing the knot invariants from affine algebras (see, however, [76]). However, in the case of finite-dimensional groups, the integrals can be defined. We do not enter the details and discuss here only the corresponding Racah matrices, postponing explanations of how to construct from them knot invariants to a separate publication.

9.3.1. *Infinite-dimensional unitary representations of $SU_q(1,1)$*

We start from the case of $0<q<1$ and discuss the Racah matrices for unitary infinite-dimensional irreducible representations of $SU_q(1,1)$ [44]. In this case, there are five different series of representations [75, 77]: positive and negative discrete series $\pi_k^{\pm}$, principal $\pi_{\rho,\epsilon}^P$, complementary $\pi_{\lambda,\epsilon}^C$ and strange series, where $k>0$, $0\le\epsilon<1$, $0\le\rho\le-\frac{\pi}{2\log q}$, and $-\frac{1}{2}<\lambda<-\epsilon(-\frac{1}{2}<\lambda<\epsilon-1)$ for $0\le\epsilon<\frac{1}{2}$ $(\frac{1}{2}\le\epsilon\le 1)$. For the sake of illustrative purposes, we write down the Racah matrices in three different cases [44]. To this end, we need to know how the tensor products of the representations decompose into irreducible irreps in these cases. These decompositions are:

$$\pi^+_{k_1} \otimes \pi^+_{k_2} = \oplus_{j=0}^{\infty} \pi^+_{k_1+k_2+j} \tag{9.22}$$

$$\pi^+_{k_1} \otimes \pi^-_{k_2} = \int_0^{-\frac{\pi}{2\log q}} d\rho\, \pi^P_{\rho,\{k_1-k_2\}} \quad \text{for } k_1+k_2 \geq \frac{1}{2}, \quad k_1-k_2 \geq -\frac{1}{2} \tag{9.23}$$

$$\pi^P_{\rho,\epsilon} \otimes \pi^-_k = \int_0^{-\frac{\pi}{2\log q}} d\sigma\, \pi^P_{\sigma,\{\epsilon-k\}} \oplus_j \pi^-_{j-\epsilon+k} \tag{9.24}$$

$$\pi^+_k \otimes \pi^P_{\rho,\epsilon} = \int_0^{-\frac{\pi}{2\log q}} d\sigma\, \pi^P_{\sigma,\{\epsilon+k\}} \oplus_j \pi^+_{j+\epsilon+k}. \tag{9.25}$$

Here the second subscript $\{\epsilon\}$ in $\pi^P_{\rho,\{\epsilon\}}$ is the fractional part of ϵ. Below, we use hat in order to denote the procedure of replacing q with q^2 and using the logarithmic parameters as in (9.7): $\widehat{f}(q^x|a) := f(q^{2x}|q^{2a})$. The Racah matrices corresponding to described products (9.22)–(9.25) are:

- for the product $\pi^+_{k_1} \otimes \pi^+_{k_2} \otimes \pi^-_{k_3}$, when the representations emerging in the decomposition of the first two factors, (9.22) are labelled by non-negative integers j and those emerging in the decomposition of the second two factors, (9.23) are labelled by the continuous parameter ρ,

$$U_{j\rho}\begin{bmatrix} k_1 & k_2 \\ k_3 & \tau \end{bmatrix} = \mathfrak{N}_1 \cdot \widehat{P}^{AW}_j\left(q^{i\rho}+q^{-i\rho}\middle| k_2-k_3+\frac{1}{2}, k_2+k_3-\frac{1}{2}, k_1+i\tau, k_1-i\tau\right)$$

$$\mathfrak{N}_1 := \frac{2\pi q^{-2jk_1} q^{j(j-1)/2}(q^{2k_1+2k_2+2k_3-1+2j+2i\tau}, q^{2k_1+2k_2+2k_3-1+2j-2i\tau}; q^2)_\infty}{(q^{4k_1};q^2)_j (q^2, q^2, q^{4k_3}, q^{4k_1+4k_2+4j}, q^{4i\tau}, q^{-4i\tau}; q^2)_\infty}$$
$$\times \frac{\theta(q^{2k_1+2k_2-2k_3+1+2j+2i\tau};q^2)\theta(q^{2k_1+2k_2-2k_3+1+2j-2i\tau};q^2)}{\sqrt{(q^{4k_1+4k_2+2j}, q^{1-2k_1-2k_2+2k_3-2j+2i\tau}, q^{1-2k_1-2k_2+2k_3-2j-2i\tau}; q^2)_j}}. \tag{9.26}$$

- for the product $\pi^+_{k_1} \otimes \pi^P_{\rho,\epsilon} \otimes \pi^-_{k_3}$, when the representations emerging in the decompositions of the both first two factors, (9.25) and the second two factors, (9.24) are labelled by the continuous parameters σ, ζ,

$$U_{\sigma\zeta}\begin{bmatrix} k_1 & (\rho,\epsilon) \\ k_3 & \tau \end{bmatrix} = \mathfrak{N}_2 \cdot \widehat{\Psi}_{q^{i\zeta}}\left(q^{i\sigma}\middle| k_1+i\rho, k_3+i\tau, k_3-i\tau, 1-k_1+i\rho\right)$$

$$\mathfrak{N}_2 := [2\pi\theta(q^{1+2k_1-2k_3+2\epsilon+2i\tau};q^2)\theta(q^{1+2k_1-2k_3+2\epsilon-2i\tau};q^2)]/[(q^2, q^{4k_1}, q^{4k_3}, q^{4i\tau},$$
$$q^{-4i\tau}, q^{1+2\epsilon+2k_1+2i\sigma}, q^{1+2\epsilon+2k_1-2i\sigma}, q^{1-2\epsilon+2k_3+2i\zeta}, q^{1-2\epsilon+2k_3-2i\zeta}; q^2)_\infty]. \tag{9.27}$$

- for the product $\pi^+_{k_1} \otimes \pi^+_{k_2} \otimes \pi^P_{\rho,\epsilon}$, when the representations emerging in the decomposition of the first two factors, (9.22) are labelled by non-negative integers j and those emerging in the decomposition of the second two factors, (9.25) are labelled by the continuous parameter σ,

$$U_{j\sigma}\begin{bmatrix} k_1 & k_2 \\ (\rho,\epsilon) & \tau \end{bmatrix} = \mathfrak{N}_3 \cdot \widehat{P}^{AW}_j\left(q^{i\sigma}+q^{-i\sigma}\middle|k_2+i\rho, k_2-i\rho, k_1+i\tau, k_1-i\tau\right),$$

$$\mathfrak{N}_3 := 2\pi q^{\frac{1}{2}j(j-4k_1-1)} \frac{\prod_{\varepsilon=\pm}(q^{1-2k_1-2k_2-2\epsilon+2i\varepsilon\tau};q^2)_\infty \prod_{\varepsilon_{1,2}=\pm}(q^{2k_1+2k_2+2j+2i\varepsilon_1\rho+2i\varepsilon_2\tau};q^2)_\infty}{(q^{4k_1};q^2)_j(q^2,q^{4k_1+4k_2+4j},q^{1-2\epsilon+2i\rho},q^{1-2\epsilon-2i\rho},q^{4i\tau},q^{-4i\tau};q^2)_\infty}$$

$$\times\sqrt{\frac{(q^{1-2k_1-2k_2-2\epsilon-2j+2i\tau},q^{1-2k_1-2k_2-2\epsilon-2j-2i\tau};q^2)_j}{(q^{4k_1+4k_2+2j};q^2)_j}}. \tag{9.28}$$

One can use these Racah matrices in order to construct knot polynomials. We will return to this problem elsewhere.

9.3.2. $|q|=1$ *and Hikami invariants*

As we observed in Sec. 7.3, there is an extension of the Askey–Wilson functions, i.e. the functions that solve the Askey–Wilson difference equation (3.30) to the case of $|q|=1$, i.e. in the case of real τ, $q=e^{2\pi i\tau}$. In fact, their modular properties (that follow from (7.32)) provide that the Askey–Wilson functions in this case also satisfy the other, modular inverted equation with $\tilde{q}:=e^{2\pi i/\tau}$. This is a consequence of the duality with respect to interchanging $q\leftrightarrow\tilde{q}$, i.e. $\tau\to 1/\tau$ [78–81] and is associated with the modular double of $SL_q(2)$ (which was introduced by Faddeev [82]), or with the quantum algebra $SL_q(2)\times SL_{\tilde{q}}(2)$ (see also [83]). In this case, a corresponding class of one-parametric representations with weights $\alpha_i^\vee$ was described in [81, 84, 85] (see also [86]), their Racah matrices [81] being expressed through the Askey–Wilson functions at $|q|=1$ (see also [87] for the non-deformed case):

$$U_{\alpha_s^\vee\alpha_t^\vee}\begin{bmatrix} \alpha_1^\vee & \alpha_2^\vee \\ \alpha_3^\vee & \alpha_4^\vee \end{bmatrix} \sim \Psi^\tau_{q^{\alpha_t}}\left(-2\alpha_s\middle|q^{\alpha_4-\alpha_3+Q/b},q^{\alpha_1-\alpha_2+Q/b},q^{\alpha_3+\alpha_4},q^{\alpha_1+\alpha_2}\right) \tag{9.29}$$

$$\text{where}\quad Q=b+b^{-1},\quad \tau=b^2,\quad \alpha_i:=\frac{\alpha_i^\vee}{b}-\frac{Q}{2b}.$$

Here $\alpha_s^\vee$ and $\alpha_t^\vee$ denote the intermediate representations in the Racah matrices.

Having constructed the Racah matrices, one can further find the R-matrix [82, 84, 88] (see also [89, 90]) and further knot invariants [22]. These knot invariants are known and called Hikami invariants, or integral state model [90–93]. For the simplest knots they look like

$$\mathbb{H}(3_1,x) = e^{\frac{3x^2}{\hbar}+\frac{i\pi x}{\hbar}}, \tag{9.30}$$

$$\mathbb{H}(4_1,x) = \int dy \frac{\Phi_b(x-y)}{\Phi_b(y)} e^{2\pi i x(2y-x)}, \tag{9.31}$$

$$\mathbb{H}(5_2, x) = \int dz \frac{e^{\pi i(z^2-x^2)}}{\Phi_b(z)\Phi_b(z-x)\Phi_b(x+z)}, \tag{9.32}$$

where Faddeev's quantum dilogarithm [94] is defined as

$$\Phi_b(x) := \frac{\left(-e^{2\pi bx}\sqrt{q}; q\right)_\infty}{\left(-e^{2\pi x/b}\sqrt{\tilde{q}}; \tilde{q}\right)_\infty}. \tag{9.33}$$

Note that the Racah matrix (9.29) is nothing but the modular kernel that gives rise to the modular transformation of the conformal blocks [81] in the two-dimensional conformal theory with central charge $c = 1 + 6Q^2$, the parameter $\alpha^\vee$ of representations being associated with the conformal dimensions of the fields in the conformal block in accordance with $\Delta = \alpha^\vee(Q - \alpha^\vee)$.

Being the Askey–Wilson function at $|q| = 1$, the Racah matrix (9.29) satisfies the Askey–Wilson difference equation (3.30) (see Sec. 7.3). This equation determines the modular kernel as a function of intermediate dimensions ($\alpha_s^\vee$ and $\alpha_t^\vee$) and coincides with the equation for the modular kernel obtained within the *pq*-duality approach [15], where the conformal blocks are treated as eigenfunctions of the monodromy check operators [95].

10. Conclusion

In this paper, we have reviewed properties of the orthogonal polynomials, mostly hypergeometric polynomials, and discussed several extensions in application to various branches of mathematical physics: to integrable theories, matrix models, two-dimensional conformal theories and knot theory. Our discussion was inevitably short and sketchy, since we tried to include all these topics into a general framework. However, the approaches described here can be immediately pushed forward.

In particular, as was explained in Sec. 9, with the knowledge of Racah matrices, one can construct knot invariants. The knot invariants made from the Racah matrices for finite-dimensional representations of $SL_q(2)$ are known, these are the Jones polynomials. However, only recently it was discovered [3, 73, 74] how to construct the knot invariants colored with symmetric representations of $SL_q(N)$, these are the corresponding HOMFLY-PT polynomials [61]. The problem of constructing the Racah matrices in non-symmetric representations, both rectangular (when there are no multiplicities) [96] and non-rectangular [97] (see also [98]) as well as their interpretation in terms of hypergeometric polynomials remains unsolved. Similarly, the knot polynomials colored with representations of the modular double of $SL_q(2)$ are also known, these are Hikami invariants [91]. However, the extension of these invariants to other representations (see, e.g., [99]) or to groups of higher ranks has not been done yet. It can be immediately done using results of Sec. 9.3.2 and the trick that allowed us to come from $SL_q(2)$ Racah matrices to $SL_q(N)$ ones in Sec. 9.2. Similarly, one could construct knot invariants from infinite-dimensional unitary representations of $SU_q(1,1)$ and higher rank groups using the results of Sec. 9.3.1. We shall return to these issues elsewhere.

Another issue related to the same Racah matrices for the modular double is related to their interpretation as modular kernels of the modular transformations of conformal blocks. This also waits for its extension to higher groups, which corresponds to special conformal blocks of the W-algebras. It would be also interesting to interpret these results within the pq-duality approach [15], where the conformal blocks are treated as eigenfunctions of the monodromy check operators [95].

Note that absolutely mysterious here remain extensions of these modular kernels to conformal blocks of the q-Virasoro algebra. This issue has to be explained within the more general context of the Ding–Iohara–Miki algebras [100, 101] and the corresponding network matrix models [101, 102].

Note that the integral representations that we discussed in Sec. 5, and the related modular transformations of Secs. 5 and 9 have a lot to do with the new emerged topic of constructing solutions to the Painléve and discrete Painléve equations in terms of the conformal blocks of the Virasoro and q-Virasoro algebras. This problem also will be addressed in a separate publication [102].

At last, there is a rather traditional question of relating the eigenvalue type matrix models with those of the Kontsevich type. So far, this correspondence included at most rectangular complex matrix models and has been established by using the Laguerre polynomials (see Sec. 8.3). The correspondence inspired by more general orthogonal polynomials is not known yet, but it probably could be constructed. Another direction is to construct a generalization of these relations to the q-deformed case even in the case of the q-Laguerre polynomials [103].

Acknowledgments

Our work was partly supported by the grant of the Foundation for the Advancement of Theoretical Physics "BASIS" (A.Mor. and A.S.), by RFBR grants 16-01-00291 (A.Mir.) and 16-02-01021 (A.Mor.), by joint grants 17-51-50051-YaF, 15-51-52031-NSC-a, 16-51-53034-GFEN, 16-51-45029-IND-a (A.M.'s and A.S.), by grant 16-31-60082-mol-a-dk (A.S.). C.T. Chan was supported by the grant of Ministry of Science and Technology of Taiwan under the grants 104-2923-M-002-003-MY3 and 106-2112-M-029-003. The discussions with M. Ismail on the Shohat-Favard theorem is greatly acknowledged.

Appendix A. Hypergeometric Series

In this Appendix, we collect all the definitions and notation related to the (q-)hypergeometric series that we use throughout the paper. The main textbooks are [1, 8, 104].

A.1. *Generalized hypergeometric functions*

The generalized hypergeometric functions of one variable are usually defined as an analytical continuation (or resummations) of the generalized hypergeometric series,

$$ {}_rF_s\left(\begin{matrix} a_1,\dots,a_r \\ b_1,\dots,b_s \end{matrix}\middle| z\right) := \sum_{n=0}^{\infty} \frac{(a_1)_n \dots (a_r)_n}{(b_1)_n \dots (b_s)_n} \cdot \frac{z^n}{n!} = \sum_{n=0}^{\infty} \frac{(a_1,\dots,a_p)_n}{(b_1,\dots,b_q)_n} \frac{z^n}{n!}, \tag{A.1} $$

with $(a)_n := \frac{\Gamma(a+n)}{\Gamma(a)} = \prod_{k=0}^{n-1}(a+k)$, $(a_1,\dots,a_i)_n := (a_1)_n \cdot \dots \cdot (a_i)_n$.

One way to obtain an integral representation of the generalized hypergeometric function is to use the Euler integral formula,

$$ {}_{r+1}F_{s+1}\left(\begin{matrix} a_1,\dots,a_r,a \\ b_1,\dots,b_s,b \end{matrix}\middle| z\right) = \frac{\Gamma(b)}{\Gamma(a)\Gamma(b-a)} \int_0^1 t^{a-1}(1-t)^{b-a-1} \, {}_rF_s\left(\begin{matrix} a_1,\dots,a_r \\ b_1,\dots,b_s \end{matrix}\middle| tz\right) dt. \tag{A.2} $$

In the special case $r = s+1$, we get

$$ {}_{q+1}F_q\left(\begin{matrix} a_0,a_1,\dots,a_q \\ b_1,\dots,b_q \end{matrix}\middle| z\right) = \frac{\Gamma(b_1)\Gamma(b_2),\dots,\Gamma(b_q)}{\Gamma(a_1)\Gamma(a_2),\dots,\Gamma(a_q)\cdot\Gamma(b_1-a_1)\Gamma(b_2-a_2),\dots,\Gamma(b_q-a_q)} \times \prod_{j=1}^{q}\left(\int_0^{t_{j+1}} dt_j \, t_j^{a_j-a_{j-1}-1}(1-t_j/t_{j+1})^{b_j-a_j-1}\right) t_1^{a_0}(1-t_1 z)^{-a_0}, \tag{A.3} $$

where $t_{q+1} := 1$.

There are also recursions which change the difference $r-s$, but they involve exponential weights:

$$ {}_{r+1}F_s\left(\begin{matrix} a_1,\dots,a_{r+1} \\ b_1,\dots,b_s \end{matrix}\middle| z\right) = \frac{1}{\Gamma(a_{r+1})} \int_0^{\infty} dt \, e^{-t} t^{a_{r+1}-1} \, {}_rF_s\left(\begin{matrix} a_1,\dots,a_r \\ b_1,\dots,b_s \end{matrix}\middle| zt\right) \tag{A.4} $$

and

$$ {}_rF_{s+1}\left(\begin{matrix} a_1,\dots,a_r \\ b_1,\dots,b_{s+1} \end{matrix}\middle| z\right) = \frac{\Gamma(b_{s+1})}{2\pi i} \int_{c-i\infty}^{c+i\infty} dt \, e^t t^{-b_{s+1}} \, {}_rF_s\left(\begin{matrix} a_1,\dots,a_r \\ b_1,\dots,b_s \end{matrix}\middle| \frac{z}{t}\right), \tag{A.5} $$

where $c > 0$. These formulas are proved by comparing the coefficients at both sides of formulas using the integral representations of the Γ-functions

$$ \Gamma(z) = \int_0^{\infty} dt \, e^{-t} t^{z-1}, \qquad \frac{1}{\Gamma(z)} = \frac{1}{2\pi i} \int_{c-i\infty}^{c+i\infty} dt \, e^t t^{-z}. \tag{A.6} $$

A.2. *Basic hypergeometric series*

The q-analogue of the generalized hypergeometric series, called Heine's basic hypergeometric series, or hypergeometric q-series is defined as,

$$
{}_r\phi_s\left(\begin{matrix} a_1,\dots,a_r \\ b_1,\dots,b_s \end{matrix}\,\middle|\, q,\, z\right) := \sum_{k=0}^{\infty} \frac{(a_1,\dots,a_r;q)_k}{(b_1,\dots,b_s;q)_k}\cdot\frac{z^k}{(q;q)_k}\left[(-1)^k q^{k(k-1)/2}\right]^{1+s-r}, \tag{A.7}
$$

where

$$
\begin{aligned}
(a_1,\dots,a_r;q)_k &:= \prod_{i=1}^{r}(a_i;q)_k, \quad (a;q)_k := \prod_{i=0}^{k-1}(1-aq^i), \\
(a;q)_0 = 1, &\quad (a;q)_\infty := \prod_{i=0}^{\infty}(1-aq^i).
\end{aligned} \tag{A.8}
$$

Note that, at non-natural k, one usually substitutes $(a;q)_k$ with

$$
(a;q)_k \to \frac{(a;q)_\infty}{(aq^k;q)_\infty}. \tag{A.9}
$$

We apply this trick in the paper in various places.

The basic hypergeometric series ${}_{r+1}\phi_r$ is called *balanced* if

$$
\prod_{i=1}^{r+1} a_i = q\prod_{j=1}^{r} b_j \tag{A.10}
$$

and *very-well-poised* if $a_2 = q\sqrt{a_1}$, $a_3 = -q\sqrt{a_1}$, $b_1 = \sqrt{a_1}$, $b_2 = -\sqrt{a_1}$, $b_j = qa_1/a_{j+1}$ for $j>2$. We denote the very-well-poised hypergeometric series

$$
{}_{r+1}W_r(a_1;a_4,a_5,\dots,a_{r+1}|q,z) := {}_{r+1}\phi_r\left(\begin{matrix} a_1, q\sqrt{a_1}, -q\sqrt{a_1}, a_4, \dots, a_{r+1} \\ \sqrt{a_1}, -\sqrt{a_1}, qa_1/a_4, \dots, qa_1/a_{r+1} \end{matrix}\,\middle|\, q, z\right). \tag{A.11}
$$

Appendix B. Very Classical Orthogonal Polynomials

In this Appendix, we list some of the very classical polynomials: the Jacobi, Laguerre and Hermite polynomials, and their properties. These are the polynomials that we discuss in the paper.

B.1. *List of the very classical orthogonal polynomials*

Jacobi Polynomials

- **The Jacobi polynomial as a hypergeometric polynomial**

$$
\begin{aligned}
P_n^{(a,b)}(x) &:= \frac{(a+1)_n}{n!}\,{}_2F_1\left(\begin{matrix} -n, & a+b+n+1 \\ & a+1 \end{matrix}\,\middle|\, \frac{1-x}{2}\right) \\
&= P_n^{(a,b)}(1)\,{}_2F_1\left(\begin{matrix} -n, & a+b+n+1 \\ & a+1 \end{matrix}\,\middle|\, \frac{1-x}{2}\right).
\end{aligned} \tag{B.1}
$$

- **weight and orthogonal relation**

$$\int_{-1}^{1} P_m^{(a,b)}(x)P_n^{(a,b)}(x)(1-x)^a(1+x)^b dx$$
$$= \frac{2^{a+b+1}}{2n+a+b+1}\frac{\Gamma(n+a+1)\Gamma(n+b+1)}{\Gamma(n+a+b+1)n!}\delta_{m,n}, \quad a,b>-1. \tag{B.2}$$

- **differential equation**

$$(1-x^2)P_n^{(a,b)''}+[b-a-x(a+b+2)]P_n^{(a,b)'}+n(n+a+b+1)P_n^{(a,b)}=0. \tag{B.3}$$

- **3-term relation**

$$\frac{2n(n+a+b)(2n+a+b-2)}{2n+a+b-1}P_n^{(a,b)}(x)$$
$$= \left[(2n+a+b)(2n+a+b-2)x+a^2-b^2\right]P_{n-1}^{(a,b)}(x)$$
$$- \frac{2(n+a-1)(n+b-1)(2n+a+b)}{2n+a+b-1}P_{n-2}^{(a,b)}(x). \tag{B.4}$$

- **Rodriguez formula**

$$P_n^{a,b}(x) = \frac{(-1)^n}{2^n n!}(1-x)^{-a}(1+x)^{-b}\frac{d^n}{dx^n}\left[(1-x)^{a+n}(1+x)^{b+n}\right]. \tag{B.5}$$

- **generating function**

$$\frac{2^{a+b}}{R(1-t+R)^a(1+t+R)^b} = \sum_{n=0}^{\infty} P_n^{(a,b)}(x)t^n. \quad R := \sqrt{1-2tx+t^2}. \tag{B.6}$$

- **specializations of the Jacobi polynomials**
  - for $a=b=\nu-1/2$ we get **Gegenbauer** polynomials,

$$C_n^{(\nu)}(x) := \frac{(2\nu)_n}{(\nu+1/2)_n}P_n^{(\nu-1/2,\nu-1/2)}(x). \tag{B.7}$$

  - For $a=b=0$ we get **Legendre** polynomials,
  - for $a=b=\pm\frac{1}{2}$ we get **Chebyshev** polynomials.

Laguerre Polynomials

- **Laguerre polynomial as a hypergeometric polynomial**

$$L_n^{(a)}(x) = \frac{(a+1)_n}{n!}\,{}_1F_1\left(\begin{matrix}-n\\ a+1\end{matrix}\middle| x\right). \tag{B.8}$$

- **weight and orthogonal relation**

$$\int_0^{\infty} L_m^{(a)}(x)L_n^{(a)}(x)x^a e^{-x}dx = \frac{\Gamma(n+a+1)}{n!}\delta_{m,n}, \quad a>-1. \tag{B.9}$$

- **differential equation**

$$xL_n^{(a)''} + (a+1-x)L_n^{(a)'} + nL_n^{(a)} = 0. \tag{B.10}$$

- **3-term relation**

$$L_n^{(a)}(x) = \left(2 - \frac{x-a+1}{n}\right) L_{n-1}^{(a)}(x) - \left(1 + \frac{a-1}{n}\right) L_{n-2}^{(a)}(x). \tag{B.11}$$

- **Rodriguez formula**

$$L_n^a(x) = \frac{e^x x^{-a}}{n!} \frac{d^n}{dx^n} \left[e^{-x} x^{n+a}\right]. \tag{B.12}$$

- **generating function**

$$\frac{1}{(1-t)^{a+1}} e^{-\frac{xt}{1-t}} = \sum_{n=0}^{\infty} L_n^{(a)}(x) t^n. \tag{B.13}$$

- **as a limit of the Jacobi polynomial**

$$\lim_{b\to\infty} P_n^{(a,b)}\left(1 - 2b^{-1}x\right) = L_n^{(a)}(x). \tag{B.14}$$

Hermite Polynomials

- **Hermite polynomial as a hypergeometric polynomial**

$$H_n(x) = (2x)^n\ {}_2F_0\left(\left.\begin{matrix} -n/2,\ -(n-1)/2 \\ - \end{matrix}\right| -\frac{1}{x^2}\right)$$

$$= \begin{cases} \dfrac{(-1)^m n!}{m!}\ {}_1F_1\left(\left.\begin{matrix} -m \\ \frac{1}{2} \end{matrix}\right| x^2\right) & n = 2m \\ \dfrac{(-1)^m n!}{m!}\ 2x\ {}_1F_1\left(\left.\begin{matrix} -m \\ \frac{3}{2} \end{matrix}\right| x^2\right) & n = 2m+1 \end{cases}. \tag{B.15}$$

We can check this by deriving the corresponding differential equation.

- **weight and orthogonal relation**

$$\int_{-\infty}^{\infty} H_m(x) H_n(x) e^{-x^2} dx = 2^n\, n! \sqrt{\pi}\, \delta_{m,n}. \tag{B.16}$$

- **differential equation**

$$H_n'' - 2xH_n' + 2nH_n = 0. \tag{B.17}$$

- **3-term relation**

$$H_{n+1}(x) - 2xH_n(x) + 2nH_{n-1}(x) = 0. \tag{B.18}$$

• **Rodriguez formula**

$$H_n(x) = (-1)^n e^{x^2} \frac{d^n}{dx^n} e^{-x^2}. \tag{B.19}$$

• **generating function**

$$e^{2xt-t^2} = \sum_{n=0}^{\infty} H_n(x) \frac{t^n}{n!}. \tag{B.20}$$

• **as a limit of Jacobi polynomial**

$$\lim_{a\to\infty} a^{-\frac{n}{2}} P_n^{(a,a)}(a^{-\frac{1}{2}}x) = \frac{H_n(x)}{2^n\, n!}. \tag{B.21}$$

B.2. *Orthogonality of the Jacobi polynomials*

For illustrative purposes, we reproduce here a proof of the orthogonality of the Jacobi polynomials (B.2).

Let us denote left-hand side of (B.2) by J, expand Jacobi polynomials as hypergeometric series (B.1) and get

$$J = \frac{(a+1)_n(a+1)_m}{n!\, m!} \sum_{k,l=0}^{\infty} \frac{(-n)_k(a+b+n+1)_k(-m)_l(a+b+m+1)_l}{(a+1)_k(a+1)_l\, k!\, l!\, 2^{k+l}} \times \int_{-1}^{1} (1-x)^{k+l+a}\, (1+x)^b\, dx. \tag{B.22}$$

To evaluate the integral we use Euler integral representation for beta function:

$$\begin{aligned}\int_{-1}^{1} (1-x)^{k+l+a}(1+x)^b dx &\overset{x=1-2t}{=} -\int_{1}^{0} 2^{k+l+a+b+1}\, t^{k+l+a}(1-t)^b dt \\ &= 2^{k+l+a+b+1} \frac{(k+l+a)!\, b!}{(a+b+k+l+1)!}.\end{aligned} \tag{B.23}$$

Then after simple algebraic manipulations Eq. (B.22) takes the form

$$\begin{aligned}J = \frac{(a+1)_n(a+1)_m}{n!\, m!} &\sum_{l} \frac{(-m)_l(a+b+m+1)_l\, a!\, b!\, 2^{a+b+1}}{(a+b+l+1)_l\, l!} \\ &\sum_{k} \frac{(-n)_k(a+b+n+1)_k(a+l+1)_k}{(a+b+l+2)_k(a+1)_k\, k!}.\end{aligned} \tag{B.24}$$

The last sum is a hypergeometric function ${}_3F_2(...|1)$. To this function we apply famous Saalschütz's identity

$${}_3F_2\left(\begin{matrix} a, b, -n \\ c, 1+a+b-c-n \end{matrix} \middle| 1\right) = \frac{(c-a)_n(c-b)_n}{(c)_n(c-a-b)_n}. \tag{B.25}$$

Therefore we get

$$J = \frac{(a+1)_n(a+1)_m}{n!\,m!} \times \sum_l \frac{(-m)_l(a+b+m+1)_l\, a!\, b!\, 2^{a+b+1}}{(a+b+l+1)_l\, l!} \times \frac{(-b-n)_n(-l)_n}{(a+1)_n(-a-b-n-l-1)_n}. \tag{B.26}$$

Factors $(-m)_l$ and $(-l)_n$ give us non-vanishing condition

$$n \le l \le m. \tag{B.27}$$

Then by simple algebraic manipulations we represent a sum over l as a hypergeometric function ${}_2F_1(...|1)$ and apply identity (B.25) once again:

$$\begin{aligned} J &= \frac{(a+1)_n(a+1)_m\,(a+b+m+1)_l\, a!\, b!\, 2^{a+b+1}\,(-1)^n\,(-b-n)_n}{n!\,m!\,(a+1)_n(a+b+2n+1!\,(a+b+m)!\,(m-n)!} \times \sum_l \frac{(n-m)_l(a+b+m+n+1)_l}{(a+b+2n+2)_l} \\ &= \frac{(a+1)_n(a+1)_m\,(a+b+m+1)_l\, a!\, b!\, 2^{a+b+1}\,(-1)^n\,(-b-n)_n}{n!\,m!\,(a+1)_n(a+b+2n+1!\,(a+b+m)!\,(m-n)!} \times \frac{(a+b+2n+1)!}{(n-m)!\,(a+b+n+m+1)!}. \end{aligned} \tag{B.28}$$

It gives us second condition

$$n \ge m. \tag{B.29}$$

Both conditions (B.27) and (B.29) imply

$$n = m. \tag{B.30}$$

It finishes our proof of the orthogonality. In order to get right-hand side of (B.2) we put $n = m$ in formula (B.28):

$$J = \frac{(a+n)!\,(b+n)!\,2^{a+b+1}}{(a+b+n)!\,n!\,(a+b+2n+1)}\delta_{n,m}. \tag{B.31}$$

References

[1] A. Erdelyi, *Higher Transcendental Functions* (*H. Bateman Manuscript Project*), Vol. 2 (McGraw Hill, New York, 1953); http://apps.nrbook.com/bateman/Vol2.pdf.

[2] A. Mironov and A. Morozov, arXiv:1708.07479.

[3] A. Mironov, A. Morozov and A. Sleptsov, arXiv:1709.02290.

[4] A. Alexandrov, A. Mironov and A. Morozov, *JHEP* **0912** (2009) 053, arXiv: 0906.3305.

[5] G. E. Andrews and R. Askey, *Classical Orthogonal Polynomials*, Lecture Notes in Math. Vol. 1171 (Springer-Verlag), pp. 36–62; J. Labelle, *Tableau d'Askey*, *ibid.* pp. xxxvi–xxxvii; R. Askey and J. Wilson, *Memo. Amer. Math. Soc.* **54** (1985) 319; R. Koekoek and R. F. Swarttouw, The Askey-scheme of hypergeometric orthogonal polynomials and its q-analogue, No. 98-17, Delft University of Technology, Faculty of Information Technology and Systems, Department of Technical Mathematics and Informatics (1998).

[6] R. Koekoek, P. A. Lesky and R. F. Swarttouw, *Hypergeometric Orthogonal Polynomials and Their q-Analogues* (Springer Science & Business Media, 2010); T. Koornwinder, arXiv:1401.0815.

[7] T. S. Chihara, *An Introduction to Orthogonal Polynomials* (Gordon and Breach, 1978); reprinted (Dover Publications, 2011).

[8] G. Gasper and M. Rahman, *Basic Hypergeometric Series* (Cambridge University Press, 1990).

[9] T. Koornwinder, *Computer Algebra in Quantum Field Theory* (Springer-Verlag, 2013), pp. 145–170, arXiv:1303.2825.

[10] J. A. Shohat and J. D. Tamarkin, *The Problem of Moments* (American Mathematical Society, 1943).

[11] J. Wilson, Hypergeometric series recurrence relations and some new orthogonal functions, Ph.D. thesis, Univ. Wisconsin, Madison (1978); R. Askey and J. Wilson, *SIAM J. Math. Anal.* **10** (1979) 1008–1016.

[12] L. Vinet and A. Zhedanov, *J. Phys. A: Math. Theor.* **44** (2011) 085201, arXiv:1011.1669.

[13] M. E. H. Ismail and M. Rahman, *Trans. Amer. Math. Soc.* **328** (1991) 201–237.

[14] D. Galakhov, A. Mironov and A. Morozov, *JHEP* **2012** (2012) 67, arXiv:1205.4998; N. Nemkov, *J. Phys. A: Math. Theor.* **47** (2014) 105401, arXiv:1307.0773.

[15] D. Galakhov, A. Mironov and A. Morozov, *JHEP* **06** (2014) 050, arXiv:1311.7069.

[16] N. Nemkov, *Theor. Math. Phys.* **189** (2016) 1574–1591, arXiv:1409.3537; *JHEP* **1510** (2015) 039, arXiv:1504.04360.

[17] A. Mironov and A. Morozov, *Phys. Lett. B* **773** (2017) 34–46, arXiv:1707.02443.

[18] M. Jimbo and H. Sakai, chao-dyn/9507010; K. Kajiwara, M. Noumi and Y. Yamada, *J. Phys. A: Math. Theor.* **50** (2017) 073001, arXiv:1509.08186; Ya. Yamada, talk at the VII Workshop on Geometric Correspondences of Gauge Theories, Trieste (2017).

[19] H. Sakai, *Nonlinearity* **11** (1998) 823–833.

[20] O. Gamayun, N. Iorgov and O. Lisovyy, *JHEP* **10** (2012) 038, arXiv:1207.0787; N. Iorgov, O. Lisovyy and Yu. Tykhyy, *JHEP* **12** (2013) 029, arXiv:1308.4092; N. Iorgov, O. Lisovyy and J. Teschner, *Comm. Math. Phys.* **336** (2015) 671–694, arXiv:1401.6104; M. Bershtein and A. Shchechkin, *Comm. Math. Phys.* **339** (2015) 1021–1061, arXiv:1406.3008; G. Bonelli, A. Grassi and A. Tanzini, arXiv:1603.01174; arXiv:1704.01517; P. Gavrylenko and O. Lisovyy, arXiv:1608.00958; H. Nagoya, arXiv:1611.08971; G. Bonelli, O. Lisovyy, K. Maruyoshi, A. Sciarappa and A. Tanzini, arXiv:1612.06235.

[21] M. Jimbo, H. Nagoya and H. Sakai, arxiv:1706.01940; M. Bershtein, P. Gavrylenko and A. Marshakov, arXiv:1711.02063.

[22] D. Galakhov, A. Mironov and A. Morozov, *Theor. Math. Phys.* **187** (2016) 678–694, arXiv:1510.05366.

[23] Vl. Dotsenko and V. Fateev, *Nucl. Phys. B* **240** (1984) 312–348.

[24] A. Marshakov, A. Mironov and A. Morozov, *Phys. Lett. B* **265** (1991) 99; A. Mironov and S. Pakuliak, *Theor. Math. Phys.* **95** (1993) 604–625 (*Teor. Mat. Fiz.* **95** (1993) 317–340), hep-th/9209100; S. Kharchev, A. Marshakov, A. Mironov, A. Morozov

and S. Pakuliak, *Nucl. Phys. B* **404** (1993) 717–750, hep-th/9208044; H. Awata, Y. Matsuo, S. Odake and J. Shiraishi, *Phys. Lett. B* **347** (1995) 49, hep-th/9411053; *Soryushiron Kenkyu* **91** (1995) A69–A75, hep-th/9503028; H. Awata, Y. Matsuo, S. Odake, and J. Shiraishi, *Nucl. Phys. B* **449** (1995) 347–374, hep-th/9503043.

[25] R. Dijkgraaf and C. Vafa, arXiv:0909.2453; H. Itoyama, K. Maruyoshi and T. Oota, *Prog. Theor. Phys.* **123** (2010) 957–987, arXiv:0911.4244; T. Eguchi and K. Maruyoshi, arXiv:0911.4797; arXiv:1006.0828; A. Mironov, A. Morozov and Sh. Shakirov, *JHEP* **02** (2010) 030, arXiv:0911.5721; *Int. J. Mod. Phys. A* **25** (2010) 3173–3207, arXiv:1001.0563; *JHEP* **1103** (2011) 102, arXiv:1011.3481; A. Mironov, A. Morozov and An. Morozov, *Nucl. Phys. B* **843** (2011) 534–557, arXiv:1003.5752.

[26] H. Awata and H. Kanno, *JHEP* **0907** (2009) 076, arXiv:0905.0184; H. Awata and Y. Yamada, *JHEP* **1001** (2010) 125, arXiv:0910.4431; *Prog. Theor. Phys.* **124** (2010) 227, arXiv:1004.5122; S. Yanagida, *J. Math. Phys.* **51** (2010) 123506, arXiv:1005.0216; A. Mironov, A. Morozov, S. Shakirov and A. Smirnov, *Nucl. Phys. B* **855** (2012) 128, arXiv:1105.0948; F. Nieri, S. Pasquetti, F. Passerini and A. Torrielli, *JHEP* **12** (2014) 040, arXiv:1312.1294; M. C. Tan, *JHEP* **12** (2013) 031, arXiv:1309.4775; arXiv:1607.08330.

[27] A. Gerasimov, S. Khoroshkin, D. Lebedev, A. Mironov and A. Morozov, *Int. J. Mod. Phys. A* **10** (1995) 2589–2614, hep-th/9405011; S. Kharchev, A. Mironov and A. Morozov, q-alg/9501013; A. Mironov, hep-th/9409190; *Theor. Math. Phys.* **114** (1998) 127, q-alg/9711006.

[28] A. Mironov and A. Morozov, *Nucl. Phys. B* **825** (2009) 1–37, arXiv:0908.2569.

[29] A. Belavin, A. Polyakov and A. Zamolodchikov, *Nucl. Phys. B* **241** (1984) 333–380.

[30] A. Zamolodchikov and Al. Zamolodchikov, Conformal field theory and critical phenomena in 2d systems (2009) (in Russian); L. Alvarez-Gaume, *Helv. Phys. Acta* **64** (1991) 361; P. Di Francesco, P. Mathieu and D. Senechal, *Conformal Field Theory* (Springer, 1996).

[31] L. Alvarez-Gaune, C. Gomez and S. Sierra, *Phys. Lett. B* **220** (1989) 142.

[32] G. Moore and N. Seiberg, *Comm. Math. Phys.* **123** (1989) 177–254.

[33] R. Askey, *SIAM J. Math. Anal.* **5** (1974) 119–124.

[34] T. Koornwinder, *Indag. Math.* **34** (1972) 188–191.

[35] B. L. J. Braaksma and B. Meuienbeld, *Indag. Math.* **30** (1968) 384–389; A. Dijksma and T. Koornwinder, *Indag. Math.* **33** (1971) 191–196.

[36] I. G. Macdonald, Orthogonal polynomials associated with root systems, *Sémi. Lothar. Combin.* **45** (2000), Article B45a, 40 pp, math/0011046.

[37] T. H. Koornwinder, Askey–Wilson polynomials for root systems of type BC, in *Hypergeometric Functions on Domains of Positivity, Jack Polynomials, and Applications*, ed. D. St. P. Richards, Contemp. Math., Vol. 138 (Amer. Math. Soc., Providence, R.I., 1992), pp. 189–204.

[38] M. V. Tratnik, *J. Math. Phys.* **32** (1991) 2065–2073.

[39] G. Gasper and M. Rahman, *Ramanujan J.* **13** (2007) 389–405, arXiv:math/0410250.

[40] P. Iliev, *Trans. Amer. Math. Soc.* **363** (2011) 1577–1598, arXiv:0801.4939.

[41] M. Rahman, *J. Math. Anal. Appl.* **164** (1992) 263–284.

[42] E. Koelink and J. V. Stokman, *Intern. Math. Res. Notices* **22** (2001) 1203–1227, math/0004053.

[43] J. V. Stokman, math/0301330.

[44] W. Groenevelt, *Acta Appl. Math.* **91** (2006) 133–191, math/0501511.

[45] V. Spiridonov and A. Zhedanov, *Duke Math. J.* **89** (1997) 283–305, q-alg/9605034.

[46] M. L. Mehta, *Random Matrices*, 2nd edn. (Academic Press, New York, 1991); E. Brézin, C. Itzykson, G. Parisi and J.-B. Zuber, *Comm. Math. Phys.* **59** (1978)

35; D. Bessis, *Comm. Math. Phys.* **69** (1979) 147; D. Bessis, C. Itzykson and J.-B. Zuber, *Adv. Appl. Math.* **1** (1980) 109; C. Itzykson and J.-B. Zuber, *J. Math. Phys.* **21** (1980) 411.

[47] A. Gerasimov, A. Marshakov, A. Mironov, A. Morozov and A. Orlov, *Nucl. Phys. B* **357** (1991) 565–618.

[48] S. Kharchev, A. Marshakov, A. Mironov, A. Orlov and A. Zabrodin, *Nucl. Phys. B* **366** (1991) 569–601.

[49] A. Morozov, *Phys. Usp.* (*UFN*) **35** (1992) 671–714; **37** (1994) 1; hep-th/9502091; hep-th/0502010; A. Mironov, *Int. J. Mod. Phys. A* **9** (1994) 4355; *Phys. Part. Nucl.* **33** (2002) 537; hep-th/9409190.

[50] S. Kharchev, A. Marshakov, A. Mironov and A. Morozov, *Nucl. Phys. B* **397** (1993) 339–378, hep-th/9203043.

[51] P. J. Hansen and D. J. Kaup, *J. Phys. Soc. Japan* **54** (1985) 4126.

[52] T. Miwa, *Proc. Japan Acad.* **58** (1982) 9–12.

[53] S. Kharchev, A. Marshakov, A. Mironov, A. Morozov and A. Zabrodin, *Nucl. Phys. B* **380** (1992) 181–240, hep-th/9201013; *Phys. Lett. B* **275** (1992) 311–314, hep-th/9111037; S. Kharchev, hep-th/9810091.

[54] A. Mironov and A. Morozov, *Phys. Lett. B* **771** (2017) 503–507, arXiv:1705.00976; *Phys. Lett. B* **774** (2017) 210–216, arXiv:1706.03667.

[55] C. Itzykson and J.-B. Zuber, *J. Math. Phys.* **21** (1980) 411.

[56] T. Morris, *Nucl. Phys. B* **356** (1991) 703–728; Yu. Makeenko, *Pis'ma v ZhETF* **52** (1990) 885–888; Yu. Makeenko, A. Marshakov, A. Mironov and A. Morozov, *Nucl. Phys. B* **356** (1991) 574–628.

[57] H. Itoyama, A. Mironov and A. Morozov, *JHEP* **06** (2017) 115, arXiv:1704.08648.

[58] F. A. Berezin and F. I. Karpelevich, *DAN SSSR* **118** (1958) 9; A. D. Jackson, M. K. Sener and J. J. M. Verbaarschot, *Phys. Lett. B* **387** (1996) 355–360, hep-th/9605183; T. Guhr and T. Wettig, *J. Math. Phys.* **37** (1996) 6395–6413, hep-th/9605110.

[59] J. W. Alexander, *Trans. Amer. Math. Soc.* **30**(2) (1928) 275–306; V. F. R. Jones, *Invent. Math.* **72** (1983) 1; *Bull. AMS* **12** (1985) 103; *Ann. Math.* **126** (1987) 335; L. Kauffman, *Topology* **26** (1987) 395.

[60] J. H. Conway, Algebraic properties, in *Computational Problems in Abstract Algebra*, ed. J. Leech, Proc. Conf. Oxford, 1967 (Pergamon Press, Oxford, New York, 1970), pp. 329–358.

[61] P. Freyd, D. Yetter, J. Hoste, W. B. R. Lickorish, K. Millet and A. Ocneanu, *Bull. AMS.* **12** (1985) 239; J. H. Przytycki and K. P. Traczyk, *Kobe J. Math.* **4** (1987) 115–139.

[62] E. Guadagnini, M. Martellini and M. Mintchev, *Clausthal 1989, Proceedings, Quantum Groups* (Elsevier Science, B. V., 2003), pp. 307–317; *Phys. Lett. B* **235** (1990) 275; N. Yu. Reshetikhin and V. G. Turaev, *Comm. Math. Phys.* **127** (1990) 1–26.

[63] E. Witten, *Comm. Math. Phys.* **121** (1989) 351–399.

[64] R. K. Kaul and T. R. Govindarajan, *Nucl. Phys. B* **380** (1992) 293–336, hep-th/9111063; P. Ramadevi, T. R. Govindarajan and R. K. Kaul, *Nucl. Phys. B* **402** (1993) 548–566, hep-th/9212110; *Nucl. Phys. B* **422** (1994) 291–306, hep-th/9312215; P. Ramadevi and T. Sarkar, *Nucl. Phys. B* **600** (2001) 487–511, hep-th/0009188.

[65] A. Mironov, A. Morozov and An. Morozov, *JHEP* **1203** (2012) 034, arXiv:1112.2654; in *Strings, Gauge Fields, and the Geometry Behind: The Legacy of Maximilian Kreuzer*, eds. A. Rebhan, L. Katzarkov, J. Knapp, R. Rashkov and E. Scheidegger (World Scientific, 2013), pp. 101–118, arXiv:1112.5754; H. Itoyama, A. Mironov, A. Morozov and An. Morozov, *Int. J. Mod. Phys. A* **27** (2012) 1250099,

arXiv:1204.4785; A. Anokhina, A. Mironov, A. Morozov and An. Morozov, *Nucl. Phys. B* **868** (2013) 271–313, arXiv:1207.0279; *Adv. High Energy Phys.* **2013** (2013) 931830, arXiv:1304.1486; A. Anokhina, arXiv:1412.8444; S. Dhara, A. Mironov, A. Morozov, An. Morozov, P. Ramadevi, V. K. Singh and A. Sleptsov, arXiv:1711.10952.

[66] https://en.wikipedia.org/wiki/Reidemeister_move
http://mathworld.wolfram.com/ReidemeisterMoves.html

[67] A. Caudron, Classification des noeuds et des enlacements, *Publ. Math. Orsay.* **82-4**, University of Paris XI, Orsay (1982); F. Bonahon and L. C. Siebenmann, http://www-bcf.usc.edu/~fbonahon/Research/Preprints/BonSieb.pdf, New geometric splittings of classical knots and the classification and symmetries of arborescent knots (2010).

[68] P. Ramadevi, T. R. Govindarajan and R. K. Kaul, *Mod. Phys. Lett. A* **9** (1994) 3205–3218, hep-th/9401095; S. Nawata, P. Ramadevi and Zodinmawia, *J. Knot Theory Ramifications* **22** (2013) 13, arXiv:1302.5144; Zodinmawia's PhD thesis (2014); D. Galakhov, D. Melnikov, A. Mironov, A. Morozov and A. Sleptsov, *Phys. Lett. B* **743** (2015) 71–74, arXiv:1412.2616; S. Nawata, P. Ramadevi and V. K. Singh, arXiv:1504.00364; A. Mironov, A. Morozov, An. Morozov, P. Ramadevi and V. K. Singh, *JHEP* **1507** (2015) 109, arXiv:1504.003719.

[69] A. Mironov and A. Morozov, *Phys. Lett. B* **755** (2016) 47–57, arXiv:1511.09077; A. Mironov, A. Morozov, An. Morozov, P. Ramadevi, V. K. Singh and A. Sleptsov, *J. Phys. A: Math. Theor.* **50** (2017) 085201, arXiv:1601.0419.

[70] G. Racah, *Phys. Rev.* **62** (1942) 438–462; E. P. Wigner, Manuscript in 1940, in *Quantum Theory of Angular Momentum* (Academic Press, New York, 1965), pp. 87–133; *Group Theory and Its Application to the Quantum Mechanics of Atomic Spectra* (Academic Press, New York, 1959).

[71] A. Kirillov and N. Reshetikhin, in *New Developments in the Theory of Knots* (World Scientific, Singapore, 1989).

[72] L. D. Landau and E. M. Lifshitz, *Quantum Mechanics: Non-Relativistic Theory*, 3rd edn. (Pergamon Press, 1977); J. Scott Carter, D. E. Flath and M. Saito, *The Classical and Quantum 6j-Symbols* (Princeton University Press, 1995).

[73] S. Nawata, P. Ramadevi and Zodinmawia, *Lett. Math. Phys.* **103** (2013) 1389–1398, arXiv:1302.5143.

[74] A. Mironov, A. Morozov and A. Sleptsov, *JHEP* **07** (2015) 069, arXiv:1412.8432.

[75] N. Vilenkin and A. Klimyk, *Representation of Lie Groups and Special Functions. Volume 1: Simplest Lie Groups, Special Functions and Integral Transforms* (Springer Science+Business Media Dordrecht, 1991).

[76] H. Awata, H. Kanno, A. Mironov, A. Morozov, An. Morozov, Y. Ohkubo and Y. Zenkevich, *Nucl. Phys. B* **918** (2017) 358–385, arXiv:1611.07304.

[77] T. Masuda, K. Mimachi, Y. Nakagami, M. Noumi, Y. Saburi and K. Ueno, *Lett. Math. Phys.* **19** (1990) 195–204; I. M. Burban and A. U. Klimyk, *J. Phys. A* **26** (1993) 2139–2151.

[78] L. Faddeev, *Lett. Math. Phys.* **34** (1995) 249–254, hep-th/9504111.

[79] L. Faddeev, R. Kashaev and A. Volkov, *Comm. Math. Phys.* **219** (2001) 199–219, hep-th/0006156.

[80] L. Faddeev and R. Kashaev, *J. Phys. A* **35** (2002) 4043–4048, hepth/0201049.

[81] B. Ponsot and J. Teschner, hep-th/9911110; *Comm. Math. Phys.* **224** (2001) 613–655, arXiv:math/0007097.

[82] L. Faddeev, *Math. Phys. Stud.* **21** (2000) 149–156, math/9912078.

[83] S. Kharchev, D. Lebedev and M. Semenov-Tian-Shansky, *Comm. Math. Phys.* **225** (2002) 573–609, hep-th/0102180.

[84] A. Bytsko and J. Teschner, *Comm. Math. Phys.* **240** (2003) 171–196, math/0208191.

[85] J. Teschner, *Class. Quant. Grav.* **18** (2001) R153–R222, hep-th/0104158; J. Teschner and G. Vartanov, arXiv:1202.4698; arXiv:1302.3778; I. Nidaiev and J. Teschner, arXiv:1302.3454.

[86] S. Derkachov and L. Faddeev, arXiv:1302.5400.

[87] R. Ismagilov, *Funk. Anal. i Prilozhen.* **40** (2006) 69–72 (*Funct. Anal. Appl.* **40** (2006) 222–224); *Mat. Sbornik* **198** (2007) 77–90 (*Sb. Math.* **198** (2007) 369–381); S. Derkachov and V. Spiridonov, arXiv:1711.07073.

[88] R. Kashaev, math/0008148.

[89] R. Kashaev, *Algebra i Analiz* **8** (1996) 63–74, q-alg/9503005.

[90] K. Hikami and R. Inoue, *J. Knot Theory Ramifications* **23** (2014) 1450006, arXiv:1212.6042; *Algebr. Geom. Topol.* **15** (2015) 2175–2194, arXiv:1304.4776; *J. Phys. A: Math. Theor.* **47** (2014) 474006, arXiv:1404.2009.

[91] K. Hikami, *Int. J. Mod. Phys. A* **16** (2001) 3309–3333, math-ph/0105039; *J. Geom. Phys.* **57** (2007) 1895–1940, math/0604094.

[92] T. Dimofte, S. Gukov, J. Lenells and D. Zagier, *Comm. Num. Theor. Phys.* **3** (2009) 363–443, arXiv:0903.2472.

[93] J. E. Andersen and R. Kashaev, arXiv:1109.6295.

[94] L. Faddeev and R. Kashaev, *Mod. Phys. Lett. A* **9** (1994) 427–434, hep-th/9310070.

[95] A. Alexandrov, A. Mironov and A. Morozov, *Int. J. Mod. Phys. A* **21** (2006) 2481–2518, hep-th/0412099; *Fortsch. Phys.* **53** (2005) 512–521, hep-th/0412205; A. Mironov and A. Morozov, *SIGMA* **13** (2017) 047, arXiv:1701.03057.

[96] A. Mironov, A. Morozov, An. Morozov and A. Sleptsov, *JETP Lett.* **104** (2016) 56–61, arXiv:1605.03098; A. Morozov, *Nucl. Phys. B* **911** (2016) 582–605, arXiv:1605.09728; *JHEP* **1609** (2016) 135, arXiv:1606.06015; arXiv:1612.00422; *Phys. Lett. B* **766** (2017) 291–300, arXiv:1701.00359; Sh. Shakirov and A. Sleptsov, arXiv:1611.03797; A. Morozov, arXiv:1711.09277.

[97] J. Gu and H. Jockers, arXiv:1407.5643; A. Mironov, A. Morozov, An. Morozov and A. Sleptsov, *J. Mod. Phys. A* **30** (2015) 1550169, arXiv:1508.02870; *JHEP* **2016** (2016) 134, arXiv:1605.02313; *Phys. Lett. B* **760** (2016) 45–58, arXiv:1605.04881.

[98] http://www.knotebook.org

[99] L. Faddeev, arXiv:0712.2747.

[100] J. Ding and K. Iohara, *Lett. Math. Phys.* **41** (1997) 181–193, q-alg/9608002; K. Miki, *J. Math. Phys.* **48** (2007) 123520.

[101] H. Awata, H. Kanno, T. Matsumoto, A. Mironov, A. Morozov, An. Morozov, Y. Ohkubo and Y. Zenkevich, *JHEP* **07** (2016) 103, arXiv:1604.08366.

[102] A. Mironov, A. Morozov and Y. Zenkevich, *Phys. Lett. B* **762** (2016) 196–208, arXiv:1603.05467.

[103] C.-T. Chan and H.-F. Liu, arXiv:1612.09039.

[104] A. Erdelyi, *Higher Transcendental Functions* (*H. Bateman Manuscript Project*), Vol. 1 (McGraw Hill, New York, 1953), http://apps.nrbook.com/bateman/Vol1.pdf.

OPE for XXX

Philippe Di Francesco[*,‡] and Fedor Smirnov[†,§,¶]

[*]IPhT, Université Paris Saclay, CNRS, CEA,
F-91191 Gif-sur-Yvette, France
and
Department of Mathematics, University of Illinois, MC-382,
Urbana, IL 61821, USA
[†]Sorbonne Université, UPMC Univ. Paris 06,
CNRS, UMR 7589, LPTHE,
F-75005, Paris, France
[‡]philippe@illinois.edu
[§]smirnov@lpthe.jussieu.fr

We explain a new method for finding the correlation functions for the XXX model which is based on the concepts of Operator Product Expansion of Quantum Field Theory on one hand and of fermionic bases for the XXX spin chain on the other. With this method, we are able to perform computations for up to 11 lattice sites. We show that these "experimental" data allow to guess exact formulae for the OPE coefficients.

In memory of Ludwig Dmitrievich Faddeev

1. Introduction

This paper deals with the isotropic Heisenberg spin chain. However, even dealing with a lattice model, it is important to have in mind certain general ideas coming from Quantum Field Theory (QFT). One of them is the operator product expansion (OPE).

Suppose for a given QFT we know a complete set of local operators $\mathcal{O}_i(x)$. Then for small x, we have an expansion

$$\mathcal{O}_i(x)\mathcal{O}_j(0) = \sum_k C_{i,j}^k(x)\mathcal{O}_k(0)\,.$$

We take the OPE quite symbolically leaving aside the necessity of time ordering, the convergence issues, etc. An important feature of the OPE is that it is defined only by the short distance, ultra-violet (UV) nature of the theory. The infrared

[¶]Membre du CNRS.

(IR) environment becomes relevant when the correlation functions are computed. In the OPE approach, it enters through the one-point functions only:

$$\langle \mathcal{O}_i(x)\mathcal{O}_j(0)\rangle_{\text{env}} = \sum_k C_{i,j}^k(x)\langle \mathcal{O}_k(0)\rangle_{\text{env}}\,. \tag{1}$$

A first serious problem concerning the application of OPE consists in finding a convenient way to enumerate the local operators. It is well known that in a two-dimensional conformal field theory (CFT) they correspond to representations of the Virasoro algebra. A lesser known example is provided by the sine(h)-Gordon model for which there exists a fermionic basis for the local operators where all the one-point functions are expressed as determinants [1, 2]. In the UV limit, the relation of the fermionic basis to the Virasoro basis is known, and this allows in principle to apply Perturbed CFT methods to the computation of the two-point functions.

On the other hand, let us imagine a situation when for a large number of IR environments the one-point functions and two-point functions are known. Then the relations (1) provide equations for the coefficients of the OPE. One can dream that these equations could be sufficient to entirely fix all the coefficients of the OPE. For the moment, this program does not look realistic for the QFT. However, we shall show that it is applicable to integrable lattice models.

Generally, the OPE does not depend on the IR environment. However, one can consider classes of IR environments possessing certain symmetries. For example, they can respect translational invariance. In this case, the spatial derivatives of one-point functions vanish, and we can drop them in the right-hand side of the OPE. We shall deal with much wider symmetries of this kind, and write the OPE modulo symmetries as

$$\mathcal{O}_i(x)\mathcal{O}_j(0) \equiv \sum_k C_{i,j}^k(x)\mathcal{O}_k(0)\,.$$

Certainly, the symmetries in question will have to be carefully implemented. They can be not only continuous, but also discrete ones, like C-symmetry for instance.

In this paper, we shall consider the celebrated isotropic Heisenberg magnet (XXX spin chain) with the Hamiltonian

$$H = \sum_{j=-\infty}^{\infty} h_{j,j+1}, \quad h_{j,j+1} = \sigma_j^a\sigma_{j+1}^a\,,$$

with usual notations for Pauli matrices σ_j^a at site j, and where summation over repeated indices is always implied. For some technical reasons which will be explained later, we consider the two-point functions of $\mathfrak{sl}_2$-invariant operators. The examples will be $\sigma_1^a\sigma_N^a$ and $h_{1,2}h_{N-1,N}$.

We consider the correlation functions with arbitrary Matsubara data which will be introduced in the next section. This is a quite general situation which includes important physical applications: zero-temperature anti-ferromagnet, (anti)-ferromagnet with non-zero temperature and magnetic field or even the case of generalized Gibbs ensemble. All these cases correspond to choosing certain special Matsubara data and taking limits for them.

The set of local operators in the lattice analogue of OPE is expressed via the fermionic basis. Crucial to our study are the cases with simple Matsubara data. For these the one-point functions are computed as determinants and there is an efficient procedure for computing the two-point functions using the quantum inverse scattering method (QISM) techniques [3–5]. This allows to obtain an over-determined system of equations for the OPE coefficients. We check that these systems have solutions up to $N = 11$ and compute all the OPE coefficients. These coefficients can now be used for any Mastubara data.

It will be very interesting to guess general formulae for the OPE coefficients from our "experimental" data. This part of the work is still under way, but we shall present here some preliminary results.

The paper is organized as follows. In Sec. 2, we formulate the problem of computing correlation functions on a cylinder (Matsubara environment). In Sec. 3, the fermionic basis and its restriction to our problem are explained. In Sec. 4, we present an efficient computation procedure for small Matsubara chains. In Sec. 5, we further restrict the fermionic basis. The results of computations and formulae for some elements of OPE are given in Sec. 6.

2. Formulation of the Problem

Consider the following square lattice wrapped onto a cylinder $\mathbb{R} \times S^1$:

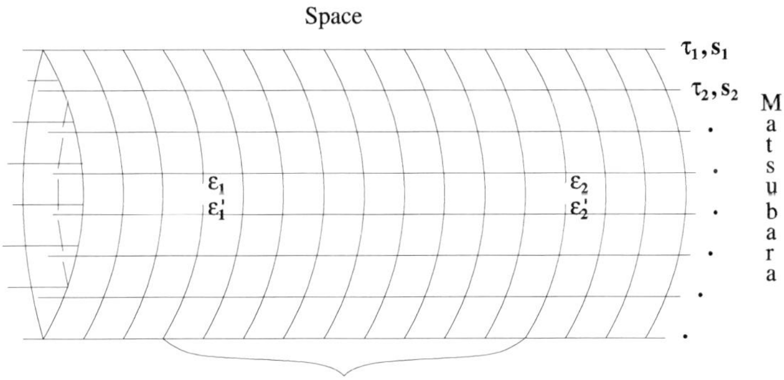

Fig. 1. Expectation value on a cylinder.

We associate the space $\mathbb{C}^2$ to each horizontal lattice site (along each of the parallel circles), and the space $\mathbb{C}^{2s_{\mathbf{m}}+1}$ together with the complex parameter $\tau_{\mathbf{m}}$ to the $\mathbf{m}$th site of the compact Matsubara lattice (along horizontal lines). Denote by π^{2s}_{λ} the evaluation representation of the Yangian with the dimension $2s+1$ and evaluation parameter λ. Every crossing corresponds to the $\mathfrak{sl}_2$ Yangian R-matrix in the tensor product of two evaluation representations: π^1_0 and $\pi^{2s_{\mathbf{m}}}_{\tau_{\mathbf{m}}-1/2}$. Summation is implied over all indices except the four fixed: $\epsilon_1, \epsilon_2, \epsilon'_1, \epsilon'_2$. We denote the partition function defined by the picture of Fig. 1 by $Z^{\epsilon'_1,\epsilon'_2}_{\epsilon_1,\epsilon_2}$, then the expectation value of $\sigma^a_1\sigma^a_N$ (correlation function of spins) with the Matsubara data $\{s_{\mathbf{m}}, \tau_{\mathbf{m}}\}$ is defined by

$$G^{\sigma\sigma}(N) = 1 + \frac{2}{Z}\left(Z^{-+}_{+-} + Z^{+-}_{-+} - 2Z^{+-}_{+-}\right),$$

where Z is the usual partition function (without insertions). A similar formula based on the partition function with four operator insertions (at the sites $1, 2, N-1, N$) is easy to write for the expectation value $G^{hh}(N)$ of $h_{1,2}h_{N-1,N}$ (correlation function of densities of the Hamiltonian). The Matsubara direction has finite size $\mathbf{n}$ while the horizontal (space) direction is infinite. The latter infinity may cause trouble, so we regularize the horizontal space by assuming it has finite length equal to $2L$ and imposing periodic boundary conditions (this amounts to putting our lattice on a torus), and then by finally sending $L \to \infty$.

Let us be more formal. Consider the representation of Yangian $\pi_{\mathbf{S}}$ which is the tensor product of $2L$ representations π^1_0. Consider in addition the representation $\pi_{\mathbf{M}}$ which is the tensor product of the $\mathbf{n}$ representations $\pi^{2s_{\mathbf{m}}}_{\tau_{\mathbf{m}}-1/2}$, $\mathbf{m} = 1, 2, \ldots, \mathbf{n}$. Define

$$T_{\mathbf{S},\mathbf{M}} = (\pi_{\mathbf{S}} \otimes \pi_{\mathbf{M}})(\mathcal{R}),$$

with $\mathcal{R}$ being the universal R-matrix. Then for any $\mathcal{O}$ acting on a small (compared to L) number of tensor components of the representation space of $\pi_{\mathbf{S}}$ define

$$Z(\mathcal{O}) = \lim_{L\to\infty} \frac{\mathrm{Tr}_{\mathbf{M}}\,\mathrm{Tr}_{\mathbf{S}}(T_{\mathbf{S},\mathbf{M}}\mathcal{O})}{\mathrm{Tr}_{\mathbf{M}}\,\mathrm{Tr}_{\mathbf{S}}(T_{\mathbf{S},\mathbf{M}})}.$$

This is the more general definition of the functional Z which was introduced earlier in a particular case.

To the left and to the right of the insertion (see Fig. 1) we have products of Matsubara transfer-matrices

$$T_{\mathbf{M}} = T_{\mathbf{M}}(0), \quad T_{\mathbf{M}}(\lambda) = \mathrm{Tr}\left(\pi^1_\lambda \otimes \pi_{\mathbf{M}}(\mathcal{R})\right),$$

where the trace is taken with respect to the first tensor component. Generically, the transfer-matrix $T_{\mathbf{M}}$ has a single eigenvector $|\mathrm{max}\rangle$ corresponding to an eigenvalue T of maximal absolute value. Suppose the operator $\mathcal{O}$ is localized on the interval $[1, N]$ (like in Fig. 1). Clearly, because of the limit $L \to \infty$

$$Z(\mathcal{O}) = \frac{\langle \mathrm{max}|\mathrm{Tr}_{[1,N]}\left(T_{[1,N],\mathbf{M}}\mathcal{O}\right)|\mathrm{max}\rangle}{T^N\langle \mathrm{max}|\mathrm{max}\rangle}.$$

This quantity was actually computed using the fermionic basis in [6]. The computation is purely algebraic and uses only the fact that $|\mathrm{max}\rangle$ is an eigenvector. For that reason the computation extends to more general quantities of the form

$$Z(\mathcal{O}, \Psi) = \frac{\langle \Psi|\mathrm{Tr}_{[1,N]}\left(T_{[1,N],\mathbf{M}}\mathcal{O}\right)|\Psi\rangle}{T^N\langle \Psi|\Psi\rangle}, \tag{2}$$

with an arbitrary eigenstate $|\Psi\rangle$ with eigenvalue T of the transfer-matrix $T_{\mathbf{M}}$.

3. Fermionic Basis

The fermionic basis is defined in [7] for the anisotropic XXZ model with the anisotropy parameter $\Delta = \frac{1}{2}(q+q^{-1})$. Taking the limit $q \to 1$ is not a difficult problem. More importantly, [7] deals with the quasi-local operators of the form $q^{\alpha \sum_{j=-\infty}^{0} \sigma_j^3} \mathcal{O}$. Taking the limit $\alpha \to 0$ in this expression is non-trivial, and requires some explanation.

In the XXZ case, the basic algebraic tool is the quantum affine algebra $U_q(\widehat{\mathfrak{sl}}_2)$ which replaces the Yangian used for the XXX case. The main result of [7] is the existence of creation operators $\mathbf{t}_j^*$ (boson), $\mathbf{b}_j^*$, $\mathbf{c}_j^*$ (fermions) for $j = 1, 2, \ldots$. By acting on the primary field $\Phi_\alpha = q^{\alpha \sum_{j=-\infty}^{0} \sigma_j^3}$, these operators create a basis in the space of quasi-local operators for which the expectation values of the type considered in the previous section are easily computed. It should be noticed that contrary to the fermions the operators $\mathbf{t}_j^*$ are of rather simple origin, however they make computations much more involved.

In the limit $\alpha \to 0$, the operators $\mathbf{b}_j^*$, $\mathbf{c}_j^*$ acting on Φ_α develop simple poles in α. As a consequence, computing expectation values involves using L'Hospital's rules. This is doable, but leads eventually to the doubling of fermions: one has to consider for every fermion its regular part and residue at $\alpha = 0$. However, there is a notable exception of this general picture. The algebra $U_q(\widehat{\mathfrak{sl}}_2)$ contains two finite-dimensional subalgebras isomorphic to $U_q(\mathfrak{sl}_2)$. Consider the operators $\mathcal{O}$ commuting with the action of one of them. In a *weak* sense (when inserted in expectation values), it can be shown that in the limit $\alpha \to 0$ these operators are created by the action of regular parts of fermions only. This is a great simplification. Another nice point is that the operators $\mathbf{t}^*$ do not contribute for $\alpha = 0$.

Passing to the XXX case with $\alpha = 0$ we assert that the local operators are created by $\mathbf{b}_j^*$, $\mathbf{c}_j^*$. We use the multi-index notations: for $I = \{i_1, \ldots, i_n\}$ we define $\mathbf{b}_I^* = \mathbf{b}_{i_1}^* \cdots \mathbf{b}_{i_n}^*$, $\mathbf{c}_I^* = \mathbf{c}_{i_n}^* \cdots \mathbf{c}_{i_1}^*$. We denote by $\#(I) = n$ the cardinality of I, and by $|I| = \sum_{\ell=1}^{n} i_\ell$. Then according to previous considerations the invariant operators are contained in the space $\mathfrak{H}_0$ of operators with the basis

$$\mathbf{b}_I^* \mathbf{c}_J^* \cdot \mathrm{I}\,, \qquad \#(I) = \#(J)\,, \tag{3}$$

I standing for the unit operator. The operator (3) is supported on an interval of length at most $|I| + |J|$.

It is important to identify among these the operators which are localized on an interval of length N. The following first two conditions follow, for instance, from [8]:

$$\#(I) \le [N/2]\,, \quad \max(I \cup J) \le N\,. \tag{4}$$

Moreover, we consider C-invariant operators (invariant under $\sigma_j^a \to -\sigma_j^a$). For them, it can be shown that the following additional condition can be imposed:

$$|I| + |J| \equiv 0 \pmod 2\,. \tag{5}$$

Let us denote the subspace of $\mathfrak{H}_0$ satisfying (4), (5) by $\mathfrak{H}_0^{(N)}$.

More sophisticated conditions result from the homogeneous limit of the results of [9]. Consider the annihilation operators $\mathbf{b}_j$, $\mathbf{c}_j$ canonically conjugated to $\mathbf{b}^*_j$, $\mathbf{c}^*_j$, and annihilating the vacuum. Introduce the operators

$$Q_m = \sum_{j=1}^{m-1} c_j b_{m-j}\,, \quad m = 1, 2, \dots .$$

We state that the operators localized on $[1, N]$ are linear combinations of (3) satisfying

$$Q_m \mathcal{O} = 0\,, \qquad m > N\,. \tag{6}$$

Due to the second condition in (4) for given N, we can restrict the values of m in the above to $m \le 2N + 1$.

The main theorem of [6] states that for $\#(I) = \#(J)$

$$Z(\mathbf{b}^*_I \mathbf{c}^*_J \cdot \mathrm{I}, \Psi) = \det\bigl(\omega_{i_p,j_q}\bigr)_{p,q=1,\dots,\#(I)}\,, \tag{7}$$

where $\omega_{i,j}$ is an infinite matrix depending on

$$\text{Matsubara data} = \{\{s_\mathbf{m}, \tau_\mathbf{m}\}^{\mathbf{n}}_{\mathbf{m}=1}\,,\ \ |\Psi\rangle\}\,.$$

The exact definition of $\omega_{i,j}$ will be given in the next section.

In our case ($\alpha = 0$), the matrix $\omega_{i,j}$ is symmetric. Since we decided to work in a weak sense, this imposes additional restrictions. The first of them is obvious: we can consider the space $\mathfrak{H}_0/\iota$, with ι being the anti-homomorphism $\mathbf{b}^*_j \leftrightarrow \mathbf{c}^*_j$. For example, we can require that $I \le J$ in lexicographical order. Another consequence is less trivial. The following simple example explains that there are linear relations between minors of any given symmetric matrix:

$$\begin{vmatrix}\omega_{1,3} & \omega_{1,4}\\ \omega_{2,3} & \omega_{2,4}\end{vmatrix} - \begin{vmatrix}\omega_{1,2} & \omega_{1,4}\\ \omega_{2,3} & \omega_{3,4}\end{vmatrix} + \begin{vmatrix}\omega_{1,2} & \omega_{1,3}\\ \omega_{2,4} & \omega_{3,4}\end{vmatrix} = 0\,.$$

This identity implies that working in a weak sense we can factor out the null-vector

$$\bigl(b^*_1 b^*_2 c^*_4 c^*_3 - b^*_1 b^*_3 c^*_4 c^*_2 + b^*_2 b^*_3 c^*_4 c^*_1\bigr) \cdot \mathrm{I}\,.$$

We look for a general formula for such null-vectors. Consider the operator

$$C = \sum_{i=1}^{\infty} c^*_i b_i\,,$$

and the space $\mathfrak{H}_2$ spanned by $\mathbf{b}^*_I \mathbf{c}^*_J \cdot \mathrm{I}$ with $\#(I) = \#(J) + 2$. Then the subspace $\mathfrak{N} = C\mathfrak{H}_2$ of the space $\mathfrak{H}_2$ consists of null-vectors. Indeed, by taking linear combinations, the formula (7) can be rewritten as

$$Z(\mathcal{O}, \Psi) = I^*\, e^{\Omega}\, \mathcal{O}\,,$$

where $\Omega = \sum_{i,j} \omega_{i,j} b_i c_j$ and the left vacuum I^* is annihilated by the right action of creation operators. The fact that $\mathfrak{N}$ consists of null-vectors follows from the commutation of C with Ω. We do not have a formal proof that belonging to $\mathfrak{N}$ is necessary for a vector to be a null-vector, but since the computer does not contradict this, we shall assume it is true.

Let us summarize what we have for the moment. In the weak sense for any local operator $\mathcal{O}$ acting on N sites, we have

$$\mathcal{O} \equiv \sum_{\alpha} C_\alpha v_\alpha \,, \tag{8}$$

where v_α is a basis of the space

$$\mathfrak{V}^{(N)} = \bigcap_{i=N+1}^{2N+1} \mathrm{Ker}\Big(Q_i\big|_{\mathfrak{H}_0^{(N)}/\iota/\mathfrak{N}}\Big) \,, \tag{9}$$

the dimension of this space is much smaller than the dimension of $\mathfrak{H}_0^{(N)}/\iota$. We shall see that it is possible to reduce the space further, but with $\mathfrak{V}^{(N)}$ we can already start working.

4. Explicit Computations for Small Matsubara Chain

The main goal of this section is to show how to produce a large number of Matsubara data, and subsequently compute the matrix $(\omega_{i,j})$ and expectation values using them.

Suppose we are given some Matsubara parameters $\tau_1, s_1, \ldots, \tau_{\mathbf{n}}, s_{\mathbf{n}}$. The eigenvectors $|\Psi\rangle$ are parametrized by Bethe roots $\beta_1, \ldots, \beta_m$. They satisfy the Bethe equations which are conveniently written in terms of the Baxter function $Q(\lambda) = \prod_{j=1}^{m}(\lambda - \beta_j)$:

$$a(\beta_j)Q(\beta_j + 1) + d(\beta_j)Q(\beta_j - 1) = 0, \quad j = 1, \ldots, m \,, \tag{10}$$

where

$$a(\lambda) = \prod_{\mathbf{m}=1}^{\mathbf{n}} (\lambda - \tau_{\mathbf{m}} - s_{\mathbf{m}}), \qquad d(\lambda) = \prod_{\mathbf{m}=1}^{\mathbf{n}} (\lambda - \tau_{\mathbf{m}} + s_{\mathbf{m}}) \,.$$

Recall $s_{\mathbf{m}}$ are given integers, however the subsequent formulae are all analytical in $s_{\mathbf{m}}$, hence we may consider them as arbitrary complex parameters. Then $a(\lambda), d(\lambda)$ will be considered as arbitrary monic polynomials of degree $\mathbf{n}$:

$$a(\lambda) = \lambda^{\mathbf{n}} + \sum_{j=1}^{\mathbf{n}} a_j \lambda^{\mathbf{n}-j}, \quad d(\lambda) = \lambda^{\mathbf{n}} + \sum_{j=1}^{\mathbf{n}} d_j \lambda^{\mathbf{n}-j} \,. \tag{11}$$

The usual problem of the theory of quantum integrable models is to find solutions to Bethe equations for the input data

$$\text{input} = \{a_1, \ldots, a_{\mathbf{n}}, d_1, \ldots, d_{\mathbf{n}}\} \,. \tag{12}$$

Now solving the Bethe equations we would obtain an equivalent set of parameters to the previously introduced

$$\text{Matsubara data} = \{\{\tau_{\mathbf{m}}\}, \{s_{\mathbf{m}}\}, \Psi\} \equiv \{\beta_1, \ldots, \beta_m, a_1, \ldots, a_{\mathbf{n}}, d_1, \ldots, d_{\mathbf{n}}\} \,. \tag{13}$$

For a certain well-known reason, we set $m \leq [\mathbf{n}/2]$.

The main problem here is that the Bethe equations constitute a complicated system of algebraic equations for $\beta_1, \dots, \beta_m$. But we have to realize that our goal is different from the usual one. We do not need to find the spectrum for a given chain. Rather we need a stock of solutions of the Bethe equations in order to obtain equations for the OPE coefficients. That is why we make the following new choice of input data

$$\text{input} = \{\beta_1, \dots, \beta_m, a_{m+1}, \dots, a_{\mathbf{n}}, d_1, \dots, d_{\mathbf{n}}\}. \tag{14}$$

The number of continuous parameters is the same, but the unknowns are now $a_1, \dots, a_m$. Note that for them, the Bethe equations are *linear*. Solving them is easy, and we obtain as a result some Matsubara data (13).

Given some Matsubara data, we need first of all to construct the infinite matrix $\omega_{i,j}$. The latter is conveniently coded by the generating function

$$\omega(\lambda, \mu) := \sum_{i,j=1}^{\infty} \lambda^{i-1}\mu^{j-1}\omega_{i.j}. \tag{15}$$

Let us define the latter in the spirit of [10]. Introduce the kernel and "half-kernel" functions:

$$K(\lambda) = -\frac{2}{\lambda^2 - 1}, \qquad H(\lambda) = \frac{1}{(\lambda - 1)\lambda},$$

and the measure

$$dm(\lambda) = \frac{d\lambda}{1 + \mathfrak{a}(\lambda)}, \qquad \mathfrak{a}(\lambda) = \frac{a(\lambda)Q(\lambda + 1)}{d(\lambda)Q(\lambda - 1)}.$$

We need an auxiliary function defined by the integral equation

$$G(\eta, \mu) = H(\eta - \mu) - \frac{1}{2\pi i}\oint_{\Gamma} K(\eta - \sigma)G(\sigma, \mu)dm(\sigma),$$

where the contour Γ goes around the Bethe roots $\beta_1, \dots, \beta_m$ and the point $\sigma = \mu$. For a finite Matsubara chain, we have a finite number of Bethe roots for which the equation above reduces to a linear system for $G(\beta_j, \mu)$. The function $\omega(\lambda, \mu)$ is given by

$$\omega(\lambda, \mu) = \frac{1}{2\pi i}\oint_{\Gamma'} H(\eta - \lambda)G(\eta, \mu)dm(\eta) - \frac{1}{4}K(\lambda - \mu),$$

with Γ' containing one more point: $\eta = \lambda$. It is easy to see that this function is symmetric. Finally, extracting the coefficients of $\omega(\lambda, \mu)$ from (15) allows to compute the right hand side of

$$\langle \mathcal{O} \rangle_{\mathrm{Md}} = \sum_{\alpha} C_\alpha \langle v_\alpha \rangle_{\mathrm{Md}}, \tag{16}$$

where Md stands for Mastubara data and $\{v_\alpha\}$ is the reduced fermion basis (8).

The left-hand side of (16) is computed by QISM techniques. Let us begin by introducing the main character of QISM, the *monodromy matrix*. We use the block decomposition with respect to the first tensor component

$$(\pi^1_\lambda \otimes \pi_{\mathbf{M}})(\mathcal{R}) = \begin{pmatrix} A(\lambda) & B(\lambda) \\ C(\lambda) & D(\lambda) \end{pmatrix},$$

with $A(\lambda), B(\lambda), C(\lambda), D(\lambda)$ acting in the Matsubara space. We shall denote $A = A(0)$, etc. We want to compute

$$\langle \Psi | \operatorname{Tr}_{[1,N]}\bigl(T_{[1,N],\mathbf{M}}\mathcal{O}\bigr)|\Psi\rangle ,$$

for $\mathcal{O}$ located on sites $[1, N]$. For given Matsubara data we write

$$\begin{aligned} \langle \Psi| &= \langle \beta_1, \dots, \beta_m| = \langle \downarrow | B(\beta_1)\cdots B(\beta_m) , \\ |\Psi\rangle &= |\beta_1, \dots, \beta_m\rangle = C(\beta_1)\cdots C(\beta_m)|\downarrow\rangle . \end{aligned}$$

The normalization is provided by the Gaudin formula below. Obviously, for given $\mathcal{O}$ the expression $\operatorname{Tr}_{[1,N]}\bigl(T_{[1,N],\mathbf{M}}\mathcal{O}\bigr)$ is certain linear combination of monomials of A, B, C, D of total degree N. So, we have to learn how to compute their averages in an efficient way.

We start with the Slavnov formula for the scalar product of the Bethe covector $\langle \beta_1, \dots, \beta_m|$ with an off-shell vector $|\mu_1, \dots, \mu_m\rangle = C(\mu_1)\cdots C(\mu_m)|\downarrow\rangle$, μ_j being arbitrary. The formula is

$$\langle \beta_1, \dots, \beta_m | \mu_1, \dots, \mu_m\rangle = \prod_{j=1}^m d(\beta_j) d(\mu_j) \frac{\prod\limits_{i,j=1}^m (\beta_j - \mu_i + 1)}{\prod\limits_{i<j} (\mu_i - \mu_j) \prod\limits_{i<j} (\beta_j - \beta_i)} \det(\mathcal{N}) , \tag{17}$$

where the matrix $\mathcal{N}$ has entries

$$N_{i,j} = \frac{1}{(\mu_i - \beta_j)} \left(\frac{1}{(\mu_i - \beta_j - 1)} - \frac{1}{(\mu_i - \beta_j + 1)} \mathfrak{a}(\mu_i) \right) .$$

The left-hand side of (17) is by definition a symmetric polynomial of $\mu_1, \dots, \mu_m$, hence the singularities in the right-hand side cancel. We shall return to this point later.

The normalization of the Bethe vectors is given by the Gaudin formula, expressible via L'Hospital rules from (17). Explicitly, we have

$$\langle \beta_1, \dots, \beta_m | \beta_1, \dots, \beta_m\rangle = \prod_{j=1}^m a(\beta_j) d(\beta_j) \prod_{i \neq j} \frac{\beta_i - \beta_j + 1}{\beta_i - \beta_j} \det(\mathcal{G}) , \tag{18}$$

where the matrix $\mathcal{G}$ has entries

$$G_{k,l} = \frac{\partial}{\partial \beta_l} \log \mathfrak{a}(\beta_k) , \quad k, l = 1, \dots, m .$$

We want to compute

$$\langle \beta_1, \dots, \beta_m | X_1 X_2 \cdots X_N | \beta_1, \dots, \beta_m\rangle ,$$

where X_j is one of A, B, C, D. We proceed as follows. Start with the Slavnov formula. Use the following formulae which are easily obtained by methods of QISM.

Consider any Bethe covector $\langle\Phi|$. Introduce the notations $v(\lambda) = 1/\lambda$, $u(\lambda) = v(\lambda) + 1$. Provided

$$f(\mu_1, \ldots, \mu_m) = \langle\Phi|\mu_1, \ldots, \mu_m\rangle\,,$$

is known for arbitrary $\mu_1, \ldots, \mu_m$ we have

$$\begin{aligned}\langle\Phi|A|\mu_1, \ldots, \mu_m\rangle = \prod_{j=1}^{m} u(-\mu_j)a(0)f(\mu_1, \ldots, \mu_m)\\ - \sum_{j=1}^{k} v(-\mu_j)\prod_{r\neq j}^{k} u(\mu_j - \mu_r)a(\mu_j)f(\mu_1, \ldots, \widehat{\mu_j}, \ldots, \mu_m, 0)\,,\end{aligned} \tag{19}$$

$$\begin{aligned}\langle\Phi|D|\mu_1, \ldots, \mu_m\rangle = \prod_{j=1}^{k} u(\mu_j)d(0)\prod_{j=1}^{k} C(\mu_j)|\downarrow\rangle\\ - \sum_{j=1}^{m} v(\mu_j)\prod_{r\neq j}^{m} u(\mu_r - \mu_j)d(\mu_j)f(\mu_1, \ldots, \widehat{\mu_j}, \ldots, \mu_m, 0)\,,\end{aligned} \tag{20}$$

$$\begin{aligned}\langle\Phi|C|\mu_1, \ldots, \mu_m\rangle = \Bigg(\sum_{j=1}^{k} v(-\mu_j)\prod_{r\neq j}^{k} u(-\mu_r)(\mu_r - \mu_j)a(0)d(\mu_j)\\ + v(\mu_j)\prod_{r\neq j}^{k} u(\mu_r)(\mu_j - \mu_r)d(0)a(\mu_j)\Bigg) f(\mu_1, \ldots, \widehat{\mu_j}, \ldots, \mu_m)\\ + \sum_{j>i}\Bigg(d(\mu_i)a(\mu_j)v(-\mu_i)v(-\lambda)u(\mu_j - \mu_i)\prod_{r\neq i,j}^{k} u(\mu_r - \mu_i)u(\mu_j - \mu_r)\\ + a(\mu_i)d(\mu_j)v(-\mu_j)v(\mu_i)u(\mu_i - \mu_j)\prod_{r\neq i,j}^{k} u(\mu_i - \mu_r)u(\mu_r - \mu_j)\Bigg)\\ \times f(\mu_1, \ldots, \widehat{\mu_i}, \ldots, \ldots, \widehat{\mu_j}, \ldots, \mu_m, 0)\,,\end{aligned} \tag{21}$$

$$\langle\Phi|B|\mu_1, \ldots, \mu_m\rangle = f(\mu_1, \ldots, \mu_m, 0)\,. \tag{22}$$

Notice that in the last two formulae the number of arguments changes. Using these formulae we compute inductively

$$\langle\beta_1, \ldots, \beta_m|X_1X_2\cdots X_N|\mu_1, \ldots, \mu_m\rangle\,,$$

and then set $\mu_j = \beta_j$, $j = 1, \ldots, m$.

Our goal is to implement the formulae above on a computer. To proceed fast, we have to avoid symbolic computations as much as possible. The input data are chosen numerically as a set of random integers from 1 to 10. Still in the procedure described above we cannot immediately exclude the symbolic variables μ_j. It has been said that $\langle\beta_1, \ldots, \beta_m|\mu_1, \ldots, \mu_m\rangle$ is a symmetric polynomial in $\mu_1, \ldots, \mu_m$. So are the expressions (19), (22), (21), (20). However, to arrive at the polynomial form we have to cancel some singularities by factorizing the expressions in the right-hand side. Factorization is a time consuming operation, it should be ultimately avoided. Actually, looking at the Slavnov formula one realizes that it can be easily rewritten as a sum of Schur polynomials. Then with some effort one rewrites the right-hand sides of (19), (22), (21), (20) as operators acting in the basis of Schur polynomials (indexed by Young diagrams). In this way, we completely eliminate time-consuming symbolic computations. In fact acting in the space of Young diagrams can be made very fast.

5. Final Restriction of the Number of Variables

Now for any given input data we construct the full Matsubara data which allow to compute, as explained in the previous section, the quantities $\langle \mathcal{O} \rangle_{\mathrm{Md}}$ and $\langle v_\alpha \rangle_{\mathrm{Md}}$ in the main formula (16).

We start with $\mathcal{O} = \sigma_1^a \sigma_N^a$. It is interesting that in order to fix C_α up to $N = 4$ we do not need to resort to non-trivial excited states Ψ in the Matsubara data to fix the coefficients. Thus for $N = 4$, absolutely trivial computations give results allowing access to the correlation functions in any environment. For $N = 5$, we need one one-particle state Ψ. Starting from $N = 8$ a few two-particle states are needed. We always take more Matsubara data than the number of unknowns, so that the systems of equations to solve for C_α are overdetermined. The compatibility of all these equations is an important check of the validity of our procedure.

We computed up to $N = 9$, and going further was very hard. However, analyzing the results we made the following important observation. Decompose the operators v_α in the fermionic basis to rewrite

$$\mathcal{O} \equiv \sum_{I,J} D_{I,J} b_I^* c_J^* \cdot \mathrm{I}\,.$$

A priori I, J are only lexicographically ordered, but we observe that the coefficients $D_{I,J}$ vanish unless

$$i_p \leq j_p, \quad p = 1, \ldots, \#(I)\,.$$

We shall write $I \preccurlyeq J$.

We want to reduce the number of unknowns using this observation. The naïve idea consists in repeating the procedure of Sec. 3 starting with the subspace $\widetilde{\mathfrak{H}}_0^{(N)}$ spanned by $b_I^* c_J^* \cdot \mathrm{I}$ satisfying (5), (4) and $I \preccurlyeq J$. However, the resulting space (9) would be too small, there would be no solutions to the equations for C_α. More careful consideration and further experiments show that the correct choice is the subspace $\widetilde{\mathfrak{V}}^{(N)}$ of $\widetilde{\mathfrak{H}}_0^{(N)}$ such that

$$\bigoplus_{i=N+1}^{2N+1} \mathrm{Im}\big(Q_i\big|_{\widetilde{\mathfrak{V}}^{(N)}}\big) \subset \mathfrak{N}\,.$$

This results in a drastic reduction of the dimension of the space. The dimensions up to $N = 12$ are given in the table below.

N	2	3	4	5	6	7	8	9	10	11	12
$\dim(\widetilde{\mathfrak{H}}_0^{(N)})$	1	3	11	26	99	253	1038	2816	12041	34062	148630
$\dim(\widetilde{\mathfrak{V}}^{(N)})$	1	2	6	12	31	79	178	434	1141	2946	7888

6. Results and Conjectures

We computed the OPE coefficients for $\sigma_1^a\sigma_N^a$ and $h_{1,2}h_{N-1,N}$ up to $N = 11$. Starting from $N = 7$ the data becomes too large to be presented here, they are available upon request.

With the OPE coefficients at hand we can compute the correlation functions for any Matsubara data, as well as the corresponding matrix $\omega_{i,j}$ or equivalently the function $\omega(\lambda, \mu)$. In particular, for the anti-ferromagnet at zero temperature and zero magnetic field we have

$$\omega(\lambda, \mu) = -\frac{1}{2} + 2\log(2) + \sum_{k=1}^{\infty}(\lambda - \mu)^{2k}\left(2\zeta(2k+1)(1 - 2^{-2k}) - \frac{1}{2}\right).$$

Let us present the numerical values for $G^{\sigma\sigma}$. Up to $N = 8$ they are known from [11], so, we use the normalization of this paper $G^{\sigma\sigma}(N) \to G^{\sigma\sigma}(N)/12$ to simplify the comparison.

N	$G^{\sigma\sigma}(N)/12$
2	−0.147715726853315103139077
3	0.0606797699564353014934941
4	−0.0502486272572352479593931
5	0.0346527769827281656556596
6	−0.0308903666476093257628751
7	0.0244467383279589065417695
8	−0.0224982227633722183770986
9	0.0189734169587321977494075
10	−0.0177751064604679461357958
11	0.0155478493216075886422179

The last two values are already in a good agreement with the asymptotic formula of [12].

We can proceed with computations for finite temperature and other Matsubara environments, however, the most interesting applications of our "experimental" data consists in trying to guess general structure of the OPE coefficients. We have some incomplete results in this direction which we expose now.

Consider the decompositions

$$\begin{aligned}\sigma_1^1\sigma_N^a &\equiv \sum_{I\preccurlyeq J} D^{\sigma\sigma}_{I,J}(N)\mathbf{b}^*_I\mathbf{c}^*_J\cdot \mathrm{I}\,,\\ h_1h_{N-1} &\equiv \sum_{I\preccurlyeq J} D^{hh}_{I,J}(N)\mathbf{b}^*_I\mathbf{c}^*_J\cdot \mathrm{I}\,.\end{aligned} \tag{23}$$

A first question which we would like to address is the behavior of the coefficients as functions of N. The experimental data show that they are polynomial in N of degrees

$$d^{\sigma\sigma}_{I,J}=\frac{1}{2}(|I|+|J|)+\#(I)-1\,,\qquad d^{hh}_{I,J}=\frac{1}{2}(|I|+|J|)+\#(I)-4\,. \tag{24}$$

Moreover, in these polynomials not all the coefficients are independent, we find that they are of the form

$$D^{\sigma\sigma}_{I,J}(N)=\sum_{s=l^{\sigma\sigma}_{I,J}}^{d^{\sigma\sigma}_{I,J}} X^{\sigma\sigma}_{s,I,J}\binom{N-2}{s}\,,\qquad D^{hh}_{I,J}(N)=\sum_{s=l^{hh}_{I,J}}^{d^{hh}_{I,J}} X^{hh}_{s,I,J}\binom{N-4}{s}\,. \tag{25}$$

The lower limits $l^{\sigma\sigma}_{I,J}$, $l^{hh}_{I,J}$ are rather subtle. Lower bounds for them are

$$l^{\sigma\sigma}_{I,J}\geq \max(J)-2\,,\qquad l^{hh}_{I,J}\geq \max(J)-4\,.$$

These estimates show that the coefficients vanish if $\max(J)$ goes far from the rest of elements of $I\cup J$. However, if two elements of J become large no vanishing happens. There are some additional simple reason for the coefficients to vanish which will be clear from examples below. In certain cases we were even able to make general conjectures for the coefficients. Let us present them.

We begin with the simple case of $X^{\sigma\sigma}_{s,I,J}$ (coefficients in (25)) for the two-fermion case with $I=\{k-p\}$ and $J=\{k+p\}$. In that case, we have $3-p\leq 3$ terms in our formula. Since the space $\mathfrak{N}$ is empty for the zero fermion case we have for any s

$$\sum_p X^{\sigma\sigma}_{s,\{k-p\},\{k+p\}}=0\,. \tag{26}$$

In particular, for $s=k-2$ there is only one term in this sum. Hence $X^{\sigma\sigma}_{k-2,\{k\},\{k\}}=0$, this is the above mentioned additional reason for coefficients to vanish. For nontrivial coefficients it is easy to guess from our data that

$$X^{\sigma\sigma}_{k,\{k-p\},\{k+p\}}=2(-1)^{p-1}\left(\binom{2}{p}+2\binom{1}{p}\right)\,,\quad X^{\sigma\sigma}_{k-1,\{k-p\},\{k+p\}}=2(-1)^{p-1}\binom{1}{p}\,.$$

Both coefficients vanish for $p>2$ in agreement with the previous explanation.

It so happens that all the two fermions contribution to $G^{hh}(N)$ vanish. The next simplest case is that of four fermions contribution. For that case we use for the sets $I=\{k-p,i\}$, $J=\{k+p,j\}$ for $i+j$ even and $I=\{k-p,i\}$, $J=\{k+p+1,j\}$ for $i+j$ odd. For fixed i,j, we shall vary k and p within the obvious limits. Non-vanishing coefficients X^{hh} for some (k,p) correspond to boxes in a triangular array (see Table 1 for $i = 1, j = 1$). The vertical and horizontal sizes of the array are equal to $S = \left[\frac{i+j}{2}\right] + 3$. Counting the rows by $p = 0, \ldots, S-1$, and the columns by $r = 0, \ldots, S-1$, then let $s = r + k - 4$ for $i + j$ even and $s = k + r - 3$ for $i + j$ odd. For given i, j we shall simplify notations: define $X_{r,p}(k) := X^{hh}_{s,\{k-p,1\},\{k+p,1\}}$ with r and s related as explained. Sums over columns equal zero for the same reason as explained above, so, we do not write down $X_{r,r}(k)$. Notice, in particular, that $X_{0,0}(k) = 0$.

We start from the simplest case $i = j = 1$. Non-vanishing coefficients correspond to boxes in the table:

Table 1.

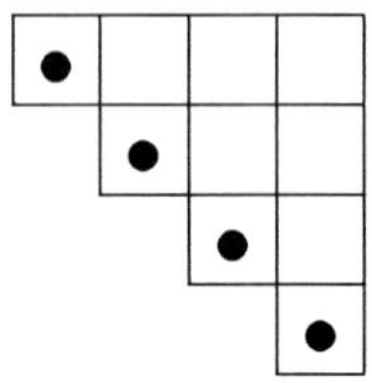

So, we shall not write formulae for the boxes with bullets for the reasons which have been explained. The coefficients grow rather fast with k, so, at the first glance there is no hope to find a formula for them. However, looking at them attentively, we observed that the simplest one (the second from the left in the first row) is just

$$X_{1,0}(k) = \frac{1}{6}(3^{k-2} - 1)\,.$$

Based on this observation and the experimental data we guessed further

$$\begin{aligned}
X_{2,0}(k) &= \frac{5}{8} \cdot 3^{k-1} + \frac{(2k^2 + 4k - 45)}{24}\,,\\
X_{3,0}(k) &= \frac{5}{4} \cdot 3^{k-1} + \frac{2k^2 + 8k - 49}{12}\,,\\
X_{2,1}(k) &= -\frac{5}{2} \cdot (3^{k-2} - 1)\,,\\
X_{3,1}(k) &= -\frac{5}{8} \cdot 3^{k} - \frac{2k^2 + 8k - 159}{24}\,,\\
X_{3,2}(k) &= \frac{27}{4} \cdot (3^{k-3} - 1) - \frac{(k-3)(k+7)}{6}\,.
\end{aligned}$$

For a certain reason, $X_{3,2}(3) = 0$. This is made explicit in the last line. We shall continue making such vanishings explicit.

Let us consider now the case $i = 1$, $j = 2$. The table is the same as for the previous case. From experimental data, we guess the formulae

$$
\begin{aligned}
X_{1,0}(k) &:= -\frac{k-2}{24}\cdot(3^{k-1}-1)-\frac{(k-1)(k+2)}{12}, \\
X_{2,0}(k) &= -\frac{5(k-1)}{16}\cdot 3^{k-1}-\frac{2k^3+18k^2+23k+21}{48}, \\
X_{3,0}(k) &:= -\frac{5k}{32}\cdot 3^{k}-\frac{6k^3+52k^2+89k+128}{96}, \\
X_{2,1}(k) &= \frac{5(k-1)}{32}\cdot 3^{k}+\frac{2k^3+38k^2+57k+63}{96}, \\
X_{3,1}(k) &:= \frac{k}{32}\cdot 3^{k+2}+\frac{2k^3+20k^2+31k+64}{32}, \\
X_{3,2}(k) &= -\frac{5}{32}k\cdot 3^{k}+\frac{2k^3+4k^2+23k-64}{96}.
\end{aligned}
$$

At this point, one can think that the general formula consists of combinations with polynomial coefficients of 1 and 3^k. Actually, this is not the case: we were able to fix several additional coefficients observing that 6^k starts to appear.

Consider $i = 2, j = 2$. In this case our data allow to define only three coefficients:

$$
\begin{aligned}
X_{1,0}(k) &= \frac{1}{125}\cdot(6^{k-1}-1)-\frac{(2k+1)}{144}\cdot(3^{k-1}-1)+\frac{(k-1)(45k-37)}{1800}, \\
X_{2,0}(k) &= \frac{49}{250}\cdot(6^{k-1}-1)+\frac{(k^2-19k-41)}{96}\cdot(3^{k-1}-1) \\
&\quad+\frac{(k-1)(25k^2+210k+179)}{1200}, \\
X_{2,1}(k) &= -\frac{98}{375}\cdot(6^{k-1}-1)+\frac{(6k+13)}{24}\cdot(3^{k-1}-1)-\frac{(45k+43)(k-1)}{300}.
\end{aligned}
$$

Similarly for $i = 1, j = 3$:

$$
\begin{aligned}
X_{1,0}(k) &= -\frac{48}{625}\cdot(6^{k-2}-1)+\frac{(7k-11)}{60}\cdot(3^{k-2}-1)-\frac{(k-2)(5k-53)}{750}, \\
X_{2,0}(k) &= -\frac{144\cdot 49}{625}\cdot(6^{k-3}-1)-\frac{27(2k^2-15k-18)}{720}\cdot(3^{k-3}-1) \\
&\quad+\frac{(k-3)(25k^2-465k+10054)}{3000}, \\
X_{2,1}(k) &= \frac{32\cdot 49}{625}\cdot 6^{k-2}-2k\cdot 3^{k-2}+\frac{75k^2+1105k-964}{1875}.
\end{aligned}
$$

Finally, we were able to guess one coefficient for each $i = 2$, $j = 3$ and $i = 1$, $j = 4$. They are respectively

$$\begin{aligned} X_{1,0}(k) &= -\frac{6(k-4)}{625}\cdot(6^{k-1}-1) - \frac{(4k^2-9k-8)}{240}\cdot(3^{k-1}-1) \\ &\quad - \frac{(k-1)(65k^2+91k+96)}{3000}, \\ X_{1,0}(k) &= \frac{12(k-4)}{625}\cdot(6^{k-1}-1) + \frac{k^2+9k-40}{120}\cdot(3^{k-1}-1) \\ &\quad + \frac{(k-1)(5k^2+47k+132)}{500}. \end{aligned}$$

An important question is that of the general structure. Is it true that the next exponent to appear will be 9^k? This should be possible to answer considering the cases $i + j = 6$ for which we observe experimentally faster growth than 6^k. Unfortunately, our experimental data do not allow to guess a general formula for this case.

For four-fermion contributions to $G^{\sigma\sigma}$ the coefficients are more complicated, we shall not present them because we know only very few. We would mention, however, that the structure looks similar, but 6^k starts to appear from the very beginning.

7. Conclusion

We have demonstrated the power of the method to fix the OPE coefficients for the XXX model based on the fermionic basis. Somebody with better computer skills than ours may probably produce some more data. For example, $N = 12$ should be reachable.

The main problem which we could not solve for the moment is finding general formulae for the coefficients even for the four-fermion contributions to G^{hh}. For example, the coefficient $X_{r,p}(k)$ is known for $\{i,j\} = \{1,1\}$, $\{1,2\}$, $\{1,3\}$, $\{1,4\}$, $\{2,2\}$, $\{2,3\}$, but still we were unable to find a conjecture for general i, j. Probably, an independent look is needed?

Acknowledgments

PDF is partially supported by the Morris and Gertrude Fine endowment. FS would like to thank H. Boos, M. Jimbo and T. Miwa for stimulating discussions.

References

[1] M. Jimbo, T. Miwa and F. Smirnov, Hidden Grassmann structure in the XXZ model V: sine-Gordon model, *Lett. Math. Phys.* **96** (2011) 325–365.

[2] S. Negro and F. Smirnov, On one-point functions for sinh-Gordon model at finite temperature, *Nucl. Phys. B* **875(1)** (2013) 166–185.

[3] L. D. Faddeev, E. K. Sklyanin and L. A. takhtajan, The quantum inverse problem method, *Theor. Math. Phys.* **40** (1980) 688–711.

[4] V. E. Korepin, N. M. Bogoliubov and A. G. Izergin, *Quantum Inverse Scattering Method and Correlation Functions* (Cambridge University Press, 1993).

[5] N. A. Slavnov, Calculation of scalar products of wave functions and form factors in the framework of the alcebraic Bethe ansatz, *Theor. Math. Phys.* **79** (1989) 502–508.

[6] M. Jimbo, T. Miwa and F. Smirnov, Hidden Grassmann structure in the XXZ model III: Introducing Matsubara direction, *J. Phys. A* **42** (2009) 304018, 31 pp.

[7] H. Boos, M. Jimbo, T. Miwa, F. Smirnov and Y. Takeyama, Hidden Grassmann structure in the XXZ model II: Creation operators, *Comm. Math. Phys.* **286** (2009) 875–932.

[8] H. Boos, M. Jimbo, T. Miwa and F. Smirnov, Completeness of a fermionic basis in the homogeneous XXZ model, *J. Math. Phys.* **50** (2009) 095206.

[9] H. Boos, M. Jimbo, T. Miwa, F. Smirnov and Y. Takeyama, Fermionic basis for space of operators in the XXZ model, *SISSA Proceedings of Science* (2007) Paper 015, 34 pp.

[10] H. Boos, F. Göhmann, A. Klümper and J. Suzuki, Factorization of the finite temperature correlation functions of the XXZ chain in a magnetic field, *J. Phys. A* **40** (2007) 10699–10727.

[11] J. Sato, M. Shiroishi and M. Takahashi, Exact evaluation of density matrix elements for the Heisenberg chain *J. Stat. Mech.* **0612** (2006) P12017.

[12] S. Lukyanov and V. Terras, Long-distance asymptotics of spin–spin correlation functions for the XXZ spin chain, *Nucl. Phys. B* **654** (2003) 323–356.

Chiral Anomaly, Topological Field Theory, and Novel States of Matter

Jürg Fröhlich
Theoretical Physics, ETH Zurich, Wolfgang Pauli Strasse, 8093 Zurich, Switzerland
juerg@phys.ethz.ch

Starting with a description of the motivation underlying the analysis presented in this paper and a brief survey of the chiral anomaly, I proceed to review some basic elements of the theory of the quantum Hall effect in 2D incompressible electron gases in an external magnetic field, ("Hall insulators"). I discuss the origin and role of anomalous chiral edge currents and of anomaly inflow in 2D insulators with explicitly or spontaneously broken time reversal, i.e. in Hall insulators and "Chern insulators". The topological Chern–Simons action yielding the large-scale response equations for the 2D bulk of such states of matter is displayed. A classification of Hall insulators featuring quasi-particles with abelian braid statistics is sketched.

Subsequently, the chiral edge spin currents encountered in some time-reversal invariant 2D topological insulators with spin-orbit interactions and the bulk response equations of such materials are described.

A short digression into the theory of 3D topological insulators, including "axionic insulators", follows next.

To conclude, some open problems are described and a problem in cosmology related to axionic insulators is mentioned.

As far as the quantum Hall effect and the spin currents in time-reversal invariant 2D topological insulators are concerned, this review is based on extensive work my collaborators and I carried out in the early 1990's.

Dedicated to the memory of Ludvig Dmitrievich Faddeev — a great scientist who will be remembered

1. General Goals of Analysis

The purpose of this paper is to review some applications of the chiral anomaly and of topological field theory to condensed matter physics, more specifically to the theoretical analysis of novel, topologically protected states of matter. This is an area that has been of much interest to me, ever since the late 1980's, and to which I have made contributions that I feel are interesting and important. For this reason, it makes sense to highlight them, one more time, trying to put the emphasis on some of the most important concepts and most elegant ideas. The results sketched in the following are based on the papers [1–6] (and references given therein) all published in the early nineties.

I am confident that the material exposed here would have interested Ludvig Faddeev. Among many other important contributions, he greatly enhanced our mathematical understanding of quantum gauge theories and of the origin and implications of the chiral anomaly. Some of his results on gauge theories and anomalies provide the theoretical underpinning for much of what I will discuss in the following.

I start by sketching the general motivation underlying the efforts of my collaborators and myself (see [4–8]) that have led to the results discussed in the following.

Key purposes of our work are:

- To classify bulk and surface states of systems of condensed matter, using concepts and results from gauge theory, current algebra and General Relativity. In particular, *effective actions* (= generating functionals of connected current Green functions yielding expressions for transport coefficients, such as conductivities, via Kubo formulae) are used extensively in our analysis; their form being constrained by *gauge-invariance*, by the cancellation of *gauge anomalies* between bulk and surface terms ("holography"), and by *locality, unitarity* and "*power counting*". It turns out that these general principles *completely determine* the most relevant terms in the effective actions of, for example, insulators. Using effective actions, one discovers new ways of characterizing novel states of matter, as indicated next.
- To complement the *Landau Theory* of Phases and Phase Transitions by a "*Gauge Theory of Phases/States of Matter*", with the purpose of describing novel "topologically protected" states of matter that cannot be characterized by a local order parameter.
- To apply our "Gauge Theory of Phases of Matter" to extend or develop:
  - The theory of the Fractional Quantum Hall Effect; (our work extended over the period of 1989–2012).
  - The theory of time-reversal invariant Topological Insulators and of Topological Superconductors; (our first paper appeared in 1993).
  - A description of higher-dimensional cousins of the Quantum Hall Effect, with applications in cosmology, e.g., towards understanding the origin of intergalactic primordial magnetic fields in the Universe, etc.; (from 1999 to the present).

The Chiral Anomaly

While vector currents, J^{μ}, are usually conserved, i.e. $\partial_{\mu} J^{\mu} = 0$, *axial* currents, J_5^{μ}, defined on space-times of even dimension fail to be conserved, i.e. tend to be *anomalous*, when coupled to gauge fields — even if they are associated with *massless* matter fields. This phenomenon is called "*chiral anomaly*"; (see [9] and references given therein).

Let $\vec{E}$ denote the electric field and $\vec{B}$ the magnetic field (or induction). In two space-time dimensions, the chiral anomaly is expressed by

$$\partial_\mu J_5^\mu = \frac{\alpha}{2\pi} E, \quad \alpha := \frac{q^2}{\hbar},$$

where q is the electric charge of a massless field that gives rise to the current J_5^μ, and $\hbar$ is Planck's constant, and by the (non-vanishing) anomalous equal-time commutator

$$[J_5^0(\vec{y},t), J^0(\vec{x},t)] = i\alpha\delta'(\vec{x}-\vec{y}).$$

In four space-time dimensions, the anomaly is expressed by

$$\partial_\mu J_5^\mu = \frac{\alpha}{\pi} \vec{E} \cdot \vec{B},$$

and the anomalous commutator

$$[J_5^0(\vec{y},t), J^0(\vec{x},t)] = i\frac{\alpha}{\pi}\big(\vec{B}(\vec{y},t) \cdot \nabla_{\vec{y}}\,\delta\big)(\vec{x}-\vec{y}).$$

For a massive matter field, there are additional terms contributing to the divergence, $\partial_\mu J_5^\mu$, of the axial current that are proportional to the mass, m, of that field.

It is then obvious that the *chiral* currents

$$J_\ell^\mu := J^\mu - J_5^\mu, \quad \text{and} \quad J_r^\mu := J^\mu + J_5^\mu$$

are anomalous, too, (the signs of their divergence being opposite to one another).

In this paper, we will not make use of anomalous commutators. But they play an important role in the study of transport in dissipationless one-dimensional and three-dimensional conductors; see [7, 8]. Faddeev and collaborators [10] contributed important insights to a clear understanding of the mathematics underlying the chiral anomaly and anomalous commutators.

Contents, Credits and Acknowledgments

This paper is devoted to reviewing efforts to understand novel states of matter that I have personally been involved in and that have spanned more than two decades. I think they have yielded a number of important and elegant results. Of course, numerous other people have contributed important insights towards clarifying issues discussed in this paper, and it is completely impossible to do justice to all their contributions in a short review like this one. Some of the results on the fractional quantum Hall effect described in Sec. 2 overlap with independent ones due to Wen [11].[a] Ideas and results concerning 2D time-reversal invariant topological insulators related to some of the ones described in Sec. 3 of this paper have also been brought forward by Shoucheng Zhang and collaborators and will be referred to

[a] I will not review work on the integer quantum Hall effect in 2D non-interacting electron gases, and I refrain from quoting papers on this effect, because there are too many of them, and they are well known.

later.[b] In Sec. 4, some results on 3D time-reversal invariant topological insulators, including axionic insulators, are reviewed. These results, too, overlap with ones described by Shoucheng Zhang and collaborators.

I am indebted to the following colleagues and friends for collaboration and/or many illuminating discussions on the subject matter covered in this review and for encouragement:

A. Alekseev, S. Bieri, A. Boyarsky, V. Cheianov, G. M. Graf, B. I. Halperin, T. Kerler, I. Levkivskyi, L. Molenkamp, G. Moore, B. Pedrini, O. Ruchayskiy, C. Schweigert, E. Sukhorukov, J. Walcher, Ph. Werner, P. Wiegmann, and A. Zee.

Very special thanks are due to R. Morf, who introduced me to the quantum Hall effect and joined me for countless discussions, and to U. M. Studer and E. Thiran, without whose devoted help and collaboration my efforts to understand aspects of the fractional quantum Hall effect would have failed.

2. Theory of the Fractional Quantum Hall Effect: Anomalous Chiral Edge Currents, Effective Actions, Classification of Hall Fluids

Setup, Basic Quantities

We consider a two-dimensional electron gas (2D EG) forming at the interface between an insulator and a semi-conductor when a gate voltage is applied; see Fig. 1. The gas is confined to a domain Ω in the xy-plane. A uniform magnetic field $\vec{B}_0 \perp \Omega$ is applied to the system. The so-called filling factor

$$\nu := \frac{n}{e|\vec{B}_0|/h\,c},$$

where n is the particle density of the 2D EG, e is the elementary electric charge and c is the speed of light, is chosen such that the longitudinal resistance, R_L, of the gas vanishes. It is known, experimentally, that such filling factors exist, and these findings are supported by numerical simulations. But there are essentially no analytical results that enable us to understand why, at certain values of ν, the longitudinal resistance vanishes (or, put differently, why there is a strictly positive mobility gap above the groundstate energy). Here we will not consider this difficult problem, but see, e.g., [12].[c] Rather, we *assume* that, at certain values of the filling factor, R_L vanishes and then derive non-trivial, mathematically precise consequences of this assumption.

---

[b]However, their work only appeared several years after the publication of the one of my collaborators and myself.

[c]I refer the reader to [13, 14], and references given there, for much useful information on the quantum Hall effect.

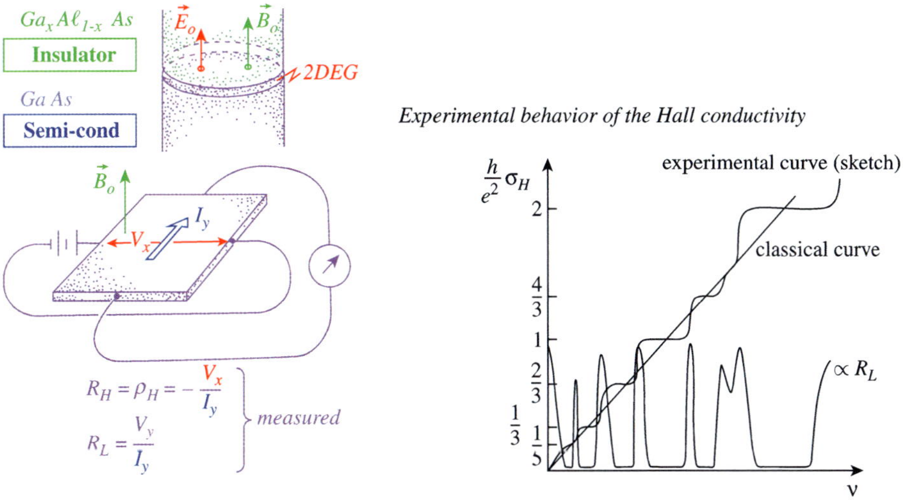

Fig. 1. Schematic representation of the experimental setup and of experimental results on the QHE.

Observations: $R_L = 0 \leftrightarrow (\nu, \sigma_H) \in$ plateau; plateau heights $\in \frac{e^2}{h}\mathbb{Q}$; the cleaner the sample, the more numerous are the observed plateaux and the narrower they are; if $\frac{h}{e^2}\sigma_H \notin \mathbb{Z}$ there appear to exist fractional electric charges.

Applications: Metrology; novel computer memories; topological quantum computing (?) using quasi-particles with non-abelian braid statistics [15].

We first study the response of the 2D EG to small perturbations in the external electromagnetic field, $\vec{E} \| \Omega$ and $\vec{B} \perp \Omega$. We set

$$\vec{B}^{tot} = \vec{B}_0 + \vec{B}, \quad B := |\vec{B}|, \quad \underline{E} := (E_1, E_2).$$

Note that, as long as electron spin is neglected, only the components, $\underline{E}$, of the electric field parallel to the sample plane and the component, B, of the magnetic field perpendicular to the sample plane enter the laws of motion of electrons in the domain Ω. The field tensor of the electromagnetic field on the three-dimensional space-time of the 2D EG is then given by

$$F := \begin{pmatrix} 0 & E_1 & E_2 \\ -E_1 & 0 & -B \\ -E_2 & B & 0 \end{pmatrix}.$$

When interpreted as a 2-form, F is the exterior derivative, $\mathrm{d}A$, of a 1-form, $A = \sum A_\mu \mathrm{d}x^\mu$, where $x = (x^0 \equiv t, \underline{x})$, $\underline{x} = (x^1, x^2) \in \mathbb{R}^2$, and A is the electromagnetic vector potential.

Electrodynamics of 2D Incompressible Electron Gases

The electric current density in the 2D EG is defined by

$$j^\mu(x) = \langle J^\mu(x) \rangle_A, \quad \mu = 0, 1, 2,$$

where $J^\mu(x)$ is the quantum-mechanical vector current density — an operator-valued distribution — and $\langle(\cdot)\rangle_A$ is the state of the EG in the presence of an external electromagnetic field with vector potential A.

We begin by summarizing the basic equations governing the electrodynamics in the bulk of a 2D EG, which, for $\underline{E} = 0$ and $B = 0$, is assumed to be *incompressible*, i.e. $R_L = 0$.

(I) **Hall's Law** (assuming that $R_L = 0$)

$$\underline{j}(x) = \sigma_H(\underline{E}(x))^*, \tag{1}$$

where σ_H is the Hall conductivity, and $\underline{E}^*$ is the vector arising from counter-clockwise rotation of $\underline{E}$ in the (xy-) plane of the sample through an angle of 90°. This equation shows that space reflections in lines, P, and time reversal, T, are *not* symmetries of the gas, i.e. are broken, which is, of course, a consequence of the presence of the external magnetic field $\vec{B}_0 \perp \Omega$.

(II) **Charge Conservation**

$$\frac{\partial}{\partial t}\rho(x) + \underline{\nabla} \cdot \underline{j}(x) = 0. \tag{2}$$

(III) **Faraday's Induction Law**

$$\frac{\partial}{\partial t}B_3^{tot} + \underline{\nabla} \wedge \underline{E}(x) = 0. \tag{3}$$

It follows that

$$\frac{\partial \rho}{\partial t} \overset{(2)}{=} -\underline{\nabla} \cdot \underline{j} \overset{(1)}{=} -\sigma_H \underline{\nabla} \wedge \underline{E} \overset{(3)}{=} \sigma_H \frac{\partial B}{\partial t}. \tag{4}$$

Integrating Eq. (4) in the time variable t, with integration constants chosen as follows

$$j^0(x) := \rho(x) + e \cdot n, \quad B(x) = \left(\vec{B}^{tot}(x) - \vec{B}_0\right) \cdot \vec{e}_3\,,$$

where $\vec{e}_3$ is the unit vector in the z-direction orthogonal to Ω, we find the

(IV) **Chern–Simons Gauss Law**

$$j^0(x) = \sigma_H B(x)\,. \tag{5}$$

Combination of laws (1) and (5) yields

$$\boxed{j^\mu(x) = \tfrac{1}{2}\sigma_H\, \varepsilon^{\mu\nu\lambda} F_{\nu\lambda}(x).} \tag{6}$$

Next, when combining laws (II), (III) and Eq. (6) we seemingly conclude that

$$0 \overset{(2)}{=} \partial_\mu j^\mu \overset{(3),(6)}{=} \frac{1}{2}\varepsilon^{\mu\nu\lambda}(\partial_\mu \sigma_H) F_{\nu\lambda} \neq 0, \tag{7}$$

wherever $\sigma_H \neq const.$, e.g., at $\partial\Omega$. We seem to arrive at a contradiction. Well, actually, there isn't any contradiction! The point is that the current density j^μ appearing in Eq. (6) is the *bulk* current density, (j^μ_{bulk}), which is *not* the conserved *total* electric current density, j^μ_{tot}:

$$j^\mu_{tot} = j^\mu_{bulk} + j^\mu_{edge}, \quad \text{with} \quad \partial_\mu j^\mu_{tot} = 0, \tag{8}$$

where j^μ_{edge} is the "edge current density", a current distribution with support on all lines in Ω across which the value of σ_H jumps; in particular, supp(j^μ_{edge}) contains the boundary/edge, $\partial\Omega$, of the gas. In general, only j^μ_{tot} is conserved, but

$$\partial_\mu j^\mu_{bulk}, \partial_\mu j^\mu_{edge} \neq 0,$$

as follows from Eqs. (7) and (8).

Anomalous Chiral Edge Currents

The edge current density is defined in such a way that

$$\text{supp } j^\mu_{edge} = \text{supp}(\underline{\nabla}\sigma_H) \supseteq \partial\Omega, \quad \underline{j}_{edge} \perp \underline{\nabla}\sigma_H.$$

One may view it as an instance of "*holography*" that, on supp($\underline{\nabla}\sigma_H$),

$$\boxed{\partial_\mu j^\mu_{edge} \overset{(8)}{=} -\partial_\mu j^\mu_{bulk}|_{\text{supp}(\underline{\nabla}\sigma_H)} \overset{(6)}{=} -\sigma_H E_\parallel|_{\text{supp}(\underline{\nabla}\sigma_H)},} \tag{9}$$

where $E_\parallel$ is the component of the electric field $\vec{E}|_{\text{supp}(\underline{\nabla}\sigma_H)}$ parallel to the plain of the sample and perpendicular to $\underline{\nabla}\sigma_H$. Equation (9) is the chiral anomaly in $1+1$ dimensions!

The edge current, $j^\mu_{edge} \equiv j^\mu_{\ell/r}$, is an anomalous chiral current in $1+1$ dimensions: At the edge of the sample, the Lorentz force acting on the electrons must be cancelled by the force confining the electrons to the interior of the domain Ω. Hence

$$\frac{e}{c}|B^{tot} v_\parallel| = |(\underline{\nabla}V_{edge})|,$$

where $v_\parallel$ is the propagation speed of an electron moving along $\partial\Omega$, and V_{edge} is the potential of the force confining the electrons to the interior of Ω. The chiral nature of j^μ_{egde} becomes obvious from Fig. 2.

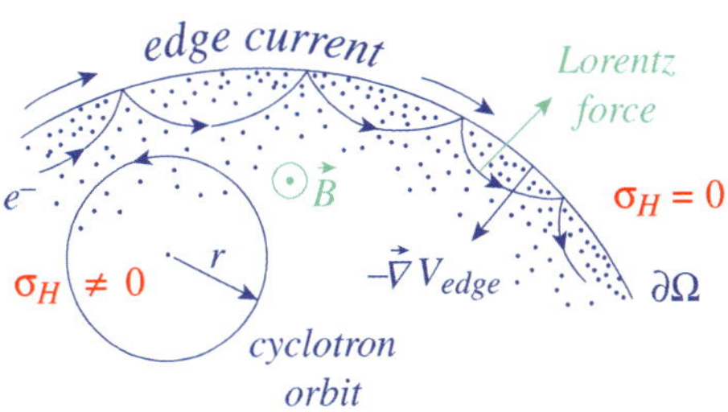

Fig. 2. Skipping orbits of electrons moving along the boundary/edge of a 2D electron gas confined to a disk — electrons near $\partial\Omega$ perform a chiral motion.

Classically, the orbits of electrons moving close to the edge $\partial\Omega$ of the sample are so-called "skipping orbits" which describe a chiral motion of electrons, the chirality being determined by the direction of $\vec{B}_0$.

There is an analogous phenomenon in classical physics, namely, in the theory of hurricanes: Under the replacements

$$\vec{B} \to \vec{\omega}_{\text{earth}} \text{ (angular velocity)}, \quad \text{Lorentz force} \to \text{Coriolis force},$$

$$V_{edge} \to \text{air pressure},$$

the theory of charge transport in a 2D EG close to the edge of the sample is mapped to the theory of air-transport in a hurricane near the surface of the Earth. Actually, one may view a hurricane as an example of a "Hall effect" in a rotating gas. It is expected that a quantized Hall effect is encountered in a rapidly rotating layer of a superfluid Bose gas; (see last subsection of Sec. 1).

Recall that the chiral anomaly in $(1+1)$D says that (for $j^\mu_{edge} = j^\mu_\ell$)

$$\partial_\mu j^\mu_{edge} = -\frac{1}{h}\left(\sum_\alpha q_\alpha^2\right) E_\| \overset{(9)}{\Rightarrow} \boxed{\sigma_H = \frac{1}{h}\sum_\alpha q_\alpha^2\,,} \tag{10}$$

where $q_\alpha = e\widetilde{Q}_\alpha$ is the electric charge of an edge current corresponding to a counterclockwise-chiral edge mode α; (similar contributions come from clockwise modes, but with a *reversed sign* on the right-hand side in Eq. (10)). Equation (10) yields a precise expression for the Hall conductivity σ_H in terms of charges of chiral currents circulating at the edge of Ω. The integer quantum Hall effect is observed if the charge quantum numbers $\widetilde{Q}_\alpha = \pm 1$, for all α. The observation of a system in a *fractional* quantum Hall state then implies that there are currents carried by chiral modes with *fractional* charge circulating at the edge of the system.

Apparently, starting from the basic equations of the electrodynamics of a 2D EG in an external electromagnetic field, we are led to rediscover Halperin's edge currents [16]; (see also [17]).

Edge and Bulk Effective Actions

Equation (10) shows that if $\sigma_H \notin \frac{e^2}{h}\mathbb{Z}$ then there must exist *fractionally charged* modes (quasi-particles) propagating along supp($\underline{\nabla}\sigma_H$) and, in particular, along $\partial\Omega$. The quantum-mechanical chiral edge current density J^μ_{edge} turns out to generate a $U(1)$-current algebra that can be described in terms of free massless chiral Bose fields, φ^α. The Green functions of J^μ_{edge} can be derived from the anomalous chiral action, $\Gamma_{\partial\Omega\times\mathbb{R}}(A_\|)$, in $1+1$ dimensions — see Eq. (12), below — where $A_\|$ is the restriction of the vector potential A to the boundary, $\partial\Omega \times \mathbb{R}$, of the space-time of the 2D EG. The action $\Gamma_{\partial\Omega\times\mathbb{R}}(A_\|)$ is anomalous in the sense that it is *not* invariant under gauge transformations of $A_\|$.

The anomaly of $\Gamma_{\partial\Omega\times\mathbb{R}}(A_\|)$ is a consequence of the fact that the chiral edge current J^μ_{edge} is *not* conserved, i.e. is anomalous. The anomaly of the edge current must be cancelled by the anomaly of the bulk current J^{μ}_{bulk}; see Eq. (9). The generating function of the Green functions of the bulk current density is the *bulk effective action*, $S_{\Omega\times\mathbb{R}}(A)$, which we will determine presently. This action is not invariant under gauge transformations of A either. Its variation under gauge transformations of A must be cancelled by the gauge variation of an appropriate multiple of the anomalous chiral action $\Gamma_{\partial\Omega\times\mathbb{R}}(A_{\|})$.

Next, we derive the expression for the bulk effective action. By the definition of effective actions, the bulk current is given by the variational derivative of $S_{\Omega\times\mathbb{R}}(A)$ with respect to A. Thus,

$$\begin{aligned} j^{\mu}_{bulk}(x) &= \langle J^{\mu}_{bulk}(x)\rangle_A \equiv \frac{\delta S_{\Omega\times\mathbb{R}}(A)}{\delta A_{\mu}(x)} \\ &\overset{(6)}{=} \frac{1}{2}\sigma_H \varepsilon^{\mu\nu\lambda}F_{\nu\lambda}(x), \quad x \notin \partial\Omega\times\mathbb{R}\,. \end{aligned}$$

It follows that

$$\boxed{S_{\Omega\times\mathbb{R}}(A) = \frac{\sigma_H}{4}\int_{\Omega\times\mathbb{R}} A\wedge \mathrm{d}A.} \tag{11}$$

Note that the action $S_{\Omega\times\mathbb{R}}(A)$ is *not* invariant under gauge transformations of A *not* vanishing at the boundary, $\partial\Omega\times\mathbb{R}$, of the space-time of the sample. The *total effective action* is obtained by adding to the anomalous Chern–Simons action in Eq. (11) the anomalous chiral action $\Gamma_{\partial\Omega\times\mathbb{R}}(A_{\|})$ (see Eq. (12), below), multiplied by a coefficient chosen in such a way that the total action is gauge-invariant. In fact, the coefficient in front of $\Gamma_{\partial\Omega\times\mathbb{R}}(A_{\|})$ must be equal to σ_H! This observation implies that the *bulk definition* of the Hall conductivity coincides with its *edge definition*; a conclusion that first appeared in [2, 4].

Finally, we present an expression for the anomalous chiral action $\Gamma_{\partial\Omega\times\mathbb{R}}$ of a single chiral matter field on $\partial\Omega\times\mathbb{R}$ coupled to a vector potential a:

$$\Gamma_{\partial\Omega\times\mathbb{R}}(a) := \frac{1}{2}\int_{\partial\Omega\times\mathbb{R}} \left[a_+a_- - 2a_{\pm}\frac{\partial^2_{\mp}}{\Box}a_{\pm}\right]\mathrm{d}^2u, \tag{12}$$

where $a = a_+\mathrm{d}u^+ + a_-\mathrm{d}u^- \equiv A_{\|}$, u^+ and u^- are "lightcone coordinates" on the boundary $\partial\Omega\times\mathbb{R}$ of the space-time of the sample, and "+" $\leftrightarrow$ "counterclockwise", "−" $\leftrightarrow$ "clockwise".

Exercise: Check that the anomaly of the (bulk) Chern–Simons action $S_{\Omega\times\mathbb{R}}(A)$ in (11), which is a boundary term, is cancelled by the one of $\sigma_H\,\Gamma_{\partial\Omega\times\mathbb{R}}(A_{\|})$.

Whatever we have said about 2D electron gases in a homogeneous external magnetic field exhibiting the quantum Hall effect can be extended to so-called *Chern insulators* [18] (see also [2]), which have the property that reflections in lines ("parity") and time reversal are broken *even* in the absence of an external magnetic field, (e.g., because of magnetic impurities in the bulk). An experimental realization of Haldane's model [18] has been described in [19]. The low-energy physics of quasi-particles in the bulk of a Chern insulator may resemble the one of two-component relativistic Dirac fermions coupled to the electromagnetic vector potential, with an effective action given by (11), with $A = A_{\mathrm{tot}}$, and $\sigma_H = e^2/2h$ (= Chern number of a certain vector bundle of Dirac fermion wave functions over the Brillouin zone $\mathbb{T}^2$).

Classification of "Abelian" Hall Fluids and Chern Insulators

Next, I sketch a general classification [5] of 2D insulators with broken time reversal and parity in topologically protected states (i.e. groundstates of many-body Hamiltonians with a strictly positive mobility gap above the groundstate energy) exhibiting quasi-particles with *abelian braid statistics* [15]; ("non-abelian states" are discussed in [20]). Explicit results will be limited to Hall fractions in the interval $(0, 1]$.

In the following, J_{tot} denotes the *total* electric current density (bulk + edge), which is conserved, i.e. $\partial_\mu J^\mu_{tot} = 0$. We consider a general ansatz:

$$J_{tot} = \sum_{\alpha=1}^{M} Q_\alpha J_\alpha,$$

where the densities J_α are *separately conserved* current densities corresponding to different quasi-particle species, and the coefficients $Q_\alpha \in \mathbb{R}$ are "charges". (In the simple case of the integer quantum Hall effect for non-interacting electrons, α labels filled Landau levels of which there are M, and $Q_\alpha = 1, \forall \alpha$.) On a three-dimensional, simply connected space-time $\Lambda = \Omega \times \mathbb{R}$, a conserved current density J can be derived from a vector potential: If $\mathcal{J}$ denotes the 2-form dual to J then conservation of J implies that $\mathrm{d}\mathcal{J} = 0$, and hence

$$\mathcal{J} = \mathrm{d}B,$$

by Poincaré's lemma, where the 1-form B is the vector potential of J and is determined by the current density up to the gradient of a scalar function β; i.e. B and $B + \mathrm{d}\beta$ yield the same current density J.

Chern–Simons Effective Action of Conserved Currents

Henceforth we use units where $\frac{e^2}{h} = 1$. For a 2D insulator (i.e. a state of matter protected by a mobility gap above the groundstate energy), the field theory of the conserved currents $(J_\alpha)_{\alpha=1}^{M}$ in the limit of very large distance scales and very low frequencies must be *topological*: If parity and time reversal are broken the "most relevant" contribution to the action of the currents J_α is the Chern–Simons action

$$S_\Lambda(\underline{B}, A) := \sum_{\alpha=1}^{M} \int_\Lambda \{B_\alpha \wedge \mathrm{d}B_\alpha + A \wedge Q_\alpha \mathrm{d}B_\alpha\} + \text{boundary terms}, \qquad (13)$$

where A is the electromagnetic vector potential, and the boundary terms must be added to cancel the anomalies of the Chern–Simons terms (1st term on right-hand side) in (13) under the gauge transformations $B \to B + \mathrm{d}\beta$ and $A \to A + \mathrm{d}\chi$.

Carrying out the Gaussian functional integrals over the fields $\left(B_\alpha\right)_{\alpha=1}^{M}$, we find that

$$\int \exp(iS_\Lambda(\underline{B}, A)) \prod_{\alpha=1}^{M} \mathcal{D}B_\alpha = \exp\left(i\frac{\sigma_H}{4}\int_\Lambda A \wedge \mathrm{d}A + \sigma_H \Gamma_\Lambda(A_\|)\right), \tag{14}$$

where

$$\sigma_H = \sum_{\alpha=1}^{M} Q_\alpha^2. \tag{15}$$

Physical excitations (states) of a topological Chern–Simons theory are labeled by networks of Wilson lines contained in the half-space $\Lambda_- = \Omega \times \mathbb{R}_-$ corresponding to negative values of the time variable and ending in points of the spatial surface $\Omega \times \{0\}$. Scalar products of physical states of the Chern–Simons theory with action given by (13) can be calculated by inserting into the Gaussian functional integral on the left-hand side of Eq. (14) networks of Wilson lines contained in Λ_- hooked up to reflected networks of Wilson lines, contained in $\Lambda_+ = \Omega \times \mathbb{R}_+$, at common intersection points in $\Omega \times \{0\}$, so as to yield a network of *gauge-invariant* Wilson networks/loops. The Hamiltonian of Chern–Simons theory vanishes; bulk excitations are static, reflecting the feature that there is a positive mobility gap above the groundstate energy.

Classification of 2D "Abelian" Topological Insulators with Broken Time Reversal — Bulk Degrees of Freedom

The operator measuring the electric charge of an excitation of the Chern–Simons theory with action given by (13) contained in a region $\mathcal{O}$ of space, $\Omega \times \{0\}$, is given by

$$Q_{\mathcal{O}} = \int_{\mathcal{O}} J^0(x)\mathrm{d}^2x = \sum_{\alpha=1}^{M} Q_\alpha \int_{\partial\mathcal{O}} B_\alpha.$$

The electric charge deposited inside the domain $\mathcal{O}$ by excitations corresponding to a network of Wilson lines can be inferred by inserting, besides the Wilson network (hooked up to its reflection at $\Omega \times \{0\}$), a Wilson loop

$$\exp\left(i\, t \sum_{\alpha=1}^{M} Q_\alpha \int_{\partial\mathcal{O}} B_\alpha\right), \quad t \in \mathbb{R},$$

into the functional integral; (see Eq. (14)). If a network of Wilson lines creates an excitation describing n electrons or holes located inside $\mathcal{O}$ from the groundstate of the insulator then its electric charge, as measured by the operator $Q_{\mathcal{O}}$, is given by $-n+2k$, $k = 1, \ldots, n$, where k is the number of holes. If n is odd this excitation must have Fermi–Dirac statistics, if n is even it must have Bose–Einstein statistics. This *relation between the electric charge of an excitation and its statistics* implies that the M-tuples of quantum numbers assigned to lines in Wilson networks creating multi-electron-hole excitations must be sites of an *odd-integral lattice*, Γ, of rank M, and that the "co-vector" $Q := (Q_1, \ldots, Q_M)$ must belong to the dual lattice Γ^*, (in fact, Q must be a so-called "visible vector" in Γ^*); see [5, 6]. Hence

$$\sigma_H = \sum_{\alpha=1,\ldots,M} Q_\alpha^2 \in \mathbb{Q},$$

i.e. the Hall conductivity σ_H is a *rational multiple* of $\frac{e^2}{h}$. It takes some more effort to show that the so-called Witt sublattice of Γ is an A-, D- or E-root lattice; see [5, 6].

Classification of 2D "Abelian" Topological Insulators with Broken Time Reversal — Edge Degrees of Freedom

The chiral anomaly in the form of Eq. (10) reflects the feature that there are several (N) species of gapless quasi-particles propagating either clockwise or counterclockwise along the edge $\partial\Omega$ of the 2D EG. The dynamics of these quasi-particles is described by N massless chiral scalar Bose fields $\{\varphi^\alpha\}_{\alpha=1}^N$ with propagation speeds $\{v_\alpha\}_{\alpha=1}^N$; (we choose a basis, $\left(\varphi^\alpha\right)_{\alpha=1}^N$, in field space such that the velocity matrix $\left(v_\alpha\right)_{\alpha=1}^N$ is *diagonal*). We then find that:

1. The chiral electric edge current operator and the Hall conductivity are given by

$$J^\mu_{edge} = e\sum_{\alpha=1}^{N} \widetilde{Q}_\alpha \partial^\mu \varphi^\alpha, \quad \widetilde{Q} = (\widetilde{Q}_1, \ldots, \widetilde{Q}_N), \quad \text{and} \quad \sigma_H = \frac{e^2}{h}\widetilde{Q}\cdot\widetilde{Q}^T,$$

   respectively.
2. Multi-electron/hole states localized along the edge of the 2D EG are created by vertex operators

$$:\exp i\left(\sum_{\alpha=1}^{N} q_\alpha \varphi^\alpha\right):, \quad q = \begin{pmatrix} q_1 \\ \vdots \\ q_N \end{pmatrix} \in \widetilde{\Gamma}, \tag{16}$$

   where the double colons indicate Wick ordering, and $\widetilde{\Gamma}$ is a lattice. The electric charge carried by the vertex operator in (16) is given by

$$e\,\widetilde{Q}\cdot q.$$

   By the same reasoning process as above, the relation between the charge of an edge excitation created by the vertex operator in (16) and its statistics implies that the "quantum numbers" q must correspond to sites in an *odd-integral lattice*, $\widetilde{\Gamma}$, of rank N. The Witt sublattice of $\widetilde{\Gamma}$ is an A-, D- or E-root lattice and gives rise to an $\hat{A}$-, $\hat{D}$- or $\hat{E}$-Kac–Moody current algebra at level 1 acting as a quantum symmetry on the edge states of the 2D electron gas. The matrix

$$q^* := \left(q_\alpha^{*\,k}\right)_{\alpha,k=1,\ldots,N}$$

where, for all k, the vector q^{*k} determines a vertex operator creating a one-electron state propagating along the edge of the 2D EG, is a close analogue of the Cabibbo–Kobayashi–Maskawa matrix of the weak sector of the Standard Model.

We conclude that:

3. The classifying data for the edge degrees of freedom of a 2D EG exhibiting the quantum Hall effect are given by
$$\{\widetilde{\Gamma};\ \widetilde{Q} \in \widetilde{\Gamma}^*(\text{“visible”});\ q^*;\ v = (v_\alpha)_{\alpha=1}^N\}. \tag{17}$$
The charged quasi-particles predicted by these data exhibit *abelian braid statistics* and transform under $\hat{A}$-, $\hat{D}$- or $\hat{E}$-Kac–Moody symmetries at level 1.

For “ideal” Hall insulators, one expects that $\widetilde{\Gamma} = \Gamma$ and $\widetilde{Q} = Q$; (see previous subsection on bulk degrees of freedom). But, in general, the two lattices Γ and $\widetilde{\Gamma}$ need not be identical. This can happen when the edge of a 2D incompressible EG has a layered structure. However, in any case, the constraint that the edge conductivity $\frac{e^2}{h}\widetilde{Q}\cdot\widetilde{Q}^T$ is equal to the bulk Hall conductivity $\frac{e^2}{h}Q\cdot Q^T$ must *always* be satisfied, as required by gauge invariance.

A (partial) classification of 2D electron gases with quasi-particles exhibiting *non-abelian* braid statistics and quantized Hall conductivity has been presented in [20].

It is an interesting, partly open problem to classify edge degrees of freedom localized on interfacial lines at which two different electron gases, with non-vanishing Hall conductivities $\sigma_H^{(1)} \neq \sigma_H^{(2)} \neq 0$, join.

Success of Classification, Comparison with Data

In [5, 6], we have given a list of “physically plausible” odd-integral lattices Γ and visible vectors $Q \subset \Gamma^*$ with the property that $0 < (\frac{e^2}{h})^{-1}\sigma_H \leq 1$. We write
$$\left(\frac{e^2}{h}\right)^{-1}\sigma_H = \frac{n_H}{d_H},$$
where n_H and d_H are relatively prime integers, with $n_H \leq d_H$. Plotting d_H vertically (downwards) and $(\frac{e^2}{h})^{-1}\sigma_H$ horizontally (to the right), we have obtained the table displayed in Fig. 3, below, comparing data on 2D electron gases exhibiting the quantum Hall effect observed in laboratories with our theoretical predictions. This figure is taken from a plenary lecture I presented at the 1994 International Congress of Mathematicians in Zurich [21].[d]

Explanation of Fig. 3: Experimentally well-established Hall fractions are indicated by a bullet “•”. These are fractions for which the longitudinal resistance, R_L, has a minimum very close to 0 and R_H shows a plateau. Fractions for which R_L has a minimum and R_H shows an inflection are indicated by a circle “∘”. If

[d] On this occasion, I was kindly introduced to the audience by Ludvig Faddeev.

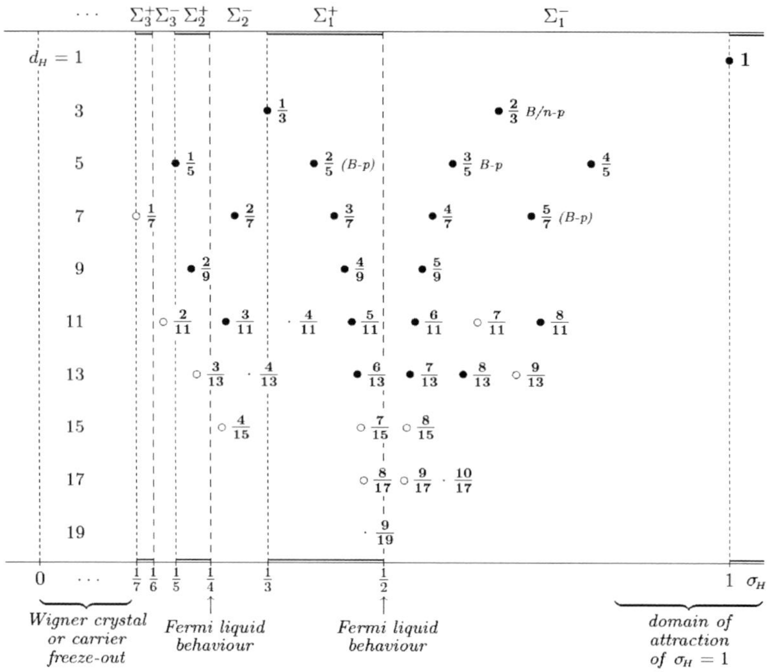

Fig. 3. Observed Hall fractions $\sigma_H = n_H/d_H$ in the interval $0 < \sigma_H \leq 1$ and their experimental status in single-layer 2D electron gases exhibiting the quantum Hall effect.

there are only weak experimental indications for a Hall insulator state or if the data are controversial the symbol "·" is used. *All* bullets and circles in this figure, and *more* (!), are predicted theoretically; see [5, 6].

Hall fractions at which a magnetic field (B) and/or density-driven transition from one insulator state to another one (at a constant value of σ_H) has been observed are marked by the letters "$B/n - p$", "$B - p$". In the interval $[\frac{1}{3}, 1] = \Sigma_1^+ \cup \Sigma_1^-$, the agreement between our theory, namely, the classification of 2D "abelian" Hall insulator states, as presented above, and experimentally observed Hall fractions, including the prediction of transitions at a constant value of σ_H, is nearly perfect. (Only the fractions $\frac{8}{13}$ and $\frac{10}{17}$ are somewhat uncertain.) Furthermore, there are (abelian and *non-abelian*) states at $\sigma_H = \frac{1}{4}, \frac{1}{2}$ in double-layer electron gases and at $\sigma_H = 1$; see [20]. Predictions of plateaux in the intervals $\Sigma_p^+ \cup \Sigma_p^-, p = 2, 3, \ldots$, are related to those in the interval $\Sigma_1^+ \cup \Sigma_1^-$ by the so-called "*shift map*"; see [6, 20]. For a detailed discussion of the shift map and of the Kac–Moody symmetries of Hall insulators alluded to above, with $\sigma_H < 1$ (as indicated in Fig. 3), see [5, 6, 20].

Rotating Bose Gases, and a Duality between Photonics and Electronics

We conclude our discussion of Hall and Chern insulators with some remarks.

Remark 1. A discussion of the relevance of the gravitational anomaly and of some new invariants for the theory of the quantum Hall effect (in a spirit somewhat related to the one of this review) can be found in recent papers by P. Wiegmann and collaborators [22].

Remark 2. A very rapidly rotating 2D layer of a superfluid Bose gas of atoms (with "hard-core repulsion" between atoms) may settle in a state analogous to a Hall insulator. The centrifugal force acting on the bosonic atoms in the gas may be precisely cancelled by a harmonic potential trapping the gas, and the Coriolis force then plays the role of the Lorentz force acting on the electrons in a 2D EG. The energy levels of a bosonic atom precisely correspond to the Landau levels of an electron in a 2D EG. This effect has first been sketched in [4, 6] and discussed in considerably more detail in [23] — among other papers. The basic formalism can be found, e.g., in [24].

Remark 3. In two space dimensions, the electromagnetic vector potential A is *"dual"* to the vector potential B ($:= \sum_\alpha Q_\alpha B_\alpha$) of the conserved electromagnetic current density J; (see Eqs. (13) and (14)). Under the replacements

$$A \mapsto B, \quad B \mapsto A,$$

conventional 2D *insulators* with time-reversal symmetry are mapped to 2D *superconductors*, and conversely, as discussed in [6]. Furthermore, electronic Hall insulators are mapped to 2D *gapped photonic wave guides* exhibiting extended electromagnetic surface waves. Assuming that time reversal is *not* a symmetry of the system, the leading terms in the action of the electromagnetic field in such a wave guide are given by

$$S_\Lambda^{WG}(A) = \frac{1}{2e^2}\int_\Lambda \{\underline{E}(x)\cdot\varepsilon\underline{E}(x) - \mu^{-1}B^2(x)\}\mathrm{d}^3x$$
$$+ \frac{\sigma_H}{4}\int_\Lambda A\wedge \mathrm{d}A + \sigma_H\Gamma_\Lambda(A_\parallel), \tag{18}$$

where ε is the dielectric tensor of the wave guide and μ is its magnetic permeability. This is the action functional of *massive* QED in 2+1 space-time dimensions; see [25]. The Chern–Simons term, $\propto \int_\Lambda A\wedge \mathrm{d}A$, on the right-hand side of Eq. (18) arises from coupling the electromagnetic vector potential A to the electric current density $\mathcal{J} = \mathrm{d}B$, as in

$$\int_\Lambda A\wedge \mathrm{d}B,$$

assuming that the action functional for the vector potential B of the current density is given, in the simplest instance, by

$$S_\Lambda(B) = \frac{1}{\sigma_H}\int_\Lambda B\wedge \mathrm{d}B + \text{boundary action} + \text{l.r. terms},$$

as appropriate for an insulator with broken time reversal, (see Eq. (13), with $N = 1$); the "boundary action" (see (12)) is there to restore gauge invariance, i.e. invariance under $B \mapsto B + \mathrm{d}\beta$, of $S_\Lambda(B)$, and "l.r. terms" stands for "less (infrared) relevant terms" in the action functional. We then find the response equations

$$\begin{aligned}\langle F_{\mu\nu}(x)\rangle_B &= \varepsilon_{\mu\nu\rho}\frac{\delta S_\Lambda(B)}{\delta B_\rho(x)} \\ &= \sigma_H^{-1}\varepsilon_{\mu\nu\rho}j^\rho(x)\,, \end{aligned} \tag{19}$$

with $\varepsilon_{\mu\nu\rho}j^\rho = (\mathrm{d}B)_{\mu\nu}$.

Using arguments similar to those presented in connection with Eqs. (7) and (9), we find that quantized electromagnetic waves can propagate chirally and essentially without dispersion along the edge of such a wave guide.

3. Chiral Spin Currents in Planar Time-Reversal Invariant Topological Insulators

So far, we have not paid attention to electron spin, although there are actually 2D EG exhibiting the fractional quantum Hall effect where electron spin must be taken into account; see [6]. But it would take us too far off the main line of this paper to discuss this issue here.

In the theory of 2D *time-reversal invariant topological insulators*, which we will sketch next, electron spin undoubtedly plays a crucial role! In order to prepare the ground for an account of this theory, we briefly review the Pauli Equation for a non-relativistic spinning electron in an arbitrary external electromagnetic field:

$$i\hbar D_0\Psi_t = -\frac{\hbar^2}{2m}\vec{D}\cdot\vec{D}\,\Psi_t, \tag{20}$$

where m is the (effective) mass of an electron,

$$\Psi_t(\vec{x}) = \begin{pmatrix}\psi_t^\uparrow(\vec{x}) \\ \psi_t^\downarrow(\vec{x})\end{pmatrix} \in L^2(\mathbb{R}^3)\otimes\mathbb{C}^2$$

is a two-component Pauli spinor describing the state of an electron located at the point $\vec{x}$ at time t, and the operators D_0 and $\vec{D} = (D_1, D_2, D_3)$ are covariant derivatives given by

$$i\hbar D_0 = i\hbar\partial_t + e\varphi - \underbrace{\vec{W}_0\cdot\vec{\sigma}}_{\text{Zeeman coupling}}\quad, \quad \vec{W}_0 = \mu c^{-1}\vec{B} + \frac{\hbar}{4}\vec{\nabla}\wedge\vec{V}, \tag{21}$$

where φ is the electrostatic potential, $\vec{\sigma} = (\sigma_1, \sigma_2, \sigma_3)$ is the vector of Pauli matrices, μ is the Bohr magneton, $\vec{B}$ is the magnetic field and $\vec{V}$ is the velocity field generating an incompressible (i.e. $\vec{\nabla}\cdot\vec{V} = 0$) classical motion/flow of the system (relative to the inertial laboratory frame). Furthermore,

$$\frac{\hbar}{i} D_k = \frac{\hbar}{i} \partial_k + eA_k - mV_k - \vec{W}_k \cdot \vec{\sigma}, \tag{22}$$

$k = 1, 2, 3$, where $\vec{A} = (A_1, A_2, A_3)$ is the electromagnetic vector potential, and $\vec{W}_k$ is given by

$$\vec{W}_k \cdot \vec{\sigma} := \underbrace{\left[\left(-\tilde{\mu}\vec{E} + \frac{\hbar}{c^2}\dot{\vec{V}}\right) \wedge \vec{\sigma}\right]_k}_{\text{spin-orbit interactions}}, \qquad \text{with } \tilde{\mu} = \mu + \frac{e\hbar}{4mc}, \tag{23}$$

describing spin-orbit interactions and Thomas precession.[e] Note that the 2×2 matrices $\vec{W}_0 \cdot \vec{\sigma}$ and $\vec{W}_k \cdot \vec{\sigma}$, $k = 1, 2, 3$, determine an $SU(2)$-gauge field/connection on the bundle of two-component Pauli spinors over space-time, and that the Pauli equation (20) obeys $U(1)_{em} \times SU(2)_{spin}$-*gauge invariance.*

For a detailed discussion of the Pauli equation and its gauge invariance, as well as plenty of physical consequences thereof, see [4, 6].

Effective Action of a 2D Time-Reversal Invariant Topological Insulator

Next, we consider a 2D gas of interacting electrons confined to a domain Ω in the xy-plane, assuming that $\vec{B} \perp \Omega$ and $\vec{E}$, $\vec{V} \| \Omega$. Then the $SU(2)$-connection, $\vec{W}_\mu$, is given by $W^3_\mu \cdot \sigma_3$, with $W^K_\mu \equiv 0$, $K = 1, 2$, for $\mu = 0, 1, 2$.

Thus the connection determining parallel transport of the component $\psi^\uparrow$ of a Pauli spinor Ψ is given by $a + w$, while parallel transport of $\psi^\downarrow$ is determined by $a - w$, with $a_0 = -e\varphi$, $a_k = -eA_k + mV_k$, and $w_\mu = W^3_\mu$. These connections are *abelian*, i.e. just affect the phases of $\psi^\uparrow$ and $\psi^\downarrow$, respectively. Under time reversal, the components of these connections transform as follows:

$$a_0 \to a_0, \quad a_k \to -a_k, \quad \text{but} \quad w_0 \to -w_0, \quad w_k \to w_k. \tag{24}$$

The term in the effective action of a 2D insulator with the smallest scaling dimension, i.e. the term dominating the large-scale physics, is a Chern–Simons term. If there were only the gauge field a, with $w \equiv 0$, or only the gauge field w, with $a \equiv 0$, a Chern–Simons term would *not* be invariant under time reversal, and the dominant term in the effective action would be given by

$$S(a) = \int_{\Omega \times \mathbb{R}} \mathrm{d}^3x \{\underline{E} \cdot \varepsilon \underline{E} - \mu^{-1} B^2\}, \tag{25}$$

with ε the dielectric tensor and μ the magnetic permeability. This is the effective action of a conventional time-reversal invariant insulator. But, in the presence of

[e] One should actually also include the "spin connection" of the sample geometry in the definition of the $SU(2)$-gauge field $\vec{W}_\mu$; see [4].

electron spin and hence of *two* gauge fields, a and w, satisfying (24), the dominant term is a time-reversal invariant combination of two Chern–Simons terms:

$$S(a,w) = \frac{\sigma}{4}\int \{(a+w)\wedge \mathrm{d}(a+w) - (a-w)\wedge \mathrm{d}(a-w)\}$$
$$= \frac{\sigma}{2}\int \{a\wedge \mathrm{d}w + w\wedge \mathrm{d}a\}. \qquad (26)$$

This action reproduces an action similar to the one in (25) if the connection w is chosen as in (21) and (22). However, the action (26) describes the response of a topological insulator for an *arbitrary* choice of the gauge field $w_\mu = W^3_\mu$. The gauge fields a and w transform *independently* under gauge transformations, and the Chern–Simons action $S(a,w)$ in Eq. (26) is *anomalous* under gauge transformations of $a \pm w$ *not* vanishing at the boundary, $\partial\Lambda$, of a 2D sample with space-time given by $\Lambda = \Omega \times \mathbb{R}$.

Edge Degrees of Freedom: Spin Currents

The anomalous chiral boundary action

$$\sigma\{\Gamma((a+w)|_{||}) - \Gamma((a-w)|_{||})\}, \qquad (27)$$

cancels the anomaly of the bulk action $S(a,w)$ in (26). The two terms in the boundary action are the generating functionals of connected Green functions of two *counter-propagating* chiral electric edge currents, one carried by electrons with "spin-up", the other one carried by electrons with "spin-down", generating two $U(1)$-current algebras. Thus, whenever the edge degrees of freedom support a current then a net *chiral edge spin current*, s^3_{edge}, circulates at the boundary of the sample.

Bulk Response Equations of a 2D Time-Reversal Invariant Topological Insulator

From the Chern–Simons action $S(a,w)$ in Eq. (26) one easily derives the following response equations in the bulk of a 2D topological insulator with unbroken time-reversal invariance:

$$\underline{j}(x) = \sigma(\underline{\nabla} B)^*, \quad \text{and} \quad s^\mu_3(x) = \frac{\delta S(a,w)}{\delta w_\mu(x)} = \sigma\varepsilon^{\mu\nu\lambda}F_{\nu\lambda}(x), \qquad (28)$$

where $\underline{j}$ is the electric current density in the bulk of the insulator, and s^μ_3 is the bulk spin current density, (more precisely its 3-component in spin space); compare to Eqs. (1) and (6). By arguments analogous to those in Eqs. (7) through (9), we see that Eq. (28) implies that the spin current s^μ_3 is anomalous and that its anomaly is cancelled by a chiral egde spin current.

The Chern–Simons effective action in (26), the response equations in (28), chiral spin currents and the role of spin-orbit interactions in connection with the presence of a non-trivial gauge field w have first been described, among many other effects, in [4], in 1993; see also [6].

Quasi-Particles in 2D Time-Reversal Invariant Topological Insulators

We should ask what kinds of quasi-particles may induce the effective Chern–Simons actions

$$S_{\pm}(a \pm w) = \pm\frac{\sigma}{4}\int (a \pm w) \wedge \mathrm{d}(a \pm w) \pm \sigma\Gamma\big((a \pm w)|_{\|}\big),$$

see Eq. (26), where "+" stands for "spin-up" and "−" stands for "spin-down". It is well known that a two-component relativistic Dirac fermion with mass $M > 0$ ($M < 0$) breaks time-reversal invariance and — when coupled to an abelian gauge field, A — induces a Chern–Simons term

$$\underset{(-)}{+}\frac{1}{4\pi}\int A \wedge \mathrm{d}A$$

in the effective action; see [25]. Thus, a 2D time-reversal invariant topological insulator with chiral edge spin currents might exhibit two species of charged quasi-particles in the bulk, with one species (spin-up) related to the other one (spin-down) by time reversal, each species having two degenerate states per wave vector and mimicking, at small wave vectors, a 2-component Dirac fermion. The value of the constant σ is then determined!

Experiments, Other Approaches, Chiral Spin Liquids

Materials with properties similar to those predicted by the theory described above have been studied experimentally in the group of L. Molenkamp in Würzburg [26]. More detailed theoretical aspects have been discussed in [28, 29]. However, because of external electromagnetic fields violating the assumption that $\vec{B} \perp \Omega$ and $\vec{E}\|\Omega$, and because of magnetic impurities in the material, the property that $\vec{W}_\mu = (0, 0, W_\mu^3)$ will usually not hold exactly, and hence the spin current s_3^μ is *not* conserved; (although $\vec{s}^\mu$ is always "*covariantly* conserved", [4, 6]). There then exist spin-umklapp processes that cause transitions between the two chiralities of edge currents and hide the quantization of the value of σ. For these reasons, one cannot expect that real materials exactly reproduce the theoretical predictions of the theory sketched above.

One may, however, expect that there *are* materials with edge currents, j_ℓ and j_r, of opposite chirality related to each other by time reversal that do *not* mix. The current j_ℓ may couple to a chiral abelian gauge field a_ℓ and the current j_r to a gauge field a_r. The effective actions of a_ℓ and a_r are given by $(\pm 1) \times$ the anomalous action in (12). The two gauge fields may have the form

$$a_{\ell/r} = a \pm w,$$

where w is a Berry connection. The transformation of a and w under time reversal is the one given in (24). The theory sketched above provides a description of such materials. This theory has the advantageous feature that it is *not* based on a single-particle picture. More recently, $\mathbb{Z}_2$-topological invariants based on a single-particle picture and characterizing non-interacting time-reversal invariant topological insulators have been studied by Kane and Mele and numerous further authors; see, e.g., [29] and references given there.

In [4, 6], various further 2D systems with somewhat exotic magnetic properties have been described. For example, it has been speculated (see also [30]) that there may be *chiral spin liquids* with broken time-reversal symmetry not exhibiting any long-range spin order. The response laws for the spin current, $\vec{s}^{\mu}$, of chiral spin liquids can be derived from a non-abelian Chern–Simons action,

$$\frac{k}{4\pi}\int_{\Lambda} \mathrm{tr}\left(w \wedge \mathrm{d}w + \frac{2}{3} w \wedge w \wedge w\right) + \text{bd. terms},$$

where k is an integer ($k = 0, 1$ for electron gases), and $w = \sum w_\mu \mathrm{d}x^\mu$ is an $SU(2)$-gauge field coupling to electron spin, with $w_\mu = \vec{W}_\mu \cdot \vec{\sigma}$.

4. 3D Topological Insulators, Axions

Encouraged by the findings of the last section, we propose to consider insulators in three dimensions confined to a region

$$\Lambda := \Omega \times \mathbb{R}, \quad \Omega \subset \mathbb{R}^3, \ \partial\Lambda \neq \emptyset$$

of space-time. We are interested in the general form of the effective action describing the response of such materials to turning on an external electromagnetic field. Until the mid nineties, it was expected that the effective action of a 3D insulator is always given by

$$S_\Lambda(A) = \frac{1}{2}\int_\Lambda \mathrm{d}t\, \mathrm{d}^3x \{\vec{E} \cdot \varepsilon \vec{E} - \vec{B} \cdot \mu^{-1}\vec{B}\}, \tag{29}$$

where A is the electromagnetic vector potential of the electromagnetic field $(\vec{E}, \vec{B})$, ε is the dielectric tensor of the material and μ is the magnetic permeability tensor. In $(3+1)$ dimensions, the action in (29) is dimensionless.

In the seventies, particle theorists have taught us that one could add another dimensionless term to the action in (29), the θ-term first introduced in QCD.

An Effective Action with a Topological Term

We replace the action $S_\Lambda(A)$ by

$$S_\Lambda(A) \to S_\Lambda^{(\theta)}(A) := S_\Lambda(A) + \theta\, I_\Lambda(A), \quad \theta \in [0, 2\pi), \tag{30}$$

where θ is the analogue of the "vacuum angle" in particle physics, and I_Λ is a topological term ("instanton number") given by

$$\boxed{(4\pi^2)I_\Lambda(A) = 2\int_\Lambda \mathrm{d}t\,\mathrm{d}^3x\,\vec{E}\cdot\vec{B} = \int_\Lambda F_A \wedge F_A \underset{\text{Stokes}}{=} \int_{\partial\Lambda} A\wedge \mathrm{d}A} \tag{31}$$

see [6, 30, 31]. The partition function of a topological insulator in three dimensions whose effective action is given by (30) and (31) is

$$\mathcal{Z}_\Lambda^{(\theta)}(A) = \exp\big(iS_\Lambda^{(\theta)}(A)\big).$$

In the thermodynamic limit, $\Omega \nearrow \mathbb{R}^3$, $\mathcal{Z}_\Lambda^{(\theta)}(A)$ is *periodic* in θ with period 2π and *invariant under time reversal* iff

$$\theta = 0, \pi.$$

Surface Degrees of Freedom

A conventional insulator corresponds to $\theta = 0$. For $\theta = \pi$, the partition function $\mathcal{Z}_\Lambda^{(\theta)}(A)$ has a boundary term given by

$$\exp\left(\frac{i}{4\pi}\int_{\partial\Lambda} A\wedge \mathrm{d}A\right), \tag{32}$$

see (31). As mentioned in the last section, this is the leading term in the partition function of $(2+1)$-dimensional charged, "relativistic", two-component Dirac fermions on $\partial\Lambda$ coupled to an external electromagnetic field $A|_{\partial\Lambda}$. Two species of charged spin-polarized quasi-particles with a conical dispersion law mimicking a two-component relativistic Dirac fermion and propagating along the surface of the system may be encountered in certain 3D topological insulators with two degenerate bands "communicating" with each other and an effective action given by (30), (31), with $\theta = \pi$. Such systems support surface Hall currents, as is evident from the form of the surface action in (31) and (32); cf. Eq. (11). They have been studied experimentally in [27]; see also [31].

Axionic Topological Insulators

One may wonder whether, following ideas of Peccei and Quinn [32], it might make sense to view the angle θ as the (groundstate) expectation value of a dynamical field, φ, and replace the term $\theta\, I_\Lambda(A)$ by

$$I_\Lambda(\varphi, A) := \frac{1}{4\pi^2}\int_\Lambda \varphi\, F_A\wedge F_A + S_0(\varphi), \tag{33}$$

where $S_0(\varphi)$ is invariant under constant shifts $\varphi \mapsto \varphi + n\pi$, $n\in\mathbb{Z}$, and φ is a pseudo-scalar field called "axion field", (see [8]). The effective action given in (33) then gives rise to Halperin's 3D Hall effect:

$$\boxed{j = -\frac{1}{2\pi^2} \mathrm{d}\varphi \wedge F_A; \quad \text{in particular} \quad \vec{j} = \frac{1}{2\pi^2} \vec{\nabla}\varphi \times \vec{E}.} \tag{34}$$

In a crystalline insulator, and for an axion linear in $\vec{x}$, $\vec{\nabla}\varphi = \text{cst.}\vec{K}$, where, by invariance under lattice translations, $\vec{K}$ must belong to the dual lattice. Hence $\vec{\nabla}\varphi$ is "quantized".[f] It is argued that dynamical axions may emerge in certain topological insulators with anti-ferromagnetic short-range order and two bands of electron states "communicating" with each other. The time derivative of φ then has the interpretation of a chemical potential conjugate to the axial charge: The chiral anomaly tells us that

$$\int \mathrm{d}\varphi \wedge j_5 = -\text{cst.} \int \varphi F_A \wedge F_A + \text{terms} \propto \text{masses}. \tag{35}$$

The action of Eq. (33) has been proposed in [8] as the effective action describing a higher-dimensional cousin of the Hall effect and has been applied to the problem of unravelling possible origins of intergalactic primordial magnetic fields in the Universe; see also [7]. Later it appeared in [34, 35].

Mott Transition and Instabilities in Axionic Topological Insulators

It is interesting to analyze physical effects observable in putative states of matter described by the effective action

$$S_\Lambda^{tot}(A) := S_\Lambda(A) + I_\Lambda(\varphi, A),$$

where $S_\Lambda(A)$ is as in (29) and $I_\Lambda(\varphi, A)$ as in (33). Since the action $S_0(\varphi)$ on the right-hand side of Eq. (33) has been assumed to be periodic under constant shifts $\varphi \mapsto \varphi + n\pi, n \in \mathbb{Z}$, one may expect that, at positive (but low) temperatures, the bulk of such a state of matter will exhibit *axion domain walls* across which the mean value of φ changes by an integer multiple of π. Recalling the insight described after (32), one predicts that gapless, charged two-component Dirac fermions propagate along domain walls across which the mean value of φ changes by an odd multiple of π. We are thus led to predict that the formation of macroscopic axion domain walls gives rise to a "Mott transition" from an insulator to a state with non-vanishing bulk conductivity, due to the extended charged surface modes propagating along such domain walls; (see [30]).

Axion-electrodynamics exhibits some tantalizing *instabilities*. One such instability, triggered by the term $I_\Lambda(\varphi, A)$, manifests itself in the growth of long-wave-length magnetic fields in the bulk of the system; see [7, 8]. A related instability has been described in [36]: Above a certain critical field strength, E_c, a constant external electric field, $E\vec{n}$ (with $|\vec{n}| = 1$ and $E > E_c$), applied to an axionic topological insulator is *screened*, with the excess electric field $(E - E_c)\vec{n}$ converted to a *magnetic field.*

[f] I thank Gregory Moore for having explained this point to me; see [33].

Expository articles covering theoretical and experimental aspects of topological insulators can be found, e.g., in [37, 38].

Conclusions and Outlook

- The Physics of 2D (and 3D) insulators is surprisingly rich and has potential to lead to promising technological applications. Interesting mathematical techniques — ranging from abstract algebra over the topology of fiber bundles all the way to hard analysis — have non-trivial applications to problems concerning these fascinating states of matter.
- 2D electron gases, Bose gases and magnetic materials represent stimulating play grounds for experimentalists and theorists alike — not least because very general principles of quantum theory, such as *braid statistics, fractional spin and fractional electric charges, anomalies and their cancellation, current algebra* and *holography* appear to manifest themselves in the arena of 2D condensed matter physics. Unfamiliar species of quasi-particles, in particular, *two-component Dirac-like fermions* and *Majorana fermions* are encountered in the physics of various exotic two- and three-dimensional systems whose exploration has begun only recently.
- It is interesting to consider higher-dimensional cousins of the QHE and of time-reversal invariant topological insulators; see [8, 39]. Some of them might be relevant in cosmology, e.g., in connection with the generation of intergalactic primordial magnetic fields in the Universe. Ideas on cosmology somewhat related to those reviewed in this paper are presently actively pursued.

Many ideas originally developed in particle physics turn out to have fruitful applications in condensed matter physics. Likewise, ideas originating in condensed matter physics and statistical mechanics promise to have very interesting applications in cosmology. But that's another story that I cannot go into here. — In any event, these fields of physics appear to feature highly stimulating exchanges and fruitful interactions, or call for such. Moreover, different mathematical techniques and styles of research have turned out to be useful and to complement each other in productive ways in particle physics, condensed matter physics and cosmology. New insights gained in pure mathematics sometimes have spectacular applications in physics, and conversely. Yet, disappointingly, interactions across certain disciplinary boundaries apparently continue to be weaker than they ought to be, and people in different communities are not seldom reluctant to profit from each other's experiences. A determination to change this belongs to the legacy Ludvig Faddeev has left us.

References

[1] J. Fröhlich and P. A. Marchetti, Quantum field theory of anyons, *Lett. Math. Phys.* **16** (1988) 347–358.

[2] J. Fröhlich and T. Kerler, Universality in quantum Hall systems, *Nucl. Phys. B* **354** (1991) 369–417.

[3] J. Fröhlich and A. Zee, Large-scale physics of the quantum Hall fluids, *Nucl. Phys. B* **364** (1991) 517–540.

[4] J. Fröhlich and U. M. Studer, Gauge invariance and current algebra in non-relativistic many-body theory, *Rev. Mod. Phys.* **65** (1993) 733–802.

[5] J. Fröhlich and E. Thiran, Integral quadratic forms, Kac–Moody algebras, and fractional quantum Hall effect. An ADE – $\mathcal{O}$ classification, *J. Stat. Phys.* **76** (1994) 209–283.

[6] J. Fröhlich, U. M. Studer and E. Thiran, Quantum theory of large systems of non-relativistic matter (Course 8, Part I), in *Géométries fluctuantes en mécanique statistique et en théorie des champs* (*Les Houches, 1994*), eds. F. David, P. Ginsparg and J. Zinn-Justin (Elsevier Science B.V., Amsterdam, 1996), pp. 771–912.

[7] A. Yu. Alekseev, V. V. Cheianov and J. Fröhlich, Universality of transport properties in equilibrium, the Goldstone theorem, and chiral anomaly, *Phys. Rev. Lett.* **81** (1998) 3503–3506.

[8] J. Fröhlich and B. Pedrini, New applications of the chiral anomaly, in *Mathematical Physics 2000*, eds. A. Fokas, A. Grigoryan, T. Kibble and B. Zegarlinski (Imperial College Press, London and Singapore, 2000), pp. 9–47; A. Boyarsky, J. Fröhlich and O. Ruchayskiy, Self-consistent evolution of magnetic fields and chiral asymmetry in the early Universe, *Phys. Rev. Lett.* **108** (2012) 031301.

[9] R. Jackiw, D. Gross and S. B. Treiman, *Current Algebra and Anomalies* (Princeton University Press, Princeton NJ, 1986).

[10] J. S. Bell and R. Jackiw, A PCAC puzzle: $\pi^0 \to yy$ in the σ-model, *Nuovo Cimento A* **60** (1969) 47–61; S. L. Adler, Axial-vector vertex in spinor electrodynamics, *Phys. Rev.* **177** (1969) 2426–2438; L. D. Faddeev, Operator anomaly for the Gauss law, *Phys. Lett. B* **145** (1984) 81–84; A. G. Reiman, M. A. Semenov-Tyan-Shanskii and L. D. Faddeev, Quantum anomalies and cocycles on gauge groups, *Funct. Anal. Appl.* **18** (1984) 319–326; L. D. Faddeev and S. L. Shatashvili, Algebraic and Hamiltonian methods in the theory of non-Abelian anomalies, *Theor. Math. Phys.* **60** (1984) 770–778; J. Lott, A proof of the axial anomaly, *Comm. Math. Phys.* **97** (1985) 371–379; A. Yu. Alekseev, Ya. Madaichik, L. D. Faddeev and S. L. Shatashvili, Derivation of anomalous commutator in the functional integral formalism, *Theor. Math. Phys.* **73** (1987) 1149–1151.

[11] X. G. Wen, Chiral Luttinger liquid and the edge excitations in the fractional quantum Hall states, *Phys. Rev. B* **41** (1990) 12838–12844; B. Blok and X. G. Wen, Effective theories of the fractional quantum Hall effect at generic filling fractions, *Phys. Rev. B* **42** (1990) 8133–8144.

[12] G. Landwehr (ed.), *High Magnetic Fields in Semiconductor Physics III — Quantum Hall Effect, Transport and Optics*, Springer Series in Solid-State Sciences, Vol. 101 (Springer Verlag, Berlin, Heidelberg, 1992); See, in particular, the contribution of R. Morf, and references given therein.

[13] R. E. Prange and S. M. Girvin (eds.), *The Quantum Hall Effect*, 2nd edn., Graduate Texts in Contemporary Physics (Springer-Verlag, New York, 1990).

[14] B. I. Halperin, Theory of the quantized Hall conductance, *Helv. Phys. Acta* **56** (1983) 75–102.

[15] J. Fröhlich, Statistics of fields, the Yang–Baxter equation, and the theory of knots and links, in *Non-Perturbative Quantum Field Theory*, eds. G. 't Hooft *et al.*, Cargèse 1987 (Plenum Press, New York, 1988), pp. 71–100.

[16] B. I. Halperin, Quantized Hall conductance, current-carrying edge states, and the

existence of extended states in a two-dimensional disordered potential, *Phys. Rev. B* **25** (1982) 2185–2190.

[17] J. Fröhlich, G. M. Graf and J. Walcher, On the extended nature of edge states of quantum Hall Hamiltonians, *Ann. Henri Poincaré* **1** (2000) 405–442.

[18] F. D. M. Haldane, Model for a quantum Hall effect without Landau levels: Condensed-matter realization of the "parity anomaly", *Phys. Rev. Lett.* **61** (1988) 2015–2018.

[19] G. Jotzu, M. Messer, R. Desbuquois, M. Lebrat, T. Uehlinger, D. Greif and T. Esslinger, Experimental realization of the topological Haldane model with ultracold fermions, *Nature* **515** (2014) 237–240.

[20] J. Fröhlich, B. Pedrini, C. Schweigert and J. Walcher, Universality in quantum Hall systems: Coset construction of incompressible states, *J. Stat. Phys.* **103** (2001) 527–567.

[21] J. Fröhlich (*RG in* $M \cup \Phi$), The fractional quantum Hall effect, Chern–Simons theory, and integral lattices, in *Proceedings of ICM '94*, ed. S. D. Chatterji (Birkhäuser Verlag, Basel, Boston, Berlin, 1995), pp. 75–105.

[22] T. Can, M. Laskin and P. Wiegmann, Fractional quantum Hall effect in a curved space: Gravitational anomaly and electromagnetic response, *Phys. Rev. Lett.* **113** (2014) 046803; S. Klevtsov and P. Wiegmann, Geometric adiabatic transport in quantum Hall states, *Phys. Rev. Lett.* **115** (2015) 086801.

[23] N. R. Cooper, N. K. Wilkin and J. M. F. Gunn, Quantum phases of vortices in rotating Bose–Einstein condensates, *Phys. Rev. Lett.* **87** (2001) 120405; N. R. Cooper, Rapidly rotating atomic gases, *Adv. Phys.* **57** (2008) 539–616.

[24] E. Schmutzer and J. Plebański, Quantum mechanics in non-inertial frames of reference, *Fortschr. Physik* **25** (1977) 37–82.

[25] S. Deser, R. Jackiw and S. Templeton, Topologically massive gauge theories, *Ann. Phys. (NY)* **140** (1982) 372–411; A. N. Redlich, Gauge noninvariance and parity nonconservation of three-dimensional fermions, *Phys. Rev. Lett.* **52** (1984) 18–21.

[26] M. König, H. Buhmann, L. W. Molenkamp, T. Hughes, C. X Liu, X. L. Qi and S. C. Zhang, The quantum spin Hall effect: Theory and experiment, *J. Phys. Soc. Jpn.* **77** (2008) 031007–031020; A. Roth, C. Brüne, H. Buhmann, L. W. Molenkamp, J. Maciejko, X. L. Qi and S. C. Zhang, Nonlocal transport in the quantum spin Hall state, *Science* **325** (2009) 294–297; C. Brüne, A. Roth, E. G. Novik, H. Buhmann, E. M. Hankiewicz, W. Hanke, J. Sinova and L. W. Molenkamp, Evidence for the ballistic intrinsic spin Hall effect in HgTe nanostructures, *Nat. Phys.* **6** (2010) 448–454; C. Brüne, A. Roth, H. Buhmann, E. M. Hankiewicz, L. W. Molenkamp, J. Maciejko, X. L. Qi and S. C. Zhang, Spin polarization of the quantum spin Hall edge, *Nat. Phys.* **8** (2012) 485–490.

[27] C. Brüne, C.-X. Liu, E. G. Novik, E. M. Hankiewicz, H. Buhmann, Y.-L. Chen, X.-L. Qi, Z.-X. Shen, S.-C. Zhang and L. W. Molenkamp, Quantum Hall effect from the topological surface states in strained bulk HgTe, *Phys. Rev. Lett.* **106** (2011) 126803–126809.

[28] S. Murakami, N. Nagaosa and S. C. Zhang, Spin-Hall insulator, *Phys. Rev. Lett.* **93** (2004) 156804; B. A. Bernevig and S. C. Zhang, Quantum spin Hall effect, *Phys. Rev. Lett.* **96** (2006) 106802–106805; B. A. Bernevig, T. L. Hughes and S. C. Zhang, Quantum spin Hall effect and topological phase transition in HgTE quantum wells, *Science* **314** (2006) 1757–1761,

[29] C. L. Kane and E. J. Mele, Z_2 topological order and the quantum spin Hall effect, *Phys. Rev. Lett.* **95** (2005) 146802; L. Fu and C. L. Kane, Topological insulators with inversion symmetry, *Phys. Rev. B* **76** (2007) 045302; A. Kitaev, Periodic table for

topological insulators and superconductors, in *Adv. Theor. Physics: Landau Memorial Conf.*, AIP Conf. Proc., Vol. 1134 (American Institute of Physics, 2009), pp. 22–30. M. Z. Hasan and C. L. Kane, Topological insulators, *Rev. Mod. Phys.* **82** (2010) 3045–3067; G. M. Graf and M. Porta, Bulk-edge correspondence for two-dimensional topological insulators, *Comm. Math. Phys.* **324** (2013) 851–895; K. Gawedzki, Bundle gerbes for topological insulators (December 2015), arXiv:1512.01028 [math-ph]; H. Schulz-Baldes, Topological insulators from the perspective of noncommutative geometry and index theory, *Jahresbericht der Deutschen Mathematiker-Vereinigung* **118** (2016) 247–273; R. Kennedy and M. R. Zirnbauer, Bott periodicity for $\mathbb{Z}_2$ symmetric ground states of gapped free-fermion systems, *Comm. Math. Phys.* **342** (2016) 909–963.

[30] J. Fröhlich and Ph. Werner, Gauge theory of topological phases of matter, *EPL* **101** (2013) 47007-p1–47007-p6.

[31] Z. Wang, X. G. Qi and S. C. Zhang, Equivalent topological invariants of topological insulators, *New J. Phys.* **12** (2010) 065007.

[32] R. D. Peccei and H. R. Quinn, *CP* conservation in the presence of pseudoparticles, *Phys. Rev. Lett.* **38** (1977) 1440–1443.

[33] G. Moore, A comment on Berry connections (6 June, 2017); arXiv:1706.01149v2 [hep-th].

[34] Yu. N. Obukhov and F. W. Hehl, Measuring a piecewise constant axion field in classical electrodynamics, *Phys. Lett.* **341** (2005) 357–365.

[35] R. Li, J. Wang, X. L. Qi and S. C. Zhang, Dynamical axion field in topological magnetic insulators, *Nat. Phys.* **6** (2010) 284–288.

[36] H. Ooguri and M. Oshikawa, Instability in magnetic materials with a dynamical axion field, *Phys. Rev. Lett.* **108** (2012) 161803–161806.

[37] J. E. Moore, The birth of topological insulators, *Nature* **464** (2010) 194–198.

[38] M. Franz and L. Molenkamp (eds.), *Topological Insulators*, Contemporary Concepts of Condensed Matter Science, Vol. 6 (Elsevier, Oxford, Amsterdam, Waltham MA and San Diego CA, 2013).

[39] H. M. Price, O. Zilberberg, T. Ozawa, I. Carusotto and N. Goldman, Four-dimensional quantum Hall effect with ultracold atoms, *Phys. Rev. Lett.* **115** (2015) 195303.

On Some Hamiltonian Properties of the Isomonodromic Tau Functions

A. R. Its[*] and A. Prokhorov[†]

Department of Mathematical Sciences, Indiana University–Purdue University, 402 N. Blackford St., Indianapolis, IN 46202-3267, USA

and

St. Petersburg State University, Universitetskaya emb. 7/9, 199034, St. Petersburg, Russia

[*] *aits@iupui.edu*

[‡] *aprokhor@iupui.edu*

We discuss some new aspects of the theory of the Jimbo–Miwa–Ueno tau function which have come to light within the recent developments in the global asymptotic analysis of the tau functions related to the Painlevé equations. Specifically, we show that up to the total differentials the logarithmic derivatives of the Painlevé tau functions coincide with the corresponding classical action differential. This fact simplifies considerably the evaluation of the constant factors in the asymptotics of tau functions, which has been a long-standing problem of the asymptotic theory of Painlevé equations. Furthermore, we believe that this observation is yet another manifestation of L. D. Faddeev's emphasis of the key role which the Hamiltonian aspects play in the theory of integrable system.

1. Introduction

Consider a system of linear ordinary differential equations with rational coefficients,

$$\frac{d\Phi}{dz} = A(z)\Phi, \tag{1}$$

where $A(z)$ is an $N \times N$, $N > 1$ matrix-valued rational function. The object of our study is the Jimbo–Miwa–Ueno tau function associated with the isomonodromy deformation of system (1). Let us remind, following [20], the general set-up associated with this notion.

Denote the poles of the matrix-valued rational function $A(z)$ on $\mathbb{C}P^1$ by $a_1, \dots, a_n, \infty$ and by $r_1, r_2, \dots, r_n, r_\infty$ the corresponding Poincaré ranks. The matrix function $A(z)$ can then be written as,

$$A(z) = \sum_{\nu=1}^{n}\sum_{k=1}^{r_\nu+1} \frac{A_{\nu,-k+1}}{(z-a_\nu)^k} + \sum_{k=0}^{r_\infty-1} z^k A_{\infty,-k-1}, \quad A_{\nu,-k+1},\ A_{\infty,-k-1} \in \mathfrak{sl}_N(\mathbb{C}),$$
$$k = 1, \dots, r_\nu + 1, \quad \nu = 1, \dots, n.$$

We are going to make the standard assumption that all highest order matrix coefficients $A_\nu \equiv A_{\nu,-r_\nu}$ are diagonalizable

$$A_{\nu,-r_\nu} = G_\nu \Theta_{\nu,-r_\nu} G_\nu^{-1}; \quad \Theta_{\nu,-r_\nu} = \operatorname{diag}\{\theta_{\nu,1}, \ldots, \theta_{\nu,N}\},$$

and that their eigenvalues are distinct and non-resonant:

$$\begin{cases} \theta_{\nu,\alpha} \neq \theta_{\nu,\beta} & \text{if } \; r_\nu \geq 1, \quad \alpha \neq \beta, \\ \theta_{\nu,\alpha} \neq \theta_{\nu,\beta} \mod \mathbb{Z} & \text{if } \; r_\nu = 0, \quad \alpha \neq \beta. \end{cases}$$

At each singular point, the system (1) admits a unique *formal* solution,

$$\Phi^{(\nu)}_{\text{form}}(z) = G^{(\nu)}(z) e^{\Theta_\nu(z)}, \qquad \nu = 1, \ldots, n, \infty, \tag{2}$$

where $G^{(\nu)}(z)$ are formal series,

$$G^{(\nu)}(z) = G_\nu \left[\mathrm{I} + \sum_{k=1}^{\infty} g_{\nu,k} (z - a_\nu)^k \right], \quad G^{(\infty)}(z) = G_\infty \left[\mathrm{I} + \sum_{k=1}^{\infty} g_{\infty,k} z^k \right],$$

and $\Theta_\nu(z)$ are diagonal matrix-valued functions,

$$\Theta_\nu(z) = \sum_{k=-r_\nu}^{-1} \frac{\Theta_{\nu,k}}{k} (z - a_\nu)^k + \Theta_{\nu,0} \ln(z - a_\nu),$$

$$\Theta_\infty(z) = -\sum_{k=1}^{r_\infty} \frac{\Theta_{\infty,-k}}{k} z^k - \Theta_{\infty,0} \ln z.$$

For every $\nu \in \{1, \ldots, n, \infty\}$, the matrix coefficients $g_{\nu,k}$ and $\Theta_{\nu,k}$ can be explicitly computed in terms of the coefficients of the matrix-valued rational function $G_\nu^{-1} A(z) G_\nu$, see [20].

The non-formal global properties of solutions of Eq. (1) are described by its *monodromy data* M which include: (i) formal monodromy exponents $\Theta_{\nu,0}$, (ii) appropriate connection matrices between canonical solutions at different singular points, and (iii) relevant Stokes matrices at irregular ($r_\nu \geq 1$) singular points. Let us denote the space of monodromy data of the system (1) by $\mathcal{M}$. It can be described in more details as follows.

Let a_ν be an irregular singular point of rank r_ν. For $j = 1, \ldots, 2r_\nu + 1$, let also

$$\Omega_{j,\nu} = \left\{ z\colon 0 < |z - a_\nu| < \epsilon,\ \theta_j^{(1)} < \arg(z - a_\nu) < \theta_j^{(2)},\ \theta_j^{(2)} - \theta_j^{(1)} = \frac{\pi}{r_\nu} + \delta \right\}$$

be the *Stokes sectors* around a_ν (see, e.g., [10, Chap. 1] or [32] for more details). According to the general theory of linear systems, in each sector $\Omega_{j,\nu}$ there exists a unique *canonical solution* $\Phi_j^{(\nu)}(z)$ of (1) which satisfies the asymptotic condition

$$\Phi_j^{(\nu)}(z) \simeq \Phi^{(\nu)}_{\text{form}}(z) \qquad \text{as } z \to a_\nu, \quad z \in \Omega_{j,\nu}, \quad j = 1, \ldots, 2r_\nu + 1.$$

Different canonical solutions are related by *Stokes matrices*, $S_j^{(\nu)}$, and *connection matrices*, C_ν:

$$\Phi_{j+1}^{(\nu)} = \Phi_j^{(\nu)} S_j^{(\nu)}, \quad j = 1, \ldots, 2r_\nu, \qquad \Phi_1^{(\nu)} = \Phi_1^{(\infty)} C_\nu, \quad \nu = 1, \ldots, n.$$

Let us assume that the irregular singular points are ∞ and the first $n_0 \leq n$ points among the singular points $a_1, \ldots, a_n$. Denote by $\mathcal{S}_\nu$ the collection of Stokes matrices at an irregular point a_ν, i.e.

$$\mathcal{S}_\nu = \{S_1^{(\nu)}, \ldots, S_{2r_\nu}^{(\nu)}\}.$$

The space $\mathcal{M}$ of monodromy data of the system (1) consists of formal monodromy exponents $\Theta_{\nu,0}$, connection matrices C_ν and Stokes matrices $S_j^{(\nu)}$, i.e.

$$\mathcal{M} = \{M \equiv (\Theta_{\nu,0},\ \nu = 1, \ldots, n, \infty;\ C_\nu,\ \nu = 1, \ldots, n;\ \mathcal{S}_\nu,\ \nu = 1, \ldots, n_0, \infty)\}.$$

We shall use the notation,

$$\vec{m} = (m_1, \ldots, m_d), \quad d = N(n+1) + nN^2 + \left(\frac{N(N-1)}{2}\right)\left(\sum_{\nu=1}^{n_0} 2r_\nu + 2r_\infty\right),$$

for the points $\vec{m} \in \mathcal{M}$. In addition, we denote by $\mathcal{T}$ the set of *times*,

$$a_1, \ldots, a_n, \quad (\Theta_{\nu,k})_{ll}, \quad k = -r_\nu, \ldots, -1, \quad \nu = 1, \ldots, n_0, \infty, \quad l = 1, \ldots, N.$$

We shall use the notation,

$$\vec{t} = (t_1, \ldots, t_L), \quad L = n + N\left(\sum_{\nu=1}^{n_0} r_\nu + r_\infty\right),$$

for the points $\vec{t} \in \mathcal{T}$. Let us also denote by $\mathcal{A}$ the variety of all rational matrix-valued functions $A(z)$ with a fixed number of poles of fixed orders. The so-called Riemann–Hilbert correspondence states that, up to submanifolds where the inverse monodromy problem for (1) is not solvable, the space $\mathcal{A}$ can be identified with the product $\widetilde{\mathcal{T}} \times \mathcal{M}$, where $\widetilde{\mathcal{T}}$ denotes the universal covering of $\mathcal{T}$. We shall loosely write,

$$\mathcal{A} \simeq \widetilde{\mathcal{T}} \times \mathcal{M}.$$

It should be mentioned that in each concrete case one has to specify the gauge normalization of the matrix $A(z)$ as well as the choice of the gauge matrices G_ν in order to make this identification well defined. In Sec. 3, we will demonstrate how these specifications can be done in the case of Painlevé equations.

The *Jimbo–Miwa–Ueno 1-form* is defined as the following differential form on $\mathcal{A}$:

$$\omega_{\text{JMU}} = -\sum_{\nu=1,\ldots,n,\infty} \operatorname*{res}_{z=a_\nu} \operatorname{Tr}\left((G^{(\nu)}(z))^{-1}\frac{dG^{(\nu)}}{dz}(z)\, d_{\mathcal{T}}\Theta_\nu(z)\right), \tag{3}$$

where $G^{(\nu)}(z)$ are the series from (2) and we put $a_\infty \equiv \infty$. The notation $d_{\mathcal{T}}\Theta_\nu(z)$ stands for

$$d_{\mathcal{T}}\Theta_\nu(z) = \sum_{k=1}^{L} \frac{\partial \Theta_\nu(z)}{\partial t_k}\, dt_k, \qquad L = n + N\left(\sum_{\nu=1}^{n_0} r_\nu + r_\infty\right).$$

The significance of this form is that, being restricted to any *isomonodromic family* in the space $\mathcal{A}$,

$$A(z) \equiv A(z;\vec{t},M), \qquad \vec{t} = (t_1,\dots,t_L), \quad M \equiv \text{const}$$

it becomes closed with respect to times $\mathcal{T}$, i.e.

$$d_{\mathcal{T}}\left(\omega_{\text{JMU}}\big|_{A(z;\vec{t},M\equiv\text{const})}\right) = 0.$$

The closedness of the 1-form ω_{JMU} with respect to $\mathcal{T}$ in turn implies that locally there is a function $\tau \equiv \tau(\vec{t},M)$ on $\mathcal{T}\times\mathcal{M}$ such that

$$d_{\mathcal{T}} \ln \tau = \omega_{\text{JMU}}\big|_{A(z;\vec{t},M)}. \tag{4}$$

A remarkable property of the *tau function* $\tau(\vec{t},M)$, which was established in [25, 27], is that it admits analytic continuation as an entire function to the whole universal covering $\widetilde{\mathcal{T}}$ of the parameter space $\mathcal{T}$. Furthermore, zeros of $\tau(\vec{t},M)$ correspond to the points in $\mathcal{T}$ where the inverse monodromy problem for (1) is not solvable for a given set M of monodromy data (or, equivalently, where the holomorphic vector bundle over $\mathbb{C}P^1$ determined by M becomes nontrivial). Hence a central role of the concept of tau function in the monodromy theory of systems of linear differential equations.

The isomonodromicity of the family $A(z;\vec{t},M)$ means that all equations from it have the same set $M\in\mathcal{M}$ of monodromy data. This implies that the corresponding solution $\Phi(z)\equiv\Phi(z,\vec{t})$ satisfies an overdetermined system

$$\begin{cases} \partial_z\Phi = A(z,\vec{t})\Phi(z,\vec{t}), \\ d_{\mathcal{T}}\Phi = B(z,\vec{t})\Phi(z,\vec{t}). \end{cases} \tag{5}$$

The coefficients of the matrix-valued differential form $B\equiv\sum_{k=1}^{L} B_k(z,\vec{t})dt_k$ are rational in z. Their explicit form may be algorithmically deduced from the expression for $A(z)$ (see again [20]). The compatibility of the system (5) yields the *monodromy preserving deformation equation*:

$$d_{\mathcal{T}}A = \partial_z B + [B,A]. \tag{6}$$

Isomonodromy equation (6) is of great interest on its own. Indeed, it includes as special cases practically all known integrable differential equations. The first nontrivial cases of (6), where the set of isomonodromic times effectively reduces to a single variable t, cover all six classical Painlevé equations. Solutions of the latter are dubbed as *nonlinear special functions*, and they indeed play this role in many areas of modern nonlinear science (see [10, 4, 7, 12, 30, 31]).

The principal analytic issue concerning the tau function, in particular from the point of view of applications, is its behavior near the critical hyperplanes, where either $a_\mu = a_\nu$ for some $\mu\neq\nu$, or $\theta_{\nu,\alpha}=\theta_{\nu,\beta}$ for some ν and some $\alpha\neq\beta$. In the case of Painlevé equations this is the behavior of respective tau functions near the $t=\infty$ (PI, II, IV), $t=\infty,0$ (PIII, V), and $t=\infty,0,1$ (PVI). A special challenge in the asymptotic analysis of the tau functions is the evaluation of the constant pre-factors in their asymptotics. In fact, it is these pre-factors which usually contain the most important information about the physical properties of the model under investigation. At the same time, they cannot be obtained directly via the Riemann–Hilbert approach. The latter method is one of the principal modern tools of the asymptotic analysis of Painlevé transcendents, and it is based on the asymptotic evaluation of the above-mentioned Riemann–Hilbert correspondence. In other words, the Riemann–Hilbert technique allows to evaluate the asymptotics of the matrix $A(z, \vec{t})$ and hence the asymptotics of the differential form $\omega_{\rm JMU}$. In view of (4), this gives the asymptotics of the logarithmic derivatives of the tau function. In order to obtain the complete asymptotic description of the tau function itself, which would include the above-mentioned pre-factors, one has to solve the "constant problem": to find the constant of integration arising from the formal integration of (4). More precisely, since the tau function is itself defined up to a multiplicative constant, we are actually talking about the evaluation, in terms of monodromy data, of the ratios of constant factors corresponding to different critical points (Painlevé III, V, VI) or to different critical directions (Painlevé I, II, IV).

The first rigorous solution of a constant problem for Painlevé equations (a special Painlevé III transcendent appearing in the Ising model) has been obtained in the work of Tracy [29]. After that, several other important special cases have been also solved. We refer the reader to [16] for a detailed history of the question. It is important to emphasize that all these works were concerned with the very special families of the Painlevé functions, and they used the techniques which could not be extended to the generic tau functions.

The means to solve the "constant problem" for tau functions corresponding to the generic solutions of Painlevé equations started to develop since the 2013–2014 works [18, 17] of Iorgov, Lisovyy, Shchechkin, and Tykhyy where a very important discovery of the conformal block interpretation of tau functions was made. For the history of the question, we refer the reader to the paper [16] where the heuristic though truly pioneering results of [18, 17] have been rigorously proven. Another conjectural pre-factor formula, this time concerning the third Painlevé equation (work [19]), was proven in [15]. Later on, the method of [16] and [15] was successfully applied to the first Painlevé equation in [24].

The method of [16, 15] is inspired by the earlier works of Malgrange [25] and Bertola [2] and it is based on an extension of the Jimbo–Miwa–Ueno differential form $\omega_{\rm JMU}$ to a 1-form, ω on the whole space $\mathcal{A} \simeq \widetilde{\mathcal{T}} \times \mathcal{M}$,

$$\omega = \sum_{k=1}^{L} P_k(\vec{t}, M)dt_k + \sum_{j=1}^{d} Q_j(\vec{t}, M)dm_j,$$

such that

$$\omega(\partial_{t_k}) = \omega_{\rm JMU}(\partial_{t_k}),$$

and the exterior differential of ω, i.e. the form,

$$\Omega_0 := d\omega \tag{7}$$

is a 2-form on $\mathcal{M}$ only. Furthermore, it is independent of isomonodromic times $\mathcal{T}$. The construction of the form ω will be described in detail in the next section.[a]

The time-independence of the 2-form Ω_0 in conjunction with the Riemann–Hilbert computability of the asymptotics of $\Phi(z)$ determines what should be added to the form ω to make it closed, i.e. to transform it into the form $\hat{\omega}$ which satisfies the two crucial properties:

$$d\hat{\omega} \equiv d_{\mathcal{T}}\hat{\omega} + d_{\mathcal{M}}\hat{\omega} = 0, \quad \text{and} \quad \hat{\omega}(\partial_{t_k}) = \omega_{\mathrm{JMU}}(\partial_{t_k}).$$

Having the form $\hat{\omega}$, the tau function can be represented as

$$\ln \tau = \int \hat{\omega}. \tag{8}$$

Equation (8) allows one to use the asymptotic behavior of $\Phi(z)$ to evaluate the asymptotics of the associated tau function up to a numerical (i.e. independent of monodromy data) constant. The latter can be calculated by applying the final formulae to trivial solutions of deformation equations. This program has been first realized in [15] for the sine-Gordon reduction of the Painlevé III equation and later in [16] for Painlevé VI and II equations and in [24] for Painlevé I equation.

We want also mention the most recent work [11] where a general Fredholm determinant formula was found for the Painlevé VI tau function which allows to produce rigorously both the evaluation of the relevant asymptotic constants and the combinatorial series expansions of the tau function at the critical points.

In the course of the asymptotic analysis performed in [15, 16], an interesting observation has been made with respect to the form ω in the Painlevé III and II cases. This observation is concerned with the Hamiltonian aspect of the theory of isomonodromy deformations which we have not yet discussed. As a matter of fact, the space $\mathcal{A}$ can be equipped with a symplectic structure — see [9, 14, 3, 23, 1], so that the isomonodromy equation (6) induces L commuting Hamiltonian flows on $\mathcal{A}$. A striking property of the tau function is that in many (though not all[b]) known special cases its logarithm serves as the *generating function* of the Hamiltonians H_k of these flows:

$$\frac{\partial \ln \tau(\vec{t}, M)}{\partial t_k} = \gamma H_k\big|_{A(z;\vec{t},M)}. \tag{9}$$

Here γ is numerical constant (in many cases, γ =1). This fact for all six Painlevé equations was first established by Okamoto [28] who also was the first who suggested the Hamiltonian formalism for all six Painlevé equations. The above-mentioned observation of [15, 16] is that in the Painlevé II and III cases the 2-form $\Omega_0 = d\omega$ is nothing else but, up to a numerical factor, the corresponding symplectic form. Hence, in these examples, the 1-form ω, up to a numerical factor and the addition of an explicit total differential, is an extension to the space $\widetilde{\mathcal{T}} \times \mathcal{M}$ of the differential

[a] The exact relation of the form ω to the original Malgrange–Bertola 1-form is explained in detail in [16] — see Remark 4.4 there.

[b] The statement actually depends on the specific choice of the symplectic structure.

of classical action; moreover, in [15, 16] these total differentials have been explicitly found. Similar relation to the classical action in the case of the Painlevé I tau function has been obtained in [24], and in the case of the Schlesinger equations — the pure Fuchsian system (1), in [26].[c]

The goal of this paper is to show that the relation between the tau function and the classical action established in [15, 16, 24] for the special cases of Painlevé III, II, and I is true for all Painlevé equations. We shall also present some arguments allowing one to expect that this relation is, most likely, a general fact of the monodromy theory of linear systems.

The detailed construction of the form ω is given, following [16], in the next section. In this section we also provide the arguments in favor of the connection between the form ω and the classical action in the general case of linear system (1) and formulate two conjectures concerning with this connection. In Sec. 3, these conjectures are justified for all six Painlevé equations and for an arbitrary Schlesinger system. In the cases of Painlevé I and Schlesinger equations we just reproduce the results of [24, 26], respectively.

2. The Extended Jimbo–Miwa–Ueno Differential and the Classical Action Functional

As in the Introduction, we shall, unless the otherwise is explicitly indicated, be treating all the objects which are defined on $\mathcal{A} \simeq \widetilde{\mathcal{T}} \times \mathcal{M}$ as functions of $(\vec{t}, M) \equiv (\vec{t}, \vec{m})$. In particular,

$$\omega_{\text{JMU}} \equiv \omega_{\text{JMU}}(\vec{t}, M) = \omega_{\text{JMU}}\big|_{A(z;\vec{t},M)},$$

and for any function F on $\mathcal{A}$, the partial derivatives with respect to t_k will mean the partial derivatives of F as a function of $(\vec{t}, M)$, i.e.

$$\frac{\partial F}{\partial t_k} \equiv \frac{\partial}{\partial t_k} F(\vec{t}, M) = \frac{\partial}{\partial t_k}\big(F\big|_{A(z;\vec{t},M)}\big).$$

We will also use the notations

$$dF \equiv dF(\vec{t}, M) = \sum_{k=1}^{L} \frac{\partial F}{\partial t_k} dt_k + \sum_{k=1}^{d} \frac{\partial F}{\partial m_k} dm_k \equiv d_{\mathcal{T}} F + d_{\mathcal{M}} F.$$

Our starting point is the following Lemma.

Lemma 1 ([20]). *The* 1*-form* (3) (*considered as a* 1*-form on* $\widetilde{\mathcal{T}} \times \mathcal{M}$) *can be alternatively written as*

$$\begin{aligned}\omega_{\text{JMU}} &= \sum_{k=1}^{L} \sum_{\nu=1,\ldots,n,\infty} \operatorname{res}_{z=a_\nu} \operatorname{Tr}\left(A\, \frac{\partial G^{(\nu)}}{\partial t_k} \left(G^{(\nu)}\right)^{-1}\right) dt_k \\ &\equiv \sum_{\nu=1,\ldots,n,\infty} \operatorname{res}_{z=a_\nu} \operatorname{Tr}\left(A\, d_{\mathcal{T}} G^{(\nu)} \left(G^{(\nu)}\right)^{-1}\right). \end{aligned} \qquad (10)$$

[c]The authors are grateful to Marta Mazzocco for pointing out this result of Malgrange to us.

We give the version [16] of the proof of this Lemma in Sec. A.1 of the Appendix.

A direct corollary of Lemma 1 is the following integral formula for the tau function,

$$\ln \tau \equiv \ln \tau(\vec{t}_1, \vec{t}_2, M) = \int_{\vec{t}_1}^{\vec{t}_2} \sum_{\nu,k} \mathrm{res}_{z=a_\nu} \mathrm{Tr}\left(A \frac{\partial G^{(\nu)}}{\partial t_k} (G^{(\nu)})^{-1} \right) dt_k. \tag{11}$$

Another consequence of the Lemma is the idea to take formula (10) as the motivation to introduce the following 1-form (cf. [16, 15])

$$\begin{aligned}\omega &= \sum_{k=1}^{L} \sum_{\nu=1,\dots,n,\infty} \mathrm{res}_{z=a_\nu} \mathrm{Tr}\left(A \frac{\partial G^{(\nu)}}{\partial t_k} (G^{(\nu)})^{-1} \right) dt_k \\ &\quad + \sum_{k=1}^{d} \sum_{\nu=1,\dots,n,\infty} \mathrm{res}_{z=a_\nu} \mathrm{Tr}\left(A \frac{\partial G^{(\nu)}}{\partial m_k} (G^{(\nu)})^{-1} \right) dm_k \\ &\equiv \sum_{\nu=1,\dots,n,\infty} \mathrm{res}_{z=a_\nu} \mathrm{Tr}(A\, d_{\mathcal{T}} G^{(\nu)} (G^{(\nu)})^{-1}) \\ &\quad + \sum_{\nu=1,\dots,n,\infty} \mathrm{res}_{z=a_\nu} \mathrm{Tr}(A\, d_{\mathcal{M}} G^{(\nu)} (G^{(\nu)})^{-1}) \\ &\equiv \sum_{\nu=1,\dots,n,\infty} \mathrm{res}_{z=a_\nu} \mathrm{Tr}(A\, dG^{(\nu)}\, (G^{(\nu)})^{-1}).\end{aligned} \tag{12}$$

Now, the key observation.

Lemma 2 ([16]). *The form $d\omega$ has no cross terms of the kind $dt_k \wedge dm_j$, $k = 1, \dots, L$, $j = 1, \dots, d$.*

We present, following [16], the proof of this Lemma in Sec. A.2 of the Appendix.

Lemma 2 plays a crucial role in the mentioned in the Introduction rigorous approach to the "constant problem". Indeed, a key issue in the determining of the monodromy dependence of the tau function is the possibility of the effective evaluation of the derivative of the integral (11) with respect to the monodromy parameters m_j. Lemma 2 implies that

$$\frac{\partial}{\partial m_j} \sum_\nu \mathrm{res}_{z=a_\nu} \mathrm{Tr}\left(A \frac{\partial G^{(\nu)}}{\partial t_k} (G^{(\nu)})^{-1} \right) = \frac{\partial}{\partial t_k} \sum_\nu \mathrm{res}_{z=a_\nu} \mathrm{Tr}\left(A \frac{\partial G^{(\nu)}}{\partial m_j} (G^{(\nu)})^{-1} \right).$$

Therefore,

$$\begin{aligned}\frac{\partial \ln \tau}{\partial m_j} &= \int_{\vec{t}_1}^{\vec{t}_2} \sum_{k=1}^{L} \frac{\partial}{\partial m_j} \sum_\nu \mathrm{res}_{z=a_\nu} \mathrm{Tr}\left(A \frac{\partial G^{(\nu)}}{\partial t_k} (G^{(\nu)})^{-1} \right) dt_k \\ &= \sum_{k=1}^{L} \int_{\vec{t}_1}^{\vec{t}_2} \frac{\partial}{\partial t_k} \sum_\nu \mathrm{res}_{z=a_\nu} \mathrm{Tr}\left(A \frac{\partial G^{(\nu)}}{\partial m_j} (G^{(\nu)})^{-1} \right) dt_k \\ &= \sum_\nu \mathrm{res}_{a_\nu} \mathrm{Tr}\left(A \frac{\partial G^{(\nu)}}{\partial m_j} (G^{(\nu)})^{-1} \right) \Bigg|_{\vec{t}_1}^{\vec{t}_2}.\end{aligned} \tag{13}$$

In other words, we conclude that in addition to the differential relation (4), i.e.

$$d_{\mathcal{T}} \ln \tau = \sum_{\nu=1,\ldots,n,\infty} \operatorname{res}_{z=a_\nu} \operatorname{Tr}(G^{(\nu)}(z)^{-1} A(z) d_{\mathcal{T}} G^{(\nu)}(z)),$$

the tau function satisfies the differential relation,

$$d_{\mathcal{M}} \ln \tau = \sum_{\nu=1,\ldots,n,\infty} \operatorname{res}_{z=a_\nu} \operatorname{Tr}(G^{(\nu)}(z)^{-1} A(z) d_{\mathcal{M}} G^{(\nu)}(z)).$$

These two differential identities allow to evaluate the asymptotic connection formulae up to the numerical constants and this is what is effectively done in [15, 16].

The arguments which led to the representation (13) for the logarithmic derivative of the tau function with respect to m_j are reminiscent to the variational equations for the classical action. In more detail, let us assume that we can identify the classical Darboux coordinates,[d] p_j, q_j on the space $\mathcal{A}$ so that the isomonodromy deformation equations (6) can be written as the commuting system of Hamiltonian dynamical equations,

$$\frac{\partial q_j}{\partial t_k} = \frac{\partial H_k}{\partial p_j}, \quad \frac{\partial p_j}{\partial t_k} = -\frac{\partial H_k}{\partial q_j}. \tag{14}$$

We remind that we still identify $\mathcal{A} \simeq \widetilde{\mathcal{T}} \times \mathcal{M}$, so that we consider p_j and q_j as the functions on $\widetilde{\mathcal{T}} \times \mathcal{M}$,

$$q_j \equiv q_j(\vec{t}, M), \quad p_j \equiv p_j(\vec{t}, M), \quad H_k \equiv H_k\big(q_j(\vec{t}, M), p_j(\vec{t}, M), \vec{t}\big).$$

The compatibility of the system (14) means (see, e.g., [1]) that all

$$c_{kl} := \{H_k, H_l\} + \frac{\partial H_k}{\partial t_l} - \frac{\partial H_l}{\partial t_k}$$

are the Casimir functions[e] (maybe depending on the times t_k). We shall assume that

$$c_{kl} = 0 \quad \forall\, k, l. \tag{15}$$

This assumption works for all example of the isomonodromy deformations equations that we know. The *classical action differential* can be defined as the differential form on $\widetilde{\mathcal{T}} \times \mathcal{M}$,

$$\omega_{\text{cla}} = \sum p_j dq_j - \sum H_k dt_k \equiv \sum_k \left(\sum_j p_j \frac{\partial q_j}{\partial t_k} - H_k \right) dt_k + \sum_k \left(\sum_j p_j \frac{\partial q_j}{\partial m_k} \right) dm_k$$

[d] The Darboux coordinates on the phase spaces $\mathcal{A}$ corresponding to Painlevé equations are introduced in [9, 13, 1]; the Darboux coordinates for more general cases of the isomonodromy deformation equations are considered in [14].

[e] Warning: here,

$$\frac{\partial H_k}{\partial t_l} = \frac{\partial}{\partial t_l} \big(H_k(\vec{p}, \vec{q}, \vec{t})\big|_{\vec{p},\vec{q} \equiv \text{const}}\big).$$

and, using assumption (15), it is easy to check that it is closed on the trajectories of the dynamical system (14), i.e.

$$d_{\mathcal{T}}\big(\omega_{\mathrm{cla}}\big|_{M\equiv\mathrm{const}}\big) = 0.$$

Note that in those cases when the logarithm of the tau function is the generating function for the Hamiltonians H_k, the Jimbo–Miwa–Ueno differential form is

$$\omega_{\mathrm{JMU}} = \sum H_k dt_k,$$

so that the integral (11) is the *truncated* action integral,

$$\ln \tau = \int_{\vec{t}_1}^{\vec{t}_2} \sum_k H_k dt_k.$$

Suppose that instead of this integral we need to study the complete action, i.e. the integral,

$$S \equiv S(\vec{t}_1, \vec{t}_2, M) = \int_{\vec{t}_1}^{\vec{t}_2} \omega_{\mathrm{cla}}(M) \equiv \int_{\vec{t}_1}^{\vec{t}_2} \sum_k \left(\sum_j p_j \frac{\partial q_j}{\partial t_k} - H_k \right) dt_k.$$

Then, the usual variational calculus arguments show that, similar to (13), in any m_j-derivative of S the integral terms would disappear. In fact, we have,

$$\begin{aligned}
\frac{\partial S}{\partial m_{jo}} &= \int_{\vec{t}_1}^{\vec{t}_2} \sum_k \left(\sum_j \frac{\partial p_j}{\partial m_{jo}} \frac{\partial q_j}{\partial t_k} + p_j \frac{\partial^2 q_j}{\partial t_k \partial m_{jo}} - \frac{\partial H_k}{\partial p_j}\frac{\partial p_j}{\partial m_{jo}} - \frac{\partial H_k}{\partial q_j}\frac{\partial q_j}{\partial m_{jo}} \right) dt_k \\
&= \sum_{j,k} p_j \frac{\partial q_j}{\partial m_{jo}}\bigg|_{\vec{t}_1}^{\vec{t}_2} + \int_{\vec{t}_1}^{\vec{t}_2} \sum_k \left(\sum_j \frac{\partial p_j}{\partial m_{jo}}\frac{\partial q_j}{\partial t_k} - \frac{\partial q_j}{\partial m_{jo}}\frac{\partial p_j}{\partial t_k} \right.\\
&\quad \left. - \frac{\partial H_k}{\partial p_j}\frac{\partial p_j}{\partial m_{jo}} - \frac{\partial H_k}{\partial q_j}\frac{\partial q_j}{\partial m_{jo}} \right) dt_k \\
&= \sum_{j,k} p_j \frac{\partial q_j}{\partial m_{jo}}\bigg|_{\vec{t}_1}^{\vec{t}_2}
\end{aligned} \tag{16}$$

and the integral term vanishes because of the equations of motion (14). Comparing (13) and (16) makes one to suspect some deep connection between the tau function and the classical action. And, indeed, taken the full exterior derivation of $\omega_{\mathrm{cla}} \equiv \omega_{\mathrm{cla}}(\vec{t}, M)$, one obtains,

$$d\omega_{\mathrm{cla}} = \sum_j d_{\mathcal{M}} p_j \wedge d_{\mathcal{M}} q_j \equiv \Omega.$$

The form Ω is the symplectic form associated with the dynamical system (14). Note, that both, the form Ω and the form Ω_0 from (7) are the closed 2-form on $\mathcal{M}$ and they do not depend on the times $\mathcal{T}$. This observation together with the similarities of the variational identities (13) and (16) allow us to formulate the following conjectures.

Conjecture 1. *Suppose that the parameter space $\mathcal{A}$ is equipped with the symplectic structure. Let Ω be a corresponding symplectic form and Ω_0 be the 2-form defined in (7). Then, there exists a numerical constant γ such that,*

$$\Omega_0 = \gamma\Omega.$$

If this conjecture is true then the two 1-forms, ω and ω_{cla}, coincide up to the total differential. Hence our next conjecture.

Conjecture 2. *There exists a rational function $G(\vec{p}, \vec{q}, \vec{t})$ of $\vec{p}, \vec{q}, \vec{t}$ such that,*

$$\omega = \gamma\omega_{\text{cla}} + dG(\vec{p}, \vec{q}, \vec{t}). \tag{17}$$

Moreover, the function $G(\vec{p}, \vec{q}, \vec{t})$ is explicitly computable.

As it has already been indicated in the Introduction, the statement of Conjecture 2 has been proven to be true in the case of the Schlesinger equations [26], in the case of the sine-Gordon reduction of Painlevé III equation [15], in the case of the (homogeneous) Painlevé II equation [16], and in the case of the Painlevé I equation [24]. In the next section of this paper we demonstrate the validity of the both conjectures for the rest of the Painlevé equations, and we also present, for completeness, the result of [26].

Remark 3. Restricting (17) to the isomonodromy family $M \equiv \text{const}$, one arrives to the identity

$$\begin{aligned}\sum_k \frac{\partial \ln \tau}{\partial t_k} dt_k = &\sum_k \left(\sum_j p_j(\vec{t}, M) \frac{\partial q_j(\vec{t}, M)}{\partial t_k} - H_k\left(\vec{p}(\vec{t}, M), \vec{q}(\vec{t}, M), \vec{t}\right) \right) dt_k \\ &+ \sum_k \frac{\partial}{\partial t_k} G(\vec{p}(\vec{t}, M), \vec{q}(\vec{t}, M), \vec{t}) dt_k\end{aligned} \tag{18}$$

and hence,

$$\ln \tau(\vec{t_1}, \vec{t_2}, M) = S(\vec{t_1}, \vec{t_2}, M) + G(\vec{p}(\vec{t}, M), \vec{q}(\vec{t}, M), \vec{t})\Big|_{\vec{t_1}}^{\vec{t_2}}. \tag{19}$$

This, in turn, would produce, taking into account (16), the following, alternative to (13), formula for the m_j-derivative of $\ln \tau$,

$$\frac{\partial \ln \tau}{\partial m_{j_0}} = \sum_{j,k} p_j \frac{\partial q_j}{\partial m_{j_0}}\Big|_{\vec{t_1}}^{\vec{t_2}} + \frac{\partial G}{\partial m_{j_0}}\Big|_{\vec{t_1}}^{\vec{t_2}}. \tag{20}$$

This version of the variational logarithmic derivatives of the tau functions turns out even more efficient then (13) in the concrete examples related to the "constant problem". Indeed, the particular cases of (20) have been used in [5] in evaluation of the constant terms in the asymptotics of the several basic distribution functions of random matrix theory expressible in terms of the Painlevé transcendents.

Remark 4. In the pioneering papers [17, 18], the evaluation of the constant prefactors has been partially based on the conjectural interpretations of these constant factors as the generating functions of the canonical transformations between the Darboux asymptotic coordinates associated with the different critical points. This generating function interpretation of the constant pre-factors, which in the case of Painlevé II, III, and I has been proven in [15, 16, 24], can be considered as a direct corollary of (19).

3. Painlevé Equations

In this section, we present the exact realization of the relations (17) and (18) between the tau function and the classical action for all six Painlevé equations and also for the case of the general Schlesinger equation. We shall start with the second Painlevé equation which we take as a case study and illustrate in detail the general constructions of Sec. 2. In particular, we will describe exactly the both spaces, $\mathcal{A}$ and $\mathcal{M}$ corresponding to the Painlevé II equation. The other Painlevé equations will be treated with less details. We will not describe the space $\mathcal{M}$ for other Painlevé equations explicitly, and instead we will refer the reader either to [21] or to [10, Chap. 5].

3.1. *Painlevé II*

According to [21], the second Painlevé equation describes the isomonodromy deformations of the 2×2 linear system having only one irregular singular point at $z=\infty$ of the Poincaré rank 3,

$$\frac{d\Phi}{dz} = A(z)\Phi, \qquad A(z) = A_2 z^2 + A_1 z + A_0. \tag{21}$$

Following again [21], we normalize the system by the conditions,

$$\operatorname{Tr} A(z) \equiv 0, \quad A_2 = \begin{pmatrix} 1 & 0 \\ 0 & -1 \end{pmatrix}, \quad A_{1,11} = A_{1,22} = 0,$$

so that the matrix coefficients A_1 and A_0 can be written in the form,

$$A_1 = \begin{pmatrix} 0 & k \\ -\frac{2p}{k} & 0 \end{pmatrix}, \quad A_0 = \begin{pmatrix} p + \frac{t}{2} & -kq \\ -\frac{2}{k}(\theta + pq) & -p - \frac{t}{2} \end{pmatrix},$$

and the complex parameters p, q, k, θ, and t can be taken as the original coordinates on the corresponding space $\mathcal{A}$,

$$\mathcal{A} = \{(p, q, k, \theta, t)\}. \tag{22}$$

The formal solution of system (21) at its only (irregular) singular point, $z=\infty$, has the structure (cf. (2)),

$$\begin{aligned} \Phi_{\text{form}}(z) &\equiv G(z)e^{\Theta(z)} \\ &= \left(I + \frac{g_1}{z} + \frac{g_2}{z^2} + \frac{g_3}{z^3} + \dots\right)e^{\Theta(z)}, \\ \Theta(z) &= \sigma_3\left(\frac{z^3}{3} + \frac{tz}{2} - \theta \ln z\right), \end{aligned} \tag{23}$$

with the first three matrix coefficients g_k, $k = 1, 2, 3$ given as functions on the space $\mathcal{A}$ (22) by the explicit formulae,

$$g_1 = \begin{pmatrix} -H & -\frac{k}{2} \\ -\frac{p}{k} & H \end{pmatrix}, \tag{24}$$

$$g_2 = \begin{pmatrix} \frac{H^2}{2} + \frac{p}{4} - \frac{t\theta}{4} & -\frac{kH}{2} + \frac{kq}{2} \\ \frac{pH}{k} - \frac{pq}{k} - \frac{\theta}{k} & \frac{H^2}{2} + \frac{p}{4} + \frac{t\theta}{4} \end{pmatrix}, \tag{25}$$

$$g_3 = \begin{pmatrix} -\frac{H^3}{6} - \frac{Hp}{4} + \frac{Ht}{6} + \frac{Ht\theta}{4} + \frac{pq}{6} + \frac{\theta^2}{6} + \frac{\theta}{3} & -\frac{kH^2}{4} + \frac{kqH}{2} + \frac{kp}{8} + \frac{kt}{4} - \frac{kt\theta}{8} \\ -\frac{pH^2}{2k} + \frac{Hpq}{k} + \frac{H\theta}{k} + \frac{p^2}{4k} + \frac{pt\theta}{4k} + \frac{pt}{2k} & \frac{H^3}{6} + \frac{Hp}{4} - \frac{Ht}{6} + \frac{Ht\theta}{4} - \frac{pq}{6} - \frac{\theta^2}{6} + \frac{\theta}{6} \end{pmatrix}, \tag{26}$$

where

$$H = \frac{p^2}{2} + pq^2 + \frac{pt}{2} + q\theta. \tag{27}$$

The system (21) has seven canonical solutions, characterized by the asymptotic condition,

$$\Phi_j(z) \simeq \Phi_{\text{form}}(z), \quad z \to \infty, \quad \frac{(2j-5)\pi}{6} < \arg z < \frac{(2j-1)\pi}{6},$$

$$\Phi_7(z) = \Phi_1(z) e^{-2\pi i \theta \sigma_3}$$

and hence it has six Stokes matrices, $S_j = \Phi_j^{-1}(z)\Phi_{j+1}(z)$, which have the following triangular structure (for more detail see [21] or [10, Chap. 5]),

$$S_{2k+1} = \begin{pmatrix} 1 & 0 \\ s_{2k+1} & 1 \end{pmatrix}, \quad S_{2k} = \begin{pmatrix} 1 & s_{2k} \\ 0 & 1 \end{pmatrix},$$

and satisfy one cyclic relation,

$$S_1 S_2 \cdots S_6 = e^{-2\pi i \theta \sigma_3}.$$

Also, as it follows from (23), the parameter θ determines the formal monodromy exponent, $\Theta_0 = \theta\sigma_3$. This means, that the space $\mathcal{M}$ in the case of system (21) can be identified with the algebraic variety of dimension 4,

$$\begin{aligned} \mathcal{M} = \{\vec{m} = (s_1, s_2, \ldots, s_6, \theta)\colon 1 + s_1 s_2 = (1 + s_4 s_5) e^{2\pi i \theta}, \\ 1 + s_2 s_3 = (1 + s_5 s_6) e^{-2\pi i \theta}, s_1 + s_3 + s_1 s_2 s_3 = -s_5 e^{2\pi i \theta}\}. \end{aligned}$$

Equation (23), also tells us that the parameter t is the only isomonodromic time, so that, in the case of system (21) we have,

$$\mathcal{T} = \{t\}.$$

The isomonodromy deformations of system (21), i.e. the conditions,

$$p = p(t, M), \quad q = q(t, M), \quad k = k(t, M), \quad M \equiv \text{const},$$

yield the second linear matrix differential equation, this time with respect to t, for the function $\Phi(z) \equiv \Phi(z; t, M)$,

$$\frac{d\Phi}{dt} = B(z)\Phi, \qquad B(z) = B_1 z + B_0, \tag{28}$$

where

$$B_1 = \frac{1}{2}\begin{pmatrix} 1 & 0 \\ 0 & -1 \end{pmatrix}, \quad B_0 = \frac{1}{2}\begin{pmatrix} 0 & k \\ -\frac{2p}{k} & 0 \end{pmatrix}.$$

Equations (21) and (28) form a *Lax pair* (cf. (5)),

$$\begin{cases} \dfrac{d\Phi}{dz} = A(z)\Phi \\ \dfrac{d\Phi}{dt} = B(z)\Phi \end{cases}$$

whose compatibility condition (6), in terms of the functional parameters p, q, k and θ reads

$$\begin{aligned} \frac{dq}{dt} &= p + q^2 + \frac{t}{2}, \\ \frac{dp}{dt} &= -2pq - \theta, \\ \frac{dk}{dt} &= -kq, \\ \frac{d\theta}{dt} &= 0. \end{aligned} \tag{29}$$

The last equation of this system is just the statement that θ, as the part of the monodromy data, is constant. The third equation gives $\ln k(t)$ as the antiderivative of $-q(t)$. The first two first-order differential equations are equivalent to one second-order differential equation, indeed, the Painlevé II equation,

$$q_{tt} = tq + 2q^3 + \alpha, \quad \alpha = \frac{1}{2} - \theta. \tag{30}$$

Assuming that

$$\theta \equiv \text{const}$$

one can easily see that the function (27) is the Hamiltonian of the second Painlevé equation (30) with respect to the symplectic form,

$$\Omega = dp \wedge dq.$$

Indeed, the first and the second equations in (29) are just

$$\frac{dq}{dt} = \frac{\partial H}{\partial p}, \quad \text{and} \quad \frac{dp}{dt} = -\frac{\partial H}{\partial q},$$

respectively.

Let us now discuss the forms ω_{JMU} and ω corresponding to (21). The linear system (21) has only ∞ as its singular point. Therefore, the general definition (3) of the form ω_{JMU} transforms to the equation,

$$\omega_{\mathrm{JMU}} = -\operatorname{res}_{z=\infty} \operatorname{Tr}\left(G^{-1}(z)\frac{dG(z)}{dz}\frac{d\Theta(z)}{dt}\right) dt.$$

Plugging (23) at the right-hand side we arrive at the formulae,

$$\omega_{\mathrm{JMU}} = -\operatorname{Tr}\left(\frac{1}{2}g_1\sigma_3\right),$$

or, taking into account (24) (cf. (9)),

$$\omega_{\mathrm{JMU}} \equiv \frac{d\ln\tau}{dt}dt = Hdt. \tag{31}$$

Similarly, the general definition (12) of the form ω transforms to the equation,

$$\omega = \operatorname{res}_{z=\infty}\operatorname{Tr}(A(z)dG(z)G(z)^{-1}). \tag{32}$$

Plugging (23) into (32) we arrive at the formula,

$$\omega = \operatorname{Tr}(A_2dg_3 - A_2dg_2g_1 - A_2dg_1g_2 + A_2dg_1g_1^2 + A_1dg_2 - A_1dg_1g_1 + A_0dg_1).$$

Now, it is more involved to plug (24)–(26) into the right-hand side of the last equation. However, after performing some algebra, the final expression comes out rather simple,

$$\omega = -\frac{1}{3}qdp + \frac{2}{3}pdq - \theta\frac{dk}{k} + \frac{2}{3}tdH - \frac{1}{3}Hdt - \frac{2\theta-1}{3}d\theta,$$

and can be in turn easily transformed to the equation,

$$\omega = pdq - Hdt + d\left(\frac{2}{3}Ht - \frac{1}{3}qp - \theta\ln k - \frac{\theta^2}{3} + \frac{\theta}{3}\right) + \ln k\, d\theta. \tag{33}$$

If we again assume that

$$\theta \equiv \mathrm{const},$$

relation (33) reduces to

$$\omega = pdq - Hdt + d\left(\frac{2}{3}Ht - \frac{1}{3}qp - \theta\ln k\right). \tag{34}$$

Equation (34) proves Conjecture 2 in the case of the 2×2 system (21) with the additional constraint, $\theta \equiv \mathrm{const}$. Indeed, this is exactly the formula (17) with the specification[f]

$$G(p,q,t) = \frac{2}{3}Ht - \frac{1}{3}qp - \theta\ln k.$$

[f] From the point of view of the asymptotic analysis of the tau functions outlined in the Introduction, the appearance of the non-local term $\ln k = \int qdt$ is not an obstacle since the functional parameter k equals $-2g_{1,12}$ (see (24) where g_1 is the first coefficient of the series (23). This means that, similar to the functions p and q, it can be recovered from the underlining Riemann–Hilbert problem, see also [6].

The corresponding equation (18) is

$$\begin{aligned}\frac{d\ln\tau}{dt} &= p\frac{dq}{dt} - H + \frac{d}{dt}\left(\frac{2}{3}Ht - \frac{1}{3}qp - \theta\ln k\right)\\ &= p\frac{dq}{dt} - H + \frac{d}{dt}\left(\frac{2}{3}Ht - \frac{1}{3}qp\right) + \theta q. \qquad (35)\end{aligned}$$

We also note that Conjecture 1 follows directly from (34).

Remark 5. Together with (31), Eq. (35) is just an identity which can be proven directly by substituting (31) into the left-hand side of (35). However, it would be quite difficult to guess, without having the concept of the form ω, the existence of such connection between the truncated and the full action differentials. We believe that the fact that in the case of the Painlevé dynamical systems the truncated action differs from the full action by a total differential is a manifestation of their Lax pair integrability.

Remark 6. Let us denote

$$p_1 = p, \quad q_1 = q, \quad p_2 = \ln k, \quad q_2 = \theta,$$

then the whole system (29) becomes Hamiltonian with the same Hamiltonian (27),

$$H = \frac{p_1^2}{2} + p_1 q_1^2 + \frac{p_1 t}{2} + q_1 q_2,$$

and with respect to the symplectic form,

$$\Omega = dp_1 \wedge dq_1 + dp_2 \wedge dq_2.$$

Moreover, (34) can be written as

$$\omega = \sum_{k=1}^{2} p_k dq_k - H dt + d\left(\frac{2}{3}Ht - \frac{1}{3}q_1 p_1 - q_2 p_2 - \frac{q_2^2}{3} + \frac{q_2}{3}\right).$$

That is, if we associate with the linear system (21) not just the second Painlevé equation (30) but the full system (29) of the isomonodromy deformation equations of (21), then the G-function in the relation (17) will be totally local and, in fact, polynomial in the Darboux coordinates,

$$G(p_1, p_2, q_1, q_2, t) = \frac{2}{3}Ht - \frac{1}{3}q_1 p_1 - q_2 p_2 - \frac{q_2^2}{3} + \frac{q_2}{3}.$$

It is also worth noticing that since H does not contain p_2, the variable q_2 is an action variable, i.e. it is constant, as it should be.

Remark 7. The Lax pair (21), (28) is not the only Lax pair for the Painlevé II equation. If we use another Lax pair, for instance the Lax pair of Flaschka and Newell [8], we would get the another tau function, another Hamiltonian and another form ω. However, the Conjectures 1 and 2 would still be true — see [16, 5]. It is an interesting issue how much the Hamiltonian aspects we are promoting depend on the concrete Lax pair realization of the Painlevé equations.

3.2. *Painlevé I*

The results of this subsections belong to O. Lisovyy and J. Roussillon, and we follow here their paper [24]. The linear system associated with the first Painlevé equation is the 2×2 matrix ODE with one irregular singular point of Poincaré rank 5 at $z = \infty$ and with one Fuchsian singular point at $z = 0$,

$$\frac{d\Phi}{dz} = A(z)\Phi, \qquad A(z) = A_4 z^4 + A_2 z^2 + A_1 z + A_0 + \frac{A_{-1}}{z}. \tag{36}$$

The matrix coefficients are,

$$A_4 = \begin{pmatrix} 4 & 0 \\ 0 & -4 \end{pmatrix}, \quad A_2 = \begin{pmatrix} 0 & -4q \\ 4q & 0 \end{pmatrix}, \quad A_1 = \begin{pmatrix} 0 & -2p \\ -2p & 0 \end{pmatrix},$$

$$A_0 = \begin{pmatrix} 2q^2 + t & -2q^2 - t \\ 2q^2 + t & -2q^2 - t \end{pmatrix}, \quad A_{-1} = -\frac{1}{2}\begin{pmatrix} 0 & 1 \\ 1 & 0 \end{pmatrix},$$

and so that the space $\mathcal{A}$ is parametrized by p, q, and t,

$$\mathcal{A} = \{(p, q, t)\}.$$

The formal solution at $z = \infty$ is given by the series,

$$\Phi_{\text{form}}(z) = \left(I + \frac{g_1}{z} + \frac{g_2}{z^2} + \frac{g_3}{z^3} + \frac{g_4}{z^4} + \frac{g_5}{z^5} + O\left(\frac{1}{z^6}\right)\right) e^{\Theta(z)}, \quad \Theta(z) = \sigma_3\left(\frac{4z^5}{5} + tz\right) \tag{37}$$

with the explicit formulae for the first five coefficient matrices g_k given by the equations,

$$g_1 = \begin{pmatrix} -H & 0 \\ 0 & H \end{pmatrix}, \quad g_2 = \begin{pmatrix} \frac{H^2}{2} & \frac{q}{2} \\ \frac{q}{2} & \frac{H^2}{2} \end{pmatrix}, \quad g_3 = \begin{pmatrix} -\frac{H^3}{6} - \frac{2p-t^2}{24} & \frac{qH}{2} + \frac{p}{4} \\ -\frac{qH}{2} - \frac{p}{4} & \frac{H^3}{6} + \frac{2p-t^2}{24} \end{pmatrix}, \tag{38}$$

$$g_4 = \begin{pmatrix} \frac{H^4}{24} + \frac{2p-t^2}{24}H + \frac{q^2}{8} & \frac{qH^2}{4} + \frac{pH}{4} + \frac{2q^2+t}{8} \\ \frac{qH^2}{4} + \frac{pH}{4} + \frac{2q^2+t}{8} & \frac{H^4}{24} + \frac{2p-t^2}{24}H + \frac{q^2}{8} \end{pmatrix}, \tag{39}$$

$$g_5 = \begin{pmatrix} -\frac{H^5}{120} - \frac{2p-t^2}{48}H^2 - \frac{5q^2-2t}{40}H - \frac{4pq+1}{160} & \frac{qH^3}{12} + \frac{pH^2}{8} + \frac{2q^2+t}{8}H + \frac{2p-t^2}{48}q + \frac{1}{16} \\ -\frac{qH^3}{12} - \frac{pH^2}{8} - \frac{2q^2+t}{8}H - \frac{2p-t^2}{48}q - \frac{1}{16} & \frac{H^5}{120} + \frac{2p-t^2}{48}H^2 + \frac{5q^2-2t}{40}H + \frac{4pq+1}{160} \end{pmatrix}, \tag{40}$$

where

$$H = \frac{p^2}{2} - 2q^3 - tq. \tag{41}$$

The Fuchsian point $z = 0$ is a resonant point and hence the generic theory outlined in the Introduction is not applicable. In fact, the behavior of the solution $\Phi(z)$ at $z = 0$ is given by the formula [24],

$$\Phi(z) = \begin{pmatrix} 1 & \frac{1}{2} \\ 1 & -\frac{1}{2} \end{pmatrix} z^{-\frac{1}{2}\sigma_3}\hat{\Phi}(z),$$

where $\hat{\Phi}(z)$ is holomorphic and invertible at $z = 0$.

The set of monodromy data $\mathcal{M}$ of system (36) consists of 10 Stokes matrices associated with the irregular singularity at $z = \infty$ out of which, due to the symmetry $z \to -z$ of the system, only two are in fact free, i.e. $\dim \mathcal{M} = 2$ (see [24] for details). The structure of the essential singularity of $\Phi(z)$ at infinity described in (37) indicates that the parameter t is the only isomonodromic time. The corresponding Lax pair is formed by Eq. (36) and the following additional matrix equation,

$$\frac{d\Phi}{dt} = B(z)\,\Phi, \qquad B(z) = B_1 z + \frac{B_{-1}}{z}, \tag{42}$$

$$B_1 = \begin{pmatrix} 1 & 0 \\ 0 & -1 \end{pmatrix}, \quad B_{-1} = \begin{pmatrix} q & -q \\ q & -q \end{pmatrix}.$$

The compatibility condition of (36) and (42) yields the system of ODEs on $q \equiv q(t)$ and $p \equiv p(t)$,

$$\begin{aligned} \frac{dq}{dt} &= p, \\ \frac{dp}{dt} &= 6q^2 + t, \end{aligned} \tag{43}$$

which is equivalent to the Painlevé I equation

$$q_{tt} = 6q^2 + t.$$

The system (43) is a Hamiltonian system with the Hamiltonian (41).

We are passing now to the forms ω_{JMU} and ω corresponding to the Painlevé I system (43). Because $z = 0$ is the resonant Fuchsian point we strictly speaking cannot use the definitions (3) and (12) for the forms ω_{JMU} and ω. However, following [24], we take (3) and (12), where the undefined contribution of the resonant point $z = 0$ simply ignored, as the definitions of these forms in the case of the Painlevé I equations, i.e. as in the case of the second Painlevé equation, we put

$$\omega_{\mathrm{JMU}} = -\operatorname{res}_{z=\infty} \operatorname{Tr}\left(G^{-1}(z) \frac{dG(z)}{dz} \frac{d\Theta(z)}{dt} \right) dt, \tag{44}$$

and

$$\omega = \operatorname{res}_{z=\infty} \operatorname{Tr}(A(z) dG(z) G(z)^{-1}), \tag{45}$$

where $G(z)$ and $\Theta(z)$ are the series and the exponent from (37). As it is shown in [24] such approach preserves the validity of Lemmas 1 and 2. Just as in the case of Painlevé II, we obtain from (44) at once that

$$\omega_{\mathrm{JMU}} \equiv \frac{d \ln \tau}{dt} dt = 2H dt.$$

The form ω needs more work.

Introduce the matrix coefficients h_j of the series inverse to the series (37),

$$\begin{aligned} &\left(\mathrm{I} + \frac{h_1}{z} + \frac{h_2}{z^2} + \frac{h_3}{z^3} + \frac{h_4}{z^4} + O\left(\frac{1}{z^5}\right)\right) \\ &\quad := \left(\mathrm{I} + \frac{g_1}{z} + \frac{g_2}{z^2} + \frac{g_3}{z^3} + \frac{g_4}{z^4} + \frac{g_5}{z^5} + O\left(\frac{1}{z^6}\right)\right)^{-1}. \end{aligned} \tag{46}$$

We have for the first four coefficients the relations,

$$h_1 = -g_1, \quad h_2 = -g_2 + g_1^2, \quad h_3 = -g_3 + g_2 g_1 + g_1 g_2 - g_1^3,$$

$$h_4 = -g_4 + g_3 g_1 + g_1 g_3 + g_2^2 - g_2 g_1^2 - g_1 g_2 g_1 - g_1^2 g_2 + g_1^4.$$

Plugging (46) and (37) into (45) we arrive at the formula,

$$\begin{aligned}\omega = \operatorname{Tr}(&-A_4(h_4 dg_1 + h_3 dg_2 + h_2 dg_3 + h_1 dg_4 + dg_5)\\ &- A_2(h_2 dg_1 + h_1 dg_2 + dg_3) - A_1(h_1 dg_1 + dg_2) - A_0 dg_1).\end{aligned}$$

Using (38)–(40) we get after (quite a lot of) simplifications

$$\omega = \frac{6}{5} p dq - \frac{4}{5} q dp - \frac{2}{5} H dt + \frac{8}{5} t dH,$$

and properly combining the terms,

$$\omega = 2\left[p dq - H dt + d\left(\frac{4Ht}{5} - \frac{2pq}{5}\right)\right]. \tag{47}$$

Equation (47) proves Conjectures 1 and 2, with $\gamma = 2$, in the case of the 2×2 system (36) and gives the explicit formula for $G(p, q, t)$,

$$G(p, q, t) = \frac{2}{5}(4Ht - 2pq).$$

The corresponding equation (18) is

$$\frac{d \ln \tau}{dt} = 2\left(p\frac{dq}{dt} - H\right) + \frac{2}{5}\frac{d}{dt}(4Ht - 2pq),$$

and, of course, can be easily checked directly.

3.3. *Painlevé III*

The linear system associated with the third Painlevé equation we take again from [21]. This is the system,

$$\frac{d\Phi}{dz} = A(z)\Phi, \qquad A(z) = \frac{A_{-2}}{z^2} + \frac{A_{-1}}{z} + A_0, \tag{48}$$

with

$$A_0 = \frac{1}{2}\begin{pmatrix} t & 0 \\ 0 & -t \end{pmatrix}, \quad A_{-1} = \begin{pmatrix} -\theta_\infty & -qkt \\ \frac{pq(t-p)}{kt} + \frac{\theta_0+\theta_\infty}{k} - \frac{2\theta_\infty p}{kt} & \theta_\infty \end{pmatrix},$$

$$A_{-2} = \begin{pmatrix} p - \frac{t}{2} & -kt \\ \frac{p(p-t)}{kt} & -p + \frac{t}{2} \end{pmatrix}.$$

The system has two irregular singular points at $z = \infty$ and $z = 0$, both of the Poincaré rank 1. The corresponding formal solutions are:

$$\Phi_{\text{form}}^{(\infty)}(z) = \left(I + \frac{g_{\infty,1}}{z} + O\left(\frac{1}{z^2}\right)\right) e^{\Theta_\infty(z)}, \quad \Theta_\infty(z) = \sigma_3\left(\frac{tz}{2} - \theta_\infty \ln z\right), \tag{49}$$

with

$$g_{\infty,1} = \begin{pmatrix} -\frac{H}{2} - \frac{pq}{2t} + \frac{\theta_\infty^2-\theta_0^2}{2t} + \frac{t}{2} & kq \\ \frac{pq(t-p)}{kt^2} + \frac{\theta_0+\theta_\infty}{kt} - \frac{2\theta_\infty p}{kt^2} & \frac{H}{2} + \frac{pq}{2t} - \frac{\theta_\infty^2-\theta_0^2}{2t} - \frac{t}{2} \end{pmatrix}, \tag{50}$$

at $z = \infty$, and

$$\Phi^{(0)}_{\text{form}}(z) = G_0(I + g_{0,1}z + O(z^2))e^{\Theta_0(z)}, \quad \Theta_0(z) = \sigma_3\left(\frac{t}{2z} + \theta_0 \ln z\right) \tag{51}$$

with

$$g_{0,1} = \begin{pmatrix} -\frac{H}{2} - \frac{pq}{2t} - \frac{\theta_\infty^2-\theta_0^2}{2t} + \frac{t}{2} & \frac{kqa}{t}(p-t) + \frac{ka}{t}(\theta_\infty - \theta_0) \\ -\frac{1}{kta}(pq + \theta_0 + \theta_\infty) & \frac{H}{2} + \frac{pq}{2t} + \frac{\theta_\infty^2-\theta_0^2}{2t} - \frac{t}{2} \end{pmatrix}, \tag{52}$$

at $z = 0$. In (50) and (52),

$$H = \frac{1}{t}(2p^2q^2 + p(2t - 2tq^2 + (4\theta_\infty - 1)q) - 2tq(\theta_0 + \theta_\infty) + \theta_\infty^2 - \theta_0^2), \tag{53}$$

and G_0 diagonalizes matrix A_{-2},

$$G_0^{-1}A_{-2}G_0 = -\frac{t\sigma_3}{2},$$

and it is chosen in the form

$$G_0 = \frac{1}{\sqrt{k}}\begin{pmatrix} k & -k \\ \frac{p}{t} & \frac{t-p}{t} \end{pmatrix} a^{-\frac{\sigma_3}{2}}, \tag{54}$$

with a being an extra gauge parameter, so that the full space $\mathcal{A}$ is seven-dimensional,

$$\mathcal{A} = \{p, q, k, a, t, \theta_0, \theta_\infty\}. \tag{55}$$

From the series (49) and (51) it follows that θ_∞ and θ_0 are the formal monodromy exponents and the parameter t is the isomonodromic time. The isomonodromicity with respect to t yields the second differential equation for $\Phi(z) \equiv \Phi(z,t)$,

$$\frac{d\Phi}{dt} = B(z)\Phi, \qquad B(z) = B_1 z + B_0 + \frac{B_{-1}}{z}, \tag{56}$$

where,

$$B_1 = \frac{1}{2}\begin{pmatrix} 1 & 0 \\ 0 & -1 \end{pmatrix}, \quad B_0 = \frac{1}{t}\begin{pmatrix} 0 & -qkt \\ \frac{pq(t-p)}{kt} + \frac{\theta_0+\theta_\infty}{k} - \frac{2\theta_\infty p}{kt} & 0 \end{pmatrix},$$

$$B_{-1} = \begin{pmatrix} \frac{t-2p}{2t} & k \\ \frac{p(t-p)}{kt^2} & \frac{2p-t}{2t} \end{pmatrix}.$$

The compatibility condition of the matrix equations (48) and (56) implies the following dynamical system on (55),

$$\begin{aligned} \frac{dq}{dt} &= \frac{4pq^2}{t} - 2q^2 + \frac{q(4\theta_\infty - 1)}{t} + 2, \\ \frac{dp}{dt} &= -\frac{4p^2q}{t} + \frac{p(4tq - 4\theta_\infty + 1)}{t} + 2\theta_0 + 2\theta_\infty, \\ \frac{dk}{dt} &= -\frac{4pqk}{t} + 2qk - \frac{2\theta_\infty k}{t}, \qquad \frac{da}{dt} = \frac{a}{t}(2qt + 2\theta_0), \\ \frac{d\theta_\infty}{dt} &= 0, \quad \frac{d\theta_0}{dt} = 0. \end{aligned} \tag{57}$$

It should be also mentioned, that the fourth equation, i.e. the equation for the function $a(t)$, follows from plugging (51) into Eq. (56) — the second equation of the Lax pair, and equating the terms of zero order with respect to z.

The last two equations of system (57) just state that θ_∞ and θ_0 as the part of the monodromy data, are constant. The third and the fourth equations give $\ln k(t)$ and $\ln a(t)$ as the antiderivatives of the simple combinations of p and q. The first two equations are equivalent to the third Painlevé equation,

$$q_{tt} = \frac{(q_t)^2}{q} - \frac{q_t}{t} + \frac{1}{t}(\alpha q^2 + \beta) + \gamma q^3 + \frac{\delta}{q}, \tag{58}$$

where

$$\alpha = 8\theta_0, \quad \beta = 4 - 8\theta_\infty, \quad \gamma = 4, \quad \delta = -4.$$

Assuming that θ_∞ and θ_0 are numerical constants, the function (53) becomes the Hamiltonian of (58) with p, q being the canonical variables. Also, if we denote

$$p_1 = p, \quad q_1 = q, \quad p_2 = \ln k, \quad q_2 = \theta_\infty, \quad p_3 = \ln a, \quad q_3 = \theta_0, \tag{59}$$

then the whole system (57) becomes Hamiltonian with the same Hamiltonian (53), i.e. with

$$H = \frac{1}{t}(2p_1^2 q_1^2 + p_1(2t - 2tq_1^2 + (4q_2 - 1)q_1) - 2tq_1(q_2 + q_3) + q_2^2 - q_3^2),$$

and with respect to the symplectic form,

$$\Omega = dp_1 \wedge dq_1 + dp_2 \wedge dq_2 + dp_3 \wedge dq_3.$$

The general formulae (3) and (12) transform, in the case of system (48), into the equations

$$\begin{aligned} \omega_{\mathrm{JMU}} = &- \operatorname{res}_{z=\infty} \operatorname{Tr}\left((G^{(\infty)}(z))^{-1} \frac{dG^{(\infty)}(z)}{dz} \frac{d\Theta_\infty(z)}{dt} \right) dt \\ &- \operatorname{res}_{z=0} \operatorname{Tr}\left((G^{(0)}(z))^{-1} \frac{dG^{(0)}(z)}{dz} \frac{d\Theta_0(z)}{dt} \right) dt \end{aligned} \tag{60}$$

and

$$\omega = \operatorname{res}_{z=\infty} \operatorname{Tr}(A(z) dG^{(\infty)}(z) G^{(\infty)}(z)^{-1}) + \operatorname{res}_{z=0} \operatorname{Tr}(A(z) dG^{(0)}(z) G^{(0)}(z)^{-1}), \tag{61}$$

respectively. Substituting the series $G^{(\infty,0)}(z)$ and the exponentials $\Theta_{\infty,0}(z)$ from (49) and (51) into (60), we obtain that

$$\omega_{\mathrm{JMU}} = -\frac{1}{2}\operatorname{Tr}(g_{\infty,1}\sigma_3)dt - \frac{1}{2}\operatorname{Tr}(g_{0,1}\sigma_3)dt$$

and using (50) and (52) we arrive at the final formula for ω_{JMU},

$$\omega_{\mathrm{JMU}} \equiv \frac{d \ln \tau}{dt} dt = H dt + \frac{pq}{t} dt - t dt. \tag{62}$$

Note the additional to Hdt terms in the right-hand side of (62). Similar substitution of $G^{(\infty,0)}(z)$ from (49) and (51) into (61) leads us to the formula,

$$\omega = \mathrm{Tr}(A_{-1}dG_0G_0^{-1} + G_0^{-1}A_{-2}G_0dg_{0,1} - A_0dg_{\infty,1}),$$

and using again (50), (52) and (54) we arrive at the equation,

$$\omega = pdq + tdH - \theta_\infty\frac{dk}{k} - \theta_0\frac{da}{a} - tdt.$$

After regrouping the terms we obtain that

$$\omega = pdq - Hdt + d\left(Ht - \theta_\infty \ln k - \theta_0 \ln a - \frac{t^2}{2}\right) + \ln k\, d\theta_\infty + \ln a\, d\theta_0,$$

or, using the definition (59) of the canonical coordinates,

$$\omega = p_1dq_1 + p_2dq_2 + p_3dq_3 - Hdt + d\left(Ht - q_2p_2 - q_3p_3 - \frac{t^2}{2}\right). \tag{63}$$

Equation (63) proves Conjectures 1 and 2, with $\gamma = 1$, in the case of the 2×2 system (48) and gives the explicit formula for $G(p_j, q_j, t)$,

$$G(p_1,p_2,p_3,q_1,q_2,q_3,t) = Ht - q_2p_2 - q_3p_3 - \frac{t^2}{2}.$$

The corresponding equation (18) is

$$\frac{d\ln\tau}{dt} = p\frac{dq}{dt} - H + \frac{d}{dt}\left(Ht - \theta_\infty \ln k - \theta_0 \ln a - \frac{t^2}{2}\right). \tag{64}$$

Remark. One can deduce from (57) that

$$\frac{pq}{t} = \frac{1}{4}\frac{d}{dt}\ln\frac{a}{k} - \frac{\theta_0+\theta_\infty}{2t}.$$

Combining this with (62) and (64), we arrive at the equation,

$$Hdt = pdq - Hdt + d\left(Ht + \frac{1-4\theta_\infty}{4}\ln k - \frac{1+4\theta_0}{4}\ln a + \frac{\theta_0+\theta_\infty}{2}\ln t\right),$$

where $df \equiv d_tf = \frac{df}{dt}$. In other words, although the truncated action, Hdt, is not in this case exactly the Jimbo–Miwa–Ueno form ω_{JMU}, it still coincides with the full classical action, up to a total differential. As we will see, this is true in all other examples when accidentally $Hdt \neq \omega_{\mathrm{JMU}}$.

3.4. *Painlevé IV*

This time (see again [21]), the linear system is the 2×2 system with one irregular singular point at $z=\infty$ with the Poincaré rank 2 and one Fuchsian point at $z=0$:

$$\frac{d\Phi}{dz} = A(z)\Phi, \qquad A(z) = \frac{A_{-1}}{z} + A_0 + A_1z, \tag{65}$$

where

$$A_1 = \begin{pmatrix} 1 & 0 \\ 0 & -1 \end{pmatrix}, \quad A_0 = \begin{pmatrix} t & k \\ -\frac{q(4p-q-2t)+4\theta_\infty}{2k} & -t \end{pmatrix},$$

$$A_{-1} = \frac{1}{2}\begin{pmatrix} \frac{q(4p-q-2t)}{2} & -kq \\ \frac{q^2(4p-q-2t)^2-16\theta_0^2}{4kq} & -\frac{q(4p-q-2t)}{2} \end{pmatrix}.$$

The corresponding formal solution at $z=\infty$ is

$$\Phi_{\text{form}}(z) = \left(\mathrm{I} + \frac{g_1}{z} + \frac{g_2}{z^2} + O\left(\frac{1}{z^3}\right)\right)e^{\Theta(z)}, \quad \Theta(z) = \sigma_3\left(\frac{z^2}{2} + tz - \theta_\infty \ln z\right) \tag{66}$$

with

$$g_1 = \frac{1}{2}\begin{pmatrix} -\frac{2H+q}{2} & -k \\ -\frac{q(4p-q-2t)+4\theta_\infty}{2k} & \frac{2H+q}{2} \end{pmatrix},$$

$$g_2 = \frac{1}{8}\begin{pmatrix} \frac{1}{4}((2H+q+2t)^2-4t^2-8\theta_0^2+8\theta_\infty^2) & -k\,(2H-q-4t) \\ \frac{1}{2k}((2H-q)(q(4p-q-2t)+4\theta_\infty+4)+8q) & \frac{1}{4}((2H+q+2t)^2-4t^2+8\theta_0^2-8\theta_\infty^2) \end{pmatrix}, \tag{67}$$

and

$$H = 2p^2q - \frac{1}{8}q^3 - \frac{1}{2}tq^2 + \frac{1}{2}(2\theta_\infty - 1 - t^2)q + 2\theta_\infty t - \frac{2\theta_0^2}{q}. \tag{68}$$

The behavior of the solution of (65) at the (non-resonant, this time) Fuchsian point $z=0$ is described by the equation,

$$\Phi^{(0)}(z) = G_0(\mathrm{I} + O(z))z^{\theta_0\sigma_3}, \quad z \to 0, \tag{69}$$

where G_0 diagonalizes the matrix A_{-1},

$$G_0^{-1}A_{-1}G_0 = \theta_0\sigma_3$$

and it is chosen in the form,

$$G_0 = \frac{1}{2\sqrt{kq\theta_0}}\begin{pmatrix} -kq & -kq \\ -\frac{q(4p-q-2t)-4\theta_0}{2} & -\frac{q(4p-q-2t)+4\theta_0}{2} \end{pmatrix} a^{-\frac{\sigma_3}{2}}. \tag{70}$$

The full parameter space,

$$\mathcal{A} = \{p, q, k, a, t, \theta_0, \theta_\infty\},$$

is again seven-dimensional with t being the isomonodromic time and θ_∞ and θ_0 serving as the formal monodromy exponents at the respective singular points. The isomonodromicity with respect to t yields the second differential equation for $\Phi(z)$,

$$\frac{d\Phi}{dt} = B(z)\Phi, \qquad B(z) = B_1 z + B_0, \tag{71}$$

where

$$B_1 = A_1, \quad B_0 = \begin{pmatrix} 0 & k \\ \frac{q(4p-q-2t)+4\theta_\infty}{2k} & 0 \end{pmatrix}$$

and the compatibility of (71) and (65) implies,

$$\begin{aligned} \frac{dq}{dt} &= 4pq, \\ \frac{dp}{dt} &= -2p^2 + \frac{3}{8}q^2 + qt + \frac{1}{2}t^2 - \theta_\infty + \frac{1}{2} - \frac{2\theta_0^2}{q^2}, \\ \frac{dk}{dt} &= -(q+2t)k, \quad \frac{da}{dt} = \frac{4\theta_0}{q}a, \\ \frac{d\theta_\infty}{dt} &= 0, \quad \frac{d\theta_0}{dt} = 0. \end{aligned} \tag{72}$$

As in the previous section, the fourth equation follows from the substitution of (69) into (71).

Similar to the previous cases, the last equations of (72) manifests the time-independence of the formal monodromy exponents, the third and the fourth equations express k and a in terms of p and q, while the first and the second equations are equivalent to a Painlevé equation, this time to the fourth Painlevé equation,

$$q_{tt} = \frac{(q_t)^2}{2q} + \frac{3}{2}q^3 + 4tq^2 + 2(t^2 - \alpha)q + \frac{\beta}{q}, \tag{73}$$

where

$$\alpha = 2\theta_\infty - 1, \quad \beta = -8\theta_0^2.$$

Assuming that θ_∞ and θ_0 are numerical constants, the function (68) becomes the Hamiltonian of (73) with p, q being the canonical variables. Also, if we again denote

$$p_1 = p, \quad q_1 = q, \quad p_2 = \ln k, \quad q_2 = \theta_\infty, \quad p_3 = \ln a, \quad q_3 = \theta_0, \tag{74}$$

then the whole system (72) becomes Hamiltonian with the same Hamiltonian (68), i.e. with

$$H = 2p_1^2 q_1 - \frac{1}{8}q_1^3 - \frac{1}{2}tq_1^2 + \frac{1}{2}(2q_2 - 1 - t^2)q_1 + 2q_2 t - \frac{2q_3^2}{q_1},$$

and with respect to the symplectic form,

$$\Omega = dp_1 \wedge dq_1 + dp_2 \wedge dq_2 + dp_3 \wedge dq_3.$$

The general formulae (3) and (12) transform, in the case of system (65), into the equations,

$$\omega_{\mathrm{JMU}} = -\operatorname{res}_{z=\infty} \operatorname{Tr}\left((G^{(\infty)}(z))^{-1} \frac{dG^{(\infty)}(z)}{dz} \frac{d\Theta_\infty(z)}{dt} \right) dt \tag{75}$$

and

$$\omega = \operatorname{res}_{z=\infty} \operatorname{Tr}(A(z)dG^{(\infty)}(z)G^{(\infty)}(z)^{-1}) + \operatorname{res}_{z=0} \operatorname{Tr}(A(z)dG^{(0)}(z)G^{(0)}(z)^{-1}), \tag{76}$$

respectively. Substituting the series $G^{(\infty)}(z)$ and the exponentials $\Theta_\infty(z)$ from (66) into (75), and using (67) we obtain that

$$\omega_{\mathrm{JMU}} \equiv \frac{d\ln\tau}{dt}dt = Hdt + \frac{1}{2}qdt. \tag{77}$$

Note again the additional to Hdt term in the right-hand side of (77). Similar substitution of $G^{(\infty,0)}(z)$ from (66) and (69) into (76) followed by the use of (67) and (70) leads us to the formulae,

$$\begin{aligned}\omega &= \operatorname{Tr}(A_{-1}dG_0G_0^{-1} - A_1dg_2 + A_1dg_1g_1 - A_0dg_1)\\ &= -\frac{1}{2}qdp + \frac{1}{2}pdq + \frac{1}{2}tdH - \frac{1}{2}Hdt - \theta_\infty\frac{dk}{k} - \theta_0\frac{da}{a} + \theta_0d\theta_0 - \frac{2\theta_\infty - 1}{2}d\theta_\infty.\end{aligned}$$

Regrouping the last equation, we arrive at the final answer for the form ω,

$$\begin{aligned}\omega = pdq - Hdt + d\left(\frac{Ht}{2} - \frac{pq}{2} - \theta_\infty\ln k - \theta_0\ln a + \frac{\theta_0^2}{2} + \frac{\theta_\infty}{2} - \frac{\theta_\infty^2}{2}\right)\\ + \ln k\, d\theta_\infty + \ln a\, d\theta_0\end{aligned}$$

or, using the definition (74) of the canonical coordinates,

$$\begin{aligned}\omega &= p_1dq_1 + p_2dq_2 + p_3dq_3 - Hdt\\ &\quad + d\left(\frac{Ht}{2} - \frac{p_1q_1}{2} - q_2p_2 - q_3p_3 + \frac{q_3^2}{2} + \frac{q_2}{2} - \frac{q_2^2}{2}\right).\end{aligned} \tag{78}$$

Equation (78) proves Conjectures 1 and 2, with $\gamma = 1$, in the case of the 2×2 system (65) and gives the explicit formula for $G(p_j, q_j, t)$,

$$G(p_1,p_2,p_3,q_1,q_2,q_3,t) = \frac{Ht}{2} - \frac{p_1q_1}{2} - q_2p_2 - q_3p_3 + \frac{q_3^2}{2} + \frac{q_2}{2} - \frac{q_2^2}{2}.$$

The corresponding equation (18) and the formula for the truncated action are

$$\frac{d\ln\tau}{dt} = p\frac{dq}{dt} - H + \frac{d}{dt}\left(\frac{Ht}{2} - \frac{pq}{2} - \theta_\infty\ln k - \theta_0\ln a\right),$$

and

$$Hdt = pdq - Hdt + d\left(\frac{Ht}{2} - \frac{pq}{2} + \frac{1-2\theta_\infty}{2}\ln k - \theta_0\ln a + \frac{t^2}{2}\right), \quad d \equiv d_t,$$

respectively.

3.5. *Painlevé V*

In [21], the following linear system is associated with the fifth Painlevé equation,

$$\frac{d\Phi}{dz} = A(z)\Phi, \qquad A(z) = A_2 + \frac{A_0}{z} + \frac{A_1}{z-1}, \tag{79}$$

where

$$A_2 = \frac{t}{2}\begin{pmatrix} 1 & 0 \\ 0 & -1 \end{pmatrix},$$

$$A_0 = \begin{pmatrix} -pq - \theta_\infty - \theta_1 & k(pq + \theta_\infty + \theta_1 - \theta_0) \\ -\frac{1}{k}(pq + \theta_\infty + \theta_1 + \theta_0) & pq + \theta_\infty + \theta_1 \end{pmatrix},$$

$$A_1 = \begin{pmatrix} pq + \theta_1 & -kq(pq + 2\theta_1) \\ \frac{p}{k} & -pq - \theta_1 \end{pmatrix}.$$

This system has one irregular singular point of Poincaré rank 1 at $z = \infty$ and two Fuchsian singular points, $z = 0$ and $z = 1$. The corresponding formal solution at $z = \infty$ is given by the formulae,

$$\Phi_{\text{form}}(z) = \left(I + \frac{g_1}{z} + O\left(\frac{1}{z^2}\right)\right) e^{\Theta(z)}, \quad z \to \infty, \quad \Theta(z) = \sigma_3\left(\frac{tz}{2} - \theta_\infty \ln z\right) \tag{80}$$

with

$$g_1 = \begin{pmatrix} -H & \frac{k(pq^2 - pq + 2\theta_1 q - \theta_\infty - \theta_1 + \theta_0)}{t} \\ -\frac{2pq - 2p + 2\theta_\infty + 2\theta_1 + 2\theta_0}{tk} & H \end{pmatrix} \tag{81}$$

and

$$H = \frac{p^2(q-1)^2 q}{t} + p\left(\frac{q^2}{t}(\theta_0 + 3\theta_1 + \theta_\infty) + \frac{q}{t}(t - 2\theta_\infty - 4\theta_1) + \frac{1}{t}(\theta_\infty + \theta_1 - \theta_0)\right)$$
$$+ \frac{q2\theta_1}{t}(\theta_\infty + \theta_1 + \theta_0) + \frac{\theta_0^2 - \theta_1^2 - \theta_\infty^2 + \theta_1 t - 2\theta_1\theta_\infty}{t}. \tag{82}$$

The behavior of the solutions of (79) at the (non-resonant) Fuchsian points $z = 0$ and $z = 1$ are described by the equations,

$$\Phi^{(0)}(z) = G_0(\mathrm{I} + O(z))z^{\theta_0\sigma_3}, \quad z \to 0, \tag{83}$$

and

$$\Phi^{(1)}(z) = G_1(\mathrm{I} + O(z-1))(z-1)^{\theta_1\sigma_3}, \quad z \to 1, \tag{84}$$

respectively. The matrices G_0 and G_1 diagonalize the matrix coefficients A_0 and A_1,

$$G_0^{-1}A_0G_0 = \theta_0\sigma_3, \quad G_1^{-1}A_1G_1 = \theta_1\sigma_3,$$

and are chosen in the form,

$$G_0 = \frac{1}{\sqrt{-4k\theta_0}}\begin{pmatrix} k(2pq + 2\theta_\infty + 2\theta_1 - 2\theta_0) & k \\ 2pq + 2\theta_\infty + 2\theta_1 + 2\theta_0 & 1 \end{pmatrix} a^{-\frac{\sigma_3}{2}}, \tag{85}$$

and

$$G_1 = \frac{1}{\sqrt{2k\theta_1}} \begin{pmatrix} k(pq+2\theta_1) & kq \\ p & 1 \end{pmatrix} b^{-\frac{\sigma_3}{2}}. \tag{86}$$

The full space

$$\mathcal{A} = \{p, q, k, a, b, t, \theta_0, \theta_1, \theta_\infty\}$$

is nine-dimensional with t being the isomonodromic time and θ_∞, θ_0, and θ_1 serving as the formal monodromy exponents at the respective singular points. The isomonodromicity with respect to t yields the second differential equation for $\Phi(z)$,

$$\frac{d\Phi}{dt} = B(z)\Phi, \qquad B(z) = B_1 z + B_0, \tag{87}$$

where

$$B_1 = \frac{A_2}{t},$$

$$B_0 = \begin{pmatrix} 0 & \frac{k}{t}(-pq^2 + pq - 2\theta_1 q + \theta_\infty + \theta_1 - \theta_0) \\ -\frac{1}{tk}(pq + \theta_\infty - p + \theta_1 + \theta_0) & 0 \end{pmatrix},$$

and the compatibility of (87) and (79) implies,

$$\begin{aligned}
\frac{dq}{dt} &= \frac{2pq(q-1)^2}{t} + \frac{q^2}{t}(\theta_0 + 3\theta_1 + \theta_\infty) + \frac{q}{t}(t - 2\theta_\infty - 4\theta_1) + \frac{1}{t}(\theta_\infty + \theta_1 - \theta_0),\\
\frac{dp}{dt} &= -\frac{p^2}{t}(3q^2 - 4q + 1) - p\left(\frac{2q}{t}(\theta_0 + 3\theta_1 + \theta_\infty) + \frac{1}{t}(t - 2\theta_\infty - 4\theta_1)\right)\\
&\quad - \frac{2\theta_1}{t}(\theta_\infty + \theta_1 + \theta_0),\\
\frac{dk}{dt} &= -\frac{k}{t}(pq^2 - 2pq + p + 2\theta_1 q - 2\theta_\infty - 2\theta_1),\\
\frac{da}{dt} &= \frac{a}{t}(p - pq^2 - 2\theta_1 q - 2\theta_0),\\
\frac{db}{dt} &= -\frac{b}{t}(3pq^2 + p - 4pq + 2\theta_\infty q + 4\theta_1 q + 2\theta_0 q - 2\theta_\infty - 2\theta_1 + t),\\
\frac{d\theta_\infty}{dt} &= 0, \quad \frac{d\theta_0}{dt} = 0, \quad \frac{d\theta_1}{dt} = 0.
\end{aligned} \tag{88}$$

As before, the equations for a and b follow from the substitution of (83) into (87) and (84) into (87), respectively, and they simply express the functions $a(t)$ and $b(t)$ in terms of p and q. The third equation in (88) is also trivial — just an expression of k in terms of p and q, and the last three equations are the manifestation of the time-independence of the formal monodromy exponents. The nontrivial first two equations are equivalent to the fifth Painlevé equation,

$$q_{tt} = \left(\frac{1}{2q} + \frac{1}{q-1}\right)(q_t)^2 - \frac{q_t}{t} + \frac{(q-1)^2}{t^2}\left(\alpha q + \frac{\beta}{q}\right) + \gamma\frac{q}{t} + \delta\frac{q(q+1)}{(q-1)}, \tag{89}$$

where

$$\alpha = \frac{(\theta_0 - \theta_1 + \theta_\infty)^2}{2}, \quad \beta = -\frac{(\theta_0 - \theta_1 - \theta_\infty)^2}{2}, \quad \gamma = (1 - 2\theta_0 - 2\theta_1), \quad \delta = -\frac{1}{2}.$$

Assuming that θ_∞, θ_0, and θ_1 are numerical constants, the function (82) becomes the Hamiltonian of (89) with p, q being the canonical variables. Also, if we denote

$$p_1 = p, \quad q_1 = q, \quad p_2 = \ln k, \quad q_2 = \theta_\infty, \quad p_3 = \ln a, \quad q_3 = \theta_0, \quad p_4 = \ln b, \quad q_4 = \theta_1, \tag{90}$$

then the whole system (88) becomes Hamiltonian with the same Hamiltonian (82), i.e. with

$$H = \frac{p_1^2(q_1-1)^2 q_1}{t} + p_1\left(\frac{q_1^2}{t}(q_3 + 3q_4 + q_2) + \frac{q_1}{t}(t - 2q_2 - 4q_4) + \frac{1}{t}(q_2 + q_4 - q_3)\right)$$
$$+ \frac{2q_1 q_4}{t}(q_2 + q_4 + q_3) + \frac{q_3^2 - q_4^2 - q_2^2 + q_4 t - 2q_4 q_2}{t},$$

and with respect to the symplectic form,

$$\Omega = dp_1 \wedge dq_1 + dp_2 \wedge dq_2 + dp_3 \wedge dq_3 + dp_4 \wedge dq_4.$$

The general formulae (3) and (12) transform, in the case of system (79), into the equations,

$$\omega_{\mathrm{JMU}} = -\operatorname{res}_{z=\infty} \operatorname{Tr}\left((G^{(\infty)}(z))^{-1} \frac{dG^{(\infty)}(z)}{dz} \frac{d\Theta_\infty(z)}{dt}\right) dt \tag{91}$$

and

$$\omega = \operatorname{res}_{z=\infty} \operatorname{Tr}(A(z) dG^{(\infty)}(z) G^{(\infty)}(z)^{-1}) + \operatorname{res}_{z=0} \operatorname{Tr}(A(z) dG^{(0)}(z) G^{(0)}(z)^{-1})$$
$$+ \operatorname{res}_{z=1} \operatorname{Tr}(A(z) dG^{(1)}(z) G^{(1)}(z)^{-1}), \tag{92}$$

respectively. Substituting the series $G^{(\infty)}(z)$ and the exponent $\Theta_\infty(z)$ from (80) into (91) and using (81), we obtain that, similar to the Painlevé II case,

$$\omega_{\mathrm{JMU}} \equiv \frac{d\ln\tau}{dt} dt = H dt. \tag{93}$$

Substituting $G^{(\infty,0,1)}(z)$ from (80), (83), and (84) into (92) and using after that (81), (85), (86) leads us to the formulae,

$$\omega = \operatorname{Tr}(A_0 dG_0 G_0^{-1} + A_1 dG_1 G_1^{-1} - A_2 dg_1)$$
$$= pdq + tdH - \theta_\infty \frac{dk}{k} - \theta_0 \frac{da}{a} - \theta_1 \frac{db}{b} + d\theta_0 + d\theta_1.$$

Regrouping the last equation, we arrive at the final answer for the form ω,

$$\omega = pdq - Hdt + d(Ht - \theta_\infty \ln k - \theta_0 \ln a - \theta_1 \ln b + \theta_0 + \theta_1)$$
$$+ \ln a \, d\theta_0 + \ln b \, d\theta_1 + \ln k \, d\theta_\infty,$$

or, using the definition (90) of the canonical coordinates,

$$\omega = p_1 dq_1 + p_2 dq_2 + p_3 dq_3 + p_4 dq_4 - Hdt + d(Ht - q_2 p_2 - q_3 p_3 - q_4 p_4 + q_3 + q_4). \tag{94}$$

Equation (94) proves Conjectures 1 and 2, with $\gamma = 1$, in the case of the 2×2 system (79) and gives the explicit formula for $G(p_j, q_j, t)$,

$$G(p_1, p_2, p_3, p_4, q_1, q_2, q_3, q_4, t) = Ht - q_2p_2 - q_3p_3 - q_4p_4 + q_3 + q_4.$$

The corresponding equation (18) is

$$\frac{d \ln \tau}{dt} = p\frac{dq}{dt} - H + \frac{d}{dt}(Ht - \theta_\infty \ln k - \theta_0 \ln a - \theta_1 \ln b).$$

Again, this equation together with (93) make an identity, which, this time, would not be so easy to check directly.

3.6. *Painlevé VI*

Consider the 2×2 Fuchsian system with four regular singularities at 0, 1, t and ∞,

$$\frac{d\Phi}{dz} = A(z)\Phi, \qquad A(z) = \frac{A_0}{z} + \frac{A_t}{z-t} + \frac{A_1}{z-1}, \tag{95}$$

where

$$A_0, A_1, A_t \in \mathfrak{sl}_2(\mathbb{C}), \quad A_0 + A_1 + A_t = -\theta_\infty \sigma_3.$$

Following to [21], we introduce the parametrization,

$$A_0 = \begin{pmatrix} x_0 + \theta_0 & -ux_0 \\ \frac{x_0+2\theta_0}{u} & -x_0 - \theta_0 \end{pmatrix}, \quad A_1 = \begin{pmatrix} x_1 + \theta_1 & -vx_1 \\ \frac{x_1+2\theta_1}{v} & -x_1 - \theta_1 \end{pmatrix},$$

$$A_t = \begin{pmatrix} x_t + \theta_t & -wx_t \\ \frac{x_t+2\theta_t}{w} & -x_t - \theta_t \end{pmatrix}.$$

Observe that, $\pm\theta_0, \pm\theta_1, \pm\theta_t$ are the eigenvalues of A_0, A_1, A_t, and that the following constraints are satisfied,

$$x_0 + \theta_0 + x_1 + \theta_1 + x_t + \theta_t = -\theta_\infty, \tag{96}$$

$$ux_0 + vx_1 + wx_t = 0, \tag{97}$$

$$\frac{x_0 + 2\theta_0}{u} + \frac{x_1 + 2\theta_1}{v} + \frac{x_t + 2\theta_t}{w} = 0. \tag{98}$$

We also introduce the parameters k and q by writing the entry $A_{12}(z)$ of the matrix $A(z)$ as,

$$A_{1,2}(z) = \frac{k(z-q)}{z(z-1)(z-t)}.$$

Notice that,

$$ux_0(1+t) + vx_1 + wx_t = k, \quad ux_0t = kq. \tag{99}$$

Finally we put

$$p = A_{11}(q) = \frac{x_0 + \theta_0}{q} + \frac{x_1 + \theta_1}{q-1} + \frac{x_t + \theta_t}{q-t}. \tag{100}$$

Solving Eq. (97) and (99) with respect to u, v, w, we get

$$u = \frac{kq}{x_0 t}, \quad v = \frac{k(q-1)}{x_1(1-t)}, \quad w = \frac{k(t-q)}{x_t t(1-t)}. \tag{101}$$

Next, we express x_1, x_t from (96), (100), and then we express x_0 from (98). The result is

$$\begin{aligned}
x_0 &= \frac{p^2q^2(q-1)(q-t)}{t2\theta_\infty} + \frac{pq(q-1)(q-t)}{t} + \frac{\theta_\infty q(q-t-1)}{2t} + \frac{\theta_1^2(t-1)}{2t\theta_\infty(q-1)} \\
&\quad - \frac{\theta_t^2 t(t-1)}{2\theta_\infty(q-t)} - \frac{\theta_1^2(1-t)}{2t\theta_\infty} - \frac{\theta_t^2 t(t-1)}{2t\theta_\infty} - \theta_0 - \frac{\theta_0^2}{2\theta_\infty}, \\
x_1 &= \frac{p^2q(q-1)^2(t-q)}{(t-1)2\theta_\infty} + \frac{pq(q-1)(t-q)}{t-1} + \frac{\theta_\infty(q-1)(t-q-1)}{2(t-1)} - \frac{\theta_0^2 t}{2q\theta_\infty(t-1)} \\
&\quad + \frac{\theta_t^2 t(t-1)}{2\theta_\infty(q-t)} + \frac{\theta_0^2 t}{2(t-1)\theta_\infty} + \frac{\theta_t^2 t(t-1)}{2(t-1)\theta_\infty} - \theta_1 - \frac{\theta_1^2}{2\theta_\infty}, \\
x_t &= \frac{p^2q(q-1)(t-q)^2}{t(t-1)2\theta_\infty} + \frac{pq(q-1)(q-t)}{t(t-1)} + \frac{\theta_\infty(q-t)(q+t-1)}{2t(t-1)} + \frac{\theta_0^2 t}{2q\theta_\infty(t-1)} \\
&\quad - \frac{\theta_1^2(t-1)}{2\theta_\infty t(q-1)} - \frac{\theta_0^2}{2(t-1)\theta_\infty} + \frac{\theta_1^2}{2t\theta_\infty} - \theta_t - \frac{\theta_t^2}{2\theta_\infty}.
\end{aligned} \tag{102}$$

Equations (101) and (102) provide parametrization of the matrices A_0, A_1, and A_t by the parameters q, p, k, θ_0, θ_1, θ_t, θ_∞, which will prove to be the Darboux coordinates, and by the parameter t which is the isomonodromic time.

Solutions of (95) have the following behavior at $z = 0, 1, t$, and ∞,

$$\begin{aligned}
\Phi^{(\infty)}(z) &= (\mathrm{I} + O(z^{-1}))z^{-\theta_\infty\sigma_3}, \quad z \to \infty. \\
\Phi^{(0)}(z) &= G_0(\mathrm{I} + O(z))z^{\theta_0\sigma_3}, \quad z \to 0, \\
\Phi^{(1)}(z) &= G_1(\mathrm{I} + O(z-1))(z-1)^{\theta_1\sigma_3}, \quad z \to 1, \\
\Phi^{(t)}(z) &= G_t(\mathrm{I} + g_1(z-t) + O((z-t)^2))(z-t)^{\theta_t\sigma_3}, \quad z \to t,
\end{aligned}$$

where

$$g_1 = \begin{pmatrix} \frac{H}{2\theta_t} - p\frac{q(q-1)}{2\theta_t t(t-1)} - \frac{\theta_\infty(q-t)}{2\theta_t t(t-1)} & \frac{Hc}{2\theta_t(1-2\theta_t)} - p\frac{q(q-1)c}{2\theta_t t(t-1)} - \frac{\theta_\infty(q-t)c}{2\theta_t t(t-1)} \\ & \quad - \frac{\theta_t(2qt-t^2-q)c}{(2\theta_t-1)t(t-1)(q-t)} \\ \frac{H}{2\theta_t(1+2\theta_t)c} + p\frac{q(q-1)}{2\theta_t t(t-1)c} + \frac{\theta_\infty(q-t)}{2\theta_t t(t-1)c} & -\frac{H}{2\theta_t} + p\frac{q(q-1)}{2\theta_t t(t-1)} + \frac{\theta_\infty(q-t)}{2\theta_t t(t-1)} \\ \quad - \frac{\theta_t(2qt-t^2-q)}{(2\theta_t+1)t(t-1)(q-t)c} & \end{pmatrix} \tag{103}$$

and

$$\begin{aligned}
H &= p^2\frac{q(q-1)(q-t)}{t(t-1)} + p\frac{q(q-1)}{t(t-1)} + \frac{\theta_\infty(1-\theta_\infty)(q-t)}{t(t-1)} + \frac{\theta_0^2(q-t)}{qt(t-1)} \\
&\quad - \frac{\theta_1^2(q-t)}{(q-1)t(t-1)} + \frac{\theta_t^2(t^2-q(2t-1))}{(q-t)t(t-1)}.
\end{aligned} \tag{104}$$

The matrices G_0, G_1, and G_t diagonalize the matrix residues A_0, A_1, and A_t,

$$G_0^{-1}A_0G_0 = \theta_0\sigma_3, \quad G_1^{-1}A_1G_1 = \theta_1\sigma_3, \quad G_t^{-1}A_tG_t = \theta_t\sigma_3$$

and they are chosen in the form,

$$G_0 = \sqrt{\frac{kq}{t}}\begin{pmatrix} 1 & 1 \\ \frac{1}{u} & \frac{x_0+2\theta_0}{ux_0} \end{pmatrix} a^{-\frac{\sigma_3}{2}}, \tag{105}$$

$$G_1 = \sqrt{\frac{k(q-1)}{1-t}}\begin{pmatrix} 1 & 1 \\ \frac{1}{v} & \frac{x_1+2\theta_1}{vx_1} \end{pmatrix} b^{-\frac{\sigma_3}{2}}, \tag{106}$$

$$G_t = \sqrt{\frac{k(t-q)}{t(1-t)}}\begin{pmatrix} 1 & 1 \\ \frac{1}{w} & \frac{x_t+2\theta_t}{wx_t} \end{pmatrix} c^{-\frac{\sigma_3}{2}}. \tag{107}$$

The whole parameter space $\mathcal{A}$ has dimension 11,

$$\mathcal{A} = \{p, q, k, a, b, c, t, \theta_0, \theta_1, \theta_t, \theta_\infty\}. \tag{108}$$

The isomonodromicity with respect to t yields the second differential equation for $\Phi(z)$,

$$\frac{d\Phi}{dt} = -\frac{A_t}{z-t}\Phi \tag{109}$$

and the equations

$$\frac{dG_0}{dt} = \frac{A_t}{t}G_0, \quad \frac{dG_1}{dt} = \frac{A_t}{t-1}G_1, \quad \frac{dG_t}{dt} = \left(\frac{A_0}{t} + \frac{A_1}{t-1}\right)G_t, \tag{110}$$

for the gauge matrices G_0, G_1, G_t. The compatibility of (109) and (95) together with the Eqs. (110) imply the following dynamical system on (108),

$$\frac{dq}{dt} = \frac{2pq(q-1)(q-t)}{t(t-1)} + \frac{q(q-1)}{t(t-1)},$$

$$\frac{dp}{dt} = \frac{1}{4t(t-1)}(4p^2(2tq-3q^2-t+2q) + 4p(1-2q) + 4\theta_\infty(\theta_\infty-1))$$

$$-\frac{\theta_0^2}{q^2(t-1)} + \frac{\theta_1^2}{t(q-1)^2} - \frac{\theta_t^2}{(q-t)^2},$$

$$\frac{dk}{dt} = \frac{k(2\theta_\infty-1)(q-t)}{t(t-1)},$$

$$\frac{da}{dt} = -\frac{2\theta_0(q-t)a}{qt(t-1)}, \quad \frac{db}{dt} = \frac{2\theta_1(q-t)b}{t(t-1)(q-1)}, \quad \frac{dc}{dt} = \frac{2\theta_t(q(2t-1)-t^2)c}{(q-t)t(t-1)},$$

$$\frac{d\theta_0}{dt} = \frac{d\theta_1}{dt} = \frac{d\theta_t}{dt} = \frac{d\theta_\infty}{dt} = 0. \tag{111}$$

As before, the only nontrivial equations are the first two, and they are equivalent to the sixth Painlevé equation for the function $q(t)$,

$$\frac{d^2q}{dt^2} = \frac{1}{2}\left(\frac{1}{q} + \frac{1}{q-1} + \frac{1}{q-t}\right)\left(\frac{dq}{dt}\right)^2 - \left(\frac{1}{t} + \frac{1}{t-1} + \frac{1}{q-t}\right)\frac{dq}{dt}$$
$$+ \frac{q(q-1)(q-t)}{t^2(t-1)^2}\left(\alpha + \beta\frac{t}{q^2} + \gamma\frac{t-1}{(q-1)^2} + \delta\frac{t(t-1)}{(q-t)^2}\right), \tag{112}$$

where

$$\alpha = \frac{(2\theta_\infty - 1)^2}{2}, \quad \beta = -2\theta_0^2, \quad \gamma = 2\theta_1^2, \quad \delta = \frac{1-4\theta_t^2}{2}.$$

Assuming that θ_j, $j = 0, 1, t, \infty$ are numerical constants, the function (104) becomes the Hamiltonian of (112) with p, q being the canonical variables. Also, if we denote

$$\begin{gathered} p_1 = p, \quad q_1 = q, \quad p_2 = \ln k, \quad q_2 = \theta_\infty, \quad p_3 = \ln a, \quad q_3 = \theta_0, \\ p_4 = \ln b, \quad q_4 = \theta_1, \quad p_5 = \ln c, \quad q_5 = \theta_t, \end{gathered} \tag{113}$$

then the whole system (111) becomes Hamiltonian with the same Hamiltonian (104), that is with,

$$H = p_1^2\frac{q_1(q_1-1)(q_1-t)}{t(t-1)} + p_1\frac{q_1(q_1-1)}{t(t-1)} + \frac{q_2(1-q_2)(q_1-t)}{t(t-1)}$$
$$+ \frac{q_3^2(q_1-t)}{q_1t(t-1)} - \frac{q_4^2(q_1-t)}{(q_1-1)t(t-1)} + \frac{q_5^2(t^2-q_1(2t-1))}{(q_1-t)t(t-1)},$$

and with respect to the symplectic form,

$$\Omega = \sum_{j=1}^{5} dp_j \wedge dq_j.$$

The general formulae (3) and (12) transform, in the case of system (95), into the equations,

$$\omega_{\mathrm{JMU}} = -\operatorname{res}_{z=t}\operatorname{Tr}\left((G^{(t)}(z))^{-1}\frac{dG^{(t)}(z)}{dz}\frac{d\Theta_t(z)}{dt}\right)dt \tag{114}$$

and

$$\omega = \operatorname{res}_{z=\infty}\operatorname{Tr}(A(z)dG^{(\infty)}(z)G^{(\infty)}(z)^{-1}) + \operatorname{res}_{z=0}\operatorname{Tr}(A(z)dG^{(0)}(z)G^{(0)}(z)^{-1}),$$
$$\operatorname{res}_{z=1}\operatorname{Tr}(A(z)dG^{(1)}(z)G^{(1)}(z)^{-1}) + \operatorname{res}_{z=t}\operatorname{Tr}(A(z)dG^{(t)}(z)G^{(t)}(z)^{-1}). \tag{115}$$

From (114) it follows that

$$\omega_{\mathrm{JMU}} = \theta_t \operatorname{Tr}(g_1\sigma_3),$$

and taking into account (103), we obtain that,

$$\omega_{\mathrm{JMU}} \equiv \frac{d\ln\tau}{dt}dt = Hdt - p\frac{q(q-1)}{t(t-1)}dt - \frac{\theta_\infty(q-t)}{t(t-1)}dt.$$

Similarly, (115) reduces to the equation,

$$\omega = \mathrm{Tr}(G_0^{-1}A_0 dG_0 + G_1^{-1}A_1 dG_1 + G_t^{-1}A_t dG_t - A_t G_t g_1 G_t^{-1} dt),$$

which after using (105)–(107) and simplifying yields the formula

$$\omega = pdq - Hdt - \theta_\infty \frac{dk}{k} - \theta_0 \frac{da}{a} - \theta_1 \frac{db}{b} - \theta_t \frac{dc}{c} + d\theta_\infty.$$

This, in turn, can be rewritten as

$$\begin{aligned}\omega = pdq - Hdt &+ d(\theta_\infty - \theta_0 \ln a - \theta_1 \ln b - \theta_t \ln c - \theta_\infty \ln k) \\ &+ \ln k d\theta_\infty + \ln a\, d\theta_0 + \ln b\, d\theta_1 + \ln c\, d\theta_t\end{aligned}$$

or, remembering the definitions (113) of the canonical coordinates,

$$\omega = \sum_{j=1}^{5} pdq - Hdt + d(q_2 - q_3 p_3 - q_4 p_4 - q_5 p_5 - q_2 p_2). \tag{116}$$

Equation (116) proves Conjectures 1 and 2, with $\gamma = 1$, in the case of the 2×2 system (95) and gives the explicit formula for $G(p_j, q_j, t)$,

$$G(p_1, p_2, p_3, p_4, p_5, q_1, q_2, q_3, q_4, q_5, t) = q_2 - q_3 p_3 - q_4 p_4 - q_5 p_5 - q_2 p_2.$$

The corresponding equation (18) and the truncated action are

$$\frac{d \ln \tau}{dt} = p\frac{dq}{dt} - H - \frac{d}{dt}(\theta_0 \ln a + \theta_1 \ln b + \theta_t \ln c + \theta_\infty \ln k)$$

and

$$Hdt = pdq - Hdt + d\left(\frac{1}{2}\ln\left(\frac{k(q-t)}{t(t-1)}\right) - \theta_0 \ln a - \theta_1 \ln b - \theta_t \ln c - \theta_\infty \ln k\right), d \equiv d_t,$$

respectively.

3.7. *Schlesinger system*

This section reproduces the result of [26, Subsection 5.6, Remark 5.5]. Once again, we are grateful to M. Mazzocco for informing us about this part of Malgrange's work.

Consider the Fuchsian system

$$\frac{d\Phi}{dz} = A(z)\Phi(z), \quad A(z) = \sum_{\nu=1}^{n} \frac{A_\nu}{z - a_\nu}, \quad A_\nu \in \mathfrak{sl}_N(\mathbb{C}). \tag{117}$$

We assume that all matrix coefficients A_ν are diagonalizable

$$A_\nu = G_\nu \Theta_\nu G_\nu^{-1}; \quad \Theta_\nu = \mathrm{diag}\{\theta_{\nu,1}, \dots, \theta_{\nu,N}\},$$

and that their eigenvalues are distinct and non-resonant:

$$\theta_{\nu,\alpha} \neq \theta_{\nu,\beta} \mod \mathbb{Z}.$$

We also assume that the residue of $A(z)$ at $z = \infty$ is diagonal, i.e.

$$A_\infty = -\sum_{\nu=1}^{n} A_\nu = \Theta_\infty.$$

Solutions of (117) have the following behavior at the singular points

$$\Phi^{(\infty)}(z) = (I + O(z^{-1}))z^{-\Theta_\infty}, \quad z \to \infty,$$

$$\Phi^{(\nu)}(z) = G_\nu(I + g_{\nu,1}(z - a_\nu) + O((z - a_\nu)^2))(z - a_\nu)^{\Theta_\nu} C_\nu, \quad z \to a_\nu.$$

The isomonodromic times are now positions of the singular points a_ν. The isomonodromy deformations with respect to these times yields the equation

$$\frac{d\Phi}{da_\nu} = B_\nu(z)\Phi(z), \quad B_\nu(z) = -\frac{A_\nu}{z - a_\nu}.$$

The compatibility conditions give the Schlesinger system

$$\frac{dA_\mu}{da_\nu} = \frac{[A_\mu, A_\nu]}{a_\mu - a_\nu}, \quad \mu \neq \nu, \qquad \frac{dA_\nu}{da_\nu} = -\sum_{\mu \neq \nu} \frac{[A_\mu, A_\nu]}{a_\mu - a_\nu}, \tag{118}$$

and also the equations (cf. Eqs. (110))

$$\frac{dG_\mu}{da_\nu} = \frac{A_\nu}{a_\nu - a_\mu} G_\mu, \quad \mu \neq \nu, \qquad \frac{dG_\nu}{da_\nu} = -\sum_{\mu \neq \nu} \frac{A_\mu}{a_\mu - a_\nu} G_\nu. \tag{119}$$

Following [22] we introduce matrix Darboux coordinates

$$Q_\nu = G_\nu \Theta_\nu, \quad P_\nu = G_\nu^{-1}$$

and Hamiltonians

$$H_\nu = \sum_{\mu \neq \nu}^{n} \frac{\mathrm{Tr}(Q_\mu P_\mu Q_\nu P_\nu)}{a_\nu - a_\mu}.$$

Notice that $A_\nu = Q_\nu P_\nu$. Then, as it is shown in [26], the isomonodromy equations (118) and (119) are equivalent to the Hamiltonian system

$$\frac{dP_{\mu,jk}}{da_\nu} = -\frac{\partial H_\nu}{\partial Q_{\mu,kj}}, \quad \frac{dQ_{\mu,jk}}{da_\nu} = \frac{\partial H_\nu}{\partial P_{\mu,kj}}.$$

Moreover, by a rather straightforward calculation, one can show that the general formulae (3) and (12) in the case of the Fuchsian system (117) produce the following expressions of the forms ω_{JMU} and ω,

$$\omega_{\mathrm{JMU}} = \sum_{\nu=1}^{n} H_\nu da_\nu$$

and

$$\omega = \sum_{\nu=1}^{n} \mathrm{Tr}(P_\nu dQ_\nu) - H_\nu da_\nu.$$

In other words, we have validity of Conjectures 1 and 2, with $\gamma = 1$, in the case of the Fuchsian system (117). Moreover, the form ω just coincides with ω_{cla}.

Acknowledgments

In addition to M. Mazzocco, we would also like to thank O. Lisovyy for many very useful discussions and comments. The present work was supported by the NSF Grant DMS-1700261 and Russian Science Foundation grant No. 17-11-01126.

Appendix

A.1. *The Proof of Lemma 1*

The proof of Lemma 1 is rather short. Indeed, noticing that

$$(G^{(\nu)})^{-1}\frac{dG^{(\nu)}}{dz} = (G^{(\nu)})^{-1}AG^{(\nu)} - \frac{d\Theta_\nu}{dz} \tag{120}$$

and plugging this into the right-hand side of (3), we have,

$$\begin{aligned}\omega_{\mathrm{JMU}} = &-\sum_{k=1}^{L}\sum_{\nu=1,\dots,n,\infty} \mathrm{res}_{z=a_\nu}\,\mathrm{Tr}\left((G^{(\nu)})^{-1}AG^{(\nu)}\frac{\partial\Theta_\nu}{\partial t_k}\right)dt_k\\ &+\sum_{k=1}^{L}\sum_{\nu=1,\dots,n,\infty} \mathrm{res}_{z=a_\nu}\,\mathrm{Tr}\left(\frac{d\Theta_\nu}{dz}\frac{\partial\Theta_\nu}{\partial t_k}\right)dt_k.\end{aligned} \tag{121}$$

The expression $\frac{d\Theta_\nu}{dz}\frac{\partial\Theta_\nu}{\partial t_k}$ has poles of order at least 2, so it does not have residues and hence the second sum in (121) vanishes. We also have,

$$\frac{\partial\Theta_\nu}{\partial t_k} = (G^{(\nu)})^{-1}B_kG^{(\nu)} - (G^{(\nu)})^{-1}\frac{\partial G^{(\nu)}}{\partial t_k}. \tag{122}$$

Substituting (122) into (121) we transform it to the equation,

$$\begin{aligned}\omega_{\mathrm{JMU}} = &-\sum_{k=1}^{L}\sum_{\nu=1,\dots,n,\infty} \mathrm{res}_{z=a_\nu}\,\mathrm{Tr}(AB_k)dt_k\\ &+\sum_{k=1}^{L}\sum_{\nu=1,\dots,n,\infty} \mathrm{res}_{z=a_\nu}\,\mathrm{Tr}\left(A\frac{\partial G^{(\nu)}}{\partial t_k}(G^{(\nu)})^{-1}\right)dt_k.\end{aligned}$$

The function AB_k is rational, therefore sum of its residues is zero. So we get (10).

A.2. *The Proof of Lemma 2*

Denote

$$I=\sum_{k=1}^{L}\sum_{j=1}^{d}\sum_{\nu=1,\dots,n,\infty} \mathrm{res}_{z=a_\nu}\,\mathrm{Tr}\left(\frac{\partial}{\partial m_j}\left(A\frac{\partial G^{(\nu)}}{\partial t_k}(G^{(\nu)})^{-1}\right)-\frac{\partial}{\partial t_k}\left(A\frac{\partial G^{(\nu)}}{\partial m_j}(G^{(\nu)})^{-1}\right)\right).$$

We have

$$I = \sum_{k=1}^{L}\sum_{j=1}^{d}\sum_{\nu=1,\dots,n,\infty} \operatorname{res}_{z=a_\nu} \operatorname{Tr}\left(\frac{\partial A}{\partial m_j}\frac{\partial G^{(\nu)}}{\partial t_k}(G^{(\nu)})^{-1}\right.$$

$$\left. - A\frac{\partial G^{(\nu)}}{\partial t_k}(G^{(\nu)})^{-1}\frac{\partial G^{(\nu)}}{\partial m_j}(G^{(\nu)})^{-1} - \frac{\partial A}{\partial t_k}\frac{\partial G^{(\nu)}}{\partial m_j}(G^{(\nu)})^{-1}\right)$$

$$+ \sum_{k=1}^{L}\sum_{j=1}^{d}\sum_{\nu=1,\dots,n,\infty} \operatorname{res}_{z=a_\nu} \operatorname{Tr}\left(A\frac{\partial G^{(\nu)}}{\partial m_j}(G^{(\nu)})^{-1}\frac{\partial G^{(\nu)}}{\partial t_k}(G^{(\nu)})^{-1}\right).$$

We use the formula (122) to get rid of $\frac{\partial G^{(\nu)}}{\partial t_k}$ and Eq. (6) to replace $\frac{\partial A}{\partial t_k}$. Omitting terms with zero residue and after some cancellations we have

$$I = \sum_{k=1}^{L}\sum_{j=1}^{d}\sum_{\nu=1,\dots,n,\infty} \operatorname{res}_{z=a_\nu} \operatorname{Tr}\left(-\frac{\partial A}{\partial m_j}G^{(\nu)}\frac{\partial \Theta_\nu}{\partial t_k}(G^{(\nu)})^{-1}\right.$$

$$\left. + AG^{(\nu)}\frac{d\partial\Theta_\nu}{\partial t_k}(G^{(\nu)})^{-1}\frac{\partial G^{(\nu)}}{\partial m_j}(G^{(\nu)})^{-1} - \frac{dB_k}{dz}\frac{\partial G^{(\nu)}}{\partial m_j}(G^{(\nu)})^{-1}\right)$$

$$+ \sum_{k=1}^{L}\sum_{j=1}^{d}\sum_{\nu=1,\dots,n,\infty} \operatorname{res}_{z=a_\nu} \operatorname{Tr}\left(-A\frac{d\partial G^{(\nu)}}{\partial m_j}\frac{\partial\Theta_\nu}{\partial t_k}(G^{(\nu)})^{-1}\right).$$

We replace B_k using again formula (122). After that, we notice that the residue of the derivative with respect to z of formal series is zero. Therefore we can "integrate by parts", moving the derivative from one term to another. We do that with the term, where we replaced B_k. Using (120), we have

$$I = \sum_{k=1}^{L}\sum_{j=1}^{d}\sum_{\nu=1,\dots,n,\infty} \operatorname{res}_{z=a_\nu} \operatorname{Tr}\left(-\frac{\partial A}{\partial m_j}G^{(\nu)}\frac{\partial \Theta_\nu}{\partial t_k}(G^{(\nu)})^{-1}\right.$$

$$\left. + AG^{(\nu)}\frac{\partial\Theta_\nu}{\partial t_k}(G^{(\nu)})^{-1}\frac{dG^{(\nu)}}{\partial m_j}(G^{(\nu)})^{-1} - A\frac{\partial G^{(\nu)}}{\partial m_j}\frac{\partial\Theta_\nu}{\partial t_k}(G^{(\nu)})^{-1}\right)$$

$$+ \sum_{k=1}^{L}\sum_{j=1}^{d}\sum_{\nu=1,\dots,n,\infty} \operatorname{res}_{z=a_\nu} \operatorname{Tr}\left(-G^{(\nu)}\frac{\partial\Theta_\nu}{\partial t_k}(G^{(\nu)})^{-1}\frac{\partial G^{(\nu)}}{\partial m_j}(G^{(\nu)})^{-1}A\right.$$

$$\left. + \frac{\partial\Theta_\nu}{\partial t_k}(G^{(\nu)})^{-1}\frac{\partial G^{(\nu)}}{\partial m_j}\frac{d\Theta_\nu}{dz}\right)$$

$$+ \sum_{k=1}^{L}\sum_{j=1}^{d}\sum_{\nu=1,\dots,n,\infty} \operatorname{res}_{z=a_\nu} \operatorname{Tr}\left(G^{(\nu)}\frac{\partial\Theta_\nu}{\partial t_k}(G^{(\nu)})^{-1}\frac{\partial^2 G^{(\nu)}}{\partial z\partial m_j}(G^{(\nu)})^{-1}\right).$$

Finally using (120) one more time we get $I = 0$.

References

[1] M. Babich, On canonical parametrization of the phase spaces of equations of isomonodromic deformations of Fuchsian systems of dimension 2×2. Derivation of the Painlevé VI equation, *Uspekhi Mat. Nauk* **64**(1) (2009) 51–134 (Russian); English translation in *Russian Math. Surveys*, **64**(1) (2009) 45–127.

[2] M. Bertola, The dependence on the monodromy data of the isomonodromic tau function, *Comm. Math. Phys.* **294** (2010) 539–579; arXiv:0902.4716 [nlin.SI].

[3] P. Boalch, Symplectic Manifolds and Isomonodromic Deformations, *Adv. Math.* **163** (2001) 137–205.

[4] E. Brézin and V. A. Kazakov, Exactly solvable field theories of closed strings, *Phys. Letts. B* **236** (1990) 144–150.

[5] T. Bothner, A. Its and A. Prokhorov, On the analysis of incomplete spectra in random matrix theory through an extension of the Jimbo–Miwa–Ueno differential, arXiv:1708.06480v1 [math-ph].

[6] J. Baik, R. Buckingham, J. DiFranco and A. Its, Total integrals of global solutions to Painlevé II, *Nonlinearity* **22** (2009) 1021–1061; arXiv:0810.2586 [math.CA].

[7] M. Douglas and S. Shenker, Strings in less than one dimension, *Nucl. Phys. B* **335** (1990) 635–654.

[8] H. Flaschka and A. C. Newell, Monodromy- and spectrum-preserving deformations I, *Comm. Math. Phys.* **76** (1980) 65–116.

[9] H. Flaschka and A. C. Newell, The inverse monodromy transform is a canonical transformation, in *Nonlinear Problems: Present and Future* (*Los Alamos, N.M., 1981*), Math. Studies, Vol. 61 (North Holland, 1982), pp. 65–91.

[10] A. S. Fokas, A. R. Its, A. A. Kapaev and V. Yu. Novokshenov, *Painlevé Transcendents: The Riemann–Hilbert Approach*, Mathematical Surveys and Monographs, Vol. 128 (AMS, Providence, RI, 2006).

[11] P. Gavrylenko and O. Lisovyy, Fredholm determinant and Nekrasov sum representations of isomonodromic tau functions (2016); arXiv:1608.00958 [math-ph].

[12] D. Gross and A. Migdal, A nonperturbative treatment of two-dimensional quantum gravity, *Nucl. Phys. B* **340** (1990) 333–365.

[13] J. Harnad and M. A. Wisse, Loop algebra moment maps and Hamiltonian models for the Painlevé transcendants, *Field's Inst. Commun.* **7** (1993) 155–169.

[14] J. Harnad and M. A. Wisse, Moment maps to loop algebras, classical R-matrix and integrable systems, in *Quantum Groups Integrable Models and Statistical Systems* (World Scientific, Singapore, 1993), pp. 105–117.

[15] A. Its and A. Prokhorov, Connection problem for the tau-function of the sine-Gordon reduction of Painlevé-III equation via the Riemann–Hilbert approach, *Int. Math. Res. Not.* **22** (2016) 6856–6883.

[16] A. Its, O. Lisovyy and A. Prokhorov, Monodromy dependence and connection formulae for isomonodromic tau functions, to appear in *Duke Math. J.*; arXiv:1604.03082v1 [math-ph].

[17] N. Iorgov, O. Lisovyy, A. Shchechkin and Yu. Tykhyy, Painlevé functions and conformal blocks, *Constr. Approx.* **39** (2014) 255–272.

[18] N. Iorgov, O. Lisovyy and Yu. Tykhyy, Painlevé VI connection problem and monodromy of $c = 1$ conformal blocks, *JHEP* **12** (2013) 029; arXiv:1308.4092v1 [hep-th].

[19] A. Its, O. Lisovyy and Yu. Tykhyy, Connection problem for the sine-Gordon/Painlevé III tau function and irregular conformal blocks, *Int. Math. Res. Not.* **18** (2015) 8903–8924; arXiv:1403.1235 [math-ph]; *Phys. D* **2** (1981) 306–352.

[20] M. Jimbo, T. Miwa and K. Ueno, Monodromy preserving deformation of linear ordinary differential equations with rational coefficients. I, *Phys. D* **2** (1981) 306–352.

[21] M. Jimbo and T. Miwa, Monodromy preserving deformation of linear ordinary differential equations with rational coefficients. II, *Phys. D* **2** (1981) 407–448.

[22] M. Jimbo, T. Miwa, Y. Môri and M. Sato, Density matrix of an impenetrable Bose gas and the fifth Painlevé transcendent, *Phys. D* **1** (1980) 80–158.

[23] I. Krichever, Isomonodromy equations on algebraic curves, cannonical transformations and Whitham equations, *Moscow Math. J.* **2** (2002) 717–752; arXiv:hep-th/0112096.

[24] O. Lisovyy and J. Roussillon, On the connection problem for Painlevé I, *J. Phys. A: Math. Theor.* **50** (2017) 255202.

[25] B. Malgrange, Sur les déformations isomonodromiques, I. Singularités régulières, in *Mathematics and Physics (Paris, 1979/1982)*, Prog. Math., Vol. 37 (Birkhäuser, Boston, MA, 1983), pp. 401–426.

[26] B. Malgrange, Déformations isomonodromiques, forme de Liouville, fonction τ, *Ann. Inst. Fourier* **54**(5) (2004) 1371–1392.

[27] T. Miwa, Painlevé property of monodromy preserving deformation equations and the analyticity of τ-functions, *Publ. Res. Inst. Math. Sci.* **17** (1981) 703–712.

[28] K. Okamoto, Polynomial Hamiltonians associated with Painlevé equations, I, *Proc. Japan Acad. Ser. A Math. Sci. Volume* **56**(6) (1980) 264–268.

[29] C. A. Tracy, Asymptotics of the τ-function arising in the two-dimensional Ising model, *Comm. Math. Phys.* **142** (1991) 297–311.

[30] C. A. Tracy and H. Widom, Level-spacing distributions and the Airy kernel, *Comm. Math. Phys.* **159** (1994) 151–174; hep-th/9211141.

[31] C. A. Tracy and H. Widom, Fredholm determinants, differential equations and matrix models, *Comm. Math. Phys.* **163** (1994) 33–72; hep-th/9306042.

[32] W. Wasow, *Asymptotic Expansions for Ordinary Differential Equations* (Dover, New York, 2002).

Spectral Equations for the Modular Oscillator

Rinat M. Kashaev*,‡ and Sergey M. Sergeev†,§

*Section de Mathématiques, Université de Genève,
2-4 rue du Lièvre, 1211 Genève 4, Switzerland

†Faculty of Education, Science, Technology, & Mathematics,
University of Canberra, Bruce ACT 2601, Australia
and
Department of Theoretical Physics, Research School of Physics and Engineering,
Australian National University, Canberra, ACT 0200, Australia

‡Rinat.Kashaev@unige.ch

§sergey.m.sergeev@gmail.com

Motivated by applications for non-perturbative topological strings in toric Calabi–Yau manifolds, we discuss the spectral problem for a pair of commuting modular conjugate (in the sense of Faddeev) Harper type operators, corresponding to a special case of the quantized mirror curve of local $\mathbb{P}^1 \times \mathbb{P}^1$ and complex values of Planck's constant. We illustrate our analytical results by numerical calculations.

In memory of Ludwig Faddeev

1. Introduction

Topological string theory is known for its prominent role in relating different subjects of mathematical physics and mathematics such as Chern–Simons theory, Gromov–Witten invariants, mirror symmetry, enumerative geometry, quantum topology, integrable systems, supersymmetric gauge theory, random matrix theory, etc. The recent progress in string theory has lead to remarkable connections between spectral theory, integrable systems and local mirror symmetry. The idea of relating topological strings in toric Calabi–Yau manifolds with quantum mechanical spectral problems originating from integrable systems was put forward in [1] and subsequently has been materialized to a powerful and numerically testable conjecture of Grassi–Hatsuda–Mariño [7]. It has been shown that quantization of mirror curves leads to trace class (one-dimensional) quantum mechanical operators, whose spectral properties not only contain the full information of the enumerative geometry of the underlying Calabi–Yau manifold, but also allow to formulate the string theory non-perturbatively, see [14] for a review and references therein. Outside the context of string theory, these results are preceded by connections of some quantum mechanical spectral problems with integrable systems and conformal field theory [3, 4].

In the case of toric Calabi–Yau threefold known as local $\mathbb{P}^1 \times \mathbb{P}^1$ or local $\mathbb{F}_0$, the corresponding operator obtained from quantization of the mirror curve is of the form

$$\boldsymbol{O}_{\mathbb{F}_0,m} = \boldsymbol{v} + \boldsymbol{v}^{-1} + \boldsymbol{u} + m\boldsymbol{u}^{-1} \tag{1}$$

with positive self-adjoint operators $\boldsymbol{u}$ and $\boldsymbol{v}$ satisfying the Heisenberg–Weyl commutation relation

$$\boldsymbol{u}\boldsymbol{v} = \mathrm{e}^{\mathrm{i}\hbar}\,\boldsymbol{v}\boldsymbol{u}, \quad \hbar \in \mathbb{R}_{>0}, \tag{2}$$

and positive real parameter m. In the special case $m = 1$, operator $\boldsymbol{O}_{\mathbb{F}_0,1}$ structurally resembles the Harper operator [8], but the fact that operators $\boldsymbol{u}$ and $\boldsymbol{v}$ are positive self-adjoint rather than unitary makes its quantum mechanical content very different: it has purely discrete positive spectrum, and its inverse $\boldsymbol{O}_{\mathbb{F}_0,m}^{-1}$ is of trace class [11, 12]. Nonetheless, recent findings in [9, 10] indicate to hidden connections of the two cases through Faddeev's modular duality [5].

The spectral problem of operator (1) has been addressed in [16] as a part of a larger class of integrable systems related to supersymmetric gauge theories, and, more recently, in [13] from the perspective of exact WKB approximation and in [15] by using a matrix integral representation of the eigenfunctions. One can approach the same problem from the standpoint of quantum integrable systems by noting that both the Bethe ansatz equations and the quantum method of separation of variables can be formulated in more general and universal framework of Baxter's $T - Q$ functional difference equations [2] which, in their turn, can be interpreted as (one-dimensional) quantum mechanical spectral problems. Complemented with Faddeev's modular duality idea, the second-named author has shown in [17] that the modular $T - Q$ equations of general form admit a particularly tractable approach to their solution in the so called "strong coupling" regime corresponding to a complexified parameter $\hbar$ in the commutation relation (2), namely, $\hbar = 2\pi\mathrm{e}^{\mathrm{i}2\theta}$ with real $\theta \in]0, \pi/2[$. In this case, operators $\boldsymbol{u}$ and $\boldsymbol{v}$ cannot be neither self-adjoint nor unitary, but one can choose them to be normal and satisfying the relation

$$\boldsymbol{u}\boldsymbol{v}^\dagger = \boldsymbol{v}^\dagger\boldsymbol{u}, \tag{3}$$

implying that operator (1) stays normal, and thus its spectral problem still makes sense.

In this paper, motivated by topological string applications, we address the spectral problem of operator (1) in the special case $m = 1$ and $\hbar = 2\pi\mathrm{e}^{\mathrm{i}2\theta}$ with $\theta \in [0, \pi/2[$. Our approach is based on solving Baxter's $T - Q$ equations in terms of analytic functions on the entire complex plane. The strong coupling regime $\theta \in]0, \pi/2[$ is treated along the approach of [17], while the case $\theta = 0$ (corresponding to $\hbar = 2\pi$) is treated by solving the functional difference equation directly. We illustrate our analytical results with numerical calculations.

The organization of the paper is as follows. In Sec. 2, we formulate the problem, fix the notation and conventions. In Secs. 3–6, we describe the solution of the problem in the strongly coupled regime. First, we introduce the main functional difference equation and describe some of its properties (Sec. 3), then we construct the eigenfunctions with minimal number of poles (Sec. 4), describe the spectral equations as conditions of cancelling the poles (Sec. 5), and give some numerical analysis in the special case $\hbar = 2\pi\mathsf{i}$ (Sec. 6). In the final Sec. 7, we give a solution of the case $\hbar = 2\pi$.

2. Background from Basic Quantum Mechanics

Let $\boldsymbol{x}$ and $\boldsymbol{p}$ be normalised self-adjoint quantum mechanical position and momentum operators in the Hilbert space $L^2(\mathbb{R})$ defined by the functional equalities

$$\boldsymbol{x}\psi(x) = x\psi(x), \quad \boldsymbol{p}\psi(x) = \frac{1}{2\pi\mathsf{i}}\frac{\partial\psi(x)}{\partial x} \tag{4}$$

so that the Heisenberg commutation relation between them takes the form

$$[\boldsymbol{p}, \boldsymbol{x}] := \boldsymbol{p}\boldsymbol{x} - \boldsymbol{x}\boldsymbol{p} = \frac{1}{2\pi\mathsf{i}}, \tag{5}$$

i.e. with specific choice of Planck's constant $h = 2\pi\hbar = 1$. As these are unbounded operators, it is assumed that the function $\psi(x)$ in (4) is taken from the respective domains $D(\boldsymbol{x})$ and $D(\boldsymbol{p})$.

For each non-zero complex parameter b, there are two pairs of exponential operators

$$\boldsymbol{u} := \mathrm{e}^{2\pi\mathsf{b}\boldsymbol{x}}, \quad \boldsymbol{v} := \mathrm{e}^{2\pi\mathsf{b}\boldsymbol{p}}, \tag{6}$$

and

$$\bar{\boldsymbol{u}} := \mathrm{e}^{2\pi\mathsf{b}^{-1}\boldsymbol{x}}, \quad \bar{\boldsymbol{v}} := \mathrm{e}^{2\pi\mathsf{b}^{-1}\boldsymbol{p}}, \tag{7}$$

which satisfy the Weyl commutations relations

$$\boldsymbol{u}\boldsymbol{v} = q^2\boldsymbol{v}\boldsymbol{u}, \quad \bar{\boldsymbol{v}}\bar{\boldsymbol{u}} = \bar{q}^2\bar{\boldsymbol{u}}\bar{\boldsymbol{v}}, \tag{8}$$

where

$$q := \mathrm{e}^{\mathsf{i}\pi\mathsf{b}^2}, \quad \bar{q} := \mathrm{e}^{-\mathsf{i}\pi\mathsf{b}^{-2}}, \tag{9}$$

and all other commutation relations are trivial. In coordinate representation (4), we have

$$\boldsymbol{u}\psi(x) = \mathrm{e}^{2\pi\mathsf{b}x}\psi(x), \quad \boldsymbol{v}\psi(x) = \psi(x - \mathsf{i}\mathsf{b}), \tag{10}$$

and

$$\bar{\boldsymbol{u}}\psi(x) = \mathrm{e}^{2\pi\mathsf{b}^{-1}x}\psi(x), \quad \bar{\boldsymbol{v}}\psi(x) = \psi(x - \mathsf{i}\mathsf{b}^{-1}). \tag{11}$$

The case where parameter b is pure imaginary corresponds to unitary exponentials, while in this paper we will mainly assume the *strong coupling* regime

$$\mathsf{b} = \mathrm{e}^{\mathsf{i}\theta}, \quad \theta \in]0, \pi/2[\subset \mathbb{R}, \tag{12}$$

which corresponds to unbounded exponentials. In this latter case, the operation bar in (7)–(9) and (11) is the complex conjugation for numbers and Hermitian conjugation for operators. We will keep that notation in a broader context of analytically continued parameters and variables always assuming that

$$|q| < 1 \Leftrightarrow \Im \mathsf{b}^2 > 0 \Leftrightarrow |\bar{q}| < 1. \tag{13}$$

As in (4), the function $\psi(x)$ in (10) and (11) is assumed to belong to respective domains of unbounded operators. For example, the domain of operator $\boldsymbol{v}$ consists of analytic functions in the open strip between the real axis $\mathbb{R} \subset \mathbb{C}$ and the line $-\mathsf{i}\mathsf{b} + \mathbb{R}$, which are continuous on the closure of the strip, and which restrict to square integrable functions on the boundary of the strip.

We remark that the simultaneous appearance in a theory of both pairs of exponential operators (6) and (7) is a manifestation of *Faddeev's modular duality* [5].

Statement of the problem. We address the spectral problem of the following pair of commuting operators (to be called *Hamiltonians*):

$$\boldsymbol{H} = \boldsymbol{v} + \boldsymbol{v}^{-1} + \boldsymbol{u} + \boldsymbol{u}^{-1}, \quad \bar{\boldsymbol{H}} = \bar{\boldsymbol{v}} + \bar{\boldsymbol{v}}^{-1} + \bar{\boldsymbol{u}} + \bar{\boldsymbol{u}}^{-1}. \tag{14}$$

In the strong coupling regime (12), $\bar{\boldsymbol{H}} = \boldsymbol{H}^\dagger$ so that commutativity of $\boldsymbol{H}$ and $\bar{\boldsymbol{H}}$ means that $\boldsymbol{H}$ is a normal operator. In the coordinate representation (4), the spectral problems

$$\boldsymbol{H}\psi = \varepsilon\psi, \quad \bar{\boldsymbol{H}}\psi = \bar{\varepsilon}\psi \tag{15}$$

correspond to the following pair of functional difference equations:

$$\psi(x + \mathsf{i}\mathsf{b}) + \psi(x - \mathsf{i}\mathsf{b}) = (\varepsilon - 2\cosh(2\pi \mathsf{b} x))\psi(x), \tag{16}$$

$$\psi(x - \mathsf{i}\mathsf{b}^{-1}) + \psi(x + \mathsf{i}\mathsf{b}^{-1}) = (\bar{\varepsilon} - 2\cosh(2\pi \mathsf{b}^{-1} x))\psi(x). \tag{17}$$

By definition, solving these spectral problems corresponds to construction of a function $\psi(x)$ and determination of specific values of ε and $\bar{\varepsilon}$ such that

(1) $\psi(x)$ is analytic in an open strip around the real axis whose closure contains $\pm\mathsf{i}\mathsf{b}$ and $\pm\mathsf{i}\mathsf{b}^{-1}$, and it is continuous on the closure of that strip;
(2) $\psi(x)$ is square integrable on the real axis and all of its translates by $\pm\mathsf{i}\mathsf{b}$ and $\pm\mathsf{i}\mathsf{b}^{-1}$;
(3) Eqs. (16), (17) are satisfied for all $x \in \mathbb{R}$.

Remark 1. As we will see, the solution of the spectral problems (16), (17) is given by analytic functions on the entire complex plane.

Conventions. Throughout the text, each time when we use the variables u, $\bar{u}$ and x, we assume that they are related as follows:

$$u = \mathrm{e}^{2\pi \mathsf{b} x}, \quad \bar{u} = \mathrm{e}^{2\pi x/\mathsf{b}}. \tag{18}$$

Similarly, as soon as we use the variables s, $\bar{s}$ and σ, we assume that they are related as follows:

$$s = \mathrm{e}^{2\pi \mathsf{b}\sigma}, \quad \bar{s} = \mathrm{e}^{2\pi\sigma/\mathsf{b}}. \tag{19}$$

Else, if $f = f(u, \varepsilon, q)$, then we write $\bar{f} = \overline{f(u, \varepsilon, q)}$ meaning the analytic continuation of the complex conjugation in the strong coupling regime.

Remark 2. In the limit $\mathsf{b} \to 0$ one has

$$\boldsymbol{H} = 4 + (2\pi\mathsf{b})^2(\boldsymbol{p}^2 + \boldsymbol{x}^2) + \mathcal{O}(\mathsf{b}^4), \tag{20}$$

what justifies the term "oscillator".

3. The Main Functional Equation

In this section, we study the following functional finite difference equation

$$f\left(\frac{u}{q^2}\right) + q^2u^2 f(q^2u) = (1 - \varepsilon u + u^2)f(u). \tag{21}$$

We can rewrite it as a first order difference matrix equation

$$\begin{pmatrix} f\left(\frac{u}{q^2}\right) \\ f(u) \end{pmatrix} = L(u) \begin{pmatrix} f(u) \\ f(q^2u) \end{pmatrix} \tag{22}$$

where

$$L(u) := \begin{pmatrix} 1 - \varepsilon u + u^2 & -q^2u^2 \\ 1 & 0 \end{pmatrix}. \tag{23}$$

Upon iteration of (22), we obtain

$$\begin{pmatrix} f\left(\frac{u}{q^2}\right) \\ f(u) \end{pmatrix} = M_n(u) \begin{pmatrix} f(q^{2n-2}u) \\ f(q^{2n}u) \end{pmatrix}, \quad \forall n \in \mathbb{Z}_{>0}, \tag{24}$$

where

$$M_n(u) := L(u)L(q^2u)\cdots L(q^{2(n-1)}u) \tag{25}$$

is a sequence of matrix valued polynomials in u (the degree of $M_n(u)$ is $2n$) which satisfies the following obvious relations

$$M_{m+n}(u) = M_m(u)M_n(q^{2m}u), \quad \forall m, n \in \mathbb{Z}_{\geq 0}. \tag{26}$$

3.1. *The regular solution* $\chi_q(u, \varepsilon)$

The limit $n \to \infty$ leads to a matrix-valued entire function $M_\infty(u)$, i.e. a holomorphic function on the entire complex plane of the variable u. This follows from the following functional analytic argument.

For any $r \in \mathbb{R}_{>0}$, let V_r be the vector space of $\mathbb{C}^2$-valued continuous functions on the closed disk

$$D_r := \{z \in \mathbb{C} \mid |z| \leq r\}. \tag{27}$$

It is a Banach space with respect to the norm

$$\|g\|_r := \max_{z\in D_r} \|g(z)\|, \quad \|g(z)\| := \max_{i\in\{1,2\}} |g_i(z)|, \quad \forall g \in V_r. \tag{28}$$

We have the following monotonicity property of the norms with different disk radii:

$$s \le r \Rightarrow \|g|_{D_s}\|_s \le \|g\|_r, \quad \forall g \in V_r. \tag{29}$$

Lemma 1. *For any $r \in \mathbb{R}_{>0}$, the sequence of restrictions $M_n|_{D_r}$, $n \in \mathbb{Z}_{n>0}$, forms a (fundamental) Cauchy sequence with respect to the induced norm in the (Banach) algebra of bounded linear operators over V_r.*

Proof. On the level of matrix valued continuous functions $A\colon D_r \to \mathrm{End}(\mathbb{C}^2)$, the induced norm is calculated as follows:

$$\|A\|_r = \max_{z\in D_r} \|A(z)\|, \quad \|A(z)\| := \max_{i\in\{1,2\}} \sum_{j=1}^{2} |A_{i,j}(z)|. \tag{30}$$

Using this formula and the explicit form (23) of the polynomial matrix $L(u)$, we have

$$\|L\|_r \ge 1, \quad \forall r \in \mathbb{R}_{\ge 0}, \tag{31}$$

and

$$\|L\|_r \le 1 + r(r+1)T, \quad T := \max_{i\in\{1,2\}} \|L_i\| \ge 1 + |q|^2, \tag{32}$$

where $L_i \in \mathrm{End}(\mathbb{C}^2)$ are the expansion coefficients of the matrix $L(u)$:

$$L(u) = \sum_{i=0}^{2} L_i u^i, \tag{33}$$

the coefficient $L_0 = L(0)$ being a projection matrix of unit norm:

$$L_0^2 = L_0, \quad \|L_0\| = 1. \tag{34}$$

That allows us to get a uniform in m upper bound for the norms of $M_m(u)$:

$$\|M_m\|_r \le \max_{z\in D_r} \prod_{i=0}^{m-1} \|L(zq^{2i})\| \le \prod_{i=0}^{m-1} \|L\|_{r|q|^{2i}} \le \prod_{i=0}^{\infty} \|L\|_{r|q|^{2i}} =: \nu(r), \tag{35}$$

where the infinite product converges due to inequality (32). On the other hand, the norms of the coefficients in the expansions

$$M_m(u) = \sum_{k=0}^{2m} M_{m,k} u^k, \quad M_{m,k} := \sum_{\substack{0\le i_1,\dots,i_m\le 2\\ i_1+\cdots+i_m=k}} L_{i_1}\cdots L_{i_m} q^{2\sum_{j=1}^{m-1} j i_{j+1}}, \tag{36}$$

are also bounded from above uniformly on m:

$$\|M_{m,k}\| \le \sum_{\substack{0\le i_1,\dots,i_m\le 2\\ i_1+\cdots+i_m=k}} \|L_{i_1}\|\cdots\|L_{i_m}\| |q|^{2\sum_{j=1}^{m-1} j i_{j+1}} \le \frac{3T^k}{(|q|^2;|q|^2)_{m-1}} \le \frac{3T^k}{(|q|^2;|q|^2)_\infty} \tag{37}$$

with the standard notation for the deformed Pochhammer symbols

$$(x;q)_n := \prod_{i=0}^{n-1}(1-xq^i). \tag{38}$$

For any $m,n \in \mathbb{Z}_{>0}$, there exists a matrix valued polynomial $A_{m,n}(u)$ of degree $2\max(m,n)-1$ such that

$$M_m(u) - M_n(u) = uA_{m,n}(u). \tag{39}$$

Explicitly, we have

$$A_{m,n}(u) = \sum_{k=0}^{2\max(m,n)-1} (M_{m,k} - M_{n,k})u^k \tag{40}$$

so that

$$\|A_{m,n}\|_r \leq \sum_{k=0}^{2\max(m,n)-1} (\|M_{m,k}\| + \|M_{n,k}\|)r^k \leq \frac{6\big(1-(rT)^{2\max(m,n)}\big)}{(|q|^2;|q|^2)_\infty(1-rT)}. \tag{41}$$

Let $N_r \in \mathbb{Z}_{>0}$ be such that $rT|q|^{2N_r} < 1$. Then for any $m \geq N_r$ and $n \geq 0$, taking into account the formula

$$M_{m+n+1}(u) - M_{m+1}(u) = M_m(u)uq^{2m}A_{n+1,1}(uq^{2m}) \tag{42}$$

and the inequalities (35) and (41), we obtain

$$\begin{aligned}\|M_{m+n+1} - M_{m+1}\|_r &\leq |q|^{2m} r \|M_m\|_r \|A_{n+1,1}\|_{r|q|^{2m}} \\ &\leq |q|^{2m} r\nu(r) \frac{6(1-(rT|q|^{2m})^{2n+2})}{(|q|^2;|q|^2)_\infty(1-rT|q|^{2m})} \\ &\leq |q|^{2m} \frac{6r\nu(r)}{(|q|^2;|q|^2)_\infty(1-rT|q|^{2N_r})}.\end{aligned} \tag{43}$$

The obtained equality implies that $(M_m|_{D_r})_{m>0}$ is a Cauchy sequence in the Banach algebra of bounded linear operators over V_r. □

Lemma 1 implies the uniform convergence of the matrix coefficients of $M_n(u)$ on all compact subsets of $\mathbb{C}$. Indeed , let $f_n(u)$ be a matrix element of $M_n(u)$ and $K \subset \mathbb{C}$ a compact set. Then there exists $r \in \mathbb{R}_{>0}$ such that $K \subset D_r$ and thus

$$|f_m(u) - f_n(u)| \leq \|M_m(u) - M_n(u)\| \leq \|M_m - M_n\|_r, \quad \forall u \in K. \tag{44}$$

Thus, $M_\infty(u)$ is an entire function on $\mathbb{C}$ as a uniformly convergent limit on all compact subsets of $\mathbb{C}$ of a sequence of polynomial functions.

By taking into account the equalities

$$M_\infty(u) = L(u)M_\infty(q^2u) = M_\infty(u)L(0), \quad M_\infty(0) = L(0), \tag{45}$$

we immediately arrive to the following structure of the matrix $M_\infty(u)$:

$$M_\infty(u) = \begin{pmatrix} \chi_q(q^{-2}u) & 0 \\ \chi_q(u) & 0 \end{pmatrix}, \tag{46}$$

where $\chi_q(u) = \chi_q(u,\varepsilon)$ is an entire function normalised so that $\chi_q(0) = 1$ and which solves the functional difference equation (21). On the other hand, if write

$$M_n(u) = \begin{pmatrix} a_n(u) & b_n(u) \\ c_n(u) & d_n(u) \end{pmatrix}, \tag{47}$$

then the matrix equality

$$L(u)M_n(q^2u) = M_n(u)L(q^{2n}u) \tag{48}$$

implies that

$$b_n(q^2u) = -q^{4n+2}u^2c_n(u), \quad d_n(q^2u) = q^{4n}(a_n(u) - (1 - \varepsilon u + u^2)c_n(u)), \tag{49}$$

and we explicitly see the rate at which $b_n(u)$ and $d_n(u)$ converge to zero functions. Furthermore, by taking into account the limits

$$\lim_{n\to\infty} a_n(u) = \chi_q(q^{-2}u), \quad \lim_{n\to\infty} c_n(u) = \chi_q(u), \tag{50}$$

we obtain from (49)

$$\lim_{n\to\infty} q^{-4n}b_n(u) = -q^{-2}u^2\chi_q(q^{-2}u), \quad \lim_{n\to\infty} q^{-4n}d_n(u) = -q^{-2}u^2\chi_q(u), \tag{51}$$

where in the last limit we have used Eq. (21) for $\chi_q(u)$.

Theorem 1 (Uniqueness of the regular solution). *In the case $|q| < 1$, let $f(u)$ be a solution of the functional equation (21) which is regular (i.e. holomorphic) at $u = 0$ and normalized so that $f(0) = 1$. Then $f(u) = \chi_q(u)$.*

Proof. Taking the limit $n \to \infty$ in the matrix equality (24) and using the regularity and normalization properties of $f(u)$ at $u = 0$, we obtain the equality

$$\begin{pmatrix} f(q^{-2}u) \\ f(u) \end{pmatrix} = M_\infty(u)\begin{pmatrix} 1 \\ 1 \end{pmatrix} = \begin{pmatrix} \chi_q(q^{-2}u) \\ \chi_q(u) \end{pmatrix}. \tag{52}$$

□

Lemma 2. *In the space of solutions of (21), there is an involution which associates to any solution $f(u)$ another solution $\check{f}(u)$ defined by*

$$\check{f}(u) = u^{-1}f(u^{-1}). \tag{53}$$

Proof. This is an easy direct verification. □

Applying Lemma 2 to our regular solution $\chi_q(u,\varepsilon)$, we obtain a second solution of the recursion equation (21)

$$\check{\chi}_q(u,\varepsilon) := u^{-1}\chi_q(u^{-1},\varepsilon) \tag{54}$$

which is holomorphic in $\mathbb{C}_{\neq 0}$.

3.2. *Orthogonal polynomials associated to $\chi_q(u,\varepsilon)$*

The function $\chi_q(u,\varepsilon)$ can be expanded as a power series

$$\chi_q(u,\varepsilon)=\sum_{n\geq 0} c_n u^n, \quad c_0=1, \tag{55}$$

with infinite radius of convergence. The coefficients here are functions of ε and q. The functional equation (21) in this case is translated into a system of recurrence equations on the coefficients

$$(1-q^{-2-2n})c_{n+1}=\varepsilon c_n-(1-q^{2n})c_{n-1}, \quad n\in\mathbb{Z}_{\geq 0}, \tag{56}$$

which, upon multiplication by $(q^{-2};q^{-2})_n$, is rewritten as a system of recurrence relations defining orthogonal polynomials $\chi_{q,n}(\varepsilon)\in\mathbb{C}[\varepsilon]$, $n\in\mathbb{Z}_{n\geq 0}$:

$$\chi_{q,0}(\varepsilon)=1, \quad \chi_{q,n+1}(\varepsilon)=\varepsilon\chi_{q,n}(\varepsilon)+(q^n-q^{-n})^2\chi_{q,n-1}(\varepsilon), \tag{57}$$

where

$$\chi_{q,n}(\varepsilon):=(q^{-2};q^{-2})_n c_n. \tag{58}$$

Notice that the polynomials $\chi_{q,n}(\varepsilon)$ also depend on the deformation parameter q in symmetric way in the sense that they are unchanged under the replacement $q\mapsto q^{-1}$. Here are the explicit forms of the first four polynomials

$$\begin{gathered}\chi_{q,0}(\varepsilon)=1, \quad \chi_{q,1}(\varepsilon)=\varepsilon, \quad \chi_{q,2}(\varepsilon)=\varepsilon^2+(q-q^{-1})^2, \\ \chi_{q,3}(\varepsilon)=\varepsilon(\varepsilon^2+(q^2-q^{-2})^2+(q-q^{-1})^2), \quad \ldots\end{gathered} \tag{59}$$

Thus, our solution $\chi_q(u,\varepsilon)$ is represented in the form of a power series

$$\chi_q(u,\varepsilon):=\sum_{n\geq 0}\frac{\chi_{q,n}(\varepsilon)}{(q^{-2};q^{-2})_n}u^n=\sum_{n\geq 0}(-1)^n q^{n(n+1)}\frac{\chi_{q,n}(\varepsilon)}{(q^2;q^2)_n}u^n. \tag{60}$$

In order to see explicitly how this series absolutely convergences on the entire complex plane of the variable u, it suffices to find the growth rate of the polynomials $\chi_{q,n}(\varepsilon)$ at large index n. To this end, we observe that the recurrence relation (57) at large n approaches the form

$$\chi_{n+1}\simeq\varepsilon\chi_n+q^{-2n}\chi_{n-1} \tag{61}$$

where we have suppressed the arguments q and ε of $\chi_{q,n}(\varepsilon)$ and have taken into account the inequality $|q|<1$. This equivalently can be rewritten as follows:

$$\frac{\chi_{n+1}}{\chi_n}\frac{\chi_n}{\chi_{n-1}}\simeq\varepsilon\frac{\chi_n}{\chi_{n-1}}+q^{-2n}\Leftrightarrow\frac{\chi_n}{\chi_{n-1}}\simeq q^{-n}\Leftrightarrow\chi_n\simeq q^{-n^2/2}. \tag{62}$$

The latter relation immediately implies the absolute convergence of the sum in (60).

Theorem 2. *The orthogonal polynomials $\chi_{q,n}(\varepsilon)$ satisfy the following multiplication rules*

$$\chi_{q,m}(\varepsilon)\chi_{q,n}(\varepsilon) = \sum_{k=0}^{\min(m,n)} \frac{(q^{2m};q^{-2})_k(q^{2n};q^{-2})_k(q^{2(k-m-n)};q^2)_k}{(q^2;q^2)_k}\chi_{q,m+n-2k}(\varepsilon) \tag{63}$$

Proof. This is a direct check by recurrence on $\max(m,n)$. □

3.3. *Non-linear first order functional difference equations*

The vector space of solutions of (21) is a module over the algebra of q^2-periodic functions, i.e. functions $g(u)$ that satisfy the functional difference equation $g(q^2u) = g(u)$. For any nontrivial solution $f(u)$ of (21), the combinations

$$R_f(u) := \frac{f(q^{-2}u)}{f(u)} \tag{64}$$

and

$$P_f(u) := \frac{q^2u^2f(q^2u)}{f(u)} \tag{65}$$

are invariant with respect to multiplication by q^2-periodic functions, so that they can be used for characterization of the equivalence classes of solutions of (21) differing by q^2-periodic functions. The two combinations are related to each other through the functional equalities

$$P_f(u)R_f(q^2u) = q^2u^2, \quad P_f(u) + R_f(u) = 1 - \varepsilon u + u^2. \tag{66}$$

The second order functional linear difference equation (21) implies first order non-linear functional difference equations for $R(u) = R_f(u)$

$$R(u)R(q^2u) = (1 - \varepsilon u + u^2)R(q^2u) - q^2u^2 \tag{67}$$

and for $P(u) = P_f(u)$

$$P(u)P(q^{-2}u) = (1 - \varepsilon u + u^2)P(q^{-2}u) - q^{-2}u^2. \tag{68}$$

Notice that two equations are related to each other by the symmetry $q \mapsto q^{-1}$.

By using (24) and (47), we easily obtain the following linear fractional transformation formulae

$$R(q^{2n}u) = \frac{d_n(u)R(u) - b_n(u)}{a_n(u) - c_n(u)R(u)}, \quad \forall n \in \mathbb{Z}_{\geq 0}. \tag{69}$$

Theorem 3. *Let $R, P\colon U \to \mathbb{C}$, $U \subset \mathbb{C}$, be solutions of (67) and (68), respectively, and $z \in U$ be such that $\chi_q(z) \neq 0$. Then the sequences $(R(q^{2n}z))_{n\geq 0}$ and $(P(q^{2n}z))_{n\geq 0}$ converge with the limits*

$$\lim_{n\to\infty} R(q^{2n}z) = \begin{cases} 0 & \textit{if } R(z) \neq R_{\chi_q}(z); \\ 1 & \textit{otherwise.} \end{cases} \tag{70}$$

and

$$\lim_{n\to\infty} P(q^{2n}z) = \begin{cases} 1 & \text{if } P(z) \neq P_{\chi_q}(z); \\ 0 & \text{otherwise.} \end{cases} \tag{71}$$

Proof. Formula (71) follows from (70) if we define $R(u) := 1 - \varepsilon u + u^2 - P(u)$.

If $R(z) = R_{\chi_q}(z)$, then formula (69) with $u = z$ implies that $R(q^{2n}z) = R_{\chi_q}(q^{2n}z)$ for all $n \in \mathbb{Z}_{\geq 0}$. Thus,

$$\lim_{n\to\infty} R(q^{2n}z) = \lim_{n\to\infty} R_{\chi_q}(q^{2n}z) = R_{\chi_q}(0) = 1 \tag{72}$$

by the regularity and the normalization properties of $\chi_q(u)$ at $u = 0$.

If we assume that $R(z) \neq R_{\chi_q}(z)$, then in equality (69) with $u = z$ the denominator is distinct from zero for sufficiently large n since it converges to a non-zero number:

$$\begin{aligned}\lim_{n\to\infty}(a_n(z) - c_n(z)R(z)) &= a_\infty(z) - c_\infty(z)R(z) \\ &= \chi_q(q^{-2}z) - \chi_q(z)R(z) \\ &= \chi_q(z)(R_{\chi_q}(z) - R(z)),\end{aligned} \tag{73}$$

while the numerator converges to $b_\infty(z)R(z) - d_\infty(z) = 0$. □

3.4. *Wronskian of two solutions of the main functional equation*

An important function associated with any two solutions of the functional equation (21) is their Wronskian or, more precisely, the functional difference analogue of the Wronskian in the theory of linear second-order ordinary differentian equations. As we will see below, any Wronskian satisfies a particularly simple difference equation of the form

$$W(q^2u) = \frac{1}{q^2u^2}W(u) \Leftrightarrow W(u/q) = u^2W(qu). \tag{74}$$

Definition–Proposition 1. *Let $f(u)$ and $g(u)$ satisfy Eq. (21). Then their Wronskian $[f,g](u)$ defined by*

$$[f,g](u) := f(q^{-2}u)g(u) - g(q^{-2}u)f(u) \tag{75}$$

satisfies Eq. (74).

Proof. Subtracting (21) for $g(u)$ multiplied by $f(u)$ from (21) for $f(u)$ multiplied by $g(u)$, we obtain

$$\begin{aligned}\big(f(q^{-2}u) + q^2u^2f(q^2u)\big)g(u) &= \big(g(q^{-2}u) + q^2u^2g(q^2u)\big)f(u) \\ &\Leftrightarrow f(q^{-2}u)g(u) - g(q^{-2}u)f(u) \\ &= q^2u^2\big(g(q^2u)f(u) - f(q^2u)g(u)\big) \\ &\Leftrightarrow [f,g](u) = q^2u^2[f,g](q^2u).\end{aligned} \tag{76}$$

□

Of particular importance for the sequel will be the Wronskian of χ and $\check{\chi}$:

$$[\chi, \check{\chi}](u) = \chi(q^{-2}u)\check{\chi}(u) - \check{\chi}(q^{-2}u)\chi(u). \tag{77}$$

Here and hereafter we shorten the notation $\chi(u) = \chi_q(u, \varepsilon)$ and so on.

Lemma 3. *The Wronskian* $[\chi, \check{\chi}]$ *is not the identically zero function.*

Proof. Using the power series expansions, we obtain a Laurent series

$$[\chi, \check{\chi}](u) = \sum_{n\in\mathbb{Z}} [\chi, \check{\chi}]_n u^n, \tag{78}$$

where the residue $[\chi, \check{\chi}]_{-1}$ is given explicitly by the absolutely convergent series

$$[\chi, \check{\chi}]_{-1} = \sum_{m\geq 0} \left(\frac{\chi_{q,m}(\varepsilon)}{(q^{-2}; q^{-2})_m} \right)^2 (q^{-2m} - q^{2m+2}) \tag{79}$$

where each term is non-negative if the variables ε and q are real with $0 < q < 1$. Thus, dropping all but the very first term, we obtain

$$[\chi, \check{\chi}]_{-1} \geq 1 - q^2 > 0. \tag{80}$$

□

3.5. *Parametrization in terms of θ-functions*

Consider Jacobi's θ-function

$$\theta_1(u, q) = \frac{1}{\mathsf{i}} \sum_{n\in\mathbb{Z}} (-1)^n q^{(n+1/2)^2} u^{n+1/2}. \tag{81}$$

It has the properties

$$\theta_1(1, q) = 0, \quad \theta_1(u^{-1}, q) = -\theta_1(u, q), \quad \theta_1(q^2 u, q) = -\frac{1}{qu}\theta_1(u, q), \tag{82}$$

and

$$\overline{\theta_1(u, q)} = \mathsf{b}\mathrm{e}^{\pi\mathsf{i}/4 - \mathsf{i}\pi x^2}\theta_1(u, q). \tag{83}$$

Any solution $W(u)$ of (74) in the form of a Laurent series in u, can be brought to the form

$$W(u) = \varrho\theta_1(su, q)\theta_1(s^{-1}u, q) \tag{84}$$

for some constants s and ϱ. The properties of the θ-function given by (82) and (83) imply that

$$\overline{W(u)} = \mathsf{i}\mathsf{b}^2 \bar{\varrho}\varrho^{-1} \mathrm{e}^{-2\pi\mathsf{i}(\sigma^2 + x^2)}\, W(u), \quad u = \mathrm{e}^{2\pi \mathsf{b} x}, \quad s = \mathrm{e}^{2\pi\mathsf{b}\sigma}. \tag{85}$$

In the case when $W = [\chi, \check{\chi}]$, the variables s and ϱ will be determined in terms of ε and q:

$$s = s(\varepsilon, q), \quad \varrho = \varrho(\varepsilon, q). \tag{86}$$

The first of these dependences is of particular importance for us, and it will be the subject of a numerical study.

3.6. *The main functional equation with q replaced by q^{-1}*

We turn now to the discussion of the functional equation (21) with q replaced by q^{-1}:

$$f(q^2u) + \frac{u^2}{q^2} f\left(\frac{u}{q^2}\right) = (1 - \varepsilon u + u^2) f(u) \,. \tag{87}$$

Let $\boldsymbol{\chi}_{q^{-1}}(u, \varepsilon)$ be the normalized solution (60) with q replaced by q^{-1}, i.e. a power series of the form

$$\boldsymbol{\chi}_{q^{-1}}(u, \varepsilon) \simeq \sum_{n \geq 0} \frac{\chi_{q,n}(\varepsilon)}{(q^2; q^2)_n} u^n \tag{88}$$

where we use the semi-equality symbol $\simeq$ to indicate the fact that the series is now only a formal power series as it has zero radius of convergence. This is an immediate consequence of the asymptotic behavior of $\chi_{q,n}(\varepsilon)$ at large n given by (62). Thus, there does not exist a regular at $u = 0$ solution of Eq. (87). Nonetheless, the formal power series solution (88) is interpreted as the asymptotic expansion of a true solution which is not analytic at $u = 0$. Before formulating this result, we do some preparatory work.

Lemma 4. *Let $f(u)$ and $W(u)$ satisfy the functional equations* (21) *and* (74), *respectively. Then the ratio $f(u)/W(u)$ satisfies* (87).

Proof. This is a direct verification. □

Lemma 5. *Let $f(u)$ and $g(u)$ satisfy the functional equations* (21) *and* (87), *respectively. Then the combination*

$$\langle f, g\rangle(u) := f\left(\frac{u}{q^2}\right) g(u) - \frac{u^2}{q^2} g\left(\frac{u}{q^2}\right) f(u) \tag{89}$$

is q^2-periodic, i.e.

$$\langle f, g\rangle(u) = \langle f, g\rangle(q^2u), \quad \forall u. \tag{90}$$

Proof. Subtracting (87) multiplied by $f(u)$ from (21) multiplied by $g(u)$, we obtain

$$\begin{aligned}\left(f\left(\frac{u}{q^2}\right) + q^2u^2 f(q^2u)\right) g(u) &= \left(g(q^2u) + \frac{u^2}{q^2} g\left(\frac{u}{q^2}\right)\right) f(u) \\ &\Leftrightarrow f\left(\frac{u}{q^2}\right) g(u) - \frac{u^2}{q^2} g\left(\frac{u}{q^2}\right) f(u) \\ &= f(u) g(q^2u) - q^2u^2 g(u) f(q^2u) \end{aligned} \tag{91}$$

where the latter equality is exactly (90). □

Theorem 4. *The formal power series* (88) *is the asymptotic expansion of the following solution of the functional equation* (87)

$$\chi_{q^{-1}}(u,\varepsilon) := \frac{\check{\chi}_q(u,\varepsilon)}{[\chi_q,\check{\chi}_q](u)} \tag{92}$$

where the Wronskian $[\chi_q,\check{\chi}_q](u)$ *is defined in* (77).

Proof. The fact that $\chi_{q^{-1}}(u,\varepsilon)$ defined in (92) satisfies (87) follows from Lemma 4. The rest of the proof is based on the following two easily verifiable functional identities

$$\langle\chi_q,\chi_{q^{-1}}\rangle(u) = 1 \Leftrightarrow \chi_q(u)\chi_{q^{-1}}(q^2u) - q^2u^2\chi_{q^{-1}}(u)\,\chi_q(q^2u) = 1 \tag{93}$$

and

$$\langle\check{\chi}_q,\chi_{q^{-1}}\rangle(u) = 0 \Leftrightarrow P_{\check{\chi}_q}(u) = \frac{\chi_{q^{-1}}(q^2u)}{\chi_{q^{-1}}(u)}. \tag{94}$$

The Wronskian $[\chi,\check{\chi}](u)$ can equivalently be rewritten in the form

$$[\chi,\check{\chi}](u) = \chi(u)\check{\chi}(u)(R_\chi(u) - R_{\check{\chi}}(u)) \tag{95}$$

which implies that if $[\chi,\check{\chi}](z) \neq 0$ then we have $R_\chi(z) \neq R_{\check{\chi}}(z)$, and by Theorem 3 combined with equality (94) we conclude that

$$\lim_{n\to\infty}\frac{\chi_{q^{-1}}(q^{2+2n}z)}{\chi_{q^{-1}}(q^{2n}z)} = \lim_{n\to\infty} P_{\check{\chi}}(q^{2n}z) = 1. \tag{96}$$

On the other hand, from equality (93) we also obtain a sequence of equalities

$$\frac{1}{\chi_{q^{-1}}(q^{2n}z)} = \frac{\chi_q(q^{2n}z)\chi_{q^{-1}}(q^{2+2n}z)}{\chi_{q^{-1}}(q^{2n}z)} - q^{2+4n}z^2\chi_q(q^{2+2n}z), \quad \forall n \in \mathbb{Z}_{\geq 0} \tag{97}$$

so that

$$\lim_{n\to\infty}\frac{1}{\chi_{q^{-1}}(q^{2n}z)} = 1 \Leftrightarrow \lim_{n\to\infty}\chi_{q^{-1}}(q^{2n}z) = 1. \tag{98}$$

Now, expanding in a formal power series $\chi_{q^{-1}}(u)$ around $u = 0$, we recover the series (88) as it is fixed uniquely by the functional difference equation (87) and the initial condition $\chi_{q^{-1}}(0) = 1$. □

4. Behavior at Infinity and Ansatz for the Eigenfunction

In the limit $x \to -\infty$, Eq. (16) is approximated by the equation

$$\psi(x + \mathsf{ib}) + \psi(x - \mathsf{ib}) = -\mathrm{e}^{-2\pi\mathsf{b}x}\psi(x), \tag{99}$$

where, in the left-hand side, any one of the two terms can be dominating giving rise to two equations

$$\psi(x + \epsilon\mathsf{ib}) = -\mathrm{e}^{-2\pi\mathsf{b}x}\psi(x), \quad \epsilon \in \{\pm 1\}, \tag{100}$$

with particular solutions

$$\psi(x) = \mathrm{e}^{\epsilon i\pi x^2 + 2\pi\eta x} \tag{101}$$

which describe the asymptotic behavior at $x \to -\infty$ of two solutions of (16) and (17)

$$\psi_\epsilon(x) = \mathrm{e}^{\epsilon i\pi x^2 + 2\pi\eta x}\phi_\epsilon(x), \quad \phi_\epsilon(x)|_{x\to-\infty} = \mathcal{O}(1), \quad \epsilon \in \{\pm 1\}. \tag{102}$$

Inspection of the equations for $\phi_\epsilon(x)$ gives solutions

$$\phi_\epsilon(x) = \boldsymbol{\chi}_{q^{-\epsilon}}(u)\overline{\boldsymbol{\chi}_{q^\epsilon}(u)}, \quad \epsilon \in \{\pm 1\}, \tag{103}$$

up to multiplication by doubly periodic functions of x. Thus, a general solution for (16) and (17) exponentially decays at $x \to -\infty$ and is given by the formula

$$\psi(x) = \mathrm{e}^{2\pi\eta x} \sum_{\epsilon\in\{\pm 1\}} A_\epsilon \mathrm{e}^{\epsilon i\pi x^2}\phi_\epsilon(x) \tag{104}$$

where, in general, A_ϵ are doubly periodic functions of x. As non-constant doubly periodic functions provide unwanted extra poles of the eigenfunction, we assume A_ϵ to be constants. Moreover, using the modular transformation formula (85) with $W(u) = [\boldsymbol{\chi}, \check{\boldsymbol{\chi}}](u)$, and choosing

$$A_1 = \mathsf{b}^{-1}\varrho\mathrm{e}^{\pi\mathsf{i}\sigma^2 - \xi\pi\mathsf{i}/4}, \quad A_{-1} = \bar{A}_1 = \mathsf{b}\bar{\varrho}\mathrm{e}^{\xi\pi\mathsf{i}/4 - \pi\mathsf{i}\sigma^2}, \tag{105}$$

where $\xi^2 = 1$, we arrive to the final form for the eigenfunction:

$$\psi(x) = \mathsf{b}^{-1}\mathrm{e}^{\pi\mathsf{i}\sigma^2 - \xi\pi\mathsf{i}/4}\mathrm{e}^{2\pi\eta x + i\pi x^2}\frac{\check{\boldsymbol{\chi}}(u)\overline{\boldsymbol{\chi}(u)} + \xi\boldsymbol{\chi}(u)\overline{\check{\boldsymbol{\chi}}(u)}}{\theta_1(su, q)\theta_1(s^{-1}u, q)}. \tag{106}$$

One can easily verify that

$$\psi(-x) = \xi\psi(x), \tag{107}$$

so that the parameter ξ is identified with the parity of the state. We also have the exponential decay at both infinities

$$|\psi(x)| \sim \mathrm{e}^{-2\pi\eta|x|}, \quad x \to \pm\infty, \tag{108}$$

and the reality property

$$\overline{\psi(x)} = \psi(x). \tag{109}$$

5. The Quantization Condition

In the previous section, we have constructed a solution (106) for (16) and (17) which exponentially decays at plus and minus infinities and has a minimal number of poles. The quantization condition is the analyticity condition for (106) in the strip

$$S_\mathsf{b} := \left\{z \in \mathbb{C} \mid |\Im z| < \max(\Re\mathsf{b}, \Re\mathsf{b}^{-1})\right\}. \tag{110}$$

As a matter of fact, this happens to be equivalent to the absence of poles in the entire complex plane $\mathbb{C}$, at least in the case when all zeros of the wronskian $[\chi, \check{\chi}](u)$ are simple.

Let us define a function

$$G_q(u,\varepsilon) := \frac{\chi_q(u,\varepsilon)}{\check{\chi}_q(u,\varepsilon)}, \tag{111}$$

which has the (anti)symmetry property

$$G_q(u,\varepsilon)G_q(1/u,\varepsilon) = 1, \quad \forall u \in \mathbb{C}. \tag{112}$$

Theorem 5. *Let $\varepsilon = \varepsilon(\sigma)$ be such that*

$$W(u) := [\chi_q, \check{\chi}_q](u) = \varrho\theta_1(su)\theta_1(s^{-1}u), \quad \forall u \in \mathbb{C}, \tag{113}$$

and assume that $s \notin \pm q^{\mathbb{Z}}$ (recall that $s = s(\sigma) = \mathrm{e}^{2\pi \mathsf{b}\sigma}$). Then the eigenfunction $\psi(x)$ defined by (106) does not have poles in the strip S_{b} if the variable σ is such that

$$G_q(s,\varepsilon) = -\xi\overline{G_q(s,\varepsilon)}. \tag{114}$$

In that case, the eigenfunction $\psi(x)$ is an entire function on the whole complex plane $\mathbb{C}$.

Proof. The zeros of the denominator in (106) are given by

$$u = q^{2n}s^{\pm 1}, \quad n \in \mathbb{Z}, \tag{115}$$

which all are simple provided $s \notin \pm q^{\mathbb{Z}}$. The numerator of (106) vanishes for any u satisfying the equation

$$G_q(u,\varepsilon) = -\xi\overline{G_q(u,\varepsilon)}. \tag{116}$$

On the other hand, the identity $W(q^{2n}s^{\pm 1}) = 0$ is equivalent to

$$G_q(q^{2n}s,\varepsilon) = G_q(s,\varepsilon) \tag{117}$$

which means that the infinite number of cancellation conditions arising from (116) at $u = q^{2n}s^{\pm 1}$ reduce to the single equation (114). □

Remark. The spectral equation (114) together with the (anti)symmetry property (112) implies that

$$\xi^2 = 1 \tag{118}$$

regardless of our previous parity and reality considerations.

6. Numerical Study for $\hbar = 2\pi\mathrm{i}$

6.1. *The structure of* $\varepsilon = \varepsilon(\sigma)$

Let us remind the definition of the Wronskian (77):

$$W(u,\varepsilon) := \chi\left(\frac{u}{q^2},\varepsilon\right)\check{\chi}(u,\varepsilon) - \check{\chi}\left(\frac{u}{q^2},\varepsilon\right)\chi(u,\varepsilon) = \varrho(\varepsilon)\theta_1(su)\theta_1(s^{-1}u). \quad (119)$$

Let σ be fixed ($s = \mathrm{e}^{2\pi\mathsf{b}\sigma}$). Consider the following equation relating σ and ε:

$$W(\mathrm{e}^{2\pi\mathsf{b}\sigma},\varepsilon) = 0. \quad (120)$$

We have used the Newton method to solve (120) with respect to ε. This method uses an initial point as input, and different initial points provide different solutions:

$$\varepsilon \in \{\varepsilon_1,\varepsilon_2,\varepsilon_3,\dots\}. \quad (121)$$

Varying σ, one obtains

$$\varepsilon(\sigma) \in \{\varepsilon_1(\sigma),\varepsilon_2(\sigma),\varepsilon_3(\sigma),\dots\}, \quad (122)$$

where $\varepsilon_k(\sigma)$ stands for k^{th} sheet of the Riemann surface of the multivalued function $\varepsilon(\sigma)$.

If ε and σ are related by (120), then one has

$$W(\mathrm{e}^{2\pi\mathsf{b}(\pm\sigma+\mathrm{i}\mathsf{b}n+\mathrm{i}\mathsf{b}^{-1}m)},\varepsilon) = 0, \quad \forall n,m\in\mathbb{Z}. \quad (123)$$

We deduce that

$$\varepsilon(\sigma) = \varepsilon(\pm\sigma + \mathrm{i}\mathsf{b}n + \mathrm{i}\mathsf{b}^{-1}m). \quad (124)$$

In general, this relation involves exchanges of sheets of the Riemann surface. However, more careful numerical studies show that

- quantization corresponds to only real σ;
- all branching points of $\varepsilon = \varepsilon(\sigma)$ lie outside the real axis.

Numerical studies show that indeed

$$\varepsilon_k(\sigma) = \varepsilon_k(-\sigma) = \varepsilon_k(2\sin\theta - \sigma), \quad \sigma\in\mathbb{R}, \quad (125)$$

for the same kth sheet of the Riemann surface. In what follows, the image of the function $\varepsilon_k(\sigma)$ with $\sigma\in\mathbb{R}$ will be called *orbit*. The principal domain for $\varepsilon_k(\sigma)$ is then

$$\sigma\in\mathbb{R}, \qquad 0\le\sigma\le\sin\theta. \quad (126)$$

In the numerical studies we use $\mathsf{b} = \mathrm{e}^{\frac{\mathrm{i}\pi}{4}}$ which corresponds to

$$\theta = \frac{\pi}{4}, \quad \sin\theta = \cos\theta = \frac{1}{\sqrt{2}}, \quad q = \mathrm{e}^{-\pi}. \quad (127)$$

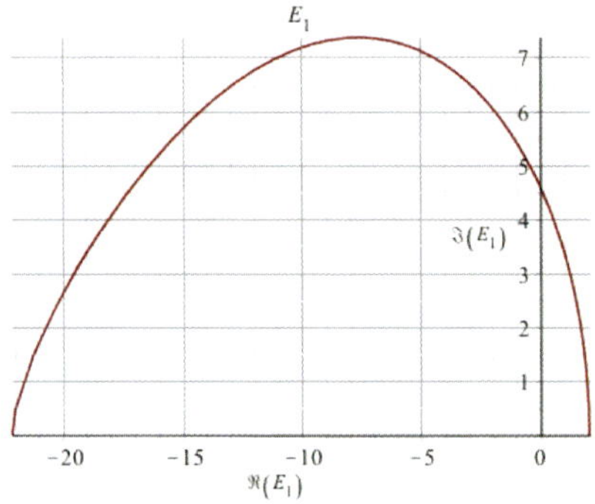

Fig. 1. The orbit of $\varepsilon = \varepsilon_1(\sigma)$. Turning points (endpoints on the plot) are $\varepsilon_1(0) = 1.9962511523$ and $\varepsilon_1(\sin\theta) = -22.1838257068$.

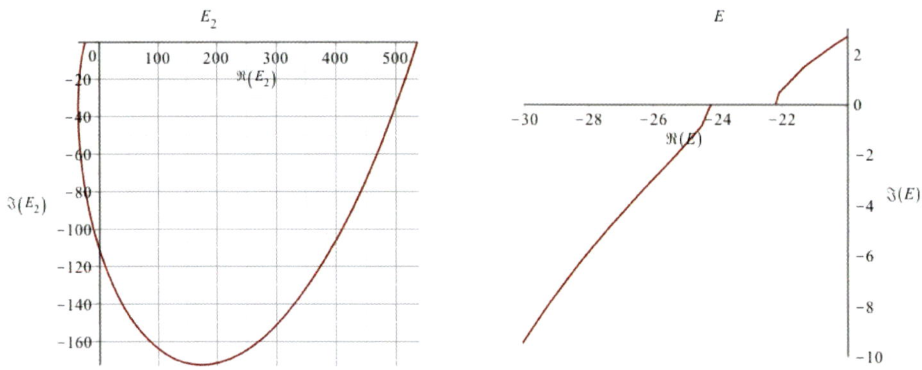

Fig. 2. Left: The orbit of $\varepsilon = \varepsilon_2(\sigma)$. Turning points are $\varepsilon_2(0) = 535.49351947$ and $\varepsilon_2(\sin\theta) = -24.183825694$. Right: The orbits of $\varepsilon_1(\sigma)$ and $\varepsilon_2(\sigma)$ in the vicinity of $\sigma = 1/\sqrt{2}$. Recall that $\varepsilon_1(1/\sqrt{2}) = -22.1838257068$ vs. $\varepsilon_2(1/\sqrt{2}) = -24.183825694$.

Figures 1 and 2 show the orbits of $\varepsilon_1(\sigma)$ and $\varepsilon_2(\sigma)$, respectively, in the complex plane. Two branching points for the branch cuts between ε_1 and ε_2 are located approximately at

$$\sigma \approx \frac{1}{\sqrt{2}} \pm 0.007\mathsf{b}. \tag{128}$$

The right-hand side of Fig. 2 shows the gap between ε_1 and ε_2. Figure 3 shows the orbit $\varepsilon_3(\sigma)$. The right-hand side of Fig. 3 shows all three orbits in the log scale.

It is worth discussing the endpoints $\sigma = 0$ and $\sigma = \sin\theta$ in general. The point $\sigma = 0$ gives (see (119))

$$W(1,\varepsilon) = \chi(1,\varepsilon)(\chi(q^{-2},\varepsilon) - q^2\chi(q^2,\varepsilon)) = 0, \tag{129}$$

and the case

$$\chi(1,\varepsilon) = 0 \tag{130}$$

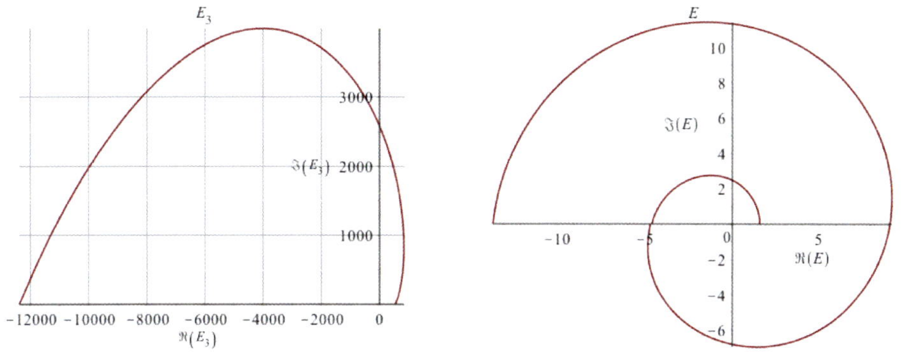

Fig. 3. Left: The orbit of $\varepsilon = \varepsilon_3(\sigma)$. Turning points are $\varepsilon_3(0) = 535.49726832$ and $\varepsilon_3(\sin\theta) = -12391.6479693$. Right: Collection of $\varepsilon_1(\sigma)$, $\varepsilon_2(\sigma)$ and $\varepsilon_3(\sigma)$ in the logarithmic scale. Low resolution does not allow one to see the gaps between the orbits. Approximation of this spiral trajectory is $\varepsilon(\sigma) \simeq \mathrm{e}^{2\pi \mathrm{b}\sigma}$, $\sigma \geq 0$.

gives $\varepsilon_k(0)$ for even sheets,[a]

$$\begin{aligned}
\varepsilon_2(0) &= \frac{1}{q^2}\left(1 + q^4 - q^6 - q^8 - q^{10} - 2q^{12} - q^{14} + q^{18} + 6q^{20} + 11q^{22} + \cdots\right),\\
\varepsilon_4(0) &= \frac{1}{q^4}\left(1 + 2q^8 + q^{14} - q^{18} - 3q^{20} - 8q^{22} - 13q^{24} + \cdots\right),\\
\varepsilon_6(0) &= \frac{1}{q^6}\left(1 + 2q^{12} + q^{22} + q^{24} + q^{26} + 2q^{32} + 2q^{34} + \cdots\right),
\end{aligned} \tag{131}$$

and so on. Another branch

$$\boldsymbol{\chi}(q^{-2},\varepsilon) - q^2\boldsymbol{\chi}(q^2,\varepsilon) = 0 \tag{132}$$

gives $\varepsilon_k(0)$ for the odd sheets:

$$\varepsilon_1(0) = 2 - 2q^2 - 4q^4 - 2q^6 + 14q^8 + 50q^{10} + 40q^{12} - 268q^{14} - 1136q^{16} + \cdots, \tag{133}$$

$$\varepsilon_3(0) = \frac{1}{q^2}(1 + 3q^4 + 3q^6 + q^8 - 15q^{10} - 52q^{12} - 43q^{14} + 264q^{16} + 1127q^{18} + \cdots), \tag{134}$$

and so on.

The point $\sigma = \sin\theta$ gives

$$W(-q^{-1},\varepsilon) \sim \left(\boldsymbol{\chi}(-q^{-1},\varepsilon) - q\boldsymbol{\chi}(-q,\varepsilon)\right)\left(\boldsymbol{\chi}(-q^{-1},\varepsilon) + q\boldsymbol{\chi}(-q,\varepsilon)\right) \tag{135}$$

so that

$$\boldsymbol{\chi}(-q^{-1},\varepsilon) - q\boldsymbol{\chi}(-q,\varepsilon) = 0 \tag{136}$$

gives the values for $\varepsilon_k(\sin\theta)$ on the odd sheets, for instance

$$\varepsilon_1(\sin\theta) = -q^{-1}\left(1 - q + q^2 - q^4 - q^6 + q^7 - 2q^8 + 2q^9 - 2q^{10} + 5q^{11} - 4q^{12} + \cdots\right). \tag{137}$$

[a]It suffices to substitute $\varepsilon = q^{-2k}\sum_{n=0}^{\infty} c_n q^{2n}$ into $\boldsymbol{\chi}(1)$ as a series with respect to q^2 and equate the coefficients of each power q^{2n} to zero.

Equation

$$\chi(-q^{-1},\varepsilon)+q\chi(-q,\varepsilon)=0 \tag{138}$$

gives the values for $\varepsilon_k(\sin\theta)$ on the even sheets, for instance

$$\varepsilon_2(\sin\theta)=-q^{-1}(1+q+q^2-q^4-q^6-q^7-2q^8-2q^9-2q^{10}-5q^{11}-4q^{12}-\cdots). \tag{139}$$

6.2. *The spectrum*

We turn now to the spectrum of the Hamiltonians (14).

The quantization conditions (114) for real σ are

$$\mathrm{Im}(G_q(s,\varepsilon))=0, \qquad \xi=-1, \quad \text{odd state}, \tag{140}$$

and

$$\mathrm{Re}(G_q(s,\varepsilon))=0, \qquad \xi=1, \quad \text{even state}. \tag{141}$$

One can see that formally there are two points $\sigma=0$ and $\sigma=\sin\theta$ on each sheet of the Riemann surface of $\varepsilon(\sigma)$ corresponding to these quantization conditions. These are the exceptional cases as denominator $W(u)$ has the second-order zeros while in the cases $\sigma=0$ for odd sheets, and $\sigma=\sin\theta$ for all sheets, the numerator of (106) has the first-order zeros. Therefore, all these cases must be eliminated from the spectrum. On the other hand, if $\sigma=0$ on even sheets, corresponding to $\chi(1)=0$, the numerator of (106) has second-order zeros at some second-order zeros of the denominator, but not at all of them and therefore the wave function (106) could not be a holomorphic function either. Theorem 5 concerns only the elimination of simple poles, and further analysis is needed to completely understand the situation with double poles.

6.2.1. *Sheet 1*

Real and imaginary parts of $G_q(s,\varepsilon_1(\sigma))$ (the first sheet) are plotted in Fig. 4.

An even state from the first sheet is given by

$$\sigma=\frac{1}{2}\sin\theta, \quad \varepsilon_1(\sigma)=4.594358809836918 94\mathrm{i}. \tag{142}$$

This is the ground state.

An odd state from the first sheet is given by

$$\begin{aligned}\sigma&=0.6121173716461672675,\\ \varepsilon_1(\sigma)&=-13.8783047780366906+6.161296243244348685\mathrm{i}.\end{aligned} \tag{143}$$

The cases corresponding to $\sigma=0$ and $\sigma=\sin\theta$ are discarded.

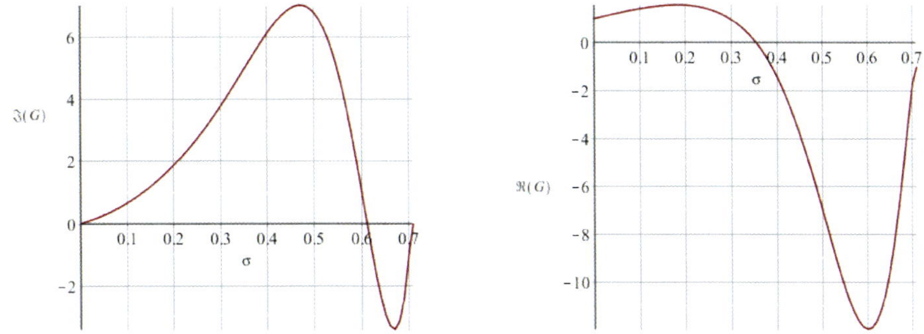

Fig. 4. Left: $\Im(G_q(s,\varepsilon_1))$ as function of σ, $0 \leq \sigma \leq \sin\theta$. Right: The same for $\Re(G_q(s,\varepsilon_1))$.

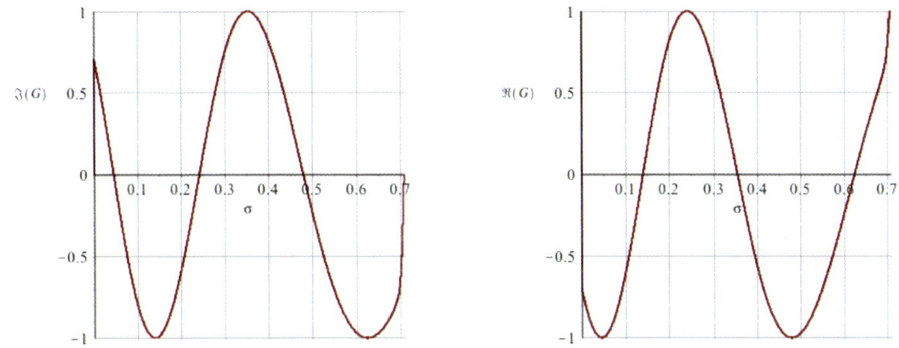

Fig. 5. Left: $\frac{\Im(G_q(s,\varepsilon_2))}{|G_q(s,\varepsilon_2)|}$ as function of σ, $0 \leq \sigma \leq \sin\theta$. Right: The same for $\frac{\Re(G_q(s,\varepsilon))}{|G_q(s,\varepsilon_2)|}$. Behavior near $\sigma = 0$ is misleading since one has an ambiguity $0/0$ there.

6.2.2. *Sheet 2*

Imaginary and real parts of the function $\frac{G_q(s,\varepsilon_2(\sigma))}{|G_q(s,\varepsilon_2(\sigma))|}$ (on the second sheet) are plotted in Fig. 5.

The odd and even states from the second sheet are given in Tables 1 and 2 respectively. The endpoint on the second orbit is given by

$$\sigma = 0, \quad \varepsilon_2(0) = 535.493519473629469, \tag{144}$$

corresponding to $\boldsymbol{\chi}(1,\varepsilon_2) = 0$, but this value does not correspond to an eigenvalue since it is not possible to cancel all second-order poles in the wave function.

Table 1. The odd states from the second sheet.

σ	$\varepsilon_2(\sigma)$
0.0449074054136668986	429.937612699070933 − 86.9352869839236228i
0.241612973133940861	87.33324987160330085 − 160.859744733070428i
0.478766031821187121	−33.7154767687408649 − 54.1710567496918622i

Table 2. The even states from the second sheet.

σ	$\varepsilon_2(\sigma)$
0.139460116715428804	234.614101715470239 − 167.345168305129794i
$\frac{1}{2}\sin\theta$	−111.300184113096796i
0.623413635048467267	−31.32504489967260473 − 12.15333894226767358i

Table 3. The states from the first sheet, $k = 1$.

σ	Parity	$\varepsilon_1(\sigma)$
0.35	even	4.594358809836918 94i
0.61	odd	−13.8783047780366906 + 6.161296243244348685i

Table 4. The states from the second sheet, $k = 2$.

σ	Parity	$\varepsilon_2(\sigma)$
0.62	even	−31.32504489967260473 − 12.15333894226767358i
0.48	odd	−33.7154767687408649 − 54.1710567496918622i
0.35	even	−111.300184113096796i
0.24	odd	87.33324987160330085 − 160.859744733070428i
0.14	even	234.614101715470239 − 167.345168305129794i
0.045	odd	429.937612699070933 − 86.9352869839236228i

6.3. *Table of the spectrum for* $\mathsf{b} = \mathsf{e}^{\mathsf{i}\pi/4}$

We summarize the spectrum from the first and second sheets in Tables 3 and 4, respectively. In our numerics we have used the precision 10^{-40}, therefore all our digits for the energies are precise.

7. The Self-Conjugate Case $\hbar = 2\pi$

Despite the fact that the special case $\mathsf{b} = 1$ corresponds to the limiting value of the strong coupling regime $\theta \to 0$, we do not know how to approach that limit on the level of eigenvalues and eigenfunctions. For this reason, we treat that case separately, independently of the strong coupling regime. This case is very special as the operators $\boldsymbol{u}$ and $\boldsymbol{v}$ are positive self-adjoint and formally commuting, while both functional equations (16), (17) reduce to one and the same but still non-trivial functional difference equation

$$\psi(x - \mathsf{i}) + \psi(x + \mathsf{i}) = (\varepsilon - 2\cosh(2\pi x))\psi(x), \tag{145}$$

which, under the substitution

$$\varphi(x) := \psi(\mathsf{i}x), \tag{146}$$

becomes an equation of the form

$$\varphi(x - 1) + \varphi(x + 1) + (2\cos(2\pi x) - \varepsilon)\varphi(x) = 0. \tag{147}$$

Despite being a special case of Harper's functional equation [8] on the real line, this equation should nonetheless be considered on the entire complex plane as its quantum mechanical content is extracted from the restriction of its solutions to the imaginary axis.

7.1. *The spectral curve*

Due to periodicity of the potential, for any solution of (147), its Weil–Gel'fand–Zak transform [6, 18, 19]

$$\tilde{\varphi}(x,y) = \sum_{m\in\mathbb{Z}} \varphi(x+m)\mathrm{e}^{-2\pi \mathrm{i} y m} \tag{148}$$

gives another solution which is quasi-periodic in x and periodic in y:

$$\tilde{\varphi}(x+1,y) = \mathrm{e}^{2\pi\mathrm{i}y}\tilde{\varphi}(x,y), \quad \tilde{\varphi}(x,y+1) = \tilde{\varphi}(x,y). \tag{149}$$

That means that the functional difference equation (147) is transformed into an algebraic functional equation

$$\left(\cos(2\pi x) + \cos(2\pi y) - \frac{\varepsilon}{2}\right)\tilde{\varphi}(x,y) = 0, \tag{150}$$

which means that the support of $\tilde{\varphi}(x,y)$ is localized to the *spectral curve*

$$\Gamma_\varepsilon := \left\{ (x,y) \in \mathbb{C}^2 \,\middle|\, \cos(2\pi x) + \cos(2\pi y) = \frac{\varepsilon}{2} \right\} \tag{151}$$

which covers the finite (affine) part of the elliptic curve

$$\mathcal{F}_\varepsilon := \left\{ (u,v) \in \mathbb{C}^2 \,|\, u + u^{-1} + v + v^{-1} = \varepsilon \right\} \tag{152}$$

through the map

$$\mathrm{q}\colon \Gamma_\varepsilon \to \mathcal{F}_\varepsilon, \quad (x,y) \mapsto \left(\mathrm{e}^{2\pi\mathrm{i}x}, \mathrm{e}^{2\pi\mathrm{i}y}\right). \tag{153}$$

Projection to the first component

$$x\colon \Gamma_\varepsilon \to \mathbb{C} \tag{154}$$

is a covering map branched at the points satisfying the condition $\sin(2\pi y) = 0$ which corresponds to a discrete infinite set of values of x:

$$(-\mathrm{i}\alpha + \mathbb{Z}) \cup (\mathrm{i}\alpha + \mathbb{Z}) \cup (-\mathrm{i}\beta + \mathbb{Z}) \cup (\mathrm{i}\beta + \mathbb{Z}) \tag{155}$$

where parameters $\alpha = \alpha(\varepsilon)$ and $\beta = \beta(\varepsilon)$ are determined through the equations

$$\varepsilon = 2 + 2\cosh(2\pi\alpha) = 4\cosh(\pi\alpha)^2, \tag{156}$$

$$\cosh(2\pi\beta) = \cosh(2\pi\alpha) + 2 \Leftrightarrow \sinh(\pi\beta) = \pm\cosh(\pi\alpha). \tag{157}$$

Since the spectrum of our operator is contained in $\mathbb{R}_{>4}$, we assume that $\varepsilon \in \mathbb{R}_{>4}$. In that case, we can choose α and β in a canonical way as real positive solutions of Eqs. (156) and (157).

The automorphism group of the spectral curve Γ_ε contains a subgroup generated by three elements

$$\rho,\sigma,\tau\colon \Gamma_\varepsilon \to \Gamma_\varepsilon, \quad \rho(x,y) = (-x,y), \quad \sigma(x,y) = (y,x), \quad \tau(x,y) = (x+1,y). \tag{158}$$

The two involutions ρ and σ cover involutions $\dot\rho$ and $\dot\sigma$ of $\mathcal{F}_\varepsilon$, while τ acts fiber-wise, i.e.

$$\mathrm{q}\circ\rho=\dot\rho\circ\mathrm{q},\quad \mathrm{q}\circ\sigma=\dot\sigma\circ\mathrm{q},\quad \mathrm{q}\circ\tau=\mathrm{q}\,. \tag{159}$$

7.2. *The deformed symplectic potential*

For the construction of a Bloch–Jost function to be defined below, we will use the restrictions to Γ_ε of a one-parameter family of *deformed* (*holomorphic*) *symplectic potentials* on $\mathbb{C}^2$

$$\theta_\lambda := \left(x+\frac{\lambda}{\sin(2\pi x)}\right)\mathrm{d}y,\quad \lambda\in\mathbb{C}, \tag{160}$$

where the non-deformed part

$$\theta_0 = x\,\mathrm{d}y \tag{161}$$

is the restriction of the standard holomorphic symplectic potential in $\mathbb{C}^2$, while the deforming part is given by the pull-back of a holomorphic 1-form on $\mathcal{F}_\varepsilon$

$$\frac{\mathrm{d}y}{\sin(2\pi x)}=-\frac{\mathrm{d}x}{\sin(2\pi y)}=\mathrm{q}^*(\omega),\quad \omega=\frac{\mathrm{d}v}{\pi v(u-u^{-1})}=-\frac{\mathrm{d}u}{\pi u(v-v^{-1})}, \tag{162}$$

which is unique, up to multiplication by a non-zero complex constant. With respect to the automorphisms (158), the form θ_λ has the following properties:

$$\rho^*(\theta_\lambda)=-\theta_\lambda,\quad \sigma^*(\theta_\lambda)=\mathrm{d}(xy)-\theta_\lambda,\quad \tau^*(\theta_\lambda)=\mathrm{d}y+\theta_\lambda. \tag{163}$$

7.3. *The Bloch–Jost function*

Definition 1. A holomorphic function $f\colon\Gamma_\varepsilon\to\mathbb{C}$ is called *Bloch–Jost function* if it satisfies the functional relations

$$f(x+1,y)\mathrm{e}^{-2\pi\mathrm{i}y}=f(x,y+1)=f(x,y),\quad \forall(x,y)\in\Gamma_\varepsilon. \tag{164}$$

For a fixed base point $P_0\in\Gamma_\epsilon$, let $\mathrm{p}\colon\tilde\Gamma_\varepsilon\to\Gamma_\varepsilon$ be the universal covering given by the homotopy classes of all paths starting from P_0. We define

$$g_{\lambda,\varepsilon}\colon\tilde\Gamma_\varepsilon\to\mathbb{C}_{\neq0},\quad \gamma\mapsto\mathrm{e}^{2\pi\mathrm{i}\int_\gamma\theta_\lambda} \tag{165}$$

Theorem 6. *Let* $(\lambda,\varepsilon)\in\mathbb{C}\times\mathbb{R}_{>4}$ *be such that*

$$g_{\lambda,\varepsilon}(\gamma)=1, \tag{166}$$

for any γ *for which* $x\circ\gamma$ *and* $\mathrm{q}\circ\gamma$ *are closed paths. Then, there exists a Bloch–Jost function* f *such that* $g_{\lambda,\varepsilon}=\mathrm{p}^*(f)$.

Proof. The fact that $g_{\lambda,\varepsilon}=\mathrm{p}^*(f)$ for some $f\colon\Gamma_\varepsilon\to\mathbb{C}$ is immediate. It suffices to remark that any closed path on Γ_ε projects down to closed paths both in $\mathbb{C}$ and $\mathcal{F}_\varepsilon$. So, we need to verify only the quasi-periodicity properties of f.

Let paths $\mu,\nu\colon[0,1]\to\Gamma_\varepsilon$ be such that

$$\mu(0)=P_0,\quad \mu(1)=\nu(0)=(x,y),\quad \nu(1)=(x,y+1). \tag{167}$$

Then, we have

$$f(x, y+1) = \mathrm{e}^{2\pi \mathrm{i} \int_{\mu \cdot \nu} \theta_\lambda} = \mathrm{e}^{2\pi \mathrm{i} \int_\mu \theta_\lambda} \mathrm{e}^{2\pi \mathrm{i} \int_\nu \theta_\lambda} = f(x, y), \tag{168}$$

since both $x \circ \nu$ and $\mathrm{q} \circ \nu$ are closed.

Redefine now $\mu, \nu \colon [0,1] \to \Gamma_\varepsilon$ so that

$$\mu(0) = P_0, \quad \mu(1) = \nu(0) = (x, y), \quad \nu(1) = (x+1, y). \tag{169}$$

Then, we have

$$f(x+1, y) = \mathrm{e}^{2\pi \mathrm{i} \int_{\mu \cdot \nu} \theta_\lambda} = \mathrm{e}^{2\pi \mathrm{i} \int_\mu \theta_\lambda} \mathrm{e}^{2\pi \mathrm{i} \int_\nu \theta_\lambda} = f(x, y) \mathrm{e}^{2\pi \mathrm{i} \int_\nu \theta_\lambda}, \tag{170}$$

where, this time, the path $x \circ \nu$ is not closed. Using the second equality in (163), we have

$$\mathrm{e}^{2\pi \mathrm{i} \int_\nu \theta_\lambda} = \mathrm{e}^{2\pi \mathrm{i} \int_{\sigma \circ \nu} \sigma^* \theta_\lambda} = \mathrm{e}^{2\pi \mathrm{i} \int_{\sigma \circ \nu} \mathrm{d}(xy)} \mathrm{e}^{-2\pi \mathrm{i} \int_{\sigma \circ \nu} \theta_\lambda} = \mathrm{e}^{2\pi \mathrm{i} y} \tag{171}$$

where we have used the fact that the paths $x \circ \sigma \circ \nu$ and $\mathrm{q} \circ \sigma \circ \nu$ are closed. □

7.4. *The spectral equations*

We call Eqs. (166) *spectral equations.* Our goal here is to reduce them to a minimal set so that any other equation is a consequence of the equations from that minimal set. By using two functions

$$r \colon \mathbb{R} \to [\alpha, \beta], \quad \cosh(2\pi r(t)) = 1 - \cos(\pi t) + \cosh(2\pi\alpha), \tag{172}$$

and

$$s \colon \mathbb{R} \to [0, \alpha], \quad \sinh(\pi s(t)) = \sinh(\pi\alpha)\sin(\pi t), \tag{173}$$

we define four paths $\xi, \check{\xi}, \zeta, \hat{\zeta} \colon [0,1] \to \Gamma_\varepsilon$ by

$$\xi(t) = (\mathrm{i}s(t+1/2), \mathrm{i}s(t)), \quad \check{\xi}(t) = (-\mathrm{i}s(t+1/2), -\mathrm{i}s(t)) \tag{174}$$

and

$$\zeta(t) = (\mathrm{i}r(t), t/2), \quad \hat{\zeta}(t) = (\mathrm{i}r(1-t), (t+1)/2). \tag{175}$$

We observe that $(\xi, \check{\xi})$ and $(\zeta, \hat{\zeta})$ are composable pairs, $\xi \cdot \check{\xi}$ being closed, while $\zeta \cdot \hat{\zeta}$ projects down to closed paths both in $\mathbb{C}$ and F_ε. Moreover, their images generate the fundamental group of F_ε, which means that the equations

$$\mathrm{e}^{2\pi \mathrm{i} \int_{\xi \cdot \check{\xi}} \theta_\lambda} = \mathrm{e}^{2\pi \mathrm{i} \int_{\zeta \cdot \hat{\zeta}} \theta_\lambda} = 1 \tag{176}$$

constitute a complete set of spectral equations. We calculate

$$\int_{\xi \cdot \check{\xi}} \theta_\lambda = 2 \int_\xi \theta_\lambda = 4 \int_0^{1/2} \left(\frac{\lambda}{\sinh(2\pi s(t+1/2))} - s(t+1/2) \right) s'(t)\, \mathrm{d}t =: A\lambda - \tilde{A} \tag{177}$$

and

$$\int_{\zeta \cdot \hat{\zeta}} \theta_\lambda = 2 \int_\zeta \theta_\lambda = \mathrm{i} \int_0^1 \left(r(t) - \frac{\lambda}{\sinh(2\pi r(t))} \right) \mathrm{d}t =: \mathrm{i}(\tilde{B} - \lambda B) \tag{178}$$

where $A, \tilde{A}, B, \tilde{B}$ are positive functions of ε satisfying the inequality

$$A\tilde{B} - B\tilde{A} > 0 \tag{179}$$

which is seen if we write out all integrals explicitly:

$$A\tilde{B} - B\tilde{A} = 4\int_{[0,1/2]\times[0,1]} \mathrm{d}(t,u)\frac{s'(t)(r(u)\sinh(2\pi r(u)) - s(t+1/2)\sinh(2\pi s(t+1/2)))}{\sinh(2\pi s(t+1/2))\sinh(2\pi r(u))} \tag{180}$$

and take into account the inequalities

$$0 \leq s(t) \leq \alpha \leq r(u) \leq \beta, \quad \forall t, u \in \mathbb{R}. \tag{181}$$

Now, the spectral equations (176) imply that

$$A\lambda - \tilde{A} =: n+1 \in \mathbb{Z}, \quad \mathsf{i}(\tilde{B} - \lambda B) \in \mathbb{Z}, \tag{182}$$

so that the second condition fixes a unique choice for λ,

$$\lambda = \tilde{B}/B, \tag{183}$$

while the first condition becomes a quantization condition for the energy:

$$A\tilde{B} - B\tilde{A} = (n+1)B, \quad n \in \mathbb{Z}_{\geq 0}, \tag{184}$$

where we have taken into account the inequality (179).

Numerical calculation at $n = 0$ for the ground state energy ε_0 gives the result

$$\log(\varepsilon_0) = 2.88181542992629678247713987172363292221616219\ldots \tag{185}$$

which is in agreement with formula (5.39) of [7].

7.5. *The eigenfunctions*

Theorem 7. *Let $(\lambda, \varepsilon) \in \mathbb{C} \times \mathbb{R}_{>4}$ satisfy the assumptions of Theorem 6, and choose the base point $P_0 = (\mathsf{i}\alpha, 0) \in \Gamma_\varepsilon$. Then, there exists a holomorphic function $\varphi\colon \mathbb{C} \to \mathbb{C}$ such that*

$$\sin(2\pi y)\varphi(x) = \sin\left(2\pi \int_\gamma \theta_\lambda\right), \quad \forall (x, y) \in \Gamma_\varepsilon, \tag{186}$$

for any path $\gamma\colon [0,1] \to \Gamma_\varepsilon$ such that $\gamma(0) = P_0$ and $\gamma(1) = (x, y)$. In particular, $\varphi(x)$ is a solution of the functional difference equation (147).

Proof. The right-hand side of (186) can be rewritten in the form

$$2\mathsf{i}\sin\left(2\pi \int_\gamma \theta_\lambda\right) = f(x, y) - f(x, -y), \tag{187}$$

where $f(x, y)$ is the Bloch–Jost function defined by

$$f(x, y) := \mathrm{e}^{2\pi\mathsf{i}\int_\gamma \theta_\lambda} \tag{188}$$

and the choice of the base point is such that

$$f(x, y)f(x, -y) = 1. \tag{189}$$

Thus, the function

$$h(x, y) := \frac{f(x, y) - f(x, -y)}{\sin(2\pi y)} \tag{190}$$

is explicitly periodic and symmetric in y:

$$h(x, y+1) = h(x, y) = h(x, -y). \tag{191}$$

On the other hand, if $(a, b) \in \Gamma_\varepsilon$ then

$$x^{-1}(a) = \{a\} \times (\{b + \mathbb{Z}\} \cup \{-b + \mathbb{Z}\}) \tag{192}$$

so that

$$h(a, b') = h(a, b), \quad \forall (a, b') \in x^{-1}(a), \tag{193}$$

i.e. there exists a function φ: $\mathbb{C} \to \mathbb{C}$ such that $h = x^*(\varphi)$. It remains to see that $\varphi(x)$ is holomorphic at any of the ramification points (155) corresponding to $\sin(2\pi y) = 0$, i.e. either $y = 0$ or $y = 1/2$. The case $y = 0$ is explicitly regular. To see the regularity at $y = 1/2$, it suffices to see that

$$f(\pm \mathsf{i}\beta, 1/2)^2 = 1. \tag{194}$$

In the case of plus sign, from the definition (175), it follows that

$$f(\mathsf{i}\beta, 1/2) = \mathrm{e}^{2\pi \mathsf{i} \int_\varsigma \theta_\lambda} = 1 \tag{195}$$

where the last equality is satisfied due to (177) and (183).

In the case of minus sign in (194), we verify it by using (195) and the symmetry property of the Bloch–Jost function under the negation of the first component:

$$f(-x, y) = f(-\mathsf{i}\alpha, 0)/f(x, y) = \mathrm{e}^{2\pi \mathsf{i} \int_\xi \theta_\lambda}/f(x, y) = (-1)^{n+1}/f(x, y) \tag{196}$$

where we have used definition (174) and the equation in the first part of (182). $\square$

Thus, we have solved the spectral problem (145), where the spectrum is determined by Eqs. (182) and the corresponding eigenfunctions are given by the formula $\psi(x) = \varphi(\mathsf{i}x)$ for all $x \in \mathbb{R}$, where $\varphi(x)$ is defined in (186). These eigenfunctions coincide with the ones constructed in [15].

Acknowledgments

The work is partially supported by Australian Research Council, and the work of the first-named author is supported in part by Swiss National Science Foundation.

We would like thank Jørgen Ellegaard Andersen, Vladimir Bazhanov, Tobias Ekholm, Ludwig Faddeev, Vladimir Mangazeev, Marcos Mariño, Szabolc Zakany for valuable discussions. Our special thanks go to Alba Grassi, Marcos Mariño and Szabolc Zakany for sharing with us their results prior to publication. The work is partially supported by Australian Research Council, and the work of R.M.K. is partially supported by Swiss National Science Foundation.

References

[1] M. Aganagic, R. Dijkgraaf, A. Klemm, M. Mariño and C. Vafa, Topological strings and integrable hierarchies, *Comm. Math. Phys.* **261**(2) (2006) 451–516.

[2] R. J. Baxter, *Exactly Solved Models in Statistical Mechanics* (Academic Press, Inc. [Harcourt Brace Jovanovich, Publishers], London, 1982).

[3] V. V. Bazhanov, S. L. Lukyanov and A. B. Zamolodchikov, Spectral determinants for Schrödinger equation and **Q**-operators of conformal field theory, in *Proceedings of the Baxter Revolution in Mathematical Physics* (*Canberra, 2000*), Volume 102 (2001), pp. 567–576.

[4] P. Dorey and R. Tateo, Anharmonic oscillators, the thermodynamic Bethe ansatz and nonlinear integral equations, *J. Phys. A* **32**(38) (1999) L419–L425.

[5] L. D. Faddeev, Discrete Heisenberg–Weyl group and modular group, *Lett. Math. Phys.* **34**(3) (1995) 249–254.

[6] I. M. Gel'fand, Expansion in characteristic functions of an equation with periodic coefficients, *Doklady Akad. Nauk SSSR* (*N.S.*) **73** (1950) 1117–1120.

[7] A. Grassi, Y. Hatsuda and M. Mariño, Topological strings from quantum mechanics, *Ann. Henri Poincaré* **17**(11) (2016) 3177–3235.

[8] P. G. Harper, Single band motion of conduction electrons in a uniform magnetic field, *Proc. Phys. Soc. A* **68**(10) (1955) 874–892.

[9] Y. Hatsuda, H. Katsura and Y. Tachikawa, Hofstadter's butterfly in quantum geometry (2016); arXiv:1606.01894.

[10] Y. Hatsuda, Y. Sugimoto and Z. Xu, Calabi–Yau geometry and electrons on 2d lattices (2017); arXiv:1701.01561.

[11] R. Kashaev and M. Mariño, Operators from mirror curves and the quantum dilogarithm, *Comm. Math. Phys.* **346**(3) (2016) 967–994.

[12] R. Kashaev, M. Mariño and S. Zakany, Matrix models from operators and topological strings, 2, *Ann. Henri Poincaré* **17**(10) (2016) 2741–2781.

[13] A.-K. Kashani-Poor, Quantization condition from exact WKB for difference equations (2016); arXiv:1604.01690.

[14] M. Mariño, Spectral theory and mirror symmetry (2015); arXiv:1506.07757.

[15] M. Mariño and S. Zakany, Exact eigenfunctions and the open topological string (2016); arXiv:1606.05297.

[16] N. A. Nekrasov and S. L. Shatashvili, Quantization of integrable systems and four dimensional gauge theories, in *XVIth International Congress on Mathematical Physics* (World Sci. Publ., Hackensack, NJ, 2010), pp. 265–289.

[17] S. M. Sergeev, A quantization scheme for modular q-difference equations, *Teoret. Mat. Fiz.* **142**(3) (2005) 500–509.

[18] A. Weil, *L'intégration dans les groupes topologiques et ses applications*, Actual. Sci. Ind., No. 869 (Hermann et Cie., Paris, 1940). [This book has been republished by the author at Princeton, N. J., 1941.]

[19] J. Zak, Finite translations in solid state physics, *Phys. Rev. Lett.* **19** (1967) 1385–1397.

Quantum Spectral Curve of γ-Twisted $\mathcal{N} = 4$ SYM Theory and Fishnet CFT

Vladimir Kazakov

Laboratoire de Physique Théorique, Ecole Normale Supérieure,
24 rue Lhomond, F-75231 Paris Cedex 05, France
and
PSL Universités, Sorbonne Universités,
UPMC Université Paris 6, CNRS, France
kazakov@lpt.ens.fr

We review the quantum spectral curve (QSC) formalism for the spectrum of anomalous dimensions of $\mathcal{N} = 4$ SYM, including its γ-deformation. Leaving aside its derivation, we concentrate on the formulation of the "final product" in its most general form: a minimal set of assumptions about the algebraic structure and the analyticity of the Q-system — the full system of Baxter Q-functions of the underlying integrable model. The algebraic structure of the Q-system is entirely based on (super)symmetry of the model and is efficiently described by Wronskian formulas for Q-functions organized into the Hasse diagram. When supplemented with analyticity conditions on Q-functions, it fixes completely the set of physical solutions for the spectrum of an integrable model. First, we demonstrate the spectral equations on the example of $gl(N)$ and $gl(K|M)$ Heisenberg (super)spin chains. Supersymmetry $gl(K|M)$ occurs as a simple "rotation" of the Hasse diagram for a $gl(K+M)$ system. Then we apply this method to the spectral problem of AdS_5/CFT_4-duality, describing the QSC formalism. The main difference with the spin chains consists in more complicated analyticity constraints on Q-functions which involve an infinitely branching Riemann surface and a set of Riemann–Hilbert conditions. As an example of application of QSC, we consider a special double scaling limit of γ-twisted $\mathcal{N} = 4$ SYM, combining weak coupling and strong imaginary twist. This leads to a new type of non-unitary CFT dominated by particular integrable, and often computable, 4D fishnet Feynman graphs. For the simplest of such models — the bi-scalar theory — the QSC degenerates into the Q-system for integrable non-compact Heisenberg spin chain with conformal, $SU(2,2)$ symmetry. We describe the QSC derivation of Baxter equation and the quantization condition for particular fishnet graphs — wheel graphs, and review numerical and analytic results for them.

1. Introduction

In the past 40 years, a multitude of supersymmetric conformal quantum field theories (CFT) in four dimensions has been discovered and studied [1]. Typically, they are various deformations of super-Yang–Mills theories, with $\mathcal{N} = 1 - 4$ supersymmetries (see [2] for modern classification). On the other hand, well-identified non-supersymmetric and/or non-gauge CFTs in four dimensions are rare species. Apart from a rather exotic Banks–Zaks theory [3] or critical Potts model [4] there are hardly known examples which are explicitly constructed and well understood.[a]

Even more rare are the integrable 4-dimensional CFT's. The $\mathcal{N} = 4$ SYM theory is the emblematic example of such a theory: it is conformal for $SU(N_c)$ gauge group for any N_c but integrable only in the large N_c, 't Hooft limit (see the review [5] and references therein).[b] It was long believed that its large global $PSU(2,2|4)$ superconformal symmetry is responsible for the integrability. However, under specific deformations breaking the supersymmetry partially or entirely, the theory seems to retain its integrability.

An important, and rather general class of such deformations is the γ-twist [1, 6, 7]. It breaks the global symmetry to $SU(2,2) \times U(1)^3$, i.e. only three Cartan subgroups are left from R-symmetry and preserves, at least on the tree level, the 4D conformal symmetry. The last one could be endangered by various conformal anomalies [8] but, remarkably, it survives when adding to the action a well defined set of double-trace counter-terms [9–13] and tuning the double-trace couplings to certain critical values. The critical double-trace couplings are, in general, complex functions of the 't Hooft coupling which parametrizes the whole family of these non-unitary CFTs.

All the quantum integrability properties known from the undeformed $\mathcal{N} = 4$ SYM seem to survive as well this γ-deformation [13]. In particular, the quantum spectral curve (QSC) [14–16] — the most advanced formalism of AdS/CFT integrability, giving a comprehensive solution of the problem of spectrum of anomalous dimensions of local (and some non-local) operators — remains valid the γ-deformation, with minor modifications, and describes this non-unitary CFT precisely at the (complex) critical line [13]. The QSC method already found numerous applications in the study of planar spectrum of $\mathcal{N} = 4$ SYM theory [16–28] (see also recent review [29] and references therein).

Recently, Gürdogan and the author proposed in [30] a special double scaling (DS) limit of the γ-deformed $\mathcal{N} = 4$ SYM, combining the weak coupling limit and large imaginary values of the γ-parameters. It gives rise to a new 4D non-unitary CFT where the gauge interactions decouple and only chiral 4-scalar and Yukawa interactions are left. They are also expected to inherit the integrability properties of their "mother" theory – the γ-deformed $\mathcal{N} = 4$ SYM. In the simplest case, when only one double-scaling coupling is kept non-zero, it becomes a simple theory of two interacting complex scalars (referred to in what follows as the "bi-scalar theory"). Nevertheless, it is still a non-trivial interacting CFT but in addition it is

[a] In 3 dimensions, a well-defined and studied example of non-supersymmetric CFT's is the Ising model, but not much beyond that.

[b] Its 3-dimensional cousin is the ABJM model, also integrable at large N_c.

integrable in planar limit! Its integrability, unlike the integrability of its "mother" theory, has a clear origin: its perturbation theory for various correlation functions is dominated by the "fishnet" Feynman graphs. This means that sufficiently large planar graphs have in the bulk the shape of regular square lattice. It was noticed long ago [31] that such a graph defines an integrable 2D statistical-mechanical spin system with $SU(2,2) \sim SO(4,2)$ symmetry, which is a 4-dimensional conformal group. Thus the integrability of the bi-scalar theory is tightly related with the integrability of the conformal, non-compact $SU(2,2)$ Heisenberg spin chain. The theory of integrable non-compact spin chains has a long history [32–38] following the fundamental works of Faddeev and the Leningrad school (see [39] and references therein). It had also a few important applications, such as BFKL approximation in high-energy, Regge limit in QCD [35, 36]. Many of these and other old results on non-compact integrable spin chains appear to be very helpful in the study of non-perturbative dynamics of the bi-scalar theory and the other CFT's from the family of chiral CFT obtained in the DS limit from γ-deformed $\mathcal{N}=4$ SYM [40].

In this work, we will review the formalism of quantum spectral curve (QSC) for the spectrum of anomalous dimensions of the planar $\mathcal{N}=4$ SYM theory, including its γ-deformed version (named below as QSCγ). We will concentrate ourselves on the minimal set of basic propositions when formulating the QSCγ equations, leaving aside its derivation. The algebraic part of QSCγ, entirely dictated by the global $PSU(2,2|4)$ symmetry (broken to $SU(2,2)\times U(1)^3$ by γ-deformation) and quantum integrability, is most conveniently formulated in terms of the Q-system — a set of 2^8 Baxter's Q-functions of spectral parameter u. Q-functions are organized into the Hasse diagram, reflecting the fact that they are Grassmannian coordinates and they obey certain Plücker relations. The analytic part of QSCγ construction consists of description of the structure of Riemann surfaces of Q-functions, where the main element is an infinite "ladder" of equally spaced "Zhukovsky" quadratic cuts at $u \in (-2g + i\mathbb{Z}, 2g + i\mathbb{Z})$, where g is the 't Hooft coupling. The large u asymptotics of Q-functions and their specific monodromy properties around the Zhukovsky cuts conclude the formulation of spectral problem. Roughly speaking, QSCγ represents a system of non-linear Riemann–Hilbert equations. We will start with Sec. 2 where we demonstrate the similar formalism and the emergence of the supersymmetric Q-system on the example of Heisenberg $GL(K|M)$ spin chain where the analyticity constraints look much simpler. Then, in Sec. 3, we will describe the QSCγ for γ-deformed $\mathcal{N}=4$ SYM (AdS_5/CFT$_4$ duality). In Sec. 4, we will describe the CFTs emerging in the DS limit of this theory, and in particular the so-called bi-scalar theory dominated by integrable "fishnet" graphs. We will review the results obtained for these models from QSCγ, such as the exact computation of certain multi-loop "wheel" graphs and discuss the equivalence of the bi-scalar theory to the conformal integrable Heisenberg spin chain. Section 6 is devoted to conclusions and unsolved problems.

2. Spectrum of Heisenberg $GL(K|M)$ Spin Chain from Baxter Q-Functions

In this section, we will give an alternative formulation of the well known solution for the spectrum of compact, Heisenberg $GL(K|M)$ spin chain which will be useful for the generalization to AdS/CFT integrability. We will avoid the direct use of standard Bethe equations since in the sigma-model on $AdS_5 \times S^5$ background, which describes the string side of the duality, the notion of Bethe roots is a tricky and not very invariant issue. We will rather rely on the full system of Baxter Q-functions, forming a Grassmannian, spectral parameter dependent structure in the $(K+M)$-dimensional space. The Q-functions are most naturally classified by the vertices of Hasse diagram — the $(K+M)$-dimensional hypercube. Specifying the analyticity properties of these Q-functions w.r.t. the spectral parameter u one can classify and efficiently study the spectrum. The supersymmetric generalization of this picture in terms of Hasse diagram, from $GL(K+M)$ bosonic group to $GL(K|M)$ supersymmetric group will be essentially a "rotation" of the hypercube when imposing specific determinant relations ("determinant" flow) and analyticity conditions. Many of the details, missing in this short overview of spin chains from the point of view of Q-functions, can be found in [16, 41–43] and in the recent review [29].

2.1. *Spectrum of $GL(N)$ spin chain via Q-functions on Hasse diagram*

Let us start from the Heisenberg spin chain with spins belonging to the bosonic group $GL(N)$ and twisted boundary conditions. This spin-chain is defined through the 1D hamiltonian

$$\hat{H} = \sum_{i=1}^{L-1} \mathcal{P}_{i,i+1} + \mathcal{P}_{L,1} G_L G_1^{-1} \tag{1}$$

where the spin at each site i takes the values $s_i = 1, 2, \ldots, N$, the permutation acts on a pair of spins as $\mathcal{P}_{i,j}(s_1, \ldots, s_i, \ldots, s_j, \ldots, s_L) = (s_1, \ldots, s_j, \ldots, s_i, \ldots, s_N)$ and the twist $G = \mathrm{diag}\{x_1, x_2, \ldots, x_N\} \in SU(N)$ is a fixed element of Cartan subgroup.[c] Explicitly, in components, various terms in (1) mean

$$\mathcal{P}_{i,i+1} = \delta_{s_1}^{s_1'} \delta_{s_2}^{s_2'} \cdots \delta_{s_i}^{s_{i+1}'} \delta_{s_{i+1}}^{s_i'} \cdots \delta_{s_L}^{s_L'}, \tag{2}$$

$$\mathcal{P}_{L,1} G_L G_1^{-1} = \delta_{s_2}^{s_2'} \delta_{s_3}^{s_3'} \cdots \delta_{s_L-1}^{s_L'-1} \delta_{s_L}^{s_1'} \delta_{s_1}^{s_L'} \frac{x_{s_L}}{x_{s_1}}. \tag{3}$$

To formulate the solution for the spectrum of this hamiltonian we introduce a set of N Baxter $Q_k(u)$-functions of spectral parameter u with a single index

$$Q_k(u) = x_k^{iu} \prod_{j=1}^{R_k} (u - u_j^{(k)}), \qquad k = 1, 2, \ldots, N, \tag{4}$$

[c] All eigenvalues are assumed to be different, and we impose the unimodularity $\prod_{j=1}^{N} x_j = 1$.

each of them being a polynomial of spectral parameter u times a twist-dependent exponential factor. The positive integers $\{R_1, R_2, \ldots, R_N\}$ are in fact the Cartan charges of the $U(1)^N$ residual symmetry left after breaking the original $GL(N)$ symmetry by twisting.

Let us also define a natural object — the multi-index Q-functions — by the following Wronskian formula:

$$Q_I(u) \equiv Q_{j_1,\ldots,j_k}(u) = \frac{\det_{1\le m,n\le k} Q_{j_m}^{[-1-k+2n]}}{\prod_{j=0}^{k-1} Q_\emptyset^{[-k+1+2j]}}, \tag{5}$$

where we denoted by capital letter $I = \{j_1, \ldots, j_k\} \in \bar{\emptyset} \equiv \{1, 2, \ldots, N\}$ a subset of the full set $\bar{\emptyset}$ of indices. By definition, all indices in this subset are different and ordered from left to right. Any permutation P of indices in (5) can only change the overall sign by factor $(-1)^P$. It was also natural to introduce the "empty set" Q-function $Q_\emptyset(u)$ in denominator of (5), but the reasons which will be clear below, when we will discuss the Plücker relations (12). In total, we have 2^N different Q-functions, but they are obviously interrelated since they are given in terms of only N single index functions.

To fix all the roots of these Q-functions, and thus to find all the eigenvalues of the above hamiltonian, it is enough to find all solutions of the following equation [44, 45]

$$Q_{12\cdots N}(u) = \Delta(x_1, \ldots, x_N)\, u^L, \qquad Q_\emptyset(u) = 1, \tag{6}$$

where, according to (5),

$$Q_{12\cdots N}(u) \equiv \det_{1\le k,j\le N} Q_k^{[-1-N+2j]} \tag{7}$$

is the Wronskian of the full set of single index Q-functions and $\Delta(x_1, \ldots, x_N) = \prod_{k>j}(x_k - x_j)$ is the Vandermonde determinant of twist eigenvalues. Here and below we use the notations for standard shifts of arguments of the functions: $f^{[n]} \equiv f(u + \frac{in}{2})$.

Once one finds a solution of (6), the corresponding energy — the eigenvalue of the hamiltonian (1) — is given by the familiar formula

$$E = L + i\partial_u \log \left.\frac{Q_{\bar{k}}(u - i/2)}{Q_{\bar{k}}(u + i/2)}\right|_{u=0}, \tag{8}$$

where we used the $N-1$ index Q-functions $Q_{\bar{k}} \equiv Q_{12\ldots\hat{k}\ldots,N}$ given by $(N-1)\times(N-1)$-determinant according to the formula (5).[d] The answer does not depend on the choice of $Q_{\bar{k}}$ (this symmetry is related to the so-called particle-hole duality).

It is natural to attach all these 2^N Q-functions to the vertices of the N-dimensional hypercube which is called in this occasion the Hasse diagram. For example, in the simplest case of $GL(2)$ spin chain ($N = 2$) we have the set of 4 Q-functions: $Q_\emptyset, Q_1, Q_2, Q_{12}$ which we place at 4 vertices of the square, as

[d]In general, the bar over a subset I, i.e. $\bar{I}$, means a subset complementary to I w.r.t. the full set $\bar{\emptyset} \equiv \{1, 2, \ldots, N\}$, and $\{\bar{k}\} \equiv \{1, 2, \ldots, \hat{k}, \ldots, N\}$ denotes the set of $N-1$ indices with k missing.

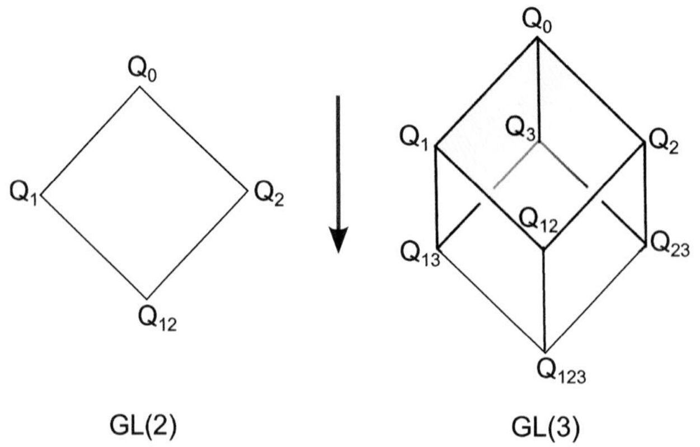

Fig. 1. Examples of Hasse diagrams for Q-system of Baxter functions of integrable models with $GL(2)$ symmetry (on the left) and $GL(3)$ symmetry (on the right). The arrow shows the direction of "determinant flow": the Q-functions with k indices are $k \times k$ determinants (5) of single index Q-functions, i.e. increasing in size with the increase of the level k.

shown on Fig. 1 (left). The upper vertex is occupied by $Q_\emptyset$, which is connected by two edges with Q_1 and Q_2, which, in turn, are connected by two edges with $Q_{\bar\emptyset} \equiv Q_{12} = Q_1^+ Q_2^- - Q_1^- Q_2^+$.[e] The spectral equation (6) takes the form

$$Q_1^+ Q_2^- - Q_1^- Q_2^+ = (x - 1/x) u^L \tag{9}$$

with the twist $G = \{1/x, x\}$. Imposing the specific analyticity condition – the "polynomiality" of Q-functions (4) — we obtain the usual Bethe equation for the roots of $Q_1(u)$:

$$\frac{Q_1^{++}(u_j^{(1)})}{Q_1^{--}(u_j^{(1)})} = -\left(\frac{u_j^{(1)} + i/2}{u_j^{(1)} - i/2}\right)^L, \qquad j = 1, 2, \ldots, R_1. \tag{10}$$

We used for that two relations (9) at the roots of Q_1, shifted from the original one by $\pm i/2$, and divided one over another. A similar equation for the roots of Q_2 leads to the same spectrum given by (8). In the formula for energy (8) we can use either $Q_{\bar 1} = Q_2$ or $Q_{\bar 2} = Q_1$.

For $N = 3$, the Hasse diagram is 3D cube Fig. 1 (right). It is convenient to orient the cube in such a way that two of the vertices connected by the main diagonal appear to be the upper and the lower ones. We place again $Q_\emptyset$ at the top vertex, the single-indexed Q_1, Q_2, Q_3 — at the vertices adjacent to it, then, say, Q_{13} — on the next level, at the vertex adjacent to both Q_1 and Q_3, etc. One could pictorially think of the Hasse diagram as of the "globe", where the level of Q-functions with given number of indices is like a "latitude", the $\emptyset$-vertex and the $\bar\emptyset$-vertex — like the "north and south poles", respectively.

The generalization to any N is straightforward. All $N + 1$ levels from top to bottom are ordered w.r.t. the number of indices $|I| \equiv \text{Coordinality}(I)$ in the cor-

[e] We used the notations $f^\pm \equiv f(u \pm \frac{i}{2})$.

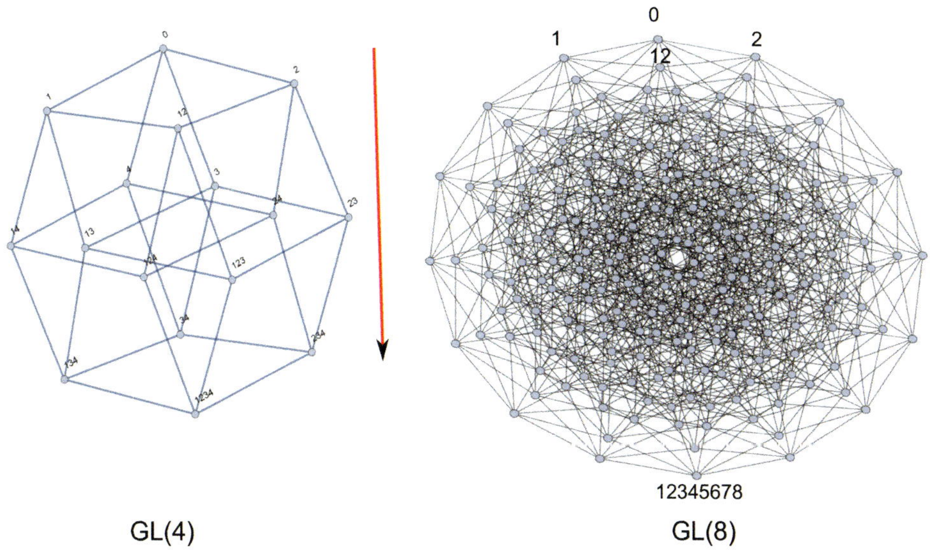

Fig. 2. Hasse diagrams for Q-system of Baxter functions of integrable models with $GL(4)$ symmetry (on the left) and $GL(8)$ symmetry (on the right) are hypercubes with the dimension of the rank of the symmetry of integrable quantum system. The arrow shows the direction of determinant flow according to (5). The $GL(8)$ Hasse diagram will be the same as for the $PSU(2,2|4)$ super-group of AdS_5/CFT_4 duality (up to a certain "rotation" of direction of the determinant flow and the details of analyticity structure).

responding Q_I functions. This induces a natural direction in parametrization (5) which we will call "determinant flow". At a given u, the collection of functions $Q_I(u)$ on a particular n-level with $|I|=n$ forms an n-dimensional linear subspace $V_{(n)}(u)$ representing the Plücker coordinates of a point on the Grassmannian $\mathbf{G}_N^n$ defined on the linear space $\mathbb{C}^N$. As was pointed out in [16], the quantum integrability, constraining the spectra of various integrable models, from spin chains to $2D$ quantum field theories, is based on the following abstract relation between these Plücker coordinates:

$$V_{(n)}(u+i/2)\cap V_{(n)}(u-i/2)=V_{(n-1)}(u), \qquad \forall n\in\{1,2,\dots,N-1\}, \tag{11}$$

following of course from (5). The Grassmannian structure of quantum integrability was first pointed out in [46] on the example of transfer-matrices and Hirota bi-linear finite difference equations.

We also present on Fig. 2 two other important examples. On Fig. 2 (left) we depicted the Hasse diagram for the $GL(4)$ system, relevant for the conformal or for the R-symmetry subgroups of $PSU(2,2|4)$. On Fig. 2 (right) the Hasse diagram for the $GL(8)$ system is presented. As we will see, the last one is closely related to the full $PSU(2,2|4)$ symmetry group of $\mathcal{N}=4$ SYM theory.

The determinant flow (5) leads to the following Plücker relation (which is also called the QQ-relation in the AdS/CFT integrability literature) between the four

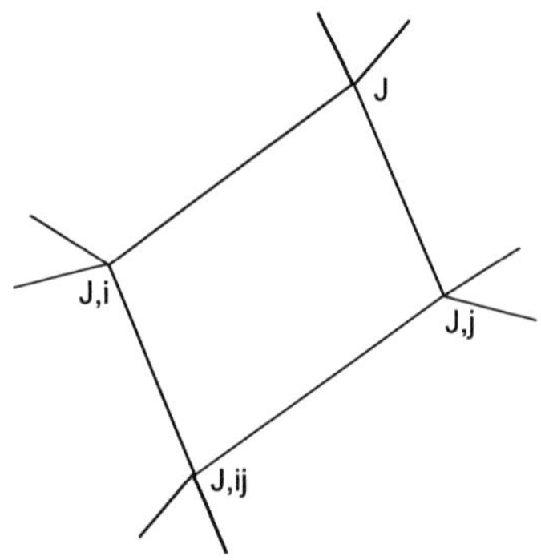

Fig. 3. QQ-relations (Plücker identities for the Grassmannian) (12). They emerge at any 2-dimensional face of the hypercube of Hasse diagram as a consequence of the determinant relations (5).

Q-functions adjacent to the same 2-dimensional face of the Hasse diagram (shown on Fig. 3):

$$Q_J Q_{J,jk} = Q^+_{J,j} Q^-_{J,k} - Q^-_{J,j} Q^+_{J,k} \tag{12}$$

where $J = \{j_1, \ldots, j_k\} \in \bar{\emptyset} \equiv \{1, 2, \ldots, N\}$ is a particular vertex on Hasse diagram. Notice, that the introduction of arbitrary function $Q_\emptyset$ in denominator of (6) was necessary for satisfying the QQ-relations (12) on the whole Hasse diagram, including the $\emptyset$-vertex.

To obtain the standard nested Bethe ansatz equations, one has to choose a set of Q-functions along a "meridian" of the Hasse diagram, say, $Q_\emptyset, Q_1, Q_{12}, Q_{123}, \ldots, Q_{123\ldots N}$. Then one can use the above Plücker relations along the faces adjacent to this "meridian" (on the same side of it) and exclude all other Q-functions at the roots of the "meridional" Q-functions by the trick similar to the one which led us to the Bethe equation (10) [42, 47, 48].

2.2. *Spectrum of supersymmetric $GL(K|M)$ spin chain via Q-functions*

The hamiltonian of the $GL(K|M)$ super-spin chain has the following form

$$\hat{H} = \sum_{i=1}^{L-1} \mathcal{P}^{K|M}_{i,i+1} + \mathcal{P}^{K|M}_{L,1} G_L G_1^{-1} \tag{13}$$

where the super-spin at each site i takes two kinds of values

$$s_i = 1, 2, \ldots, K|K+1, \ldots, N. \tag{14}$$

They correspond to two different gradings: $p_s = 1$, if $s \le K$, $p_s = 0$, if $s > K$. The super-permutation acts on a pair of spins as $\mathcal{P}_{i,j}(s_1, \ldots, s_i, \ldots, s_j, \ldots, s_L) = (-1)^{p_{s_i}+p_{s_j}}(s_1, \ldots, s_j, \ldots, s_i, \ldots, s_N)$ and the twist $G = \mathrm{diag}\{x_1, \ldots, x_K| y_1, \ldots, x_M\} \in gl(K|M)$ is a fixed group element.[f]

[f] All eigenvalues are assumed distinct here.

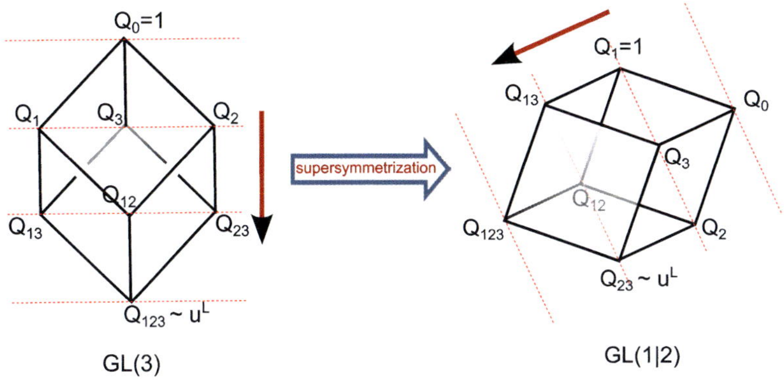

Fig. 4. Supersymmetrization of Hasse diagram of a QQ system of rank-3 on the example of rational Heisenberg spin chains: we pass from $GL(3)$ spin chain to $SU(2|1)$ spin chain by rotating the direction of the determinant flow, i.e. imposing different analyticity (polynomiality) conditions ((15) and (16) instead of (4)) and fixing a different pair of diametrally opposed Q-functions different in each case. The determinant flow is the same in both cases.

The supersymmetric generalization of the above picture in terms of Q-functions can be nicely and easily presented as a specific "rotation" of Hasse diagram, when imposing the analyticity ("polynomiality") conditions. Namely, for $GL(K|M)$ case we can preserve the same determinant flow (5) along the Hasse diagram as for $GL(K+M)$ bosonic case. But to fix the analyticity conditions we choose, instead of $Q_\emptyset, Q_{\bar{\emptyset}}$, a pair of Q-functions at the extremes of a different main diagonal of the hypercube, one on the level K, another on the level M. For example, we can pick $Q_{12\cdots K}$ and $Q_{K+1,K+2,\ldots,M}$.[g] This supersymmetrization procedure is shown for the example $GL(3) \to GL(1|2)$ on Fig. 4.

To find the spectrum of the Heisenberg super-spin chain we impose the following analyticity (polynomial times exponential for twist) conditions on the Q-functions:

$$Q_{1,2\ldots,\hat{k},\ldots,K}(u) = x_k^{iu} \prod_{j=1}^{R_k} (u - u_j^{(k)}), \qquad k = 1, 2, \ldots, K; \tag{15}$$

$$Q_{1,\ldots,K,K+m}(u) = y_m^{iu} \prod_{j=1}^{\hat{R}_m} (u - u_j^{(m)}), \qquad m = 1, 2, \ldots, M, \tag{16}$$

where the hat over $\hat{k}$ in the left-hand side of the first equation means that the corresponding index is missing from the sequence. For the example $GL(2|1)$ of Fig. 4 such three Q-functions in (15), (16) are $Q_\emptyset$ and Q_{12}, Q_{13}, respectively. The positive integers $\{R_1, R_2, \ldots, R_K | \hat{R}_1, \hat{R}_2, \ldots, \hat{R}_M\}$ are the Cartan super-charges of the $U(1)^{K+M}$ residual symmetry left after breaking the original $SU(K|M)$ symmetry by twisting. To fix all the roots of these Q-functions, and thus to find all the eigenvalues of the above hamiltonian of super-spin chain of length L, it is enough to find

[g] Other choices of two such functions at the same levels are simple re-labelings.

all solutions of the Q-system with the following conditions imposed

$$Q_{K+1,K+2,\dots,K+M}(u) = 1, \qquad Q_{12\cdots K}(u) = \text{const} \times u^L. \tag{17}$$

Once one finds a solution of (17), the corresponding energy — an eigenvalue of the hamiltonian (13) — is given by the familiar formula, through the Q-functions neighboring the "momentum-carrying" Q-function $Q_{12\cdots K}$ on Hasse diagram

$$E = L + i\partial_u \log \frac{Q_{k,k+1,\dots,K+M}(u+i/2)}{Q_{k,k+1,\dots,K+M}(u-i/2)}\Bigg|_{u=0} \tag{18}$$

$$= L + i\,\partial_u \log \frac{Q_{K+1,K+2,\dots,K+\hat{k},\dots,K+M}(u-i/2)}{Q_{K+1,K+2,\dots,K+\hat{k},\dots,K+M}(u+i/2)}\Bigg|_{u=0}, \tag{19}$$

where the answer does not depend on the choice of k or $\hat{k}$.

We see that the scheme of solution for supersymmetric case is almost identical to the previous, bosonic case. But the "rotation" of the Hasse diagram in such a way leads to the dramatic change of analyticity properties. For example, the known function $Q_{12\cdots K}(u)$ of Eq. (17) cannot be expressed through the basic functions (15), (16) as a simple determinant, as in the bosonic case, but rather has to be found by solving a chain of Plücker relations (12), which leads to more complicated formulas. One can also derive the corresponding supersymmetric Bethe ansatz equations [49, 50] directly from the QQ-relations (12) as it was done in Sec. 5 of [47] in less invariant notations.[h]

The Q-functional approach has a long history [42, 46, 47, 51–60] and it has been developed in the form described above in the series of papers [15, 16, 42, 48, 61–63] where the reader can find many more details. This approach is not only aesthetically attractive, it also appeared to be more efficient in certain explicit computations compared to more conventional Bethe equations for rational spin chains [26, 27]. The construction in terms of Q-system, based on the Hasse diagram, presented above, can be applied for more complicated quantum integrable systems, such as non-compact (super)-spin chains and 2D sigma models in a finite volume. In this case, the analyticity conditions should be modified, since the Q-functions, or at least a part of them, cannot be parametrized by polynomials anymore. However, the solutions for the spectrum in such problems can be still formulated in terms of certain analyticity conditions on the set of Q-functions for which the algebraic structure of Q-system is entirely defined by the symmetry group. This approach was successfully applied for example for the study of spectrum of the principal chiral field model on a finite space-circle [64, 65]. In the next section, we will use the Q-system approach to formulate, in the most concise and general way, the Quantum Spectral

[h]For the sake of a unified description of Q-systems of bosonic and supersymmetric models, we avoid here the use of supersymmetric notations of the original papers [15, 16] where we would denote $Q_{I|J} \equiv Q_{I,\bar{\emptyset}_M \backslash J+K}$, where $I \subset \{1,2,\dots,K\}$, $J \subset \bar{\emptyset}_M \equiv \{1,2,\dots,M\}$ and by $\bar{\emptyset}_M \backslash J + K$ we denoted the subset, complementary to the set $\bar{\emptyset}_M$, with every index shifted by $+K$. We mean that on the r.h.s. of this definition we have the Q-function in our current notations (5).

Curve (QSC) equations [14, 15] — a system of non-linear functional equations for computation of anomalous dimensions of arbitrary local operators, at any coupling, in the planar limit of $\mathcal{N}=4$ Super-Yang–Mills (SYM) theory.

3. Quantum Spectral Curve for Twisted $\mathcal{N}=4$ SYM

In this section we will give a concise formulation of the quantum spectral curve (QSC) for the spectrum of anomalous dimensions of local operators in planar $\mathcal{N}=4$ Super-Yang–Mills (SYM) theory, first introduced in [14, 15], including its twisted version [16, 20]. It will be based on the Q-system[i] approach described above. We will first make precise the algebraic structure of this AdS_5/CFT$_4$ Q-system, based on the super-conformal $PSU(2,2|4)$ symmetry of the model. Then we will describe the analyticity properties of the underlying Q-functions and the Riemann–Hilbert "sewing" conditions allowing to completely fix the system of equations for the physical solutions.

Let us stress that we do not give here any derivation of the AdS_5/CFT$_4$ QSC. We only formulate the final mathematical formalism, ready for further applications. Until the last chapter devoted to a particular application of QSC to the chiral double limit of γ-twisted $\mathcal{N}=4$ SYM, we avoid, on purpose, the discussion of any consequences of QSC equations and of secondary details, concentrating only on the basic foundations of QSC construction. For the derivation, details and numerous consequences, the reader can turn to the original papers [14–16], to the recent review [29] as well as to the already rich literature of its generalizations and applications [16–28].

3.1. *Algebraic structure of the AdS_5/CFT_4 Q-system*

The Hasse diagram for the AdS/CFT Q-system is similar to the one for the $gl(4|4)$ super-spin chain described in the previous section. It represents an 8-dimensional hypercube with the $2^8 = 256$ Q-functions attached to its vertices, as shown on Fig. 2 (right). The Q-functions have the same determinant flow as described by Eq. (5). Let us note that the Q-system obeys a certain residual $sl(4)\times sl(4)$ symmetry corresponding to two bosonic subgroups of the $PSU(2,2|4)$ symmetry. This algebraic symmetry refers to the linear transformations of, separately, 4 functions with 3 indices $Q_{1,\ldots,\hat{k},\ldots,4}(u)$, $k=1,2,3,4$ and 4 functions with 5 indices $Q_{1,\ldots,4,j}(u)$, $j=1,2,3,4$. Another, "gauge" symmetry of the Q-system, due to the homogeneity of QQ-relations, consists of the rescalings of Q-functions (there are two such rescaling parameters, see [15]).

But we should demand for QSC even more: we impose two to unit value

$$Q_{1234}(u) = 1 = Q_{5678}(u) \tag{20}$$

[i]Called also AdS/CFT Q-system, referring to the duality between $\mathcal{N}=4$ SYM and the string sigma model on $AdS_5\times S^5$ background. For other recent reviews on integrability methods for this system see [5, 29].

at any spectral parameter u. The first of these conditions can be achieved by rescalings. But the second one, the Q-function diametrally opposed on Hasse diagram (i.e. Q_{1234} and Q_{5678} are Hodge dual to each other),[j] this is an additional condition which replaces (17) for the super-spin chain.[k] It actually reflects the projectivity and super-unimodularity of the $PSU(2,2|4)$ symmetry of the system.

This gauge appears to be the most suitable for the formulation of analyticity properties of the whole Q-system. But these properties are more complicated then the polynomial ansatz (15) since we deal not with the rational super-spin chain (which however occurs to be the case in the weak-coupling, one-loop limit of $\mathcal{N}=4$ SYM [66–68]) but with the integrable string sigma-model on $AdS_5 \times S^5$ coset.

We introduce special notations for the most useful "near-equator" Q-functions mentioned above:

$$\mathbf{P}_a \equiv Q_{1,\dots,\hat{a},\dots,4}(u), \qquad a = 1,2,3,4; \tag{21}$$

$$\mathbf{Q}_j \equiv Q_{1,2,3,4,j+4}(u), \qquad j = 1,2,3,4, \tag{22}$$

where by $\hat{a}$ we denote again the missing index from the set $\{1,2,3,4\}$. For example, $\mathbf{P}_2 = Q_{134}$ and $\mathbf{Q}_3 = Q_{12347}$. Hence $\mathbf{P}_a$ have 3 indices and $\mathbf{Q}_j$ has 5 indices in the standard notations for Q functions, as in (5).

Their Hodge dual Q-functions have the same, but upper indices:

$$\mathbf{P}^a \equiv Q_{a,5,6,7,8}(u), \qquad a = 1,2,3,4; \tag{23}$$

$$\mathbf{Q}^j \equiv Q_{5,\dots,\widehat{j+4},\dots,8}(u), \qquad j = 1,2,3,4. \tag{24}$$

For example, $\mathbf{P}^2 = Q_{25678}$ and $\mathbf{Q}^3 = Q_{568}$. The positions of these functions on $(4|4)$ Hasse diagram are pictorially presented on Fig. 5.

The $\mathbf{P}$ and $\mathbf{Q}$ functions are, roughly, responsible for the dynamics of string fields on S^5 (related to R-symmetry) and on AdS_5 (related to the conformal symmetry) projections of the dual string sigma-model, as will be seen from their large u asymptotics.

Another useful set of 16 Q-functions and of their 16 Hodge duals deserves a special notation:

$$\mathbf{Q}_{a|j} \equiv Q_{1,\dots,\hat{a},\dots,4,j+4}(u), \qquad a = 1,2,3,4; \quad j = 1,2,3,4, \tag{25}$$

$$\mathbf{Q}^{a|j} \equiv Q_{a,5,\dots,\widehat{j+4},\dots,8}(u), \qquad a = 1,2,3,4; \quad j = 1,2,3,4, \tag{26}$$

where by "hat" we denote again the missing member from consecutive integers. For example, $\mathbf{Q}_{2|3} = Q_{1347}$ and $\mathbf{Q}^{2|3} = Q_{2568}$.

[j] A Q-function diametrally opposed to a given Q-function is usually called its Hodge dual. Their index sets are complementary to each other w.r.t. the full set. All the algebraic relations of AdS/CFT Q-system are invariant w.r.t. the Hodge transformation, i.e. w.r.t. exchange of upper and lower indices, once (20) is imposed.

[k] We could have imposed the condition (20) also for the super-spin chain. However, it would have spoiled the polynomial ansatz (16) and significantly complicate the analyticity properties. The 2nd condition (20) is not a simple normalization, but a dynamical restriction, for the case of $(N|N)$ Q-systems [63].

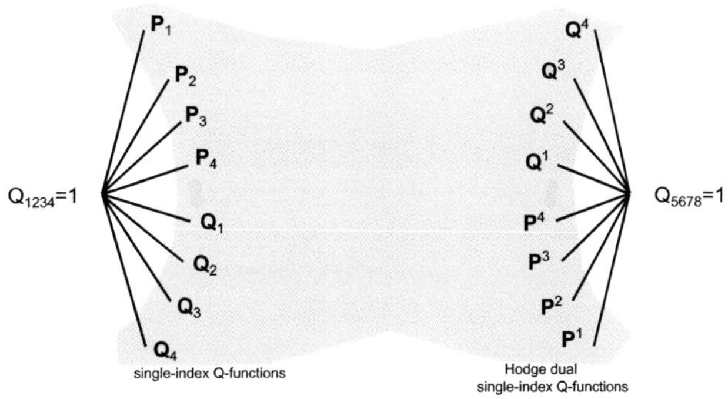

Fig. 5. Schematic presentation of positions of $\mathbf{P}$ functions and $\mathbf{Q}$ functions (21)–(24), within the (4|4) Hasse diagram. They are neighboring the two "poles" of Hasse diagram corresponding to empty-set and full-set labels. Each pair of functions $\mathbf{P}_a, \mathbf{P}^a$ or $\mathbf{Q}_j, \mathbf{Q}^j$, with the same label, are placed at the diametrally opposite vertices of Hasse diagram, i.e. they are Hodge dual to each other w.r.t. the Grassmannian structure of the Q-system. These 16 functions have the simplest analytic structure on the physical sheet. The gray "cloud" signifies all the elements of 8D hypercube missing on the picture.

Due to the determinant flow (5), together with the gauge conditions (20), these functions satisfy a useful set of algebraic identities:

$$\mathbf{Q}_j = -\mathbf{Q}^{\pm}_{a|j}\mathbf{P}^a, \quad \mathbf{P}_a = -\mathbf{Q}^{\pm}_{a|j}\mathbf{Q}^j \qquad \text{("metric" property)}, \tag{27}$$

$$\mathbf{Q}^{+}_{a|j} - \mathbf{Q}^{-}_{a|j} = \mathbf{P}_a\mathbf{Q}_j \qquad \text{(special } QQ\text{-relation)}. \tag{28}$$

Notice that we can raise and lower the indices by the rules similar to the standard tensor algebra. A useful automatic consequence of the Grassmannian structure of Q-system and of the gauge (20) is the orthogonality relations: $\mathbf{P}_a\mathbf{P}^a = \mathbf{Q}_j\mathbf{Q}^j = 0$.

The QSC formalism is based on a set of 256 Q-functions, out of which only a few are algebraically independent. The rest of them can be deduced from the determinant flow or from the Plücker QQ-relations (12). The choice of the most convenient algebraically independent subset of Q-functions depends on the problem being solved, i.e. on the type of studied operators and on the chosen approximations (weak coupling, strong coupling, numerics, etc). Thus there exist many useful forms of QSC equations. Let us mention one particularly important, especially for various weak coupling limits — the coupled system of 4th order Baxter equations on the functions (21)–(24). Namely, excluding the functions $\mathbf{Q}_{a|j}$ from the relations (27) and (28)[1] one gets the following linear 4th order finite difference Baxter equation [69]

$$a\mathbf{Q}^{[+4]} - b\mathbf{Q}^{[+2]} + c\mathbf{Q}^{[0]} - \bar{b}\mathbf{Q}^{[-2]} + \bar{a}\mathbf{Q}^{[-4]} = 0 \tag{29}$$

where the coefficients are explicit functionals of $\mathbf{P}_a, \mathbf{P}^a$-functions:

$$a(u) = d_0, \qquad b(u) = d_1 - \mathbf{P}_a^{[+2]}\mathbf{P}^{a[+4]}d_0,$$
$$c(u) = d_3 + \mathbf{P}_a\mathbf{P}^{a[+4]}d_0 + \mathbf{P}_a\mathbf{P}^{a[+2]}d_1$$

[1]One uses for that a linear system of 5 dependent equations on the same $\mathbf{Q}_{a|j}$ obtained from (27) by 5 consecutive shifts of u by integers$\times i$ and a multiple use of (28) for bringing various functions $\mathbf{Q}_{a|j}$ to the same argument u (see [69] for details).

where $d_m = \det_{1\le a,k\le 4}(\mathbf{P}^a)^{[4-2k+2\theta_{k,m}]}$, $(\theta_{m,k} = 1$ if $m \ge k$ and $\theta_{m,k} = 0$ if $m < k)$. (30)

Four solutions of this equation give the functions $\mathbf{Q}_j$. Of course, any independent linear combinations of these 4 $\mathbf{Q}_j$-functions with i-periodic coefficients[m] are also algebraically admissible $\mathbf{Q}_j$-functions.

For a general state/operator, this Baxter equation should be supplemented by three similar equations. One of them, for $\mathbf{Q}^j$-functions, uses the Hodge duality of the Q-system, obtained from the above equation by exchange of all upper$\Leftrightarrow$lower indices of the $\mathbf{P}_a$-functions in the coefficients. Two other 4th order Baxter equations, on $\mathbf{P}_a$- and $\mathbf{P}^a$-functions, can be obtained from the previous two by simply exchanging all $\mathbf{P}$- and $\mathbf{Q}$-functions. The existence of the last two equations is a simple consequence of the algebraic symmetry within the full Q-system between $\mathbf{P}$- and $\mathbf{Q}$-functions.

Let us note that the most frequent cases of the $\mathcal{N} = 4$ SYM operators studied in the literature are those which obey the so-called left–right (LR) symmetry w.r.t. to the exchange of two subgroups of the full superconformal group: $SU(2|2)_L \times SU(2|2)_R \in PSU(2,2|4)$. This symmetry has direct algebraic consequence for the underlying AdS/CFT Q-system. Namely, due this symmetry we can raise and lower the indices of $\mathbf{P}$-functions and $\mathbf{Q}$-functions, i.e. a-type or j-type, by means of a "metric" whose role is played by a fixed constant matrix [15, 70]:

$$\mathbf{P}_a = \chi_{ab}\mathbf{P}^b, \quad \mathbf{Q}_j = \chi_{jk}\mathbf{Q}^k, \quad \text{where } \chi = \begin{pmatrix} 0 & 0 & 0 & 1 \\ 0 & 0 & -1 & 0 \\ 0 & 1 & 0 & 0 \\ -1 & 0 & 0 & 0 \end{pmatrix}. \tag{31}$$

Obviously, in this case only two 4th order Baxter equations are algebraically independent: one for $\mathbf{Q}_j$ and one for $\mathbf{P}_a$, which significantly simplifies the problem. In addition, in various weak coupling limits, such as one-loop [68, 71] or BFKL [69] approximations, or the double scaling (DS) limit of γ-deformed $\mathcal{N} = 4$ SYM [30] described in the next section, the analytic properties of $\mathbf{P}$-functions simplify even further: they can have only finite order poles at $u = 0$ and $u = \infty$ and thus they can be parametrized by a finite number of coefficients in the corresponding polynomials. Then the Baxter equation (29) on $\mathbf{Q}$-functions starts to remind the one for the integrable $SU(2,2)$ spin chain reflecting the 4D conformal symmetry of the problem.

3.2. *Analyticity: quantum spectral curve as a Riemann–Hilbert problem*

The QSC formalism is based on two fundamental ingredients: the first is the algebraic structure of the underlying Q-system, entirely based on the superconformal symmetry $PSU(2,2|4)$ of the model, and the second is the analyticity properties

[m]That is, the coefficients are functions of u periodic w.r.t. the shift $u \to u + i$.

of the underlying Q-functions. The analyticity is greatly, but not completely dictated by the algebraic structure of Q-system. It was established in the original papers [14, 15]. It was extracted from the exact solution of the AdS/CFT spectral problem, first proposed in the form of the AdS/CFT Y-system [72] and then via the TBA approach [73–75]. The papers [70, 76] have been important steps towards the discovery of the QSC formalism.

In the rest of this section, we will describe the analytic properties of Q-functions. The main ingredients of their analytic structure are (i) Infinitely branching Riemann surface due to branch cuts at fixed positions — "Zhukovsky cuts"[n]; (ii) Asymptotics at large values of spectral parameter fixing the $SU(2,2|4)$ representation of state/operator; (iii) Riemann–Hilbert "sewing" conditions relating various Q-functions via monodromies around Zhukovsky cuts; (iv) Absence of any other singularities anywhere on the Riemann surface of any Q-function, except mentioned above. Let us inspect these properties in detail.

3.2.1. *Zhukovsky branch cuts and Riemann surface for Q-functions*

The main analyticity observation in QSC formalism is about the particular subset of 16 Q-functions, precisely the ones listed in (21)–(24) and shown on Fig. 5. Namely, the 8 functions $\mathbf{P}_a$ and $\mathbf{P}^a$ have, each, a special sheet of the Riemann surface (which will be called physical) where it has only one Zhukovsky cut for the range of spectral parameters $u \in (-2g, 2g)$ and $g^2 = \frac{1}{16\pi^2} N_c g_{YM}^2$ is the 't Hooft coupling.[o] The physical sheet is depicted on Fig. 6 (left). Similarly, the other 8 Q-functions, $\check{\mathbf{Q}}_j$ and $\check{\mathbf{Q}}^j$, have a special, physical sheet where they have only one cut with the same branch-points but passing through $u = \infty$, i.e. for $u \in (-\infty, -2g) \cup (2g, \infty)$. The physical sheet is depicted on Fig. 6 (top right). It is natural call the first type of cuts as "short cuts" and the second one as "long cuts". The positions of the branchpoints of these cuts are actually the only place in the QSC formalism where the 't Hooft–Yang–Mills coupling constant is encoded.[p]

There are no other singularities on the physical sheets of these Q-functions except the one at $u = \infty$, described in the next subsection.

Next, we want to know what happens under the cut, on the next sheet of the Riemann surface. In fact, the structure of the Q-system, and in particular of the QQ-relations (12), dictates that for $\mathbf{P}_a$ and $\mathbf{P}^a$ functions, apart from the same short cut $u \in (-2g, 2g)$, we find on the second sheet an infinite "ladder" of its periodically[q] repeating replicas at $u \in (-2g + i\mathbb{Z}, 2g + i\mathbb{Z})$, as shown on the right of

[n] The name originating from the Zhukovsky conformal map, inverse of $u = g(x + 1/x)$ function.

[o] g_{YM} is the original Yang–Mills coupling and N_c is the number of colors of the $U(N_c)$ gauge group. From now on, for brevity, we will rather call g the 't Hooft coupling.

[p] From now on, we will distinguish, by "check" or its absence, two functions $\check{\mathbf{Q}}_j$ and $\mathbf{Q}_j$. In fact, it is the same function but the Riemann sheets are organized differently in two cases: $\check{\mathbf{Q}}_j$ has only long cuts and the notation $\mathbf{Q}_j$ will be reserved for the same function with Riemann sheets re-glued in such a way that all its cuts are short. See the text below and the Fig. 6 for more explanations.

[q] We call this periodicity the i-periodicity since the cuts are spaced by the distance i in complex plane.

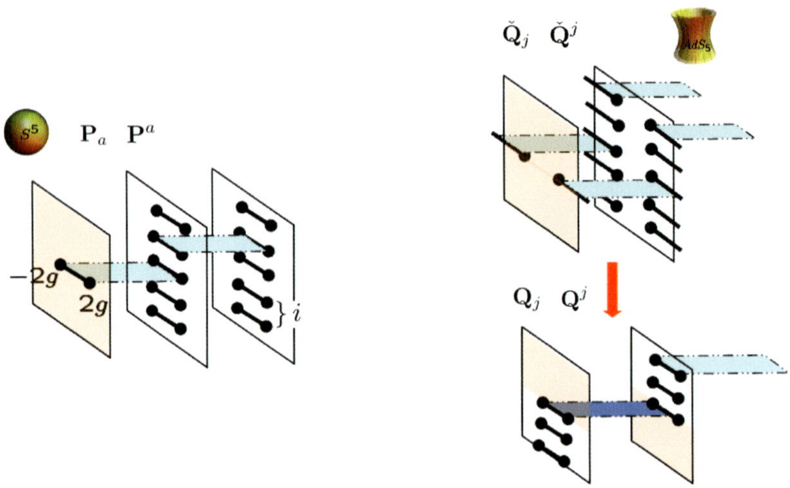

Fig. 6. Schematic depiction of analytic structure of Riemann surfaces of $\mathbf{P}$ and $\mathbf{Q}$ functions defined by (21)–(24). On the left, $\mathbf{P}_a$ and $\mathbf{P}^a$ have, each, a special, physical sheet of the Riemann surface where it has only one Zhukovsky cut for the range of spectral parameters $u \in (-2g, 2g)$, where $g^2 = \frac{1}{16\pi^2} N_c g_{YM}^2$ is the 't Hooft coupling. This cut is connected to the next sheet which has a ladder of equidistant Zhukovsky cuts spaced by i, at positions $u \in (-2g + i\mathbb{Z}, 2g + i\mathbb{Z})$, along the imaginary axis. On the top right, the same picture of Riemann surface is true for the $\check{\mathbf{Q}}_j$ and $\check{\mathbf{Q}}^j$ functions, with an important difference: short Zhukovsky cuts should be replaced by long Zhukovsky cut, passing through $u = \infty$, i.e. for $u \in (-\infty, -2g) \cup (2g, \infty)$. On the next sheets we have an infinite ladder of such cuts spaced by i. On the bottom right, we show the rearrangement of Riemann surface, by re-gluing the upper half-plane of the physical sheet with the lower half-plane of the next sheet, and vice versa for the other two halves. This flips the long Zhukovsky cut on the real axis to short cut, but also creates a sequence of cuts in the lower half-plane (which can be made short by the same re-gluing procedure for the next sheets).

Fig. 6 (left). If we pass through any of these cuts we will encounter another sheet, with the same infinite ladder of short cuts repeating periodically along the whole imaginary axis. Passing through any of these cuts we discover the other sheets with the same infinite ladder of cuts. Consequently, each $\mathbf{P}$-function lives on an infinitely branching Riemann surface of the topology of sphere with a puncture at $u = \infty$.[r]

As for the functions $\check{\mathbf{Q}}_j$ and $\check{\mathbf{Q}}^j$, the picture of cuts on the sheets next to the physical one is exactly the same as for the functions $\mathbf{P}_a$ and $\mathbf{P}^a$, except that all cuts are long, i.e. they are i-periodic, at positions $u \in (-\infty, -2g + i\mathbb{Z}) \cup (2g + i\mathbb{Z}, \infty)$, as shown on the right of the Fig. 6 (top right). Of course the fact that these cuts are accumulated at $u = \infty$ leads to an infinite branching at infinity and allows for asymptotics with arbitrary power law w.r.t. spectral parameter. As we will see below, this is the way how the parameter $\gamma(g)$ — the anomalous dimension of the operator — arises in the QSC formalism as a power in the large u asymptotics of $\mathbf{Q}$-functions.

[r]We conjecture the spherical topology of the Riemann surface since we see no obvious reasons for the existence of any non-contractible closed paths on this surface (except those encircling $u = \infty$).

3.2.2. *Large u asymptotics*

Now we describe the behavior of $\mathbf{P}$- and $\check{\mathbf{Q}}$-functions at $u\to\infty$ on the physical sheet. To avoid complications with degeneracy of solutions we first consider the case of the totaly deformed superconformal symmetry of the model: $PSU(2,2|4)\to[U(1)^2\times\mathbb{R}]_{conf.}\times[U(1)^3]_R$. This is done by introduction of special twist on the $AdS_5\times\mathrm{CFT}_4$ duality parametrized by a fixed Cartan group element: $\mathrm{diag}\{x_1,x_2,x_3,x_4|y_1,y_2,y_3,y_4\}\in PSU(2,2|4)$, with the group constraint $\prod_{a=1}^4 x_a=1=\prod_{j=1}^4 y_j$. This deformation is easy to perform directly in the $\mathcal{N}=4$ SYM action [1, 6, 8, 77] for the case of so-called γ-twist, when the conformal part of the superconformal symmetry is not twisted $y_1=y_2=y_3=y_4=1$ and the x_a are parameters of $SU(4)$ R-symmetry deformation.[s]

The picture for $\mathbf{P}$-functions is very simple: since the only singularity at the finite part of the u-plane is a short Zhukovsky cut, we can approach the $u=\infty$ by any path, and the asymptotics is completely fixed by the global R-symmetry charges $\{J_1,J_2,J_3\}\in so(6)\sim su(4)$ and the value of the twist:

$$\mathbf{P}_b\sim x_b^{i\,u}\,u^{-\lambda_b}\left(1+\frac{p_1^{(b)}}{u}+\frac{p_2^{(b)}}{u^2}+\cdots\right),\qquad \text{where } \{x_1,x_2,x_3,x_4\}\in SU(4), \tag{32}$$

$$\lambda_b=\frac{1}{2}\{+J_1+J_2-J_3,+J_1-J_2+J_3,-J_1+J_2+J_3,-J_1-J_2-J_3\}. \tag{33}$$

So we see that these asymptotics can have only integer or half-integer powers.[t]

The situation with large u asymptotics on the physical sheet of $\mathbf{Q}$-functions is slightly more involved: due to the presence of the long Zhukovsky cut we should speak in principle separately of the large u asymptotics in the upper half-plane (UHP) and in the lower half-plane (LHP). However one can easily argue that those two asymptotics can be different only by an overall constant (see [15] for the calculation of this constant). We can thus impose that, for example in the UHP, far away from the real axis (to avoid the vicinity of the long cut) their exponential and power-like parts are defined, respectively, by $SU(2,2)$ twists and the Cartan charges of conformal group $\{\Delta,S_1,S_2\}\in so(4,2)\sim su(2,2)$, as follows

$$\check{\mathbf{Q}}_j\sim y_j^{-iu}u^{-\nu_j}\left(1+\frac{q_1^{(j)}}{u}+\frac{q_2^{(j)}}{u^2}+\cdots\right),\qquad \text{where } \{y_1,y_2,y_3,y_4\}\in SU(2,2), \tag{34}$$

$$\nu_j=\frac{1}{2}\{+\Delta-S_1-S_2,\quad +\Delta+S_1+S_2,\quad -\Delta-S_1+S_2,\quad -\Delta+S_1-S_2\}. \tag{35}$$

---

[s]Twisting the conformal part of the supergroup leads to a non-commutative generalization of $\mathcal{N}=4$ SYM theory [77].

[t]In fact, in more "physical" quantities, such as transfer-matrix eigenvalues, the $\mathbf{P}$-functions enter bi-linearly, so that the asymptotics becomes single-valued at $u=\infty$. So in some sense the quadratic branch-point of $\mathbf{P}$ at $u=\infty$ is not an important issue for the analyticity properties of the Q-system, see [15] for the details.

Here S_1, S_2 are integer conformal spins and $\Delta(g) \equiv \Delta(0) + \gamma(g)$ is the dimension of the studied operator (energy of the state on the string side of duality) which is the main quantity under study in QSC formalism. Generically, $\Delta(g)$ is a complicated function of the 't Hooft coupling g, of conserved charges $J_1, J_2, J_3|S_1, S_2$ and of the twist parameters $\{x_1, x_2, x_3, x_4|y_1, y_2, y_3, y_4\}$. With all these parameters fixed, we should have a finite or infinite discrete set of operators/states with different anomalous dimensions fixed by the values of the other conserved charges present in this integrable model. The presence of an arbitrary (if we vary g) power $\pm\Delta/2$ in the asymptotics means the presence of, in general, infinite branching at $u = \infty$. This is a natural consequence of the presence of a long cut passing through $u = \infty$ point. Notice that on the next sheets of $\check{\mathbf{Q}}$-functions it is hardly possible to speak about such power-like $\times$ exponential asymptotics due to the accumulation of long cuts forming an infinite ladder. On the contrary, one can define this kind of asymptotics at large u for the $\mathbf{P}$-functions if we avoid approaching $u = \infty$ along the imaginary axis, in the vicinity of infinite ladder of short cuts.

Notice that we did not impose separately the asymptotics of Hodge dual $\mathbf{P}^a$ and $\mathbf{Q}^j$-functions since those are not independent of $\mathbf{P}_a$ and $\mathbf{Q}_j$. They are completely constrained by the structure of the Q-system (with an important role of the gauge condition (20)) and the leading asymptotics are inverse powers w.r.t. the original $\mathbf{P}_a$ and $\mathbf{Q}_j$, namely,

$$\begin{aligned} \mathbf{P}^b &\sim x_b^{-iu} u^{\lambda_b} \left(1 + \frac{p_1''^{(b)}}{u} + \frac{p_2''^{(b)}}{u^2} + \cdots \right), \\ \check{\mathbf{Q}}^j &\sim y_j^{iu} u^{\nu_j} \left(1 + \frac{q_1''^{(j)}}{u} + \frac{q_2''^{(j)}}{u^2} + \cdots \right). \end{aligned} \tag{36}$$

It is important to notice that all these asymptotics are multiplied by the expansion in integer powers w.r.t. $1/u$. This is a special choice of the Q-functions, since any linear combination of them would spoil this property and mix up different combinations of twists and charges. We call our choice "pure" asymptotics, and this choice will be important for the rest of analyticity properties given below in the form of Riemann–Hilbert conditions.

We can partially remove the deformations by making some of the twist parameters $\{x_1, x_2, x_3, x_4|y_1, y_2, y_3, y_4\}$ equal to each other, i.e. restoring some subgroups of $PSU(2, 2|4)$ symmetry. Then we have to modify the asymptotics (32), (34) by shifting the exponent by certain integers because, asymptotically, certain determinant formulas for Q-functions will become ambiguous and will not render the right asymptotics. The whole classification of various twist configurations and of the corresponding asymptotics is given in [16]. We will discuss a couple of the most important cases. One of them, used in the next section, is the so-called γ-deformation which preserves the entire conformal subgroup $SU(2, 2)$, i.e. $y_1 = y_2 = y_3 = y_4 = 1$, and leaves arbitrary twists $\{x_1, x_2, x_3, x_4\}$, $\prod_j x_j = 1$, thus breaking R-symmetry

$SU(4) \to U(1)^3$. Then the asymptotics (34), should be modified as follows

$$\check{\mathbf{Q}}_j \sim u^{-\nu'_j}\left(1+\frac{q_1^{(j)}}{u}+\frac{q_2^{(j)}}{u^2}+\cdots\right) \tag{37}$$

where

$$\nu'_j = \frac{1}{2}\{+\Delta - S_1 - S_2, +\Delta + S_1 + S_2 - 2, -\Delta - S_1 + S_2 - 4, -\Delta + S_1 - S_2 - 6\}, \tag{38}$$

where as the leading asymptotics of $\mathbf{P}_b$ remain as given by (32), (33).

Finally, the most studied case is of course the fully untwisted, completely $PSU(2,2|4)$ symmetric SYM theory (or the equivalent dual superstring sigma-model on $AdS_5 \times S^5$ background). In this case, the above asymptotics of $\check{\mathbf{Q}}$ are the same as in (34), but for $\mathbf{P}$ they look now as follows [14, 15]

$$\mathbf{P}_b \sim u^{-\lambda'_b}\left(1+\frac{p_1^{(b)}}{u}+\frac{p_2^{(b)}}{u^2}+\cdots\right), \tag{39}$$

with

$$\lambda'_b = \frac{1}{2}\{+J_1 + J_2 - J_3, +J_1 - J_2 + J_3 - 2, -J_1 + J_2 + J_3 - 4, -J_1 - J_2 - J_3 - 6\}. \tag{40}$$

Using these asymptotics and the Grassmannian structure of the QQ system we can even compute a few leading coefficients of all these asymptotics, which appear to depend only on the global charges, not on particular solutions [14, 15]. The classification of the coefficients of the leading asymptotics can be found in [16]. We do not give here explicit formulas since we limit ourselves only to the formulation of basic rules of QSC construction, leaving aside its consequences.

3.2.3. *Riemann–Hilbert sewing conditions*

Finally, we have to describe how one can move among the sheets of the Riemann surface for the $\mathbf{P}$ and $\check{\mathbf{Q}}$-functions. In other words, one should detail the properties of monodromy around the branch-points of Zhukovsky cuts of these functions.

It was noticed in [22] that, after having imposed the "purity" of asymptotics, as discussed after Eq. (36), one can fix completely the system of spectral equations, by demanding within the QSC formalism the following Riemann–Hilbert sewing conditions [15][u]

$$\begin{pmatrix} \bar{\check{\mathbf{Q}}}_1 \\ \bar{\check{\mathbf{Q}}}_2 \\ \bar{\check{\mathbf{Q}}}_3 \\ \bar{\check{\mathbf{Q}}}_4 \end{pmatrix} \simeq \begin{pmatrix} 0 & \beta_1 & 0 & 0 \\ \bar{\beta}_1 & 0 & 0 & 0 \\ 0 & 0 & 0 & \beta_2 \\ 0 & 0 & \bar{\beta}_2 & 0 \end{pmatrix} \begin{pmatrix} \check{\mathbf{Q}}^1 \\ \check{\mathbf{Q}}^2 \\ \check{\mathbf{Q}}^3 \\ \check{\mathbf{Q}}^4 \end{pmatrix}, \tag{41}$$

[u]For the origin and explanations of these sawing conditions, see [15, Sec. 4.4.2], and in particular Eq. (4.63). In [22] these relations are called "gluing conditions", but we call them here "sewing conditions" which seems to be a more frequent terminology for Riemann–Hilbert problems.

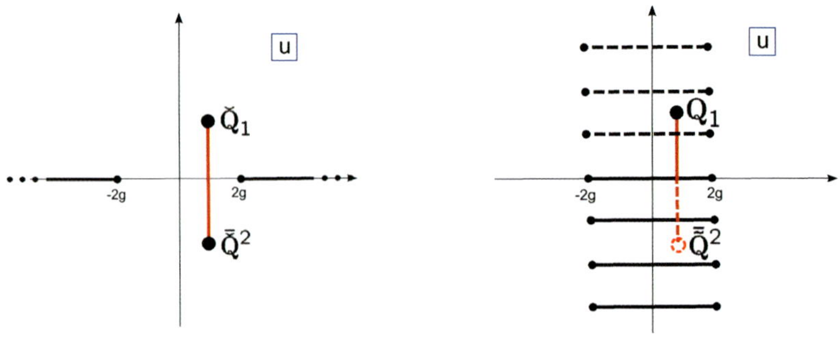

Fig. 7. Demonstration of the Riemann–Hilbert sewing relations: on the left, the complex conjugation relation (41) between a pair of functions $\check{\mathbf{Q}}_1 = \bar{\beta}_1 \bar{\check{\mathbf{Q}}}^2$ on the physical sheet with a single long cut is presented. Notice that the path connecting them should go between the branch-points. On the right, the same relation is demonstrated on the sheet with short cuts. It takes the form $\tilde{\mathbf{Q}}_1 = \bar{\beta}_1 \bar{\tilde{\mathbf{Q}}}^2$. Dotted cuts, are situated on the second sheet. The conjugation path is now passing through the short cut at the real axis, i.e. the conjugation involves now the monodromy (denoted by tilde) as well.

where $\bar{\check{\mathbf{Q}}}_j$ means the complex conjugation, i.e. reflection of the main sheet w.r.t. the real axis where the long cut is present, see Fig. 7 (left), and β_1, β_2 are constants non-trivially depending on the parameters of the operator/state, to be defined self-consistently in the process of solution of QSC equations. The origins of this sewing condition originate already from the properties of quasi-momenta of classical finite gap solution of the string dual — the sigma model on $AdS_5 \times S_5$ coset [78–80].[v]

These conditions mean that the $\check{\mathbf{Q}}$-functions are not all independent but rather glued together into a smaller number of analytic functions. This sewing condition is the finite element of QSC construction which locks completely the QSC relations into a closed system of equations for spectrum. Their solution renders a discrete set of dimensions/energies of all the operators/states with the given 't Hooft coupling g, the global charges $J_1, J_2, J_3|S_1, S_2$ and twist parameters $\{x_1, x_2, x_3, x_4|y_1, y_2, y_3, y_4\}$.

For various applications, especially related to the weak coupling approximations $g \to 0$, it is very convenient to reshuffle the sheets of the Riemann surface of each $\check{\mathbf{Q}}$ function in such a way that it would have only short cuts on the whole Riemann surface. For example, the function

$$x(u) = \frac{1}{2g}\left(u + \sqrt{u-2g}\sqrt{u+2g}\right), \tag{42}$$

inverse to Zhukovsky map $\frac{u}{g} = x + 1/x$, has by definition a short cut. But its analytic continuation, the function $\check{x}(u) = \frac{1}{2g}\left(u + i\sqrt{4g^2 - u^2}\right)$, already has a long cut[w] and it corresponds to regluing halves of the two sheets of the first functions

[v]See Sec. 2.5 and [80, Eqs. (2.59) and (3.3)]. The parameter x is related to the spectral parameter used here by Zhukovsky map $u = x + 1/x$ (the 't Hooft coupling g plays the role of the "Planck constant" for the string sigma model and can be scaled out in classical limit).

[w]What can be immediately seen if one plots real and imaginary parts of these two functions on Mathematica using Plot3D function.

along the real axis. As was already mentioned, we will reserve from now on the notation $\mathbf{Q}_j$ (without "check") for the configuration of Riemann surface with short cuts. This transition $\check{\mathbf{Q}} \to \mathbf{Q}$ is shown on Fig. 6 (bottom right). Re-gluing in this way the halves of the first and second sheets we obtain the first, upper sheet of the new Riemann surface which is free of singularities in the UHP, inheriting this analyticity from the UHP of the original physical sheet with the long cut; but on the real axis and below it we will have now a half-infinite ladder of Zhukovsky cuts inherited from the second sheet of the original Riemann surface. All these cuts can be made short by the same operation, involving the sheets next to the second one. So we can always work only with the short cuts, but the analyticity on the main sheet becomes more involved.

The sewing Riemann–Hilbert relations (41) now look as follows

$$\begin{pmatrix} \tilde{\mathbf{Q}}_1 \\ \tilde{\mathbf{Q}}_2 \\ \tilde{\mathbf{Q}}_3 \\ \tilde{\mathbf{Q}}_4 \end{pmatrix} \simeq \begin{pmatrix} 0 & \bar{\beta}_1 & 0 & 0 \\ \beta_1 & 0 & 0 & 0 \\ 0 & 0 & 0 & \bar{\beta}_2 \\ 0 & 0 & \beta_2 & 0 \end{pmatrix} \begin{pmatrix} \bar{\mathbf{Q}}^1 \\ \bar{\mathbf{Q}}^2 \\ \bar{\mathbf{Q}}^3 \\ \bar{\mathbf{Q}}^4 \end{pmatrix}, \tag{43}$$

where tilde sign corresponds to the monodromy of a function around the Zhukovsky branchpoint on the real axis. That means that $\tilde{\mathbf{Q}}_j$ is identical to $\check{\mathbf{Q}}_j$ on the second sheet on Fig. 6. Notice that on both sides of (43) the sequence of short cuts goes from the real axis upwards.

In fact, it turns out that any particular one of the conditions (43)

$$\tilde{\mathbf{Q}}_1 \sim \bar{\mathbf{Q}}^2, \quad \tilde{\mathbf{Q}}_2 \sim \bar{\mathbf{Q}}^1, \quad \tilde{\mathbf{Q}}_3 \sim \bar{\mathbf{Q}}^4, \quad \tilde{\mathbf{Q}}_4 \sim \bar{\mathbf{Q}}^3, \tag{44}$$

imposed on functions with pure asymptotics is enough to fix completely the set of physical solutions for energies/dimensions of states with given global charges of the superconformal symmetry [21]. The first of them is demonstrated on Fig. 7 (right). In fact, it was also observed in [21] that, in the case of complete twisting, as in (36), only one of these conditions is enough to fix completely such a set of solutions. The other three will follow from it. If we have only a partial twisting, or coinciding twist parameters, the situation is more complex since the asymptotics for some groups of Q-functions will be different only by integer powers and the choice of pure solutions becomes ambiguous.[x] This is the case for example in the case of γ-deformation (36) where there is no twisting for $\mathbf{Q}$-functions. In this case, a pair of relations (44), containing asymptotics with different sign of Δ is enough (the first and third, or the second and fourth).

To conclude, in this and preceding sections, we gave the general scheme of the QSC formalism, concentrating on its universal and the most general defining features. We specially avoided so far any secondary details and consequences of this construction. In the next section, we will discuss an interesting physical application

[x]The discussion and classification of asymptotics of all possible generations of twists can be found in [16].

of QSCγ, related to conformal filed theory following from the γ-twisted $\mathcal{N} = 4$ SYM in a specific double scaling (DS) limit of large (imaginary) γ twists and weak coupling.

4. Double Scaling Limit of γ-Twisted $\mathcal{N} = 4$ SYM and Fishnet Feynman Graphs

In this section, we demonstrate the force of QSC on a particular example of the study of the double scaling (DS) limit of γ-deformed $\mathcal{N} = 4$ SYM theory proposed in [30]. The resulting non-unitary chiral CFTs inherit the integrability properties of the full γ-twisted $\mathcal{N} = 4$ SYM. However it demystifies to some extent the, still hypothetic though always properly working, AdS/CFT integrability: at least in its simplest, bi-scalar version the theory in DS limit is dominated by so-called "fishnet" Feynman graphs, explicitly related to the integrable conformal, $SU(2,2)$ Heisenberg spin chain [24, 30, 31].

4.1. *Lagrangian and conformal properties of γ-deformed $\mathcal{N} = 4$ SYM*

The QSC formalism described in the previous section is deeply rooted in the bootstrap solution of the string σ-model on γ-deformed $AdS_5 \times S^5$ background. It is the result of a long development of integrability methods, such as Y-system, TBA and Destri–De Vega-type equations, applied to this 2-dimensional string σ-model. However, the AdS/CFT correspondence, as applied to the γ-deformed case [6, 7, 10, 77], states that the energy spectrum of this σ-model is in one-to-one correspondence with the spectrum of anomalous dimensions of the γ-deformed $\mathcal{N} = 4$ SYM theory with the Lagrangian given in the following explicit form (see [12, 30])

$$\begin{aligned}\mathcal{L} = N_c \operatorname{tr}\Big[&-\frac{1}{4}F_{\mu\nu}F^{\mu\nu} - \frac{1}{2}D^\mu\phi_i^\dagger D_\mu\phi^i + i\bar\psi_A^{\dot\alpha} D_{\dot\alpha}^\alpha \psi_\alpha^A \\ &+ g^2\left(\frac{1}{4}\{\phi_i^\dagger,\phi^i\}\{\phi_j^\dagger,\phi^j\} - e^{-i\epsilon^{ijk}\gamma_k}\phi_i^\dagger\phi_j^\dagger\phi^i\phi^j\right) \\ &+ g(-e^{-\frac{i}{2}\gamma_j^-}\bar\psi_j\phi^j\bar\psi_4 + e^{+\frac{i}{2}\gamma_j^-}\bar\psi_4\phi^j\bar\psi_j + i\epsilon_{ijk}e^{\frac{i}{2}\epsilon_{jkm}\gamma_m^+}\bar\psi^k\phi^i\bar\psi^j \\ &+ \text{conjugate terms})\Big], \end{aligned} \tag{45}$$

where $A = 1,2,3,4$, and we sum up over all doubly repeated or (abusing the standard tensorial notations) triply repeated indices $i,j,k,m = 1,2,3$.[y] Here $\phi_{j=1,2,3}$ are complex scalar fields and ψ_j^α are Majorana–Weyl fermions.[z] We also used the

[y]Such a summation over triply repeated indices occurs since the $SU(4) \sim SO(6)$ symmetry is broken by γ-twists.

[z]We also suppressed the spinorial indices in the second and third lines in the above formula, always assuming that they are contracted in the standard way: $(\cdots\psi\cdots\psi) \to (\cdots\psi^\alpha\cdots\psi_\alpha)$ and $(\cdots\bar\psi\cdots\bar\psi) \to (\cdots\bar\psi^{\dot\alpha}\cdots\bar\psi_{\dot\alpha})$.

shorthand notations $\gamma_1^\pm = \frac{\gamma_3\pm\gamma_2}{2}$, $\gamma_2^\pm = \frac{\gamma_1\pm\gamma_3}{2}$, $\gamma_3^\pm = \frac{\gamma_2\pm\gamma_1}{2}$. The three parameters γ_j are related to the twist parameters x_k of (37) by the following formulas [16]

$$\begin{aligned} x_1 &= e^{\frac{i}{2}[(\gamma_2-\gamma_1)J_3-(\gamma_1+\gamma_3)J_2+(\gamma_2+\gamma_3)J_1]}, \\ x_2 &= e^{\frac{i}{2}[(\gamma_1+\gamma_2)J_3-(\gamma_2+\gamma_3)J_1+(\gamma_1-\gamma_3)J_2]}, \\ x_3 &= e^{\frac{i}{2}[-(\gamma_1+\gamma_2)J_3+(\gamma_1+\gamma_3)J_2+(\gamma_3-\gamma_2)J_1]}, \\ x_4 &= e^{\frac{i}{2}[(\gamma_3-\gamma_1)J_2+(\gamma_1-\gamma_2)J_3+(\gamma_2-\gamma_3)J_1]}. \end{aligned} \tag{46}$$

Notice that if we put $\gamma_1 = \gamma_2 = \gamma_3 = 0$ we obtain the standard case of the superconformal, $\mathcal{N}=4$ SYM — a CFT with the unbroken $PSU(2,2|4)$ symmetry.[aa] The twisting corresponds to the following rule. In each term of the Lagrangian (45), the deformation factors depend on the order of the fields under the trace. For two arbitrary fields A and B the matrix product AB is replaced by a star product:

$$AB \to A \star B \equiv q_{A,B}AB, \quad \text{where} \quad q_{A,B} = e^{-\frac{i}{2}\epsilon^{mjk}\gamma_m J_j^A J_k^B} = (q_{B,A})^{-1} \tag{47}$$

and J_1^A, J_2^A, $J_3^A \in SO(6)$ are the three Cartan charges of A-field. Using this rule, it is easy to recover the γ-deformed Lagrangian (45) from the undeformed one. On the classical level, the full superconformal symmetry appears to be explicitly broken: $PSU(2,2|4) \to SU(2,2)\times U(1)^3$ but the conformal symmetry remains.

Strictly speaking, on the quantum level the γ-deformation breaks the conformal symmetry, even in the large N_c limit [8]: although the 't Hooft coupling g does not run with RG flow, a few new, double-trace terms of the type $\mathrm{tr}(\phi_j\phi_k^\dagger)\,\mathrm{tr}(\phi_k\phi_j^\dagger)$ or $\mathrm{tr}(\phi_j\phi_k)\,\mathrm{tr}(\phi_j^\dagger\phi_k^\dagger)$ [8, 10, 11] are generated, whose couplings do run. In particular, for the double-trace interaction term $\alpha_{jj}^2\,\mathrm{tr}(\phi_j\phi_j)\,\mathrm{tr}(\phi_j^\dagger\phi_j^\dagger)$ the one-loop beta-function is given by [81]

$$\beta_{\alpha_{jj}^2} = \frac{g^4}{\pi^2}\sin^2\gamma_j^+\sin^2\gamma_j^- + \frac{\alpha_{jj}^4}{4\pi^2} + \mathcal{O}(g^6)), \qquad (\text{no sum over } j), \tag{48}$$

so that at a complex fixed points

$$\alpha_{jj}^2 = \pm 2ig^2\sin\gamma_j^+\sin\gamma_j^- + O(g^4) \tag{49}$$

the γ-deformed $\mathcal{N}=4$ SYM theory becomes again a true, though non-unitary, CFT! [12, 13]. This statement was demonstrated in [13] in a few orders of perturbation theory in the specific double scaling limit described below. It was also claimed in [13], and checked in many different ways, that the γ-deformed QSC of the previous section solves the problem of planar spectrum of the γ-deformed $\mathcal{N}=4$ SYM precisely at this fixed point.

4.2. *Double scaling limit and fishnet graphs*

The γ-deformed $\mathcal{N}=4$ SYM theory admits an interesting DS limit [30] which significantly clarifies the origins of integrability of $\mathcal{N}=4$ SYM itself. The DS limit

[aa] Unless it is spontaneously broken in Coulomb branch.

can be explicitly performed on the level of Lagrangian (45). Namely, it combines the weak coupling limit and big imaginary γ_j parameters:

$$g \to 0, \qquad e^{-i\gamma_j/2} \to \infty, \qquad \xi_j = g e^{-i\gamma_j/2} \; - \text{fixed}, \qquad (j = 1, 2, 3). \tag{50}$$

The resulting non-unitary CFT directly follows from (45). It depends on three DS couplings ξ_j and is defined[ab] by the following Lagrangian [30]

$$\begin{aligned} \mathcal{L} = N_c \operatorname{tr}\Big[&-\frac{1}{2}\partial^\mu \phi_i^\dagger \partial_\mu \phi^i + i\bar{\psi}_A^{\dot\alpha} \partial_{\dot\alpha}^{\alpha} \psi_\alpha^A + \xi_1^2 \phi_2^\dagger \phi_3^\dagger \phi_2 \phi_3 + \xi_2^2 \phi_3^\dagger \phi_1^\dagger \phi_3 \phi_1 \\ &+ \xi_3^2 \phi_1^\dagger \phi_2^\dagger \phi_1 \phi_2 + i\sqrt{\xi_2\xi_3}(\psi^3 \phi^1 \psi^2 + \bar\psi_3 \phi_1^\dagger \bar\psi_2) + i\sqrt{\xi_1\xi_3}(\psi^1 \phi^2 \psi^3 + \bar\psi_1 \phi_2^\dagger \bar\psi_3) \\ &+ i\sqrt{\xi_1\xi_2}(\psi^2 \phi^3 \psi^1 + \bar\psi_2 \phi_3^\dagger \bar\psi_1)\Big]. \end{aligned} \tag{51}$$

Notice that the gauge field and the 4th component of fermion are completely decoupled in DS limit.

This theory obeys a certain chirality property which shows up if we consider the Feynman perturbation theory for graphs with fixed topology appearing in $1/N_c$ expansion. Namely, each term in the last two lines of (51), with quartic scalar or Yukawa interaction, does not have its Hermitian conjugate counterpart. This means that, only a certain order of propagators around each vertex of a planar graph is possible, and the vertex with opposite order, which would correspond to the Hermitian conjugated term in the Lagrangian, does not appear. We will show in the next subsection, along the lines of [30, 40], that this chiral property leads to the RG independence of ξ_j couplings and the absence of mass generation, since the corresponding Feynman graphs would necessarily include the vertices of both chiralities.

4.3. *Bi-scalar model and integrable fishnet graphs*

Let us discuss a particular case of the model (51), taking $\xi_1 = \xi_2 = 0$ and keeping only $\xi \equiv \xi_3 \neq 0$. Its Lagrangian appears to be extremely simple [30]

$$\mathcal{L}_{\text{bi-scalar}}[\phi_1, \phi_2] = \frac{N_c}{2} \operatorname{tr}\left(\partial^\mu \phi_1^\dagger \partial_\mu \phi_1 + \partial^\mu \phi_2^\dagger \partial_\mu \phi_2 + 2\xi^2\, \phi_1^\dagger \phi_2^\dagger \phi_1 \phi_2\right). \tag{52}$$

The planar Feynman graphs are built of two types of scalar propagators:

$$\left\langle \phi_1^{*ij}(y) \phi_1^{kl}(x) \right\rangle_0 = \left\langle \phi_2^{*ij}(y) \phi_2^{kl}(x) \right\rangle_0 = \frac{1}{N} \delta^{ik} \delta^{jl} \frac{1}{(x-y)^2} \tag{53}$$

and a scalar vertex

$$V = \xi^2 \operatorname{tr}(\phi_1^\dagger \phi_2^\dagger \phi_1 \phi_2). \tag{54}$$

These elements are presented in double-line notations on Fig. 8.

The perturbative expansion for this theory appears to contain very limited set of Feynman graphs, with very specific structure. Notice for example that, in the lowest order of perturbation theory, the Feynman diagrams renormalizing the coupling ξ

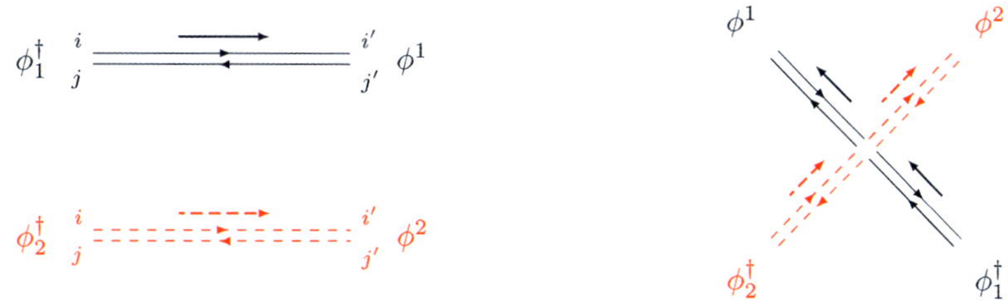

Fig. 8. Propagators and the vertex of bi-scalar model in double-line 't Hooft notations. The external arrows show the direction from a field to its Hermitian conjugate. The model has a particular orientation of this arrows for two different fields (solid lines for ϕ_1 and dotted lines for ϕ_2) around the vertex, fixing its chirality. The vertex of opposite chirality, corresponding to the complex conjugate interaction term, is absent.

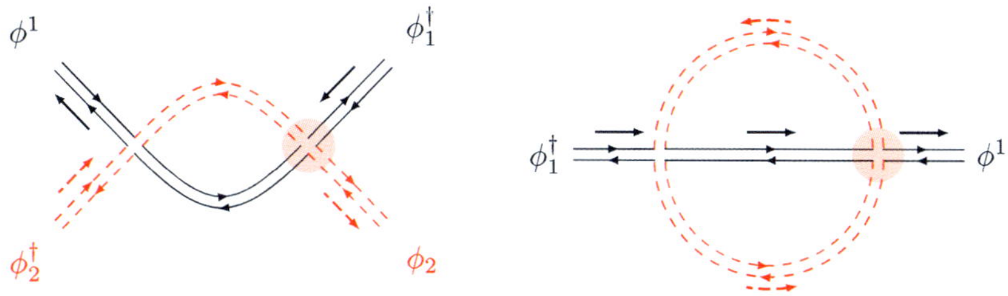

Fig. 9. These elements of graphs, renormalizing the vertex (on the left) and the mass (on the right), are absent from the planar Feynman diagrams of bi-scalar model due to the wrong chirality of vertices marked by a spot, absent from the action. This is an illustration of a general phenomenon leading to the all-loop conformality of the model.

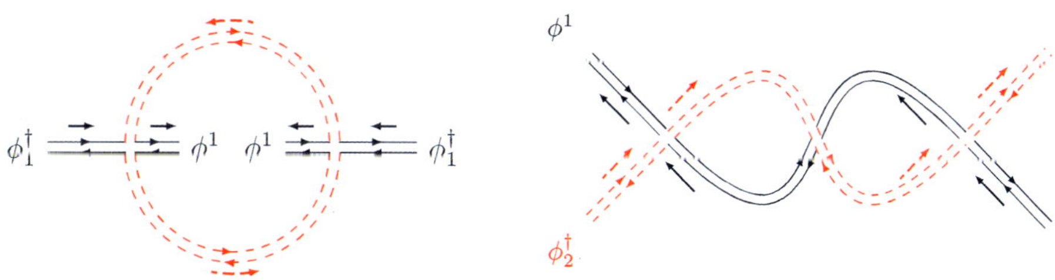

Fig. 10. Feynman diagrams generating the double-trace interactions in γ-deformed $\mathcal{N} = 4$ SYM theory in general, and in the bi-scalar model in particular. They survive even in the planar limit, but with the appropriate fine-tuning of these (complex) couplings the theory remains conformal.

(on the left of Fig. 9) and the mass of scalars (on the right of Fig. 9) are absent since they can be built only from two vertices of opposite chirality. This property persists in higher orders as well.

However, double-trace vertices will still be generated by RG, similarly to the full γ-deformed $\mathcal{N} = 4$ SYM. For example, the diagrams of the types depicted on Fig. 10 will generate the double-trace terms

[ab]Up to the already discussed double-trace scalar interactions, tuned to the conformal point.

$$\alpha_1^2 \sum_{i=1}^{2} \operatorname{tr}(\phi_i\phi_i)\operatorname{tr}(\phi_i^\dagger\phi_i^\dagger) - \alpha_2^2 \operatorname{tr}(\phi_1\phi_2)\operatorname{tr}(\phi_2^\dagger\phi_1^\dagger) - \alpha_3^2 \operatorname{tr}(\phi_1\phi_2^\dagger)\operatorname{tr}(\phi_2\phi_1^\dagger). \tag{55}$$

The couplings α_i can be again adjusted, as functions of the non-renormalized coupling ξ, to the fixed point where the model becomes a non-unitary CFT [12, 13]. The last two couplings appear to be only one-loop renormalizable, so that at the critical point they are fixed to $\alpha_2^2 = \alpha_3^2 = \xi^2$, whether as the critical coupling $\alpha_1(\xi)$ is more complicated and can have two complex conjugate values [13], given in dimensional regularization scheme by

$$\alpha_{1,\pm}^2 = \pm\frac{i\xi^2}{2} - \frac{\xi^4}{2} \mp \frac{3i\xi^6}{4} + \xi^8 \pm \frac{65i\xi^{10}}{48} - \frac{19\xi^{12}}{10} + O(\xi^{14}). \tag{56}$$

To study the spectrum of anomalous dimensions of local operators in such a theory, we have to be able to compute the mixing matrix among various operators which is given in terms of two-point correlation functions. Generically, such operators are of single trace type (in the planar limit) and can be presented as linear combinations of "words" (with cyclic symmetry) built out of four fields $\phi_1, \phi_1^\dagger, \phi_2, \phi_2^\dagger$ (taken at the same space-time point x) and two light-cone derivatives $\partial_\pm$ applied any number of times to any of the scalar fields constituting a local operator. Not all of these operators are independent: some of them are descendants (full derivatives) of simpler ones and some can be excluded by equations of motion. Still their amount is quickly increasing with the number of constituent fields. Diagonalizing the mixing matrix with particular linear combinations of such operators we obtain the anomalous dimensions as its eigenvalues.

Let us consider one type of such operators — the multi-magnon operators built only from $L-M$ fields ϕ_1 and M fields ϕ_2 (with no derivatives): [ac]

$$\mathcal{O}_{L,M}(x) = \operatorname{tr}\left(\underbrace{\phi_2\phi_1\phi_1\phi_1\phi_2\dots\phi_1}_{L\ \text{fields}}\right)(x) + \text{permutations}. \tag{57}$$

There exist linear combinations of such operators with given L, M, diagonalizing the mixing matrix, i.e. leading to the standard conformal two-point correlation functions

$$\langle \mathcal{O}(x)\mathcal{O}(0)\rangle \sim |x|^{-2L-2\gamma(\xi)} \tag{58}$$

where $\gamma(\xi)$ is the anomalous dimension of such an operator — typically a complicated function of ξ which we want to compute. The simplest of such operators is the "vacuum" operator[ad] without magnons ($M = 0$)

$$\mathcal{O}_L(x) = \operatorname{tr}[\phi_1(x)]^L. \tag{59}$$

[ac] In the description of these operators and the related Feynman graphs we closely follow the paper [40] and most of the figures are also borrowed from there.

[ad] This operator is protected in the undeformed $\mathcal{N} = 4$ SYM theory, i.e. it has the dimension $\Delta = L$ for any value of coupling. The corresponding string state is usually called BMN vacuum. We keep calling it the "vacuum" operator, though it gets non-trivial corrections in γ-deformed case, produced by so-called wrapped Feynman graphs.

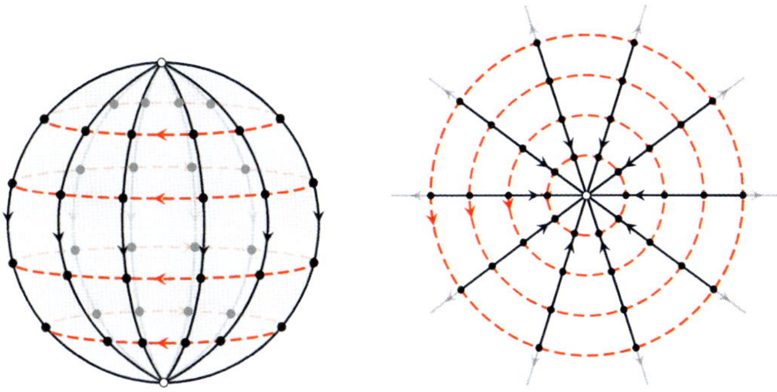

Fig. 11. The "globe" graphs (on the left) are the only type of Feynman graphs (for $L > 2$) contributing to the pair correlation function of the vacuum operator $\mathrm{tr}(\phi_1)^L(x)$. In the bulk, such a Feynman diagram has the structure of fishnet: the regular square lattice of propagators with ϕ^4 interactions in the vertices. Due to the conformal invariance of these graphs we can send the coordinate at one of the poles to infinity and, using the fact that we compute UV divergent expression, factor out the propagators around this pole as $|x|^{-2L}$. We will be thus left with the calculation of the "wheel" graph, on the right of the picture. To compute the corresponding anomalous dimension we need to know only the coefficient of the simple pole $1/\epsilon$ of the wheel graph. The wheel graphs are integrable. They can be studied by means of QSCγ (see the next section) or using its $SU(2,2)$ Heisenberg spin chain interpretation.

It does not mix with any other operator and hence it has a particular anomalous dimension $\gamma_L(\xi)$. If we try to compute the pair correlation function of such operator $\langle \mathcal{O}_L^\dagger(x)\mathcal{O}_L(0)\rangle$ by the Feynman perturbation technique we quickly realize that a single non-zero Feynman graph contributes at each ξ^{2L} order of perturbation theory and it has the shape of a "globe" with meridians consisting of only ϕ_1-type propagators and the parallels consisting of only ϕ_2-type propagators, as depicted on Fig. 11 (left). Due to the conformal invariance, we can send in this two-point correlator $x \to \infty$ without the loss of information. In this case, the trivial divergent contribution $|x|^{-2L}$ will factor out from the correlator (58) and we can chop off L propagators adjacent to the "north pole" of the globe responsible for that contribution and reduce the globe graph to a wheel graph shown on Fig. 11 (right). We will call this procedure a UV reduction. That means that if we were able to compute such a Feynman integral at any loop order at a given L we would calculate a very non-trivial quantity in this CFT — the anomalous dimension of the "vacuum" operator. Remarkably, the bulk of this graph looks like a regular square lattice — "fishnet" — and is known to define an integrable statistical-mechanical lattice model [31]. The problem of vacuum anomalous dimension is exactly solvable due to integrability and QSC is a very efficient approach for that. In the next subsection we will describe these results.

Let us also consider a more general case of the operators (57), in the presence of magnons, i.e. $M \neq 0$. They are also dominated by very particular Feynman graphs of a spiral type, as shown on Fig. 12 (left). The UV reduction, similar to the previous globe/wheel case, brings us to a graph on Fig. 12 (right) which can

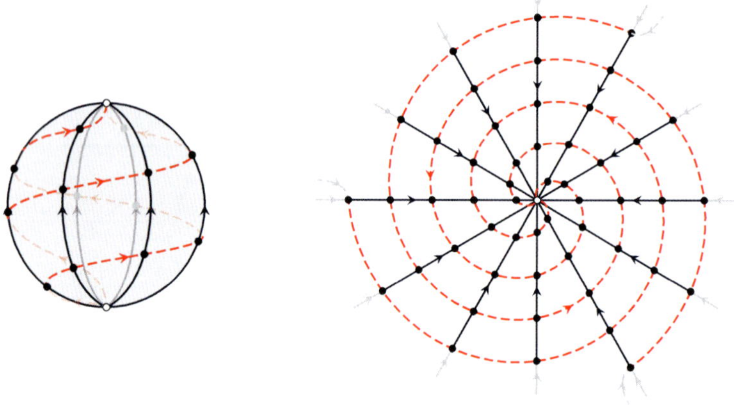

Fig. 12. The spiral graphs, such as one on the left — an example with 2 magnons, dominate the operators with magnons of the type (57). The UV reduction, similar to the one leading from globe graphs to wheel graphs for vacuum operators, gives a spiderweb-type graph on the right, obtained from the spiral graph by chopping off the propagators around one of the poles, such as the 3-magnon graph on the right. All those graphs are integrable and thus, in principal calculable, at least for the simple pole contributions in dimensional regularization.

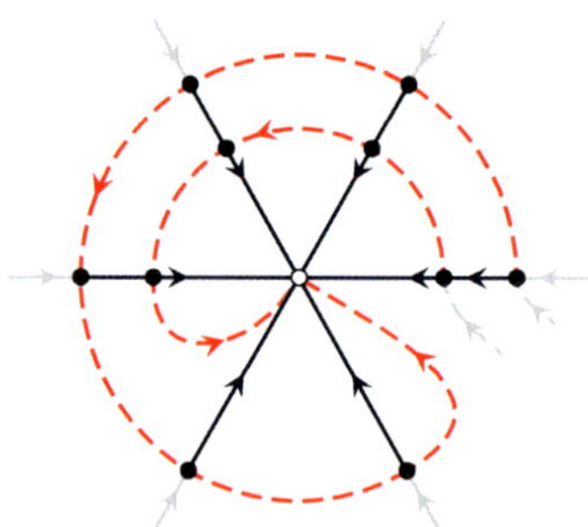

Fig. 13. An unwrapped magnon graph (spiderweb), calculable by the asymptotic Bethe ansatz methods developed in [40]. The absence of wrapping concerns the topology of planar graph. It can be seen as existence of a path connecting the center (fixed) node with a point outside ("infinity") without crossing any propagators.

be called a "spiderweb" graph.[ae] These graphs also have a fishnet structure in their bulk. However, on the boundary of this fishnet the structure of the spiderweb graph is very different from the wheel graph. We can see that these are the same integrable lattice systems but with different boundary conditions. QSC allows to compute the anomalous dimensions of such multi-magnon operators as well. Such a calculation at arbitrary coupling, or arbitrary loop order, is yet to be done. But for low orders of perturbation theory, corresponding to "unwrapped" magnon graphs, such as shown on Fig. 13 can be computed by means of the asymptotic Bethe ansatz (ABA) [40] — a doubly scaled version of Beisert–Staudacher equations [82].

[ae]This is precisely the way the spiders weave their web.

Let us note that, unlike the vacuum operators, the multi-magnon operators mix with each other for $M > 1$ magnons. QSC approach automatically solves the problem of finding the true anomalous dimensions — the eigenvalues of the mixing matrix.

The limited Feynman graph content of correlation functions in bi-scalar theory brings us to the idea of using integrability for exact computation of fishnet graphs (or at least of their specific UV singularity) with various integrable boundaries, specified by the appropriate conformal operators. This program was significantly advanced in [30] where the double wheel graphs (with double wrapping) have been computed using the TBA results of [83], in [40][af] where the 5-loop unwrapped 2-magnon graphs have been computed using the doubly scaled ABA equations, and finally in [24] where the problem of $L = 3$ wheel graphs (i.e. with 3 spokes) is reduced, by the double scaling procedure applied to QSC formalism of the previous section, to a Baxter equation with specific quantization conditions described in the next subsection. The last result gives essentially the full solution of the problem since it is very easy to generate from this equation the results for UV $1/\epsilon$ divergency of such a graph at very high loop orders (12 loops are reached by a laptop mathematica program), in terms of explicit multiple ζ-value (MZV) expressions. Numerical solution for the anomalous dimension at finite couplings, with very high precision, is available as well.

We will describe the results of [24] for the vacuum operator in the next section. We will conclude this section by an interesting observation which can have important consequences for the non-perturbative study of this model.

4.4. *Feynman graphs of bi-scalar model and $SU(2,2)$ conformal Heisenberg spin chain*

It was noticed in [30] that the problem of computation of the wheel graphs can be formulated in terms of the "graph building" operator in the space of space-time coordinates x_l

$$\hat{t}_L = \xi^{2L} \prod_{l=1}^{L} \frac{1}{(x_{l+1} - x_l)^2} \prod_{l=1}^{L} \Delta_{x_l}^{-1} \tag{60}$$

which represents one row of a wheel graph, as depicted in Fig. 14 (left). Here the first factor represents propagators of field ϕ_2, placed along the circular frame in angular direction, and the second factor uses the standard property of inverse Laplacian

$$\Delta_x^{-1}\, \delta^{(4)}(x-y) = \frac{1}{(x-y)^2}$$

to insert the propagators in radial direction.

[af] To make it precise, the explicit results for each 2-magnon 5-loop graph come about from a combination of ABA approach of [40] and explicit computation of one of these graphs in [84].

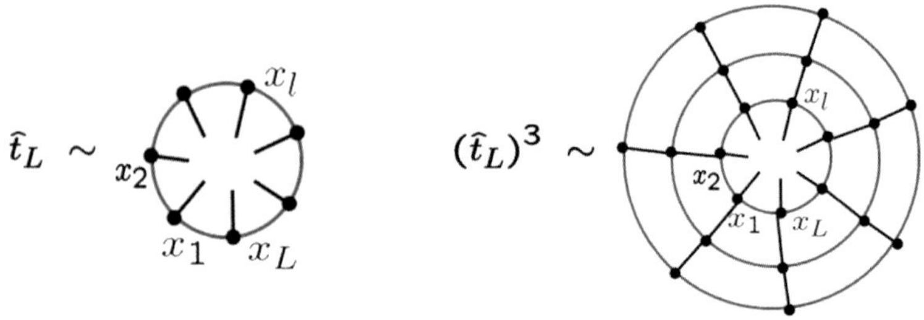

Fig. 14. On the left, we give a schematic representation of the graph-building operator $\hat{t}_L$ (60) (transfer-matrix), with a row of propagators around the circle and the radial propagators given by inverse Laplacians. On the right, the 3rd power of this operator is depicted as a consecutive action of the kernel of this operator, producing the integrations w.r.t. the 4D variables in the intermediate vertices. These graphs give the main contribution to a certain $2L$-point correlation function of the bi-scalar model.

The wheel graph with n frames is given by the following formal expression:

$$W_{L,n} = \int d^4x_1 \cdots \int d^4x_L \langle x_1, \ldots, x_L | (\hat{t}_L)^n | 0, \ldots, 0 \rangle. \tag{61}$$

A power of graph-building operator $\hat{t}_L$ is illustrated by the Fig. 14 (on the right we show the example of $\hat{(t_L)}^3$). Summing up matrix elements of powers of $\hat{t}_L$ we obtain a certain $2L$-point correlation function

$$K_L(x_1, \ldots, x_L | y_1, \ldots, y_L) = \prod_{l=1}^{L} (x_{l+1} - x_l)^2 \times \sum_{n=1}^{\infty} \langle x_1, \ldots, x_L | (\hat{t}_L)^n | y_1, \ldots, y_L \rangle$$

$$= \prod_{l=1}^{L} (x_{l+1} - x_l)^2 \times \langle x_1, \ldots, x_L | \frac{\hat{t}_L}{1 - \hat{t}_L} | y_1, \ldots, y_L \rangle \tag{62}$$

given by the sum of cylindric fishnet graphs, as the one on Fig. 14 (right).

Of course these expressions are UV divergent and need to be regularized. The standard dimensional regularization introduces the dimension as the regularization parameter $\epsilon = 4 - D$. To extract the anomalous dimension $\gamma_L(\xi)$ of the vacuum operator $\mathrm{tr}(\phi_1)^L$, we only need to know the residue of the lowest, $1/\epsilon$ pole of $W_{L,n}$ for each n.

We will now argue that the graph-building operator (60) is a nontrivial conserved charge of the noncompact Heisenberg spin chain based on the conformal group $su(2,2)$. Indeed, let us define a Lax operator

$$\hat{L}_{\alpha\beta}(u) = u\delta_{\alpha\beta} + \frac{1}{2} s^{ab}_{\alpha\beta} \rho_{ab} \tag{63}$$

where u is the spectral parameter, $s^{ab}_{\alpha\beta} = \delta^a_\alpha \delta^b_\beta - \frac{1}{4}\delta^{ab}\delta_{\alpha\beta}$ are the standard adjoint $su(4)$ generators and $\rho_{ab} = \{P_{i,\mu}, L_{i,\mu\nu}, D_i, K_{i,\mu}\} \in su(2,2)$ is a 4×4 matrix of generators of conformal group in representation $(S_1 = 0, h = 1, S_2 = 0)$, i.e. with zero conformal spins and a unit weight (dimension) corresponding to that of the scalar field. This Lax operator satisfies the Yang–Baxter equations graphically

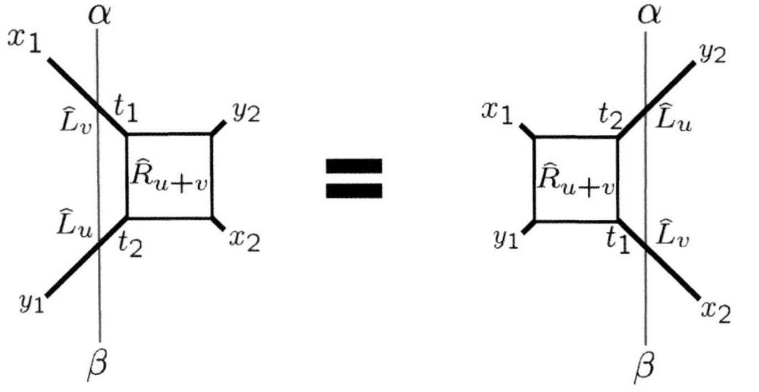

Fig. 15. Yang–Baxter relations for the Lax operators (63), including the intertwiner R-matrix (64) leaving in the product of principal series representations of the conformal group $SU(2,2)$ realized by 4D coordinates x_j, y_j, t_j. Lax operator leaves in the product of principal series representation (thick black lines) and fundamental representation (thin red lines), with indices $\alpha, \beta = 1,2,3,4$. The integrations over coordinates t_1, t_2 are assumed.

represented on Fig. 15, with the intertwining R-matrix living in the direct product of principal series representations and given by the following expression [38]

$$R_u(z_1, x_1|y_1, z_2) = \frac{c(u)}{[(x_1-z_1)^2]^{-u-1}[(x_1-z_2)^2(z_1-y_1)^2]^{u+2}[(y_1-z_2)^2]^{-u+1}}, \tag{64}$$

where $c(u) = \frac{2^{4u}}{\pi^4}\frac{\Gamma^2(u+2)}{\Gamma^2(-u)}$ is a useful normalization factor. This R-matrix can be used to construct a transfer-matrix by taking the trace of their matrix product in auxiliary space (with the 4D variables $z_1, \ldots, z_L$ as labels):

$$\begin{aligned}\hat{T}_L(u) = \int d^4 z_1 \int d^4 z_2 \cdots \int d^4 z_L R_u(z_1, x_1|y_1, z_2)\\ \times R_u(z_2, x_2|y_2, z_3)\cdots R_u(z_L, x_L|y_L, z_1).\end{aligned} \tag{65}$$

This quantity is depicted on Fig. 16 (upper part).

Remarkably, when tuned to a particular value of the spectral parameter $u = -1+\epsilon$, $(\epsilon \to 0)$ this transfer-matrix becomes exactly the graph-building operator (60)! [24] Namely,

$$\begin{aligned}\langle x_1, \ldots, x_L|T_L(-1+\epsilon)|y_1, \ldots, y_L\rangle &= \frac{1}{(16\pi^2\epsilon)^L}\prod_{i=1}^{L}\frac{1}{(y_i-y_{i+1})^2(y_i-x_i)^2}\\ &\sim \langle x_1, \ldots, x_L|\hat{t}_L|y_1, \ldots, y_L\rangle,\end{aligned}$$

which can be also easily seen from the Fig. 16 (lower part).

With the help of Lax operator (63) we can also construct another u-dependent transfer-matrix,

$$\hat{T}_L(u) = \hat{L}_{\alpha_1\alpha_2}(u)\hat{L}_{\alpha_2\alpha_3}(u)\cdots\hat{L}_{\alpha_L\alpha_1}(u). \tag{66}$$

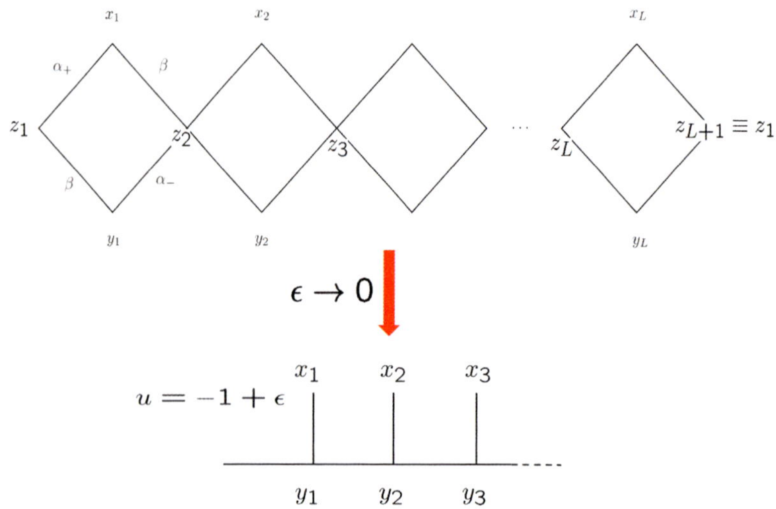

Fig. 16. Graphical demonstration of relation between the R-matrix (64) (upper chain of squares, with integrations over variables z_j) and the graph-building operator (60). The latter one emerges from the former for a particular limit of spectral parameter $u \to -1$, when $\alpha_+ \to 0, \beta, \alpha_- \to 1$, and $\delta^{(4)}(z_i - y_i)$ occur. This demonstrates the fact that the graph-building operator is in involution with all the conserved charges of the integrable conformal $su(2,2)$ spin chain.

Both transfer-matrices represent generating functions of quantum integrals of motion for the conformal, $SU(2,2)$ Heisenberg spin chain. In virtue of the Yang–Baxter relations they should commute:

$$[\hat{\mathcal{T}}_L(u), \hat{T}_L(u')] = 0, \tag{67}$$

and, consequently, they both commute at any u with the graph-building transfer matrix (60):

$$[\hat{T}_L(u), \hat{t}_L] = [\hat{\mathcal{T}}_L(u), \hat{t}_L] = 0. \tag{68}$$

This means that many problems of computation of physical quantities in the bi-scalar theory, such as OPE data, correlation functions, etc., given by fishnet Feynman graphs, can be formulated and studied within the relatively well-developed formalism of integrable non-compact Heisenberg spin chains [32–38]. This promising approach to the study of bi-scalar model and its generalizations, based on the conformal spin chain, is still at its very early stage, though some important observations have been done on this way in [24]. This Lax approach appeared to be very fruitful in application to the scattering amplitudes of the bi-scalar model where the Yangian symmetry has been discovered and explicitly demonstrated [85, 86], though in the original $\mathcal{N} = 4$ SYM this symmetry remains still an open issue.

We will turn now to a more developed QSC approach which is however limited so far to the computation of spectra of local operators.[ag]

[ag] And some non-local operators as well, such as a cusped Wilson loop [20].

5. QSC Solution for Wheel Graphs

In this section, we will consider, as an application of QSC method, the calculation of anomalous dimensions for the vacuum operators $\mathrm{tr}(\phi_1)^L$ in the bi-scalar CFT (52), with a particular stress on the $L = 3$ case. We will give only the main ideas of calculations. The interested reader can find all the details in the original papers [13, 24] where these results have been obtained.[ah]

Our computation will be based on three main ingredients.

- Analyticity properties of $\mathbf{P}$ and $\mathbf{Q}$ functions in the DS limit, as described in the previous section.
- The DS limit of general QSCγ Baxter equation (29) for $\mathbf{Q}$-functions.
- Quantization condition for solutions of Baxter equation as a consequence of RH sewing relations (43).

5.1. *DS Baxter equations*

Let us first notice that, the operators $\mathrm{tr}(\phi_1)^L$ belongs to representation ($J_1 = L$, $J_2 = 0, J_3 = 0$) of $SO(6) \to U(1)^3$ broken R-symmetry group of the full γ-deformed SYM theory. This state, as well as more general multi-magnon operators (57), obey the LR-symmetry and hence the upper index $\mathbf{P}^a$ and $\mathbf{Q}^j$ functions are trivially expressed through lower index $\mathbf{P}_a$ and $\mathbf{Q}_j$ functions the relations (31), thus greatly simplifying the algebraic structure of QSC Baxter equation (29), even before the DS limit.

The asymptotic $1/u$ expansions of $\mathbf{P}_a$ and $\mathbf{Q}_j$ functions follow from the values of Cartan charges for this state: $\{J_1, J_2, J_3|\Delta, S_1, S_2\} = \{L, 0, 0|\Delta, 0, 0\}$. We have from (32), (33), (37), (38)

$$\mathbf{P}_a \sim A_a x_a^{iu} u^{-\lambda_a}(1 + \mathcal{O}(1/u)), \qquad \mathbf{Q}_i \sim B_i u^{-\nu'_i}(1 + \mathcal{O}(1/u)), \tag{69}$$

where

$$\lambda_a = \left\{\frac{L}{2}, \frac{L}{2}, -\frac{L}{2}, -\frac{L}{2}\right\}, \qquad \nu'_i = \left\{-\frac{\Delta}{2}, -1 - \frac{\Delta}{2}, -2 + \frac{\Delta}{2}, -3 + \frac{\Delta}{2}\right\}. \tag{70}$$

The leading coefficients of asymptotics A_a, B_j, as well as the subleading ones, can be fixed, up to the normalization conventions, by plugging them into the QSC Baxter equation (29).

We can also make some precisions on the 't Hooft coupling g dependence of $\mathbf{P}$ functions, using their important analyticity property — the presence of a single short Zhukovsky cut for $x \in (-2g, 2g)$ on its physical sheet. That is why we can uniformize $\mathbf{P}$ functions by expanding them in powers of variable $x(u)$ instead of the spectral parameter u itself. Namely, we can write

$$\mathbf{P}_a(u) = x_a^{iu}(gx(u))^{-\lambda_a}\mathbf{p}_a(u), \tag{71}$$

[ah] Most of the figures of this section are also borrowed from [24].

where

$$\mathbf{p}_a = \{A_1 f_1(u), A_2 f_1(-u), A_3 f_2(u), A_4 f_2(-u)\} \tag{72}$$

and

$$f_1 = 1 + g^{2L} \sum_{n=1}^{\infty} \frac{g^{2n-2} c_{1,n}}{(gx)^n}, \tag{73}$$

$$f_2 = (gx)^{-L} \left(u^L + \sum_{k=0}^{L-1} c_{2,-k} u^k + \sum_{n=1}^{\infty} \frac{g^{2n} c_{2,n}}{(gx)^n} \right), \tag{74}$$

where the coefficients $c_{1,n}(g), c_{2,n}(g)$ are functions of g and γ_j, yet to be defined. We use the Zhukovsky variable which can be also expanded in powers of g/u

$$gx(u) = \frac{1}{2}\left(u + \sqrt{u^2 - 4g^2}\right) = u\left(1 - \frac{g^2}{u^2} - \frac{g^4}{u^4} - \frac{2g^6}{u^6} + O(g^8)\right). \tag{75}$$

The asymptotics (69) are already incorporated into this expansion and we used here natural assumptions about the symmetry of $\mathbf{P}$ functions of the vacuum state $\mathrm{tr}(\phi_1)^L$ w.r.t. the reflection $u \leftrightarrow -u$.

In the DS limit, we take[ai]

$$\kappa = e^{-\frac{i}{2}(\gamma_3+\gamma_2)} \to \infty, \quad \hat{\kappa} = e^{-\frac{i}{2}(\gamma_3-\gamma_2)} \to \infty, \quad g \to 0, \quad (\xi = g\,\kappa \to \text{fixed}), \tag{76}$$

we have to make some natural assumptions about the g-dependence of coefficients $c_{1,n}(g), c_{2,n}(g)$. We will assume that all of them have regular expansion around $g = 0$. Notice that $f_1 = 1 + \mathcal{O}(g^{2L})$, which reflects the fact that the non-trivial, wrapping contributions to the main asymptotics start here from g^{2L} terms [16]. The power g^{2n-2} in each term of expansion in (73) is needed to make each of these terms regular in the weak coupling limit $g \to 0$ in the function $\tilde{\mathbf{P}}_a(u)$ obtained from $\mathbf{P}_a(u)$ by monodromy around the branchpoint. This monodromy $\mathbf{P}_a \to \tilde{\mathbf{P}}_a$ is achieved by simply flipping everywhere $x(u) \to 1/x(u)$, so that, e.g. $\frac{g^{2n}}{(gx)^n} \to \frac{g^{2n}x^n}{g^n} \sim \mathcal{O}(g^0)$. A similar reasoning applies to f_2. Notice that positive powers of u can be always converted to expansion in x using the inverse map $u = g(x + 1/x)$.

Plugging the large u asymptotics of expansion (71)–(74) into the coefficients (30) of QSC Baxter equation we can immediately fix the following relation between the leading coefficients A_a of the asymptotics: [16]

$$A_1 = -A_2 = \frac{\hat{\kappa}^L(\kappa^L - 1)^3}{(1+\kappa^L)(\kappa^L - \hat{\kappa}^L)((\kappa\hat{\kappa})^L - 1)}, \tag{77}$$

$$A_3 = -A_4 = -\frac{\kappa^L(\hat{\kappa}^L - 1)^3}{(1+\hat{\kappa}^L)(\kappa^L - \hat{\kappa}^L)((\kappa\hat{\kappa})^L - 1)}, \tag{78}$$

where we chose the twist parameters as $x_a = \{\kappa^L, \kappa^{-L}, \hat{\kappa}^L, \hat{\kappa}^{-L}\}$. Notice that for the bi-scalar limit $\hat{\kappa} \to \kappa$.[aj]

[ai] There is no γ_1 dependence for this particular state.

[aj] We consider the case $\hat{\kappa} \neq \kappa$ for regularization of certain divergencies. Then the LR-symmetry relation (31) between upper and lower index $\mathbf{P}$'s should be accompanied by the simultaneous exchange $\hat{\kappa} \leftrightarrow \kappa$. [16] At the end we will put them equal.

We can use all this information to compute the coefficients (30) of QSC Baxter equation in the DS limit but for finite u. To this end, we expand the coefficients $c_{1,n}(g), c_{2,n}(g)$, as well as the Zhukovsky variable $gx(u)$, in regular series in g^2:

$$c_{m,n}(g) = c_{m,n}(0) + c'_{m,n}(0)g^2 + \frac{1}{2}c''_{m,n}(0)g^4 + O(g^6) \quad (m = 1, 2), \tag{79}$$

plug these expansions into (73), (74) and use the resulting expansions of f_1, f_2 to compute the coefficients (30) of QSC Baxter equation. Notice that we assume the right DS scaling regime for these functions to be $g \ll u \sim 1$. As we mentioned in formulating the basic "axioms" of QSC, we assumed that the short Zhukovsky cut is the only singularity of $\mathbf{P}$ functions at finite part of the physical sheet. Hence the expansion (75) is the only source of poles at the origin in the DS limit of $\mathbf{P}$ functions.

Now we can perform the DS limit in coefficients (30) by grouping the powers of g with the powers of twist parameters $\hat{\kappa}, \kappa$ (appearing in coefficients A_a in (77) and (78)) into finite DS couplings $\xi = g\kappa, \hat{\xi} = g\hat{\kappa}$ and dropping all the subleading terms $\mathcal{O}(g, 1/\kappa, 1/\hat{\kappa})$. Remarkably, no terms which blow up in this limit occur during this DS procedure, which perfectly confirms our assumptions for the ansatz (73), (74). Of course, only a finite number of expansion coefficients is retained, though their number increases with the length of operator L, as is obvious from the form of the leading coefficients (77) and (78).[ak] Another positive sign is that all the coefficients (30) of QSC Baxter equation appear to be of the same order in g and hence we have at the end a perfectly defined Baxter equation in DS limit for the "vacuum" state, valid for the full chiral CFT (51). If we want to limit ourselves to the bi-scalar case we simply put $\hat{\xi} = \xi$ which appears to be a smooth limit in the DS Baxter equation.

The resulting Baxter equation, for slightly modified definition of $\mathbf{Q}$-function

$$\mathbf{Q}_j = u^{L/2} q_j, \tag{80}$$

takes a rather symmetric form [24]

$$\begin{aligned} A(u+i)q(u+2i) - B\left(u + \frac{i}{2}\right)q(u+i) + C(u)q(u) \\ - B\left(u - \frac{i}{2}\right)q(u-i) + A(u-i)q(u-2i) = 0, \end{aligned} \tag{81}$$

where for $L = 3$

$$A(u) = u^3, \qquad B(u) = u\left(4u^2 - \frac{\alpha+5}{2}\right), \tag{82}$$

$$C(u) = 6u^3 - (\alpha+5)u + \frac{(\alpha-1)^2}{16u} + \frac{-m^2}{u^3} \tag{83}$$

[ak]Practically, extracting the DS limit from such expansions was done by using Mathematica program. Increasing L can be rather time-consuming, so that we have done the actual calculations only for $L = 2, 3, 4$. [24].

and $\alpha = (\Delta - 2)^2$. The higher conserved charge $m(\xi)$, as well as the value of dimension $\Delta(\xi)$ for this state — our main goal — will be fixed from the auxiliary quantization condition following essentially from the QSC Riemann–Hilbert sewing conditions (41).

The Baxter equation (81) is very suggestive as concerns already mentioned direct relation of the current problem to the periodic $SU(2,2)$ Heisenberg spin chain described in [38] (see Appendix A of [24]). One can even justify its form (81), with $A(u) = u^L$, which is completely fixed by the choice of spin representation $\{0,1,0\}$, whereas $B(u)$ and $u^L C(u)$ are polynomials in u of degree L and $2L$, respectively, obeying the symmetry (proper to this state): $B(u) = (-1)^L B(-u), u^L C(u) = (-u)^L C(-u)$. This information, as well as the asymptotic behavior of q functions:

$$q \sim u^\delta, \qquad \delta = \left\{\frac{\Delta - L}{2}, \frac{\Delta - L}{2} + 1, 2 - \frac{\Delta + L}{2}, 3 - \frac{\Delta + L}{2}\right\} \tag{84}$$

following from (69) partially fix the coefficients of (81) at any L [24]:

$$B(u) = 4u^L - \frac{1}{2}(\alpha + 3L - 4)u^{L-2} + bu^{L-4} + \sum_{k=3}^{[L/2]} d_k u^{L-2k}, \tag{85}$$

$$C(u) = 6u^L - (\alpha + 3L - 4)u^{L-2} + \frac{(\alpha - 4)^2 + 32b + 3L^2 + 2(\alpha - 7)L}{16} u^{L-4}$$
$$+ \sum_{k=3}^{L} c_k u^{L-2k}, \tag{86}$$

where $\alpha = (\Delta - 2)^2$. These expressions depend on $1 + (L-2) + ([L/2] - 2) = L + [L/2] - 3$ arbitrary constants b, c_k, d_k, to be fixed by additional, yet to be derived, quantization conditions.

5.2. *DS quantization condition*

The Baxter equation (81) has a few unfixed coefficients, including the main quantity under study — the dimension $\Delta(\xi)$. In addition, any linear combination, with i-periodic coefficients, of 4 independent solutions of this equation is again a solution. We have to find such a set of 4 solutions, corresponding to $q_i = u^{-L/2} Q_i$ that they are pure functions. But even the condition of purity does not fix them completely. We have to find an additional condition which fixes the solutions, as well as the yet unfixed coefficients in (85), completely. Such a quantization condition should be based on the RH sewing relations (41) which we did not use so far.

Let us now concentrate on the case $L = 3$ for the bi-scalar model. First of all, we notice that the 4th order Baxter equation (81) can be factorized in this case to two 2nd order equations: $(\hat{L}_+ u^3 \hat{L}_-)q \equiv (\hat{L}_- u^3 \hat{L}_+)q = 0$, where

$$\hat{L}_\pm q(u) \equiv \left(\frac{(\Delta - 1)(\Delta - 3)}{4u^2} \pm \frac{m}{u^3} - 2\right) q(u) + q(u+i) + q(u-i) = 0. \tag{87}$$

The asymptotics (70) suggest that the pure solutions of the first of these equations $\hat{L}_- q(u,m)=0$ are

$$q_2(u,m) = u^{\Delta/2-1/2}\left(1+\frac{a_1}{u}+O\left(\frac{1}{u^2}\right)\right), \tag{88}$$

$$q_4(u,m) = u^{-\Delta/2+3/2}\left(1+\frac{b_1}{u}+O\left(\frac{1}{u^2}\right)\right). \tag{89}$$

Then $q_1(u,m)=q_2(u,-m)$, $q_3(u,m)=q_4(u,-m)$.

From solutions to the Baxter equation (87), we can now construct four $\mathbf{Q}$-functions. In the double scaling limit they are linear combinations of q functions

$$\begin{aligned}
\mathbf{Q}_1(u) &= \frac{-is^6}{2m(\Delta-2)}u^{3/2}[q_2(u,m)-q_2(u,-m)],\\
\mathbf{Q}_2(u) &= \frac{u^{3/2}}{2}[q_2(u,m)+q_2(u,-m)],\\
\mathbf{Q}_3(u) &= \frac{is^6}{2m(\Delta-2)}u^{3/2}[q_4(u,m)-q_4(u,-m)],\\
\mathbf{Q}_4(u) &= \frac{u^{3/2}}{2}[q_4(u,m)+q_4(u,-m)],
\end{aligned} \tag{90}$$

where the coefficients on the left are chosen as a normalization and the other two are fixed from compatibility with the QSC Baxter equation (29). These functions obey a fixed parity w.r.t. $u\to -u$, as assumed in the initial ansatz, since the Baxter equations are invariant w.r.t. the simultaneous change $u\to -u, m\to -m$. Notice that this also means that

$$\mathbf{Q}_j(u)\sim \mathbf{Q}_j(-u) \tag{91}$$

since we will see below that for physical solutions the constant m is purely imaginary.

The RH conditions (41), together with the LR-symmetry (31), lead to an additional constraint $\beta_1 = 1/\bar{\beta}_2$ and thus to the following analyticity constraints

$$\tilde{\mathbf{Q}}_1(u)=\bar{\beta}_1\bar{\mathbf{Q}}_3(u), \quad \tilde{\mathbf{Q}}_2(u) = -\beta_1\bar{\mathbf{Q}}_4(u). \tag{92}$$

Notice that if we take the argument at one of the branch-points $u^*=\pm 2g$ then we have an obvious equality $\tilde{\mathbf{Q}}_j(u^*)\sim \mathbf{Q}_j(u^*)$ since, by assumption, the only singularities of $\mathbf{Q}_j$ at finite u are the Zhukovsky cuts. That means that keeping $u=u^*$ we can extract from (92), (88), (89) two expressions for β_1:

$$\beta_1 = \frac{\mathbf{Q}_1(u^*)}{\mathbf{Q}_3(-u^*)} = -\frac{\mathbf{Q}_2(u^*)}{\mathbf{Q}_4(-u^*)}. \tag{93}$$

Since in the DS limit $u^*=\pm 2g\to 0$, plugging here (90) we obtain the final quantization condition[al]:

[al]I thank N. Gromov for sharing with me this and the next shortcuts for deriving the quantization condition and the formula for m^2. See more rigorous original derivation in our paper [24].

$$q_2(0,-m)q_4(0,m) + q_2(0,m)q_4(0,-m) = 0 \tag{94}$$

which fixes unambiguously the physical solution of Baxter equation (87).[am]

Recall that the $\mathbf{Q}$ functions, by the assumptions of QSC formalism, can only have the short cuts in the lower half-plane, starting from the real axis. In the DS limit, these cuts can give only the poles at $u = -in$, $n = 0, 1, 2, \ldots$. Inspecting Eq. (87) around these poles we realize that, first of all, they cannot be of the order higher than $\frac{1}{u^3}$, and second, $q_2(u,m), q_4(u,m)$ are regular functions at $u = 0$.[an] Similar argument applies to $\bar{\mathbf{Q}}$, except that the poles can be now found only in the upper half-plane, and hence $q_2(u,-m), q_4(u,-m)$ are also regular at $u = 0$.

This quantization condition suffices to establish the relation between Δ and m. However, we do not know yet how any of them are related to the coupling constant ξ. The dependence of m on ξ was derived in [24] from the QSC formalism and appeared to be very simple

$$m^2 = -\xi^6 \tag{95}$$

implying, as we already mentioned, that m is imaginary. We will not reproduce here these arguments but we note that it can be obtained entirely within the $SU(2,2)$ Heisenberg spin chain formalism [24, 38] by showing that the coefficient m^2, an eigenvalue of the conserved charge $\hat{m}^2$, can be directly related to the graph-building hamiltonian $\hat{h}_3$ defined by (60), namely

$$\hat{m}^2 = -\xi^6(\hat{h}_3)^{-1}. \tag{96}$$

Then we notice that the energy of this state should be defined by the position of the pole in the expansion (62), namely at $\hat{h}_3 \to 1$. For the eigenvalue of $\hat{m}$ this means precisely the relation (95).[ao]

In conclusion, we obtained the following formulation of solution of the problem of computation of anomalous dimension of the operator $\mathrm{tr}(\phi_1)^3$ in the bi-scalar model: to fix $\Delta(\xi)$, we find a pair of pure solutions of Baxter equation (81) (choosing the lower sign "−" there and fixing $m = \pm i\xi^3$) satisfying the quantization condition (94), the asymptotic expansions (88) and (89) and having the poles up to the third order for $u = -in, n = 1, 2, 3, \ldots$. This will fix a discrete set of dimensions $\Delta(\xi)$, the lowest of them corresponding to the operator $\mathrm{tr}(\phi_1)^3$.

5.3. *Some results for $J = 3$ spectrum*

Let us briefly present some results of the analysis, perturbative and numerical, of equations obtained in the previous section.

[am]This relation implies the cancellation of the proportionality constants appearing in (91). These constants can be computed by following the asymptotic behavior of functions in (90) when going from $u \to \infty$ to $u \to -\infty$ along a big semicircle in the upper half-plane.

[an]A pole at $u = 0$ would immediately produce a pole at $u = i$, which is absent by assumption.

[ao]The last argument will appear in the forthcoming work [87].

Weak coupling solution for the operator tr$(\phi_1)^3$: At zero order we have $m = -i\xi^3 = 0$ and $\Delta(0) = 3$. We find thus a pair of independent solutions of (81): $q_I(u,0) = u, q_{II}(u,0) = 1$. Expanding $\Delta = L + \sum_k \xi^{6k}\Delta^{(k)}$ and, for each function, $q(u,m) = w_1(u) + mw_2(u) + m^2w_3(u) + \cdots$ and solving (81) iteratively by means of varying coefficients, we can find the next term of $m = \pm i\xi^3$ expansion with the right pole structure:

$$\begin{aligned} q_I &= u - im(\eta_1 - \eta_2 u) \\ &\quad + m^2\left(-\eta_{1,2} + \eta_{2,1} + u\eta_{1,3} - u\eta_{2,2} - \frac{i\delta}{2} + \frac{1}{2}i\eta_1 u\delta + \frac{u\delta}{2}\right) + \mathcal{O}(m^3), \\ q_{II} &= 1 - im(\eta_2 - \eta_3 u) \\ &\quad + m^2\left(-\eta_{2,2} + \eta_{3,1} + u\eta_{2,3} - u\eta_{3,2} - \frac{1}{2}i\eta_1\delta + \frac{1}{2}i\eta_2 u\delta\right) + \mathcal{O}(m^3), \end{aligned}$$

where $\Delta = 3 - m^2\delta + O(m^4)$ and we introduced a standard set of functions [88]

$$\eta_{s_1,\ldots,s_k}(u) = \sum_{n_1>n_2>\cdots>n_k\geq 0} \frac{1}{(u+in_1)^{s_1}\cdots(u+in_k)^{s_k}}. \tag{97}$$

Using these two q-functions we can find, order by order, the right linear combinations of them corresponding to two pure solutions q_2 and q_4 which fit the asymptotics (88) and (89). Leaving aside the details, which can be found in [24], we give here the result for the dimension of tr$(\phi_1)^3$ up to 12 loops:

$$\begin{aligned} \Delta_3 - 3 &= -12\zeta_3\xi^6 + \xi^{12}(189\zeta_7 - 144\zeta_3{}^2) + \xi^{18}\Big(-1944\zeta_{8,2,1} - 3024\zeta_3{}^3 - 3024\zeta_5\zeta_3{}^2 \\ &\quad + 6804\zeta_7\zeta_3 + \frac{198\pi^8\zeta_3}{175} + \frac{612\pi^6\zeta_5}{35} + 270\pi^4\zeta_7 + 5994\pi^2\zeta_9 - \frac{925911\zeta_{11}}{8}\Big) \\ &\quad + \xi^{24}\Big(-93312\zeta_3\zeta_{8,2,1} + \frac{10368}{5}\pi^4\zeta_{8,2,1} + 5184\pi^2\zeta_{9,3,1} + 51840\pi^2\zeta_{10,2,1} \\ &\quad - 148716\zeta_{11,3,1} - 1061910\zeta_{12,2,1} + 62208\zeta_{10,2,1,1,1} - 77760\zeta_3{}^4 \\ &\quad - 145152\zeta_5\zeta_3{}^3 - \frac{576}{7}\pi^6\zeta_3{}^3 - 864\pi^4\zeta_5\zeta_3{}^2 - 2592\pi^2\zeta_7\zeta_3{}^2 + 244944\zeta_7\zeta_3{}^2 \\ &\quad + 186588\zeta_9\zeta_3{}^2 + \frac{9504}{175}\pi^8\zeta_3{}^2 - 2592\pi^2\zeta_5{}^2\zeta_3 + \frac{29376}{35}\pi^6\zeta_5\zeta_3 + 298404\zeta_5\zeta_7\zeta_3 \\ &\quad + 12960\pi^4\zeta_7\zeta_3 + 287712\pi^2\zeta_9\zeta_3 - 5555466\zeta_{11}\zeta_3 + \frac{2910394\pi^{12}\zeta_3}{2627625} \\ &\quad + 57672\zeta_5{}^3 - 71442\zeta_7{}^2 + \frac{13953\pi^{10}\zeta_5}{1925} + \frac{7293\pi^8\zeta_7}{175} - \frac{19959\pi^6\zeta_9}{5} \\ &\quad + \frac{119979\pi^4\zeta_{11}}{2} + \frac{10738413\pi^2\zeta_{13}}{2} - \frac{4607294013\zeta_{15}}{80}\Big) + O(\xi^{30}), \end{aligned} \tag{98}$$

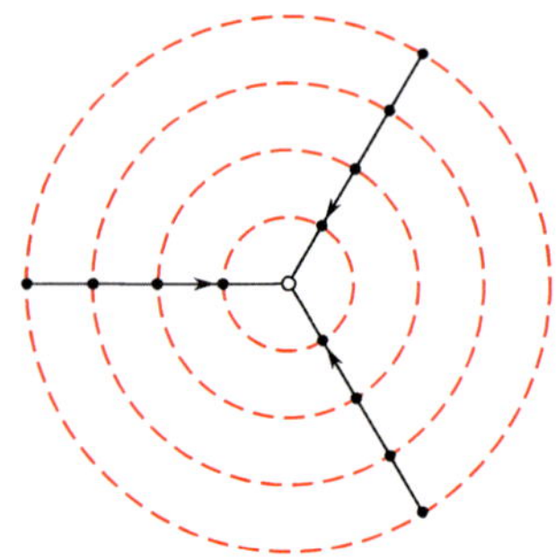

Fig. 17. The "wheel" Feynman graph corresponding to $O(\xi^{24})$ term in the weak coupling expansion (98) of anomalous dimension of the operator $\mathrm{Tr}(\phi_1^3)$.

where $\zeta_{i_1,\ldots,i_k} = \sum_{n_1>\cdots>n_k>0} 1/(n_1^{i_1}\cdots n_k^{i_k})$ are multiple Riemann ζ-values. Here the coefficients in front of ξ^{6M} give the residues at simple pole $1/\epsilon$ in dimensionally regularized Feynman integrals corresponding to the $L=3$ wheel graphs with $M = 1,2,3,4,\ldots$ frames (see Fig. 17). The first two terms of (98) (one and two wrappings) coincide with the known results [89, 90].

Numerical solution: The equations of the previous subsections can be also solved, very efficiently and with virtually unlimited accuracy, numerically at all interesting finite values of the coupling, using the methods developed in [20]. The results for the dimension $\Delta_3(\xi)$ are presented on Fig. 18. This dimension is real for sufficiently small values of ξ, but it becomes imaginary starting from a certain value $\xi^3 \simeq 0.21$, which is not an abnormal behavior for a non-unitary theory, such as our bi-scalar model.

Higher twist solutions and Jordan cells for mixing matrix: Actually, our Baxter equation (81), together with the quantization conditions (94), describes

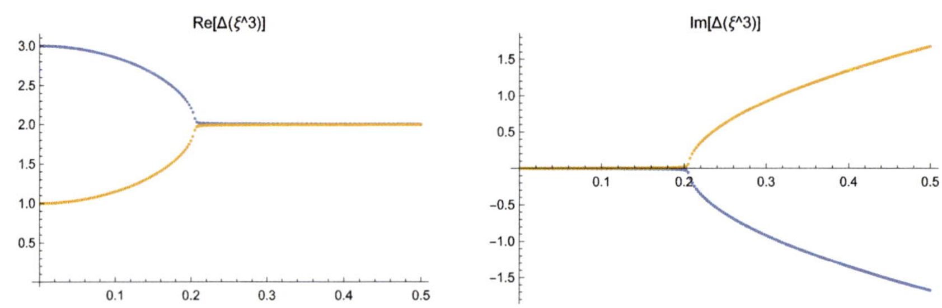

Fig. 18. Numerical results for the scaling dimension of the operator $\mathrm{tr}(\phi_1^3)$ as a function of the coupling ξ^3. At $\xi^3 \simeq 0.21$ the scaling dimension hits the value $\Delta = 2$ and becomes imaginary. This point defines the radius of convergency of the weak coupling expansion. The second branch, starting from $\Delta(0) = 1$, arises due to the symmetry of the Baxter equation (81) under $\Delta \to 4-\Delta$. It corresponds to a non-local, "shadow" operator.

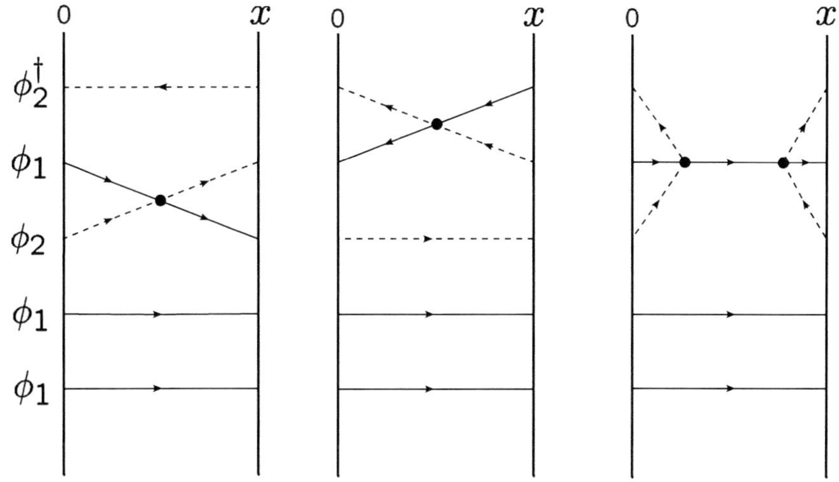

Fig. 19. The lowest order Feynman graphs contributing to the mixing matrix (101) of operators of charge $J_1 = 3, J_2 = J_3 = 0$ and the length $L = 5$. Bringing this matrix to the Jordan form we encounter two complex conjugate anomalous dimensions (99) as well a (2×2)-Jordan cell leading to logarithmic conformal correlator (102), typical for non-unitary CFTs.

not only the operator $\mathrm{tr}(\phi_1)^3$ but also all operators with the same R-charge $J_1 = L, J_2 = J_3 = 0$. They can be represented as linear combinations of operators of length $= 3+2n$, of the type: $\mathrm{tr}[(\phi_1)^3(\phi_2^\dagger\phi_2)^n]$ with all possible permutations of fields there.[ap] The weak coupling expansion, similar to the described above, leads to the following, complex conjugate values of two dimensions of length-5 operators:

$$\Delta_5^{\pm} = 5 \mp 2i\xi^3 + 3\xi^6 \pm \frac{31i\xi^9}{4} + \xi^{12}\left(3\zeta_3 - \frac{97}{4}\right) \pm i\xi^{15}\left(\frac{27\zeta_3}{2} - \frac{5359}{64}\right) + \cdots. \quad (99)$$

In fact, they correspond to a multiplet formed by four operators

$$\begin{aligned} O_1 &= \mathrm{tr}(\phi_1^3\phi_2\phi_2^\dagger), & O_2 &= \mathrm{tr}(\phi_1^2\phi_2\phi_1\phi_2^\dagger), \\ O_3 &= \mathrm{tr}(\phi_1\phi_2\phi_1^2\phi_2^\dagger), & O_4 &= \mathrm{tr}(\phi_2\phi_1^3\phi_2^\dagger). \end{aligned} \quad (100)$$

Their mixing matrix is not Hermitian, reflecting the non-unitarity of the theory. Computing it in the lowest order of perturbation theory by means of Feynman graphs presented on Fig. 19 and bringing it to Jordan form we obtain

$$V = \frac{1}{4\pi^2}\begin{bmatrix} 0 & \xi^2 & 0 & 0 \\ 0 & 0 & \xi^2 & 0 \\ 0 & -\xi^4 & 0 & \xi^2 \\ 0 & 0 & 0 & 0 \end{bmatrix} = U\cdot\begin{pmatrix} 0 & 1 & 0 & 0 \\ 0 & 0 & 0 & 0 \\ 0 & 0 & -i\xi^3 & 0 \\ 0 & 0 & 0 & i\xi^3 \end{pmatrix}\cdot U^{-1}. \quad (101)$$

[ap]The operators with insertions of $\phi_1^\dagger$ appear to be protected.

Plugging it into Callan–Symanzik equation $\mu\frac{d}{d\mu}O_i(x) = -V_{ij}O_j(x)$ and solving it for pair correlation functions we obtain a correlation matrix consisting of two standard conformal correlation functions on the diagonal, corresponding to the dimensions (99) (at two lowest orders), as well as a Jordan block of correlators with the logarithmic behavior characteristic for the non-unitary CFTs[aq]:

$$\langle \tilde{\mathcal{O}}^\dagger_\alpha(x)\tilde{\mathcal{O}}_\beta(0)\rangle_{\text{ren}} = \frac{1}{x^{10}}\begin{bmatrix} 0 & 1 \\ 1 & \log x^2\mu^2 \end{bmatrix}_{\alpha\beta}. \tag{102}$$

We also performed in [24] numerical calculations for this and a few higher multiplets, revealing a rich structure of the spectrum of this conformal CFT (see Fig. 20).

The behavior of these dimensions was also analyzed in [24] analytically in the strong coupling limit $\xi \to \infty$. The results suggested an interesting interpretation of the system in terms of the classical dynamics of three non-compact spins. It is an interesting step in the direction of understanding whether the bi-scalar theory has a string dual description. Probably for a better understanding of this problem one has to analyze the operators of the type $\text{tr}(\phi_1)^L$ at $L \to \infty$, in analogy with the Frolov–Tseytlin limit in standard $\mathcal{N} = 4$ SYM theory.

6. Prospects and Unsolved Problems

This review has two main purposes. Firstly, to give the most general, albeit minimalist formulation of the quantum spectral curve (QSC) formalism for the spectrum of dimensions of local operators of the γ-deformed $\mathcal{N} = 4$ SYM theory. Secondly, to demonstrate the power of QSC on the example of analysis of the spectrum of certain operators in the bi-scalar CFT, dominated by "fishnet" graphs in planar limit, emerging from the γ-deformed $\mathcal{N} = 4$ in a certain double scaling limit combining weak coupling and strong imaginary twist.

The QSC formalism already permitted to obtain outstanding new results in $\mathcal{N} = 4$ SYM theory (see references in introduction and in the review [29]). The QSC construction is also known for the ABJM model [18, 92, 93] and even for the Hubbard model [94]. The QSC formalism is designed first of all for the spectral problem for local operators. However, with appropriate modifications it was also used for computing the dimensions of non-local quantities, such as the cusped Wilson loop, the quark–anti-quark potential or the BFKL limit for twist-2 operators in planar $\mathcal{N} = 4$ SYM [20, 22, 23, 69]. So far, the generalizations of QSC to more complicated OPE data, such as the structure constants, is not known, though the underlying physical quantities — the three-point correlation functions — obey remarkable integrability properties [95, 96]. Another desirable generalization of QSC would be the $1/N_c^2$ corrections where the integrability seems to be also helpful [97]. The QSC is also already constructed for the other deformations of SYM related to quantum groups [25].

[aq]The presence of such logarithmic correlation functions in the bi-scalar theory, omnipresent in non-unitary CFTs [91], was first noticed by J. Caetano (unpublished).

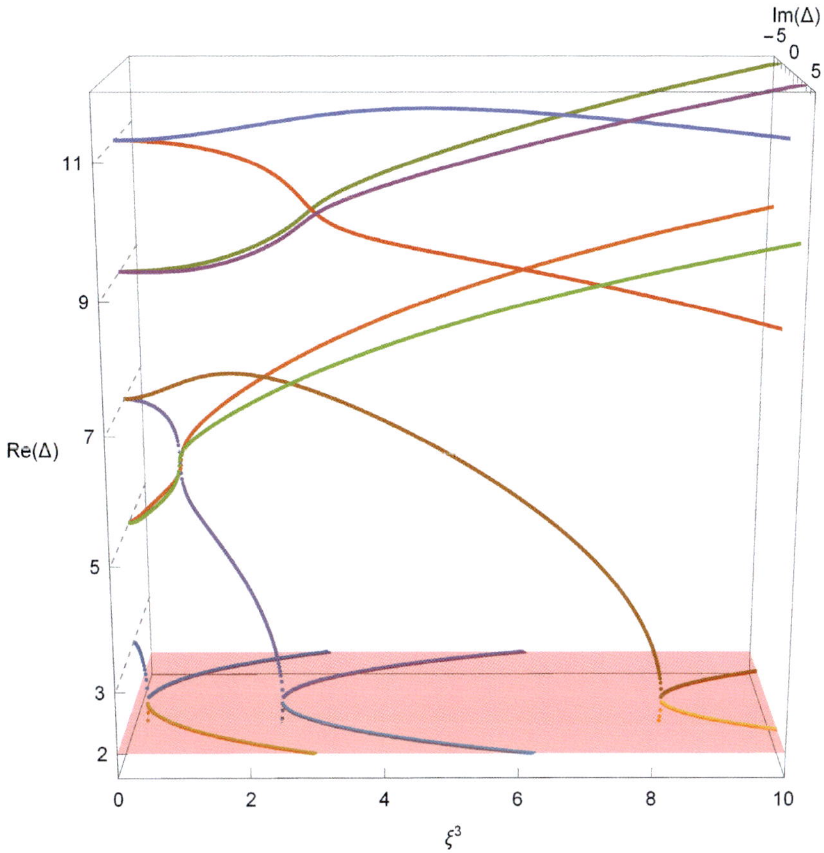

Fig. 20. Real and imaginary part of the scaling dimension of the nine lowest lying states with $L = 3$. The curve that starts at $\Delta(0) = 3$ corresponds to the operator $\mathrm{tr}(\phi_1^3)$. The pair of states that start at $\Delta(0) = 3 + 2k$ with $k = 1, 2, 3, 4$ correspond to the operators of the form $\mathrm{tr}[(\phi_1)^3(\phi_2^\dagger\phi_2)^k]$ + permutations.

The bi-scalar CFT (52) and its generalizations (51) [30, 98], obtained from γ-deformed $\mathcal{N} = 4$ SYM theory in special double scaling (DS) limit, apart from being new interesting examples of *integrable planar 4-dimensional CFTs*, also play an important conceptual role. Namely, they are dominated by very particular sets of *integrable* planar Feynman graphs, such as "fishnet" graphs (of the shape of regular square lattice in the bulk of graph) of bi-scalar model, or the "brick wall" graphs of the case $\xi_3 = 0$, $\xi_1 = \xi_2 \neq 0$ of the model (51) formed by regular hexagonal lattice of Yukawa-type vertices [40, 86]. Thus these CFTs explicitly demonstrate, for the first time, the all-loop integrability of the original $\mathcal{N} = 4$ SYM theory, at least in this specific DS limit and it might be the key of understanding of the origins of the full AdS_5/CFT_4 integrability and of the gauge-string duality. Similar considerations are applicable to the 3-dimensional regular triangular planar graphs emerging from the ABJM theory [98]. A similar chiral CFT dominated by hexagonal graphs can be constructed in six dimensions [99] but, curiously, its 6D "mother" SYM theory is unknown. It is also worth asking whether a similar chiral CFT dominated by regular planar graphs could be found in two dimensions and whether it could be an analogous DS limit of the twisted AdS_3/CFT_2 duality.

The chiral CFT's emerging in the double scaling are dominated by very few graphs at each order (sometimes only a single one, such as the "wheel" (Fig. 11) and "spiral" (Fig. 12) graphs for the vacuum and one magnon operators of bi-scalar CFT). This opens an opportunity to compute these graphs exactly, at any number of loops, as we demonstrated it here for the wheel graphs. Another interesting case is the fishnet amplitudes defined and studied in [85, 86]: each of them is dominated by a single fishnet diagram with specific boundary — a disc cut out from a piece of regular square lattice. Some particular graphs of this kind have been recently computed [100, 101] and it would be interesting to understand and generalize these results from the point of view of the Yangian symmetry discovered in [85, 86].

The question of existence of a string dual for the double scaling limit considered here remains open. Naively, the classical string picture is gone since in the weak coupling limit the AdS radius goes to zero. On the other hand, we deal with multi-loop Feynman graphs which might provide, at high loop orders, at least for long operators, a new dual string description. The results of [24] in the strong coupling limit, already with respect to the DS coupling ξ, are encouraging in this sense since they show that the bi-scalar model can be described by a classical model of a few non-compact spins. If we increase the R-charge of operator, and thus the number of spins, we could reach a classical string picture. The study of wheel graphs of higher lengths, certainly possible by integrability, would be an important step in this direction.

Many physical quantities which seem prohibitively difficult to compute in the full $\mathcal{N}=4$ SYM theory appear to be accessible in the DS limit. In particular, the exact dimension $\Delta_2(\xi, S)$ of $L=2$ wheel operator and of similar operators with conformal spin S, of the form $\text{tr}(\phi_1 \partial_\pm^S \phi_1)$, was computed in [13] in explicit form and is given in very simple explicit form, as solutions of

$$\frac{1}{16}(\Delta+S-2)(\Delta+S)(\Delta-S-2)(\Delta-S-4)=\xi^4. \tag{103}$$

Moreover, an exact four-point function for specific scalar operators, the $L=2$ case of the correlator (62), given by the cylindric graphs of the type drawn on Fig. 14, was explicitly computed in [13] in all loops — the only example, to our knowledge, of explicit all-loop calculation of a non-trivial four-point function in a 4D CFT.[ar] The formula (103) was obtained in [13] by analyzing the divergencies of this four-point correlation function, as well as from the QSC formalism. This four-point correlation

[ar] A similar computation was done in the D-dimensional analogue of the bi-scalar fishnet model, as proposed in [102]. Its 2D version looks particularly interesting since it is directly related to the $SU(2,C)$ spin chain describing a 2D CFT for BFKL physics [35, 36, 103]. Its 1D version describes, in the case of a similar 4-point correlation function, the summation of ladder graphs appearing in the scalar sSYK model [104].

function appears to have a nice OPE expansion generating infinitely many exact structure constants involving the above operators of $L = 2$.

The computation of more complicated correlators, involving longer operators and multi-point correlators, is a complicated but very promising enterprise. To do it efficiently, we have to learn how to efficiently taylor such quantities from the integrable conformal spin chains, in the spirit of the one-loop procedure of [105].

Acknowledgments

We thank V. Bazhanov, B. Basso, D. Grabner, N. Gromov, J. Caetano, S. Derkachov, G. Korchemsky, I. Kostov, S. Leurent, F. Levkovich-Maslyuk, E. Olivucci, D. Serban, Z. Tsuboi, D. Volin, D. Zhong for reading the manuscript and providing very useful comments. The work was supported by the European Research Council (Programme "Ideas" ERC-2012-AdG 320769 "AdS-CFT-solvable").

References

[1] R. G. Leigh and M. J. Strassler, Exactly marginal operators and duality in four-dimensional $N = 1$ supersymmetric gauge theory, *Nucl. Phys. B* **447** (1995) 95–136.
[2] C. Cordova, T. T. Dumitrescu and K. Intriligator, Multiplets of superconformal symmetry in diverse dimensions, arXiv:1612.00809; Deformations of superconformal theories, *JHEP* **11** (2016) 135; arXiv:1602.01217 [hep-th].
[3] T. Banks and A. Zaks, On the phase structure of vector-like gauge theories with massless fermions, *Nucl. Phys. B* **196** (1982) 189–204.
[4] R. Vasseur and J. L. Jacobsen, Operator content of the critical Potts model in d dimensions and logarithmic correlations, *Nucl. Phys. B* **880** (2014) 435–475; arXiv:1311.6143 [cond-mat.stat-mech].
[5] N. Beisert *et al.*, Review of AdS/CFT integrability: An overview, *Lett. Math. Phys.* **99** (2012) 3–32; arXiv:1012.3982 [hep-th].
[6] O. Lunin and J. M. Maldacena, Deforming field theories with $U(1) \times U(1)$ global symmetry and their gravity duals, *JHEP* **0505** (2005) 033; arXiv:hep-th/0502086 [hep-th].
[7] S. Frolov, Lax pair for strings in Lunin–Maldacena background, *JHEP* **05** (2005) 069; arXiv:hep-th/0503201 [hep-th].
[8] J. Fokken, C. Sieg and M. Wilhelm, Non-conformality of γ_i-deformed $N = 4$ SYM theory, *J. Phys. A* **47** (2014) 455401; arXiv:1308.4420 [hep-th].
[9] E. Witten, Multitrace operators, boundary conditions, and AdS/CFT correspondence (2001); arXiv:hep-th/0112258 [hep-th].
[10] A. Dymarsky, I. R. Klebanov and R. Roiban, Perturbative search for fixed lines in large N gauge theories, *JHEP* **08** (2005) 011; arXiv:hep-th/0505099 [hep-th].
[11] A. A. Tseytlin and K. Zarembo, Effective potential in nonsupersymmetric $SU(N) \times SU(N)$ gauge theory and interactions of type 0 D3-branes, *Phys. Lett. B* **457** (1999) 77–86; arXiv:hep-th/9902095 [hep-th].
[12] C. Sieg and M. Wilhelm, On a CFT limit of planar γ_i-deformed $\mathcal{N} = 4$ SYM theory, *Phys. Lett. B* **756** (2016) 118–120; arXiv:1602.05817 [hep-th].
[13] D. Grabner, N. Gromov, V. Kazakov and G. Korchemsky, Strongly γ-deformed $N = 4$ SYM as an integrable CFT (2017); arXiv:1711.04786 [hep-th].

[14] N. Gromov, V. Kazakov, S. Leurent and D. Volin, Quantum spectral curve for planar $\mathcal{N}$ = super-Yang–Mills theory, *Phys. Rev. Lett.* **112**(1) (2014) 011602.
[15] N. Gromov, V. Kazakov, S. Leurent and D. Volin, Quantum spectral curve for arbitrary state/operator in AdS_5/CFT_4, *JHEP* **09** (2015) 187; arXiv:1405.4857 [hep-th].
[16] V. Kazakov, S. Leurent and D. Volin, T-system on T-hook: Grassmannian solution and twisted quantum spectral curve (2015); arXiv:1510.02100 [hep-th].
[17] D. Bombardelli, A. Cavaglià, D. Fioravanti, N. Gromov and R. Tateo, The full quantum spectral curve for AdS_4/CFT_3, *JHEP* **09** (2017) 140; arXiv:1701.00473 [hep-th].
[18] C. Andrea, D. Fioravanti, N. Gromov and R. Tateo, Quantum spectral curve of the $\mathcal{N} = 6$ supersymmetric Chern–Simons theory, *Phys. Rev. Lett.* **113**(2) (2014) 021601; arXiv:1403.1859 [hep-th].
[19] A. Cavaglia, N. Gromov and F. Levkovich-Maslyuk, On the exact interpolating function in ABJ theory, *JHEP* **12** (2016) 086; arXiv:1605.04888 [hep-th].
[20] N. Gromov and F. Levkovich-Maslyuk, Quantum spectral curve for a cusped Wilson line in $\mathcal{N} = 4$ SYM, *JHEP* **04** (2016) 134; arXiv:1510.02098 [hep-th].
[21] N. Gromov, F. Levkovich-Maslyuk and G. Sizov, Pomeron eigenvalue at three loops in $\mathcal{N} = 4$ supersymmetric Yang–Mills theory, *Phys. Rev. Lett.* **115**(25) (2015) 251601; arXiv:1507.04010 [hep-th].
[22] N. Gromov, F. Levkovich-Maslyuk and G. Sizov, Quantum spectral curve and the numerical solution of the spectral problem in AdS_5/CFT_4, *JHEP* **06** (2016) 036; arXiv:1504.06640 [hep-th].
[23] N. Gromov and F. Levkovich-Maslyuk, Quark–anti-quark potential in $\mathcal{N} = 4$ SYM, *JHEP* **12** (2016) 122; arXiv:1601.05679 [hep-th].
[24] N. Gromov, V. Kazakov, G. Korchemsky, S. Negro and G. Sizov, Integrability of conformal fishnet theory, arXiv:1706.04167 [hep-th].
[25] R. Klabbers and S. J. van Tongeren, Quantum spectral curve for the eta-deformed $AdS_5 \times S^5$ superstring, *Nucl. Phys. B* **925** (2017) 252–318; arXiv:1708.02894 [hep-th].
[26] C. Marboe and D. Volin, Quantum spectral curve as a tool for a perturbative quantum field theory, *Nucl. Phys. B* **899** (2015) 810–847; arXiv:1411.4758 [hep-th].
[27] C. Marboe, V. Velizhanin and D. Volin, Six-loop anomalous dimension of twist-two operators in planar $\mathcal{N} = 4$ SYM theory, *JHEP* **07** (2015) 084; arXiv:1412.4762 [hep-th].
[28] C. Marboe and D. Volin, The full spectrum of AdS5/CFT4 I: Representation theory and one-loop Q-system (2017); arXiv:1701.03704 [hep-th].
[29] N. Gromov, Introduction to the spectrum of $N = 4$ SYM and the quantum spectral curve (2017); arXiv:1708.03648 [hep-th].
[30] O. Gurdogan and V. Kazakov, New integrable 4D quantum field theories from strongly deformed planar $\mathcal{N} = 4$ supersymmetric Yang–Mills theory, *Phys. Rev. Lett.* **117**(20) (2016) 201602; Addendum, *ibid.* **117**(25) (2016) 259903; arXiv:1512.06704 [hep-th].
[31] A. B. Zamolodchikov, 'Fishnet' diagrams as a completely integrable system, *Phys. Lett. B* **97** (1980) 63–66.
[32] A. G. Izergin and V. E. Korepin, Lattice versions of quantum field theory in two dimensions, *Nucl. Phys. B* **205** (1982) 401–413.
[33] A. Volkov and L. Faddeev, Yang-baxterization of the quantum dilogarithm, *Zapiski Nauchnykh Seminarov POMI* **224** (1995) 146–154; *J. Math. Sci.* **88** (1998) 202–207.

[34] V. V. Bazhanov, V. V. Mangazeev and S. M. Sergeev, Faddeev–Volkov solution of the Yang–Baxter equation and discrete conformal symmetry, *Nucl. Phys. B* **784** (2007) 234–258; arXiv:hep-th/0703041 [hep-th].

[35] L. N. Lipatov, High-energy asymptotics of multicolor QCD and two-dimensional conformal field theories, *Phys. Lett. B* **309** (1993) 394–396.

[36] L. D. Faddeev and G. P. Korchemsky, High-energy QCD as a completely integrable model, *Phys. Lett. B* **342** (1995) 311–322; arXiv:hep-th/9404173 [hep-th].

[37] S. E. Derkachov, G. P. Korchemsky and A. N. Manashov, Noncompact Heisenberg spin magnets from high-energy QCD: 1. Baxter Q operator and separation of variables, *Nucl. Phys. B* **617** (2001) 375–440; arXiv:hep-th/0107193 [hep-th].

[38] D. Chicherin, S. Derkachov and A. P. Isaev, Conformal group: R-matrix and star-triangle relation, *JHEP* **04** (2013) 020; arXiv:1206.4150 [math-ph].

[39] L. D. Faddeev, How algebraic Bethe ansatz works for integrable model, in *Relativistic Gravitation and Gravitational Radiation*, Proceedings, School of Physics, Les Houches, France, September 26–October 6, 1995 (1996), pp. 149–219; arXiv: hep-th/9605187 [hep th].

[40] J. Caetano, O. Gurdogan and V. Kazakov, Chiral limit of $N = 4$ SYM and ABJM and integrable Feynman graphs (2016); arXiv:1612.05895 [hep-th].

[41] V. Kazakov and P. Vieira, From characters to quantum (super)spin chains via fusion, *JHEP* **10** (2008) 050; arXiv:0711.2470 [hep-th].

[42] V. Kazakov, S. Leurent and Z. Tsuboi, Baxter's Q-operators and operatorial Backlund flow for quantum (super)-spin chains, *Comm. Math. Phys.* **311** (2012) 787–814; arXiv:1010.4022 [math-ph].

[43] A. Alexandrov, V. Kazakov, S. Leurent, Z. Tsuboi and A. Zabrodin, Classical tau-function for quantum spin chains, *JHEP* **09** (2013) 064; arXiv:1112.3310 [math-ph].

[44] G. P. Pronko and Yu. G. Stroganov, Bethe equations 'on the wrong side of equator', *J. Phys. A* **32** (1999) 2333–2340; arXiv:hep-th/9808153 [hep-th].

[45] G. P. Pronko and Yu. G. Stroganov, The complex of solutions of the nested Bethe ansatz. The A_2 spin chain, *J. Phys. A* **33** (2000) 8267; arXiv:hep-th/9902085 [hep-th].

[46] I. Krichever, O. Lipan, P. Wiegmann and A. Zabrodin, Quantum integrable systems and elliptic solutions of classical discrete nonlinear equations, *Comm. Math. Phys.* **188** (1997) 267–304; arXiv:hep-th/9604080 [hep-th].

[47] V. Kazakov, A. S. Sorin and A. Zabrodin, Supersymmetric Bethe ansatz and Baxter equations from discrete Hirota dynamics, *Nucl. Phys. B* **790** (2008) 345–413; arXiv:hep-th/0703147 [hep-th].

[48] S. Leurent, Integrable systems and AdS/CFT duality, Ph.D. thesis, Paris U., IV (2012); arXiv:1206.4061 [hep-th]; http://inspirehep.net/record/1118754/files/arXiv: 1206.4061.pdf.

[49] P. P. Kulish and E. K. Sklyanin, On the solution of the Yang–Baxter equation, *J. Sov. Math.* **19** (1982) 1596–1620; *Zap. Nauchn. Semin.* **95** (1980) 129–160.

[50] P. P. Kulish, Integrable graded magnets, *J. Sov. Math.* **35** (1986) 2648–2662; *Zap. Nauchn. Semin.* **145** (1985) 140–163.

[51] R. J. Baxter, *Exactly Solved Models in Statistical Mechanics* (Courier Corporation, 2007).

[52] A. Kluemper and P. Pearce, Conformal weights of RSOS lattice models and their fusion hierarchies, *Physica A* **183** (1992) 304–350.

[53] V. V. Bazhanov, S. L. Lukyanov and A. B. Zamolodchikov, Integrable structure of conformal field theory. 2. Q operator and DDV equation, *Comm. Math. Phys.* **190** (1997) 247–278; arXiv:hep-th/9604044 [hep-th].

[54] Z. Tsuboi, Analytic Bethe ansatz and functional equations for Lie superalgebra $sl(r+1|s+1)$, *J. Phys. A* **30** (1997) 7975–7991; arXiv:0911.5386 [math-ph].
[55] Z. Tsuboi, From the quantum Jacobi–Trudi and Giambelli formula to a nonlinear integral equation for thermodynamics of the higher spin Heisenberg model, *J. Phys. A* **37** (2004) 1747–1758; arXiv:cond-mat/0308333 [cond-mat].
[56] V. V. Bazhanov, A. N. Hibberd and S. M. Khoroshkin, Integrable structure of $W(3)$ conformal field theory, quantum Boussinesq theory and boundary affine Toda theory, *Nucl. Phys. B* **622** (2002) 475–547; arXiv:hep-th/0105177 [hep-th].
[57] V. V. Bazhanov, R. Frassek, T. Lukowski, C. Meneghelli and M. Staudacher, Baxter Q-operators and representations of Yangians, *Nucl. Phys. B* **850** (2011) 148–174; arXiv:1010.3699 [math-ph].
[58] R. Frassek, T. Lukowski, C. Meneghelli and M. Staudacher, Baxter operators and Hamiltonians for 'nearly all' integrable closed $\mathfrak{gl}(n)$ spin chains, *Nucl. Phys. B* **874** (2013) 620–646; arXiv:1112.3600 [math-ph].
[59] N. Gromov and P. Vieira, Complete 1-loop test of AdS/CFT, *JHEP* **04** (2008) 046; arXiv:0709.3487 [hep-th].
[60] A. V. Belitsky, S. E. Derkachov, G. P. Korchemsky and A. N. Manashov, Baxter Q-operator for graded SL(2|1) spin chain, *J. Stat. Mech.* **0701** (2007) P01005; arXiv:hep-th/0610332 [hep-th].
[61] N. Gromov, V. Kazakov, S. Leurent and Z. Tsuboi, Wronskian solution for AdS/CFT Y-system, *JHEP* **01** (2011) 155; arXiv:1010.2720 [hep-th].
[62] Z. Tsuboi, Solutions of the T-system and Baxter equations for supersymmetric spin chains, *Nucl. Phys. B* **826** (2010) 399–455; arXiv:0906.2039 [math-ph].
[63] Z. Tsuboi, Wronskian solutions of the T, Q and Y-systems related to infinite dimensional unitarizable modules of the general linear superalgebra $gl(M|N)$, *Nucl. Phys. B* **870** (2013) 92–137; arXiv:1109.5524 [hep-th].
[64] N. Gromov, V. Kazakov and P. Vieira, Finite volume spectrum of 2D field theories from Hirota dynamics, *JHEP* **12** (2009) 060; arXiv:0812.5091 [hep-th].
[65] V. Kazakov and S. Leurent, Finite size spectrum of $SU(N)$ principal chiral field from discrete Hirota dynamics, *Nucl. Phys. B* **902** (2016) 354–386; arXiv:1007.1770 [hep-th].
[66] J. A. Minahan and K. Zarembo, The Bethe ansatz for $N = 4$ super Yang–Mills, *JHEP* **03** (2003) 013; arXiv:hep-th/0212208 [hep-th].
[67] N. Beisert and M. Staudacher, The $N = 4$ SYM integrable super spin chain, *Nucl. Phys. B* **670** (2003) 439–463; arXiv:hep-th/0307042 [hep-th].
[68] N. Beisert, V. A. Kazakov, K. Sakai and K. Zarembo, Complete spectrum of long operators in $N = 4$ SYM at one loop, *JHEP* **07** (2005) 030; arXiv:hep-th/0503200 [hep-th].
[69] M. Alfimov, N. Gromov and V. Kazakov, QCD Pomeron from AdS/CFT quantum spectral curve, *JHEP* **07** (2015) 164; arXiv:1408.2530 [hep-th].
[70] N. Gromov, V. Kazakov, S. Leurent and D. Volin, Solving the AdS/CFT Y-system, *JHEP* **07** (2012) 023; arXiv:1110.0562 [hep-th].
[71] N. Beisert, The complete one loop dilatation operator of $N = 4$ super Yang–Mills theory, *Nucl. Phys. B* **676** (2004) 3–42; arXiv:hep-th/0307015 [hep-th].
[72] N. Gromov, V. Kazakov and P. Vieira, Exact spectrum of anomalous dimensions of planar $N = 4$ supersymmetric Yang–Mills theory, *Phys. Rev. Lett.* **103** (2009) 131601; arXiv:0901.3753 [hep-th].
[73] N. Gromov, V. Kazakov, A. Kozak and P. Vieira, Exact spectrum of anomalous dimensions of planar $N = 4$ supersymmetric Yang–Mills theory: TBA and excited states, *Lett. Math. Phys.* **91** (2010) 265–287; arXiv:0902.4458 [hep-th].

[74] D. Bombardelli, D. Fioravanti and R. Tateo, Thermodynamic Bethe ansatz for planar AdS/CFT: A proposal, *J. Phys. A* **42** (2009) 375401; arXiv:0902.3930 [hep-th].
[75] G. Arutyunov and S. Frolov, Thermodynamic Bethe ansatz for the $AdS_5 \times S^5$ mirror model, *JHEP* **05** (2009) 068; arXiv:0903.0141 [hep-th].
[76] A. Cavaglia, D. Fioravanti and R. Tateo, Extended Y-system for the AdS_5/CFT_4 correspondence, *Nucl. Phys. B* **843** (2011) 302–343; arXiv:1005.3016 [hep-th].
[77] N. Beisert and R. Roiban, Beauty and the twist: The Bethe ansatz for twisted $N = 4$ SYM, *JHEP* **0508** (2005) 039.
[78] N. Gromov and V. Kazakov, Review of AdS/CFT integrability, Chapter III.7: Hirota dynamics for quantum integrability, *Lett. Math. Phys.* **99** (2012) 321–347; arXiv:1012.3996 [hep-th].
[79] V. A. Kazakov, A. Marshakov, J. A. Minahan and K. Zarembo, Classical/quantum integrability in AdS/CFT, *JHEP* **05** (2004) 024; arXiv:hep-th/0402207 [hep-th].
[80] N. Beisert, V. A. Kazakov, K. Sakai and K. Zarembo, The algebraic curve of classical superstrings on $AdS_5 \times S^5$, *Comm. Math. Phys.* **263** (2006) 659–710; arXiv: hep-th/0502226 [hep-th].
[81] J. Fokken, C. Sieg and M. Wilhelm, A piece of cake: The ground-state energies in γ_i-deformed $\mathcal{N} = 4$ SYM theory at leading wrapping order, *JHEP* **09** (2014) 078; arXiv:1405.6712 [hep-th].
[82] N. Beisert, B. Eden and M. Staudacher, Transcendentality and crossing, *J. Stat. Mech.* **0701** (2007) P01021; arXiv:hep-th/0610251 [hep-th].
[83] C. Ahn, Z. Bajnok, D. Bombardelli and R. I. Nepomechie, TBA, NLO Luscher correction and double wrapping in twisted AdS/CFT, *JHEP* **12** (2011) 059; arXiv:1108.4914 [hep-th].
[84] A. Georgoudis, V. Gonçalves, E. Panzer and R. Pereira, Five-loop massless propagator integrals (2018); arXiv:1802.00803 [hep-th].
[85] D. Chicherin, V. Kazakov, F. Loebbert, D. Müller and D. Zhong, Yangian symmetry for bi-scalar loop amplitudes (2017); arXiv:1704.01967 [hep-th].
[86] D. Chicherin, V. Kazakov, F. Loebbert, D. Müller and D.-l. Zhong, Yangian symmetry for fishnet Feynman graphs, *Phys. Rev. D* **96**(12) (2017) 121901; arXiv:1708.00007 [hep-th].
[87] D. Grabner, N. Gromov, V. Kazakov and G. Korchemsky, in preparation.
[88] S. Leurent and D. Volin, Multiple zeta functions and double wrapping in planar $N = 4$ SYM, *Nucl. Phys. B* **875** (2013) 757–789; arXiv:1302.1135 [hep-th].
[89] D. J. Broadhurst, Evaluation of a Class of Feynman diagrams for all numbers of loops and dimensions, *Phys. Lett. B* **164** (1985) 356–360.
[90] E. Panzer, On the analytic computation of massless propagators in dimensional regularization, *Nucl. Phys. B* **874** (2013) 567–593; arXiv:1305.2161 [hep-th].
[91] V. Gurarie, Logarithmic operators in conformal field theory, *Nucl. Phys. B* **410** (1993) 535–549; arXiv:hep-th/9303160 [hep-th].
[92] L. Anselmetti, D. Bombardelli, A. Cavaglià and R. Tateo, 12 loops and triple wrapping in ABJM theory from integrability, arXiv:1506.09089v2.
[93] R. N. Lee and A. I. Onishchenko, ABJM quantum spectral curve and Mellin transform, arXiv:1712.00412v1.
[94] A. Cavaglia, M. Cornagliotto, M. Mattelliano and R. Tateo, A Riemann-Hilbert formulation for the finite temperature Hubbard model, arXiv:1501.04651v3.
[95] B. Basso, S. Komatsu and P. Vieira, Structure constants and integrable bootstrap in planar $N = 4$ SYM theory (2015); arXiv:1505.06745 [hep-th].

[96] T. Fleury and S. Komatsu, Hexagonalization of correlation functions, *JHEP* **01** (2017) 130; arXiv:1611.05577 [hep-th].

[97] T. Bargheer, J. Caetano, T. Fleury, S. Komatsu and P. Vieira, Handling handles I: Nonplanar integrability (2017); arXiv:1711.05326 [hep-th].

[98] J. Caetano, unpublished.

[99] O. Mamroud and G. Torrents, RG stability of integrable fishnet models, *JHEP* **06** (2017) 012; arXiv:1703.04152 [hep-th].

[100] B. Basso and L. J. Dixon, Gluing ladder Feynman diagrams into fishnets, *Phys. Rev. Lett.* **119**(7) (2017) 071601; arXiv:1705.03545 [hep-th].

[101] J. L. Bourjaily, A. J. McLeod, M. Spradlin, M. von Hippel and M. Wilhelm, The elliptic double-box integral, arXiv:1712.02785.

[102] V. Kazakov and E. Olivucci, Bi-scalar integrable CFT at any dimension (2018); arXiv:1801.09844 [hep-th].

[103] G. P. Korchemsky, Conformal bootstrap for the BFKL Pomeron, *Nucl. Phys. B* **550** (1999) 397–423; arXiv:hep-ph/9711277 [hep-ph].

[104] D. J. Gross and V. Rosenhaus, All point correlation functions in SYK, *JHEP* **2017** (2017) 148; arXiv:1710.08113 [hep-th].

[105] J. Escobedo, N. Gromov, A. Sever and P. Vieira, Tailoring three-point functions and integrability II. Weak/strong coupling match, *JHEP* **09** (2011) 029; arXiv:1104.5501 [hep-th].

3D Current Algebra and Twisted K Theory

Jouko Mickelsson
Department of Mathematics and Statistics, University of Helsinki, Finland
jouko@kth.se

Equivariant twisted K theory classes on compact Lie groups G can be realized as families of Fredholm operators acting in a tensor product of a fermionic Fock space and a representation space of a central extension of the loop algebra LG using a supersymmetric Wess–Zumino–Witten model. The aim of the present paper is to extend the construction to higher loop algebras using an abelian extension of a 3D current algebra. We have only partial success: Instead of true Fredholm operators we have formal algebraic expressions in terms of the generators of the current algebra and an infinite dimensional Clifford algebra. These give rise to sesquilinear forms in a Hilbert bundle which transform in the expected way with respect to 3D gauge transformations but do not define true Hilbert space operators.

1. Introduction

I have warm memories from many occasions where I had the opportunity to meet Ludwig Faddeev, mostly in Helsinki and Stockholm. At one point, we had very close mathematical interests although we did not have direct cooperation on the subject. This contribution to Faddeev's memorial volume is closely related to that topic. I was working during 1983 on the problem of how an additional Chern–Simons term in the Yang–Mills action affects the current algebra commutation relations through an abelian extension of the current algebra; the paper came out as a preprint in the fall 1983, and published later as [5]. I met Ludwig next spring in Helsinki and I understood that he had been thinking about the same problem and in fact he had just completed a paper with Samson Shatasvili [8] with similar results. In this contribution, the abelian extensions of current algebras which we found at that time, play a central role.

Twisted K theory on compact Lie groups has been discussed from several points of view in the past. First, let us recall some basic definitions. Let $PU(H) = U(H)/S^1$ be the projective unitary group of a complex Hilbert space H. A principal bundle over a space X with fiber $PU(H)$ is defined up to equivalence by a cohomology class $\Omega \in \mathrm{H}^3(X, \mathbb{Z})$, called the Dixmier–Douady class. The group $PU(H)$ acts by conjugation on the space of Fredholm operators in H. This action defines a bundle over X with fiber given by the space of Fredholm operators. The homotopy classes of sections of this fiber bundle define the twisted K theory $\mathrm{K}^*(X,\Omega)$. This cohomology is $\mathbb{Z}_2$ graded, the even part corresponds to all Fredholm operators on H and the odd part to self-adjoint Fredholm operators with essential positive and negative spectrum.

In the case when $X = G$ is a compact Lie group, one is also interested in the equivariant twisted classes, $\mathrm{K}_G^*(G,\mathbb{Z})$, where the G action on itself is given by conjugation. The non-equivariant case is solved for compact simple Lie groups in [4]. The equivariant classes can be explicitly constructed using a supersymmetric conformal field theory model (Wess–Zumino–Witten model) in $1+1$ space time dimensions [3]. More detailed discussion can be found in [2] including the proof of completeness of the construction.

The construction in the equivariant case is related to the fact that the compact Lie group is actually the moduli space of gauge connections on the unit circle S^1, $G = \mathcal{A}/\mathcal{G}_0$ where $\mathcal{A}$ is the space of smooth gauge connections over S^1 and $\mathcal{G}_0$ denotes the group of based gauge transformations. Thus it is natural to ask if the construction can be generalized to higher dimensions, to the case of gauge potentials on a vector bundle over a compact spin manifold. In this paper, we shall take the modest step and study the case of the 3-dimensional torus T^3. This case is convenient since we can work algebraically with functions which are finite linear combinations of the Fourier modes. But let us first briefly recall the construction in the case of $X = G$ based on loop group representations.

Fix a unitary highest weight representation of the central extension $\widehat{LG}$ of the smooth loop group LG in a Hilbert space V. Use the standard 2-cocycle kc with

$$c(X,Y) = \frac{i}{2\pi}\int_{S^1}\langle X, dY\rangle$$

where $\langle\cdot,\cdot\rangle$ is an invariant bilinear form on the Lie algebra $\mathfrak{g}$ such that the level k is a non-negative integer in an irreducible highest weight representation of the loop group. In addition, one needs an irreducible representation of the canonical anticommutation relations algebra

$$\psi_a^n\psi_b^m + \psi_b^m\psi_a^n = 2\delta_{ab}\delta_{n,-m}\mathbf{1}$$

in a Fock space $\mathcal{F}$. Here $n,m\in\mathbb{Z}$ and $a,b = 1,2,\ldots\dim G$ relate to an orthonormal basis of $\mathfrak{g}$ with respect to $\langle\cdot,\cdot\rangle$. The representation is characterized by the existence of a vacuum vector with the property $\psi_a^n|0\rangle = 0$ for $n<0$. In addition, we have the hermiticity relation $(\psi_a^n)^* = \psi_a^{-n}$.

Following the notation in [3] we define the skew symmetric operator

$$Q = \sum \psi_a^n\left(T_a^{-n} + \frac{1}{3}K_a^{-n}\right)$$

acting in the tensor product $V\otimes\mathcal{F}$. Here $S_a^n = T_a^n\otimes\mathbf{1}+\mathbf{1}\otimes K_a^n$ are the generators of the loop algebra acting in the tensor product. The commutators of the generators

K_a^n are determined by the central extension of level $k = \kappa$, where κ is the dual Coxeter number of the simple Lie algebra $\mathfrak{g}$,

$$[K_a^n, K_b^m] = \sum_c \lambda_{ab}^c K_c^{n+m} + \kappa m \delta_{ab} \delta_{n,-m},$$

where the numbers λ_{ab}^c are the structure constants of the Lie algebra $\mathfrak{g}$. In a unitary representation of the loop algebra $(T_a^n)^* = -T_a^{-n}$ and $(K_a^n)^* = -K_a^{-n}$.

Next, define the family of operators $Q_A = Q + \tilde{k} \sum \psi_a^n A_a^{-n}$ where the A_a^n's are the Fourier components of the gauge potential A in the Lie algebra basis labeled by a and $\tilde{k} = k + \kappa$. One then shows that the family of operators Q_A transforms covariantly under the central extension $\widehat{LG}$,

$$\hat{g}^{-1} Q_A \hat{g} = Q_{A^g}$$

with $A^g = g^{-1}Ag + g^{-1}dg$ for $g \in LG$. Because of the central extension, this family cannot be pushed down to a family of operators parametrized by $\mathcal{A}/LG$ but it defines a twisted equivariant K theory element on G. The Dixmier–Douady class of the twist is given by $\tilde{k}$ times the generator $\omega \in \mathrm{H}^3(G, \mathbb{Z}) \simeq \mathbb{Z}$.

2. Gauge Current Algebra in Three Dimension

Let M be a compact connected spin 3-manifold, G a compact simple Lie group and $\mathfrak{g}$ its Lie algebra. Let E be a trivial vector bundle over M with fiber the tensor product $\mathbb{C}^2 \otimes \mathbb{C}^N$ where the first factor is the space of Weyl spinors and the group G is unitarily represented in the second factor. The Weyl–Dirac operator D_A coupled to a gauge potential A, with values in $\mathfrak{g}$, acts in a dense domain of the Hilbert space $H = L_2(M, E)$ of square integrable sections of the bundle E.

For a real number λ and a given vector potential A, one has a polarization

$$H = H_+(A, \lambda) \oplus H_-(A, \lambda)$$

of the Hilbert space H using the spectral projection associated to D_A with a spectral cut at λ. This polarization defines a representation of the canonical anticommutation relations algebra (CAR) in a fermionic Fock space with a vacuum $|A, \lambda\rangle$. The canonical anticommutation relations are generated by elements $a^*(v)$ (linear in $v \in H$) and $a(v)$ (antilinear in v) with relations

$$a^*(u)a(v) + a(v)a^*(u) = 2\langle v, u\rangle \cdot \mathbf{1}$$

where $\langle \cdot, \cdot \rangle$ is the L_2 inner product in H. The vacuum is a vector in the Fock space $\mathcal{H}$ such that

$$a(v)|A, \lambda\rangle = 0 = a^*(u)|A, \lambda\rangle \text{ for } v \in H_+(A, \lambda) \text{ and } u \in H_-(A, \lambda).$$

However, there is a problem related to the spectral flow as A varies which obstructs a construction of the Fock vacua as continuous functions of the argument A. To circumvent this problem, one has to modify the construction by a certain family of local complex line bundles over open sets in the space $\mathcal{A}$ of smooth vector potentials. The group $\mathcal{G}$ of smooth gauge transformations does not act in these line bundles; instead, one has to construct an abelian extension $\hat{\mathcal{G}}$ which acts in the line bundles. The Lie algebra of the extension is defined by the 2-cocycle

$$c(A; X, Y) = \frac{1}{24\pi^2} \int_M \operatorname{tr} A[dX, dY],$$

where $X, Y\colon M \to \mathfrak{g}$ and the trace is evaluated in the representation in $\mathbb{C}^N$. There are several ways to derive this formula but the index theory derivation in [1] describes best the topological and geometric aspects related to the construction of gerbes over the moduli space of gauge potentials.

As a consequence of the line bundle modifications is that the gauge group $\mathcal{G}$ acts through an extension $\hat{\mathcal{G}}$, infinitesimally determined by the cocycle c, on a vector bundle V over $\mathcal{A}$ with model fiber $\mathcal{H}$.

Since c is a function of A the 2-cocycle property reads as

$$\begin{aligned} &c(A; X, [Y, Z]) + c(A; Y, [Z.X]) + c(A; Z, [X, Y]) \\ &\quad + L_X c(A; Y, Z) + L_Y c(A; Z, X) + L_Z c(A, X, Y) = 0, \end{aligned}$$

where L_X denotes the Lie derivative acting on functions of A through the infinitesimal gauge transformations $L_X A = [A, X] + dX$. Alternatively, the cocycle c comes from a central extension of the gauge groupoid $(\mathcal{A}, \mathcal{G})$ with sources and targets in $\mathcal{A}$ and arrows $A \to A^g = g^{-1}Ag + g^{-1}dg$.

Remark. In a real representation of $\mathfrak{g}$, the cocycle c vanishes identically. The reason is that in a real unitary representation the Lie algebra elements are antisymmetric and therefore $\operatorname{tr} A(BC + CB) = \operatorname{tr}(A(BC + CB))^t = -\operatorname{tr}(BC + CB)A = -\operatorname{tr} A(BC + CB) = 0$. In particular, this is always the case for the **adjoint representation** of $\mathfrak{g}$.

The above remark is important in the next section. We shall consider Fock representations of the current algebra in the tensor product of two Fock spaces. The first one comes from the quantization of fermions in the adjoint representation and the second in an arbitrary complex representation. It follows that the 2-cocycle c has a contribution only from the complex representation.

3. Family of Formal Supersymmetric Hamiltonians

Using the symmetric bilinear form

$$\langle X, Y \rangle = \frac{1}{(2\pi)^3} \int \operatorname{tr} X(z) Y(z) dz$$

on the algebra of smooth functions $X, Y\colon T^3 \to \mathfrak{g}$ we associate to the Fourier modes T_a^n ($n \in \mathbb{Z}^3$ and $a = 1, 2, \ldots, N = \dim \mathfrak{g}$) the dual basis ψ_a^n and define the formal series

$$Q = \sum_{n \in \mathbb{Z}^3} \sum_{a=1}^{N} \left(\psi_a^n T_a^{-n} + \frac{1}{3} \psi_a^n K_a^{-n} \right). \tag{3.1}$$

The vectors ψ_a^n are interpreted as elements in an infinite dimensional Clifford algebra with anticommutation relations

$$\psi_a^n \psi_b^m + \psi_b^m \psi_a^n = 2\delta_{ab}\delta_{n,-m}.$$

Here the K_a^n's are derivations of the CAR algebra defined as

$$[K_a^n, \psi_b^m] = \sum_c \lambda_{ab}^c \psi_c^{n+m}. \tag{3.2}$$

The commutators of the derivations and are then given as

$$[K_a^n, K_b^m] = \sum_c \lambda_{ab}^c K_c^{n+m}.$$

As infinite formal sums,

$$K_a^n = -\sum_{b,c,m} \frac{1}{4} \lambda_{ab}^c \psi_b^m \psi_c^{n-m}.$$

More concretely, in the case of Weyl fermions in the adjoint representation of the gauge group G, after quantization in the bundle of fermionic Fock spaces $\mathcal{F}$ over $\mathcal{A}$ the current algebra generated by the K_a^n's acts without the abelian extension $(c = 0)$ by the remark in the end of the previous section. For this reason, it is actually convenient to deal with massive 4-component Dirac fermions instead of the 2-component Weyl fermions. This simplifies some things later on. For massive fermions, we have a mass gap in the spectrum of the Dirac Hamiltonians and we can define the vacuum with respect to the polarization defined by the spectral cut at zero for all Dirac Hamiltonians coupled to vector potentials. The fermionic modes ψ_a^n can stand for any of the four components of the Dirac spinor. For notational simplicity, we shall suppress the spinor label $\alpha = 1, 2, 3, 4$ in the following discussion.

The current algebra extension defined by the cocycle c comes entirely from the action of the T_a^n's on the sections of the bundle V over $\mathcal{A}$.

Since the gauge group acts in the fermionic Fock spaces $\mathcal{F}$ without the extension term, the Dixmier–Douady class of the projective Fock bundle vanishes and thus can be pushed to a true Hilbert bundle over the moduli space $\mathcal{A}/\mathcal{G}_0$, the sections of this bundle corresponding to gauge invariant sections of the pull-back bundle over $\mathcal{A}$. In particular, we are free to define a gauge covariant vacuum section $A \mapsto |A\rangle$ with $g|A\rangle = |g \cdot A\rangle$. For an arbitrary polynomial P in the generators ψ_a^n, the action of a gauge transformation on the state $P|A\rangle$ is then given by the canonical action on the generators of the Clifford algebra (infinitesimally by the formula (3.2)) and by $g|A\rangle) = |g \cdot A\rangle$.

As it stands, the infinite sum in (3.1) is ill-defined as an operator but Q defines a derivation of the CAR algebra and the gauge current algebra as

$$[Q, \psi_b^m]_+ = 2(T_b^m + K_b^m) \equiv 2S_b^m, \tag{3.3}$$

$$[Q, S_b^m] = \sum_{n,a} c(A; T_a^n, T_b^m)\psi_a^{-n}. \tag{3.4}$$

For a smooth potential A, the infinite sum in the second equation is actually converging in the L^2 norm in the CAR algebra generated by the ψ_a^n's since the Fourier coefficients of a smooth function form a rapidly decreasing sequence.

Remark. An $N \times N$ matrix X defines an element $\psi(X)$ in the dual $\mathfrak{g}^*$ by $\langle \psi(X), z\rangle = \operatorname{tr} Xz$ where z stands for the matrix representing $z \in \mathfrak{g}$ in the representation space $\mathbf{C}^N$. In the same vein, a matrix valued 3-form ω on T^3 defines an element in the dual of the current algebra $Map(T^3, \mathfrak{g})$ by

$$\langle \psi(\omega), X\rangle = \int \operatorname{tr} \omega X.$$

It follows that for a linear combination $X = \sum X_m^a S_a^m$ we have

$$[Q, X] = \psi(dX \wedge dA + dA \wedge dX).$$

From (3.3) and (3.4), one can reduce that the commutators with the formal square Q^2 are given as

$$[Q^2, \psi_a^n] = 2\psi_b^m c(A; T_b^{-m}, T_a^n), \tag{3.5}$$

$$[Q^2, S_a^n] = 2S_b^m c(A; T_b^{-m}, T_a^n) - \psi_b^m \psi_c^p \mathcal{L}_c^{-p} c(A; T_b^{-m}, T_a^n). \tag{3.6}$$

We have used the fact that

$$\sum_b \mathcal{L}_b^p c(A; T_b^{-p}, T_a^n) = 0$$

which follows from

$$\mathcal{L}_Z c(A; X, Y) = \int \operatorname{tr}[A, Z](dXdY - dYdX)$$

and the algebraic relation

$$\sum_b \operatorname{tr}[A, T_b](T_bT_a + T_aT_b) = \sum_b \operatorname{tr}[T_a, A](T_b)^2 = \sum_b \operatorname{tr} T_a[A, (T_b)^2] = 0$$

since the element $\sum_b (T_b)^2$ is the Casimir invariant in the semi simple Lie algebra $\mathfrak{g}$.

The formal square of Q can be written as

$$Q^2 = \sum T_a^n T_a^{-n} + \frac{1}{2}\sum c(A; T_a^n, T_b^m)\psi_a^n \psi_b^m$$

and one can check by direct computation from the commutation relations that the commutators of this formal series matches the commutators (3.5) and (3.6). This expression can be compared with the corresponding formula [3] in the loop group case. The important difference is that the contribution involving the dual Coxeter number is missing. The reason is that the algebra of the K_a^n's does not contain the anomaly proportional to the dual Coxeter number.

From (3.3) follows immediately

Theorem 3.1. *Let $B = B(A)$ be a matrix-value 3-form on T^3 defined as*

$$B(A) = \omega \wedge dA + dA \wedge \omega + \alpha \wedge A \wedge \beta + \gamma \wedge A + A \wedge \phi + \theta$$

where the parameters ω, α, β *are matrix-valued* 1*-forms,* γ, ϕ *are matrix-valued* 2*-forms and* θ *is a matrix-valued* 3*-form. Thus* $B(A)$ *is an affine function in the variable* A*. Define the formal sum* $Q_B = Q + \psi(B(A))$*. Then for* $X = \sum X^a_m S^m_a$

$$[Q_B, X] = \psi(B'(A))$$

where $B'(A)$ *is again a matrix-valued* 3*-form with new parameters* (ω', α', β', γ', ϕ', θ')*. In particular,*

$$\omega' = [\omega, X] + dX.$$

Thus ω transforms like a matrix-valued vector potential under the infinitesimal gauge transformations X. In particular, the action of based gauge transformations is free on the parameter ω and thus also on the space of functions $A \mapsto B(A)$,

Corollary 3.2. *The moduli space* $\mathcal{B}/\mathcal{G}_0$ *of the functions* $B(A)$ *modulo the group* $\mathcal{G}_0$ *of based gauge transformations is the classifying space for the group* $\mathcal{G}_0$ *and thus homotopy equivalent to the space* $\mathcal{A}/\mathcal{G}_0$ *of gauge equivalence classes of vector potentials* A*.*

The forms Q_B are only formal algebraic expressions but they can be used to define sesquilinear forms in dense domains of the Fock spaces. Let ξ be an arbitrary smooth section of V and $\eta(A) = |A\rangle$ the vacuum section of $\mathcal{F}$. The domain D for Q_B is then defined to consists of sections of $V \otimes \mathcal{F}$ of the form $\xi \otimes p(\psi)\eta$ where p is a polynomial in the generators ψ^n_a of the Clifford algebra.

Proposition 3.3. *With the notation above* $\langle \xi' \otimes p'(\psi)\eta, Q_B(\xi \otimes p(\psi)\eta\rangle$ *is well-defined in the fiberwise inner product for the sections.*

Proof. The vacuum section η is invariant under gauge transformations so $K^n_a \eta = 0$ and thus also $\sum \psi^n_a K^{-n}_a \eta = 0$ and so $Q(\xi \otimes \eta) = \sum T^n_a \xi \otimes \psi^{-n}_a \eta$. Using the algebra relations (3.3), (3.4), one sees that it is sufficient to show that the inner products $\langle \xi' \otimes p'\eta, T^n_a \xi \otimes \psi^{-n}_a p\eta\rangle$ exist fiberwise. First computing this at the fiber over $A = 0$, the claim follows from the fact that for large enough absolute value of n the inner product of ψ^n_a between finite polynomials in the ψ's vanish. This is a consequence of the fact that the free vacuum is an eigenvector for the quantized momentum vectors P_k with eigenvalue zero and thus the vacuum expectation value of a polynomial not commuting with the P_k's is equal to zero.

When $A \neq 0$, the above argument needs a modification.

To be more precise on the definition of the momentum operators: These are not the free momentum operators but momenta defined with respect to the vacuum $|A\rangle$. That is, choosing a unitary operator T_A in the 1-particle Hilbert space such that $T_A^{-1} \epsilon_0 T_A = \epsilon_A$ where $\epsilon_A = D_A/|D_A|$ then $p_k = T_A^{-1} i \frac{\partial}{\partial x_k} T_A$ in the 1-particle space and P_k is the second quantized operator corresponding to p_k. Here T_A is a pseudodifferential operator with an asymptotic expansion $T_A = 1 + t_{-1} + t_{-2} + \cdots$ in inverse powers of momenta; the first non-trivial term t_{-1} is linear in A, [7]. It follows that $p_k = i\frac{\partial}{\partial x_k} +$ terms of order -1 in momenta. Now the commutator $[P_k, \psi_a^n]$ is not any more equal to $n_k \psi_a^n$ but there is correction which is proportional to the inverse of $|n|$. It follows that the expectation value of ψ_a^n between polynomial states in the Clifford algebra is not exactly zero for large $|n|$ but there is a correction proportional to $1/|n|^2$ (since the extra terms in P_k have relative magnitude of order -2). But the factor $1/|n|^2$ guarantees that the sum over n in $\sum \psi_a^n T_a^{-n}$ gives a finite fiberwise expectation value between sections of the form $\xi \otimes p(\psi)\eta$. The second term $\psi_a^n K^{-n}$ gives a finite contribution by $K^n{}_a \eta = 0$ and the relation (3.3). The interaction term $\psi(B(A))$ is in fact a fiberwise well-defined operator since it is a linear combination of the ψ_a^n's with L_2 coefficients coming from a smooth form $B(A)$. □

References

[1] A. Carey, J. Mickelsson and M. Murray, Index theory, gerbes, and Hamiltonian quantization, *Comm. Math. Phys.* **183**(3) (1997) 707–722.

[2] D. S. Freed, M. J. Hopkins and C. Teleman, Loop groups and twisted K-theory III, *Ann. of Math. (2)* **174**(2) (2011) 947–1007.

[3] J. Mickelsson, Gerbes, (twisted) K-theory, and the supersymmetric WZW model, in *Infinite Dimensional Groups and Manifolds*, IRMA Lect. Math. Theor. Phys., Vol. 5 (de Gruyter, Berlin, 2004), pp. 93–107.

[4] C. L. Douglas, On the twisted K-homology of simple Lie groups, *Topology* **45**(6) (2006) 955–988.

[5] J. Mickelsson, Chiral anomalies in even and odd dimensions, *Comm. Math. Phys.* **97**(3) (1985) 361–370.

[6] J. Mickelsson, Kac–Moody groups, topology of the Dirac determinant bundle, and fermionization, *Comm. Math. Phys.* **110**(2) (1987) 173–183.

[7] J. Mickelsson, Wodizicki residue and anomalies of current algebras, in *Integrable Models and Strings*, Springer Lecture Notes in Physics, Vol. 436 (Springer, 1994), pp. 123–135.

[8] L. D. Faddeev and S. L. Shatashvili, Algebraic and Hamiltonian methods in the theory of nonabelian anomalies, *Teoret. Mat. Fiz.* **60**(2) (1984) 206–217 (in Russian).

Tying Up Instantons with Anti-Instantons

Nikita Nekrasov*

Simons Center for Geometry and Physics,
Stony Brook University, Stony Brook NY 11794-3636, USA
nnekrasov@scgp.stonybrook.edu

In quantizing classical mechanical systems, one often sums over the classical trajectories as in localization formulas, but also takes into account the contributions of the "instanton gas": a set of approximate solutions of the equations of motion. This paper attempts to alleviate some of the frustrations of this 40+ year-old approach by finding the honest solutions of equations of motion of the complexified classical mechanical system. These ideas originated in the Bethe/gauge correspondence. The examples include algebraic integrable systems, from the abstract Hitchin systems to the well-studied anharmonic oscillator. We also speculate on the applications to the black hole radiation. We elucidate the relation between Lefschetz thimbles and the Ω-deformed B-model. We propose the notion of the topological renormalization group.

In memory of L. D. Faddeev

Contents

*On leave of absence from IPPI, ITEP, Moscow, Russia.

1. Preface

I first met L. D. Faddeev in Leningrad in 1991 during a conference on mathematical physics and, more specifically, quantum groups. I only started my undergraduate studies in Moscow and was very much interested in quantum field theory, string theory and topology. The Moscow and Leningrad schools of theoretical and mathematical physics were competing, so I came from the rival's camp. In fact, one of the physicists at the ITEP with whom I discussed my first undergraduate project: black hole creation in high-energy collisions, told me not to discuss this with L. D. So, understandably, my first impression of L. D. was that of pure fear. I was not alone: all students and almost all professors (I can only think of one person who seemed not to care) who gave talks at the meeting would look at L. D. in the audience for his approval, fearing his disagreement (and there was some). He was the lion, watching over his pride. His remarks were usually aimed not at the speaker, but at the audience. My last meeting with L. D. was in 2016, in the same city, which by that time became Saint Petersburg. I lectured at the Chebyshev laboratory on non-perturbative calculations in gauge theory (which we will discuss below), the lecture was aimed at students, but L. D. nevertheless came by (on his way to meet the President of Russian Federation, to talk him into continuing the support of fundamental research in Russia), and made several remarks, aimed mostly at the audience.

Twenty five years which passed in between were more or less the second half of L. D.'s active career. His interests in physics and mathematics shaped mine, both through the direct interaction and via the interaction and collaboration with his former students, the members of Faddeev's school. Several favorite topics of L. D.'s: quantum mechanics, quantum integrable systems, quantum gauge theory, the "art of quantization" are now my favorite topics as well. This paper is a novel (or so it seemed at the time of writing) approach to the old problem which is at the interface of these themes.

2. Introduction

In the summer of 1995, I was a student at the Ettore Majorana Center summer school on sub-atomic physics in Erice. This place and the school are famous for many reasons. One of them is the lecture course taught by S. Coleman in 1977 [21] (see [78, 17] for the works on which this course was based). I studied the instanton methods using these lecture notes, and of course the book [79].

The textbook quantum mechanical problem, which shows the violation of classical intuition, is the non-perturbative splitting of the ground energy level of the quantum mechanical double well potential

$$U(x) = \frac{1}{4}\lambda(x^2 - v^2)^2. \tag{1}$$

It can be derived using the semi-classical calculation of the tunneling amplitude using the instanton–anti-instanton gas picture. In this picture, the ground state energy is extracted using the small temperature $\beta \to \infty$ limit of the matrix elements of the Euclidean evolution operators:

$$M_{lr} = \langle -v|e^{-\beta\widehat{H}}|+v\rangle\,, \qquad M_{rr} = \langle +v|e^{-\beta\widehat{H}}|+v\rangle \tag{2}$$

which are represented by the path integrals:

$$M_{lr} = \int_{x(0)=v,x(\beta)=-v} e^{-\mathbf{S}[x(t)]}Dx(t)\,, \qquad M_{rr} = \int_{x(0)=x(\beta)=v} e^{-\mathbf{S}[x(t)]}Dx(t) \tag{3}$$

with the Euclidean action

$$\mathbf{S}[x(t)] = \int_0^\beta \frac{\dot{x}^2}{2} + U(x)\,. \tag{4}$$

The critical points of $\mathbf{S}[x(t)]$ are the classical trajectories $x_{cl}(t)$ in the potential $-U(x)$. For $\beta \to \infty$, says the textbook, one has the solution $x_{\text{inst}}(t)$ with finite critical value $\mathbf{S}[x_{\text{inst}}(t)] = I$ which describes the particle, which starts at $x_{\text{inst}}(-\infty) = +v$ and ends at $x_{\text{inst}}(+\infty) = -v$, specifically:

$$\dot{x}_{\text{inst}} = -\sqrt{2U(x_{\text{inst}})} = \sqrt{\frac{\lambda}{2}}(x^2 - v^2) \le 0 \Rightarrow x_{\text{inst}}(t) = \mathbf{x}(t_0 - t) \tag{5}$$

where t_0 is arbitrary,

$$\mathbf{x}(t) = v\tanh(\omega_0 t) \tag{6}$$

and

$$2\omega_0 = v\sqrt{2\lambda} = \sqrt{U''(v)} \tag{7}$$

is the frequency of the harmonic oscillator which approximates $U(x)$ near its critical points $x = \pm v$. The critical value

$$I_0 = \mathbf{S}[\mathbf{x}(t)] = \sqrt{8\lambda}\frac{v^3}{3}\,. \tag{8}$$

In quantum field theory language, ω_0 is the mass of the perturbative quanta of the field $x(t)$ in the $(0+1)$-dimensional theory described by the action

$$S[x(t)] = \int dt \left(\frac{\dot{x}^2}{2} - U(x) \right) \tag{9}$$

in Minkowski time.

The textbook approach, then, is to note that outside the small interval $|t| \leq \omega_0^{-1}$ the solution can be approximated by the sign function with exponential accuracy:

$$\mathbf{x}(t) = v\,\mathrm{sign}(t)\,(1 - e^{-\omega_0|t|} + \cdots) \approx \mathbf{y}(t) = \begin{cases} -v\,, & t < -2\omega_0^{-1} \\ \frac{1}{2} v\omega_0 t\,, & -2\omega_0^{-1} < t < 2\omega_0^{-1} \\ +v\,, & t > 2\omega_0^{-1} \end{cases} \tag{10}$$

and then define the n-instanton/anti-instanton ($\mathcal{I}\bar{\mathcal{I}}$, in the notations of [7]) configuration by splicing the trajectory defined by:

$$x(t) = \begin{cases} \mathbf{y}(t_i^- - t)\,, & t_{i-1}^+ \leq t < t_i^- \\ \mathbf{y}(t - t_i^+)\,, & t_{i+1}^- \geq t > t_i^+ \end{cases} \tag{11}$$

with $t_0^+ = -\infty$,

$$t_1^- \ll t_1^+ \ll t_2^- \ll t_2^+ \ll \cdots \ll t_n^- \ll t_n^+\,, \tag{12}$$

and $t_{n+1}^- = +\infty$ (for M_{rr} estimates) and $t_n^- = +\infty$ (for M_{lr} estimates).

The textbook then argues, that the contribution of such configurations is, approximately,

$$M_{rr,n}(\beta) = \int dt_1^- dt_1^+ dt_2^- dt_2^+ \cdots dt_n^- dt_n^+\, K_{rr,n}[x(t)] e^{-\mathbf{S}[x(t)]} \tag{13}$$

and

$$M_{lr,n}(\beta) = \int dt_1^- dt_1^+ dt_2^- dt_2^+ \cdots dt_n^-\, K_{lr,n}[x(t)] e^{-\mathbf{S}[x(t)]} \tag{14}$$

where the integrals are taken over the moduli space (12) with lower and upper limits taken to be $\mp\beta/2$, with K the one-loop contribution times the normalization of the "zero-modes" $t_i^\pm$:

$$K_{\ldots}[x(t)] = \frac{1}{\sqrt{\mathrm{Det}'(\delta^2 \mathbf{S})}} \sqrt{\mathbf{g}}\,. \tag{15}$$

With further ingenious approximation, one arrives at the estimates $K_{lr,n} \sim K^{2n-1}$, $K_{rr,n} \sim K^{2n}$, making the n-$\mathcal{I}\bar{\mathcal{I}}$ contributions, roughly

$$M_{rr,n}(\beta) \sim \frac{\beta^{2n} K^{2n}}{(2n)!} e^{-2nI_0}\,, \qquad M_{lr,n}(\beta) \sim \frac{\beta^{2n-1} K^{2n-1}}{(2n-1)!} e^{-(2n-1)I_0}\,, \tag{16}$$

and, as a result:

$$M_{rr} \approx \frac{1}{2}(e^{\beta K e^{-I_0}} + e^{-\beta K e^{-I_0}})\,, \qquad M_{lr} \approx \frac{1}{2}(e^{\beta K e^{-I_0}} - e^{-\beta K e^{-I_0}})\,. \tag{17}$$

The trouble is, of course, that the field configurations (11) are not, for finite β, the solutions of $\delta\mathbf{S} = 0$. In fact, expanding $\mathbf{S}[\mathbf{x}(t) + \xi(t)]$ for small $\xi(t)$ one finds the non-vanishing tadpoles, i.e. the linear terms in $\xi(t)$. One can try to remedy this problem by expanding further [14]:

$$\mathbf{S}[\mathbf{x}(t) + \xi(t)] = \mathbf{S}[\mathbf{x}(t)] + \int dt\delta^{(1)}\mathbf{S}(t)\xi(t) + \frac{1}{2}\int dt_1 dt_2 \delta^{(2)}\mathbf{S}(t_1,t_2)\xi(t_1)\xi(t_2) + \cdots \tag{18}$$

and then solving for $\xi(t)$ in the quadratic approximation, to produce another reference configuration

$$\mathbf{x}_1(t) = \mathbf{x}(t) - \int dt' G(t,t')\delta^{(1)}\mathbf{S}(t') \tag{19}$$

where $G(t',t'')$ is Green's function for the symmetric operator $\delta^{(2)}\mathbf{S}$. According to [14], this procedure is suggested by A. Schwarz to be the functional analogue of the well-known Newton's algorithm for finding a true critical point, i.e. the solution to $\delta\mathbf{S} = 0$.

What does this algorithm converges to, if it converges at all? In [21], it is suggested that the true critical point of $\mathbf{S}$ is to be found "at infinity".

***

In this paper, we propose another answer to this question, in a class of quantum systems. This class of systems can be loosely described as the "quantization of the real slices of the algebraic integrable systems". An example of such a system is the anharmonic oscillator above. Another example is the many-body elliptic Calogero system:

$$\widehat{H} = -\frac{\hbar^2}{2}\sum_{i=1}^{N}\frac{\partial^2}{\partial x_i^2} + \nu(\nu-1)\sum_{i<j}\wp(x_i - x_i;\tau) \tag{20}$$

which is the model of N pair-wise repelling particles on a circle of circumference 1 with the potential having an imaginary (for $\tau \in \mathrm{i}\mathbb{R}$) second period. Yet another model would be a spin chain, of XXX, XXZ, XYZ or Gaudin type.[a] We shall classify the critical points of the analytically continued action $\mathbf{S}$, where the fields are allowed to take complex values (but the time is kept intact, be it for the Minkowski or Euclidean path integrals).

The paper is organized as follows. In Sec. 2, we review the Bethe/gauge-correspondence. In Sec. 3, we describe the supersymmetric gauge theoretic computation of the thermal partition function $\mathsf{Tr}e^{-\beta\widehat{H}}$ of the quantum mechanical system

[a]L. D. used to approach me with the remark "You seem to be a nice fellow, but your taste in integrable models is terrible — here in Leningrad we prefer spin chains to many-body systems". Had I taken his advice earlier than I actually had I would have probably missed the connection of integrable systems to gauge theories I will be exploiting below, but would have probably found something else. At any rate, without his work and the work of his school, the Bethe/gauge correspondence would not have been discovered. And in the grand picture, both many-body systems and the spin chains are on equal footing [27, 101].

dual to gauge theory on the Bethe side. In Sec. 4, we interpret this computation using the topological renormalization group, which treats path integral as a period. In this way, we arrive at the problem of finding the solutions of complexified equations of motion, which we identify with the results of the Sec. 3. The reader, not interested in the relations to supersymmetry and gauge theory, can proceed directly to Sec. 5 which treats examples. In Sec. 6, we summarize and sketch the directions for future research.

3. Bethe/Gauge-Correspondence

The Bethe/gauge-correspondence connects the gauge theories with $\mathcal{N} = (2,2)$ $d = 2$ super-Poincare invariance (for example, two-dimensional gauge theories with four supersymmetries, but also four-dimensional $\mathcal{N} = 1$ theories or four-dimensional $\mathcal{N} = 2$ theories subject to the two-dimensional Ω-deformation) and quantum integrable systems. It identifies the twisted chiral ring (the cohomology of certain nilpotent supercharge $\mathcal{Q}$) with the set of quantum integrals of motion, the space of supersymmetric vacua with the space of states, the joint eigenvalues of quantum integrals with the vacuum expectation values of the corresponding twisted chiral operators. In the gauge theory world, we do not care much about the reality of these operators. The space of all states of gauge theory is a Hilbert space with the physical Hamiltonian being Hermitian, however the states we are talking about are vacua, i.e. they are annihilated by the Hamiltonian.

Both the first evidence and the main use of this correspondence is the mathematical coincidence of the equations describing the supersymmetric vacua and the equations describing the spectrum of quantum Hamiltonians, in case those are known.

3.1. *Bethe equations*

In this case, they are known as Bethe equations, after H. Bethe who proposed in 1931 [10] an ansatz to describe the eigenstates of the Heisenberg spin chain:

$$\widehat{H} = \sum_{a=1}^{L} \vec{\sigma}_a \otimes \vec{\sigma}_{a+1} \,. \tag{21}$$

His ansatz reduces the diagonalizing the $(2^L \times 2^L)$-Hermitian matrix (for the spin chain with L sites, each occupied by a spin-$\frac{1}{2}$ system) to solving a system of N algebraic equations on N unknowns $\lambda_1, \ldots, \lambda_N$, known now as Bethe roots, or magnon rapidities:

$$\prod_{a=1}^{L} \frac{\lambda_i - \mu_a + \frac{\mathrm{i}}{2}}{\lambda_i - \mu_a - \frac{\mathrm{i}}{2}} = e^{\mathrm{i}\vartheta} \prod_{j \neq i} \frac{\lambda_i - \lambda_j + \mathrm{i}}{\lambda_i - \lambda_j - \mathrm{i}} \tag{22}$$

where $\mu_1, \ldots, \mu_L$ are the so-called inhomogeneities of the spin chain (some sort of displacement of the spin sites), and ϑ describes the quasi-periodic boundary conditions:

$$\vec{\sigma}_{a+L} = e^{-\frac{i\sigma_3\vartheta}{2}}\vec{\sigma}_a e^{\frac{i\sigma_3\vartheta}{2}}\,. \tag{23}$$

The remarkable achievement of Faddeev's school (see [25] for the pedagogical introduction) was the invention of the algebraic Bethe ansatz, in which Eqs. (22) ensure that the state

$$B(\lambda_1)\cdots B(\lambda_N)|\Omega\rangle \tag{24}$$

obtained by acting on the (quasi-)vacuum (all spins point down) with the analogues of creation operators is actually the eigenvector of the Hamiltonian and all quantum integrals of motion, generated by the (twisted) transfer-matrix

$$A(\lambda) + e^{i\theta}D(\lambda)\,. \tag{25}$$

The operators $A(\lambda)$, $B(\lambda)$, $C(\lambda)$, $D(\lambda)$ form the so-called Yangian (more precisely the Yangian of $sl(2)$), which is one of the first examples of quantum groups.

3.2. *Supersymmetric vacua*

Equations (22) were recognized [72, 73] as the equations describing the supersymmetric vacua of a two-dimensional supersymmetric gauge theory, more specifically a softly broken $\mathcal{N} = (4,4)$ theory with the gauge group $U(N)$, and L fundamental hypermultiplets. The inhomogeneities are the ratios of the masses of these fundamentals to the specific twisted mass responsible for the breaking $\mathcal{N} = 4 \to \mathcal{N} = 2$ (see [72] for details), the twist parameter is the (complexified) gauge theta angle (corresponding to the $U(1)$ factor in the gauge group). Equivalently, Eqs. (22) determine the spectrum of the so-called twisted chiral ring, or the A-type topological ring. The connection between the topological gauge theories and quantum integrable (many-body) systems has been observed earlier [36, 57], and further explored in [31, 32], culminating in [74].

The gauge theory description follows the minimization of the effective potential, which due to the low-energy supersymmetry is the norm squared of the derivatives of the so-called twisted superpotential $\widetilde{\mathcal{W}}$ so that the vacua are its generalized critical points:

$$\frac{\partial\widetilde{\mathcal{W}}(\sigma)}{\partial\sigma_i} = 2\pi i n_i\,, \qquad i = 1,\ldots,r \tag{26}$$

with σ_i being the flat coordinates on the complexified Cartan subalgebra of the gauge group, and n_i integers. Obviously, Eqs. (26) require a special coordinate system, unlike the more familiar equations describing the vacua of $\mathcal{N} = 2$ Landau–Ginzburg theories. On the quantum integrable side, the function $\widetilde{\mathcal{W}}$ is identified with the Yang–Yang potential [100], which plays an important role both in the proof of the completeness of Bethe ansatz (where it is complete) and in the understanding the normalizations of Bethe wavefunction [51] and the properties of the correlation functions [46].

3.3. *Quantum mechanics from four-dimensional gauge theory*

Some quantum integrable systems are sporadic, such as the spin chains of fixed finite spin at each site. Some come in deformation families, such as the many-body systems. In the latter case, the system can be viewed as the quantization of a classical mechanical system (we shall not discuss here to what extent the finite spin systems can be accurately described in terms of the classical mechanics, see the classical work [1]).

We are interested in the systems with variable Planck constant $\hbar$. It turns out that a natural mechanism of getting such systems within the framework of Bethe/gauge correspondence is to start with $\mathcal{N} = 2$ supersymmetric gauge theory in four dimensions. It can be still viewed as the two-dimensional theory, by choosing a $2+1+1$ decomposition $\mathbb{R}^{1,3} \to \mathbb{R}^{1,1} \times \mathbb{R}^2$ of the four-dimensional spacetime. The rotations of $\mathbb{R}^2$ become the two-dimensional $SO(2)$ R-symmetry. Now turn on the twisted mass for this symmetry [74]. This is equivalent to the two-dimensional Ω-deformation of the theory [64]. More specifically, let us take the $(3+1)$-dimensional spacetime to be the product $\mathcal{D}_\hbar \times \mathbb{S}^1 \times \mathbb{R}^1$ = cigar geometry $\times$ a circle $\times$ the time axis. The cigar is the two-dimensional disk with the rotationally invariant metric $f(r)^2(dr^2 + r^2 d\theta^2)$, with $f(r) \to 1$ as $r \to 0$. Let $\mathbb{S}^1$ have the circumference $2\pi R_2$. The Ω-deformation amounts to replacing the complex adjoint scalar ϕ in the vector multiplet by $\phi - \hbar D_\theta$, so that in particular $D_A\phi - \hbar \iota_{\partial_\theta} F_A$. Note that $\hbar$ is a complex parameter, and $\phi^* \mapsto \phi^* - \hbar^* D_\theta$. By choosing $f(r)$ so that $f(r) \to \frac{R_1}{r}$ at $r \to \infty$, and then taking $R_{1,2}$ sufficiently small, we will make the four-dimensional theory look like two-dimensional sigma model with the worldsheet $\mathbb{R}_+ \times \mathbb{R}^1$ [75]. The boundary conditions in this sigma model at $r \to 0$ coming from the absence of singularities in the four-dimensional theory happen to be that of the canonical coisotropic brane introduced in [47] and used in [48]. Morally speaking [39], the topological sigma model with the canonical coisotropic brane boundary conditions computes the same three point function as the topological open string introduced in [49] and studied in [19] (building on the work in [3]) with the purpose of giving an explicit formula for the deformation quantization of a Poisson manifold. The concept of deformation quantization is introduced in [9].

3.3.1. *Many quantum mechanics from four dimensions*

Actually, deformation quantization is not quite a quantization. Deformation quantization produces an associative algebra $\mathcal{A}_\hbar$ over formal (i.e. not necessarily converging) power series over $\hbar$, which in the $\hbar \to 0$ limit becomes the commutative algebra of functions on the Poisson manifold (the Poisson structure is the first derivative in $\hbar$ of the commutator at $\hbar = 0$). Even if this algebra happens to be well-defined for finite $\hbar$, it need not be representable. However, the construction above produces not only the algebra, it also gives its representation, in fact, a category of representations. These come from the boundary conditions at $r \to \infty$. The full classification is still lacking so we shall be informal.

3.4. *Partition function of the quantum system*

Suppose we do know the supersymmetric gauge theory which is Bethe/gauge dual to our quantum system. Using the Bethe/gauge correspondence, we can write the following identity for the partition function of the generalized Gibbs ensemble:

$$\mathcal{Z}(\bar{\beta}) = \mathrm{Tr}_{\mathcal{H}_{\mathrm{qis}}} e^{-\frac{1}{\hbar}\sum_k \beta_k \widehat{H}_k} = \mathrm{Tr}_{\mathcal{H}_{\mathrm{vac}}} e^{-\frac{1}{\hbar}\sum_k \beta_k \mathcal{O}_k}, \tag{27}$$

with $\bar{\beta} = (\beta_k)_{k=1}^r$ the set of generalized inverse temperatures, and $\mathcal{O}_k$ the basis of the twisted chiral ring, which we can further express as:

$$\mathrm{Tr}_{\mathcal{H}_{\mathrm{qis}}} e^{-\frac{1}{\hbar}\sum_k \beta_k \widehat{H}_k} = \mathrm{Tr}_{\mathcal{H}_{\mathrm{vac}}} e^{-\frac{1}{\hbar}\sum_k \beta_k \mathcal{O}_k} = \mathrm{Tr}_{\mathcal{H}_{\mathrm{vac}}} (-1)^F e^{-\frac{1}{\hbar}\sum_k \beta_k \mathcal{O}_k}, \tag{28}$$

assuming all vacua are bosonic, and as

$$\begin{aligned}\mathrm{Tr}_{\mathcal{H}_{\mathrm{qis}}} e^{-\frac{1}{\hbar}\sum_k \beta_k \widehat{H}_k} &= \mathrm{Tr}_{\mathcal{H}_{\mathrm{vac}}} (-1)^F e^{-\frac{1}{\hbar}\sum_k \beta_k \mathcal{O}_k} \\ &= \mathrm{Tr}_{\mathcal{H}_{\mathrm{gauge}}} (-1)^F e^{-\frac{1}{\hbar}\sum_k \beta_k \mathcal{O}_k},\end{aligned} \tag{29}$$

using $[\mathcal{Q}, \mathcal{O}_k] = 0$ and the usual Witten index argument.

Now assume further that the gauge theory in question is the Ω-deformed four-dimensional $\mathcal{N} = 2$ gauge theory on $\mathbb{R}^{1,1} \times \mathcal{D}$, with cigar-type geometry $\mathcal{D}$, the Ω-deformation with the parameter $\hbar$ on $\mathcal{D}$.

The corresponding Witten index is now given by the path integral on the four-dimensional space-time of the form $\mathbb{T}^2 \times \mathcal{D}$, with the Ω-deformation along $\mathcal{D}$. Now, the twisted chiral ring observables $\mathcal{O}_k$ come from the local observables $\mathcal{O}_k^{(0)}$ in the four-dimensional theory (the gauge invariant polynomials of the complex adjoint scalar ϕ). Their cohomological descendents $\mathcal{O}_k^{(2)}$ can be integrated over $pt \times \mathcal{D}$, with $pt \in \mathbb{T}^2$. In the Ω-deformed theory

$$\frac{1}{\hbar}\mathcal{O}_k(pt \times 0) = \int_{pt\times\mathcal{D}} \mathcal{O}_k^{(2)} \tag{30}$$

where 0 is the tip of the cigar. The latter description makes sense even when $\hbar \to 0$.

Thus, the quantum-mechanical system partition function is equal to the susy partition function of the $\mathcal{N} = 2$ gauge theory on $\mathbb{T}^2 \times \mathcal{D}$.

$$\mathcal{Z}(\bar{\beta}) = \int_{4d \text{ superfields}} e^{-\int_{\mathbb{T}^2\times\mathcal{D}} \mathcal{L}^{\hbar}_{\mathrm{SYM}}} \cdot e^{\sum_k \beta_k \int_{\mathcal{D}} \mathcal{O}_k^{(2)}} \tag{31}$$

with the perturbation terms $\beta_k \int_{\mathcal{D}} \mathcal{O}_k^{(2)}$ being essentially the Donaldson surface-observables integrated along the cigar $\mathcal{D}$, and $\mathcal{L}^{\hbar}_{\mathrm{SYM}}$ the Lagrangian of the Ω-deformed four-dimensional $\mathcal{N} = 2$ theory.

What is implicit in formula (31) is that on the left-hand side, one has a specific realization of the noncommutative algebra of observables in the (Hilbert) space of states $\mathcal{H}_{\mathrm{qis}}$, while on the right-hand side, one fixes some supersymmetric boundary conditions on the infinite end $\partial\mathcal{D}$ of the cigar $\mathcal{D}$.

4. Landau–Ginzburg from Gauge Theory

4.1. *Effective superpotential*

Now suppose the boundary conditions on $\partial\mathcal{D}$ are matched with $\mathcal{H}_{\mathrm{qis}}$. How can we exploit (31)? We can use the knowledge of the low-energy effective theory to get an estimate of the partition function for small $\hbar$ (and large enough β_k's).

Indeed, let us deform the metric on $\mathbb{T}^2$. The usual Witten index arguments guarantee the independence of the partition function on the size of $\mathbb{T}^2$. So let us make it very large, much larger than $\Lambda_{\mathrm{SYM}}^{-1}$ scale. Integrate out the high energy modes. We end up with the effective low-energy theory on $\mathbb{T}^2 \times \mathcal{D}$. The presence of the compact torus (it could be replaced by any compact Riemann surface at the expense of inserting an additional twisted chiral ring operator under the trace) implies the existence of the topological sectors in the effective abelian theory, namely, the electric and magnetic fluxes through $\mathbb{T}^2$:

$$m^i = \frac{1}{2\pi \mathrm{i}} \int_{\mathbb{T}^2} F^{(i)}, \qquad n_i = \frac{1}{2\pi \mathrm{i}} \sum_{j=1}^{r} \tau_{ij} \int_{\mathbb{T}^2} \star F^{(j)} \tag{32}$$

where $F^{(i)}$ are the abelian gauge curvatures, $i = 1, \ldots, r$, r being the number of the abelian vector multiplets in the effective theory,

$$\tau_{ij} = \frac{\partial^2 \mathcal{F}}{\partial a^i \partial a^j} \tag{33}$$

being the matrix of effective gauge couplings and theta angles, and $\mathcal{F}$ the prepotential of the effective theory.

Now let us perform the Kaluza–Klein compactification on this, admittedly, large $\mathbb{T}^2$, by blowing up the size of $\mathcal{D}$ to even larger proportions. It was understood in [54, 55] that the resulting two-dimensional theory on $\mathcal{D}$ is the $\mathcal{N} = 2$ supersymmetric Landau–Ginzburg theory (if instead of $\mathbb{T}^2$ one takes a Riemann surface Σ the original theory was to be partially twisted along Σ). The target space is the (disconnected) space consisting of the choices of the fluxes $\mathbf{n}, \mathbf{m}$ as in (33) and the moduli $(u_1, \ldots, u_r)$ of vacua of the four-dimensional theory. The additional degrees of freedom coming from flat abelian connections on $\mathbb{T}^2$ are effectively squeezed in the effective metric and can be neglected for the discussion in this section.

The superpotential (not to be confused with the twisted superpotential) of that theory is [54, 55]:

$$\mathcal{W}_{\mathbf{n},\mathbf{m}} = 2\pi \mathrm{i} \sum_{j=1}^{r} (n_j a^j + m^j a_{D,j}) - \sum_{k=1}^{r} \beta_k u_k \tag{34}$$

where a^j and

$$a_{D,j} = \frac{\partial \mathcal{F}}{\partial a^j} \tag{35}$$

are the electric and magnetic special coordinates on the Coulomb branch of the moduli space of vacua. Below we recall the more invariant definition of this superpotential in the language of the algebraic integrable systems, which are always behind the special geometry of the Coulomb branch of $\mathcal{N} = 2$ supersymmetric gauge theory in four dimensions [23] (see also [35, 56]).

Note that the similar flux-induced superpotential, for $\bar{\beta} = 0$ arises in the context of $\mathcal{N} = 1$, $d = 4$ theories obtained by engineering using non-compact Calabi–Yau threefold compactifications [90] and (also for $\bar{\beta} = 0$) in the context of Type II Calabi–Yau fourfold compactifications producing $\mathcal{N} = 2$ two-dimensional supergravity theories [40].

4.2. *Algebraic integrable systems*

Let $(\mathcal{X}^{2r}_{\mathbb{C}}, B^r_{\mathbb{C}}, \varpi_{\mathbb{C}}, \pi)$ be an algebraic integrable system [43], with π: $\mathcal{X}^{2r}_{\mathbb{C}} \to B^r_{\mathbb{C}}$ defining a Lagrangian fibration with the fibers $J_u = \pi^{-1}(u)$, $u \in B^r_{\mathbb{C}}$ being principally polarized abelian varieties. Let $(u_1, \ldots, u_r)$ denote the basis of global functions on $B^r_{\mathbb{C}}$, the Hamiltonians of the integrable system. Let $\beta_1, \ldots, \beta_r$ denote the corresponding Gibbs chemical potentials (generalized inverse temperatures).

Let $\Xi \subset B^r_{\mathbb{C}}$ denote the discriminant, the set of singular fibers. It is a stratified variety, with the maximal dimension strata of dimension $r - 1$. Let $u_* \in B^r_{\mathbb{C}} \backslash \Xi$ denote the generic point, and $\gamma \in H_1(J_{u_*}, \mathbb{Z})$ a cycle in the fiber over that point. Let (u, p) be a pair consisting of a point $u \in B^r_{\mathbb{C}}$ and a homotopy class p of a path connecting u_* to u. We can transport γ over u (the corresponding flat connection on the associated vector bundle $\bigcup_{u \in B^r_{\mathbb{C}}} H_1(J_u, \mathbb{C})$ is called the Gauss–Manin connection), producing a two-dimensional chain $S_p \subset \mathcal{X}^{2r}_{\mathbb{C}}$. Define:

$$\mathcal{W}_{\gamma; u_*}(u, p) = \mathrm{i} \int_{S_p} \varpi_{\mathbb{C}} - \sum_{k=1}^{r} \beta_k u_k \,. \tag{36}$$

In any integral symplectic basis A_i, B^i, $i = 1, \ldots, r$ in $H_1(J_{b_*}, \mathbb{Z}) \approx \mathbb{Z}^{2r}$, obeying

$$A_i \cap A_j = B^i \cap B^j = 0, \qquad A_i \cap B^j = \delta_i^j \tag{37}$$

with the intersection form $\cap$ given by the polarization, we can rewrite (36) as:

$$\mathcal{W}_{\gamma; u_*}(u, p) = 2\pi \mathrm{i} \sum_{i=1}^{r} (n_i a^i(u) + m^i a_{D,i}(u)) - \sum_{k=1}^{r} \beta_k u_k \,, \tag{38}$$

where $a^i(u)$, $a_{D,i}(u)$ are the special coordinates on $B^r_{\mathbb{C}}$, $m^i, n_i \in \mathbb{Z}$,

$$a^i(u) = \frac{1}{2\pi \mathrm{i}} \oint_{A_i} \lambda \,, \qquad a_{D,i}(u) = \frac{1}{2\pi \mathrm{i}} \oint_{B^i} \lambda \tag{39}$$

and $\lambda = d^{-1} \varpi_{\mathbb{C}}$. The ambiguity in the definition of λ corresponds to the choice of u_*, p in (38), shifting $\mathcal{W}_{\gamma; u_*}(u, p)$ by a constant. The fundamental group $\pi_1(B^r_{\mathbb{C}} \backslash \Xi)$ of the complement $B^r_{\mathbb{C}} \backslash \Xi$ acts on $H_1(J_{u_*}, \mathbb{Z})$ by symplectic transformations via Gauss–Manin monodromy representation.

A celebrated example of such a system is Hitchin's system [43], for which $\mathcal{X}^{2r}$ is the moduli space of stable Higgs pairs $(\mathcal{P}, \Phi)$ where $\mathcal{P}$ is a holomorphic principal $G_{\mathbb{C}}$ bundle over a complex curve C, and Φ is a holomorphic section of the bundle $K_C \otimes \mathfrak{g}_{\mathcal{P}}$. Here $r = (g - 1) \dim(G)$, where g is the genus of C.

The abovementioned elliptic Calogero–Moser system is a version of Hitchin system for $G = SL(N)$, for a punctured elliptic curve (see [37, 61], and implicitly in [22]).

The importance of this example is both the explicit nature of the Hamiltonians and the plethora of simple choices of real slices of the phase space which can be conceivably quantized. Then, it was this example (and its degeneration to the periodic Toda chain) which for the first time connected the supersymmetric gauge theory and the algebraic integrable system: namely it is the $\mathcal{N} = 2^*$ theory whose low-energy dynamics is governed by the elliptic Calogero–Moser system (this statement is now proven for the $SU(n)$ gauge theories [68, 69]).

The general correspondence is the following [88]: the phase space $\mathcal{X}_{\mathbb{C}}^{2r}$ is the moduli space of vacua of the $\mathcal{N} = 2$ theory on $\mathbb{S}^1 \times \mathbb{R}^{1,2}$ with the inverse size of $\mathbb{S}^1$ determining the Kähler class of the fibers J_b. The base $B_{\mathbb{C}}^r$ is the moduli space of vacua of the four-dimensional theory. The compactification on $\mathbb{S}^1$ and the compactification on $\mathbb{T}^2 = \mathbb{S}^1 \times \mathbb{S}^1$ are qualitatively similar: in both cases one ends up with the sigma model on $\mathcal{X}_{\mathbb{C}}^{2r}$.

From the point of view of the classical integrability, the special coordinates a^i (or $a_{D,i}$, or some integral linear combinations of those) are the action variables. They are distinguished by the fact that the canonically conjugate angle variables φ_i, such that

$$\varpi_{\mathbb{C}} = \sum_{i=1}^{r} da^i \wedge d\varphi_i = \sum_{i=1}^{r} da_{D,i} \wedge d\tilde{\varphi}^i \tag{40}$$

have the rigid periodicity:

$$\varphi_i \sim \varphi_i + 2\pi(n_i + \tau_{ij} m^j) \,, \qquad \tilde{\varphi}^i \sim \tilde{\varphi}^i + 2\pi(m^j - (\tau^{-1})^{ij} n_i) \tag{41}$$

with $n_i, m^i \in \mathbb{Z}$. They are the linear coordinates on the abelian fibers J_b, which are isomorphic to $\mathbb{C}^r/\mathbb{Z}^r \oplus \tau\mathbb{Z}^r$. If we denote by Γ the rank r lattice $\mathbb{Z}^r$, in the first description, and by $\tilde{\Gamma} = \tau\Gamma$, then the coordinates φ correspond to the isomorphism $\mathbb{C}^r \approx \Gamma \otimes \mathbb{C}$, while $\tilde{\varphi}$ correspond to $\mathbb{C}^r \approx \tilde{\Gamma} \otimes \mathbb{C}$.

4.3. *Landau–Ginzburg vacua*

Let us now look for the ground states of that effective Landau–Ginzburg theory, i.e. the critical points u_c of $\mathcal{W}_{\gamma;u_*}$. In fact, the derivative $d\mathcal{W}_{\gamma;u_*}$ is u_*-independent, so the set of critical points is u_*-independent. Let us assume that $(a^i)_{i=1}^r$ are good local coordinates on $B_{\mathbb{C}}^r$ near some critical point $u_c \in B_{\mathbb{C}}^r$. Then $d\mathcal{W}_{\gamma;u_*} = 0$ is equivalent to the system of equations:

$$n_i + \sum_{j=1}^{r} \tau_{ij} m^j = \frac{1}{2\pi \mathrm{i}} \sum_{k=1}^{r} \beta_k \frac{\partial u_k}{\partial a^i} \,, \qquad i = 1, \ldots, r \,. \tag{42}$$

What do these equations mean?

4.3.1. *Periodic orbit interpretation*

One obvious interpretation of (42) comes from the classical mechanics. The holomorphic function

$$H = -\mathrm{i}\sum_{k=1}^{r} \beta_k u_k \colon \mathcal{X}_{\mathbb{C}}^{2r} \to \mathbb{C} \tag{43}$$

generates the Hamiltonian flow, which is linear in the action-angle variables:

$$\dot{\varphi}_i = 2\pi \frac{\partial H}{\partial a^i}\,, \qquad \dot{a}^i = 0\,. \tag{44}$$

Thus, (42) means that the orbit of the flow (44) closes with the period 1.

4.3.2. *Spectral curve interpretation*

An algebro-geometric interpretation of (42) is obtained by using the functions (u_k) as local coordinates on $B^r_{\mathbb{C}}$:

$$d\mathcal{W}_{\gamma;b_*} = 0 \Leftrightarrow 2\pi\mathrm{i}\sum_{i=1}^{r}\left(n_i \frac{\partial a^i}{\partial u_k} + m^i \frac{\partial a_{D,i}}{\partial u_k}\right) = \beta_k\,, \qquad k = 1,\ldots,r\,. \tag{45}$$

Now, suppose the abelian variety J_b is a Jacobian of a (spectral) curve $\mathcal{C}_b$ (as it is the case for the $GL(n)$ Hitchin systems), or its Prym subvariety.

Then the special coordinates a^i, $a_{D,i}$ can be defined also by the periods of a meromorphic 1-differential $\tilde{\lambda}$ on $\mathcal{C}_b$:

$$a^i = \frac{1}{2\pi\mathrm{i}}\oint_{A_i}\tilde{\lambda}\,, \quad a_{D,i} = \frac{1}{2\pi\mathrm{i}}\oint_{B^i}\tilde{\lambda} \tag{46}$$

for some cycles A_i, $B^i \in H_1(\mathcal{C}_b, \mathbb{Z})$. The variations ϖ_k of $\tilde{\lambda}$

$$\varpi_k = \frac{\partial}{\partial u_k}\tilde{\lambda} \tag{47}$$

are holomorphic differentials on $\mathcal{C}_b$. Then (45) states that ϖ_k has the $C_{\mathbf{m},\mathbf{n}} = \sum_i n_i A^i + m^i B_i$ period equal to $-2\pi\mathrm{i}\beta_k$.

4.4. *Ω-deformed Landau–Ginzburg theory*

We now need to review a few facts about the Ω-deformation. It is usually discussed in the context of A-type models, starting with I and J-functions in two dimensions [33, 50] and the Z-function of the four-dimensional gauge theory [64]. In this paper, we will discuss the less familiar B-model case. Let us start by recalling the field content and supersymmetry of the topological B-model in two dimensions [92, 96].

4.4.1. *B-model off-shell*

Let $\mathcal{Y}$ be a complex manifold with the local holomorphic coordinates Y^i, $i = 1, \ldots, n = \dim \mathcal{Y}$. The fields of the model are $\mathbf{Y}^i = Y^i + \psi^i + F^i$, where $\psi^i \in \Gamma(\Omega^1_\Sigma \otimes Y^*\mathcal{T}_\mathcal{Y})$, $F^i \in \Gamma(\Omega^2_\Sigma \otimes Y^*\mathcal{T}_\mathcal{Y})$, and two pairs of boson-fermion scalars valued in $Y^*\overline{\mathcal{T}}_\mathcal{Y}$, $Y^{\bar{i}}$, $\eta^{\bar{i}}$, and $\chi^{\bar{i}}$, $H^{\bar{i}}$, (with Y, H being bosons, η, χ being fermions). Geometrically, $Y = (Y^i, Y^{\bar{i}})$: $\Sigma \to \mathcal{Y}$ is a map of a Riemann surface Σ to $\mathcal{Y}$. Define the fermionic nilpotent symmetry (twisted supersymmetry):

$$\begin{aligned} \delta Y^i &= 0\,, & \delta\psi^i &= dY^i\,, & \delta F^i &= d\psi^i\,, \\ \delta Y^{\bar{i}} &= \eta^{\bar{i}}\,, & \delta\chi^{\bar{i}} &= H^{\bar{i}}\,, & & \\ \delta\eta^{\bar{i}} &= 0\,, & \delta H^{\bar{i}} &= 0\,, & & \end{aligned} \tag{48}$$

where d stands for the de Rham operator on Σ. The action of the model is defined with the help of a Hermitian metric $\mathbf{g}_{i\bar{j}} dY^i dY^{\bar{j}}$ on $\mathcal{Y}$, the volume two-form ϖ_Σ on Σ, and the complex structure on Σ (one can trade those for a choice of a metric on Σ):

$$\mathbf{S} = \int_\Sigma \mathcal{O}_W^{(2)} + \delta \int_\Sigma [\mathbf{g}_{i\bar{j}}\psi^i \wedge \star dY^{\bar{j}} + \chi^{\bar{j}}(\mathbf{g}_{i\bar{j}}F^i - \varpi_\Sigma\, \bar{\partial}_{\bar{j}}\bar{W})] \tag{49}$$

where W: $\mathcal{Y} \to \mathbb{C}$ is a holomorphic function,

$$\begin{aligned} W(\mathbf{Y}^i) &= \sum_{i=0}^{2} \mathcal{O}_W^{(i)}\,, \qquad \mathcal{O}_W^{(i)} \in \Omega^i_\Sigma\,, \\ \mathcal{O}_W^{(1)} &= \psi^i \partial_i W\,, \\ \mathcal{O}_W^{(2)} &= F^i \partial_i W + \frac{1}{2}\psi^i \wedge \psi^j\, \partial^2_{ij} W \end{aligned} \tag{50}$$

and $\bar{W}$ is a complex conjugate of W in the twisted version of the $\mathcal{N} = (2,2)$ theory. If we only care about the δ-supersymmetry, i.e. the topological theory, we are free to deform $\bar{W}$ into anything we like, as long as the action is sufficiently non-degenerate.

When Σ has a boundary, the action (53) is not δ-invariant:

$$\delta\mathbf{S} = \int_{\partial\Sigma} \mathcal{O}_W^{(1)}\,. \tag{51}$$

4.4.2. *The Ω-deformation of the B-model*

Now suppose we have a vector field $V \in Vect(\Sigma)$ which preserves both the complex structure of Σ and ϖ_Σ (i.e. it is an isometry of the corresponding metric). The supersymmetry δ can be deformed into the equivariant supersymmetry $\delta_\hbar$ which acts as follows:

$$\begin{aligned} \delta_\hbar Y^i &= \iota_V \psi^i\,, & \delta_\hbar\psi^i &= dY^i + \iota_V F^i\,, & \delta_\hbar F^i &= d\psi^i\,, \\ \delta_\hbar Y^{\bar{i}} &= \eta^{\bar{i}}\,, & \delta_\hbar\chi^{\bar{i}} &= H^{\bar{i}}\,, & & \\ \delta\eta^{\bar{i}} &= \iota_V dY^{\bar{i}}\,, & \delta_\hbar H^{\bar{i}} &= \iota_V d\chi^{\bar{i}}\,. & & \end{aligned} \tag{52}$$

The action (53) is modified to

$$\mathbf{S}_\hbar = \int_\Sigma \mathcal{O}_W^{(2)} + \delta_\hbar \int_\Sigma [\mathbf{g}_{i\bar{j}}\psi^i \wedge \star dY^{\bar{j}} + \chi^{\bar{j}}(\mathbf{g}_{i\bar{j}}F^i - \varpi_\Sigma \bar{\partial}_{\bar{j}}\bar{W})] \tag{53}$$

where we now assume both ϖ_Σ and $\star$ to be V-invariant. In the presence of boundary, cf. (51),

$$\delta_\hbar \mathbf{S}_\hbar = \int_\Sigma d\psi^i\partial_i W + F^i\partial^2_{ij}W\iota_V\psi^j + \partial^2_{ij}W(dY^i + \iota_V F^i)\psi^j$$
$$+ \frac{1}{2}\partial^3_{ijk}W\psi^i\psi^j\iota_V\psi^k = \int_{\partial\Sigma}\psi^i\partial_i W\,. \tag{54}$$

Now let us assume $\partial\Sigma$ to be V-invariant and add a term

$$\int_{\partial\Sigma} W(Y)V^\vee \tag{55}$$

with one-form $V^\vee$ obeying $\iota_V V^\vee = 1$, $L_V V^\vee = 0$ on $\partial\Sigma$, so that its $\delta_\hbar$-variation cancels (54). The correction (55) is known as Warner term in other contexts.

Let us now analyze the localization locus for the path integral

$$\int e^{-\mathcal{S}_\hbar} \tag{56}$$

with

$$\mathcal{S}_\hbar = \mathbf{S}_\hbar - \int_{\partial\Sigma} W(Y)V^\vee\,. \tag{57}$$

We want to set the right-hand side of (52) to zero:

$$dY^i = \iota_V F^i\,,\quad d\psi^i = 0\,,\quad \iota_V\psi^i = \eta^{\bar{i}} = \iota_V d\chi^{\bar{i}} = 0\,,\quad H^{\bar{i}} = \iota_V dY^{\bar{i}} = 0\,. \tag{58}$$

The field H^i enters $\mathcal{S}_\hbar$ linearly. Thus we can integrate it out, taking into account the appropriate boundary conditions. This will lead to the constraint:

$$F^i = \varpi_\Sigma\, \mathbf{g}^{i\bar{j}}\bar{\partial}_{\bar{j}}\bar{W}\,. \tag{59}$$

This constraint implies the following gradient-like equations:

$$dY^i = \mathbf{g}^{i\bar{j}}\bar{\partial}_{\bar{j}}\bar{W}\iota_V\varpi_\Sigma\,. \tag{60}$$

Let us take $\Sigma = \mathcal{D}$ to be the disk (also known as the cigar), with the radial coordinates r, θ, $0 \le r \le R$, with the vector field $V = \hbar\partial_\theta$, and the "flat" volume form:

$$\varpi_\Sigma = rdr \wedge d\theta\,. \tag{61}$$

Equations (60) now take the form:

$$\begin{aligned} \partial_\theta Y^i &= 0 \\ \partial_r Y^i &= \hbar r \mathbf{g}^{i\bar{j}}\bar{\partial}_{\bar{j}}\bar{W}\,. \end{aligned} \tag{62}$$

Let us denote by Y_R the boundary value of (Y^i) (it is θ-independent thanks to (62)) and by Y_0 its value at the center of the disk (the tip of the cigar). The boundary term (55) now reads

$$\frac{2\pi}{\hbar} W(Y_R) \tag{63}$$

whereas the bulk action evaluates to

$$\int_\Sigma \varpi_\Sigma \, \mathbf{g}^{i\bar{j}} \partial_i W \bar{\partial}_{\bar{j}} \bar{W} = \frac{2\pi}{\hbar} \int_0^R dr \partial_r Y^i \partial_i W = \frac{2\pi}{\hbar} (W(Y_R) - W(Y_0)) \tag{64}$$

so that the path integral reduces to

$$\int_{\Gamma_R} e^{-\frac{\mathcal{W}}{\hbar}} \tilde{\Omega}_\hbar \tag{65}$$

where $\mathcal{W} = 2\pi W$, and Γ_R is a submanifold in $\mathcal{Y}$ spanned by the finite-time trajectories of the gradient vector field, i.e. the solutions to the equations

$$\partial_t Y^i = \hbar \mathbf{g}^{i\bar{j}} \bar{\partial}_{\bar{j}} \bar{W} \,, \qquad 0 \le t \le R^2 \,. \tag{66}$$

Now let us take the limit $R \to \infty$. Then the only trajectories contributing to the path integral would be those for which the bosonic action (64) is finite, which, among other things, implies that $\|\nabla W\|^2 \to 0$ as $R^2 \to \infty$. Thus, the boundary value Y_R of the map $Y\colon \Sigma \to \mathcal{Y}$ must land at one of the critical points Y_* of W, i.e. $dW|_{Y_*} = 0$. The center-value Y_0, by (64) is such, that

$$\begin{aligned} \mathrm{Im}(W(Y_0)/\hbar) &= \mathrm{Im}(W(Y_R)/\hbar) \\ \mathrm{Re}(W(Y_0)/\hbar) &\le \mathrm{Re}(W(Y_R)/\hbar) \,. \end{aligned} \tag{67}$$

Thus, Y_0 belongs to the Lefschetz timble Γ_{Y_*} of Y_*, i.e. the union of the gradient trajectories, emanating from Y_*.[b] See [44] for the earlier work on two-dimensional $\mathcal{N} = 2$ theories with boundaries. Of course, Eqs. (62) have appeared in the earlier work [20] where they describe the solitons in the Landau–Ginzburg theory. To map our setup to that of Cecotti and Vafa we need to treat the angular coordinate as the time direction, while $r^2/2$ becomes the spatial coordinate. The difference with the [20] case is that in our story the spatial interval is only semi-infinite. So only one critical point of W is involved.

The measure $\tilde{\Omega}_\hbar$ in (65) comes from the holomorphic top degree form on $\mathcal{Y}$. However, for non-compact $\mathcal{Y}$ it may have a non-trivial $\hbar$-dependence. We shall now get some idea as to what this dependence might be.

5. Topological Renormalization Group and Periods

Now let us take a break from the gauge theory and try to understand the appearance of Lefschetz thimbles in the problem in a more conventional way.

---

[b] I am grateful to E. Witten for discussions about these matters during the work on [75]. In [98, 99], the related ideas were developed without the use of the Ω-deformation. The advantage of our approach is the possibility of including the Hamiltonians u_k into the picture.

5.1. *Steepest descent deformations*

The general idea is well-known [6]. Let us view the (path) integral

$$Z = \int_{\mathcal{F}} [D\phi] e^{-\frac{S(\phi)}{\hbar}} \tag{68}$$

of some Euclidean field theory, with $\mathcal{F}$ the space of fields obeying the appropriate boundary conditions as a period integral, i.e. as an integral of the holomorphic top degree form

$$\Omega_\hbar = [D\phi] e^{-\frac{S(\phi)}{\hbar}} \tag{69}$$

along a middle-dimensional contour $\Gamma \subset \mathcal{F}^{\mathbb{C}}$ in the space of complexified fields φ. In (68) we have the starting contour $\Gamma_0 = \mathcal{F} \subset \mathcal{F}^{\mathbb{C}}$. Now, the value of the integral would not change if we moved the contour while keeping it in the region where the integral converges. In other words, we should not change the large-field, large-momentum asymptotics, but we are free to move the field space contour otherwise. For example, if V is any vector field on $\mathcal{F}^{\mathbb{C}}$, $V \in Vect(\mathcal{F}^{\mathbb{C}})$, and g_t the one-parametric family of diffeomorphisms of $\mathcal{F}^{\mathbb{C}}$ it generates, then

$$\int_{\Gamma_t} g_t^* \Omega_\hbar = Z \tag{70}$$

is t-independent.

Proof.

$$\frac{d}{dt} Z = \int_{\Gamma_t} g_t^* Lie_V \Omega_\hbar = \int_{\Gamma_t} g_t^* d(\iota_V \Omega_\hbar) = \int_{\Gamma_t} d(g_t^* \iota_V \Omega_\hbar) = 0\,. \tag{71}$$

Now, the actual form of the integrand changes with t. A popular choice is to take V to be the gradient vector field for $\mathrm{Re}(S/\hbar)$:

$$V = h^{ij} \frac{\partial}{\partial x^j} (\mathrm{Re}(S/\hbar)) \frac{\partial}{\partial x^i} \tag{72}$$

for some metric h on $\mathcal{F}^{\mathbb{C}}$. Outside the set of critical points of S the gradient flow decreases the absolute value of the measure factor $e^{-S/\hbar}$, making their contribution less and less important. As a result, asymptotically, the integral will be dominated by the contribution of the critical points of S. Note that the choice of the metric h is arbitrary. Moreover, we can iterate the procedure, by using one vector field V_1 for some "time" t_1, then another vector field V_2, e.g., the gradient vector field corresponding to another metric h_2, for some "time" t_2, or, more generally, make V "time"-dependent, e.g., of the form:

$$V(t) = h^{ij}(t) \frac{\partial}{\partial x^j} (\mathrm{Re}(S/\hbar)) \frac{\partial}{\partial x^i}\,. \tag{73}$$

Moreover, by taking h to be Hermitian, i.e. of the $(1,1)$ type in the complex structure of $\mathcal{F}^{\mathbb{C}}$, we get an additional bonus in that the imaginary part of $S/\hbar$ is preserved along the flow:

$$\mathrm{Im}(S/\hbar) = \mathrm{const}\,. \tag{74}$$

In this way, we map the problem of computing Z given by (68) to the problem of computing the partition function of the theory with the t-dependent action:

$$S_t = g_t^* S \tag{75}$$

and the measure given by the restriction of the holomorphic form Ω onto $\Gamma_t = g_t(\mathcal{F})$. We shall call the t-flow on the space of theories the *topological renormalization flow*.

Let us now discuss to what extent the critical points of S determine the universality classes of the theories. Let us from now on fix a Hermitian metric h (possibly t-dependent). Let Γ_* be the union of the trajectories, *emanating* from the critical point x_*, $dS|_{x_*}= 0$. More generally,[c] let F be a connected component of the set $Crit(S)$ of critical points of S,

$$dS|_f = 0\,, \quad f \in F\,. \tag{76}$$

We assume F is compact (otherwise our theory might be ill-defined). Let γ_{f}, $\mathsf{f} = 1, \ldots, r_F$ be a basis of the middle-dimensional homology of F. Let us choose some representatives $\widehat{\gamma}_{\mathsf{f}} \subset F$ of the corresponding cycles γ_{f}. Let Γ_{f} be the union of all trajectories emanating from the representative $\widehat{\gamma}_{\mathsf{f}}$. Now, define

$$Z_{\mathsf{f}} = \int_{\Gamma_{\mathsf{f}}} \Omega_\hbar\,. \tag{77}$$

The original integral can be expanded:

$$Z = \sum_{\mathsf{f}} n_{\mathsf{f}}\, Z_{\mathsf{f}}\,, \tag{78}$$

according to the decomposition of the homology class

$$[\Gamma] = \sum_{\mathsf{f}} n_{\mathsf{f}} [\Gamma_{\mathsf{f}}] \tag{79}$$

in the basis of Lefschetz thimbles. The integers n_{f}, in the finite dimensional case, can be computed by the intersection index

$$n_{\mathsf{f}} = \#\Gamma \cap \Gamma_{\mathsf{f}}^{\vee} \tag{80}$$

with the dual Lefschetz thimbles $\Gamma_{\mathsf{f}}^{\vee}$ (defined with the help of the gradient flow of $-\mathrm{Re}(S/\hbar)$, i.e. the negative of (72)).

In the infinite-dimensional case the intersection theory is subtle. However, one can sometimes argue for vanishing of n_{f} using (74). We should stress that the topological renormalization group flow makes the original contour Γ asymptotically approach the union of Γ_{f}'s non-uniformly. Namely, Γ itself is split into the parts Γ_f, which are individually attracted to different Lefschetz thimbles. The levels of the imaginary part $\mathrm{Im}(S/\hbar)$ evaluated at the critical points separate these parts Γ_f.

[c] I thank S. Gukov for a conversation about such situation.

5.2. *Lefschetz thimbles in quantum mechanics*

Let us now apply this formalism to the path integral computing the Euclidean partition function

$$\mathcal{Z}(\beta) = \mathsf{Tr}_{\mathcal{H}}\, e^{-\frac{\beta}{\hbar}\widehat{H}} . \tag{81}$$

It can be formally represented as an integral over the space of loops $L\mathcal{X}^{2r}$ in the classical phase space. The points in $L\mathcal{X}^{2r}$ are the maps $x\colon \mathbb{S}^1 \to \mathcal{X}^{2r}$, which we shall view as 1-periodic, $x(s+1) = x(s)$, $x(s) \in \mathcal{X}^{2r}$

$$\mathcal{Z}(\beta) = \int [Dp(s)Dq(s)]\, e^{\frac{i}{\hbar}\int pdq - \frac{\beta}{\hbar}\int H(x(s))ds} \tag{82}$$

where we introduced the Darboux coordinates for the symplectic form ϖ on $\mathcal{X}^{2r}$:

$$\varpi = \sum_{a=1}^{r} dp_a \wedge dq^a . \tag{83}$$

The details of the definition of (82) for non-exact ϖ are discussed in numerous sources so will not be addressed here. It is also well-known [26] that the domain of integration in (82) is not really the loop space of $\mathcal{X}^{2r}$, rather it is the ill-defined space of loops valued in the space of leaves of some polarization. Of course, as integrals go, the same quantity can be obtained by integration of different measures over different spaces. In fact, this ambiguity leads to the novel symmetries of quantum field theories, Nakajima's algebras [58–60] and their generalizations [65]. In this paper, however, we are only interested in the enumeration of the possible Lefschetz thimbles, without going into their internal details. So we shall continue as if we were indeed integrating over $L\mathcal{X}^{2r}$.

Let us assume, for simplicity, that the classical phase space $(\mathcal{X}^{2r}, \varpi)$ is a real slice of a complex symplectic manifold $(\mathcal{X}^{2r}_{\mathbb{C}}, \varpi_{\mathbb{C}})$. Then $L\mathcal{X}^{2r}$ is a real slice of $L\mathcal{X}^{2r}_{\mathbb{C}}$.

The action $S = i\oint \sum_{a=1}^{r} p_a dq^a \quad \beta \int H(p,q)ds$ is now a holomorphic function on $L\mathcal{X}^{2r}_{\mathbb{C}}$. Its critical points are the 1-periodic solutions to the Hamilton equations with complex p and q, with the Hamiltonian $-i\beta H(p,q)$:

$$\dot{q} = -i\beta\partial_p H , \quad \dot{p} = i\beta\partial_q H . \tag{84}$$

Now the circle closes: assume $(\mathcal{X}^{2r}_{\mathbb{C}}, \varpi_{\mathbb{C}})$ is an algebraic integrable system. One can generalize the problem (82) to include several Hamiltonians $\beta H \to \sum_{k=1}^{r} \beta_k u_k$. The 1-periodic complex loops solving (84) are exactly the solutions to (42).

Thus, the Lefschetz thimbles of quantum mechanical partition function are in one-to-one correspondence with the critical points of the superpotential $\mathcal{W}_{\mathbf{m},\mathbf{n}}$! Well...almost. The critical point of $\mathcal{W}_{\mathbf{m},\mathbf{n}}$ defines not just one periodic solution of (84), but the whole torus J_b ($b \in B^r$ is specified by fixing the values of all Hamiltonians u_k) of those solutions: the $s = 0$ value of the angle variables $(\varphi_i)_{i=1}^{r}$ can be chosen arbitrarily. Of these r complex moduli one real modulus is simply the translation in s.

5.3. *Twists*

Suppose $(\mathcal{X}^{2r}, \varpi, H)$ has a symmetry group G, i.e. is invariant under the action of some group G. Suppose the symmetry extends to the quantization, as the unitary representation of G in $\mathcal{H}$, commuting with $\widehat{H}$. The partition function (81) generalizes to the character

$$\mathcal{Z}_g(\beta) = \mathrm{Tr}_{\mathcal{H}}(g e^{-\frac{\beta}{\hbar}\widehat{H}}) . \tag{85}$$

The character-valued partition function (85) has the path integral representation, where one integrates (again, in the [26] sense) over the space of twisted loops:

$$x(s+1) = g \cdot x(s) . \tag{86}$$

The integral only depends on the conjugacy class of g in G (the transformation $g \mapsto h^{-1}gh$ can be undone by $x(s) \mapsto h^{-1} \cdot x(s)$). What are the Lefschetz thimbles for this problem?

Let us assume, again, that $(\mathcal{X}^{2r}_{\mathbb{C}}, \varpi_{\mathbb{C}})$ is an algebraic integrable system, and that the action of G on $\mathcal{X}^{2r}$ extends to $(\mathcal{X}^{2r}_{\mathbb{C}}, \varpi_{\mathbb{C}})$, where it acts by holomorphic symplectic transformations preserving the integrals.

Suppose G is finite. The Hamiltonians u_k, the action variables a^i and their duals $a_{D,i}$ are G-invariant. The G-action on the angle variables is highly constrained. Let us assume that G acts by shifts:

$$g\colon \varphi_i \mapsto \varphi_i + \frac{2\pi}{N(g)} \left(n_i(g) + \sum_{j=1}^{r} \tau_{ij} m^j(g) \right) \tag{87}$$

where $N(g)$ is the order of $g \in G$ (recall that by Cayley's theorem $G \subset S(|G|)$ so every conjugacy class belongs to a cyclic subgroup $\mathbb{Z}/N(g)\mathbb{Z}$ of sufficiently large order), $n_i(g), m^j(g)$ are integers defined modulo $N(g)\mathbb{Z}$. Then (86) condition translates to the following modification of (42):

$$n_i + \frac{n_i(g)}{N(g)} + \sum_{j=1}^{r} \tau_{ij} \left(m^j + \frac{m^j(g)}{N(g)} \right) = \frac{1}{2\pi \mathrm{i}} \sum_{k=1}^{r} \beta_k \frac{\partial u_k}{\partial a^i} , \qquad i = 1, \dots, r . \tag{88}$$

Of course, the solutions to (88) define closed loops on the quotient $\mathcal{X}^{2r}_{\mathbb{C}}/G$.

5.4. *The rest of the computation*

To compare with [21, 78], we need to classify the critical points $u_c = u_{i,\mathbf{m},\mathbf{n}}$ of $\mathcal{W}_{\mathbf{m},\mathbf{n}}$, compute the multiplicities $n_{i,\mathbf{m},\mathbf{n}}$, evaluate the critical value $\mathcal{W}_{\mathbf{m},\mathbf{n}}(u_{i,\mathbf{m},\mathbf{n}})$ of the action, estimate the one-loop corrections, and perform the sum over $i, \mathbf{m}, \mathbf{n}$. We leave this for future work.

Let us make a couple of comments.

(1) For the solution of (42) the values of $\mathbf{m}, \mathbf{n}$ are basis-dependent. However, their greater common divisor (g.c.d.) N is basis-independent. Its invariant meaning

is the multiplicity of the 1-cycle $[C_{i,\mathbf{m},\mathbf{n}}] \in H_1(J_{u_c}, \mathbb{Z})$ on the abelian variety, which is represented by the periodic orbit $C_{i,\mathbf{m},\mathbf{n}}$: $S^1 \to J_{u_c}$ of the Hamiltonian vector field. In other words, $[C_{i,\mathbf{m},\mathbf{n}}] = N[C_0]$, where $[C_0] \in H_1(J_{u_c}, \mathbb{Z})$ is a primitive class.

Another invariant integral data, apart from the g.c.d. N, comes from the set of vanishing cycles. Let us assume β_k are so large, that u_c is close to the discriminant locus $\Xi_\alpha \subset \Xi$, where the cycle $[C_\alpha]$ vanishes. For the details of the construction of the basis of vanishing cycles see [6]. The intersection number (it is defined using the polarization of the abelian variety) $n_\alpha = \#[C] \cap [C_\alpha]$ is well-defined. This is what we might call the number of instanton–anti-instanton pairs of type α. The "dual" cycle $[C_\alpha]^\vee$ is defined up to the addition of $[C_\alpha]$ (Picard–Lefschetz theory) [6], so that the dual numbers m_α are defined modulo n_α.

(2) The critical locus of S containing the orbit $C_{i,\mathbf{m},\mathbf{n}}$ is the abelian variety J_{u_c}. The middle-dimensional homology $H_r(J_{u_{i,\mathbf{m},\mathbf{n}}}, \mathbb{Z})$ group has rank $(2r)!\,/r!^2$. It is a representation of the monodromy group, which is a subgroup of $Sp(2r, \mathbb{Z})$. The choice of the cycles $\gamma_{i,\mathbf{m},\mathbf{n}} \in H_r(J_{u_{i,\mathbf{m},\mathbf{n}}}, \mathbb{Z})$ is constrained by the monodromy equivariance property. In the $r = 1$ case, the choice is simple: the 1-cycle is represented by the orbit itself.

(3) In the instanton gas prescription the enthropy prefactor $\beta^N/N!$ of the configuration of N instantons and anti-instantons, which comes from the integration over the "zero modes", is important in converting the series into the exponential of the energy splitting. However, as we argued above, the number of zero modes of a true critical point of the analytic continued action is N-independent. Thus, we must get the $\beta^N/N!$ from the one-loop fluctuation determinant. Somehow the N zero modes of the approximate solution should flow to N low-lying eigenvalues of the linearization of the equations of motion around the true solution. Since the linearized equations are the second order differential equation on a circle of periodicity β while the (matrix) coefficients of the equation have the periodicity β/N, one can construct the eigenvectors by taking the Bloch solutions with the Bloch phases given by the Nth roots of unity.

(4) The one-loop correction is computed by the regularized determinant of the second quadratic form of the action S expanded about the critical point. As we have complexified the space of fields, we should view it as an infinite-dimensional quadratic form. The convergent Gaussian integral along the Lefschetz thimble $\Gamma_{i,\mathbf{m},\mathbf{n}}$ has a phase, which comes from the restriction of the holomorphic top form $\Omega_\hbar$ onto a real subspace of the tangent space to $L\mathcal{X}_{\mathbb{C}}^{2r}$ at C, spanned by the tangent vectors to the representative $\widehat{\gamma}_{i,\mathbf{m},\mathbf{n}}$, as well as the eigenvectors of $\delta^2\,\mathrm{Re}(S/\hbar)$ with positive eigenvalues. Of course, the multiplication by i maps such an eigenvector to the eigenvector with the opposite eigenvalue.

We should however remember that the integration domain is not precisely $L\mathcal{X}^{2r}$. If we approximate the loop by an K-gon, with a very large K, the action

S is approximated by [26]:

$$S = -\mathrm{i}\sum_{k=1}^{K} p_{a,k}(q_{k+1}^a - q_k^a) + \frac{\beta}{K}\sum_{k=1}^{K} H(p_k, q_k)\,, \tag{89}$$

with $q_{K+1} = q_1$. This is to be compared, e.g., to the integral of $-\mathrm{i}\varpi$ over the polygon in the $\mathbb{R}^{2r}$ with the vertices (p_k, q_k). In computing the $\delta^2 S$ one should keep in mind the choice of the polarization.

6. Examples of the Models

In this section, we consider a few specific examples. We start with the systems with one degree of freedom: Heun system (Gaudin on 4-points), Lamé system, and the anharmonic oscillator.

6.1. *One degree of freedom*

Let us start with the systems describing a one-dimensional particle (or a spin degree of freedom): let us assume the Hamiltonian is quadratic in momentum:

$$H(p,x) = \frac{1}{2}A(x)p^2 + B(x)p + C(x) \tag{90}$$

with some functions A, B, C. Such system is an algebraic integrable system if the energy level set $H(p,q) = E$ is an elliptic curve (perhaps missing a few points). For example, one can choose A to be degree 4 polynomial [91].

By going to the coordinate $\tilde{x} \sim \int dx/\sqrt{A(x)}$ and performing a similarity transformation of the wavefunction the Hamiltonian (90) can be mapped to the standard non-relativistic form

$$\tilde{H}(\tilde{p},\tilde{x}) = \frac{1}{2}\tilde{p}^2 + U(\tilde{x}) \tag{91}$$

for some function $U(\tilde{x})$.

6.1.1. *Gaudin model*

Let us start with a $n = 4$-point $SL(2)$ Gaudin model: the phase space $\mathcal{X}^2_{\mathbb{C}}$ is the complex symplectic quotient:

$$\mathcal{X}^2_{\mathbb{C}} = (\mathcal{O}_{\nu_1} \times \cdots \times \mathcal{O}_{\nu_n})//SL(2,\mathbb{C}) \tag{92}$$

where

$$\mathcal{O}_\nu = \{(x,y,z)\,|\,x^2 + y^2 + z^2 = \nu^2\} \tag{93}$$

with the symplectic form

$$\varpi_\nu = \frac{dx \wedge dy}{2z}\,. \tag{94}$$

The space $\mathcal{X}^2_{\mathbb{C}}$ can be identified with the space of meromorphic $\mathfrak{sl}(2,\mathbb{C})$-valued 1-forms

$$\phi(w) = \sum_{a=1}^{n} \phi_a \frac{dw}{w - w_a} \tag{95}$$

where

$$\phi_a = \begin{pmatrix} x_a & y_a + \mathrm{i}z_a \\ y_a - \mathrm{i}z_a & -x_a \end{pmatrix} \tag{96}$$

obey

$$\sum_{a=1}^{n} \phi_a = 0\,, \tag{97}$$

and we identify $(\phi_1, \ldots, \phi_n) \sim (g^{-1}\phi_1 g, \ldots, g^{-1}\phi_n g)$, for $g \in SL(2,\mathbb{C})$. The positions w_1, w_2, w_3, w_4 of the poles are the parameters of the model. Actually, only the cross-ratio

$$\mathfrak{q} = \frac{w_2 - w_1}{w_3 - w_1}\frac{w_3 - w_4}{w_2 - w_4} \tag{98}$$

is relevant (for $n > 4$ there will be $n-3$ parameters). However, sometimes it is convenient to use the redundant parametrization, i.e. in exploring various degenerations, including the one to the anharmonic oscillator. For now, however, we will choose $w_1 = 0$, $w_2 = \mathfrak{q}$, $w_3 = 1$, $w_4 = \infty$. Define the spectral curve $\mathcal{C}$:

$$\mathtt{Det}(\rho - \phi(w)) = 0 \tag{99}$$

which sits in the cotangent bundle $T^*\mathbb{CP}^1$ to $\mathbb{CP}^1 \ni z$ with the fibers over 0, $\mathfrak{q}$, 1, ∞ deleted. Explicitly:

$$\begin{aligned}\rho^2 &= \sum_{a=1}^{4} \frac{\nu_a^2}{(w - w_a)^2} + \frac{H_a}{w - w_a} \\ &= \frac{\nu_1^2}{w^2} + \frac{\nu_2^2}{(w-\mathfrak{q})^2} + \frac{\nu_3^2}{(w-1)^2} + \frac{\nu_4^2 - \nu_1^2 - \nu_2^2 - \nu_3^2}{(w-1)(w-\mathfrak{q})} + \frac{H\mathfrak{q}(\mathfrak{q}-1)}{w(w-1)(w-\mathfrak{q})}\end{aligned} \tag{100}$$

where H is the Hamiltonian of our system. There is a nice geometric picture of the phase space $\mathcal{X}^2$ as the complexification of the space of closed polygons (quadrangles) with fixed (complex) lengths of the edges (ν_a's) in the three-dimensional Euclidean space, viewed up to the isometries. The advantage of the complex situation is that there is no need for the triangle inequalities: all complex lengths are good. Let us label the vertices by $p_a \in \mathbb{C}^3$, $a = 1, \ldots, 4$, so that

$$p_{a+1} - p_a = (x_a, y_a, z_a)\,, \quad a = 1, \ldots, 4\,, \quad p_{a+4} = p_a\,. \tag{101}$$

The Darboux coordinates on this space can be chosen to be the length $p = \ell_{12}$ of the diagonal connecting p_1 and p_3, and the dihedral angle $x = \vartheta_{12}$ between the triangles $p_1p_2p_3$ and $p_3p_4p_1$. The Hamiltonian H (100) is given by:

$$H = \frac{\mathsf{Tr}(\phi_1\phi_2)}{\mathfrak{q}} + \frac{\mathsf{Tr}(\phi_2\phi_3)}{\mathfrak{q}-1}\,. \tag{102}$$

Using

$$\ell_{ab}^2 = (x_a + x_b)^2 + (y_a + y_b)^2 + (z_a + z_b)^2\,, \tag{103}$$

and

$$\cos(\vartheta_{12}) = \frac{\mathsf{Tr}[\phi_1,\phi_2][\phi_3,\phi_4]}{\sqrt{(\mathsf{Tr}[\phi_1,\phi_2]^2)(\mathsf{Tr}[\phi_3,\phi_4]^2)}} \tag{104}$$

we can rewrite (102) as:

$$\begin{aligned}\mathfrak{q}(1-\mathfrak{q})H &= (1-\mathfrak{q})(p^2-\nu_1^2-\nu_2^2) + \frac{\mathfrak{q}}{2p^2}(p^2-\nu_1^2+\nu_2^2)(p^2+\nu_3^2-\nu_4^2)\\ &\quad + \frac{\mathfrak{q}\cos(x)}{2p^2}\sqrt{((p^2-\nu_1^2-\nu_2^2)^2-4\nu_1^2\nu_2^2)((p^2-\nu_3^2-\nu_4^2)^2-4\nu_3^2\nu_4^2)}\,.\end{aligned} \tag{105}$$

See [91] for other realizations of this model.

The simplest presentation of the Hamiltonian (102) is in terms of the separated variables. The specification of that approach to the system with one degree of freedom will produce, upon the usual quantization, the Hamiltonian

$$\widehat{H} = -\hbar^2\, x(x-1)(x-\mathfrak{q})\frac{\partial^2}{\partial x^2} + \frac{\Delta_1\mathfrak{q}}{x} + \frac{\Delta_2\mathfrak{q}(\mathfrak{q}-1)}{x-\mathfrak{q}} + \frac{\Delta_3(1-\mathfrak{q})}{x-1} + \Delta_4 x \tag{106}$$

which can be mapped to the standard non-relativistic form (91) by choosing $\tilde{x}$ to live on the elliptic curve:

$$y^2 = 4x(x-1)(x-\mathfrak{q})\,, \qquad d\tilde{x} = dx/y\,. \tag{107}$$

The model (106) is important since it is Bethe/gauge dual to the four-dimensional $SU(2)$ gauge theory with $N_f = 4$ fundamental hypermultiplets. The parameters $\Delta_i = \nu_i(\nu_i - \hbar)$ are related to the masses of four quarks. The parameter $\mathfrak{q}$ is determined by the microscopic gauge coupling.

6.2. *Anharmonic oscillator*

This is the model we shall discuss in more detail: a particle in the one-dimensional potential

$$U(q) = \frac{\lambda}{4}(q^2 - v^2)^2\,. \tag{108}$$

It can be viewed as an irregular limit of the general Gaudin model. It is not Bethe/gauge dual to any four-dimensional gauge theory with Lagrangian description. However it can be obtained by a mass deformation of one of the non-trivial rank 1 $\mathcal{N}=2$ superconformal fixed points [4].

The complexified (and compactified) energy level set

$$H(p,q) = \frac{1}{2}p^2 + U(q) = E \tag{109}$$

is an elliptic curve. Let $U_0 = U(0) = \lambda v^4/4$. Let us choose the parametrization:

$$p = \mathrm{i}\sqrt{2U_0}\,\rho\,, \qquad q = v\,\xi\,, \qquad E = U_0\,\epsilon \tag{110}$$

so that the curve (109) becomes:

$$\rho^2 = (\xi^2 - 1)^2 - \epsilon\,. \tag{111}$$

The complex phase space $\mathcal{X}^2_{\mathbb{C}}$ is a partial compactification of the naive space $\mathbb{C}^2$ of complex coordinates q and momenta p, where we add the locus (two copies of $\mathbb{C}$) where q and p go to infinity, as:

$$p \sim \pm\mathrm{i}\sqrt{\frac{\lambda}{2}}\left(q^2 - v^2 - \frac{2E}{\lambda q^2} + \cdots\right). \tag{112}$$

This compactification is physically natural, since it takes only a finite time to reach infinity while moving in the inverted potential. Let us denote the compactified curve by $\mathcal{E}_\epsilon$. The symmetry $G = \mathbb{Z}_2$ acting by

$$(p,q) \mapsto (-p,-q) \tag{113}$$

preserves both the symplectic form and the Hamiltonian $H(p,x)$.

The elliptic curve $\mathcal{E}_\epsilon$ (111) maps 2:1 to another elliptic curve E_ϵ, which sits in the quotient $\mathcal{X}^2_{\mathbb{C}}/\mathbb{Z}_2$:

$$y^2 = 4x((x-1)^2 - \epsilon) \tag{114}$$

where

$$y = 2\rho\xi\,, \qquad x = \xi^2\,. \tag{115}$$

Such map is called an isogeny. As long as $\epsilon \neq 1$ the fixed points of the involution (113) do not belong to (111). Thus, (113) acts by a half-period shift, as can be seen explicitly below.

We thus have represented both $\mathcal{X}^2_{\mathbb{C}}$ and $\mathcal{X}^2_{\mathbb{C}}/\mathbb{Z}_2$ as the algebraic integrable systems, with the same base $B^1_{\mathbb{C}} \approx \mathbb{C}$, with the coordinate ϵ. The fiber J_ϵ is the curve (111) $\mathcal{E}_\epsilon \subset \mathcal{X}^2_{\mathbb{C}}$ and the curve $E_\epsilon \subset \mathcal{X}^2_{\mathbb{C}}/\mathbb{Z}_2$, respectively.

We can now analyze our problem of finding the periodic and $\mathbb{Z}_2$-twisted periodic orbits on $\mathcal{X}^2_{\mathbb{C}}$. First, let us choose a basis in $H_1(\mathcal{E}_\epsilon, \mathbb{Z})$. Suppose $|\epsilon| \ll 1$. Let

$$\xi_\pm = (1 \pm \sqrt{\epsilon})^{\frac{1}{2}} \approx 1 \pm \frac{1}{2}\sqrt{\epsilon}\,. \tag{116}$$

We choose the A-cycle to be the cycle which circles around the cut $[\xi_-, \xi_+]$ on the ξ-plane, while the B-cycle is the double cover of the interval connecting the two cuts: $[-\xi_-, \xi_-]$. The A-cycle vanishes when $\epsilon \to 0$. The monodromy under $\epsilon \mapsto e^{+2\pi\mathrm{i}}\epsilon$ takes ξ_- to ξ_+, the cycle A to itself, and

$$B \mapsto B + 2A \tag{117}$$

since the end-points of both small cuts $[\pm\xi_-, \pm\xi_+]$ get exchanged, so that $[-\xi_-, \xi_-]$ turns into

$$[-\xi_+, \xi_+] = [-\xi_+, -\xi_-] \cup [-\xi_-, \xi_-] \cup [\xi_-, \xi_+] . \tag{118}$$

The corresponding periods $\frac{1}{2\pi}\oint p dx$ are given by (with $\tilde{\epsilon} = \epsilon/64$):

$$\begin{aligned} a &= \frac{\mathrm{i}v\sqrt{2U_0}}{\pi} \int_{\xi_-}^{\xi_+} \rho d\xi \\ &= -16v\sqrt{2U_0}\tilde{\epsilon}(1 + 6\tilde{\epsilon} + 140\tilde{\epsilon}^2 + 4620\tilde{\epsilon}^3 + \cdots) , \\ a_D &= \frac{\mathrm{i}v\sqrt{2U_0}}{\pi} \int_{-\xi_-}^{\xi_-} \rho d\xi \\ &= \frac{2\mathrm{i}S_0}{2\pi} + \frac{2a}{2\pi\mathrm{i}}(\log(\tilde{\epsilon}) - 1 + 23\tilde{\epsilon} + 612\tilde{\epsilon}^2 + \cdots) \end{aligned} \tag{119}$$

where

$$2S_0 = \frac{8v\sqrt{2U_0}}{3} \tag{120}$$

is twice the instanton action. Note the 2 in the numerator of $\frac{2a}{2\pi\mathrm{i}}$ in (119) is the same 2 as in the monodromy transformation (117), and the same 2 as in (120). The period of motion along the orbit represented by the cycle $mA + nB$ with $m, n \in \mathbb{Z}$, $m \sim m + 2n$ is given by:

$$\begin{aligned} &2\pi\mathrm{i}\frac{\partial}{\partial E}(ma + na_D) \\ &\quad = \frac{\mathrm{i}\pi m + n(\log(\tilde{\epsilon}) + 40\tilde{\epsilon} + 1076\tilde{\epsilon}^2 + \cdots)}{\omega} \cdot (1 + 12\tilde{\epsilon} + 420\tilde{\epsilon}^2 + \cdots) , \end{aligned} \tag{121}$$

where

$$\omega = \sqrt{U''(v)} = \sqrt{2\lambda v^2} = 2\frac{\sqrt{2U_0}}{v} \tag{122}$$

is the frequency of small oscillations near the minimum of the potential (108). In quantum field theory language ω is the mass of the perturbative quantum of the field $x(t)$. We see that for real ϵ the contribution of the B-cycle to the period is real (it corresponds to the classically allowed motion in the inverted potential) while the contribution of the A-cycle is imaginary. But ϵ does not have to be real: in fact, by setting (121) to be equal to β we get, for $\beta \to +\infty$:

$$\tilde{\epsilon} = e^{\frac{2\pi\mathrm{i}m}{N}} \exp\left(-\frac{2\beta\omega}{N}\right) + \cdots \tag{123}$$

for $n = -|n| < 0$, $N = 2|n|$. The neglected terms are exponentially suppressed. The critical value of the action on such a solution is equal to

$$\mathcal{W}_{m,n} = 2\pi\mathrm{i}\left(1 - E\frac{\partial}{\partial E}\right)(ma + na_D) = NS_0 \cdot \left(1 - 12\, e^{\frac{2\pi\mathrm{i}m}{N}} \exp\left(-\frac{2\beta\omega}{N}\right) + \cdots\right). \tag{124}$$

Now let us discuss the modification of this result for the $\mathbb{Z}_2$ twisted trajectories. They project to the closed loops on E_ϵ which are not in the image of closed loops on $\mathcal{E}_\epsilon$ under the isogeny (115). The first homology group of E_ϵ is generated by the $\mathcal{A}$-cycle, which circles around the cut $[\xi_-^2, \xi_+^2]$ on the x-plane, and the $\mathcal{B}$-cycle, which is the double cover of the interval $[0, \xi_-^2]$. The isogeny (115) maps A to $\mathcal{A}$ and B to $2\mathcal{B}$. Accordingly the monodromy $\epsilon \mapsto e^{2\pi i}\epsilon$ maps $\mathcal{B} \mapsto \mathcal{B} + \mathcal{A}$. An analogous analysis will give the result (123), (124) with odd N.

Which pairs (m, N) with $m = 0, \ldots, N-1$ actually contribute to the path integral? In the phase space formulation (82) of the path integral the exponential has, for real $\hbar$, an imaginary part $\propto \frac{1}{\hbar}\oint pdq$. The real contour $L\mathcal{X}^{2r}$ is stratified by the value of $I = \frac{1}{\hbar}\oint pdq$, so that the flow lines of the anti-gradient $-\nabla\,\mathrm{Re}(S/\hbar)$ emanating from the critical point (m, N) may cross it at the stratum with $I \approx -12NS_0 e^{-\frac{2\beta\omega}{N}}\sin(2\pi m/N)$. So, at first sight all the critical points should contribute. However, if we integrate out p and represent the thermal partition function as the integral over real periodic trajectories $q(t) = q(t+\beta)$ with the real Euclidean action

$$\int_0^\beta dt \left(\frac{1}{2}\dot{q}^2 + U(q)\right).$$

The gradient flow deforming the contour of integration over real loops will not change the imaginary part of the action (74). Then, as only the pairs $(0, n)$ and $(n, 2n)$ give real critical values (at least in the approximation we are working), so it appears only the $\Gamma_{0,n}$ and $\Gamma_{n,2n}$ Lefschetz thimbles will contribute to the thermal partition function, at real β.

At any rate, once we take the parameters of the problem (slightly) complex, e.g., in comparing our formalism with [12], then all (m, n) pairs play a role.

6.3. *Generalization to several degrees of freedom*

6.3.1. *g identical particles in a polynomial potential*

In our previous discussion the fact that the potential $U(q)$ is quartic seems to play a crucial role. If $U(q)$ is degree $2g+1$ or degree $2g+2$ polynomial, with $g > 1$, then the curve $H(p, q) = E$ is hyperelliptic of genus g. The classical motion in the complexified phase space is still a linear flow in the "coordinate"

$$\int \frac{dq}{p} \tag{125}$$

which has $2g$ periods (the differential dq/p is holomorphic). Fixing the period of motion to be equal to β is not enough to fix all the moduli of the curve $H(p, q) = E$, but should be enough to fix E. The motion of one particle in such a potential may still be amenable to the analysis like we did for the anharmonic oscillator, but it is not an algebraic integrable system, and is not likely to correspond to some gauge theory. However, the motion of g identical particles is [85, 38]. The corresponding abelian variety is the Jacobian of the curve. The integrals of motion are the g parameters $u_1, \dots, u_g$ of the curve, which must be chosen in such a way that the derivatives of the Liouville one-form pdq with respect to u_k are holomorphic differentials. This is precisely the setup of Seiberg–Witten theory [86, 87].

6.3.2. *N-particle Toda chain*

The system of N-particles governed by the Hamiltonian

$$\widehat{H} = -\frac{\hbar^2}{2}\sum_{i=1}^{N}\frac{\partial^2}{\partial q_i^2} + \Lambda^2\sum_{i=1}^{N} e^{q_i - q_{i+1}} \tag{126}$$

where $q_{N+1} = q_1$, is famously dual to the pure $\mathcal{N} = 2$ super-Yang–Mills with gauge group $SU(N)$. It was observed in the classical limit in [35, 56], in [15, 16] in the formal quantization, in [74] in the actual L^2-quantization on the real line $q_i \in \mathbb{R}$. It is also possible to study the eigenvalue problem (126) where $q_i \in \mathrm{i}\mathbb{R}/2\pi\mathbb{Z}$, however new phenomena arise in this case, notably the non-perturbative splitting of the levels for which our formalism is being developed (see [45] for the current status of the problem for $N = 2, 3$, and [34] for new developments).

We expect to see our non-linear superpositions of instantons and anti-instantons in the quantum mechanical model in the limit where all Gibbs potentials $\beta_k \to \infty$. The corresponding spectral curve

$$Y + \frac{\Lambda^{2N}}{Y} = P_N(x) = x^N + u_1 x^{N-1} + \cdots + u_N \tag{127}$$

becomes maximally degenerate. These are the equilibrium points of Toda chain. The corresponding values of u_k's correspond to the vacua of $\mathcal{N} = 1$ theory which is obtained from $\mathcal{N} = 2$ by soft supersymmetry breaking via a superpotential deformation. At such a point $n - 1$ cycle vanishes. So we expect the critical points to be enumerated by the collections of integers $\mathbf{m}, \mathbf{n}$ with $\mathbf{n} = (n_1, \dots, n_{N-1})$ and $\mathbf{m} = (m_1, \dots, m_{N-1})$ with m_i's defined modulo some lattice generated by n_i's.

6.3.3. *n-point Gaudin system*

The maximal degeneration, which should occur when all $\beta_k \to \infty$ for $k = 1, \dots, n - 3$:

$$\rho^2 = \frac{B^2(w)(\alpha w^2 + \beta w + \gamma)}{R(w)^2} \tag{128}$$

where

$$R(z) = \prod_{i=1}^{n}(w - w_i) \tag{129}$$

with degree $n - 3$ monic polynomial $B(w)$. The n parameters α, β, γ and the coefficients of $B(w)$ are fixed by:

$$B(w_i)^2(\alpha w_i^2 + \beta w_i + \gamma) = \nu_i^2\,, \qquad i = 1, \dots, n\,. \tag{130}$$

Of course, solving the Eqs. (130) explicitly is not possible. But the connection to gauge theory gives us a good idea as to what these solutions are. For example, in the $n = 4$ case the corresponding theory is $SU(2)$ with $N_f = 4$ flavors. So the six solutions to (130) in this case are the vacua where one of the quarks, or a monopole, or a dyon becomes massless. The analogous analysis in the $n > 4$ case is simple using the correspondence described in detail in [29].

6.4. *Black holes*

Consider the free motion of a probe particle in the background of Schwarzschild black hole of mass M (we set $GN = 1$):

$$
\begin{aligned}
ds^2 &= \left(1 - \frac{2M}{r}\right) dt^2 - \left(1 - \frac{2M}{r}\right)^{-1} dr^2 - r^2 d\Omega_2^2 \\
&= \frac{32M^3 e^{-\frac{r}{2M}}}{r} du_+ du_- - r^2 d\Omega_2^2
\end{aligned}
\tag{131}
$$

where

$$
u_\pm = \pm\sqrt{\frac{r}{2M} - 1}\exp\frac{r \pm t}{4M} \tag{132}
$$

are the Kruskal–Szekeres coordinates, expressed through the Schwarzschild asymptotic time t and the radial variable r. The line element on the two-sphere $d\Omega_2^2 = d\theta^2 + \sin^2(\theta)d\phi^2$. The geodesic equation, written in the Hamilton–Jacobi form, reads as:

$$
\left(\frac{r}{2M}\right)^3 e^{\frac{r}{2M}} \partial_{u_+} S \partial_{u_-} S - 4\left((\partial_\theta S)^2 + \frac{1}{\sin^2(\theta)}(\partial_\phi S)^2\right) = (2mr)^2 \tag{133}
$$

for the probe mass m. Separating the variables via:

$$
S(u_+, u_-; \theta, \phi) = n\phi + s(\theta) + E\log(u_+/u_-) + \Sigma(\log(u_+ u_-)) \tag{134}
$$

we obtain, for the radial generating function Σ:

$$
\Sigma(x) = 4Mp \int^x w \frac{dz}{z(z-1)}\,, \tag{135}
$$

with

$$
w^2 = z(z^3 + \mu^2 z^2 + \nu^2(1 - z)) \tag{136}
$$

where $z = r/2M$, $\mu = m/p$, $\nu = L/(2Mp)$, $p^2 = \epsilon^2 - m^2$, $\epsilon = E/(4M)$, E is the energy measured by the outside observer, L is the angular momentum[d] and

$$
x = \log(u_+ u_-) = z + \log(z - 1)\,, \qquad dx = \frac{dz}{(1 - z^{-1})} \tag{138}
$$

is the tortoise coordinate.

[d]The angular potential $s(\theta)$ is given by

$$
s(\theta) = \int \sqrt{L^2 \sin^2(\theta) - n^2} \frac{d\theta}{\sin(\theta)}\,, \tag{137}
$$

and its explicit form is not important for us.

We observe that (136) is, again, an elliptic curve, when the variables p and z are extended to the complex domain. We thus expect that the tunneling under the horizon picture [77] of the Hawking radiation [42] should be properly formulated in terms of the periodic orbits on such a curve, as in the previous quantum mechanical examples. It would be interesting to explore this issue further, also in the context of charged and rotating black holes.

7. Summary and Future Directions

This paper is a result of an attempt to write an introduction to the book on modern instanton calculations. We found true critical points of the classical mechanical actions in a variety of examples. Even though these examples are special in that the complexification of the classical system is an algebraic integrable system, we believe the qualitative picture of these solutions can be used in more general problems. One qualitative feature of our solutions is that they are the non-linear superpositions of the instantons, antiinstantons, and perturbative quanta. The former correspond to the B-cycle wrappings, while the latter correspond to the A-cycle wrappings.

7.1. *Tropical instantons*

The algebraic integrability matches the perturbative modes of the system to the tunneling modes. The instanton–anti-instanton configurations then arise in the tropical limit of the algebraic integrable system, where the underlying spectral curve degenerates to the rational one, while its Jacobian becomes a combinatorial object.

7.2. *Lefschetz thimbles and Dyson–Schwinger equations*

The analytic properties of quantum mechanical correlators can be understood by exploring the Lefschetz thimbles of the complexified action functional. Imagine the correlation functions obey some differential-difference equations in the coupling constants, such as the Dyson–Schwinger equations,[e] which come from the invariance of the integral under the small perturbations of the contour of integration. Then the same equation is obeyed by the integrals over all Lefschetz thimbles. The order of such equation is equal to the number of those thimbles. Of course, it is infinite. But it is interesting how infinite it is. It appears that in the system with r degrees of freedom one gets a cone in a lattice of rank $2r$. Each point in this cone corresponds to a solution of the system of those Dyson–Schwinger equations.

Already for one degree of freedom, we found many more solutions than we could find in the literature. It appears that the bions of [5, 7, 8] are the particular examples of our solutions for special values of (m, n).

We found that the critical points of the action functional on the complexified loop space are the critical points of a flux-type superpotential defined on the (cover

[e]Unfortunately the DS equations do not form a closed system unless one introduces the couplings for all irrelevant operators, or takes some limit.

of the) base of an algebraic integrable system. It would be interesting to study Lefschetz thimbles of this superpotential, and compare them to the infinite-dimensional Lefschetz thimbles of the path integral.

7.3. *Topological renormalization group and hyperkähler metrics*

Explicit description of the topological renormalization group flow may prove very difficult. However, we have a lot of freedom in the choice of the metric on the space of fields. For example, one may choose a metric on the space of loops of $L\mathcal{X}_{\mathbb{C}}^{2r}$ which is induced from the hyperkähler metric on $\mathcal{X}_{\mathbb{C}}^{2r}$. The latter, in turn, can be simplified by taking a limit of small Kähler class of the fibers, the semi-flat metric discussed in [30]. This metric is singular, but the non-singular nearly semi-flat one might work as well. In this way, we will get the tropical limit of the Lefschetz timble.

It would be interesting to understand the connection of our solutions to the BPS states of supersymmetric gauge theory whose moduli space of vacua is the complexified phase space (it is the four-dimensional theory compactified on a circle).

7.4. *Rademacher for Gutzwiller*

It is tempting to speculate that Gutzwiller's formula [41] can be improved (a la Rademacher) by the contributions of our $\mathbf{m}, \mathbf{n}$-solutions, so as to produce an exact formula. What are the implications for the theory of quantum chaos?

7.5. *Towards quantum field theory: Finite-gap vs. lattice*

Of course, the main goal of this project is to improve on the instanton gas ansatz in quantum field theory, as it fails already in two dimensions [78]. In going from quantum mechanics to quantum field theory one may approximate the infinite dimensional system by a finite-dimensional one. The naive approach where we put the fields on a lattice, may prove too difficult, as all the symmetries would be lost. It would be great to find a finite-dimensional approximation which preserves, say, an algebraic integrability of the complexified model. Of course, for some models, such as the sine-Gordon in two dimensions, one can find an integrable lattice approximation [25, 46], but such examples are rather exceptional. However, one can find other finite-dimensional approximations, namely the finite-gap subspaces. It appears that such an approach works for the $O(n)$-sigma models in two dimensions [52]. In fact, it works better for $n > 3$ where the proper two-dimensional instantons are simply absent, since $\pi_2(S^{n-1}) = 0$ in this case!

Recall that the $O(3)$-model can be mapped to the sine-Gordon theory using Pohlmeyer reduction. Since the sine-Gordon equation is an infinite-dimensional integrable system, whose classical (complexified) evolution linearizes on the Jacobian of an infinite-dimensional complex curve, one could jump at the conclusion that the analogue of our $\mathbf{m}, \mathbf{n}$-solutions is easy to produce: simply look for the rational windings on that Jacobian variety. Unfortunately, Pohlmeyer reduction works simply only in the infinite volume systems, while our systematic approach requires taking both space and time compact. By imposing periodic or twisted periodic boundary conditions on the sigma model side, we arrive at a complicated problem on the sine-Gordon side (the infinite genus curve approach is simple when it is the sine-Gordon equation which is studied with periodic spatial boundary conditions). Fortunately, there is another approach, which uses the linear sigma model realization of the $O(n)$ model (which is usually studied in the large n limit), in which one directly gets the solutions of the complexified sigma model, in terms of analytic curves, called the Fermi curves in [52]. The finite-gap approximation is then the ansatz in which the curve in question has a finite genus component. The motion linearizes on the Prymian of that curve [52]. In fact, even more explicit solutions can be found using the "winding ansatz", in which the spatial slice of the worldsheet winds along an orbit of a $U(1)$-subgroup of $O(n)$ (or, for the twisted boundary conditions, one may choose a generic one-parametric subgroup). In this way the $(1+1)$-dimensional problem reduces to the $(0+1)$-dimensional one: that of Neumann system, which, in turn, can be reduced to the Gaudin model we discussed above [76]. Remarkably, despite the common lore that the $\mathbb{CP}^n$ model is not integrable, a (smaller) class of similar complex solutions can be found there as well.

Encouraged by the success in two dimensions, we may hope to find the analogous solutions in the four-dimensional (non-supersymmetric) Yang–Mills theory. The instanton methods in QCD have a very long history [18]. However, they dilute instanton gas is not a good approximation to the Yang–Mills path integral. We should, therefore, look for the analogues of our non-linear superpositions of the instantons, anti-instantons, and perturbative quanta (gluons). As an initial approach, again, we take the Euclidean space-time with the geometry $S^3 \times S^1$. Let R be the radius of S^3 and β the circumference of S^1. We then look for solutions of the Yang–Mills equations

$$D_A \star F_A = 0 \tag{139}$$

with the complex gauge field A (say $\mathfrak{sl}(2,\mathbb{C})$-valued for the $SU(2)$ gauge theory). The equations (139) are still conformally invariant, so we may study them on $\mathbb{R}^4\backslash 0$ then impose the invariance with respect to the discrete scaling $r \mapsto re^{-\beta/R}$. The equations on $\mathbb{R}^4\backslash 0$ can be studied as in [94] by imposing an $SO(3)$-invariance on the gauge fields (with $SO(3)$ acting both as the space rotations $SO(3) \subset SO(4)$ and as the gauge rotations). In this way we arrive at the abelian Higgs model on the periodic AdS_2, i.e. the strip τ, $\mathrm{Im}(\tau) > 0$, $\tau \sim \tau + 1$, with the metric

$$\frac{d\tau d\bar{\tau}}{\mathrm{Im}(\tau)^2}\,. \tag{140}$$

Perhaps this is not far from the sigma models which can be analyzed using the Fermi curve approach of [52]. Hopefully in this way the instanton liquid picture of [84] will be eventually justified. With more symmetry the $(3+1)$-dimensional problem reduces to the $(0+1)$-dimensional one, in fact, again to the anharmonic oscillator [76].

7.6. *Wilder speculations*

Note that in order to get an insight on the quantum mechanical model we lifted it to the two-dimensional theory and then to the four-dimensional supersymmetric gauge theory in a special background. If we naively adopt the $d \mapsto d+1+2$ point of view we would conclude that the non-perturbative treatment of a non-supersymmetric four-dimensional quantum field theory should reveal a seven-dimensional supersymmetric theory? There are many supersymmetric theories in seven dimensions: essentially all of them are given by compactifications of M-theory on four-manifolds possibly with fluxes (in the absence of fluxes the four-manifolds are hyperkähler). Of course, there are many local K3's but probably not as many as four-dimensional quantum field theories. Is the existence of the seven-dimensional completion another constraint [93] on the landscape of consistent four-dimensional quantum field theories?

Acknowledgments

I thank S. Gukov, I. Krichever, and E. Witten for discussions. The story presented above has been reported on at several conferences and workshops in the past few years, e.g. "Current Themes in High Energy Physics and Cosmology" at the NBI (Copenhagen, Aug 2016),[f] lectures at the Simons Center for Geometry and Physics (Stony Brook, Sept 2016),[g,h] Northern California Symplectic Geometry Seminar/UC Berkeley String-Math Seminar (Berkeley, Nov 2016), Caltech colloquium (Pasadena, Jan 2017), IGST-2017 in ENS (Paris, Jul 2017).[i] I thank the organizers of those events for their hospitality and the participants for interesting questions and comments. I would especially like to thank M. Shaposhnikov for bringing to my attention[j] the works [80–82] where the anharmonic oscillator example which we discussed in some detail above has been worked out for the first time.

The paper was completed during the visits to the Institut des Hautes Études Scientifiques (Bures-sur-Yvette) and CERN Theory Division. I thank IHES and CERN for their hospitality.

[f]https://indico.nbi.ku.dk/event/851/
[g]http://scgp.stonybrook.edu/video_portal/video.php?id=2772
[h]http://scgp.stonybrook.edu/video_portal/video.php?id=3028
[i]https://www.phys.ens.fr/ igst17/
[j]Albeit during the presentation of this work at CERN.

References

[1] A. Alekseev, L. D. Faddeev and S. L. Shatashvili, Quantization of symplectic orbits of compact Lie groups by means of the functional integral, *J. Geom. Phys.* **5** (1988) 391; doi:10.1016/0393-0440(88)90031-9.

[2] L. F. Alday, D. Gaiotto and Y. Tachikawa, Liouville correlation functions from four-dimensional gauge theories, *Lett. Math. Phys.* **91** (2010) 167; arXiv:0906.3219 [hep-th].

[3] M. Alexandrov, M. Kontsevich, A. Schwartz and O. Zaboronsky, The geometry of the master equation and topological quantum field theory, *Int. J. Mod. Phys. A* **12** (1997) 1405; doi:10.1142/S0217751X97001031; hep-th/9502010.

[4] P. C. Argyres and J. Wittig, Mass deformations of four-dimensional, rank 1, $\mathcal{N} = 2$ superconformal field theories, *J. Phys. Conf. Ser.* **462**(1) (2013) 012001; doi:10.1088/1742-6596/462/1/012001; arXiv:1007.5026 [hep-th].

[5] P. C. Argyres and M. Unsal, The semi-classical expansion and resurgence in gauge theories: New perturbative, instanton, bion, and renormalon effects, *JHEP* **1208** (2012) 063; doi:10.1007/JHEP08(2012)063; arXiv:1206.1890 [hep-th].

[6] V. I. Arnold, A. N. Varchenko and S. M. Gusein-Zade, *Singularities of Differentiable Maps. Volume II: Monodromy and Asymptotic Integrals* (Birkhäuser, 1988).

[7] G. Basar, G. V. Dunne and M. Unsal, Resurgence theory, ghost-instantons, and analytic continuation of path integrals, *JHEP* **1310** (2013) 041; doi:10.1007/JHEP10(2013)041; arXiv:1308.1108 [hep-th].

[8] G. Basar and G. V. Dunne, Resurgence and the Nekrasov–Shatashvili limit: Connecting weak and strong coupling in the Mathieu and Lamé systems, *JHEP* **1502** (2015) 160; doi:10.1007/JHEP02(2015)160; arXiv:1501.05671 [hep-th].

[9] F. Bayen, M. Flato, C. Fronsdal, A. Lichnerowicz and D. Sternheimer, Deformation theory and quantization. 1. Deformations of symplectic structures, *Ann. Phys.* **111** (1978) 61; doi:10.1016/0003-4916(78)90224-5.

[10] H. Z. Bethe, The theory of metals I. Eigenvalues and eigenfunctions of the linear atomic chain, *Z. Phys.* **71** (1931) 205; https://doi.org/10.1007/BF01341708.

[11] A. Behtash, G. V. Dunne, T. Schaefer, T. Sulejmanpasic and M. Unsal, Toward Picard–Lefschetz theory of path integrals, complex saddles and resurgence, *Ann. Math. Sci. Appl.* **2**(1) (2017) 95; doi:10.4310/AMSA.2017.v2.n1.a3; arXiv:1510.03435 [hep-th].

[12] C. Bender and T. T. Wu, Anharmonic oscillator, *Phys. Rev.* **184**(5) (1969) 1231–1260; doi:10.1103/PhysRev.184.1231.

[13] A. Bilal, V. Fock and I. Kogan, On the origin of W algebras, *Nucl. Phys. B* **359** (1991) 635; doi:10.1016/0550-3213(91)90075-9.

[14] E. Bogomolny, Calculation of instanton–anti-instanton contributions in quantum mechanics, *Phys. Lett. B* **91**(3, 4) (1980) 431–435.

[15] A. Braverman, Instanton counting via affine Lie algebras. 1. Equivariant J functions of (affine) flag manifolds and Whittaker vectors, math/0401409 [math-ag].

[16] A. Braverman and P. Etingof, Instanton counting via affine Lie algebras II: From Whittaker vectors to the Seiberg–Witten prepotential, math/0409441 [math-ag].

[17] C. G. Callan, Jr. and S. R. Coleman, The fate of the false vacuum. 2. First quantum corrections, *Phys. Rev. D* **16** (1977) 1762; doi:10.1103/PhysRevD.16.1762.

[18] C. G. Callan, Jr., R. F. Dashen and D. J. Gross, Toward a theory of the strong interactions, *Phys. Rev. D* **17** (1978) 2717; doi:10.1103/PhysRevD.17.2717.

[19] A. S. Cattaneo and G. Felder, A path integral approach to the Kontsevich quantization formula, *Comm. Math. Phys.* **212** (2000) 591; doi:10.1007/s002200000229; math/9902090.

[20] S. Cecotti and C. Vafa, On classification of $N = 2$ supersymmetric theories, *Comm. Math. Phys.* **158** (1993) 569; doi:10.1007/BF02096804; hep-th/9211097.

[21] S. R. Coleman, The uses of instantons, *Subnucl. Ser.* **15** (1979) 805.

[22] R. Donagi and E. Markman, Spectral curves, algebraically completely integrable Hamiltonian systems, and moduli of bundles, arXiv:alg-geom/9507017.

[23] R. Donagi and E. Witten, Supersymmetric Yang–Mills theory and integrable systems, *Nucl. Phys. B* **460** (1996) 299; doi:10.1016/0550-3213(95)00609-5; hep-th/9510101.

[24] G. V. Dunne and M. Unsal, WKB and resurgence in the Mathieu equation, arXiv:1603.04924 [math-ph].

[25] L. D. Faddeev, Algebraic aspects of Bethe Ansatz, *Int. J. Mod. Phys. A* **10** (1995) 1845; doi:10.1142/S0217751X95000905; hep-th/9404013.

[26] L. D. Faddeev and A. A. Slavnov, *Gauge Fields. Introduction to Quantum Theory*, Frontiers in Physics, Vol. 50 (Benjamin/Cummings Publishing Co., 1980); Frontiers in Physics, Vol. 83 (Addison-Wesley Publishing Co., 1991).

[27] V. A. Fateev, A. B. Zamolodchikov, Operator algebra and correlation functions in the two-dimensional Wess–Zumino $SU(2)xSU(2)$ chiral model, *Sov. J. Nucl. Phys.* **43** (1986) 657–664; *Yad. Fiz.* **43** (1986) 1031–1044.

[28] I. B. Frenkel and V. G. Kac, Basic representations of affine Lie algebras and dual resonance models, *Inv. Math.* **62** (1980) 23.

[29] D. Gaiotto, $\mathcal{N} = 2$ dualities, *JHEP* **1208** (2012) 034; arXiv:0904.2715 [hep-th].

[30] D. Gaiotto, G. W. Moore and A. Neitzke, Four-dimensional wall-crossing via three-dimensional field theory, *Comm. Math. Phys.* **299** (2010) 163; doi:10.1007/s00220-010-1071-2; arXiv:0807.4723 [hep-th].

[31] A. A. Gerasimov and S. L. Shatashvili, Higgs bundles, gauge theories and quantum groups, *Comm. Math. Phys.* **277** (2008) 323; doi:10.1007/s00220-007-0369-1; hep-th/0609024.

[32] A. A. Gerasimov and S. L. Shatashvili, Two-dimensional gauge theories and quantum integrable systems, in *Proceedings of Symposia in Pure Mathematics*, University of Augsburg, Germany (25–29 May, 2007), 29 pp; arXiv:0711.1472 [hep-th].

[33] A. Givental, A mirror theorem for toric complete intersections, arXiv:alg-geom/9701016.

[34] A. Gorsky, A. Milekhin and N. Sopenko, Bands and gaps in Nekrasov partition function, *JHEP* **1801** (2018) 133; doi:10.1007/JHEP01(2018)133; arXiv:1712.02936 [hep-th].

[35] A. Gorsky, I. Krichever, A. Marshakov, A. Mironov and A. Morozov, Integrability and Seiberg–Witten exact solution, *Phys. Lett. B* **355** (1995) 466; doi:10.1016/0370-2693(95)00723-X; hep-th/9505035.

[36] A. Gorsky and N. Nekrasov, Hamiltonian systems of Calogero type and two-dimensional Yang–Mills theory, *Nucl. Phys. B* **414** (1994) 213; doi:10.1016/0550-3213(94)90429-4; hep-th/9304047.

[37] A. Gorsky and N. Nekrasov, Elliptic Calogero–Moser system from two-dimensional current algebra, hep-th/9401021.

[38] A. Gorsky, N. Nekrasov and V. Rubtsov, Hilbert schemes, separated variables, and D-branes, *Comm. Math. Phys.* **222** (2001) 299; doi:10.1007/s002200100503; hep-th/9901089.

[39] S. Gukov and E. Witten, Branes and quantization, *Adv. Theor. Math. Phys.* **13**(5) (2009) 1445; doi:10.4310/ATMP.2009.v13.n5.a5; arXiv:0809.0305 [hep-th].

[40] S. Gukov, C. Vafa and E. Witten, CFT's from Calabi–Yau four folds, *Nucl. Phys. B* **584** (2000) 69; Erratum, *ibid.* **608** (2001) 477; doi:10.1016/S0550-3213(01)00289-9, 10.1016/S0550-3213(00)00373-4; hep-th/9906070.

[41] M. C. Gutzwiller, Periodic orbits and classical quantization conditions, *J. Math. Phys.* **12**(3) (1971) 343; doi:10.1063/1.1665596.
[42] S. W. Hawking, Particle creation by black holes, *Comm. Math. Phys.* **43** (1975) 199.
[43] N. J. Hitchin, Stable bundles and integrable systems, *Duke Math. J.* **54** (1987) 91; doi:10.1215/S0012-7094-87-05408-1.
[44] K. Hori, A. Iqbal and C. Vafa, D-branes and mirror symmetry, hep-th/0005247.
[45] S. Jeong, Splitting of surface defect partition functions and integrable systems, arXiv:1709.04926 [hep-th].
[46] M. Jimbo, T. Miwa and F. Smirnov, Hidden Grassmann structure in the XXZ model III: Introducing Matsubara direction, *J. Phys. A* **42** (2009) 304018; doi:10.1088/1751-8113/42/30/304018; arXiv:0811.0439 [math-ph].
[47] A. Kapustin and D. Orlov, Remarks on A branes, mirror symmetry, and the Fukaya category, *J. Geom. Phys.* **48** (2003) 84; doi:10.1016/S0393-0440(03)00026-3; hep-th/0109098.
[48] A. Kapustin and E. Witten, Electric-magnetic duality and the geometric Langlands program, *Comm. Num. Theor. Phys.* **1** (2007) 1; doi:10.4310/CNTP.2007.v1.n1.a1; hep-th/0604151.
[49] M. Kontsevich, Deformation quantization of Poisson manifolds 1, *Lett. Math. Phys.* **66** (2003) 157; doi:10.1023/B:MATH.0000027508.00421.bf; arXiv:q-alg/9709040 [q-alg].
[50] M. Kontsevich, Enumeration of rational curves via toric actions, in *The Moduli Spaces of Curves*, eds. R. Dijkgraaf, C. Faber and G. van der Geer, Progress in Math., Vol. 129 (Birkhäuser, 1995), pp. 335–368.
[51] V. E. Korepin, Norm of Bethe wave function as a determinant, arXiv:0911.1881 [math.CO].
[52] I. Krichever and N. Nekrasov, in progress.
[53] A. Losev, A. Marshakov and N. A. Nekrasov, Small instantons, little strings and free fermions, in *From Fields to Strings*, Vol. 1 (World Scientific Publishing, 2005), pp. 581–621; hep-th/0302191.
[54] A. Losev, N. Nekrasov and S. L. Shatashvili, Issues in topological gauge theory, *Nucl. Phys. B* **534** (1998) 549; doi:10.1016/S0550-3213(98)00628-2; hep-th/9711108.
[55] A. Losev, N. Nekrasov and S. L. Shatashvili, Testing Seiberg–Witten solution, in *Strings, Branes and Dualities (Cargese, 1997)* (Kluwer Acad. Publ., 1999), pp. 359–372; hep-th/9801061.
[56] E. J. Martinec and N. P. Warner, Integrable systems and supersymmetric gauge theory, *Nucl. Phys. B* **459** (1996) 97; doi:10.1016/0550-3213(95)00588-9; hep-th/9509161.
[57] G. W. Moore, N. Nekrasov and S. Shatashvili, Integrating over Higgs branches, *Comm. Math. Phys.* **209** (2000) 97; hep-th/9712241.
[58] H. Nakajima, Instantons on ALE spaces, quiver varieties, and Kac–Moody algebras, *Duke Math. J.* **76**(2) (1994) 365–416.
[59] H. Nakajima, Instantons and affine Lie algebras, *Nucl. Phys. B Proc. Suppl.* **46** (1996) 154–161.
[60] H. Nakajima, Quiver varieties and Kac–Moody algebras, *Duke Math. J.* **91**(3) (1998) 515–560; doi:10.1215/S0012-7094-98-09120-7.
[61] N. Nekrasov, Holomorphic bundles and many-body systems, *Comm. Math. Phys.* **180** (1996) 587; doi:10.1007/BF02099624; hep-th/9503157.
[62] N. Nekrasov, Four dimensional holomorphic theories, PhD. thesis, Princeton University (1996); `http://scgp.stonybrook.edu/people/faculty/bios/nikita-nekrasov`.

[63] N. Nekrasov, On the BPS/CFT correspondence, Lecture at the University of Amsterdam string theory group seminar (February 3, 2004).
[64] N. A. Nekrasov, Seiberg–Witten prepotential from instanton counting, *Adv. Theor. Math. Phys.* **7**(5) (2003) 831; hep-th/0206161.
[65] N. Nekrasov, BPS/CFT correspondence: non-perturbative Dyson–Schwinger equations and qq-characters, *JHEP* **1603** (2016) 181; arxiv:1512.05388 [hep-th].
[66] N. Nekrasov, BPS/CFT correspondence II: Instantons at crossroads, moduli and compactness theorem, arXiv:1608.07272 [hep-th].
[67] N. Nekrasov, BPS/CFT correspondence III: Gauge origami partition function and qq-characters, arXiv:1701.00189 [hep-th].
[68] N. Nekrasov and A. Okounkov, Seiberg–Witten theory and random partitions, *Prog. Math.* **244** (2006) 525; hep-th/0306238.
[69] N. Nekrasov and V. Pestun, Seiberg–Witten geometry of four dimensional $N = 2$ quiver gauge theories, arXiv:1211.2240 [hep-th].
[70] N. Nekrasov, V. Pestun and S. Shatashvili, Quantum geometry and quiver gauge theories, arXiv:1312.6689 [hep-th].
[71] N. Nekrasov and N. S. Prabhakar, Spiked instantons from intersecting D-branes, *Nucl. Phys. B* **914** (2017) 257; doi:10.1016/j.nuclphysb.2016.11.014; arXiv:1611.03478 [hep-th].
[72] N. A. Nekrasov and S. L. Shatashvili, Supersymmetric vacua and Bethe ansatz, *Nucl. Phys. Proc. Suppl.* **192/193** (2009) 91; doi:10.1016/j.nuclphysbps.2009.07.047; arXiv:0901.4744 [hep-th].
[73] N. A. Nekrasov and S. L. Shatashvili, Quantum integrability and supersymmetric vacua, *Prog. Theor. Phys. Suppl.* **177** (2009) 105; doi:10.1143/PTPS.177.105; arXiv:0901.4748 [hep-th].
[74] N. A. Nekrasov and S. L. Shatashvili, Quantization of integrable systems and four dimensional gauge theories, arXiv:0908.4052 [hep-th].
[75] N. Nekrasov and E. Witten, The omega deformation, branes, integrability, and Liouville theory, *JHEP* **1009** (2010) 092; doi:10.1007/JHEP09(2010)092; arXiv:1002.0888 [hep-th].
[76] N. Nekrasov, Tying up instantons and anti-instantons II: Sigma models and gauge theory, to appear.
[77] M. K. Parikh and F. Wilczek, Hawking radiation as tunneling, *Phys. Rev. Lett.* **85** (2000) 5042; doi:10.1103/PhysRevLett.85.5042; hep-th/9907001.
[78] A. M. Polyakov, Quark confinement and topology of gauge groups, *Nucl. Phys. B* **120** (1977) 429; doi:10.1016/0550-3213(77)90086-4.
[79] A. M. Polyakov, Gauge fields and strings, *Contemp. Concepts Phys.* **3** (1987) 1.
[80] J. L. Richard and A. Rouet, Complex saddle points in the double well oscillator, *Nucl. Phys. B* **183** (1981) 251; doi:10.1016/0550-3213(81)90555-1.
[81] J. L. Richard and A. Rouet, The saddle point method for the double well anharmonic oscillator, *Phys. Lett. B* **98** (1981) 305; doi:10.1016/0370-2693(81)90021-6.
[82] J. L. Richard and A. Rouet, Complex saddle points versus dilute gas approximation in the double well anharmonic oscillator, *Nucl. Phys. B* **185** (1981) 47; doi:10.1016/0550-3213(81)90363-1.
[83] M. Rocek and E. P. Verlinde, Duality, quotients, and currents, *Nucl. Phys. B* **373** (1992) 630; doi:10.1016/0550-3213(92)90269-H; hep-th/9110053.
[84] T. Schäfer and E. V. Shuryak, Instantons in QCD, *Rev. Mod. Phys.* **70** (1998) 323; doi:10.1103/RevModPhys.70.323; hep-ph/9610451.
[85] E. K. Sklyanin, Quantum inverse scattering method. Selected topics, in *Quantum Group and Quantum Integrable Systems*, Proceedings of the Nankai Lectures

on Mathematical Physics (World Scientific Publishing, 1992), pp. 63–97; hep-th/9211111.

[86] N. Seiberg and E. Witten, Electric-magnetic duality, monopole condensation, and confinement in $N = 2$ supersymmetric Yang–Mills theory, *Nucl. Phys. B* **426** (1994) 19; Erratum, *ibid.* **430** (1994) 485; hep-th/9407087.

[87] N. Seiberg and E. Witten, Monopoles, duality and chiral symmetry breaking in $N = 2$ supersymmetric QCD, *Nucl. Phys. B* **431** (1994) 484; hep-th/9408099.

[88] N. Seiberg and E. Witten, Gauge dynamics and compactification to three-dimensions, in *The Mathematical Beauty of Physics (Saclay, 1996)*, Adv. Ser. Math. Phys., Vol. 24 (World Scientific Publishing, 1997), pp. 333–366; hep-th/9607163.

[89] N. Seiberg and E. Witten, String theory and noncommutative geometry, *JHEP* **9909** (1999) 032; hep-th/9908142.

[90] T. R. Taylor and C. Vafa, R-R flux on Calabi–Yau and partial supersymmetry breaking, *Phys. Lett. B* **474** (2000) 130; doi:10.1016/S0370-2693(00)00005-8; hep-th/9912152.

[91] A. V. Turbiner, One-dimensional quasi-exactly solvable Schrödinger equations, *Phys. Rep.* **642** (2016) 1; doi:10.1016/j.physrep.2016.06.002; arXiv:1603.02992 [quant-ph].

[92] C. Vafa, Topological mirrors and quantum rings, in *Mirror Symmetry I*, AMS/IP Stud. Adv. Math., Vol. 9 (Amer. Math. Soc./Int. Press, 1998), pp. 97–120; hep-th/9111017.

[93] C. Vafa, The string landscape and the swampland, hep-th/0509212.

[94] E. Witten, Some exact multi-instanton solutions of classical Yang–Mills theory, *Phys. Rev. Lett.* **38** (1977) 121; doi:10.1103/PhysRevLett.38.121.

[95] E. Witten, Solutions of four-dimensional field theories via M theory, *Nucl. Phys. B* **500** (1997) 3; hep-th/9703166.

[96] E. Witten, Phases of $N = 2$ theories in two dimensions, *Nucl. Phys. B* **403** (1993) 159; hep-th/9301042.

[97] E. Witten, Integrable lattice models from gauge theory, arXiv:1611.00592 [hep-th].

[98] E. Witten, A new look at the path integral of quantum mechanics, arXiv:1009.6032 [hep-th].

[99] E. Witten, Analytic continuation of Chern–Simons theory, in *Chern–Simons Gauge Theory: 20 Years After*, AMS/IP Stud. Adv. Math., Vol. 50 (Amer. Math. Soc., 2011), pp. 347–446; arXiv:1001.2933 [hep-th].

[100] C. N. Yang and C. P. Yang, One-dimensional chain of anisotropic spin-spin interactions. 1. Proof of Bethe's hypothesis for ground state in a finite system, *Phys. Rev.* **150** (1966) 321; doi:10.1103/PhysRev.150.321.

[101] A. B. Zamolodchikov, Infinite additional symmetries in two-dimensional conformal quantum field theory, *Theor. Math. Phys.* **65** (1985) 1205; *Teor. Mat. Fiz.* **65** (1985) 347; doi:10.1007/BF01036128.

[102] J. Zinn-Justin, Quantum field theory and critical phenomena, *Int. Ser. Monogr. Phys.* **77** (1989) 1.

A Bäcklund Transformation for Elliptic Four-Point Conformal Blocks

André Neveu
L2C, CNRS, University of Montpellier,
Montpellier, France
andre.neveu@umontpellier.fr

We apply an integral transformation to solutions of a partial differential equation for the five-point correlation functions in Liouville theory on a sphere with one degenerate field $V_{-\frac{1}{2b}}$. By repeating this transformation, we can reach a whole lattice of values for the conformal dimensions of the four other operators. Factorizing out the degenerate field leads to integral representations of the corresponding four-point conformal blocks. We illustrate this procedure on the elliptic conformal blocks discovered in a previous publication.

1. Introduction

Liouville field theory on a two-dimensional surface is defined by the Lagrangian density

$$\mathcal{L} = \frac{1}{4\pi}\hat{g}^{ab}\partial_a\varphi\partial_b\varphi + \mu e^{2b\varphi} + \frac{Q}{4\pi}\hat{R}\varphi, \tag{1.1}$$

where $\hat{g}_{ab}$ is the metric and $\hat{R}$ the associated curvature. In this paper, we shall exclusively consider the case of the sphere. The theory is conformally invariant if the coupling constant b is related to the background charge Q by

$$Q = b + \frac{1}{b}. \tag{1.2}$$

The central charge of the Virasoro algebra is

$$c_L = 1 + 6Q^2. \tag{1.3}$$

The primary fields of the theory are associated with the exponential fields $e^{2\alpha\varphi}$ and their conformal dimension is

$$\Delta(\alpha) = \alpha(Q - \alpha). \tag{1.4}$$

By conformal invariance, one can in principle find the correlation functions of all local fields from the structure constants of the operator product expansion, starting from the structure constants of three primary fields $C^{\alpha_3}_{\alpha_1\alpha_2}$ which appear in the three-point correlation function [1–4]

$$\langle V_{\alpha_1}(z_1)V_{\alpha_2}(z_2)V_{\alpha_3}(z_3)\rangle = \frac{C(\alpha_1,\alpha_2,\alpha_3)}{|z_{12}|^{2\Delta_1+2\Delta_2-2\Delta_3}|z_{13}|^{2\Delta_1+2\Delta_3-2\Delta_2}|z_{23}|^{2\Delta_2+2\Delta_3-2\Delta_1}}, \tag{1.5}$$

with

$$C^{\alpha_3}_{\alpha_1\alpha_2} = C(\alpha_1,\alpha_2,Q-\alpha_3). \tag{1.6}$$

The four-point conformal block can then be represented as a sum over intermediate states

$$\begin{aligned}&\langle V_{\alpha_1}(z_1,\bar z_1)V_{\alpha_2}(z_2,\bar z_2)V_{\alpha_3}(z_3,\bar z_3)V_{\alpha_4}(z_4,\bar z_4)\rangle\\ &\quad= \prod_{i<j}|z_i-z_j|^{2\gamma_{ij}}\\ &\quad\times \frac{1}{2}\int_C C\left(\alpha_1,\alpha_2,\frac{Q}{2}+iP\right)C\left(\frac{Q}{2}-iP,\alpha_3,\alpha_4\right)\left|\mathfrak{F}_P\begin{pmatrix}\alpha_2 & \alpha_3\\ \alpha_1 & \alpha_4\end{pmatrix}\!\middle| x\right)\right|^2 dP,\end{aligned} \tag{1.7}$$

x is the anharmonic ration of the four points z_i

$$x = \frac{(z_1-z_2)(z_3-z_4)}{(z_1-z_3)(z_2-z_4)} \tag{1.8}$$

and the exponents γ_{ij} which are standard combinations of the α's can be found in [5].

The conformal block $\mathfrak{F}_P\left(\begin{smallmatrix}\alpha_2 & \alpha_3\\ \alpha_1 & \alpha_4\end{smallmatrix}\middle| x\right)$ sums up all the intermediate descendant states of a given primary one with conformal dimension $\Delta = \frac{Q^2}{4} - p^2$ in the operator product expansion (1.7).

In general, this function of x is not known in closed form. However, when one of the external fields is degenerate it satisfies a linear differential equation [5] which in the simplest case of the degenerate field $V_{-\frac{1}{2b}}$ can be solved explicitely in terms of hypergeometric functions. In [6], it was explained how to find more explicit expressions for conformal blocks for α_4, one of the parameters in Eq. (1.7), taking values of the form

$$\alpha_4 = \frac{Q}{2} - \frac{bm_4}{2} + \frac{n_4}{b} - \frac{b}{4}, \tag{1.9}$$

for positive integers m_4 and n_4 with the other three α's equal to $Q/2 - b/4$. In Sec. 2, we shall review this procedure, before explaining in Sec. 3 the extensions of these values of the α's by Bäcklund transformations. This paper can be considered a natural extension of [6], on which we rely heavily for notations and from which we reproduce several equations for completeness.

2. Elliptic Conformal Blocks

As in [6], let us consider the five-point correlation function with one degenerate field

$$\langle V_{-\frac{1}{2b}}(z)V_{\alpha_1}(0)V_{\alpha_2}(1)V_{\alpha_3}(\infty)V_{\alpha_4}(x)\rangle. \tag{2.1}$$

Due to the presence of the degenerate field, this correlation function satisfies a partial differential equation, second order in z, first order in x. For the purposes of this section, it is convenient to define the function $\Psi(u|q)$ as

$$\begin{aligned}&\langle V_{-\frac{1}{2b}}(z)V_{\alpha_1}(0)V_{\alpha_2}(1)V_{\alpha_3}(\infty)V_{\alpha_4}(x)\rangle\\ &\quad = z^{\frac{1}{2b^2}}(z-1)^{\frac{1}{2b^2}}\frac{\big(z(z-1)(z-x)\big)^{\frac{1}{4}}}{\big(x(x-1)\big)^{\frac{2\Delta(\alpha_4)}{3}+\frac{1}{12}}}\frac{\Theta_1(u)^{b^{-2}}}{\Theta_1'(0)^{\frac{b^{-2}+1}{3}}}\Psi(u|q),\end{aligned} \tag{2.2}$$

where the variable u is related with the variables z and x as

$$u = \frac{\pi}{4K(x)}\int_0^{\frac{z-x}{x(z-1)}}\frac{dt}{\sqrt{t(1-t)(1-xt)}}, \tag{2.3}$$

$K(x)$ is the elliptic integral of the first kind

$$K(x) = \frac{1}{2}\int_0^1\frac{dt}{\sqrt{t(1-t)(1-xt)}}, \tag{2.4}$$

$$\tau = \mathrm{i}\,\frac{K(1-x)}{K(x)} = \frac{\log q}{\mathrm{i}\pi}, \tag{2.5}$$

and $\Theta_1(u)$ is the Jacobi theta function (see definitions in the Appendix).

The function $\Psi(u|q)$ defined by Eq. (2.2) satisfies the non-stationary Schrödinger equation with doubly periodic potential

$$\left[\partial_u^2 - \mathbb{V}(u) + \frac{4\mathrm{i}}{\pi b^2}\partial_\tau\right]\Psi(u|q) = 0, \tag{2.6}$$

where the potential $\mathbb{V}(u)$ is given by

$$\mathbb{V}(u) = \sum_{j=1}^{4} s_j(s_j+1)\wp(u-\omega_j) \tag{2.7}$$

and the parameters s_k are related to the parameters α_k as

$$\alpha_k = \frac{Q}{2} - \frac{b}{2}\left(s_k + \frac{1}{2}\right). \tag{2.8}$$

In Eq. (2.7) $\wp(u)$ is the Weierstraß elliptic function with periods π and $\pi\tau$ defined by the infinite sum

$$\wp(u) = \frac{1}{u^2} + \sum_{k^2+l^2\neq 0}\left(\frac{1}{(u-\pi k-\pi\tau l)^2} - \frac{1}{(\pi k+\pi\tau l)^2}\right), \tag{2.9}$$

and ω_j are the half periods

$$\omega_1 = \frac{\pi}{2}, \qquad \omega_2 = \frac{\pi\tau}{2}, \qquad \omega_3 = \omega_1 + \omega_2, \qquad \omega_4 = 0. \tag{2.10}$$

In [6], explicit integral representations for the function $\Psi(u|q)$ are given for $s_1 = s_2 = s_3 = 0$, $s_4 = m_4 + \frac{2n_4}{b^2}$, m_4 and n_4 being integers. For example, from the general solution to the heat equation:

$$\partial_u^2 \Psi_0(u|q) + \frac{4\mathrm{i}}{\pi b^2} \partial_\tau \Psi_0(u|q) = 0 \tag{2.11}$$

$$\Psi_0(u|q) = q^{\frac{b^2\lambda^2}{4}} e^{-\lambda}, \tag{2.12}$$

for $s_1 = s_2 = s_3 = 0$ and $s_4 = 1$, the general solution to the equation

$$\left(-\partial_u^2 + 2\wp(u)\right)\Psi(u|q) = \frac{4\mathrm{i}}{\pi b^2} \partial_\tau \Psi(u|q), \tag{2.13}$$

can be obtained from the general solution $\Psi_0(u|q)$ of the heat equation (2.11) as follows:

$$\Psi(u|q) = \int_0^\pi \left(\frac{\Theta_1(v)}{\Theta_1'(0)^{\frac{1}{3}}}\right)^{b^2} \frac{E(u+v)}{E(u)E(v)} \Psi_0(u + b^2 v|q)\, dv, \tag{2.14}$$

where we introduced the notation

$$E(u) = \frac{\Theta_1(u)}{\Theta_1'(0)}. \tag{2.15}$$

In the limit of fixed λ at $b \to \infty$ the integral in (2.14) is governed by the saddle point ν, which is the solution of the equation

$$\frac{\Theta_1'(\nu)}{\Theta_1(\nu)} = \lambda \tag{2.16}$$

and $\Psi(u|q)$ has as limit (up to irrelevant normalization factors):

$$\Psi(u|q) \to \frac{\Theta_1(u+\nu)}{\Theta_1(u)} \mathrm{e}^{-\lambda u}, \tag{2.17}$$

which is the solution of the stationary Lamé equation with energy $\wp(\nu)$. So, the solution (2.14) can be viewed as a "quantization" of the solution of the stationary equation (2.17).

In [6], we explained how to obtain the four-point elliptic conformal blocks from the solutions $\Psi(u|q)$ by taking the limit $u \to 0$.

3. Bäcklund Transformations

In this section, we find it more convenient to use the following parametrization for the five-point correlation function of Eq. (2.2)

$$\begin{aligned}&\langle V_{\alpha_1}(0)V_{-\frac{1}{2b}}(zx)V_{\alpha_2}(x)V_{\alpha_3}(1)V_{\alpha_4}(\infty)\rangle\\&\quad= z^{(-\frac{1}{4b^2}+\frac{\alpha_1}{b}-\frac{Q}{2b})}x^{(\Delta-\Delta_1)}\\&\qquad\times(1-z)^{\alpha_2/b}(1-x)^{-2\alpha_2\alpha_3}(1-zx)^{\alpha_3/b}g(z,x).\end{aligned}\tag{3.1}$$

This parametrization extracts the short distance behaviors at $z = 1$, $x = 1$, $zx = 1$ which one obtains for the V's being exponentials of free fields and taking for the dimension of the intermediate state between $V_{-\frac{1}{2b}}(zx)$ and $V_{\alpha_2}(x)$ the value $(\alpha_1 - \frac{1}{2b})(Q - \alpha_1 + \frac{1}{2b})$. $\Delta = \alpha(Q-\alpha) = Q^2/4 - p^2$ is the dimension, which should be integrated over, of the intermediate state between $V_{\alpha_2}(x)$ and $V_{\alpha_3}(1)$. Then, $g(z, x)$ should be a series in z and x with only integer powers, whose coefficients can be computed recursively.

As in the previous section, one obtains the following partial differential equation for $g(z, x)$:

$$\begin{aligned}&z(1-z)(1-zx)\partial_z{}^2g(z,x)+(A+Bz+Czx+Dz^2x)\partial_zg(z,x)\\&\quad-x(1-x)\partial_xg/b^2+(E+Fx+Hzx)g=0,\end{aligned}\tag{3.2}$$

with

$$\begin{aligned}A&=2\alpha_1/b-1/b^2,\\B&=-2\alpha_1/b-2\alpha_2/b+2/b^2,\\C&=-2\alpha_1/b-2\alpha_3/b+1/b^2,\\D&=2\alpha_1/b+2\alpha_2/b+2\alpha_3/b-2/b^2,\\E&=\big(\alpha+\alpha_1+\alpha_2-1/(2b)-Q\big)\big(\alpha-\alpha_1-\alpha_2+1/(2b)\big)/b^2,\\F&=\big(-2\alpha_3(\alpha+\alpha_1+\alpha_2-1/(2b)-Q)-(\alpha_3+\alpha_4-\alpha)(\alpha_3-\alpha_4-\alpha+Q)\big)/b^2,\\H&=\big(\Delta_4-Q^2/4+(\alpha_3+\alpha_2+\alpha_1-1/(2b)-Q/2)^2\big)/b^2.\end{aligned}\tag{3.3}$$

Expanding $g(z, x)$ in powers of x:

$$g(z,x)=g_0(z)+xg_1(z)+x^2g_2(z)\ldots,\tag{3.4}$$

one obtains for g_0 a hypergeometric equation, and for the successive g_i's hypergeometric equations with non-zero right-hand sides which depend on the lower g_i's. For generic values of the α's and b, these equations have unique solutions with a power series expansion at $z = 0$ and irrational power behaviors at $z = 1$ and $z = \infty$. The conformal block

$$\langle V_{\alpha_1-1/(2b)}(0)V_{\alpha_2}(x)V_{\alpha_3}(1)V_{\alpha_4}(\infty)\rangle\tag{3.5}$$

is then given by the function $x^{(\Delta-\Delta_1')}(1-x)^{-2\alpha_2\alpha_3}g(0,x)$ with $\Delta_1' = (\alpha_1 - \frac{1}{2b})(Q - \alpha_1 + \frac{1}{2b})$. To study the properties of this conformal block, for example its bootstrap properties, one thus needs an explicit solution of Eq. (3.2) to all orders in x. Some non-trivial such solutions in terms of elliptic functions have been found in [6]. In this section, we now wish to substantially expand this set of solutions.

Consider the integral transformation

$$g_I(z,x) = \int \frac{du}{(z-u)^y}\, g(u,x). \tag{3.6}$$

This is inspired by formula (2) of subchapter (2.4) of [7] which transforms one hypergeometric function into another with different values of the parameters. We leave the integration limits unspecified. The only requirement is that boundary terms in partial integrations can be neglected: For this, one may have a closed contour or end points where the integrand exhibits exclusively non-integer power behaviors.

Look for values of y such that g_I is a solution of the same equation, but with new values of its parameters:

$$\begin{aligned} z(1-z)(1-zx)\partial_z^2 g_I(z,x) + (A_I + B_I z + C_I zx + D_I z^2 x)\partial_z g_I \\ - x(1-x)\partial_x g_I/b^2 + (E_I + F_I x + H_I zx) g_I = 0. \end{aligned} \tag{3.7}$$

This is straightforward. One finds that y must satisfy

$$y^2 + y(1-D_I) + H_I = 0$$

together with the following relations with the initial coefficients:

$$\begin{aligned} A &= A_I + 1 - y \\ B &= B_I - 2(1-y) \\ C &= C_I - 2(1-y) \\ D &= D_I + 3(1-y) \\ E &= E_I + (y + B_I)(1-y) \\ F &= F_I + (y + C_I)(1-y) \\ H &= H_I + (2D_I - 3y)(1-y). \end{aligned} \tag{3.8}$$

One can of course invert these relations, and see how the external α_i's are transformed into $\alpha_{I,i}$. One finds (the signs are correlated, they correspond to the two solutions for the quadratic equation for x, and obviously to the fact that α_4 appears in the original equation only through Δ_4):

$$y = 1 - 1/(2b^2) + (-\alpha_1 + Q/2 - \alpha_2 + Q/2 - \alpha_3 + Q/2 \mp \alpha_4 \pm Q/2)/b.$$

By analogy with string theory, introducing momenta p_i for the vertex operators through the definition $p_i = \alpha_i - Q/2$, this reads

$$y = 1 - 1/(2b^2) - (p_1 + p_2 + p_3 \pm p_4)/b,$$

$$\begin{aligned}\alpha_{I1} - Q/2 &= -1/(4b) + (\alpha_1 - Q/2 - \alpha_2 + Q/2 - \alpha_3 + Q/2 \mp \alpha_4 \pm Q/2)/2, \\ \alpha_{I2} - Q/2 &= -1/(4b) + (-\alpha_1 + Q/2 + \alpha_2 - Q/2 - \alpha_3 + Q/2 \mp \alpha_4 \pm Q/2)/2, \\ \alpha_{I3} - Q/2 &= -1/(4b) + (-\alpha_1 + Q/2 - \alpha_2 + Q/2 + \alpha_3 - Q/2 \mp \alpha_4 \pm Q/2)/2, \\ \alpha_{I4} - Q/2 &= -1/(4b) + (-\alpha_1 + Q/2 - \alpha_2 + Q/2 - \alpha_3 + Q/2 \pm \alpha_4 \mp Q/2)/2.\end{aligned} \tag{3.9}$$

The intermediate dimension is unchanged:

$$\alpha_I = \alpha.$$

In the following, we shall take the upper signs in these equations, simply remembering that the sign of p_4 is arbitrary; changing it is just choosing the other signs.

One introduces the matrix

$$M = \begin{pmatrix} \frac{1}{2} & -\frac{1}{2} & -\frac{1}{2} & -\frac{1}{2} \\ -\frac{1}{2} & \frac{1}{2} & -\frac{1}{2} & -\frac{1}{2} \\ -\frac{1}{2} & -\frac{1}{2} & \frac{1}{2} & -\frac{1}{2} \\ -\frac{1}{2} & -\frac{1}{2} & -\frac{1}{2} & \frac{1}{2} \end{pmatrix} \tag{3.10}$$

which is a symmetry with respect to the hyperplane perpendicular to the vector $(1,1,1,1)$ in the four-dimensional space of the values of the $p_i = \alpha_i - Q/2$. The integral transformation from g to g_I is thus a reflection with respect to that hyperplane, followed by a translation by $-1/(4b)$ along the vector $(1,1,1,1)$.

Now, one can change the signs of the $p_i = \alpha_i - Q/2$, $i = 1,2,3$: By the change of function $g = z^{(Q-2\alpha_1)/b} g_1$ one finds that g_1 satisfies Eq. (3.2) with p_1 changed into $-p_1$, p_2, p_3, Δ_4 and p unchanged; in the same fashion, by the change $g = ((1-z)/(1-zx))^{(Q-2\alpha_2)/b} g_2$ one finds that g_2 satisfies Eq. (3.2) with p_2 changed into $-p_2, p_1, p_3$, Δ_4 and p unchanged, and finally the change $g = (1-zx)^{(Q-2\alpha_3)/b} g_3$ changes the sign of p_3. These changes of signs correspond to symmetries with respect to the coordinate hyperplanes.

By successive applications of these geometrical transformations, one generates a lattice in the space of the values of the $\alpha_i - Q/2$. Calling R_i the change of sign of $p_i = \alpha_i - Q/2$, and I the integral transformation above, the sequence $R_1 \cdot I \cdot R_2 \cdot R_3 \cdot R_4 \cdot I \cdot R_2 \cdot R_3 \cdot R_4 \cdot I$ transforms any vector

$$K_0 = \begin{pmatrix} p_1 \\ p_2 \\ p_3 \\ p_4 \end{pmatrix} \quad \text{into} \quad K_1 = \begin{pmatrix} p_1 + 1/b \\ p_2 \\ p_3 \\ p_4 \end{pmatrix}. \tag{3.11}$$

By repeating this set of manipulations, one can clearly in principle reach an arbitrary point of the lattice:

$$K_0 + (n_1 e_{01} + n_2 e_{02} + n_3 e_{03} + n_4 e_{04})/b \tag{3.12}$$

where the vectors e_{0i} are defined by:

$$e_{01} = \begin{pmatrix} 1 \\ 0 \\ 0 \\ 0 \end{pmatrix} \quad e_{02} = \begin{pmatrix} 0 \\ 1 \\ 0 \\ 0 \end{pmatrix} \quad e_{03} = \begin{pmatrix} 0 \\ 0 \\ 1 \\ 0 \end{pmatrix} \quad e_{04} = \begin{pmatrix} 0 \\ 0 \\ 0 \\ 1 \end{pmatrix} \tag{3.13}$$

and the n_i are arbitrary positive or negative integers. This partly answers the conjecture put forward in [6] on the values of the external dimensions for which an explicit representation could be obtained.

By applying only once the integral transform on K_0, one generates the solutions for the vector

$$K_2 = \begin{pmatrix} (p_1 - p_2 - p_3 - p_4)/2 - 1/(4b) \\ (-p_1 + p_2 - p_3 - p_4)/2 - 1/(4b) \\ (-p_1 - p_2 + p_3 - p_4)/2 - 1/(4b) \\ (-p_1 - p_2 - p_3 + p_4)/2 - 1/(4b) \end{pmatrix} \tag{3.14}$$

which is another starting point from which one can reach the points of another lattice as for K_0 above.

Still another starting point for the construction of a lattice is the vector K_3 obtained by applying the sequence $I \cdot R_3 \cdot R_4 \cdot I$ to K_0. One finds

$$K_3 = \begin{pmatrix} -p_2 - 1/(2b) \\ -p_1 - 1/(2b) \\ p_4 \\ p_3 \end{pmatrix}. \tag{3.15}$$

Choosing the values $p_i = -b/4$ in K_2, applying the integral transformation, then the R_1 symmetry, one obtains the solution of Eq. (3.7) for the following vector of momenta

$$K_4 = \begin{pmatrix} -b/4 + 1/(4b) \\ b/4 - 1/(4b) \\ b/4 - 1/(4b) \\ b/4 - 1/(4b) \end{pmatrix}. \tag{3.16}$$

Then, the value p_1' which appears in the conformal block (3.5) is $-b/4 - 1/(4b) = -Q/4$. Together with the value of the other p_i, $b/4 - 1/(4b)$, we thus obtain an explicit integral representation of a conformal block for $0 < Q < 4$ $(1 < c_L < 25)$ where b is unimodular, p_1' real, the other p_i pure imaginary, all conformal dimensions of external operators real and positive and the dimension Δ of the intermediate state arbitrary. If we start from the opposite vector $p_i = b/4$, the same transformation gives us $p_1' = b/4 - 1/(4b)$, pure imaginary for $1 < c_L < 25$ while the three other $p_i = -b/4 - 1/(4b) = -Q/4$ are real for the same values of c_L. As far as we know, these two cases are the first explicit examples of four-point conformal blocks with positive dimensions at $1 < c_L < 25$ where the bootstrap could be explored analytically as was done in [6], but this would go beyond the scope of this paper.

4. Conclusion

By applying integral transformations we have extended the cases of [6] where explicit expressions for four-point conformal blocks could be found in terms of integrals involving elliptic functions. There remains the problem of obtaining still further explicit representations in the cases where the parameters s_k of Eq. (2.8) take arbitrary integer values. For the moment, such more general expressions are only known in the "classical" case $b \to \infty$. From [6], we only have explicit expressions at finite b for the cases of only one s_k non-vanishing, or, via duplication formulas for elliptic functions, two of them equal with the other two vanishing, or, still, all four equal.

Appendix

The theta function $\Theta_1(u|\tau)$ defined by

$$\Theta_1(u|\tau) = \sum_{n=-\infty}^{\infty} (-1)^{n-\frac{1}{2}} q^{(n+\frac{1}{2})^2} \mathrm{e}^{(2n+1)\mathrm{i}u}$$

is solution to the differential equation

$$\partial_u^2 \Theta_1(u|\tau) - \frac{4\mathrm{i}}{\pi} \partial_\tau \Theta_1(u|\tau) = 0.$$

It satisfies the following quasi-periodicity relations

$$\Theta_1(u+\pi) = -\Theta_1(u),$$
$$\Theta_1(u+\pi\tau) = -q^{-1}\mathrm{e}^{-2\mathrm{i}u}\Theta_1(u),$$

and transforms as follows under the action of the modular group

$$\Theta_1(u|\tau+1) = \mathrm{i}^{\frac{1}{2}}\Theta_1(u|\tau),$$
$$\Theta_1\left(\frac{u}{\tau}\middle| -\frac{1}{\tau}\right) = (\mathrm{i}\tau)^{\frac{1}{2}} \mathrm{e}^{\frac{iu^2}{\pi\tau}} \Theta_1(u|\tau).$$

Other theta-functions can be expressed through $\Theta_1(u)$ as

$$\Theta_2(u) = \Theta_1\left(u + \frac{\pi}{2}\right),$$

$$\Theta_3(u) = \mathrm{e}^{\mathrm{i}u} q^{\frac{1}{4}} \Theta_1\left(u + \frac{\pi}{2} + \frac{\pi\tau}{2}\right),$$

$$\Theta_4(u) = -\mathrm{i}\mathrm{e}^{\mathrm{i}u} q^{\frac{1}{4}} \Theta_1\left(u + \frac{\pi\tau}{2}\right).$$

References

[1] H. Dorn and H. J. Otto, On correlation functions for noncritical strings with $c \leq 1$ $d \geq 1$, *Phys. Lett. B* **291** (1992) 39–43; [hep-th/9206053].

[2] H. Dorn and H. J. Otto, Two and three point functions in Liouville theory, *Nucl. Phys. B* **429** (1994) 375–388; [hep-th/9403141].

[3] A. B. Zamolodchikov and Al. B. Zamolodchikov, Structure constants and conformal bootstrap in Liouville field theory, *Nucl. Phys. B* **477** (1996) 577–605; [hep-th/9506136].

[4] J. Teschner, On the Liouville three point function, *Phys. Lett. B* **363** (1995) 65–70; [hep-th/9507109].

[5] A. A. Belavin, A. M. Polyakov and A. B. Zamolodchikov, Infinite conformal symmetry in two-dimensional quantum field theory, *Nucl. Phys. B* **241** (1984) 333–380.

[6] V. A. Fateev, A. V. Litvinov, A. Neveu and E. Onofri, Differential equations for four-point correlation function in Liouville theory and elliptic four-point conformal blocks, *J. Phys. A* **42** (2009) 304011; [arXiv:0902.1331].

[7] A. Erdélyi, W. Magnus, F. Oberhettinger and F. G. Tricomi, *Higher Transcendental Functions* (McGraw-Hill, 1953).

Vacua on the Brink of Decay

Guilherme L. Pimentel[*,§], Alexander M. Polyakov[†,¶] and Grigory M. Tarnopolsky[‡,||]

[*]Institute of Theoretical Physics, University of Amsterdam, Science Park 904, Amsterdam, 1098 XH, The Netherlands
[†]Joseph Henry Laboratories, Princeton University, Princeton, NJ 08544, USA
[‡]Department of Physics, Harvard University, Cambridge, MA 02138, USA
[§]G.LeitePimentel@uva.nl
[¶]polyakov@Princeton.EDU
[||]grtarnop@g.harvard.edu

We consider free massive matter fields in static scalar, electric and gravitational backgrounds. Tuning these backgrounds to the brink of vacuum decay, we identify a term in their effective action that is singular. This singular term is universal, being independent of the features of the background configuration. In the case of gravitational backgrounds, it can be interpreted as a quantum mechanical analog of Choptuik scaling. If the background is tuned slightly above the instability threshold, this singular term gives the leading contribution to the vacuum decay rate.

Dedicated to the memory of Ludwig Faddeev

Contents

1. Introduction

In this paper, we study strong background fields which may be able to destroy their own environment. This happens when the mass gap of the Quantum Field Theory (QFT) in question, due to the external field, tends to zero and eventually becomes negative. We identify a universal singularity in the effective action of the background field, which signals instability of the vacuum, as the mass gap vanishes.

Background field configurations which lead to particle production are associated to the formation of a "horizon", i.e. a length scale in which it becomes energetically more favorable to produce particles than to sustain the field configuration. This definition of the horizon is more general than the one usually discussed in the literature. For example, it implies the existence of electromagnetic horizons. As a typical example of an electric horizon, consider electrically charged particles of mass m in a background electrostatic potential $a_t(x)$. It is clear that if the "voltage" $A \equiv a_t(+\infty) - a_t(-\infty)$ satisfies $A > 2m$ particles will be produced and not much will remain of the vacuum.[a] For a gravitational background, the location of the horizon defined in this new way agrees with that of the causal horizon (see, e.g., [1]).

Technically, the phenomenon of particle production can be diagnosed by calculating the vacuum decay rate, given by the imaginary part of the one-loop effective action, obtained after integrating out massive matter fields. Many results are available for the effective action for a constant background field; for an incomplete list of references, see [2–8] and [9, 10] for reviews. In these examples, the horizon is always present.[b] In order to study particle production for field configurations near the threshold, we must consider a gapped matter sector coupled to background fields. The mass gap acts as a barrier, preventing particle production for weak backgrounds. We would like to find singular terms in the effective action as we approach the particle production threshold. In this regime, the effective action is real. If we dial the background field strength above the threshold, the effective action acquires an imaginary piece coming from the singular term. This imaginary part gives the vacuum decay rate.

In this paper, we consider different scenarios of strong background fields. We consider background scalar, electric and gravitational fields. We couple these backgrounds to free massive scalar matter, and determine the singular terms in the one-loop determinant of the matter fields in the background geometry tuned to the vicinity of the threshold. The fixed backgrounds are not necessarily a solution of the source-free equations of motion; we assume that there are suitable sources that sustain the static background configuration, and focus on the quantum mechanical response of matter fields to the background.

---

[a] Similar considerations apply to magnetically charged particles in a background magnetostatic field.

[b] The critical field strength associated to Schwinger e^+e^- pair production, $E_c = m_e^2/e$, gives the electric field value for which the pair production rate becomes non-exponentially suppressed. The pair production rate is, however, non-zero for any value of the background constant electric field, $\Gamma \sim (e^2 E)\exp(-E_c/E)$.

The electromagnetic threshold singularity might be experimentally testable in the near future by producing strong electric pulses with lasers. The gravitational threshold singularity is a quantum analogue of Choptuik scaling [11]. Choptuik numerically simulated the gravitational collapse of a distribution of dust particles. If the initial data is tuned above a critical value, the final state has a black hole. Choptuik recognized a remarkable scaling law in the mass of the black hole, as a function of how much above criticality the initial data is. For various setups of initial data, he found scaling laws for the black hole mass with the same exponent. Our result shares the same robustness to the shape of the gravitational potential.

While the exact result for the effective action depends on the details of the background field, we argue that the threshold singularity is universal. The physical reason for that is the following: right above threshold, the first pair production event will be very soft, and the pair will have a very long wavelength. As the wavelength of the excitation is very long, it is not sensitive to the fine features of the background.

For an early discussion of these ideas, see [12]. For recent work that partially overlaps with the results presented here, see [13], where universality of the particle production rate is found for electric fields slightly above threshold. Our below-threshold singularity, when extrapolated above the threshold, agrees with the result reported in [13].

1.1. *Outline*

In Sec. 2, we compute the threshold singularity for three different backgrounds — scalar, electric and gravitational fields. In Sec. 3, we present our conclusions. In Appendix A, we derive in detail a formula for the gravitational effective action in terms of the transmission coefficient; all other cases follow a similar derivation. In Appendix B, we discuss the behavior of the transmission coefficient when the mass gap is small. Finally, in Appendix C, we quote the exact transmission coefficients for some electric and gravitational backgrounds. The exact results agree with our general considerations in the main text.

2. Threshold Singularities

Having set the stage, let us study the threshold singularity for various quantum field theories in background fields. In this section, we determine the piece in the effective action which becomes singular as a parameter in the external field configuration reaches the threshold. At the threshold, a very long wavelength pair is produced, which can only probe the rough features of the external field configuration. This allows us to find a universal answer for the singularity, regardless of the precise shape of the external field. The nature of the singularity is slightly different for scalar, vector and gravitational external fields.

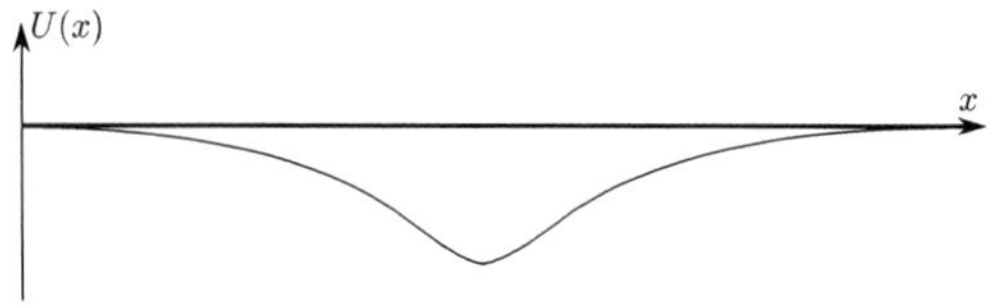

Fig. 1. Plot of a schematic form of the potential $U(x)$. As its parameters are tuned, the potential becomes deeper. At threshold, the background is too strong, and a bound state of the quantum field is spontaneously produced.

2.1. *Scalar fields*

We first consider a $1+1$ quantum field theory of a free massive scalar field in a static, position dependent background $U(x)$. We want to compute the one-loop effective action

$$\langle \mathrm{out}|\mathrm{in}\rangle = e^{iS_{\mathrm{eff}}(U)} = \int D\phi \, \exp\left(\frac{i}{2}\int \mathrm{d}t\, \mathrm{d}x((\partial_t\phi)^2 - (\partial_x\phi)^2 - (m^2+U(x))\phi^2)\right), \tag{2.1}$$

where we take $U(x)$ to be an arbitrary smooth function with asymptotic values $U(\pm\infty)\to 0$ (see Fig. 1). A very similar model was studied from a different point of view in [14].

If we consider a family of potentials controlled by some parameters, and tune these parameters in $U(x)$ to a certain threshold value, the effective action acquires an imaginary part. In this case, there is a simple way to argue that the threshold singularity will be of square root type and related to the lowest bound state in the potential $U(x)$. The effective action is proportional to the logarithm of the determinant of the Schrödinger operator

$$iS_{\mathrm{eff}}(U) = -\frac{T}{2}\mathrm{Tr}\log(\partial_x^2+\omega^2-m^2-U(x))\,, \tag{2.2}$$

where we used the Fourier representation for the time coordinate and, as we work in the approximation that the background is static, we obtain a factor of T from the amount of time that the background has been switched on. Assuming that $E_n(U)$ is the spectrum of the operator $-\partial_x^2+U(x)$, we find

$$iS_{\mathrm{eff}}(U) = -\frac{T}{2}\sum_n\int_{\mathcal{C}}\frac{\mathrm{d}\omega}{2\pi}\log(\omega^2-m^2-E_n(U))\,, \tag{2.3}$$

where the index n labels discrete and continuous eigenstates. The contour $\mathcal{C}$ in the complex ω-plane is chosen according to the Feynman $i\epsilon$-prescription $m\to m-i\epsilon$, $\epsilon>0$. On the real axis, we have multiple branch points $\omega=\pm\sqrt{m^2+E_n(U)}$, where we assume that the lowest bound state $E_0(U)>-m^2$ and the contour $\mathcal{C}$ goes above the branch cuts for $\omega>0$ and below the branch cuts for $\omega<0$ as shown in Fig. 2. Because we assumed that $E_0(U)>-m^2$, we can Wick rotate the contour $\mathcal{C}$ along the complex axis, so $\omega\to i\omega$ and we obtain a manifestly real expression for

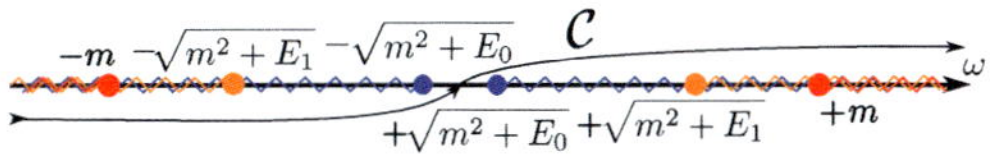

Fig. 2. The integration contour $\mathcal{C}$ in the complex ω-plane. At the threshold when $E_0(U) \to -m^2$, the contour $\mathcal{C}$ is pinched by two branch points $\omega = \pm\sqrt{m^2+E_0}$.

the effective action

$$iS_{\text{eff}}(U) = -i\frac{T}{2}\sum_n \int_{-\infty}^{+\infty} \frac{d\omega}{2\pi} \log(\omega^2 + m^2 + E_n(U)) \,. \tag{2.4}$$

We see that the possibility to Wick rotate is related to the vacuum stability. When the lowest bound state $E_0(U)$ approaches $-m^2$, the branch points start pinching the contour $\mathcal{C}$, and this leads to the appearance of the singular terms in the effective action. So, one can easily compute

$$S_{\text{eff}}(U) = -\frac{T}{2}\sqrt{m^2 + E_0(U)} + \cdots \,, \tag{2.5}$$

where we omitted less singular and non-singular terms.

It is instructive to see how this singularity arises when we express the effective action through the scattering data related to the potential $U(x)$. Namely, below we are going to show that the effective action can be expressed in terms of a logarithm of the transmission coefficient of a wave passing the potential $U(x)$. For the electric and gravitational cases, this method will be more convenient.

We begin by differentiating the effective action with respect to the mass, and obtain

$$i\frac{\partial S_{\text{eff}}(U)}{\partial m^2} = -\frac{i}{2}\int dt\, dx\, G_F(t, x; t, x) \,, \tag{2.6}$$

where $G_F(t, x; t', x') \equiv \langle \phi(x, t)\phi(x', t')\rangle$ is the Feynman Green's function.[c] We can use the Fourier representation for the time components of the Green's function, which then satisfies

$$(\partial_x^2 + \omega^2 - m^2 - U(x))G_F(\omega; x, x') = i\delta(x - x') \,. \tag{2.7}$$

By defining mode functions $f_{\text{in}}(x)$ and $f_{\text{out}}(x)$, which are annihilated by the Schrödinger operator

$$(\partial_x^2 + \omega^2 - m^2 - U(x))f_{\text{in/out}}(x) = 0 \tag{2.8}$$

and satisfy the following boundary conditions

$$f_{\text{in}}(x) \xrightarrow{x\to-\infty} \frac{e^{-ipx}}{\sqrt{2p}}, \quad f_{\text{out}}(x) \xrightarrow{x\to+\infty} \frac{e^{-ipx}}{\sqrt{2p}}, \quad p = \sqrt{\omega^2 - m^2} \,, \tag{2.9}$$

we can express the Green's function as

$$G_F(\omega; x, x') = i\frac{f_{\text{in}}(x)f^*_{\text{out}}(x')\theta(x' - x) + (x \leftrightarrow x')}{W(f_{\text{in}}, f^*_{\text{out}})} \,, \tag{2.10}$$

[c] We omit the time ordering symbol of the Feynman Green's function to avoid confusion with the time T that the background is switched on.

where the Wronskian is $W(f_{\text{in}}, f^*_{\text{out}}) \equiv f_{\text{in}}(x)\partial_x f^*_{\text{out}}(x) - f^*_{\text{out}}(x)\partial_x f_{\text{in}}(x)$. The functions f_{in} and f_{out} are related by Bogoliubov coefficients α and β

$$f_{\text{in}}(x) = \alpha f_{\text{out}}(x) + \beta f^*_{\text{out}}(x)\,, \tag{2.11}$$

where $|\alpha|^2 - |\beta|^2 = 1$, and a simple computation gives $W(f_{\text{in}}, f^*_{\text{out}}) = i\alpha$. We see that $1/\alpha$ is the transmission coefficient, which depends on ω, m and $U(x)$; it can be obtained by solving the quantum mechanical scattering problem (2.8). It is possible to show that the effective action $S_{\text{eff}}(U)$ is controlled entirely by the coefficient α [15] (see also [8, 16]).

In order to evaluate the effective action $S_{\text{eff}}(U)$, we see from (2.6) and (2.10) that we must compute $\int \mathrm{d}x\, f_{\text{in}}(x) f^*_{\text{out}}(x)$. It is possible to express this integral through the coefficient α. For this, we write the left-hand side of (2.8) for $f_{\text{in}}(x)$ with m^2, and multiply the equation by $f^*_{\text{out}}(x)$ which is solution of the same equation but with $m^2 + \delta m^2$. Analogously, we multiply the equation for $f^*_{\text{out}}(x)$ with $m^2 + \delta m^2$ by $f_{\text{in}}(x)$ with m^2. We subtract both expressions, integrate the result over x, and keep the first non-trivial terms in δm^2. This gives $\int \mathrm{d}x\, f_{\text{in}}(x) f^*_{\text{out}}(x) = -i\partial\alpha/\partial m^2$, where we used the Feynman $i\epsilon$-prescription $m \to m - i\epsilon$, $\epsilon > 0$. We finally obtain

$$S_{\text{eff}}(U) = \frac{1}{2} iT \int_C \frac{\mathrm{d}\omega}{2\pi} \log \alpha(\omega)\,, \tag{2.12}$$

where the choice of the contour C is explained above and shown in Fig. 2. In Appendix A, we give a detailed derivation of this formula for the gravitational case, which is technically the most complicated.

As we see from (2.12), finding $S_{\text{eff}}(U)$ has now been reduced to a 1-D scattering problem. Again the singularity arises in the integral (2.12) when the contour C is pinched by branch points. The result (2.12) is not so surprising, as indeed in the scattering theory it is well known that the transmission coefficient $1/\alpha$ is an analytic function of energy E on the physical sheet $\sqrt{E}(\text{Im}\sqrt{E} > 0)$, except for the points of discrete spectrum $E = E_n$, in which the amplitude has simple poles. Thus the coefficient $\alpha \sim (p - i\sqrt{|E_0|})/(p + i\sqrt{|E_0|})$ near the pinching branch points and computing the integral (2.12) for $E_0(U) \to -m^2$, one recovers the result (2.5). So we see that in the scattering approach the singularity mechanism is similar. More generally, the relation between scattering data and the determinant of the Schrödinger operator is well known and has been thoroughly investigated [17, 18].

2.2. *Electric fields*

Let us now consider the case of a free massive complex scalar in a strong electric field. We work in $1+1$ dimensions, but some of our results can be generalized to higher dimensions. The one-loop effective action is given by

$$\langle \text{out}|\text{in}\rangle = e^{iS_{\text{eff}}(a)} = \int D\phi D\bar{\phi} \exp\left(i \int \mathrm{d}t\, \mathrm{d}x (|\partial_t \phi + i a_t \phi|^2 - |\partial_x \phi|^2 - m^2|\phi|^2)\right), \tag{2.13}$$

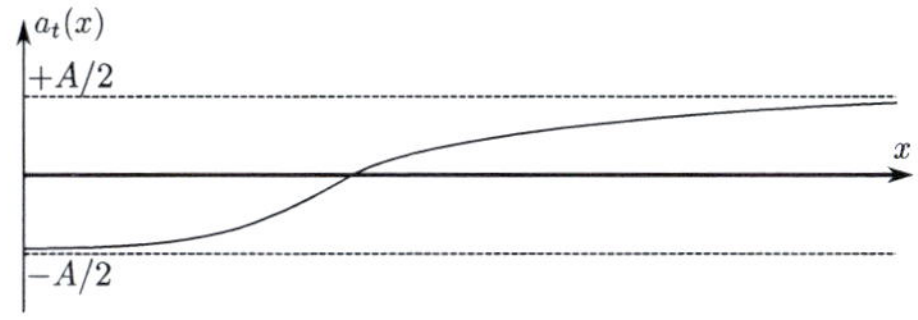

Fig. 3. Plot of a schematic form of the potential $a_t(x)$. We assume that $A < 2m$.

where we picked the static gauge $a = a_t(x)\mathrm{d}t$ for our background configuration, and chose $a_t(x)$ to be a smooth and monotonic function with asymptotic values $a_t(-\infty) = -A/2$ and $a_t(+\infty) = +A/2$ (see Fig. 3). The asymptotic values $\pm A/2$ are symmetric without loss of generality, by a simple shift of the potential. Other than monotonicity, we do not require the curve $a_t(x)$ to have any special property. We will see that particle production becomes favorable if $A > 2m$. If $A < 2m$, the effective action will be purely real and the vacuum is stable. As $A \to 2m$ from below, we will show that the effective action acquires a logarithmic singularity.

To gain some intuition of the pair production threshold in the electric case, we analyze the classical equations of motion, using band theory, in the asymptotic regimes $x \to \pm\infty$. The energies of excitations are given by

$$\omega_\pm = a_t(x) \pm \sqrt{p^2 + m^2}\,. \tag{2.14}$$

In Fig. 4, we see that the maximum and minimum points of the energy move as one goes from $x = -\infty$ to $x = +\infty$. When the bottom of the valence band comes up to the top of the conduction band, it becomes energetically favorable to disrupt the vacuum by pair production. This implies that threshold is reached when:

$$\min(\omega_+) - \max(\omega_-) = A - 2m > 0\,. \tag{2.15}$$

Now we proceed with the calculation of the effective action. Once again, it is convenient to differentiate it by mass

$$i\frac{\partial S_{\text{eff}}(a)}{\partial m^2} = -i\int \mathrm{d}t\,\mathrm{d}x\, G_F(t,x;t,x), \tag{2.16}$$

where $G_F(t,x;t',x') \equiv \langle \phi^*(x,t)\phi(x',t')\rangle$ is the Feynman Green's function. We can use the Fourier representation for the time components of the Green's function,

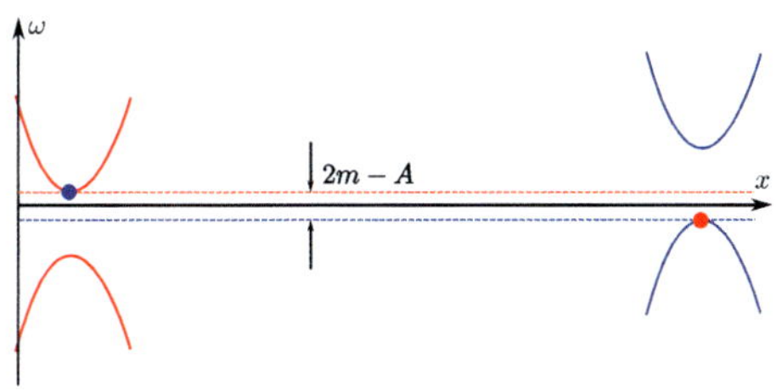

Fig. 4. Plot of the "bands" of the matter field. When the background is too strong, the bottom of the valence band comes up to the top of the conduction band and it becomes energetically favorable to trigger the tunneling and disrupt the vacuum by pair production.

which then satisfies

$$(\partial_x^2 + (\omega - a_t(x))^2 - m^2)G_F(\omega; x, x') = i\delta(x - x'). \tag{2.17}$$

Finding $S_{\text{eff}}(a)$ has now been reduced to a 1-D scattering problem similar to the scalar case we treated above. So we define mode functions f_{in} and f_{out} which are annihilated by the operator in the left-hand side of (2.17). In terms of f_{in} and f_{out}, the Green's function is given by

$$G_F(\omega; x, x') = i\frac{f_{\text{in}}(x)f^*_{\text{out}}(x')\theta(x' - x) + (x \leftrightarrow x')}{W(f_{\text{in}}, f^*_{\text{out}})}. \tag{2.18}$$

The functions f_{in} and f_{out} satisfy the following boundary conditions

$$f_{\text{in}}(x) \xrightarrow{x\to-\infty} \frac{e^{-ip_- x}}{\sqrt{2p_-}}, \quad f_{\text{out}}(x) \xrightarrow{x\to+\infty} \frac{e^{-ip_+ x}}{\sqrt{2p_+}}, \tag{2.19}$$

where $p_\pm = \sqrt{(\omega \mp A/2)^2 - m^2}$. The two solutions are related by Bogoliubov coefficients, $f_{\text{in}}(x) = \alpha f_{\text{out}}(x) + \beta f^*_{\text{out}}(x)$. Using the same method as in the scalar case, we obtain for the effective action

$$S_{\text{eff}}(a) = iT\int_C \frac{d\omega}{2\pi}\log\alpha(\omega). \tag{2.20}$$

The contour C must be chosen according to the Feynman $i\epsilon$-prescription $m \to m - i\epsilon$, $\epsilon > 0$, which gives $p_\pm \to p_\pm + i\epsilon$. There are multiple branch cuts on the real ω-axis. They start at the points corresponding to zeros of p_- and p_+ and also α. Therefore the contour C should go below the branch cuts for $\omega \to -\infty$ and above the branch cuts for $\omega \to +\infty$ and pass between the left and right branch cuts near $\omega = 0$ (see Fig. 5). In general, we may have a branch point which corresponds to $\alpha = 0$ but when A is very close to $2m$, the branch cuts corresponding to $p_- = p_+ = 0$ will pinch the contour C first. We see that this mechanism is different from the scalar case, where the effect is due to branch cuts corresponding to $\alpha = 0$.

So as $A \to 2m$ the threshold is reached when the contour C is pinched by the branch points at $\omega = +m - A/2$ and $\omega = -m + A/2$. This already hints at some universality, meaning that the most singular piece in the effective action will be largely agnostic about the particular shape of $a_t(x)$, but only depends on how close to threshold its maximum value is. Since the singularity appears as the branch points pinch the contour C, we need to look at $\alpha(\omega \approx \pm(m - A/2))$. As the band gap closes, we can use an argument which gives a general form of the coefficient α. Leaving the details to Appendix B, when $\omega \approx \pm(m - A/2)$ and $A \approx 2m$, the

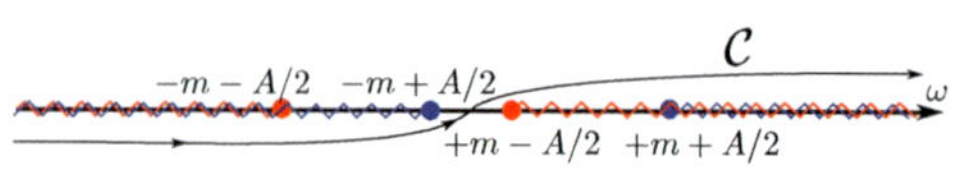

Fig. 5. The integration contour $\mathcal{C}$ in the complex ω-plane. The branch cuts here correspond to points where $p_- = p_+ = 0$. When the electric background near the threshold $A \to 2m$, the branch points $\omega = \pm(m - A/2)$ pinch the contour $\mathcal{C}$.

coefficient α is given by an infinite series in small p_+ and p_- and has the following form

$$\alpha = \frac{-ic_0 + c_-p_- + c_+p_+ + ic_{+-}p_+p_- + \cdots}{2\sqrt{p_+p_-}}, \tag{2.21}$$

with $p_\pm \equiv \sqrt{(\omega \mp A/2)^2 - m^2}$, and (c_0, c_-, c_+, c_{+-}) being shape-of-$a_t(x)$-dependent, but mass and frequency-independent real numbers. Other coefficients in this expansion are not important for the singular terms in the effective action. The conservation of current implies $c_+c_- - c_0c_{+-} = 1$.

Having an expression for α, we can evaluate the effective action. It is convenient to differentiate $\log \alpha(\omega)$ once by m^2, thus obtaining

$$\begin{aligned}\frac{\partial \log \alpha(\omega)}{\partial m^2} &= -\frac{1}{2}\left(\frac{c_+ + ic_{+-}p_-}{p_+} + \frac{c_- + ic_{+-}p_+}{p_-} + \cdots\right) \\ &\quad \times \frac{1}{-ic_0 + c_-p_- + c_+p_+ + \cdots} + \frac{1}{4}\left(\frac{1}{p_-^2} + \frac{1}{p_+^2}\right).\end{aligned} \tag{2.22}$$

At this point, the integral $\int_C \mathrm{d}\omega \frac{\partial}{\partial m^2} \log \alpha(\omega)$ is convergent and well defined. The last term in the right-hand side of (2.22) has no branch cuts and can be evaluated in closed form; it is an uninteresting, non-singular piece of $S_{\text{eff}}(a)$. So let us consider the first term in (2.22). We expect to obtain singular terms from vicinity of the points $\omega \approx \pm(m - A/2)$. It is possible to extract the non-analytic part from various integrals contributing to $\partial S_{\text{eff}}(a)/\partial m^2$. For instance, it is not difficult to show that

$$\begin{aligned}&\int_{m-A/2}^{2m} \frac{\mathrm{d}\omega}{p_-(-ic_0 + c_+p_+ + c_-p_- + ic_{+-}p_+p_- + \cdots)} \\ &\quad = k_0 - \frac{c_+}{c_0^2}(2m - A)\log\left(\frac{2m - A}{2m}\right) + k_1(2m - A) + \cdots,\end{aligned} \tag{2.23}$$

where the coefficients k_0 and k_1 depend on $c_0, c_+, c_-, c_{+-}, \ldots$ and on the upper limit of the integral, but the singular term depends only on c_0 and c_+. Analyzing various types of integrals arising from (2.22) and similar to (2.23), we finally obtain the most-singular non-analytic term of the effective action

$$\frac{\partial S_{\text{eff}}(a)}{\partial m^2} = -\frac{T}{2\pi}\frac{c_+c_- - c_0c_{+-}}{2c_0^2}\left(\frac{2m - A}{2m}\right)\log\left(\frac{2m - A}{2m}\right) + \cdots, \tag{2.24}$$

and so it follows that

$$S_{\text{eff}}(a) = \frac{Tm^3}{2\pi c_0^2}\left(\frac{2m - A}{2m}\right)^2 \log\left(\frac{2m - A}{2m}\right) + \cdots, \tag{2.25}$$

where we have omitted less singular and non-singular terms.

Let us make a few comments about (2.25):

- The term $1/c_0^2$ is proportional to the transmission amplitude of the effective potential; thus for long smooth gauge fields, it is exponentially damped.

- This term in the effective action is neither local in space (as in the usual derivative expansion) or in momentum space (as in the Euler–Heisenberg effective action). Neither of these representations can capture the threshold singularity, as we are always below threshold in the former case, and always above threshold in the latter.
- Despite depending on A, the effective action is gauge invariant, as $A = \int_{-\infty}^{+\infty} \mathrm{d}x\, E(x)$.
- $\operatorname{Im} S_{\text{eff}}$ can be reliably obtained by analytic continuation from (2.25), once we go slightly above the threshold, with $(A - 2m) \ll m$. The amount of phase space available to pair produce depends on the dimension of the spacetime. A quick estimate gives

$$\operatorname{Im} S_{\text{eff}}(a) \sim \int_0^{k_{\max}} \mathrm{d}^{d-2}k(A - 2m_{\text{eff}}(k))^2 \sim (A - 2m)^{\frac{d+2}{2}}, \tag{2.26}$$

  where $m_{\text{eff}}(k) \equiv \sqrt{m^2 + k^2}$ is the effective mass of the produced particles, and the integral over transverse momenta runs over a finite range, determined by the condition $A - 2m_{\text{eff}}(k_{\max}) = 0$. For $d = 4$, $\operatorname{Im} S_{\text{eff}}(a) \sim (A - 2m)^3$, as argued in [13]. Notice that $S_{\text{eff}}(a)$ will contain a factor of V_{d-2}, the volume of the transverse directions, in higher dimensions.
- The expression (2.25) is clearly invalid if $c_0 = 0$. In the regime $c_0 \ll (2m - A) \ll m$, one finds a different type of singularity

$$S_{\text{eff}}(a) = -\frac{Tm}{2\pi} \frac{c_+ c_-}{c_+^2 + c_-^2} \left(\frac{2m - A}{2m}\right) \log\left(\frac{2m - A}{2m}\right) + \cdots. \tag{2.27}$$

  We recover the expression above for a "quenching" electric field, with $a_t(x) = A/2\,\text{sgn}(x)$, where one has exactly $c_0 = 0$.

As a particular example of the formulas presented above, we can determine the precise form of the Bogoliubov coefficient for a family of potentials $a_t(x) = A/2\tanh(x/l)$, parametrized by the width of the potential l. The result is presented in Appendix C. This family of potentials includes the quenching example when $l \to 0$, and we can show that the singularity is different in that case, confirming our last bullet point above. Expanding the Bogoliubov coefficient around small p_+ and p_-, we find the same structure argued for in this section, given by (2.21).

2.3. *Gravitational fields*

Finally, we consider the case of a massive free scalar field in a strong gravitational background. Again, we work in $1 + 1$ dimensions and we would like to determine the one-loop effective action

$$\langle \text{out}|\text{in}\rangle = e^{iS_{\text{eff}}(v)} = \int D\phi \exp\left(-\frac{i}{2}\int \mathrm{d}t\,\mathrm{d}x\sqrt{-g}(g^{\mu\nu}\partial_\mu\phi\partial_\nu\phi + m^2\phi^2)\right). \tag{2.28}$$

For our purposes, it is convenient to describe the metric $g_{\mu\nu}$ in Painlevé–Gullstrand coordinates,

$$\mathrm{d}s^2 = g_{\mu\nu}\mathrm{d}x^\mu \mathrm{d}x^\nu = -\mathrm{d}t^2 + (\mathrm{d}x - v(x)\mathrm{d}t)^2, \tag{2.29}$$

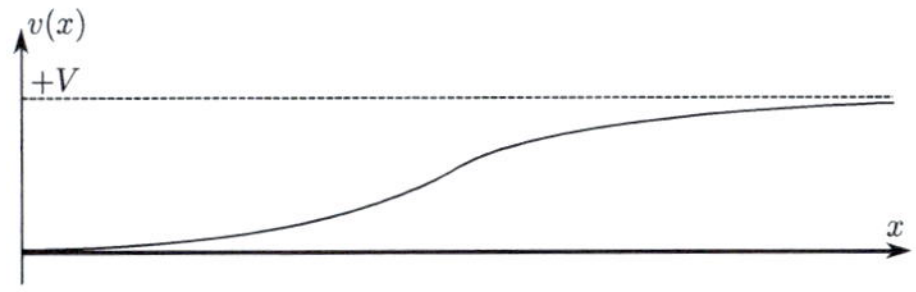

Fig. 6. Plot of a schematic form of the gravitational potential $v(x)$. We assume that $V < 1$.

where we choose $v(-\infty) = 0$ and let $v(x)$ increase smoothly for growing x up to some value $v(+\infty) = V$, similarly to the electrostatic potential $a_t(x)$ (see Fig. 6). This geometry would have a horizon at x_h if $v(x_h) = 1$ and thus $g_{tt} = 0$. As we will show later, the vanishing of g_{tt} coincides with the criterion for vacuum decay. For x at which $v(x) < 1$, we can interpret $v(x)$ as the escape velocity from the position x [19]. So our criterion for vacuum stability is that $v(x) < 1$ for all x. Therefore, we assume that $V < 1$ but we tune V to the threshold value, i.e. $V \to 1$. This case is mathematically closer to the equipotential planes with fixed asymptotics, in the electric case of the previous section. We will show that the effective action acquires a square root singularity when $V \to 1$.

Let us consider the semiclassical analysis for the gravitational case. The nature of the gap is slightly different than in the electric case. This is due to the different structure of the single-particle Hamiltonian [12, 20, 21]. The energies of excitations are given by

$$\omega_\pm(p, x) = pv(x) \pm \sqrt{p^2 + m^2}\,, \tag{2.30}$$

and threshold corresponds to $V \to 1$, as

$$\min(\omega_+) - \max(\omega_-) = 2m\sqrt{1 - V^2}\,. \tag{2.31}$$

The purpose of the mass term is just to open a gap between positive and negative energy bands (see Fig. 7), as particle production can occur for gravitational fields without a horizon when the matter sector is gapless [22].

Notice that when $v(x) = 1$, $g_{tt} = 0$, which, in these coordinates, is the usual definition of the horizon. In other words, our criterion for the location of the horizon being the distance at which it becomes energetically favorable to pair produce coincides with the definition coming from the causal structure of spacetime. Also,

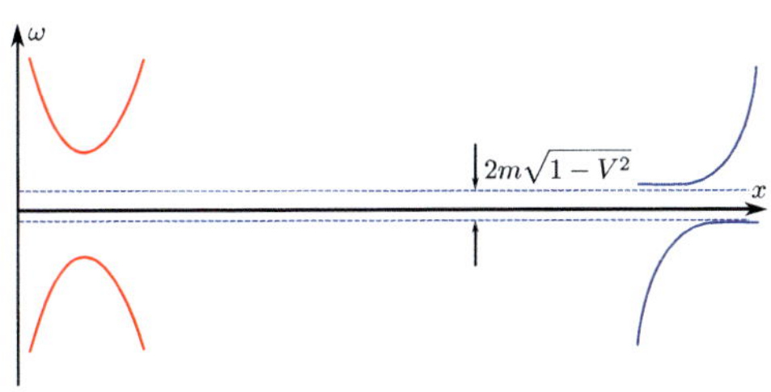

Fig. 7. Here we show the bands and the dominant tunneling event, which comes from the top of the lower blue band touching the bottom of the upper blue band.

notice that a shift $v \to v + C$ is not unphysical, for the case of a static metric. In order to remove the constant C, one could use a Galilean transformation, but this would imply a redefinition of the time coordinate. In other words, the criterion for vacuum instability depends not just on the difference between the asymptotic values of the velocity field, like in the electric field example, but also on its absolute values as $x \to \pm\infty$.

As usual, to compute the effective action, we differentiate it by the mass term

$$i\frac{\partial S_{\text{eff}}(v)}{\partial m^2} = -\frac{i}{2}\int \mathrm{d}t\,\mathrm{d}x\, G_F(x,t;x,t)\,, \tag{2.32}$$

where $G_F(x,t;x',t') \equiv \langle \phi(x,t)\phi(x',t')\rangle$ is the Feynman Green's function and we used that for our metric $\sqrt{-g} = 1$. The Green's function obeys the equation

$$(\partial_x((1-v^2)\partial_x) - 2v\partial^2_{xt} - (\partial_x v)\partial_t - \partial_t^2 - m^2)G_F(x,t;x',t') = i\delta(x-x')\delta(t-t')\,. \tag{2.33}$$

This equation contains a term with a first-order derivative in x, which naively makes it difficult to apply the previous strategy of expressing the effective action as an integral over the logarithm of the transmission coefficient. Nevertheless, by properly changing variables, we are able to obtain a similar $\log\alpha$ formula for the effective action.

To proceed, one can check that the Green's function $G_F(x,t;x',t')$ can be written in the form

$$G_F(x,t;x',t') = \int \frac{\mathrm{d}\omega}{2\pi}\frac{e^{i\omega(t-t')}e^{i(\chi(x)-\chi(x'))}}{\sqrt{(1-v^2(x))(1-v^2(x'))}}G_\omega(x,x')\,, \tag{2.34}$$

where the new Green's function $G_\omega(x,x')$ obeys a Schrödinger-like equation

$$\left(\partial_x^2 + \frac{\omega^2 - m^2(1-v^2) + (\partial_x v)^2}{(1-v^2)^2} + \frac{v\,\partial_x^2 v}{1-v^2}\right) G_\omega(x,x') = i\delta(x-x') \tag{2.35}$$

and the function χ is defined as $\partial_x\chi = \omega v/(1-v^2)$. It is easy to check that (2.34) indeed satisfies (2.33). Therefore, for the effective action, we obtain

$$\frac{\partial S_{\text{eff}}(v)}{\partial m^2} = -\frac{T}{2}\int\frac{\mathrm{d}\omega}{2\pi}\int \mathrm{d}x\frac{G_\omega(x,x)}{1-v^2(x)}\,. \tag{2.36}$$

The Green's function $G_\omega(x,x')$ as usual can be expressed through f_{in} and f_{out} functions

$$G_\omega(x,x') = i\frac{f_{\text{in}}(x)f^*_{\text{out}}(x')\theta(x'-x) + (x \leftrightarrow x')}{W(f_{\text{in}}, f^*_{\text{out}})}\,, \tag{2.37}$$

where f_{in} and f_{out} are annihilated by the Schrödinger-like operator in (2.35) and have asymptotics

$$f_{\text{in}}(x) \xrightarrow{x\to-\infty} \frac{1}{\sqrt{2p_-}}e^{-ip_-x}\,, \quad f_{\text{out}}(x) \xrightarrow{x\to+\infty} \frac{1}{\sqrt{2p_+}}e^{-ip_+x}\,, \tag{2.38}$$

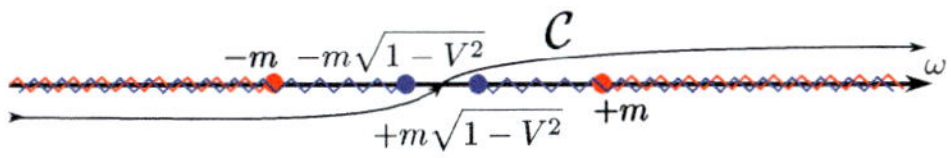

Fig. 8. The integration contour $\mathcal{C}$ in the complex ω-plane. Branch points $\omega = \pm m\sqrt{1-V^2}$ corresponding to $p_+ = 0$ pinch the contour $\mathcal{C}$ near $\omega \approx 0$ when $V \to 1$. This "pinching" region determines the singular piece of the effective action.

where we denoted

$$p_- = \sqrt{\omega^2 - m^2}, \quad p_+ = \frac{1}{1-V^2}\sqrt{\omega^2 - m^2(1-V^2)}, \tag{2.39}$$

and, as usual, we define α and β as $f_{\text{in}}(x) = \alpha f_{\text{out}}(x) + \beta f^*_{\text{out}}(x)$. As shown in Appendix A, we can bring the formula (2.36) to our usual form

$$S_{\text{eff}}(v) = \frac{1}{2} iT \int_{\mathcal{C}} \frac{\mathrm{d}\omega}{2\pi} \log \alpha(\omega). \tag{2.40}$$

The pinching singularity comes again from very small frequencies near the points where $p_+ = 0$, and we need to determine α for $\omega \sim m\sqrt{1-V^2}$ (see Fig. 8). Notice that the relevant feature is the behavior of the mode functions as $x \to \infty$, and for $\omega \sim m\sqrt{1-V^2}$, the mode function has very small modulation with ω. This means that the transmission coefficient behaves like

$$\alpha \approx \frac{-id_0 + d_+ p_+ + \cdots}{2\sqrt{p_+}}, \tag{2.41}$$

as $p_- \approx m$, and the model dependence is encoded in the coefficients (d_0, d_+). We can once again take derivatives of the effective action to isolate its singular piece. It turns out that differentiating once with respect to m^2 is enough to isolate the singular term. Following the same steps as in the electric case, we arrive at

$$S_{\text{eff}}(v) = mT\sqrt{1-V^2} + \cdots. \tag{2.42}$$

In Appendix C, we find an exact $\alpha(\omega)$ for the step potential $v(x) = V\theta(x)$. In this case, one can calculate the integral over ω in (2.40) exactly and obtain the result (2.42).

This singular part of the effective action can be written in a "local" form[d] due to the different tunneling pattern when the vacuum breaks down — pairs are produced at large x, rather than at both small and large x. There is also an analogous term to the electric threshold result, $\sim (1-V^2)\log(1-V^2)$, but it is subleading to (2.42). Notice that even in the special case $d_0 = 0$, we obtain the same singularity, albeit with different overall coefficient. Another interesting thing is that the leading term (2.42) does not care about the detailed coefficients $d_{0,+}$ — as long as they are non-zero, the only relevant data from the metric is the value of V. This is unlike the electric case, where the leading singular term depends on $2m - A$ but also on c_0.

This threshold singularity is a quantum analog of Choptuik scaling. Choptuik considered a family of initial data labeled by a parameter p. Under time evolution

[d]The volume form on the time-like surface $x \to +\infty$ is given by $\sqrt{1-V^2} = \sqrt{g_{\text{ind}}}$.

using Einstein's equations, he found [11] that the final state had a black hole of mass $M \sim (p - p_{\rm cr})^\gamma$, for $p > p_{\rm cr}$. The exponent γ is largely independent on the details of the family of initial data.

Above criticality, the formation of a black hole indicates the appearance of a horizon. Our critical exponent is entirely analogous, but is a quantum diagnostic of the appearance of the horizon. In our case, we look at $S_{\rm eff}(v)$ rather than M, criticality is reached when $V = 1$, and the critical exponent is $\gamma = 1/2$.[e]

3. Conclusions

In this paper, we argued that the crossover between the quantum mechanical stability and instability of background fields has certain universal features. This is largely due to the first unstable process triggered right above threshold having very long wavelength and low energy. This soft emission process probes only the roughest features of the external background, and the threshold singularity can be easily expressed in terms of rough background data. There are many avenues for further investigation:

- Our analysis was restricted to gaussian matter fields. How would interactions in the matter sector change the critical exponents in the threshold singularity?
- Can we connect our results to existing methods for treating backreaction in black holes [24, 25]? It would be nice to incorporate our threshold singularity to the problem of formation of a black hole, in order to see if vacuum polarization delays its formation, or prevents formation whatsoever for initial data close enough to threshold.
- Finally, it would also be interesting to find the threshold singularity for more realistic field configurations: for example, a spherically symmetric configuration, like a star, where we take a mass shell to be very close to its Schwarzschild radius.

We leave such fascinating problems to the near future.

Acknowledgments

We would like to thank Daniel Baumann, Garrett Goon, Diego Hofman, Viatcheslav Mukhanov, Andrew Strominger, Leon Takhtajan and Grigory Volovik for useful discussions, and Daniel Baumann, Garrett Goon and Diego Hofman for comments on a draft. The work of G.T. was supported by the MURI grant W911NF-14-1-0003 from ARO and by DOE grant de-sc0007870. G.P. acknowledges funding from the European Union's Horizon 2020 research and innovation programme under the Marie-Skłodowska Curie grant agreement number 751778. The work of G.P. is part of the Delta ITP consortium, a program of the Netherlands Organisation for Scientific Research (NWO) that is funded by the Dutch Ministry of Education,

[e]Interestingly, an analysis of black hole formation in a different context gives the same critical exponent [23].

Culture and Science (OCW). G.P. also acknowledges support from a Starting Grant of the European Research Council (ERC STG Grant 279617).

Appendix A. Derivation of $S_{\text{eff}} = i \int d\omega \log \alpha$

In this Appendix, we derive the formula for the gravitational effective action in terms of the Bogoliubov coefficient α in detail. The electromagnetic and scalar cases simply follow from this.

Using the expression (2.37) for $G_\omega(x, x')$ and formula (2.36), we obtain for the effective action

$$\frac{\partial S_{\text{eff}}(v)}{\partial m^2} = -i\frac{T}{2}\int \frac{d\omega}{2\pi}\int dx \frac{f_{\text{in}}(x) f^*_{\text{out}}(x)}{(1 - v^2(x))W(f_{\text{in}}, f^*_{\text{out}})}, \tag{A.1}$$

where one can easily compute $W(f_{\text{in}}, f^*_{\text{out}}) = i\alpha$. We must now calculate the integral over x in (A.1). In order to proceed, we do the following: consider the equations

$$\begin{gathered}\partial_x^2 f_{\text{in},m^2}(x) + \left(\frac{\omega^2 - m^2(1-v^2) + (\partial_x v)^2}{(1-v^2)^2} + \frac{v\,\partial_x^2 v}{1-v^2}\right) f_{\text{in},m^2}(x) = 0\,,\\ \partial_x^2 f^*_{\text{out},m^2+\delta m^2}(x) + \left(\frac{\omega^2 - (m^2+\delta m^2)(1-v^2) + (\partial_x v)^2}{(1-v^2)^2} + \frac{v\,\partial_x^2 v}{1-v^2}\right) f^*_{\text{out},m^2+\delta m^2}(x) = 0\,.\end{gathered} \tag{A.2}$$

Multiplying the first equation by $f^*_{\text{out},m^2+\delta m^2}(x)$ and the second by $f_{\text{in},m^2}(x)$ and subtracting them, we obtain

$$\begin{aligned}&\partial_x(f_{\text{in},m^2}\partial_x f^*_{\text{out},m^2+\delta m^2} - f^*_{\text{out},m^2+\delta m^2}\partial_x f_{\text{in},m^2})\\ &\quad = \delta m^2 \frac{f_{\text{in},m^2}(x) f^*_{\text{out},m^2+\delta m^2}(x)}{1 - v^2(x)}\,.\end{aligned} \tag{A.3}$$

Integrating over x the left and the right parts from $-L$ to L, where $L \to +\infty$, we get

$$\begin{aligned}&\delta m^2 \int_{-L}^{+L} dx \frac{f_{\text{in},m^2}(x) f^*_{\text{out},m^2+\delta m^2}(x)}{1 - v^2(x)}\\ &\quad = (f_{\text{in},m^2}\partial_x f^*_{\text{out},m^2+\delta m^2} - f^*_{\text{out},m^2+\delta m^2}\partial_x f_{\text{in},m^2})|_{-L}^{+L}\,.\end{aligned} \tag{A.4}$$

Because we take $L \to \infty$, we can use the asymptotic expressions for $f_{\text{in,out}}$ (2.38) and find

$$\begin{aligned}&(f_{\text{in},m^2}\partial_x f^*_{\text{out},m^2+\delta m^2} - f^*_{\text{out},m^2+\delta m^2}\partial_x f_{\text{in},m^2})|_{-L}^{+L}\\ &\quad = \delta m^2 \left(-i\frac{\partial \alpha}{\partial m^2} - L\alpha \frac{\partial(p_- + p_+)}{\partial m^2}\right.\\ &\qquad \left. - \frac{1}{2}i\left(\beta^* \frac{\partial \log p_-}{\partial m^2} e^{2iLp_-} - \beta \frac{\partial \log p_+}{\partial m^2} e^{2iLp_+}\right)\right) + \cdots\,.\end{aligned} \tag{A.5}$$

We use the Feynman $i\epsilon$-prescription $p_\pm \to p_\pm + i\epsilon$ with infinitesimal $\epsilon > 0$, so the oscillating terms above are zero for large L. Finally, we obtain

$$\begin{aligned}(f_{\mathrm{in},m^2}\partial_x f^*_{\mathrm{out},m^2+\delta m^2} &- f^*_{\mathrm{out},m^2+\delta m^2}\partial_x f_{\mathrm{in},m^2})|^{+L}_{-L} \\ &= -i\delta m^2 \alpha \frac{\partial}{\partial m^2}(\log\alpha + L(p_- + p_+))\,. \end{aligned} \tag{A.6}$$

Putting this together, we find

$$\frac{\partial S_{\mathrm{eff}}(v)}{\partial m^2} = i\frac{T}{2}\int_C \frac{\mathrm{d}\omega}{2\pi}\frac{\partial}{\partial m^2}(\log\alpha + L(p_- + p_+)). \tag{A.7}$$

Our last task is to argue that the terms proportional to L are unimportant. By that we mean that they only carry uninteresting dependence on the background. The term proportional to Lp_- is harmless, depending only on m, but the term proportional to Lp_+ seems to have non-trivial dependence on V. Let us write it more explicitly

$$\int_C \mathrm{d}\omega\, Lp_+ = \int_C \mathrm{d}\omega \frac{1}{1-V^2}(\omega^2 - m^2(1-V^2))^{1/2}\,. \tag{A.8}$$

If we change variables $\omega = \omega'(1-V^2)^{1/2}$, then the V dependence drops out of the integral and it is exactly equal to the Lp_- integral. This argument is too fast, as the integral is UV divergent. The correct argument is that the cutoff is background dependent. For the p_- integral, we are at $x \to -\infty$ so we choose some hard cutoff Λ in frequency space. At $x \to +\infty$, the metric is $\mathrm{d}t^2(1-V^2)$ so we must choose the cutoff $\Lambda/(1-V^2)^{1/2}$ in frequency space, to take into account the warping of time intervals. This renders the Lp_+ integral to have no interesting dependence on V. In summary, up to non-important terms, we find

$$S_{\mathrm{eff}}(v) = \frac{1}{2}iT\int_C \frac{\mathrm{d}\omega}{2\pi}\log\alpha(\omega)\,. \tag{A.9}$$

Appendix B. Behavior of α Near Threshold

In this Appendix, we derive the behavior of the Bogoliubov coefficient α when the effective mass gap is very small. In other words, we find the first few terms in an expansion for α around vanishing mass gap. To start, let us consider the Schrödinger equation

$$(\partial_x^2 + U(x) - m^2)f = 0\,, \tag{B.1}$$

where $U(x)$ either switches off or asymptotes to some fixed values $U(\pm\infty)$ in a smooth way. We are interested in the cases where the effective mass gap at infinity

$$p_\pm^2 \equiv U(\pm\infty) - m^2 \tag{B.2}$$

is very small, namely, $p_\pm^2 \ll m^2$. We consider here the case in which both $p_\pm^2$ are small. In the main text, the gravitational background is such that only in one extreme the mass gap vanishes. Applying our formulas to that example is straightforward.

In the region outside of which $U(x)$ is varying, the mass term is either p_+^2 or p_-^2, which we assume are small. Neglecting those terms, we get

$$\partial_x^2 f = 0\,, \tag{B.3}$$

therefore the solutions of the equations of motion are

$$f = a_1 + b_1 x, \quad x \ll 0, \qquad f = a_2 + b_2 x, \quad x \gg 0\,, \tag{B.4}$$

where the potential varies significantly close to $x = 0$. The coefficients (a_1, b_1) and (a_2, b_2) are linearly dependent

$$a_1 = c_- a_2 + c_{+-} b_2, \quad b_1 = c_0 a_2 + c_+ b_2\,, \tag{B.5}$$

where the coefficients (c_0, c_+, c_-, c_{+-}) are independent of $p_\pm$ (as $p_\pm$ do not appear in the differential equation with linear functions as solutions), and, from the conservation of current, it follows that we can choose (c_0, c_+, c_-, c_{+-}) to be real, with $c_+c_- - c_0 c_{+-} = 1$. Then matching the solutions (B.4) with the asymptotic solutions in terms of plane waves, we obtain

$$a_1 = \frac{1}{\sqrt{2p_-}}, \quad b_1 = \frac{-ip_-}{\sqrt{2p_-}}, \quad a_2 = \frac{\alpha+\beta}{\sqrt{2p_+}}, \quad b_2 = \frac{-ip_+(\alpha-\beta)}{\sqrt{2p_+}}\,, \tag{B.6}$$

and solving Eqs. (B.5), we get

$$\alpha = \frac{-ic_0 + c_- p_- + c_+ p_+ + ic_{+-} p_- p_+}{2\sqrt{p_- p_+}}, \quad \beta = \frac{ic_0 - c_- p_- + c_+ p_+ + ic_{+-} p_- p_+}{2\sqrt{p_- p_+}}\,. \tag{B.7}$$

Now, having α, we can evaluate the effective action. These expressions are only valid for $|p_\pm| \ll m$.

Appendix C. Exact Solutions

C.1. *Electric example*

The exact solution in the electric case is available for the gauge field profile $a_t(x) = A/2\tanh(x/l)$. In this case, one can obtain an exact Bogoliubov coefficient α; it is given by

$$\alpha = \frac{i/l}{\sqrt{p_+ p_-}} \frac{\Gamma(1 - ip_- l)\Gamma(1 - ip_+ l)}{\Gamma(\rho - \frac{i}{2} l(p_- + p_+))\Gamma(1 - \rho - \frac{i}{2} l(p_- + p_+))}\,, \tag{C.1}$$

where $\rho = \frac{1}{2} + \frac{1}{2}\sqrt{1 - A^2 l^2}$ and p_-, p_+ are defined below (2.19). If we first tune $l \to 0$, we obtain the step potential $a_t(x) = A/2\,\mathrm{sgn}(x)$ and the Bogoliubov coefficient (C.1) simplifies to $\alpha = (p_- + p_+)/2\sqrt{p_+ p_-}$. On the other hand, if l is fixed and we are in the regime where p_+ and p_- are small, we find

$$\alpha = \frac{\sin \pi\rho}{\pi l} \frac{(2i - l(\psi(\rho) + \psi(1-\rho) + 2\gamma_E)(p_- + p_+) + \cdots)}{2\sqrt{p_+ p_-}}\,, \tag{C.2}$$

where $\psi(x)$ is the digamma function, and γ_E is the Euler–Mascheroni constant. This form of the α coefficient agrees with (B.7).

C.2. *Gravitational example*

In this subsection, we are going to find the coefficient α in the case of a step potential $v(x) = V\theta(x)$. To proceed, it is convenient to write the Schrödinger equation for the operator (2.35) as

$$\partial_x \left((1-v^2)\partial_x \left(\frac{f_{\rm in}(x)}{\sqrt{1-v^2}} \right) \right) + \frac{\omega^2 - m^2(1-v^2)}{(1-v^2)^{3/2}} f_{\rm in}(x) = 0\,. \tag{C.3}$$

Now integrating this equation from $x = -\delta$ to $x = \delta$ with $\delta \to 0$, we find the boundary conditions for $f_{\rm in}(x)$ and $\partial_x f_{\rm in}(x)$ at $x = 0$:

$$f_{\rm in}(0^+) = \sqrt{1-V^2} f_{\rm in}(0^-), \quad \sqrt{1-V^2}\partial_x f_{\rm in}(0^+) = \partial_x f_{\rm in}(0^-)\,. \tag{C.4}$$

Using these boundary conditions, one can find

$$\alpha = \frac{p_- + (1-V^2)p_+}{2\sqrt{(1-V^2)p_- p_+}}\,, \tag{C.5}$$

where $p_- = \sqrt{\omega^2 - m^2}$ and $p_+ = \sqrt{\omega^2 - m^2(1-V^2)}/(1-V^2)$.

References

[1] R. M. Wald, *General Relativity* (The University of Chicago Press, 1984).

[2] J. S. Schwinger, On gauge invariance and vacuum polarization, *Phys. Rev.* **82** (1951) 664–679.

[3] L. Parker, Quantized fields and particle creation in expanding universes. 1., *Phys. Rev.* **183** (1969) 1057–1068.

[4] S. W. Hawking, Particle creation by black holes, *Comm. Math. Phys.* **43** (1975) 199–220.

[5] G. W. Gibbons and S. W. Hawking, Cosmological event horizons, thermodynamics, and particle creation, *Phys. Rev. D* **15** (1977) 2738–2751.

[6] G. K. Savvidy, Infrared instability of the vacuum state of gauge theories and asymptotic freedom, *Phys. Lett. B* **71** (1977) 133–134.

[7] N. K. Nielsen and P. Olesen, An unstable Yang–Mills field mode, *Nucl. Phys. B* **144** (1978) 376–396.

[8] A. M. Polyakov, De Sitter space and eternity, *Nucl. Phys. B* **797** (2008) 199–217; `arXiv:0709.2899 [hep-th]`.

[9] N. D. Birrell and P. C. W. Davies, *Quantum Fields in Curved Space*, Cambridge Monographs on Mathematical Physics (Cambridge University Press, Cambridge, UK, 1984).

[10] G. V. Dunne, Heisenberg–Euler effective Lagrangians: Basics and extensions, in *From Fields to Strings: Circumnavigating Theoretical Physics*, Vol. 1 (World Scientific, 2005), pp. 445–522; `arXiv:hep-th/0406216 [hep-th]`.

[11] M. W. Choptuik, Universality and scaling in gravitational collapse of a massless scalar field, *Phys. Rev. Lett.* **70** (1993) 9–12.

[12] A. M. Polyakov, Self-destructive manifolds and the cosmological constant, PiTP 2014, Institute for Advanced Study (2014); https://static.ias.edu/pitp/2014/schedule.html.

[13] H. Gies and G. Torgrimsson, Critical Schwinger pair production II — Universality in the deeply critical regime, *Phys. Rev. D* **95**(1) (2017) 016001; arXiv:1612.00635 [hep-th].

[14] A. Migdal, *Fermions and Bosons in Strong Fields* (Nauka, Moscow, 1978).

[15] A. I. Nikishov, On vacuum-vacuum amplitude and Bogolyubov coefficients, *J. Exp. Theor. Phys.* **96**(2) (2003) 180–192; *Zh. Eksp. Teor. Fiz.* **123**(2) (2003) 211–223; arXiv:hep-th/0207085 [hep-th].

[16] D. Krotov and A. M. Polyakov, Infrared sensitivity of unstable vacua, *Nucl. Phys. B* **849** (2011) 410–432; arXiv:1012.2107 [hep-th].

[17] L. D. Faddeev, The Inverse problem in the quantum theory of scattering, *J. Math. Phys.* **4** (1963) 72–104; *Usp. Mat. Nauk* **14** (1959) 57–119.

[18] L. D. Faddeev, Inverse problem of quantum scattering theory. II, *J. Soviet Math.* **5** (1976) 334–396.

[19] A. J. S. Hamilton and J. P. Lisle, The River model of black holes, *Am. J. Phys.* **76** (2008) 519–532; arXiv:gr-qc/0411060 [gr-qc].

[20] G. E. Volovik, On de Sitter radiation via quantum tunneling, *Int. J. Mod. Phys. D* **18** (2009) 1227–1241; arXiv:0803.3367 [gr-qc].

[21] G. E. Volovik, Particle decay in de Sitter spacetime via quantum tunneling, *JETP Lett.* **90** (2009) 1–4; arXiv:0905.4639 [gr-qc].

[22] G. L. Pimentel, A. M. Polyakov, and G. M. Tarnopolsky, Vacuum decay in CFT and the Riemann–Hilbert problem, *Nucl. Phys. B* **907** (2016) 617–632; arXiv:1512.06721 [hep-th].

[23] A. Strominger and L. Thorlacius, Universality and scaling at the onset of quantum black hole formation, *Phys. Rev. Lett.* **72** (1994) 1584–1587; arXiv:hep-th/9312017 [hep-th].

[24] G. E. Volovik, Simulation of Painleve–Gullstrand black hole in thin He-3 — A film, *JETP Lett.* **69** (1999) 705–713; *Pisma Zh. Eksp. Teor. Fiz.* **69** (1999) 662–668; arXiv:gr-qc/9901077 [gr-qc].

[25] M. K. Parikh and F. Wilczek, Hawking radiation as tunneling, *Phys. Rev. Lett.* **85** (2000) 5042–5045; arXiv:hep-th/9907001 [hep-th].

Wilson Loops in Antisymmetric Representations from Localization in Supersymmetric Gauge Theories

Jorge G. Russo[*,†,¶] and Konstantin Zarembo[‡,§,||,**]

[*]*Institució Catalana de Recerca i Estudis Avançats (ICREA), Pg. Lluis Companys, 23, 08010 Barcelona, Spain*

[†]*Departament de Física Cuántica i Astrofísica and ICCUB, Universitat de Barcelona, Martí Franquès, 1, 08028 Barcelona, Spain*

[‡]*Nordita, KTH Royal Institute of Technology and Stockholm University, Roslagstullsbacken 23, SE-106 91 Stockholm, Sweden*

[§]*Department of Physics and Astronomy, Uppsala University, SE-751 08 Uppsala, Sweden*

[¶]*jorge.russo@icrea.cat*

[||]*zarembo@nordita.org*

Large-N phase transitions occurring in massive $\mathcal{N} = 2$ theories can be probed by Wilson loops in large antisymmetric representations. The logarithm of the Wilson loop is effectively described by the free energy of a Fermi distribution and exhibits second-order phase transitions (discontinuities in the second derivatives) as the size of representation varies. We illustrate the general features of antisymmetric Wilson loops on a number of examples where the phase transitions are known to occur: $\mathcal{N} = 2$ SQCD with various mass arrangements and $\mathcal{N} = 2^*$ theory. As a byproduct, we solve planar $\mathcal{N} = 2$ SQCD with three independent mass parameters. This model has two effective mass scales and undergoes two phase transitions.

In memory of Ludvig Dmitrievich Faddeev

Contents

[**]Also at ITEP, Moscow, Russia.

1. Introduction

Localization is a powerful tool to explore supersymmetric gauge theories in the non-perturbative domain, and in particular in the large-N limit. Exact results obtained in the latter case bear direct links to the holographic duality at strong coupling.

The partition function and select observables of any $\mathcal{N} = 2$ gauge theory on S^4 localize to an effective matrix model [1], that can be studied at large-N by the standard methods of random matrix theory [2]. A somewhat unexpected outcome of this analysis is the appearance of large-N phase transitions in a variety of massive $\mathcal{N} = 2$ gauge theories [3, 4]. Their holographic description remains an interesting open problem. The difficulty is related to the strong–weak coupling nature of the holographic duality: transitions occur upon varying a coupling constant, and this is difficult to achieve in holography where an infinite-coupling regime is normally considered. It is desirable, in this respect, to devise observables that undergo phase transitions at fixed coupling while some other auxiliary parameter is being varied. Particularly promising probes of this kind are Wilson loops in large antisymmetric representations of the gauge group, which were shown to undergo phase transitions once the size of representation is dialed to a critical value [5].

Methods to calculate expectation values of Wilson loops in large representations, both holographically and from localization, have been devised in the context of the $\mathcal{N} = 4$ super-Yang–Mills (SYM) theory [6–8], which localizes to the Gaussian matrix model [1, 9]. These methods have been transplanted to $\mathcal{N} = 2$ theories, both conformal [10] and massive [5, 11], and in the massive case, antisymmetric Wilson loops were shown to undergo phase transitions in the rank of the gauge-group representation [5].

Here we discuss general features of phase transitions in antisymmetric Wilson loops and then apply the general framework to a variety of theories where the transitions are known to occur, namely to $\mathcal{N} = 2$ super-QCD for different combinations of mass parameters and to $\mathcal{N} = 2^*$ SYM, extending in the latter case the results of [5, 11].

2. Antisymmetric Wilson Loops

By localization, the expectation value of the circular Wilson loop in any $\mathcal{N} = 2$ theory on S^4 can be expressed as a matrix model correlator:

$$W_{\mathcal{R}}(C) = \left\langle \mathrm{tr}_{\mathcal{R}}\, \mathrm{e}^{L\Phi} \right\rangle_{\mathrm{MM}}, \tag{2.1}$$

where $L = 2\pi R_{S^4}$ is the circumference of the circle, which we consider to be large compared to any other scale in the problem. In the decompactification limit

$R_{S^4} \to \infty$, the circular loop should obey the same universal scaling law as any sufficiently large contour. We thus expect that (2.1) describes the universal behavior of large Wilson loops of arbitrary shape in $\mathcal{N} = 2$ gauge theories.

We concentrate on the Wilson loops in antisymmetric representations. The generating function for characters in the rank-k antisymmetric representation $\mathcal{A}_k$ is given by

$$\sum_{k=0}^{N} \mathrm{e}^{-L\nu k} \operatorname{tr}_{\mathcal{A}_k} \mathrm{e}^{L\Phi} = \det\left(1 + \mathrm{e}^{L(\Phi-\nu)}\right). \tag{2.2}$$

An expectation value in any particular representation can be computed as

$$W_{\mathcal{A}_k}(C) = L \int_{C-\frac{i\pi}{L}}^{C+\frac{i\pi}{L}} \frac{d\nu}{2\pi i}\, \mathrm{e}^{Lk\nu} \left\langle \det\left(1 + \mathrm{e}^{L(\Phi-\nu)}\right)\right\rangle_{\mathrm{MM}}. \tag{2.3}$$

In the 't Hooft limit, with

$$N \to \infty, \qquad k \to \infty, \qquad f = \frac{k}{N} \ - \text{ fixed}, \tag{2.4}$$

the saddle point of the matrix model is not affected by the Wilson loop insertion, and is characterized by the eigenvalue density

$$\rho(x) = \frac{1}{N} \operatorname{tr} \delta(x - \Phi), \tag{2.5}$$

obtained by solving the localization matrix model in the large-N limit. In the same scaling limit, the saddle-point approximation for the integral over ν in (2.3) becomes exact and the Wilson loop expectation value takes on an exponential, perimeter-law form:

$$W_{\mathcal{A}_k}(C) \simeq \mathrm{e}^{NLF(f)}. \tag{2.6}$$

The function $F(f)$ is determined by the saddle-point equation, which can written as follows. Define the function

$$\mathcal{F}(f, \nu) = f\nu + \frac{1}{L} \int dx\, \rho(x) \ln\left(1 + \mathrm{e}^{L(x-\nu)}\right). \tag{2.7}$$

Then $F(f) = \mathcal{F}(f, \nu(f))$, where $\nu(f)$ is the value of ν that maximizes $\mathcal{F}(f, \nu)$ for a given f:

$$f = \int \frac{dx\, \rho(x)}{\mathrm{e}^{L(x-\nu(f))} + 1}. \tag{2.8}$$

Once the eigenvalue distribution is known, the expectation value of the antisymmetric Wilson loop can be computed from the above two equations [7].

The distribution of eigenvalues in massive $\mathcal{N} = 2$ theories is confined to an interval $(-\mu, \mu)$, where μ is the characteristic mass scale of the underlying gauge theory, which we assume to be such that $\mu L \gg 1$. This corresponds to the low-temperature regime of the effective statistical model, and the Fermi distribution in the above formulas can be replaced by the step function:

$$\begin{aligned} \mathcal{F}(f,\nu) &= f\nu + \int_{\nu}^{\mu} dx\,\rho(x)(x-\nu), \\ f &= \int_{\nu(f)}^{\mu} dx\,\rho(x). \end{aligned} \tag{2.9}$$

It follows from these equations that

$$\begin{aligned} \frac{dF}{df} &= \nu, \\ \frac{d^2F}{df^2} &= -\frac{1}{\rho(\nu)}, \end{aligned} \tag{2.10}$$

where $\nu \equiv \nu(f)$. While the first of these equations is exact, the second one is only valid in the limit $\mu L \to \infty$.

2.1. *Phase transitions*

The eigenvalue distribution in massive theories may develop specific singularities in a certain range of parameters, which lead to phase transitions as the parameters change. The fundamental Wilson loops are not very good probes of the phase transitions, because the influence of any given singularity in the eigenvalue density is washed out by averaging over the whole eigenvalue distribution. On the contrary, the singularities are very pronounced in large antisymmetric Wilson loops due to the sharp form of the Fermi distribution at zero temperature.

Large-N solutions of the localization matrix model are known for a number of $D = 4$ supersymmetric gauge theories [4]. It was observed that in the decompactification limit, when the size of the four-sphere becomes infinite, $R_{S^4} \to \infty$, the density may develop singularities. The features encountered so far come in two types: the delta functions and the one-sided cusps. The first type of singularity arises in theories with fundamental matter, such as super-QCD, while the second type is characteristic for adjoint matter and arises, for instance, in the $\mathcal{N} = 2^*$ theory. It is these singularities which are responsible for the phase transitions. We consider the two cases in turn, first at a general, model-independent level and then on concrete examples.

2.1.1. *Delta-function singularity*

Suppose that the density has a delta-functional peak in the middle of the eigenvalue distribution:

$$\rho(x) \overset{x \to x_c}{=} \rho_0 + p\delta(x - x_c), \tag{2.11}$$

where p is the fraction of eigenvalues concentrated in the peak and ρ_0 is a constant. It is not hard to see that this structure translates into two singularities in the Wilson loop expectation value, at $f = f_{c_+}$ and $f = f_{c_-}$, where

$$f_{c_+} = \int_{x_c+0}^{\mu} dx\, \rho(x), \qquad f_{c_-} = f_{c_+} + p. \tag{2.12}$$

The function $\nu(f)$ stays flat, $\nu(f) = x_c$, for $f_{c_+} < f < f_{c_-}$. From (2.10), we see that the free energy F is continuous across the transitions together with its first derivative, while the second derivative experiences a finite jump:

$$\left.\frac{d^2F}{df^2}\right|_{f_{c_\pm}-0}^{f_{c_\pm}+0} = \pm\frac{1}{\rho_0}\,. \tag{2.13}$$

2.1.2. *Cusp*

Strictly speaking, there are two types of cusps, the left cusp and the right cusp. On one side of the cusp the density approaches a finite value, while on the other side it has an inverse square root singularity:

$$\rho(x) \overset{x\to x_{c_\pm}}{=} \begin{cases} \dfrac{C}{\sqrt{|x - x_{c_\pm}|}} & \text{at } x \to x_{c_\pm} \mp 0 \\ \rho_0 & \text{at } x \to x_{c_\pm} \pm 0. \end{cases} \tag{2.14}$$

As a result, the antisymmetric Wilson loop develops a singularity at $f = f_{c_\pm}$, where

$$f_{c_\pm} = \int_{x_{c_\pm}}^{\mu} dx\, \rho(x). \tag{2.15}$$

The free energy stays continuous together with its first derivative across the critical point, while the second derivative experiences a finite jump, given by the same formula (2.13) as for the delta-function singularity.

These are the two types of singularities encountered in the large-N solutions of the localization matrix models in $\mathcal{N} = 2$ theories in four dimensions. The delta-function singularities arise in theories fundamental matter, while cusps are characteristic of the adjoint matter. Below we consider a few examples of each type.

3. Pure $\mathcal{N} = 2$ SYM Theory

We begin with a case where there is no large-N phase transition, $\mathcal{N} = 2$ $SU(N)$ SYM theory without matter. The large-N solution of this theory was first obtained from the Seiberg–Witten theory [12]. The localization matrix model for pure $\mathcal{N} = 2$ SYM was studied in [13, 14]. In the decompactification limit $R \to \infty$, one finds [12, 13]:

$$\rho(x) = \frac{1}{\pi\sqrt{\mu^2 - x^2}}\,, \qquad \mu = c_0\Lambda R\,, \quad c_0 = 2\mathrm{e}^{-1-\gamma}, \tag{3.1}$$

where R is the radius of S^4 and Λ is the dynamically generated scale.[a]

[a]In terms of the original, renormalized 't Hooft coupling, $\Lambda R = \exp[-4\pi^2/\lambda]$.

The saddle-point equation (2.9) gives

$$f = \int_{\nu}^{\mu} dx \, \frac{1}{\pi\sqrt{\mu^2 - x^2}} = \frac{1}{\pi} \arccos\frac{\nu}{\mu}, \tag{3.2}$$

i.e.

$$\nu = \mu \cos(f\pi). \tag{3.3}$$

The Wilson loop (2.6) is given by

$$\ln W_{\mathcal{A}_k} = N \int_{\nu}^{\mu} dx \, \rho(x) x = \frac{N}{\pi}\sqrt{\mu^2 - \nu^2}. \tag{3.4}$$

Thus

$$\ln W_{\mathcal{A}_k} = NRF(f), \qquad F(f) = \frac{c_0 \Lambda}{\pi} \sin(f\pi), \qquad 0 < f < 1. \tag{3.5}$$

It is a smooth function of f, as expected, since the theory does not have any phase transitions.

4. $\mathcal{N} = 2$ Super QCD

As a first example where phase transitions occur, we consider the case of $\mathcal{N} = 2$ $SU(N)$ super-Yang–Mills theory with N_f pairs of fundamental and anti-fundamental hypermultiplets of masses $(M, -M)$, $N_f < N$.[b]

This model was studied in [4, 14]. We consider the $N \to \infty$ limit with fixed Veneziano parameter $\zeta \equiv N_f/N$ and fixed (renormalized) 't Hooft coupling $\lambda = g_{\rm YM}^2 N$, which in turn is traded by the dynamically generated scale Λ,

$$\Lambda R = \mathrm{e}^{-\frac{4\pi^2}{\lambda(1-\zeta)}}. \tag{4.1}$$

In the decompactification limit, the theory undergoes a phase transition at $\Lambda = M/2$. The different features of the model were also reproduced in [15] by taking the large-N limit on the Seiberg–Witten curve. Another interesting limit is the decompactification limit at finite N. The partition function can then be computed by incorporating instantons through the Seiberg–Witten curve [16].

It is possible to generalize the theory by considering fundamental hypermultiplets of different masses, $M_1, \ldots, M_{N_f}$. The eigenvalue density in the large-N limit can still be determined in terms of analytic formulas. The resulting theory describes multiple phase transitions occurring whenever the largest eigenvalue μ (or the lowest eigenvalue $-\mu$) crosses a new mass scale. An explicit example with three mass scales, Λ, m, M is given in Appendix A.

The present theory depends on two parameters ΛR and MR. In the large-N limit, the eigenvalue density is determined by a saddle-point equation. This

[b]The theory with $N_f = N$ with massless hypermultiplets corresponds to $\mathcal{N} = 2$ superconformal SQCD. In this case, there are no phase transitions [4].

simplifies in the decompactification limit $R \to \infty$. By differentiating the saddle-point equation once, one obtains, for $R \to \infty$,

$$2 \int_{-\mu}^{\mu} dy\, \rho(y) \ln \frac{(x-y)^2}{\Lambda^2} = \zeta \ln \frac{\left(x^2 - M^2\right)^2}{\Lambda^4}. \tag{4.2}$$

Differentiating once more, we find the equation:

$$2 \fint_{-\mu}^{\mu} dy\, \frac{\rho(y)}{x-y} = \frac{\zeta}{x+M} + \frac{\zeta}{x-M}. \tag{4.3}$$

This (singular) integral equation has two different solutions, which depends on whether the points $x = \pm M$ lie, or do not lie, within the eigenvalue distribution. The two solutions describe the two phases of the theory:

- Phase I. Strong coupling phase, $\mu > M$, with

$$\rho(x) = \frac{1-\zeta}{\pi\sqrt{\mu^2 - x^2}} + \frac{\zeta}{2}\delta(x+M) + \frac{\zeta}{2}\delta(x-M) \tag{4.4}$$

- Phase II. Weak coupling phase, $\mu < M$, with

$$\rho(x) = \frac{1-\zeta}{\pi\sqrt{\mu^2 - x^2}} + \zeta M \sqrt{M^2 - \mu^2} \frac{1}{\pi\sqrt{\mu^2 - x^2}(M^2 - x^2)}. \tag{4.5}$$

The non-analytic behavior in the Wilson loop $\ln W_{\mathcal{A}_k}$ appears when $\nu(f)$ crosses a singular point in the eigenvalue density. Since in phase II the eigenvalue density is smooth, in the weak coupling phase the $\ln W_{\mathcal{A}_k}$ will be smooth. We thus focus on the more interesting strong coupling phase, where we have a delta-function singularity of the type described in Sec. 2.1.1. In this phase, $\mu = 2\Lambda$, so the phase transition takes place at $\Lambda_c = M/2$ (see Appendix A).

The saddle-point equation (2.9) now leads to

$$f = (1-\zeta)\frac{1}{\pi}\arccos\frac{\nu}{\mu} + \frac{\zeta}{2}\theta(M-\nu) + \frac{\zeta}{2}\theta(-M-\nu). \tag{4.6}$$

The solution has several critical points f_1, f_2, f_3, f_4, where

$$\pi f_1 = (1-\zeta)\arccos\frac{M}{\mu}, \quad f_2 = f_1 + \frac{\zeta}{2}, \quad f_3 = 1 - f_2, \quad f_4 = 1 - f_1. \tag{4.7}$$

The critical points (f_1, f_2) corresponds to the points (f_{c_+}, f_{c_-}) described in Sec. 2.1.1, when ν meets the delta-function singularity at $x = M$. In the interval (f_1, f_2), ν remains constant, $\nu = M$. Similarly, (f_3, f_4) are the two critical points corresponding to the delta-function singularity at $x = -M$. Explicitly, the

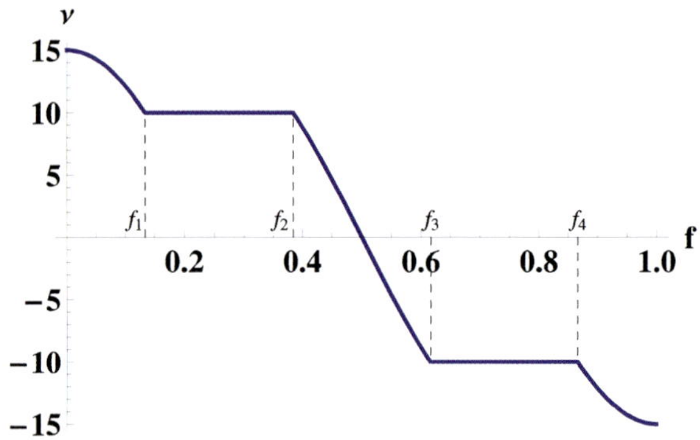

Fig. 1. ν as a function of f ($M = 10$, $\Lambda = 3M/4$, $\zeta = 1/2$). As $\nu = dF/df$, the figure also shows the behavior of the first derivative of $\ln W_{\mathcal{A}_k}$.

complete solution is

$$
\begin{aligned}
&\nu = \mu \cos \frac{\pi f}{1-\zeta}, && 0 < f < f_1 \\
&\nu = M, && f_1 < f < f_2 \\
&\nu = \mu \cos \frac{\pi (f - \zeta/2)}{1-\zeta}, && f_3 < f < f_2 \\
&\nu = -M, && f_3 < f < f_4 \\
&\nu = \mu \cos \frac{\pi (f-\zeta)}{1-\zeta}, && f_4 < f < 1.
\end{aligned}
\tag{4.8}
$$

Figure 1 shows a plot of the saddle-point ν as a function of $f = k/N$ for $\zeta = 1/2$ (i.e. $N_f = N/2$).

Let us now compute the Wilson loop. We have $\ln W_{\mathcal{A}_k} = NRF(f)$, with

$$
\begin{aligned}
F &= f\nu + \int_\nu^\mu dx\, \rho(x)(x-\nu) \\
&= f\nu + \frac{1}{\pi}(1-\zeta)\sqrt{\mu^2-\nu^2} - \frac{\nu}{\pi}(1-\zeta)\arccos\frac{\nu}{\mu} \\
&\quad + \frac{\zeta}{2}(M-\nu)\theta(M-\nu) - \frac{\zeta}{2}(M+\nu)\theta(-M-\nu).
\end{aligned}
\tag{4.9}
$$

It exhibits discontinuities in the second derivative with respect to f at the critical points. This is readily seen from (2.10), since the first derivative of $\ln W_{\mathcal{A}_k}$ is proportional to $\nu(f)$ which, as can be seen from Fig. 1, has discontinuities in its first derivatives. More directly, from (2.13), we have

$$
\left. \frac{d^2F}{df^2} \right|_{f_{c\pm}-0}^{f_{c\pm}+0} = \pm \frac{\pi\sqrt{\mu^2-M^2}}{1-\zeta}, \tag{4.10}
$$

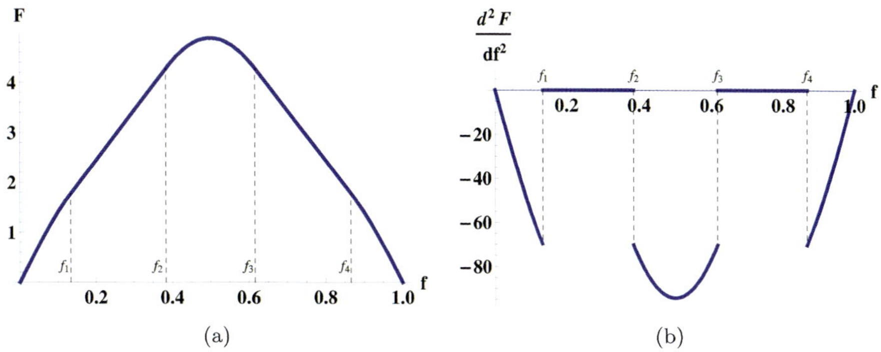

Fig. 2. (a) $F = \frac{1}{NR} \ln W_k$ as a function of f. (b) $\frac{d^2F}{df^2}$ as a function of f. Here $R = 1$, with the same values for M, Λ, ζ as in Fig. 1.

where f_{c_+}, f_{c_-} are either f_1, f_2 or f_3, f_4. Figures 2(a) and 2(b), respectively, show plots of $F = \frac{1}{NR} \ln W_{\mathcal{A}_k}$ and d^2F/df^2 as functions of f. The first derivative, dF/df, being equal to $\nu(f)$, can be seen from Fig. 1. One can check that the jumps in the second derivatives in Fig. 2(b) are in precise agreement with (4.10).

5. $\mathcal{N} = 2^*$ Theory

The field content of $\mathcal{N} = 2^*$ SYM consists of the vector multiplet and an adjoint hypermultiplet of mass M. The 't Hooft coupling $\lambda = g_{\rm YM}^2 N$ does not run and is just a parameter that characterizes the theory. The theory undergoes a fourth-order phase transition each time the width of the eigenvalue distribution 2μ passes an integer multiple of M: $\mu(\lambda_c^{(n)}) = nM/2$. It is convenient to introduce the parameters

$$n = \left[\frac{2\mu}{M}\right], \qquad \Delta = \left\{\frac{2\mu}{M}\right\}, \tag{5.1}$$

where the square (curly) brackets denote the integer (fractional) part of a real number. The eigenvalue density has left (right) cusps at

$$\begin{aligned} x_{c+}^{(k)} &= \mu - kM, \qquad & k = 1, \ldots, n \\ x_{c-}^{(k)} &= \mu - kM - \Delta, \qquad & k = 0, \ldots, n-1. \end{aligned} \tag{5.2}$$

In what follows we concentrate on the strong-coupling regime of large λ.

The width of the eigenvalue distribution 2μ grows with λ, and the number of cusps accordingly multiplies. At strong coupling the density gets a rather complicated short-scale structure with a large number of cusps, while the average density is very simple and is described by the Wigner distribution, as shown in Fig. 3(a).

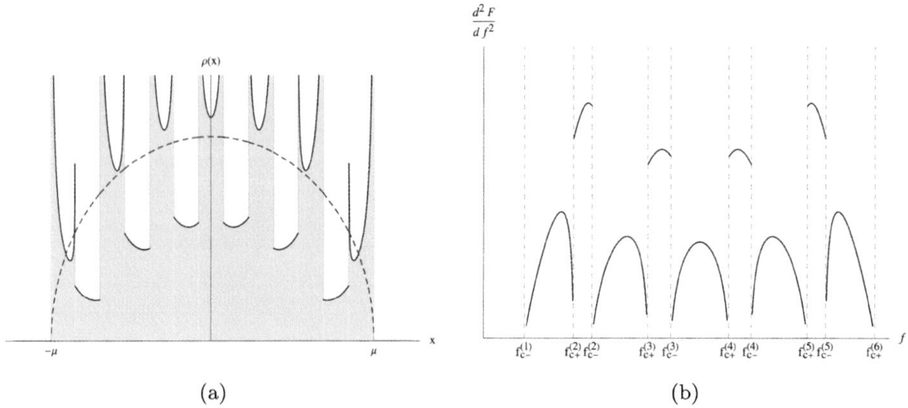

Fig. 3. (a) The eigenvalue density, for $n = 6$, along with the average Wigner distribution shown as the dashed line. (b) The second derivative of the free energy, for the same set of parameters.

To the first two orders of the strong-coupling expansion [17],

$$\rho(x) \simeq \frac{2}{\pi\mu^2}\sqrt{\mu^2 - x^2} + \frac{1}{\pi}\sqrt{\frac{M}{2\mu^5}}\left[(\mu - x)\zeta\left(\frac{1}{2}, \left\{\frac{\mu + x}{M}\right\}\right) + (\mu + x)\zeta\left(\frac{1}{2}, \left\{\frac{\mu - x}{M}\right\}\right)\right]. \quad (5.3)$$

The width of the eigenvalue distribution, to the same accuracy, is given by

$$\mu \simeq \frac{\sqrt{\lambda}M}{2\pi}, \quad (5.4)$$

but it is more convenient to regard M and μ as independent variables (rather than M and λ).

Taking into account that the zeta function has the square-root branch point at zero: $\zeta(1/2, z) \simeq 1/\sqrt{z}$, we find that the singular points (5.2) are left or right cusps:

$$\rho(x) \overset{x \to x_{c\pm}^{(k)}}{=} \begin{cases} \dfrac{C_{\pm}^{(k)}}{\sqrt{|x - x_{c\pm}^{(k)}|}} & \text{at } x \to x_{c\pm} \mp 0 \\ \rho_0^{(k)} & \text{at } x \to x_{c\pm} \pm 0, \end{cases} \quad (5.5)$$

where

$$\rho_0^{(k)} = \frac{2M\sqrt{k\,(n - k)}}{\pi\mu^2}, \quad (5.6)$$

and

$$C_{+}^{(k)} = \frac{M^2(n - k)}{\sqrt{2\pi^2\mu^5}}, \qquad C_{-}^{(k)} = \frac{M^2 k}{\sqrt{2\pi^2\mu^5}}. \quad (5.7)$$

Notice that $\rho_0^{(k)} \sim 1/\mu \gg C_\pm^{(k)}/\sqrt{\mu} \sim M/\mu^2$, so the singularities are parametrically weak.

The oscillating structure in the density (5.3) is a small correction on top of the regular Wigner distribution, and in addition the irregular part of the eigenvalue density integrates to zero upon averaging over sufficiently large interval, up to $O(M^{3/2}/\mu^{3/2})$ which is beyond the accuracy of (5.3). Therefore, to the leading order, the expectation value of the antisymmetric Wilson loop is given by the formulas for the Gaussian matrix model [7]:

$$F = \frac{2\mu}{3\pi}\sin^3\theta, \tag{5.8}$$

where $\cos\theta = \nu/\mu$, with ν from (2.9) related to the representation variable f by

$$\pi f = \theta - \frac{1}{2}\sin 2\theta. \tag{5.9}$$

Taking into account the next order introduces the oscillating peak structure in the eigenvalue density and leads to the phase transitions in the Wilson loop expectation value. According to (5.2), the phase transitions happen at $f = f_{c\pm}^{(k)}$, where the critical representation labels $f_{c+}^{(k)}$ are given by (5.9) with

$$\cos\theta_{c+}^{(k)} = 1 - \frac{2k}{n+\Delta}, \qquad k = 1, \dots, n \tag{5.10}$$

and

$$f_{c-}^{(k)} = 1 - f_{c+}^{(n-k+1)}. \tag{5.11}$$

Although $n \gg \Delta$, it is important to keep Δ in this equation to break the degeneracy under $\theta \to \pi - \theta$ and to get right the positions of the critical points.

The second derivative of the free energy experiences a jump (2.13) at the kth critical point of an amplitude

$$\left.\frac{d^2F}{df^2}\right|_{f_{c\pm}^{(k)}-0}^{f_{c\pm}^{(k)}+0} = \pm\frac{\pi\mu^2}{2M\sqrt{k(n-k)}}. \tag{5.12}$$

Since d^2F/df^2 at each critical point jumps to zero, its discontinuity is of the same order $\mathcal{O}(\mu)$ as its average value. This leads to the structure displayed in Fig. 3(b).

The above calculations apply to $f = O(1)$. For parametrically smaller $f = O(\lambda^{-3/4})$, the endpoint region of the eigenvalue distribution, where (5.3) is no longer accurate, becomes more important. The phase transitions for $f = O(\lambda^{-3/4})$ were analyzed in detail in [5]. In both cases, λ is assumed to be large, and these results are potentially relevant for the holographic description [18, 19] of the $\mathcal{N} = 2^*$ theory.

6. Conclusions

Wilson loops in large antisymmetric representations undergo phase transitions which mirror quantum phase transitions in the underlying gauge theories. Both are caused by features in the eigenvalue density, delta functions or cusps, known to occur in localization matrix models for $\mathcal{N} = 2$ gauge theories. In four dimensions only these singularities can arise. It would be interesting to extend the analysis to localization matrix models in other dimensions where different types of singularities may occur.

Holographically, antisymmetric Wilson loops of rank $k \sim N$ correspond to probe D5-branes in the dual geometry [7]. The classical solutions for D5-branes are known only for $AdS_5 \times S^5$ [7]. It would be extremely interesting to construct D5-brane configurations for the backgrounds that are dual to massive models with phase transitions, for instance in the Pilch–Warner background [18] dual to $\mathcal{N} = 2^*$ SYM. So far only D3-brane solutions, which correspond to Wilson loops in large symmetric representations, have been constructed for this background [20]. The latter are affected by singularities in the eigenvalue density to much lesser degree compared to antisymmetric representations and are not very sensitive to the phase transitions.

Acknowledgments

J.G.R. acknowledges financial support from a MINECO grant FPA2016-76005-C2-1-P and MDM-2014-0369 of ICCUB (Unidad de Excelencia 'María de Maeztu'). The work of K.Z. was supported by the grant "Exact Results in Gauge and String Theories" from the Knut and Alice Wallenberg Foundation, by the ERC advanced grant No 341222, by the Swedish Research Council (VR) grant 2013-4329 and by RFBR grant 15-01-99504.

Appendix A. SQCD with Several Mass Scales

Consider $\mathcal{N} = 2$ supersymmetric $SU(N)$ QCD on $\mathbb{S}^4$, with n_f pairs of fundamental hypermultiplets of mass $(m, -m)$, and N_f pairs of fundamental hypermultiplets of mass $(M, -M)$, with $M > m$. The case $m = M$ is the theory studied in Sec. 4.

The partition function computed by localization is given by

$$Z_{N_f}^{\mathrm{SQCD}} = \int d^{N-1}a \, \frac{\mathrm{e}^{2(N-N_f-n_f)(\ln \Lambda R+\gamma+1)\sum\limits_i a_i^2 R^2} \prod_{i<j} (a_i - a_j)^2 H^2(a_i - a_j)}{\prod_i H^{N_f}(a_i + M) H^{N_f}(a_i - M) H^{n_f}(a_i + m) H^{N_f}(a_i - m)}, \tag{A.1}$$

where R is the radius of $\mathbb{S}^4$ and Λ is the dynamically generated scale

$$\Lambda R = \mathrm{e}^{-\frac{4\pi^2}{\lambda(1-\zeta)}}, \tag{A.2}$$

with Veneziano parameters

$$\zeta = \zeta_M + \zeta_m, \qquad \zeta_M = \frac{N_f}{N}, \qquad \zeta_m = \frac{n_f}{N}. \tag{A.3}$$

We shall consider $\zeta_M + \zeta_m < 1$, in which case the theory is asymptotically free. The partition function is thus expressed in terms of a finite dimensional integral, which is still difficult to compute. Here, as Sec. 4 (see also [4, 14]), we shall compute this integral in the planar, large-N limit at fixed λ, ζ_m, ζ_M, by the saddle-point method.

Let us first summarize the physical picture of the $m = M$ case considered in Sec. 4. The theory has two phases, weak coupling $2\Lambda < M$ and strong coupling $2\Lambda > M$, separated by a third-order phase transition at $2\Lambda_c = M$. In this new theory with three mass scales Λ, M, m, the question we would like to address is whether there are more phase transitions and at which value of Λ they occur. A naive guess would be that a new phase transition should occur at $\Lambda \sim m$. However, when $\Lambda \ll M$, the hypermultiplets of mass M can be integrated out and the new dynamical scale of the theory is not Λ: it is replaced by a new effective scale that we shall compute. For example, in the theory of [4, 14] with a single mass scale, at $\Lambda \ll M$ the theory flows to pure $\mathcal{N} = 2$ SYM with $\Lambda_{\text{eff}} = \Lambda^{1-\zeta_M} M^{\zeta_M}$. In the new theory with two mass scales M, m, the next phase transition is expected not when $\Lambda = O(m)$, but, perhaps, when the new dynamical scale of the theory (left behind after the hypermultiplets of mass M have been integrated out) is of order m. The present exact calculation will determine which are the effective scales in each phase, and at which precise coupling Λ the different phase transitions occur.

In the large-N limit, the saddle-point equation becomes the following integral equation:

$$\begin{aligned} 2\fint_{-\mu}^{\mu} dy\, \rho(y) \left(\frac{1}{x-y} - \mathcal{K}(x-y) \right) &= -4(1-\zeta_M-\zeta_m)(\ln\Lambda + \gamma + 1)x \\ &\quad - \zeta_M \mathcal{K}(x+M) - \zeta_M \mathcal{K}(x-M) \\ &\quad - \zeta_m \mathcal{K}(x+m) - \zeta_m \mathcal{K}(x-m), \end{aligned} \tag{A.4}$$

where $\mathcal{K} = x(\psi(1+ix) + \psi(1-ix) + 2\gamma)$, ψ represents as usual the logarithmic derivative of the Γ-function and γ is the Euler constant, $\gamma = -\psi(1)$ (see [4] for the relation of $\mathcal{K}$ to the Barnes G-function and other properties). Here we have set $R = 1$ for the sake of clarity.

The model depends on three parameters ΛR, MR and mR. In the decompactification limit, these parameters go to infinity and the saddle-point equations simplify. In this limit, the eigenvalue distribution extends to large eigenvalues and one can use the approximation $\mathcal{K} = x \ln x^2 + 2\gamma x + O(x^{-1})$. The first derivative of the saddle-point equations is given by

$$2\int_{-\mu}^{\mu} dy\, \rho(y) \ln \frac{(x-y)^2}{\Lambda^2} = \ln \frac{\left(x^2 - M^2\right)^{2\zeta_M} \left(x^2 - m^2\right)^{2\zeta_m}}{\Lambda^{4\zeta}}. \tag{A.5}$$

Differentiating once more we obtain a singular integral equation that can be easily solved:

$$2\fint_{-\mu}^{\mu} dy\frac{\rho(y)}{x-y}=\frac{\zeta_M}{x+M}+\frac{\zeta_M}{x-M}+\frac{\zeta_m}{x+m}+\frac{\zeta_m}{x-m}. \tag{A.6}$$

Just like in the model of Sec. 4, corresponding to the $m = M$ case (or to the $\zeta_m = 0$ case), the solution to this Eq. (A.6) depends on whether $x = \pm m$ or $x = \pm M$ lie inside or outside the cut $(-\mu, \mu)$. The system has therefore three phases:

(i) Strong coupling, $\mu > M > m$. The poles at $x = \pm M$ and $x = \pm m$ lie inside the cut. The eigenvalue density $\rho(x)$ has delta function terms $\frac{1}{2}\zeta_M\delta(x\pm M)$ and $\frac{1}{2}\zeta_m\delta(x \pm m)$.
(ii) Intermediate coupling, $M > \mu > m$. The poles at $x = m$ and at $x = -m$ lie on the cut. The eigenvalue density $\rho(x)$ now contains delta-function terms $\frac{1}{2}\zeta_m\delta(x \pm m)$.
(iii) Weak coupling phase, $M > m > \mu$. All poles lie outside the eigenvalue distribution.

In what follows we describe the analytic solution in each phase.

Strong-Coupling Phase ($\mu > M > m$)

When $\mu > M > m$, the eigenvalue density that solves (A.6) is given by

$$\rho(x)=\frac{1-\zeta}{\pi\sqrt{\mu^2-x^2}}+\frac{\zeta_M}{2}\,\delta(x+M)+\frac{\zeta_M}{2}\,\delta(x-M)+\frac{\zeta_m}{2}\,\delta(x+m)+\frac{\zeta_m}{2}\,\delta(x-m). \tag{A.7}$$

The parameter μ is determined in terms of M, m and Λ by substituting the solution into (A.5). This gives

$$\mu = 2\Lambda. \tag{A.8}$$

Note that in this phase the endpoint of the eigenvalue distribution is therefore independent of the hypermultiplet masses.

The solution (A.7) holds as long as $M < \mu$. As Λ is gradually decreased, there is a critical point Λ_{c1} where $\mu = M$, which thus occurs at

$$\Lambda_{c1}=\frac{M}{2}. \tag{A.9}$$

For $\Lambda < \Lambda_{c1}$, the delta-functions $\delta(x \pm M)$ lie outside the interval $[-\mu, \mu]$ and the density (A.7) does not solve (A.6) anymore. This happens just like in the $m = M$ case discussed in Sec. 4.

Intermediate-Coupling Phase ($M > \mu > m$)

In the regime $M > \mu > m$, the solution is given by

$$\rho(x) = \frac{1-\zeta}{\pi\sqrt{\mu^2 - x^2}} + \frac{\zeta_M M\sqrt{M^2-\mu^2}}{\pi\sqrt{\mu^2-x^2}(M^2-x^2)} + \frac{\zeta_m}{2}\,\delta(x+m) + \frac{\zeta_m}{2}\,\delta(x-m). \quad \text{(A.10)}$$

Substituting into (A.5) we find the following transcendental equation for μ:

$$(1-\zeta)\ln\frac{\mu}{2M} + \zeta_M \ln\frac{\mu}{M+\sqrt{M^2-\mu^2}} = (1-\zeta)\ln\frac{\Lambda}{M}\,. \quad \text{(A.11)}$$

In the particular case $\zeta_m = 0$, corresponding to $n_f = 0$ and giving $\zeta_M = \zeta$, the resulting equation was studied in [4, 14]. Like in this case, the solution can be expressed in a parametric form:

$$\mu = M\sqrt{1-u^2}, \quad \text{(A.12)}$$

$$\left(\frac{2\Lambda}{M}\right)^{2-2\zeta_m-2\zeta_M} = (1+u)^{1-\zeta_m-2\zeta_M}(1-u)^{1-\zeta_m}. \quad \text{(A.13)}$$

Here we can see the emergence of the first relevant effective scale. When $M \gg \Lambda$, the hypermultiplets can be integrated out. From the above equations we obtain

$$\left(\frac{\mu}{2}\right)^{1-\zeta_m} \approx M^{\zeta_M}\Lambda^{1-\zeta}, \qquad M \gg \Lambda. \quad \text{(A.14)}$$

The theory left behind should be SQCD with n_f flavors and dynamical QCD scale $\Lambda_{\text{eff}1} \sim \mu$.

Note that (A.11), (A.13) simplify in the particular case $\zeta_m + 2\zeta_M = 1$. Then we can explicitly solve for μ

$$\mu = 2\sqrt{\Lambda(M-\Lambda)}. \quad \text{(A.15)}$$

There is another interesting case, $\zeta_M = \frac{3}{4}(1-\zeta_m)$, where Eq. (A.11) simplifies and can be solved explicitly. Then we find

$$\mu = \frac{\Lambda}{\sqrt{2}}\left(\sqrt{1+\frac{4M}{\Lambda}} - 1\right)^{\frac{3}{2}}. \quad \text{(A.16)}$$

For $\Lambda \ll M$, one has

$$\mu \approx 2M^{\frac{3}{4}}\Lambda^{\frac{1}{4}} = \Lambda_{\text{eff}\,1}, \quad \text{(A.17)}$$

in agreement with (A.14).

Weak-Coupling Phase ($M > m > \mu$)

As Λ is further decreased, there is a critical point Λ_{c2} where $\mu = m$. For lower values of Λ, the delta-functions $\delta(x \pm m)$ move outside the eigenvalue region $[-\mu, \mu]$ and the density (A.10) does no longer solve (A.6). In this regime $M > m > \mu$ we find the solution

$$\rho(x) = \frac{1-\zeta}{\pi\sqrt{\mu^2-x^2}} + \frac{\zeta_M M\sqrt{M^2-\mu^2}}{\pi\sqrt{\mu^2-x^2}(M^2-x^2)} + \frac{\zeta_m m\sqrt{m^2-\mu^2}}{\pi\sqrt{\mu^2-x^2}(m^2-x^2)}. \tag{A.18}$$

Substituting into the integrated form of the saddle-point equation (A.5), we now find the following transcendental equation for μ:

$$(1-\zeta)\ln\frac{\mu}{2} + \zeta_M \ln\frac{\mu}{M+\sqrt{M^2-\mu^2}} + \zeta_m \ln\frac{\mu}{m+\sqrt{m^2-\mu^2}} = (1-\zeta)\ln\Lambda. \tag{A.19}$$

This defines $\mu = \Lambda f(m/\Lambda, M/\Lambda)$. In particular, consider the case $\Lambda \ll m, M$. In this case, all hypermultiplets can be integrated out and the theory should be pure $\mathcal{N}=2$ SYM with an effective dynamical scale. The question is which the effective dynamical scale is. From (A.19), in this limit we find

$$\mu \approx 2m^{\zeta_m} M^{\zeta_M} \Lambda^{1-\zeta_M-\zeta_m} \equiv \Lambda_{\text{eff }2}. \tag{A.20}$$

This can also be written in the form

$$\Lambda_{\text{eff }2} = \bar{M}\left(\frac{\Lambda}{\bar{M}}\right)^{1-\zeta}, \qquad \bar{M}^{\zeta_m+\zeta_M} \equiv m^{\zeta_m} M^{\zeta_M}. \tag{A.21}$$

Concerning the interpretation of $\Lambda_{\text{eff }2}$ as the effective dynamical scale, one can make the following consistency check. We note that $\Lambda_{\text{eff }2} \approx \Lambda_{\text{eff }1}^{1-\zeta_m} m^{\zeta_m}$. This is indeed the dynamical scale that one should expect when $n_f = \zeta_m N$ flavors are integrated out starting from a theory with scale $\Lambda_{\text{eff }1}$.

Let us now determine at which value of the original coupling Λ_{c2} the phase transition from the intermediate to the weak-coupling phase occur. This is computed by setting $\mu = m$ in (A.19) (or, equivalently, in (A.11)). This gives

$$(1-\zeta)\ln 2\Lambda_{c2} = (1-\zeta_m)\ln m - \zeta_M \ln(M+\sqrt{M^2-m^2}). \tag{A.22}$$

Note that $M+\sqrt{M^2-m^2} = \alpha M$, where $1<\alpha<2$. Therefore, ignoring a factor of order 1, we can substitute $M+\sqrt{M^2-m^2}$ by M. Then

$$\Lambda_{c2}^{1-\zeta} \approx m^{1-\zeta_m} M^{-\zeta_M}, \tag{A.23}$$

or

$$\Lambda_{c2} \approx m\left(\frac{m}{M}\right)^{\frac{\zeta_M}{1-\zeta}}. \tag{A.24}$$

This last expression shows, in particular, that $\Lambda_{c2} \ll m$ if $m \ll M$.

Substituting $\Lambda = \Lambda_{c2}$ into $\Lambda_{\text{eff}2}$, we find that the new phase transition occurs at

$$\Lambda_{\text{eff }2} \approx m. \tag{A.25}$$

Note also that, for this Λ, one also has $\Lambda_{\text{eff }1} \approx \Lambda_{\text{eff }2}$.

Following [4, 14], it is easy to show that the two first transitions occurring at Λ_{c1}, Λ_{c2} are third order. To show this, one computes the different derivatives of the free energy at each phase using

$$\Lambda\frac{\partial F}{\partial \Lambda} = -2(1-\zeta)\int_{-\mu}^{\mu} dx\,\rho(x)x^2. \tag{A.26}$$

[Here $F = -\ln Z$ is the standard free energy of the theory — it should not be confused with the $F(f)$ appearing in (2.6), which is related to the free energy of the effective Fermi distribution.] Likewise, one can show that the Wilson loop in the fundamental representation is discontinuous in its first derivatives: recall that $\ln W_f \sim 2\pi\mu$, therefore $\ln W_f$ inherits the first-derivative discontinuities of μ at Λ_{c1}, Λ_{c2}.

Summarizing, starting from strong coupling $\Lambda \gg M$, m, as Λ is gradually decreased, the theory undergoes a first phase transition when $\Lambda = M/2$. As Λ is further reduced, one finds that nothing happens when $\Lambda = O(m)$; the theory has a smooth behavior with the coupling Λ at this scale. The new phase transition to the weak-coupling phase occurs at a lower scale Λ_{c2} (which becomes much lower if $m \ll M$). Once the hypermultiplets of mass M have been integrated out, the theory left behind has a new dynamical scale $\Lambda_{\text{eff}\,1}$. As one may expect, if a new phase transition occurs, this one will take place when $\Lambda_{\text{eff}\,1}$ is of order m and this is indeed what happens. It is reassuring that all field theory expectations are realized in a precise way through the exact large-N solution of the system.

Appendix B. Useful Integrals

The formulas for the integrals used in Sec. 2 can be derived by residue integration. One chooses a contour surrounding the cut from $(-\mu,\mu)$ and evaluates the residue of the poles outside the cut (including possible poles at infinity) by the change of variable $z = 1/x$. One finds

$$\int_{-\mu}^{\mu} dx\frac{1}{\pi\sqrt{\mu^2-x^2}} = 1, \tag{B.1}$$

$$\int_{-\mu}^{\mu} dx\frac{1}{\pi\sqrt{\mu^2-x^2}}\frac{1}{x-y} = 0, \tag{B.2}$$

$$\int_{-\mu}^{\mu} dx\frac{1}{\pi\sqrt{\mu^2-x^2}}\ln x^2 = \ln\frac{\mu^2}{4}. \tag{B.3}$$

For $\mu < M$, by residue integration, one finds

$$M\sqrt{M^2-\mu^2}\int_{-\mu}^{\mu} dx\frac{1}{\pi\sqrt{\mu^2-x^2}}\frac{1}{M^2-x^2} = 1, \tag{B.4}$$

$$M\sqrt{M^2-\mu^2}\int_{-\mu}^{\mu}dx\frac{1}{\pi\sqrt{\mu^2-x^2}}\frac{1}{M^2-x^2}\frac{1}{x-y}=\frac{1}{x+M}+\frac{1}{x-M}, \tag{B.5}$$

$$M\sqrt{M^2-\mu^2}\int_{-\mu}^{\mu}dx\frac{1}{\pi\sqrt{\mu^2-x^2}}\frac{1}{M^2-x^2}\ln\frac{x^2}{M^2}=2\ln\frac{\mu}{M+\sqrt{M^2-\mu^2}}. \tag{B.6}$$

References

[1] V. Pestun, Localization of gauge theory on a four-sphere and supersymmetric Wilson loops, *Comm. Math. Phys.* **313** (2012) 71, 0712.2824.

[2] E. Brezin, C. Itzykson, G. Parisi and J. B. Zuber, Planar diagrams, *Comm. Math. Phys.* **59** (1978) 35.

[3] J. G. Russo and K. Zarembo, Evidence for large-N phase transitions in $N = 2^*$ theory, *JHEP* **1304** (2013) 065, 1302.6968.

[4] J. Russo and K. Zarembo, Massive $N = 2$ gauge theories at large N, *JHEP* **1311** (2013) 130, 1309.1004.

[5] X. Chen-Lin and K. Zarembo, Higher rank Wilson loops in $N = 2^*$ super-Yang–Mills theory, *JHEP* **1503** (2015) 147, 1502.01942.

[6] N. Drukker and B. Fiol, All-genus calculation of Wilson loops using D-branes, *JHEP* **0502** (2005) 010, hep-th/0501109.

[7] S. Yamaguchi, Wilson loops of anti-symmetric representation and D5-branes, *JHEP* **0605** (2006) 037, hep-th/0603208.

[8] J. Gomis and F. Passerini, Holographic Wilson loops, *JHEP* **0608** 074 (2006) 074, hep-th/0604007; D. Rodriguez-Gomez, Computing Wilson lines with dielectric branes, *Nucl. Phys. B* **752** (2006) 316, hep-th/0604031; S. A. Hartnoll and S. P. Kumar, Higher rank Wilson loops from a matrix model, *JHEP* **0608** (2006) 026, hep-th/0605027; J. Gomis and F. Passerini, Wilson loops as D3-branes, *JHEP* **0701** (2007) 097, hep-th/0612022; S. Yamaguchi, Semi-classical open string corrections and symmetric Wilson loops, *JHEP* **0706** (2007) 073, hep-th/0701052; T. Okuda and D. Trancanelli, Spectral curves, emergent geometry, and bubbling solutions for Wilson loops, *JHEP* **0809** (2008) 050, 0806.4191; A. Faraggi, W. Mueck and L. A. Pando Zayas, One-loop effective action of the holographic antisymmetric Wilson loop, *Phys. Rev. D* **85** (2012) 106015, 1112.5028; B. Fiol and G. Torrents, Exact results for Wilson loops in arbitrary representations, *JHEP* **1401** (2014) 020, 1311.2058; E. Buchbinder and A. Tseytlin, The $1/N$ correction in the D3-brane description of circular Wilson loop at strong coupling, *Phys. Rev. D* **89** (2014) 126008, 1404.4952; A. Faraggi, J. T. Liu, L. A. Pando Zayas and G. Zhang, One-loop structure of higher rank Wilson loops in AdS/CFT, *Phys. Lett. B* **740** (2015) 218, 1409.3187; B. Fiol, A. Güijosa and J. F. Pedraza, Branes from light: Embeddings and energetics for symmetric k-quarks in $\mathcal{N} = 4$ SYM, *JHEP* **1501** (2015) 149, 1410.0692; M. Horikoshi and K. Okuyama, α'-expansion of anti-symmetric Wilson loops in $\mathcal{N} = 4$ SYM from Fermi gas, *PTEP* **2016** (2016) 113B05, 1607.01498; K. Zarembo, Localization and AdS/CFT correspondence, *J. Phys. A* **50** (2017) 443011, 1608.02963; X. Chen-Lin, Symmetric Wilson loops beyond leading order, *SciPost Phys.* **1** (2016) 013, 1610.02914; J. Gordon, Antisymmetric Wilson loops in $\mathcal{N} = 4$ SYM beyond the planar limit, 1708.05778.

[9] J. K. Erickson, G. W. Semenoff and K. Zarembo, Wilson loops in $N = 4$ supersymmetric Yang–Mills theory, *Nucl. Phys. B* **582** (2000) 155, hep-th/0003055; N. Drukker and D. J. Gross, An exact prediction of $N = 4$ SUSYM theory for string theory, *J. Math. Phys.* **42** (2001) 2896, hep-th/0010274.

[10] B. Fraser and S. P. Kumar, Large rank Wilson loops in $N = 2$ superconformal QCD at strong coupling, *JHEP* **1203** (2012) 077, 1112.5182; B. Fraser, Higher rank Wilson loops in the $\mathcal{N} = 2SU(N) \times SU(N)$ conformal quiver, *J. Phys. A* **49** (2016) 02LT03, 1503.05634.

[11] J. T. Liu, L. A. Pando Zayas and S. Zhou, Comments on higher rank Wilson loops in $\mathcal{N} = 2^*$, 1708.06288.

[12] M. R. Douglas and S. H. Shenker, Dynamics of $SU(N)$ supersymmetric gauge theory, *Nucl. Phys. B* **447** (1995) 271, hep-th/9503163.

[13] J. Russo and K. Zarembo, Large N limit of $N = 2$ $SU(N)$ gauge theories from localization, *JHEP* **1210** (2012) 082, 1207.3806.

[14] J. G. Russo and K. Zarembo, Localization at large N, 1312.1214.

[15] J. G. Russo, Large N_c from Seiberg–Witten curve and localization, *Phys. Lett. B* **748** (2015) 19, 1504.02958.

[16] J. G. Russo, $\mathcal{N} = 2$ gauge theories and quantum phases, *JHEP* **1412** (2014) 169, 1411.2602.

[17] X. Chen-Lin, J. Gordon and K. Zarembo, $\mathcal{N} = 2^*$ super Yang–Mills theory at strong coupling, *JHEP* **1411** (2014) 057, 1408.6040; K. Zarembo, Strong-coupling phases of planar $N = 2^*$ super-Yang–Mills theory, *Theor. Math. Phys.* **181** (2014) 1522, 1410.6114.

[18] K. Pilch and N. P. Warner, $N = 2$ supersymmetric RG flows and the IIB dilaton, *Nucl. Phys. B* **594** (2001) 209, hep-th/0004063.

[19] A. Buchel, A. W. Peet and J. Polchinski, Gauge dual and noncommutative extension of an $N = 2$ supergravity solution, *Phys. Rev. D* **63** (2001) 044009, hep-th/0008076.

[20] X. Chen-Lin, A. Dekel and K. Zarembo, Holographic Wilson loops in symmetric representations in $\mathcal{N} = 2^*$ super-Yang–Mills theory, *JHEP* **1602** (2016) 109, 1512.06420.

Fredkin Spin Chain

Olof Salberger* and Vladimir Korepin†

C. N. Yang Institute for Theoretical Physics, Stony Brook University, NY 11794, USA

**olof.salberger@stonybrook.edu*

†korepin@gmail.com

The most famous example of a spin chain is the XXX Heisenberg model. The density of the Hamiltonian can be expressed as a permutation operator [swap gate]. Faddeev represented the eigenfunctions in terms of the algebraic Bethe Ansatz. It corresponds to a matrix product state representation. We introduce a new model of interacting spin 1/2. It describes the interactions of three nearest neighbors. The density of the Hamiltonian is a controlled permutation operator [Fredkin gate]. Our construction generalizes the Shor–Movassagh chain to half-integer spins. The Fredkin chain can be solved using combinatorics related to Catalan numbers, where we consider random walks in the upper half plane of a square lattice [Dyck walks]. Each Dyck path can be mapped to a wave function of spins. The ground state is an equally weighted superposition of Dyck walks. We can also express it as a matrix product state. We further construct a model of interacting spins 3/2 and greater half-integer spins. The models with higher spins correspond to colorings of the Dyck walks. We construct a $SU(k)$ symmetric model [where k is the number of colors]. The Fredkin chain with spins 3/2 and higher has stronger quantum fluctuations than XXX.

1. Introduction

Exactly solvable spin chains have become important recently. The XXX model can be solved by the Bethe Ansatz, which yields all the eigenfunctions. We are looking for generalizations, which are solvable in a weaker sense: only the ground state wave function is known analytically [similarly to valence bond solid models]. We are looking for chains with strong quantum fluctuations. For models with a unique ground state, the entropy of the whole ground state is zero, but there might be an entropy in a subsystem [half of the chain] as we trace over the rest of the system. This is the entanglement entropy [1], it measures the strength of quantum fluctuations. Quantum fluctuations in spin models are enhanced as the number of spatial dimensions decreases. Spin chains provide good models for the experimental realization of quantum devices such as quantum computers [4, 6–8].

Entanglement entropy in spin chains has been studied intensely. Shor and Movassagh constructed a spin chain describing local interaction of integer spins in [2, 5]. For high spins, it displays abnormally high levels of quantum fluctuations [entanglement]. We generalize their results to half-integer spins.

2. The Hamiltonian and Ground State

We shall consider an open lattice with N cites. The bulk Hamiltonian describes interaction between singlet pairs of neighboring sites. It can be written as

$$H = H_{bulk} + H_{\partial} = H_{\partial} + \sum_j H_j \tag{1}$$

where we have the boundary term $H_{\partial} = |{\downarrow_1}\rangle\langle{\downarrow_1}| + |{\uparrow_N}\rangle\langle{\uparrow_N}|$, and where the bulk term is given by the following three-site interaction between singlet pairs and neighboring sites:

$$H_j = |{\uparrow_j}\rangle\langle{\uparrow_j}| \otimes |S_{j+1,j+2}\rangle\langle S_{j+1,j+2}| + |S_{j,j+1}\rangle\langle S_{j,j+1}| \otimes |{\downarrow_{j+2}}\rangle\langle{\downarrow_{j+2}}| \tag{2}$$

where $|S_{i,j}\rangle$ is the familiar singlet state $\frac{1}{\sqrt{2}}(|{\uparrow_i}\rangle|{\downarrow_j}\rangle - |{\downarrow_i}\rangle|{\uparrow_j}\rangle)$. In other words, we have an interaction term for a spin up on the left of a singlet, and a term for a spin down on the right of a singlet.

Despite its simplicity, this seemingly innocent Hamiltonian turns out to have a highly entangled ground state with a very rich combinatorial structure, as we will see in the following section which introduces the necessary formalism to write down the ground state.

Before we move on however, we allow ourselves to rewrite the bulk Hamiltonian in a few alternate ways. The first is in terms of Pauli spin operators, where it can be written as:

$$H_j = (1 + \sigma_j^Z)(1 - \vec{\sigma}_{j+1} \cdot \vec{\sigma}_{j+2}) + (1 - \vec{\sigma}_j \cdot \vec{\sigma}_{j+1})(1 - \sigma_{j+2}^Z). \tag{3}$$

Finally, we have yet another way to write down the bulk terms of the Hamiltonian, by noting their relationship to Fredkin (Controlled-Swap) gates:

$$H_j = (1 - F_{j,j+1,j+2}) + (1 - \sigma_{j+2}^x F_{j+2,j+1,j} \sigma_{j+2}^x). \tag{4}$$

Where $F_{i,j,k}$ is the Fredkin gate acting on three qubits, which is both Hermitian and unitary. It swaps j, k if site i is in state $|{\uparrow}\rangle$ and does nothing if it is in state $|{\downarrow}\rangle$.

2.1. *Combinatorial prerequisites: Dyck words*

In this section, we concisely describe the combinatorics necessary to describe the ground state and its relation to the Hamiltonian.

The central idea for defining our model is to think of the basis states of our length N chain of the two-state systems as the set of N-symbol words in a two-letter alphabet. Depending on context, the letters would then be called $\uparrow$ and $\downarrow$ for the basis states $|{\uparrow}\rangle$ and $|{\downarrow}\rangle$, 0 and 1 for $|0\rangle$ and $|1\rangle$, or more suggestively, for our purposes, (and) for $|(\rangle$ and $|)\rangle$.

Among generic two-letter words, we may single out the set of Dyck words. *These are precisely the set of words consisting of an equal number of left and right parentheses, such that every left parenthesis has a matching right parenthesis further right along the word.* That is, there are no right or left parentheses that are unmatched on either side of the word.

For example, ()(()) and (()()) are both valid six-letter Dyck words, while)(()() or ())(() are not.

Clearly, a word can only be Dyck if $N = 2n$ is even. However since splitting up a word into pieces that may be odd is a useful operation, we will still consider words of odd length.

The Dyck-ness of a word is thus a global feature which forces symbols at different places in the word to be correlated. For example, if we slice up the word along the middle, the number of unmatched left parentheses in the left half must be equal to the number of unmatched right parentheses in the right half. These correlations will be the key source of entanglement in the physical model that we propose below.

2.1.1. *Dyck words as an equivalence class*

Having defined Dyck words and observed that their definition is global, we then observe that they do admit a local description.

To do this, we consider the larger set of arbitrary words and introduce the following moves, which in analogy to Reidemeister moves we call the Fredkin moves:

$$()) \leftrightarrow)() \tag{5}$$

$$(() \leftrightarrow ()(\tag{6}$$

where we allow a matched pair of adjacent parentheses to be moved anywhere. The reason why we call these the Fredkin moves is that when viewed as bit strings, the involution that takes one side to the other is a three-bit reversible logic gate known as the Fredkin gate, up to a flipped control bit. We will sometimes also refer to these as Fredkin relations.

We now consider how these relations divide up the set of words of length N into equivalence classes. We immediately observe that the relations preserve the matchings of the parentheses, and as such, map Dyck words to Dyck words only and non-Dyck words to non-Dyck words only.

A full classification of the equivalence classes may be performed fairly straightforwardly by observing that:

(1) If any given word has at least one pair of adjacent matched parentheses, we may use the Fredkin relation to move it to the front of the word. This allows us to focus on a subword of length $N - 2$.

(2) The only words that have no matched pair of adjacent parentheses are of the form that consists of a left parentheses followed by b right parentheses.

Together, these observations imply that every word of length N is equivalent to exactly one word consisting of $2k$ matched parentheses, followed by a unmatched right parenthese, followed by b left parentheses. That is, of the form ()()...k times () ())))... a times ...)) ((((... b times...(((, where $a + b + 2k = N$. We call this the *standard form* of the word. Since clearly none of these words are equivalent to each other this completes our classification. We may denote these equivalence classes $C_{a,b}(N)$, and observe that $C_{0,0}(N)$, which is non-empty for even N, is exactly the set of Dyck words of length $N = 2n$.

We finish off by stating that the number of words in each equivalence class $C_{a,b}(N)$ is given by:

$$|C_{a,b}(N)| = \begin{cases} \binom{N}{\frac{N+a+b}{2}} - \binom{N}{\frac{N+a+b}{2}+1} & \text{if } N-a-b \text{ is even and positive} \\ 1 & \text{if } N = a+b \\ 0 & \text{otherwise} \end{cases} \tag{7}$$

which in the non-zero cases match the numbers in the Catalan triangle (where each number is the sum of the numbers above it and the one to its left). We note that in the special case where $N = 2n$ and $a = b = 0$, we recover the famous Catalan numbers $C_N = \frac{1}{n+1}\binom{2n}{n}$.

We may prove this by constructing an explicit bijection between $C_{a,b}$ and $C_{0,a+b}$, which incidentally is also a symmetry of the bulk Hamiltonian. This simply consists of exchanging each unmatched right parenthesis in a member of $C_{a,b}$ with an unmatched left parenthesis. Since the subword between one unmatched parenthesis of a word and the next must be a Dyck word, this must map matched parentheses to matched parentheses and unmatched parentheses to unmatched parentheses, which makes the inverse of the map trivial to find.

The number of elements in $C_{0,a+b}$ can then be obtained directly from Bertrand's ballot theorem with ties allowed [9], or by using the combinatorial definition of the numbers in Catalan's triangle.

2.1.2. *Dyck words reinterpreted as Dyck paths*

In this section, we introduce the path notation to describe Dyck words. This notation is particularly useful in handwritten work and has the advantage of being easier to parse. The path formalism is also generally the most convenient to use for counting the number of elements in a given Fredkin equivalence class.

The key observation is that two-letter words can also be interpreted as paths on the upper half plane of a lattice. Specifically, all words in the equivalence class $C_{a,b}(N)$ can be viewed as a path from the lattice point $(0, a)$ to the lattice point (N, b) such that on each step the path can only travel diagonally from (x, y) to $(x + 1, y \pm 1)$, and such that it is always stays in the upper half plane $y \geq 0$. The bijection is given simply by associating left parentheses with a step up and right parentheses with a step down (see Fig. 1).

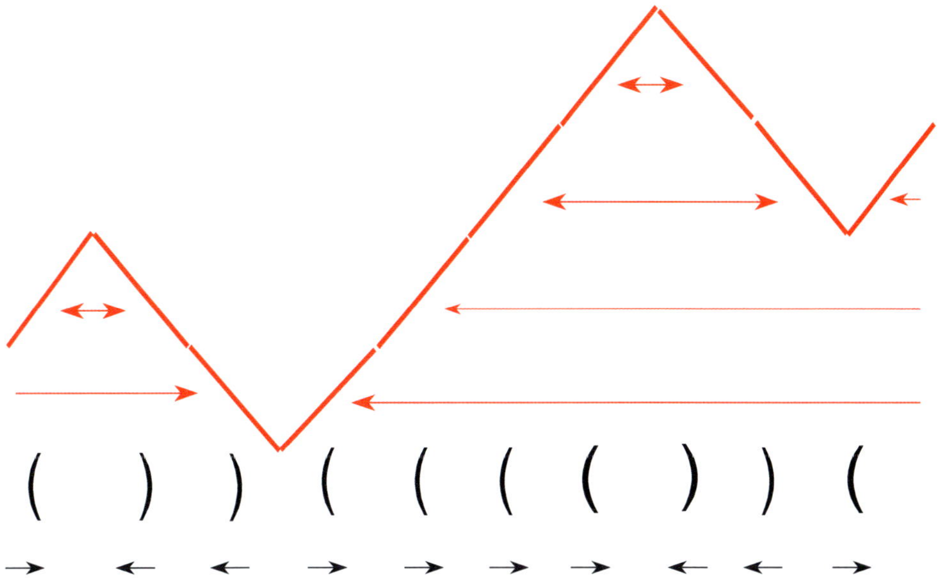

Fig. 1. In this figure, we show the different ways in which our basis states can be expressed: as a sequence of spins up or down, as Dyck words, or as paths. It also shows the relation between matched parentheses and matched steps.

Parenthesis matching can be obtained in the following way: for any word of the form $W = w_1(w_2)w_3$, we have that w_2 must be a Dyck word if and only if the two parentheses we have written out are matched, since otherwise, either the left or right parentheses would instead be matched with an unmatched parenthesis of w_2.

As a result, the path corresponding to w_2 never goes further down than its starting point. The matching of parentheses can then be translated into matching steps in the following way: for a given up step from $y = y_0$ to $y = y_0 + 1$, the first down step after it goes from $y = y_0 + 1$ to $y - y_0$, if it exists, is the *matching down step* which corresponds to the matching right parenthesis. The two-dimensional nature of paths make matchings significantly easier to parse, at the expense of using a less compact notation.

2.2. *The ground state*

The ground state of our model with $2n$ sites $|D_n\rangle$ can be written as a uniform superposition of the basis states corresponding to Dyck words(or paths).

To give some low-n examples, using the bracket notation where $|\uparrow\rangle = |(\rangle$ and $|\downarrow\rangle = |)\rangle$, we have:

$$|D_1\rangle = |(\,)\rangle$$
$$|D_2\rangle = |(\,)(\,)\rangle + |((\,))\rangle$$
$$|D_3\rangle = |(((\,)))\rangle + |((\,)(\,))\rangle + |(\,)((\,))\rangle + |((\,))(\,)\rangle + |(\,)(\,)(\,)\rangle$$

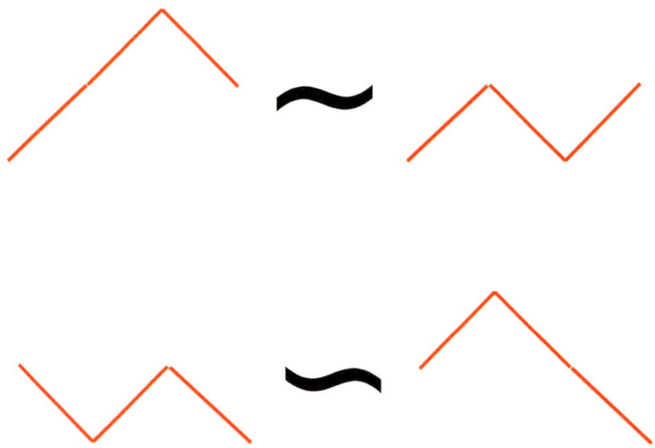

Fig. 2. The Fredkin moves in path notation.

and

$$
\begin{aligned}
|D_4\rangle = {} & |((()()))\rangle \\
& + |((()())))\rangle + |((())()))\rangle + |(()(())))\rangle \\
& + |()((()))\rangle + |((()))()\rangle + |(())(())\rangle \\
& + |(()()())\rangle + |)((()()))\rangle + |(()())()\rangle \\
& + |(())()()\rangle + |()(())()\rangle + |()()(())\rangle \\
& + |()()()()\rangle
\end{aligned}
$$

which have 1, 2, 5, and 14 terms, respectively.

To see that it is the unique ground state, we do the following: first, we show that $|D_n\rangle$ is indeed a zero eigenstate of all individual terms in the Hamiltonian. Next, we observe that this implies frustration freeness: each term is positive semi-definite. Therefore any other eigenstate with zero eigenvalues must be a zero eigenstate of all individual terms. Finally, we show that any state with that last property must be equal to $|D_n\rangle$. Alternatively, one can simply apply the orbit theorem in the appendix to the bulk Hamiltonian, and check which ground states are also ground states of the boundary terms.

We start with the boundary terms. Since every Dyck word must start with an opening parenthesis and end with a closing parenthesis (for walks, start with a step up and end with a step down), $|D_n\rangle$ is indeed a ground state of the boundary terms.

Next, we consider the bulk terms. Consider the term

$$|\uparrow_j\rangle\langle\uparrow_j| \otimes |S_{j+1,j+2}\rangle\langle S_{j+1,j+2}| = \left[|(()\rangle_j - |()(\rangle_j\right]\left[\langle(()|_j - \langle()(|_j\right],$$

where the index j denotes the position of the first site.

This term is a projector onto the state $|(()\rangle_j - |()(\rangle_j$. We then observe that we may generically write the Dyck state as $|\psi_1\rangle\otimes|(()\rangle_j\otimes|\psi_2\rangle + |\chi_1\rangle\otimes|()(\rangle_j\otimes|\chi_2\rangle + |\psi\rangle$, where $|\psi\rangle$ is orthogonal to both $|(()\rangle_j$ and $|()(\rangle_j$. By definition, the last term is annihilated by our projector, so we may focus on the first two.

However, we note that the invariance of the set of Dyck words under the Fredkin moves (5) give us that if w_1w_2 is a Dyck word, so is $w_1()(w_2$ and vice versa. Therefore, in our expression above we must have $|\psi_1\rangle = |\chi_1\rangle$ and $|\psi_2\rangle = |\chi_2\rangle$, and we have $|D_n\rangle = |\psi_1\rangle \otimes \left[|(())\rangle_j + |()()\rangle_j\right] \otimes |\psi_2\rangle + |\psi\rangle$. Both terms are orthogonal to $|(())\rangle_j - |()()\rangle_j$ and will be annihilated by our projector. The process is analogous for the second bulk term.

Therefore, the Dyck state is indeed a ground state of all individual terms of our Hamiltonian. Since such a state exists, any other ground state of the Hamiltonian must also be a ground state of all individual terms.

Now, we consider the possible ground states of the bulk terms. We now know that it must be a ground state of each individual term. We may then perform the above calculation in reverse: we break up an arbitrary ground state candidate $|\psi\rangle$ into $|\Psi\rangle = |\psi_1\rangle \otimes |(())\rangle_j \otimes |\psi_2\rangle + |\chi_1\rangle \otimes |()()\rangle_j \otimes |\chi_2\rangle + |\psi\rangle$, where $|\psi\rangle$ is orthogonal to both $|(())\rangle_j$ and $|()()\rangle_j$. If this is annihilated by the projector $\left[|(())\rangle_j - |()()\rangle_j\right]\left[\langle(())|_j - \langle()()|_j\right]$, we must have $\left[\langle(())|_j - \langle()()|_j\right]|\Psi\rangle = 0$. Expanding this, we get $|\psi_1\rangle \otimes \langle(())|_j|(())\rangle_j \otimes |\psi_2\rangle - |\chi_1\rangle \otimes \langle()()|_j|()()\rangle_j \otimes |\chi_2\rangle$, so we come to the conclusion that for any such ground state $|\chi_1\rangle = |\chi_2\rangle$ and $|\psi_1\rangle = |\psi_2\rangle$. In other words, the parts of the ground state containing $|()()\rangle_j$ and $|(())\rangle_j$ are exactly equal. We can then simply repeat this for all bulk terms in the Hamiltonian.

This is the key observation we needed to make. Writing out the ground state as $|\Psi\rangle = \sum_i c_i|w_i\rangle$, we have just proven that if the words w_i and w_j are related by a Fredkin move, then $c_i = c_j$.

To formulate this in terms of an explicit orthogonal basis for our ground states, we can use our discussion of equivalence classes under Fredkin moves in Sec. 2.1.1 to define the states $|C_{a,b}(N)\rangle = \sum_{w_i \in C_{a,b}(N)} |w_i\rangle$ which are direct superpositions of all states in the equivalence class $C_{a,b}(N)$. Then, we see that any ground state $|\Psi\rangle$ of the bulk terms can be written as $|\Psi\rangle = \sum_{a,b} |C_{a,b}(N)\rangle$, so the states $|C_{a,b}(N)\rangle$ form a complete basis for the space of bulk term ground states.

Among these, for even $N = 2n$, only $|C_{0,0}(N)\rangle = |D_n\rangle$ which has no unpaired parentheses/steps is also a ground state of the boundary terms, which finally proves that the Dyck state is indeed the unique ground state of our Hamiltonian when the boundary terms are introduced.

2.2.1. *Schmidt decomposition and measures of entanglement*

The Dyck state is highly entangled. If we break up the chain into two halves with lengths L and $N - L$, where without loss of generalization we may assume that $L < N - L$. The Schmidt rank will then be $\xi = \left\lfloor \frac{L}{2} \right\rfloor$, while the entanglement entropy will scale as

$$S = \frac{1}{2}\log(L) + O(1).$$

Our primary tool to study the entanglement between blocks will be the Schmidt decomposition. The Dyck state can be decomposed into the states

$$|D_n\rangle = \sum_{m=0}^{L} \sqrt{p_m}|C_{0,m}(L)\rangle \otimes |C_{m,0}(N-L)\rangle \tag{8}$$

where we define the states $|C_{a,b}(L)\rangle$ as a direct generalization of the Dyck state, which is a direct superposition of the basis states corresponding to the equivalence classes $C_{a,b}(N)$ that we defined in Sec. 2.1.1.

In particular, we note that $C_{a,b}(L)$ is non-empty if and only if $a+b<L$ and the parity of $a+b$ matches that of L. As a result, the non-zero terms $|C_{0,m}(L)\rangle \otimes |C_{m,0}(L)\rangle$ will be those with even $m<L$ if L is even, or odd $m<L$ if L is odd.

This leads to our above expression for the Schmidt rank, since the number of even integers $\leq L$ if L is even is given by $\frac{L}{2}$, and the number of odd integers less than L if L is odd is given by $\frac{L-1}{2}$. The two expressions may be combined concisely with the floor function.

Now we may move on to compute the entanglement entropy. We start by deriving a useful expression for p_m, which must be the product of the normalization factors of $|C_{0,m}(L)\rangle$ and $|C_{m,0}(L)\rangle$ divided by the normalization factor of $|D_n\rangle$. This gives us $p_m = \frac{|C_{0,m}(L)||C_{0,m}(N-L)|}{|C_{0,0}(N)|}$.

To find a useful expression for the p_m, we will start by cleaning up our notation a bit. We have already written $N=2n$. We will split the expression into two cases where L is odd or even.

For the even case, we write $L=2l$, and $m=2h, h\in[0,l]$ as the p_m will be zero for odd m. Then for the $h<l$ case we may simplify:

$$|C_{0,m}(L)| = \binom{2l}{l+h} - \binom{2l}{l+h+1} = \frac{2h+1}{l+h+1}\binom{2l}{l+h}$$

$$|C_{m,0}(N-L)| = \binom{2(n-l)}{n-l+h} - \binom{2(n-l)}{n-l+h+1} = \frac{2h+1}{n-l+h+1}\binom{2(n-l)}{n-l+h}$$

while in the odd case, we write $L=2l+1$ and $m=2h+1, h\in[0,l]$ which gives us

$$|C_{0,m}(L)| = \binom{2l+1}{l+h+1} - \binom{2l+1}{l+h+2} = \frac{2h+2}{l+h+2}\binom{2l+1}{l+h+1}$$

$$|C_{m,0}(N-L)| = \binom{2(n-l)-1}{(n-l)+h} - \binom{2(n-l)-1}{(n-l)+h+1}$$

which in the even case, gives us:

$$p_h = \frac{1}{Cat(n)}\frac{(2h+1)^2}{(h+1+\frac{n}{2})^2-(l-\frac{n}{2})^2}\binom{2l}{l+h}\binom{2(n-l)}{n-l+h}.$$

Our next step is then simply to use the Gaussian approximation for binomials $\binom{2n}{n+k} \approx \frac{4^n}{\sqrt{\pi n}} \exp(-\frac{k^2}{n})$, which gives us:

$$p_h \approx \frac{4^n}{\pi Cat(n)\sqrt{l(n-l)}} \frac{(2h+1)^2}{(h+1+\frac{n}{2})^2 - (l-\frac{n}{2})^2} \exp\left[-h^2\left(\frac{1}{l}+\frac{1}{n-l}\right)\right],$$

which is a good approximation when n is large, and peaks when $\frac{h^2 n}{l(n-l)} \approx 1$. Since the Gaussian factor will suppress the parts where h is comparable to $\frac{n}{2}$, our final approximation becomes

$$p_h \approx \frac{h^2}{Z} \exp\left[-h^2 \frac{n}{l(n-l)}\right]$$

where Z is a normalization factor.

This expression is largely analogous to the one in [2, 5], and from here on we may proceed in a manner fully analogous to the method outlined in these publications. The key idea is to approximate sums with integrals, and performing a change of variables to isolate the $L(N-L)$ dependence. In particular, we have:

$$\begin{aligned} S &= -\sum_h p_h \log(h) \approx -\int_0^\infty p_x \log(p_x) dx = \left[\sqrt{\frac{l(n-l)}{l}} d\alpha = dx\right] \\ &= -\int_0^\infty \sigma\rho_\alpha \log(\sigma\rho_\alpha)\sigma^{-1} d\alpha = -\int_0^\infty \rho_\alpha[\log(\sigma) + \log(\rho_\alpha)] d\alpha \\ &= -\log(\sigma) - \int_0^\infty \rho_\alpha \log(\rho_\alpha) d\alpha = \frac{1}{2}\log\left[\frac{l(n-l)}{l}\right] - \int_0^\infty \rho_\alpha \log(\rho_\alpha) d\alpha, \end{aligned}$$

where $p_x = \frac{x^2}{\tilde{Z}} \exp\left[-x^2 \frac{n}{l(n-l)}\right]$, $\tilde{Z}$ is a normalization factor so that $\int p_x dx = 1$ and $\alpha = \sigma x$ and $\sigma = \sqrt{\frac{N}{L(N-L)}}$. Furthermore, we define $\rho_\alpha = \frac{1}{Z'}\alpha^2 \exp[-\alpha^2]$ where Z' is a constant such that $\int \rho_\alpha = 1$, which gives us $p_x = \sigma\rho_\alpha$ (with a factor σ^3 from the normalization factor and a σ^{-2} from $x^2 = \sigma^{-2}\alpha^2$), which we used to insert the correct sigma factors. Thus, when N and L are large enough for the integral approximation to be valid, we have:

$$S = \frac{1}{2}\log\left[\frac{L(N-L)}{N}\right] + O(1).$$

2.2.2. *Matrix product state definition*

To find the MPS formulation, we let ourselves be guided by the form of the Fredkin moves (5). We would like the product $A^{(}A^{)}$ to commute with $A^{(}$ and $A^{)}$, or at least have a commutator which can easily be projected out, while having a more non-trivial expression for $A^{)}A^{(}$. Here A is a local matrix in matrix state representation of the ground state. It can be compared with L operator of quantum inverse scattering method.

We can do this by setting $A^{(}_{ij} = \delta^{i+1,j}$ and $A^{)}_{ij} = \delta i, j+1$ for $\infty > i, j \geq 1$ and 0 otherwise. In that case, before we truncate to a finite bond dimension, we have $A^{(}A^{)} = 1$ while $A^{)}A^{(}$ is a projector that does not commute with $A^{(}$ or $A^{)}$.

To obtain the Dyck state, we introduce the boundary (basis) vector $v_i = \delta^{i,1}$. Then we have that a the matrix element $v^T A^{k_1} A^{k_2} ... A^{k_{N-1}} A^{k_N} v$ of the product will be non-zero if and only if the indices $k_1 k_2 ... k_{N-1} k_N$, which each are) or (-valued, together form a Dyck word.

This is true because for any index values, we can remove $A^{(}A^{)} = 1$ without changing the value of the product until we are left with an expression of the form $v^T [A^{)}]^a [A^{(}]^b v$. Hence the matrix product depends only on to which equivalence class of the word formed by the indices belongs to under the Fredkin moves, and using the boundary vector $v_i = \delta^{i,1}$ the relevant matrix element will yield zero if the word belongs to any equivalence class other than that of Dyck words. Hence, the matrices above do indeed give a matrix product state description of the Dyck state.

We now only have to check whether we can truncate our infinite-dimensional matrices to finite-dimensional matrices to obtain an MPS description that is actually useful. This can be done by observing that the entries of A_{ij} with i or j larger than N do not affect the product [ground state]

$$v^T A^{i_1} A^{i_2} ... A^{i_{N-1}} A^{i_N} v,$$

meaning that we can simply truncate it to an N-dimensional MPS (which can be refined to $N/2$ exactly, and to $O(\sqrt{N})$ with arbitrarily small errors as N becomes large).

2.3. *Spin wave solutions*

The Fredkin model is closely related to the XXX Heisenberg model. In particular, since it commutes with z-component of total spin $Z = \sum_j \sigma^z_j$, we can straightforwardly apply the coordinate Bethe Ansatz to analyze the 1-magnon sector of the model.

In fact, since our model has two invariants a and b, we can obtain a slightly stronger result, namely, that all sectors with $a + b = N - 2$ have a spectrum fully isomorphic to the 1-magnon sector of a ferromagnetic Heisenberg XXX chain of length $N - 1$.

The key insight is that when viewed as a path, these equivalence classes consist of a long dip with a single /\ peak that may be moved around. The $N - 1$ basis states of this class, which we call $|\psi_j\rangle$, correspond to the path where the single peak has an up-step at site j and the down-step at site $j + 1$.

For each j, the only terms in the Fredkin Hamiltonian that can affect $|\psi_j\rangle$ non-trivially will be the three-site terms that act on both site j and $j + 1$, which leads to four candidates. By brute force consideration of all cases, we will see that up to constants there will be exactly two terms remaining, one which maps

$|\psi_j\rangle$ to $-|\psi_{j+1}\rangle$ and one which maps $|\psi_j\rangle$ to $-|\psi_{j-1}\rangle$. Therefore, the restriction of the Fredkin model Hamiltonian with open boundary conditions to this sector is isomorphic to the ferromagnetic Heisenberg XXX Hamiltonian with open boundary conditions restricted to the 1-magnon sector, on a chain of length $N-1$.

2.4. *Ground states with alternative boundary conditions*

While we have been focusing on the ground states with boundary terms $H_\partial = |\downarrow_1\rangle\langle\downarrow_1| + |\uparrow_N\rangle\langle\uparrow_N|$, our model does exhibit some rather surprising properties with other boundary conditions.

The first generalization we could make is to vary the boundary terms in the following way:

$$H_\partial(\alpha,\beta) = (1-\alpha\sigma_0^z) + (1+\beta\sigma_N^z),$$

which yields our previous boundary terms for $\alpha=\beta=1$. For $\alpha=-\beta$, the bulk ground states $|C_{N,0}\rangle$ and $|C_{0,N}\rangle$ will be ground states of the case where $\alpha=1$ and $\alpha=-1$, respectively. For $\alpha=\beta=-1$, we have an $N-2$ degeneracy as all the domain wall states $|C_{a,b}\rangle$ with $a+b=N$ and $a,b\geq 1$ are ground states. Since in all cases we have bulk ground states which are also boundary ground states, the model stays frustration-free meaning these are the only possible ground states for these boundary conditions.

Furthermore, since the model is frustration-free for all possible sign combinations of α, β, the ground state depends *only* on the signs of α, β and not on their magnitude. In other words, phase transitions may happen only when α or β is zero.

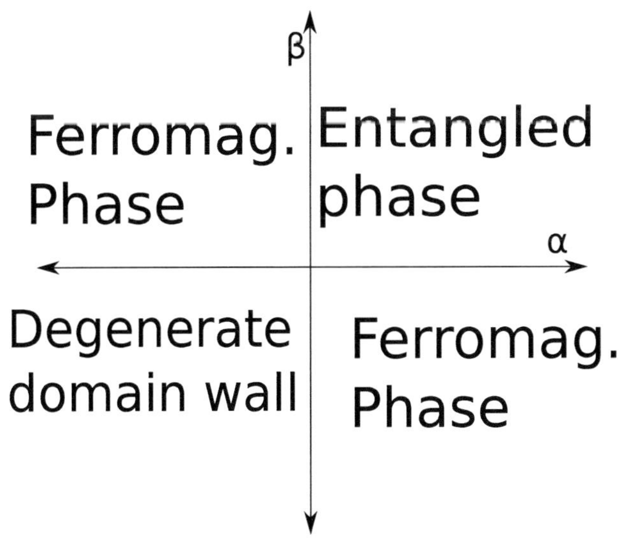

Fig. 3. A phase diagram describing the different phases of the model as we vary the boundary conditions.

These results may be combined into the phase diagram in Fig. 3, which features four different phases separated by first-order phase transitions where the energy levels become degenerate.

2.4.1. *Periodic boundary conditions*

We get a somewhat more surprising result with periodic boundary conditions. Here, we note that the Hamiltonian commutes with $Z = \sum_j \sigma_j^z$ and with the cyclic shift operator Π which maps the site j to the site $j+1$. We may use the eigenvalues of these two operators to index our ground states in the model with periodic boundary conditions.

For odd N, all ground states are eigenstates with eigenvalue 1 of Π, and can be indexed solely using the Z operator. In this case, all ground states also happened to be eigenstates of the Heisenberg XXX model and are totally symmetric. The degeneracy in this case is N.

For even $N = 2n$, the situation is different, and we have a ground state degeneracy $N+1$. N of these are eigenvectors with unit eigenvalues of the shift operator Π and are also Heisenberg XXX ground states. However, there is one additional state, which we call the *anomalous state* $|D_{an}\rangle$, which is an eigenstate of Π with eigenvalue -1, and a zero eigenstate of Z. It is given by:

$$|D_{an}\rangle = \frac{1}{\sqrt{\binom{2n}{n}}} \sum_{m=0}^{n} \sqrt{|C_{m,m}|}(-1)^m |C_{m,m}\rangle.$$

To see why this is the case, we consider what happens to the equivalence classes as we add Fredkin moves that continue periodically across the edge.

We may define a periodic matching across the edge, which is most easily done by going back to the Dyck word formalism: two parentheses are periodically matched if they are matched in the usual sense for at least one cyclic permutation of the Dyck word. Two steps are periodically matched if their corresponding parentheses in the Dyck word formalism are periodically matched.

The key idea is then that we can use our previous algorithm that reduces any path to a series of down-steps (right parentheses), followed by a series of /\ peaks (matched pairs), followed by a series of up-steps (left parentheses). However, whenever we have a down-step at site 1 and an up-step at site N, we have a peak wound across the edge.

If we have any unmatched step adjacent to it, we can move this peak away from the edge to join the other peaks. By repeating this process, we can then always reduce to the case where all matched steps are adjacent to their match, and we are left with Z unmatched up-steps and no unmatched down-steps if Z is positive, or $|Z|$ unmatched down-steps and no unmatched up-steps if Z is negative.

At this point, we are left with two cases: Either there is a peak wound across the edge, or there isn't. If $Z > 0$, we can move an unmatched up-step to site $N-1$, move the peak past it, and we will be left with no peaks wound across the edge. For $Z < 0$, we may perform the mirror image of this algorithm. However, for $Z = 0$, there is no way to transform one case into the other.

Thus, for $Z \neq 0$, we are left with exactly one equivalence class for each value of Z:

$$C_Z = \begin{cases} \bigcup_k C_{Z+k,k} & \text{if } Z > 0 \\ \bigcup_k C_{k,Z+k} & \text{if } Z < 0 \end{cases}$$

which all just contain all basis states that satisfy $\sum_j \sigma_j^z = Z$. Using the orbit theorem, we see that these all generate the same ground states as the Heisenberg XXX Hamiltonian.

In the $Z = 0$ case, we have two equivalence classes that classify whether the number of pairs matched across the edge is odd or even. Each of these gives us a ground state. The symmetric sum of these is the $Z = 0$ ground state of Heisenberg XXX, while the difference yields the anomalous state above.

3. Colored $SU(k)$-Symmetric Model with a Square Root Entanglement Entropy

It is possible to generalize the Fredkin model to exhibit a colored Dyck state as its ground state in a manner largely analogous to [5], which yields an exponentially increasing Schmidt rank and entanglement entropy. However, since we consider Dyck paths rather than Motzkin paths, the Hilbert space for each site will be of even dimension, which allows us to naturally write it as a tensor product of a two-dimensional Hilbert space and a k-dimensional "color" Hilbert space.

In particular, our colored model has a manifest $U(1) \times SU(k)$ symmetry which generalizes the $U(1)$ symmetry of the uncolored Fredkin model, where k is a positive integer giving the number of colors.

The expression for the Fredkin Hamiltonian above in terms of Pauli matrices can be generalized to an expression in terms of $SU(2) \otimes SU(c)$ matrices for each site j of the form $T_j^a = \sigma_j^z \otimes t_j^a$ where the t^a are the generators of $SU(k)$ in the fundamental representation. For convenience, we will also use the $SU(2)$ matrices that by abuse of notation we write $\sigma_j^i = \sigma_j^i \otimes 1_j$.

Finally, we define the projection operators $P_j^{\pm} = \frac{1 \pm \sigma_j^z}{2}$ and the cyclic permutation operator C_{j_1,j_2,j_3} which cyclically permutes the sites j_1, j_2, j_3.

This allows us to conveniently write the Hamiltonian in a manifestly $SU(k)$-invariant way as:

$$H = H_F + H_X + H_\partial \tag{9}$$

where

$$H_F = \sum_{j=1}^{N-2} P_j^+ P_{j+1}^+ P_{j+2}^- + P_j^+ P_{j+1}^- P_{j+2}^+ - P_j^+ P_{j+1}^+ P_{j+2}^- C_{j,j+1,j+2}$$

$$- C_{j,j+1,j+2}^\dagger P_j^+ P_{j+1}^+ P_{j+2}^- + P_j^+ P_{j+1}^- P_{j+2}^- + P_j^- P_{j+1}^+ P_{j+2}^-$$

$$- P_j^+ P_{j+1}^- P_{j+2}^- C_{j,j+1,j+2}^\dagger - C_{j,j+1,j+2} P_j^+ P_{j+1}^- P_{j+2}^-$$

$$H_X = \sum_{j=1}^{N-1} P_j^+ P_{j+1}^- \left[\sum_a (T_j^a + T_{j+1}^a)^2\right]$$

$$H_\partial = P_1^- + P_N^+.$$

All three of these operators manifestly commute with the $SU(k)$ generators $T^a = \sum_{j=1}^N T_j^a$, and with the operator $Z = \sum_{j=1}^N \sigma_j^z$, which together generate the symmetry group $SU(k) \times U(1)$ of the model, which can naturally be viewed as a full $U(k)$ symmetry. The Hilbert space at each site is then most naturally viewed as a sum of two mutually conjugate fundamental representations of $U(k)$.

To illustrate the Hamiltonian more clearly, we can expand it in the product state basis as a sum of projectors. In that case, we can naturally write:

$$H_F = \sum_{j=1}^{N-2} \sum_{c_1,c_2,c_3} \left(|\downarrow_j^{c_1}\uparrow_{j+1}^{c_2}\downarrow_{j+2}^{c_3}\rangle - |\uparrow_j^{c_2}\downarrow_{j+1}^{c_3}\downarrow_{j+2}^{c_1}\rangle\right)\left(\langle\downarrow_j^{c_1}\uparrow_{j+1}^{c_2}\downarrow_{j+2}^{c_3}| - \langle\uparrow_j^{c_2}\downarrow_{j+1}^{c_3}\downarrow_{j+2}^{c_1}|\right)$$

$$+ \left(|\uparrow_j^{c_1}\uparrow_{j+1}^{c_2}\downarrow_{j+2}^{c_3}\rangle - |\uparrow_j^{c_2}\downarrow_{j+1}^{c_3}\uparrow_{j+2}^{c_1}\rangle\right)\left(\langle\uparrow_j^{c_1}\uparrow_{j+1}^{c_2}\downarrow_{j+2}^{c_3}| - \langle\uparrow_j^{c_2}\downarrow_{j+1}^{c_3}\uparrow_{j+2}^{c_1}|\right)$$

$$H_X = \sum_{j=1}^{N-1}\left[1 - \frac{1}{k}\left(\sum_{c_1}|\uparrow_j^{c_1}\downarrow_{j+1}^{c_1}\rangle\right)\left(\sum_{c_2}\langle\uparrow_j^{c_2}\downarrow_{j+1}^{c_2}|\right)\right] P_j^+ P_{j+1}^-$$

$$H_\partial = P_1^- + P_N^+ = \sum_{c_1} |\downarrow_1^{c_1}\rangle\langle\downarrow_1^{c_1}| + |\uparrow_N^{c_1}\rangle\langle\uparrow_N^{c_1}|$$

where c_1, c_2, c_3 are color indices running from 1 to k.

This model can be shown to reduce to the Fredkin model when $k = 1$, and in general, the ground states can be obtained in a fully analogous manner and described in terms of combinatorial objects. The extension we will need is colored Dyck paths.

3.1. *Invariants of the Hamiltonian and characterization of the bulk ground states*

The basic idea here is the same as before: identifying basis states with paths, where we generalize the $k = 1$ case by considering colored paths (see Fig. 5).

Our first move should then be to identify invariants of H_F that may allow us to block diagonalize it. In the uncolored case, the only invariants were the number of

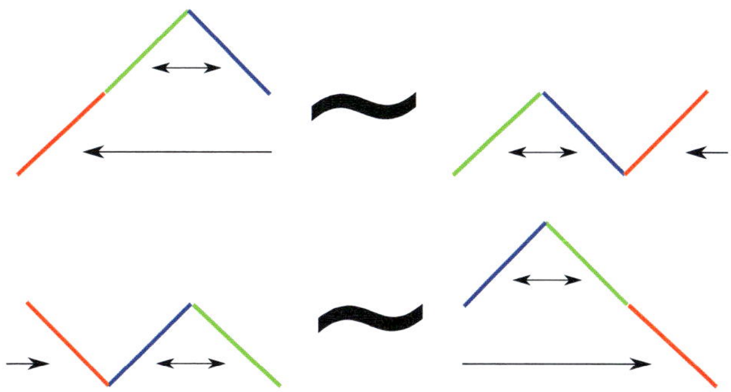

Fig. 4. The colored Fredkin moves in path notation. Note that the color of the matched pair and of the unmatched step is preserved, as well as the height of the unmatched step.

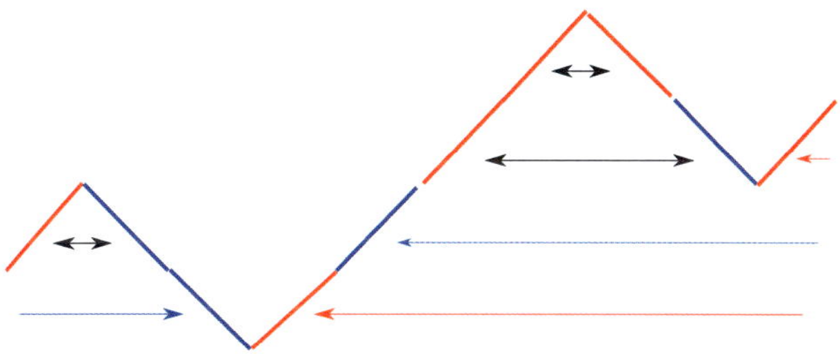

Fig. 5. An example of a path corresponding to a product state basis element.

unmatched up/down-steps in the path notation or unmatched left/right parentheses in the word notation.

In the colored case, the Fredkin moves (see Fig. 4) have more invariants, as we have already seen from the fact that the Hamiltonian commutes with the T^a operators.

The key observation here is that the Hamiltonian generates the colored Fredkin moves, which we have presented in path notation in Fig. 4. We see that these must preserve the color and order of the unmatched steps, as well as the color of the matched pairs.

Since we can move any /\ peak to any point in the path, we may use our previous algorithm where we repeatedly move the last peak to the beginning of the path until we are left with a series of single peaks followed by a dip, which up to permutations of the peaks will give the same result for any member of an equivalence class. This allows us to clearly see what equivalence classes we will be left with.

Specifically, the equivalence classes are fully specified by an ordered list of the colors of the unmatched up and down-steps, along with the number of matched pairs corresponding to each color combination along the path.

To keep our approach $SU(k)$-covariant, it is convenient to work with basis states that are not quite product states when trying to use the orbit theorem. Since the colors of up and down-steps form mutually conjugate fundamental representations of $SU(k)$, they can combine into an $SU(k)$ singlet or a member of the adjoint representation of $SU(k)$. Our choice of basis states will thus correspond to paths (σ_j^z eigenstates), where the colors of each matched pair combine to either a color singlet $|0\rangle$, or to an element $|a\rangle$, $a \in [1, k^2 - 1]$ in the adjoint representation of $SU(k)$.

Therefore, we may define the equivalence classes $C_{a,c_a,b,c_b,N^0,N^1\ldots N^{k^2-1}}$, where a, b are analogous to in the uncolored case, c_a, c_b are lists of length a, b, of the colors of the unmatched steps, N^0 is the number of matched pairs whose colors combine to singlets, and N^a is the number of matched pairs that combine into the adjoint representation state a, and where we have the additional relation $2(N^0 + \sum_{a=1}^{k^2-1} N^a) + a + b = N$.

3.1.1. *The bulk ground states*

The colored Fredkin moves generate a group that permute our chosen $SU(k)$-covariant basis vectors, preserve the equivalence classes, and can map any element of an equivalence class to any other. Therefore, as a consequence of the orbit theorem each equivalence class will have a unique ground state $|C_{a,c_a,b,c_b,N^0,N^1\ldots N^{N^2-1}}\rangle$ which is just the symmetric sum of all basis states in the class. These provide a nice orthogonal basis for all ground states of H_F (see Fig. 6).

This characterization is exactly what we will need to find out the combined ground state of $H_F + H_X + H_\partial$. The condition that the combined ground state is also a ground state of H_X leaves us with $N^a = 0$ for all a, and clearly the entire subspace that does not fulfill this cannot have any common $H_F + H_X$ zero eigenstate. Since in that case $N^0 = N - (a + b)$, we can simply call the bulk $H_F + H_X$ ground states $|C_{a,c_a,b,c_b}\rangle$.

For an even number of sites $N = 2n$, this leaves us with the unique ground state $|C_{0,\emptyset,0,\emptyset}\rangle$ as we add in the boundary terms, which we call the colored Dyck state $|D_n^k\rangle$.

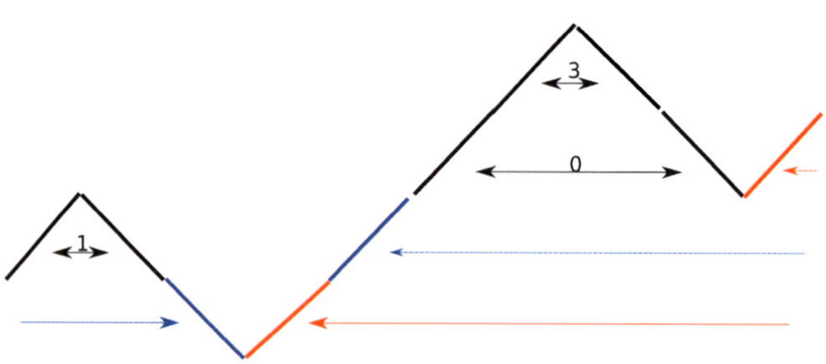

Fig. 6. An example of a path corresponding to a covariant basis element, with one $SU(k)$ singlet pair and two pairs in the $SU(k)$ adjoint representation.

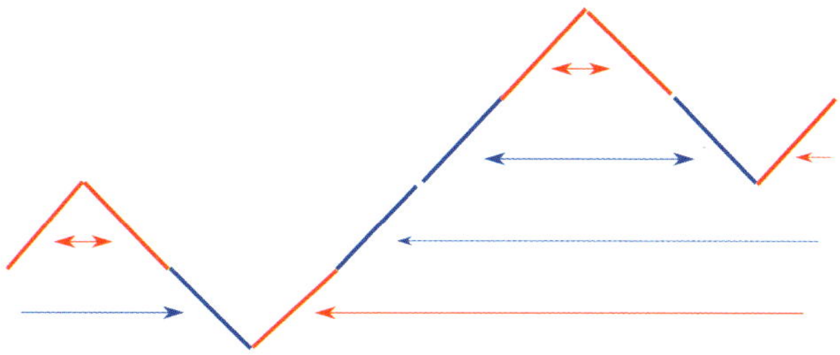

Fig. 7. An example of a properly colored path.

3.2. *The colored Dyck state*

We have now fully introduced the ground state $|D_n^k\rangle$. Depending on whether we use the covariant (see Fig. 6) or the product state (see Fig. 5) basis. In the non-local covariant basis that we used in the previous section to simultaneously block-diagonalize H_F and H_X, $|D_n^k\rangle$ is the symmetric sum of all paths colored such that all matched pairs form $SU(k)$ singlets.

We may expand this expression as a sum of product states. In the local product state basis, $|D_n^k\rangle$ is instead the sum of all basis states corresponding to *properly colored paths*, i.e. colored paths such that matching parentheses have the same color. The number of basis states in the sum is given by $k^n Cat(n)$. For $k = 2$, we will give some low-n examples in bracket notation:

$$|D_2^2\rangle = \frac{1}{\sqrt{2}}\left(|()\rangle + |[]\rangle\right)$$

$$|D_2^2\rangle = \frac{1}{\sqrt{8}}(|()()\rangle + |[]()\rangle + |()[]\rangle + |[][]\rangle$$
$$+ |(())\rangle + |([])\rangle + |[()]\rangle + |[[]]\rangle).$$

3.2.1. *The Schmidt decomposition and measures of entanglement*

Just as before, our primary tool to analyze the entanglement of the colored Dyck state will be the Schmidt decomposition, which is given by

$$|D_n^k\rangle = \sum_{m=0}^{L} \sum_{c\in\mathbb{Z}_k^m} \sqrt{p_{m,c}}\,|C_{0,\emptyset,m,c}(L)\rangle \otimes |C_{m,c,0,\emptyset}(N-L)\rangle \tag{10}$$

where c is a list of length m consisting of the colors of the unmatched parentheses of $|C_{0,\emptyset,m,c}(L)\rangle$ and $|C_{m,c,0,\emptyset}(N-L)\rangle$. Since as before

$$p_{m,c} = \frac{|C_{0,\emptyset,m,c}(L)||C_{m,c,0,\emptyset}(N-L)|}{|C_{0,0,0,0}(N)|},$$

and since $|D_n^k\rangle$ is an equally weighted sum of every coloring of each path, we get $p_{m,c}$ is independent of c. By inserting the correct powers of k to account for every possible color combination we obtain

$$p_{m,c} = p_m \frac{k^{\frac{L-m}{2}} k^{\frac{N-L-m}{2}}}{k^{\frac{N}{2}}} = k^{-m} p_m$$

where p_m are the Schmidt coefficients from the uncolored case.

This allows us to compute the Schmidt rank and the Entanglement entropy in the colored case. We will start with the Schmidt rank. The number of non-zero $p_{m,c}$ coefficient is a geometric sum. In the case where L is even and $L = 2l$, the Schmidt rank is $\sum_{h=0}^{l} k^{2h} = \frac{k^{L+2}-1}{k^2-1}$. In the case where L is odd and equal to $2l+1$, we have $\sum_{h=0}^{l} k^{2h+1} = k\frac{k^{L+1}-1}{k^2-1}$. The two may be combined into:

$$\xi = \begin{cases} \dfrac{k^{L+2}-1}{k^2-1} & \text{if } L \text{ is even} \\ k\dfrac{k^{L+1}-1}{k^2-1} & \text{if } L \text{ is odd.} \end{cases}$$

Now we may move on to the entanglement entropy. By definition, we have:

$$S = \sum_{m,c} -p_{m,c}\log(p_{m,c}) = \sum_{m} -p_m \log(k^{-m} p_m) = \sum_{m} m p_m \log(k) - p_m \log(p_m)$$

which gives us

$$S = S_{1/2} + \log(k) \sum_{m=0}^{L} p_m m \tag{11}$$

where the first term is just the log-scaling entanglement entropy from the uncolored case. The second term however, is the expectation value of the height at L for a random Dyck walk, which scales as $O(\sqrt{L})$ and is thus the key term leading to our violation of the area law.

3.2.2. *Computing the coefficient of the leading square root term in the entanglement entropy*

We are interested in computing the expression $\sum_{m=0}^{L} p_m m$. We proceed much like in the uncolored case. We consider the even case $L = 2l$ with non-zero p_m when $m = 2h, h \in [0, l]$, and consider the coefficient $p_h \approx \frac{h^2}{Z}\exp\left[-h^2 \frac{n}{l(n-l)}\right]$, so that we are left with $2\sum_h h p_h$. Replacing the sum with an integral gives us:

$$\sum_h h p_h = \int_0^\infty x p_x dx = \left[\sqrt{\frac{l(n-l)}{l}} d\alpha = dx\right] = \int_0^\infty \sigma \rho_\alpha \sigma^{-1} \alpha \sigma^{-1} d\alpha$$

$$= \sigma^{-1} \int_0^\infty \alpha \rho_\alpha d\alpha = \sqrt{\frac{l(n-l)}{n}} \int_0^\infty \frac{\alpha^3}{Z} \exp[-\alpha^2] d\alpha$$

where just like before, $p_x = \frac{x^2}{Z'} \exp\left[-x^2 \frac{n}{l(n-l)}\right]$, $\int p_x dx = 1$, $\alpha = \sigma x$ and $\sigma = \sqrt{\frac{N}{L(N-L)}}$. We move to $\rho_\alpha = \frac{1}{Z'}\alpha^2 \exp[-\alpha^2]$ where Z is a constant such that $\int \rho_\alpha = 1$, which gives us $p_x = \sigma\rho_\alpha$ (with a factor σ^3 from the normalization factor and a σ^{-2} from $x^2 = \sigma^{-2}\alpha^2$), which we used to insert the correct sigma factors.

The integral $\int_0^\infty \frac{\alpha^3}{Z} \exp[-\alpha^2] d\alpha$ is the third absolute central moment of the normal distribution, while in this formalism the constant $\frac{1}{Z}$ is chosen such that the second moment is one. Thus, we have:

$$S \approx S_{1/2} + (\log(k)) \frac{M_3}{M_2} \sqrt{2\frac{L(N-L)}{N}}$$

where M_2 and M_3 are the second and third absolute moments of the normal distribution. Computing these gives us our final approximate expression:

$$S \approx S_{1/2} + (\log(k)) \frac{2}{\sqrt{\pi}} \sqrt{2\frac{L(N-L)}{N}}.$$

Appendix A. The Orbit Theorem

In this appendix, we will prove a useful theorem that we (implicitly or explicitly) repeatedly make use of throughout this paper. It is our main tool for classifying ground states by looking at equivalence classes. We chose to call it the orbit theorem. It reads as follows:

Theorem 1. *Let V be a Hilbert space of finite dimension, and let $v_1 \cdots v_n$ be an orthonormal basis of V. Let G be a finite group which acts on $B = \{v_1, v_2 \cdots v_n\}$ by permutations. Suppose that G is generated by the generators $g_1, g_2 \cdots g_n$ and let $P_1, P_2 \cdots P_n$ be the corresponding permutation matrices. Let H be a Hamiltonian defined by $H = \sum_{i=1}^m (1 - P_i^\dagger)(1 - P_i) = \sum_{i=1}^m 2 - P_i^\dagger - P_i$ and let $W \subset V$ be the zero eigenspace of H.*

Then H is positive semidefinite, is a frustration free Hamiltonian, and the dimension of the ground state space W is equal to the number of orbits under the action of G on B. Furthermore, the Hamiltonian can be block diagonalized into the different vector spaces V_i spanned by the elements of B_i, and each block has a unique ground state which is a zero eigenstate of H. The ground states of the different blocks form an orthogonal basis for W.

Proof 1. H is clearly positive semidefinite since it is a sum of the form $\sum_i A_i^\dagger A_i$. Let $B_1, B_2 \cdots B_k$ be the orbits of G and V_i the spaces spanned by the elements of B_i. Then the linear representation of G on V decomposes into representations of G on the subspaces $V_1 \cdots V_k$ such that W is a direct sum of the spaces $W_i := W \cap V_i$ and dim W = dim W_1 + dim W_2 + $\cdots$ + dim W_k. It is thus enough to show that dim $W_i = 1$ for all i, or after change of notation that dim $W = 1$ when G acts transitively on V (if G can map any element to any element). Suppose that

$|v\rangle \in W$. Then we have $\sum_i [\langle v|(1-P_i^\dagger)][(1-P_i)|v\rangle] = \langle v|H|v\rangle = 0$ which implies that $(1-P_i)|v\rangle = 0$, which finally gives us $P_i|v\rangle = |v\rangle$ for all i.

Any $v \in W$ will therefore be fixed by all permutations generated by the P_i matrices, and since they can map any basis state to any other basis state (act transitively) the coordinates of v corresponding to any two basis states must coincide. So dim $W = 1$ as the equally weighted sum of all basis states is the only ground state of H in V, as desired. The last part of the theorem then simply follows by noticing that the sets B_i of basis vectors do not intersect, therefore their spanned vector spaces V_i are orthogonal to each other. □

We note that in all cases where we apply this theorem, the permutation matrices (corresponding to Fredkin moves) happen to be involutions(and as such have eigenvalues ± 1), so $(1-P_i^\dagger)(1-P_i) = 2(1-P_i)$, which we for convenience normalize as $1 - P_i$.

References

[1] C. H. Bennett, H. J. Bernstein, S. Popescu and B. Schumacher, Concentrating partial entanglement by local operations, *Phys. Rev. A* **53**(4) (1996) 2046–2052.

[2] S. Bravyi, L. Caha, R. Movassagh, D. Nagaj and P. W. Shor, Criticality without frustration for quantum spin-1 chains, *Phys. Rev. Lett.* **109**(20) (2012) 207202.

[3] J. Chen, X. Chen, R. Duan, Z. Ji and B. Zeng, No-go theorem for one-way quantum computing on naturally occurring two-level systems, *Phys. Rev. A* **83**(5) (2011) 050301.

[4] S. Hild, T. Fukuhara, P. Schauß, J. Zeiher, M. Knap, E. Demler, I. Bloch and C. Gross, Far-from-equilibrium spin transport in heisenberg quantum magnets, *Phys. Rev. Lett.* **113**(14) (2014) 147205.

[5] R. Movassagh and P. W. Shor, Power law violation of the area law in quantum spin chains, (2015); arXiv:1408.1657.

[6] S. Nascimbène, Y.-A. Chen, M. Atala, M. Aidelsburger, S. Trotzky, B. Paredes and I. Bloch, Experimental realization of plaquette resonating valence-bond states with ultracold atoms in optical superlattices, *Phys. Rev. Lett.* **108**(20) (2012) 205301.

[7] D. Porras and J. I. Cirac, Effective quantum spin systems with trapped ions, *Phys. Rev. Lett.* **92**(20) (2004) 207901.

[8] P. Schauß, J. Zeiher, T. Fukuhara, S. Hild, M. Cheneau, T. Macrì, T. Pohl, I. Bloch and C. Groß, Crystallization in Ising quantum magnets, *Science* **347**(6229) (2015) 1455–1458.

[9] Wikipedia, Bertrand's ballot theorem (2016) [Online accessed 10 May 2016].

Scattering on Riemannian Symmetric Spaces and Huygens Principle

Michael Semenov-Tian-Shansky

Institut de Mathématiques de Bourgogne, UMR 5584 CNRS,
Université de Bourgogne, B.P. 47870, 21078 Dijon Cedex, France
and
Steklov Mathematical Institute, St. Petersburg, Russia
semenov@u-bourgogne.fr

The famous paper by L. D. Faddeev and B. S. Pavlov (1972) on automorphic wave equation explored a highly romantic link between Scattering Theory (in the sense of Lax and Phillips) and Riemann hypothesis. An attempt to generalize this approach to general semisimple Lie groups leads to an interesting evolution system with multidimensional time explored by the author in 1976. In the present paper, we compare this system with a simpler one defined for zero curvature symmetric spaces and show that the Huygens principle for this system in the curved space holds if and only if it holds in the zero curvature limit.

To the memory of my teachers Professor Faddeev and Professor Pavlov

1. Introduction

The famous paper of Faddeev and Pavlov [1], which was published exactly 45 years ago, marked a highly romantic attempt to apply the ideas coming from the Quantum Mechanics to fundamental problems of Analytic Number Theory. By that time, the work of Selberg had established the link between Number Theory and Representation Theory of semisimple Lie groups; this link was explored in the famous monograph of Gelfand and Piatetsky-Shapiro [2]. Simultaneously, Faddeev had developed the powerful technique of Scattering Theory and Perturbation Theory for operators with continuous spectrum which he applied to problems of Quantum Mechanics. Gelfand suggested to Faddeev to use the same technique for study of the automorphic Laplace operator. This approach exposed in Faddeev's paper [3] was followed by a non-arithmetic proof of the famous Selberg formula which he obtained jointly with his two Ph.D. students, Venkov and Kalinin [4]. The almost simultaneous paper of Faddeev and Pavlov introduced another idea into the study of the automorphic Laplace operator, the non-stationary Scattering Theory in the version of Lax and Phillips [5]. The very possibility to apply the technique of Lax and Phillips depends crucially on the preliminary choice of an appropriate hyperbolic equation, the *non-Euclidean wave equation* in the Poincaré upper half-plane $\mathcal{H}$,

$$u_{tt} = \left(\Delta + \frac{1}{4}\right)u, \quad \text{where } \Delta = y^2\left(\frac{\partial^2}{\partial x^2} + \frac{\partial^2}{\partial y^2}\right), \quad z = x + iy, \ y > 0. \tag{1.1}$$

The very important constant term in (1.1) accounts for the fact that a free Laplacian in curved space does not describe a massless particle; it subtracts the "effective mass" which is due to curvature. It is easy to see that the operator $-\Delta - 1/4$ is positive and its continuous spectrum is the half-line $[0, \infty)$; the "naive" wave equation $u_{tt} = \Delta u$ describes a massive particle with mass $m^2 = 1/4$. Equation (1.1) is particularly interesting in the automorphic case when we consider the quotient $\Gamma\backslash\mathcal{H}$ over a discrete subgroup $\Gamma \subset SL(2, \mathbb{R})$. The relevant case is when $\Gamma\backslash\mathcal{H}$ is non-compact with finite volume, for instance, $\Gamma = SL(2, \mathbb{Z})$. One of the key conclusions in [1] is that in this case the scattering operator is explicitly expressed in terms of Riemann's zeta function; its non-trivial zeros are directly related to the resonances in the automorphic scattering, alias the eigenvalues of the associated contraction semigroup. While this interpretation unfortunately (and contrary to the initial hopes) did not bring us any closer to the proof of the Riemann hypothesis, the link between Riemann's zeta and scattering remains highly fascinating.[a] For many of us, at the time young Ph.D. students working in the newly created Faddeev's laboratory at the Steklov Institute in Leningrad, the work of Faddeev and Pavlov was a striking example of unity of mathematics and a challenge for the future. My personal involvement in this subject, which goes back to the early 1970's, was connected with an attempt to get a more thorough and intrinsic understanding of the role of Scattering Theory in Harmonic analysis on semisimple Lie groups. I believe the outcome is still interesting; although quite a few questions remain open, the subject certainly deserves further study.

The upper half-plane is a *rank 1 symmetric space*; a natural question, posed to me by Professors Faddeev and Pavlov, was to generalize the entire theory for arbitrary symmetric spaces. The ultimate question is to study the spaces $\Gamma\backslash X = \Gamma\backslash G/K$ (here and below G is a real semisimple Lie group, K its maximal compact subgroup and $\Gamma \subset G$ a discrete subgroup), but the question is already non-trivial for global symmetric spaces $X = G/K$, i.e. when $\Gamma = \{id\}$. It is easy to see that for arbitrary rank 1 symmetric spaces the non-Euclidean wave equation has the form

$$u_{tt} = (\Delta + \langle\rho, \rho\rangle)u, \tag{1.2}$$

where Δ is the Laplace–Beltrami operator on X and ρ is the half-sum of positive roots with their multiplicities. The question becomes much less trivial for symmetric spaces of higher rank. The point is that in the rank 1 case the algebra $D^G(X)$ of G-invariant differential operators on X is freely generated by a single element, the Laplace–Beltrami operator. When $\operatorname{rank} G = l > 1$, the algebra $D^G(X)$

[a] In 1976, Lax and Philips wrote a special monograph [7] dedicated to the Faddeev–Popov approach to automorphic forms.

is isomorphic to a polynomial algebra in l variables, and the wave equation (1.2) is replaced by an overdetermined system of equations with multidimensional time [6]. Very unexpectedly, the properties of this system are similar to those of the ordinary wave equation. In particular, it admits a natural generalization of the Lax–Phillips scattering theory; the scattering operators for this system coincide with the (properly normalized) intertwining operators for the principal series representations of G. On the heuristic level, an analogy between the poles of the intertwining operators (alias, the operators of horospheric automorphisms) and the poles of scattering matrix in Quantum Mechanics was already mentioned in Gelfand's address to the 1962 International Congress of Mathematicians [8]. The results of [6] (which justify the heuristic analogy observed by Gelfand) suggest a relation between the poles and zeros of intertwining operators (associated with the exceptional points where the principal series representations become reducible) and the resonances in the scattering problem. However, this attractive picture is obscured by the important difficulties associated with the violation of the Huygens principle. This difficulty is already present for the ordinary wave equation in Euclidean space of even dimension [9]. The violation of the Huygens principle in this case implies that the natural incoming and outgoing subspaces are not orthogonal to each other (in the Hilbert space of initial data); this makes much less efficient the subtle part of the Lax–Phillips theory which explores the contraction semigroup associated with the scattering problem. The geometric criterion which assures the validity of the Huygens principle was established in [6]: it holds if and only if all roots of (G, K) have even multiplicities. At the time, the validity of the Huygens principle for evolution equations in curved space (i.e. with variable coefficients) was rather nontrivial. The validity of the Huygens principle for wave equation in symmetric spaces was later discussed in [16, 23, 24]. In the present paper, we give a proof of another criterion: the Huygens principle holds simultaneously for the Riemannian symmetric space X and for the associated space of zero curvature. It is interesting to notice that the analytic properties of the scattering operators in the zero curvature case are actually worse than in the case of curved space: when the root multiplicities are odd, the scattering operators in the flat case contain sign flips (or Hilbert transforms), while in the curved space they remain meromorphic with poles in the "wrong" half-plane. The possibility to associate a natural contraction semigroup with the scattering problem in the global symmetric space remains unexplored to this day.

In 1970's and 1980's, the main research axis in the Faddeev laboratory has shifted towards the study of Classical and then Quantum Integrable Systems. Already in 1978, when the fascinating development of the Quantum Inverse Scattering Method was started, it became clear that there should exist a deep connection between this new technique and the Representation theory of semisimple Lie groups. Remarkably, many important quantum integrable systems, such as the Calogero–Sutherland system and the quantum Toda lattice are related to radial parts of

Laplace operators on semisimple Lie groups. However, for several decades two different trends in the theory of Quantum Integrable Systems, represented by the work on Quantum Inverse Scattering Method [14] on one hand and by the seminal observations of Kostant [15] on the theory of quantum Toda lattices on the other, remained almost disjoint. The link between the two approaches is provided be the fundamental concept of Quantum separation of variables, due to Sklyanin [12]. Their application to Representation Theory of semisimple Lie groups [13] represents a far-reaching generalization of the famous factorization formulae discovered by Gindikin and Karpelevich [10] in early 1960's in the aftermath of the fundamental works of Harish-Chandra [11]. These results are still not really understood by the experts in Representation Theory; the author believes that they will eventually lead to its complete reshaping. The intrinsic link between Representation Theory and Scattering Theory remains an important idea very closely related with these recent developments.

2. Multitemporal Wave Equation

In this section, we give a brief outline of the main results of [6]. The proof of the main result on the validity of the Huygens principle will be given in Sec. 3.

We shall start with fixing the notation. Let G be a non-compact semisimple Lie group with finite center, such as $G = SL(n, \mathbb{R})$, K its maximal compact subgroup which is stable with respect to a Cartan involution θ: $G \to G$, A its split Cartan subgroup, $\mathfrak{g}, \mathfrak{k}, \mathfrak{a}$ the corresponding Lie algebras. We denote the Cartan involution on $\mathfrak{g}$ by the same letter; let $\mathfrak{g} = \mathfrak{k} \oplus \mathfrak{p}$ be the corresponding Cartan decomposition. Let $\Sigma \subset \mathfrak{a}^*$ be the restricted root system. We fix an order in Σ and denote by Σ_+ (respectively, P) the set of positive (respectively, simple) roots of $(\mathfrak{g}, \mathfrak{a})$. For $\alpha \in \Sigma$ we denote by $\mathfrak{g}_\alpha$ the corresponding root subspace and by $m_\alpha = \dim \mathfrak{g}_\alpha$ the root multiplicity. Put $\mathfrak{n} = \bigoplus_{\alpha \in \Sigma_+} \mathfrak{g}_\alpha$, $N = \exp \mathfrak{n}$. We have $G = KAN$ and $G = KAK$ (the Iwasawa and Cartan decompositions). Let $M, M' \subset K$ be the centralizer and the normalizer of A in K; the quotient $W = M'/M$ is the restricted Weyl group of (G, K) which acts naturally on $\mathfrak{a}$ and on its dual space $\mathfrak{a}^*$. Let $X = G/K$ be the symmetric space. The ring of G-invariant differential operators on X is described by the fundamental Harish-Chandra theorem. We denote by $P^W(\mathfrak{a}^*)$ the algebra of W-invariant polynomials on $\mathfrak{a}^*$.

Theorem 2.1 (Harish-Chandra). *There exists a canonical isomorphism of algebras*

$$\Delta\colon P^W(\mathfrak{a}^*) \to D^G(X)\colon \sigma \mapsto \Delta_\sigma$$

which assigns to a symmetric polynomial σ a differential operator Δ_σ.

Remark 1. It is quite common to associate a polynomial (called *principal symbol*) to a differential operator; conversely, one can associate a differential operator to its principal symbol using the symmetrization map. In the context of symmetric spaces, we get a map $D^G(X) \to P^K(\mathfrak{p}^*)$, where $\mathfrak{g} = \mathfrak{k} \oplus \mathfrak{p}$ is the Cartan decomposition; it is easy to see that restriction to $\mathfrak{a}^* \subset \mathfrak{p}^*$ induces an isomorphism $i^*\colon P^K(\mathfrak{p}^*) \simeq P^W(\mathfrak{a}^*)$. However, the Harish-Chandra is much smarter than the simple symmetrization map, since it respects product. When σ is a homogeneous polynomial, the principal symbol of the operator Δ_σ coincides with σ (up to the isomorphism i^*), but Δ_σ contains also terms of various orders $d < \deg \sigma$. In particular, if σ is the quadratic polynomial associated with the inner product in $\mathfrak{a}$, we have

$$\Delta_\sigma = \Delta + \langle \rho, \rho \rangle,$$

where ρ is the half-sum of all positive roots. For $G = SL(2, \mathbb{R})$ this term gives precisely the 1/4-shift from (1.1).

The structure of $P^W(\mathfrak{a}^*)$ is described by the classical Chevalley theorem:

Theorem 2.2 (Chevalley). (i) *Let* $l = \operatorname{rank} G/K = \dim \mathfrak{a}$. *Then* P^W *is freely generated by* l *homogenous polynomials.*
(ii) $P(\mathfrak{a}^*) \supset P^W(\mathfrak{a}^*)$ *is a free* P^W*-module of rank* $N = \operatorname{card} W$.

We are now in a position to describe a generalization of the wave equation (1.1) which makes sense for all higher-rank spaces. We choose the time variable t to take values in $\mathfrak{a}$, so our space-time is $\mathfrak{a} \times X$. A symmetric polynomial $\sigma \in P^W(\mathfrak{a}^*)$ may be regarded, via the ordinary Fourier transform, as the symbol of a differential operator on $\mathfrak{a}$ with constant coefficients. Let $\sigma \mapsto {}'\sigma$ be the automorphism defined by

$${}'\sigma(\lambda) = \sigma(-\lambda).$$

Definition 2.3. The generalized wave equation on $\mathfrak{a} \times X$ is the compatible overdetermined system

$${}'\sigma\left(-\imath \frac{\partial}{\partial t}\right) u(t,x) = \Delta_\sigma u(t,x), \quad \sigma \in P^W(\mathfrak{a}^*). \tag{2.1}$$

Remark 2. The automorphism $\sigma \mapsto {}'\sigma$ is introduced for technical reasons to account for more natural sign conventions; if the sign flip $\lambda \mapsto -\lambda$ belongs to the Weyl group it becomes redundant.

One of the equations in (2.1) is ultra-hyperbolic (for $l = \operatorname{rank} X > 1$):

$$\left(\frac{\partial^2}{\partial t_1^2} + \cdots + \frac{\partial^2}{\partial t_l^2}\right) u = (\Delta + \langle \rho, \rho \rangle) u.$$

Surprisingly, though, it is a perfectly good evolution system which possesses:

- A well-posed Cauchy problem.
- Conserved positive energy and an associated energy inner product.
- Finite propagation speed.
- Possibility to define scattering in the sense of Lax–Phillips.

The key definitions of the space of Cauchy data and of the energy form for the system (2.1) is based on the properties of the algebraic extension $P^W(\mathfrak{a}^*) \subset P(\mathfrak{a}^*)$ (together with some elements of Galois theory). In the spirit of the Cauchy–Kowalevski theorem, a solution of an evolution equation is completely determined if we know all its time derivatives for $t = 0$. Let us denote by $\mathcal{S}$ the linear space of solutions of (2.1). We define a coupling $\mathcal{S} \times P(\mathfrak{a})^* \to C^\infty(X)$ by

$$(u, p) \mapsto p\left(-i\frac{\partial}{\partial t}\right) u(t, \cdot)|_{t=0} = \langle i_0(u), p\rangle. \tag{2.2}$$

Of course, time derivatives are not independent, the set of relations being imposed by (2.1). We may regard the space $C^\infty(X)$ as a $P^W(\mathfrak{a}^*)$-module via the Harish-Chandra isomorphism; then (2.2) means precisely that the set $i_0(u)$ of initial data is an element of the space

$$\mathcal{H} = \mathrm{Hom}_{P^W}(\mathcal{S}, C^\infty(X)) \simeq C^\infty(X) \otimes_{P^W} \mathrm{Hom}(P, P^W).$$

Let Q, Q^W be the fields of fractions of P, P^W.

Proposition 2.4. (i) *$Q^W \subset Q$ is a Galois extension with Galois group W.*
(ii) *We may identify the dual space $\mathrm{Hom}_{Q^W}(Q, Q^W)$ with Q by means of the trace form,*

$$\langle u, v\rangle = \mathrm{tr}_{Q/Q^W}(u \cdot v) := \sum_{w\in W} u^w v^w, \quad u, v \in Q. \tag{2.3}$$

Put

$$D = \left\{s \in Q;\ \mathrm{tr}\, s\cdot\sigma \in P^W \textit{ for all } \sigma \in P\right\}.$$

Then $\mathcal{H} \simeq C^\infty(X) \otimes_{P^W} D$. For complex-valued polynomials, we replace (2.3) with the corresponding Hermitean form.

It is well known that $D = \pi^{-1} \cdot P$, where π is the "product of all positive roots",

$$\pi(\lambda) = \prod_{\alpha\in\Delta_+} \alpha(\lambda);$$

up to a factor, π is the only W-anti-invariant element in P. It is much more common, of course, to define the initial data as a simple vector function, i.e. to use tensor products over $\mathbb{C}$ rather than P^W. Fix a basis $\{p_1, \ldots, p_N\}$, $N = \mathrm{card}\, W$, of P over P^W, and let $\{p^1, \ldots, p^N\}, p^i \in D \subset Q$, be the dual basis with respect to the inner product (2.3). We assume that p_i, p^i are real. We may identify elements of $\mathcal{H}$ with vector functions on X with components

$$u_k = p_k\left(-i\frac{\partial}{\partial t}\right) u(t, \cdot)|_{t=0}, \quad k = 1, \ldots, N. \tag{2.4}$$

Put $\sigma_{mn} = \pi^2 \cdot \mathrm{tr}_{Q/Q^W} p^m p^n$; obviously, the factor π^2 eliminates the denominators and hence $\sigma_{mn} \in P^W$. The energy norm of $u \in \mathcal{H}$ is given by

$$\|u\|^2 = \sum_{k,l} \langle \Delta_{\sigma_{kl}} u_k, u_l \rangle_{L_2(X)}. \tag{2.5}$$

For $\chi \in P$, let L_χ: $\mathrm{Hom}(P, P^W) \to \mathrm{Hom}(P, P^W)$ be the linear operator defined by

$$\langle L_\chi u, p \rangle = \langle u, \chi \cdot p \rangle.$$

We shall denote the operators induced by L_χ in $\mathcal{H} = C^\infty(X) \otimes_{P^W} \mathrm{Hom}(P, P^W)$ by the same letter. In realization (2.4) operator L_χ is acting as a matrix,

$$(L_\chi u)_k = \sum_m \Delta_{\chi_k^m} u_m, \quad \text{where } \chi_k^m = \mathrm{tr}_{Q/Q^W}(\chi p_k p^m). \tag{2.6}$$

Fix a basis $e_1, \ldots, e_l$ in $\mathfrak{a}$; we may regard $e_1, \ldots, e_l$ as linear functions on $\mathfrak{a}^*$, i.e. as elements of $P(\mathfrak{a}^*)$.

Theorem 2.5 ([6]). (i) *The energy norm* (2.5) *is positive and defines the structure of the pre-Hilbert space in* $\mathcal{H}$.

(ii) *System* (2.1) *is equivalent to the system of first order linear differential equations in* $\mathcal{H}$,

$$i \frac{\partial u_k}{\partial t_s} = \sum_m \Delta_{{}_s\chi_k^m} u_m, \quad \textit{where } {}_s\chi_k^m = \mathrm{tr}_{Q/Q^W}(e_s p_k p^m), \quad s = 1, \ldots, l.$$

(iii) *The associated evolution group in* $\mathcal{H}$ *is unitary, i.e. preserves energy.*

As for the ordinary wave equation, the propagation speed for system (2.1) is finite. Recall that the Cartan subgroup A acts freely on X; for each $H \in \mathfrak{a}$ the curve $e^{tH} \cdot x_0$, $t \in \mathbb{R}$, is a geodesic in X which starts from $x_0 = eK \in G/K$.

Theorem 2.6. *If the initial data for a solution* u *for* $t = 0$ *are supported in a geodesic ball* $D_R = \{x = ke^H \cdot x_0\,;\, k \in K, H \in \mathfrak{a}, |H| < R\}$, *the solution* $u(t, \cdot)$ *for any* $t \in \mathfrak{a}$ *is supported in the ball* $D_{R+|t|}$.

Thus all solutions with compact support have sharp forward front. The behavior of the back front is much more subtle. For the ordinary wave equation in $\mathbb{R}^{n+1}$ with odd n the following well-known result holds true:

Huygens principle: Solutions with compact support have sharp back front.

In curved space-time this may be false for at least two reasons:

- because of mass gap in curved space.

  (In other words, a massless particle propagating in the curved space may acquire a non-zero mass, and hence its velocity is strictly smaller than the light speed.)
- because of wave diffusion.

With our choice of the evolution system based on the Harish-Chandra isomorphism the mass gap is automatically excluded; the second reason is much more fundamental; as we shall see, it is common to the flat space of even dimension and to the corresponding curved space.

In order to control the behavior of solutions of the wave system (2.1) we need the notion of the *radiation asymptotics*. For any solution u, we set

$$\mathcal{W}_+u(\tau,k) = \lim_{t\to\infty} \pi(-i\partial/\partial\tau)e^{\rho(t)}u(t+\tau, ke^t\cdot x_0), \quad \text{where } \tau\in\mathfrak{a}, k\in K. \tag{2.7}$$

It is obvious from the definition that $\mathcal{W}_+u(\tau,k)$ depends only on the coset class kM, i.e. it is a function on the sphere K/M.

Theorem 2.7. *This limit exists when $t\to\infty$ along any ray which is strictly inside the positive Weyl chamber in $\mathfrak{a}$ (and is independent of its choice). Moreover,*

$$\|u\|_E = \|\mathcal{W}_+u\|_{L_2(\mathfrak{a}\times K/M)}.$$

Remark. "Intermediate" asymptotics when $t\to\infty$ along some wall of the Weyl chamber are also interesting; their study leads to various new Plancherel theorems closely related with the concept of factorized scattering and with the Gindikin–Karpelevich type formulae [6].

The mapping $\mathcal{W}_+\colon \mathcal{H}\to L_2(\mathfrak{a}\times K/M)$ defined by (2.7) is called the (outgoing) wave operator. Taking a different Weyl chamber we can define generalized wave operators $\mathcal{W}_w$ for any $w\in W$. We set

$$\mathcal{W}_wu(\tau,k) = \lim_{t\to\infty} \pi(-i\partial/\partial\tau)e^{\rho(t)}u(w^{-1}t+\tau, ke^t\cdot x_0), \text{ where } \tau\in\mathfrak{a}, k\in K. \tag{2.8}$$

Proposition 2.8. *For all $w\in W$ and all $s\in\mathfrak{a}$ we have*

$$(\mathcal{W}_wU_su)(\tau,k) = (\mathcal{W}_wu)(\tau+s,k).$$

In other words, wave operators provide translation representations for the evolution group associated with the wave system (2.1).

As usual, the existence of different translation representations allows to define scattering operators. We set

$$\mathcal{S}^w = \mathcal{W}_+\mathcal{W}_w^{-1}, \quad \mathbf{S}^w = \Phi\mathcal{S}^w\Phi^{-1}, \tag{2.9}$$

where Φ is the classical Fourier transform. We regard Φ as a unitary operator $\Phi\colon L_2(\mathfrak{a}\times K/M)\to L_2(\mathfrak{a}^*\times K/M)$ (acting on the first variable),

$$\Phi a(\lambda,k) = \frac{1}{(2\pi)^l}\int_{\mathfrak{a}} e^{i\lambda(t)}a(t,k)\,dt.$$

The composition $\mathbf{W}_w := \Phi\circ\mathcal{W}_w$ intertwines the evolution group U_s with multiplications by exponentials $e^{-i\lambda(s)}$, i.e. it is a *spectral representation* for the evolution group. We shall return to the properties of the wave operators in connection with the scattering theory in Theorem 2.12 below.

It is easy to compute the radiation asymptotics of the solutions represented in form of the *wave packets* constructed as linear combinations of matrix coefficients of the principle series representations of G. Before describing this computation let us briefly describe the construction of these representations. Put $B = MAN$ (recall that by definition $M \subset K$ the centralizer of A in K); the subgroup B is a minimal parabolic subgroup of G. For $\mu \in \mathfrak{a}_{\mathbb{C}}^*$ let χ_μ be the 1-dimensional representation of B defined by

$$\xi_\mu(man) = e^{\mu(\log a)}.$$

By definition, the (class one) principle series representations of G are the induced representations $\operatorname{ind}_B^G(\xi_\mu)$; these representations are unitary when $\mu = i\lambda - \rho$, where $i\lambda \in \mathfrak{a}_{\mathbb{C}}^*$ is purely imaginary and

$$\rho = \frac{1}{2}\sum_{\alpha\in\Delta_+} m_\alpha \alpha.$$

We shall write $\operatorname{ind}_B^G \xi_{i\lambda-\rho} = T_\lambda$. In one of the standard realizations, the representation space $\mathfrak{H}$ of T_λ is identified with $L_2(K/M)$; we have

$$T_\lambda(g)\psi(k) = e^{\langle i\lambda-\rho, H(g^{-1}k)\rangle}\psi(\kappa(g^{-1}k)), \tag{2.10}$$

where $e^{H(x)} \in A$, $\kappa(x) \in K$ give the Iwasawa decomposition of $x = kan \in G$. In this realization the unique (up to normalization) K-invariant vector is the constant function on K/M. Representations T_λ, T_μ are equivalent if and only if $\mu = w \cdot \lambda$ for some $w \in W$; the intertwining operators which realize this equivalence are integral operators given by the explicit formulae. It is convenient to normalize the intertwining operators by the condition $S_\lambda^w \cdot 1 = 1$; with this condition we have

$$S_\lambda^w a(k) = \frac{1}{c_w(\lambda)}\int_{\bar{N}_w} e^{-\langle i\lambda+\rho, H(v)\rangle} a(km_w\kappa(v))\, dv, \tag{2.11}$$

where $m_w \in M'$ is a representative of $w \in W = M'/M$, $\bar{N} = \theta(N)$, $\bar{N}_w = \bar{N}\bigcap m_w{}^1 N m_w$, and $c_w(\lambda)$ is given by

$$c_w(\lambda) = \int_{\bar{N}_w} e^{-\langle i\lambda+\rho, H(v)\rangle}\, dv. \tag{2.12}$$

Integrals in (2.11) and (2.12) are absolutely convergent if $\operatorname{Re} i\lambda \in \mathfrak{a}_+^*$ and admit a meromorphic continuation to the whole space $\mathfrak{a}_{\mathbb{C}}^*$. The function $c_w(\lambda)$ is explicitly given by the famous Gindikin–Karpelevich formula [10]; up to a constant factor which depends on the normalization of the Haar measures on $\bar{N}_w$ it is given by

$$c_w(\lambda) = \prod_{\alpha\in\Delta_+\cap w^{-1}\Delta_-} c_\alpha(i\lambda), \quad c_\alpha(\mu) = B\left(\frac{1}{2}m_\alpha, \frac{1}{4}m_{\alpha/2} + \frac{\langle\mu,\alpha\rangle}{\langle\alpha,\alpha\rangle}\right). \tag{2.13}$$

When $w = w_0$ is the longest Weyl group element the function $c_{w_0}(\lambda) := c(\lambda)$ is the famous *Harish-Chandra c-function* [11]. Operators S_λ^w are unitary for real λ and satisfy the relations

$$S_\lambda^w \circ T_\lambda(g) = T_{w\lambda}(g) \circ S_\lambda^w, \quad S_\lambda^{vw} = S_\lambda^w \circ S_{w\lambda}^v. \tag{2.14}$$

Let us fix $\lambda \in \mathfrak{a}^*$, $k \in K/M$, $w \in W$ and define the plane wave $\Phi(\cdot;\lambda,k)$ by

$$\Phi(t,x;\lambda,k) = e^{-i\lambda(t)} e^{\langle iw\lambda-\rho,\, H(g^{-1}k)\rangle}, \quad \text{where } x = gK. \tag{2.15}$$

The plane wave (2.15) is proportional to the image $T_{w\lambda}(g)\cdot 1$ of the spherical vector in the principal series representation. It is easy to see that it satisfies the system (2.1). For any $w \in W$, we define the Fourier transform in the space $\mathcal{H}$ of initial data by

$$F_w f(\lambda,k) = \langle f, i_0(\Phi(\cdot,\lambda,k))\rangle_E. \tag{2.16}$$

Let T be the natural action of G in $\mathcal{H}$ induced by the action of G on X; denote by T_w the natural representation of G in $L_2(\mathfrak{a}^*,\mathfrak{H})$ defined by

$$(T_w(g)a)(\lambda) = (T_{w\lambda}a(\lambda)).$$

For all $g \in G$ we have $F_w \circ T(g) = T_w(g) \circ F_w$, $S^w \circ F_w = F_1$.

Theorem 2.9 ([6]). *F_w extends to an isometry $F_w\colon \mathcal{H} \to L_2(\mathfrak{a}^* \times K/M;\mu)$ with measure $\mu(d\lambda\, dk) = |b(\lambda)|^2\, d\lambda\, dk$, where $b(\lambda) = \pi(\lambda)c(\lambda)$ is the Harish-Chandra function.*

Let us note in passing that for real λ we have $|b(\lambda)|^2 = b(\lambda)b(-\lambda)$; this provides a natural extension of the Plancherel measure to the complex domain.

Let $\widetilde{\mathcal{H}}_R \subset \mathcal{H}$ be the subspace of smooth initial data such that $f \in \widetilde{\mathcal{H}}_R$ is smooth with support in a geodesic ball B_R. The corresponding Fourier amplitudes $F_1 f := a^+, F_w f := a^w$ satisfy the following conditions:

- *For all $w \in W$ functions*

$$\lambda \mapsto \frac{1}{\pi(\lambda)}(S^w_\lambda)^* a^+_\lambda = \frac{1}{\pi(\lambda)} a^w_\lambda \tag{2.17}$$

  extend to entire functions on $\mathfrak{a}^*_{\mathbb{C}}$ with values in $\mathfrak{H} = L_2(K/M)$.
- *For all $\eta \in \mathfrak{a}^*$, $N \in \mathbb{Z}_+$, $w \in W$ we have*

$$\sup_{\xi\in\mathfrak{a}^*,\, \kappa\in K/M} (1 + \|\xi + i\eta\|)^N |a^w_{\xi+i\eta}(\kappa)| \le C_N e^{R\|\eta\|}. \tag{2.18}$$

Let $\mathcal{K}_R$ be the space of such functions, $\mathcal{K} = \bigcup_{R>0} \mathcal{K}_R$. The following theorem is a version of the Paley–Wiener theorem for symmetric spaces first proved by Helgason [19].

Theorem 2.10 ([6]). *The Fourier transforms F_w map $\widetilde{\mathcal{H}}_R$ onto $\mathcal{K}_R$.*

The inverse Fourier transform provides a large family of solutions of the system (2.1) which are realized as wave packets of plane waves,

$$u(t, g\cdot x_0) = \int_{\mathfrak{a}^*\times K/M} e^{-i\lambda(t)} e^{\langle -(i\lambda+\rho),\, H(g^{-1}\kappa)\rangle} a(\lambda,k)\, \mu(d\lambda\, d\kappa). \tag{2.19}$$

Equivalently, we may write

$$u(t, g\cdot x_0) = \int_{\mathfrak{a}^*} e^{-i\lambda(t)} \langle a,\, T_\lambda(g)1\rangle_{\mathfrak{H}}\, \mu(d\lambda). \tag{2.20}$$

Using the definition of the intertwining operators we derive from (2.20) the equivalent formulae

$$u(t, g \cdot x_0) = \int_{\mathfrak{a}^*} e^{-i\lambda(t)} \langle S^w a,\, T_{w\lambda}(g)1\rangle_{\mathfrak{H}}\, \mu(d\lambda) \tag{2.21}$$

which are valid for any $w \in W$.

According to a well-known theorem of Kostant [18], the set

$$C_T = \{H(e^T k); k \in K\}$$

coincides with the convex hull of the W-orbit of $T \in \mathfrak{a}$. In combination with the Paley–Wiener theorem for the ordinary Fourier transform this immediately implies that if $a \in \mathcal{K}_R$, the solution (2.20) vanishes outside the geodesic ball $B_{R+\|T\|+\|t\|}$, thus proving Theorem 2.6.

Let us now return to the study of the radiation asymptotics of solutions defined above in Theorem 2.7. Let us assume that u is given by (2.19) with amplitude $a \in \mathcal{K}$ and that moreover $\pi(\lambda)^{-1}a(\lambda)$ is holomorphic in some neighborhood of $\mathfrak{a}^* \subset \mathfrak{a}^*{}_{\mathbb{C}}$ (this condition allows to differentiate the solution under the integral sign).

Proposition 2.11. *We have*

$$\lim_{T\to\infty, T\in\mathfrak{a}_+} \pi\left(-i\frac{\partial}{\partial\tau}\right) u(T+\tau, ke^T \cdot x_0) = \int_{\mathfrak{a}^*} e^{-i\lambda(\tau)} a(\lambda, k)\, \frac{d\lambda}{b(-\lambda)}. \tag{2.22}$$

Sketch of a Proof. Put $g = ke^T$, where $T \in \mathfrak{a}_+$, in (2.19). We have

$$H(e^{-s}\kappa(v)) = H(e^{-s}ve^s) - H(v) - s \quad \text{for all} \quad s \in a, v \in \bar{N}.$$

The Harish-Chandra integration formula

$$\int_{K/M} \phi(k)\, dk = \int_{\bar{N}} e^{-2\rho(H(v))} \phi(\kappa(v))\, dv$$

allows to rewrite formula (2.19) as

$$u(t, ke^{\alpha T} \cdot x_0) = \int_{(\mathfrak{a}^* - i\eta)\times\bar{N}} e^{-i\lambda(t)} e^{\langle i\lambda-\rho, T\rangle} e^{-\langle i\lambda+\rho, H(v)\rangle} e^{\langle i\lambda-\rho, H(v^T)\rangle} a(\lambda, k\kappa(v^T))$$
$$\times \frac{d\lambda\, dv}{b(\lambda)b(-\lambda)}, \tag{2.23}$$

where $v^T := e^T v e^{-T}$ and $\eta \in \mathfrak{a}^*_+$ is a fixed shift of the integration contour. We have $n^T \to 1$ when $T \to \infty$ along a fixed ray in $\mathfrak{a}_+$. Passing to the limit under the sign of integral and shifting the integration contour back to the real subspace, we get (2.22), where we used the value and the properties of the Harish-Chandra integral (2.12) and (2.13) and the definition $b(\lambda) = \pi(\lambda)c(\lambda)$. The summable upper bound for the integrand in (3.19) was indicated by Harish-Chandra [11]. It is based on the well known convexity properties of the Iwasawa decomposition:

$$H(v) \in {}^+\mathfrak{a},\ H(v) - H(v^T) \in {}^+\mathfrak{a} \quad \text{for all} \quad v \in \bar{N}, T \in \mathfrak{a}_+ \tag{2.24}$$

(we denote by ${}^+\mathfrak{a}$ the dual cone of $\mathfrak{a}_+$). Choose $\epsilon \in (0,1)$ in such a way that $-\epsilon\eta + \rho \in \mathfrak{a}_+^*$; using (2.24) it is easy to check that

$$\left| e^{-\langle i\lambda+\rho, H(v)\rangle} e^{\langle i\lambda-\rho, H(v^T)\rangle} \right| \leq e^{-\langle \epsilon\eta+\rho, H(v)\rangle}$$

uniformly in $T \in \mathfrak{a}_+$. As shown by Harish-Chandra, the integral

$$\int_{\bar{N}} e^{-\langle \eta+\rho, H(v)\rangle} dv$$

is absolutely convergent for any $\eta \in \mathfrak{a}_+^*$.

Remark. For $g = e^T, T \to \infty, T \in \mathfrak{a}_+$, the integral in (2.19) is rapidly oscillating. A naive idea is to compute its asymptotics via the stationary phase approximation. The correct method applied above, which also yields a kind of localization formula, uses a more subtle technique. (The key relation we used is $n^T \to 1$ when $T \to \infty$, which means geometrically the contraction of horospheres under the action of geodesic flow on X.) The presence of Gamma functions in the final formulae means of course that the stationary phase approximation does not apply in this case; one can check that it gives the correct result only if the curvature of our symmetric space is rescaled and tends to zero simultaneously. There is one notable exception: for complex semisimple groups the stationary phase approximation applies well; in particular, the stationary phase approximation for the Harish-Chandra integral for the spherical function

$$\phi_\lambda(g \cdot x_0) = \int_{K/M} e^{-\langle i\lambda+\rho, H(g^{-1}k)\rangle} dk$$

gives the *exact* formula, which is essentially equivalent to the Weyl character formula [19]. (In this case of course all root multiplicities m_α are equal to 2 and hence the Harish-Chandra b-function reduces to a constant.) This phenomenon (closely linked to the fact that in the complex case the space K/M is symplectic) may be traced down to the Duistermaat–Heckmann theorem [20, 21] (see also the earlier paper [22]). I would like to mention that one of the favorite ideas of Professor Faddeev was the maxime: Quasiclassical approximation should be exact for problems with high degree of symmetry. Although this is of course not strictly true, this heuristic principle has proved to be most valuable; in different guises it has worked in various contexts ranging from Representation Theory to the Quantum Inverse Scattering Method. The rôle of Quasiclassics in Representation Theory of semisimple Lie groups still remains somewhat mysterious and has never been fully clarified.

To compute the radiation asymptotics in other Weyl chambers we must use formulae (2.21); together with Proposition 2.11 this yields:

$$\lim_{T\to\infty, T\in\mathfrak{a}_+} \pi\left(-i\frac{\partial}{\partial\tau}\right) u(w^{-1}T+\tau, ke^T \cdot x_0) = \int_{\mathfrak{a}^*} e^{-i\lambda(\tau)}(S^w a)(\lambda, k)\frac{d\lambda}{b(-w\lambda)}. \tag{2.25}$$

This implies that the associated scattering operator in Fourier representation (2.9) is given by

$$\mathbf{S}^w_\lambda = \left[\frac{b(-w\lambda)}{b(-\lambda)}\right] S^w_\lambda . \tag{2.26}$$

As already mentioned, wave and scattering operators defined by (2.7), (2.8), (2.26) match the definitions of the Lax–Phillips scattering theory. Let $\mathcal{H}$ be a Hilbert space, U_t: $\mathcal{H} \to \mathcal{H}$ a one-parameter group of unitary operators. A subspace $\mathcal{D}_+ \subset \mathcal{H}$ is called an *outgoing subspace* for U_t if

(i) $U_t \mathcal{D}_+ \subset \mathcal{D}_+$ for all $t > 0$.
(ii) $\bigcap_{t>0} U_t \mathcal{D}_+ = 0$.
(iii) The closure of $\bigcup_{t<0} U_t \mathcal{D}_+$ coincides with $\mathcal{H}$.

If U_t admits an outgoing subspace, it has a Lebesgue spectrum of uniform multiplicity and there exists a Hilbert space $\mathfrak{H}$ and a unitary equivalence W_+: $\mathcal{H} \to L_2(\mathbb{R}, \mathfrak{H})$ which maps $\mathcal{D}_+ \subset \mathcal{H}$ onto $L_2(\mathbb{R}_+, \mathfrak{H})$ and transforms U_t into the standard translation group acting in $L_2(\mathbb{R}, \mathfrak{H})$. For the ordinary wave equation, the construction of the outgoing subspace is based on its standard causal structure: the outgoing solutions are, by definition, those which vanish in the *forward light cone*. The causal structure for the multitemporal wave equation is defined in a similar way. We set

$$C_+ = \left\{(t, x) \in \mathfrak{a}_+ \times X; x \in B_{|t|}\right\} . \tag{2.27}$$

By definition, the outgoing subspace $\mathcal{D}_+$ for the multitemporal wave equation (2.1) consists of the initial data of solutions which vanish in C_+. Let $\mathcal{W}_+$ be the wave operator defined by (2.7) and $\mathbf{W}_+ = \Phi \circ \mathcal{W}_+$. We denote by $H_2^-(\mathfrak{a}^*; \mathfrak{H})$ the Hardy space of functions on $\mathfrak{a}^*$ with values in $\mathfrak{H} = L_2(K/M)$ which extend to analytic functions in the tube domain $\operatorname{Re} i\lambda \in \mathfrak{a}^*_+$ with the usual Hardy space property,

$$\sup_{\eta \in \mathfrak{a}^*_+} \int_{\mathfrak{a}^*} \|a(\sigma - i\eta)\|^2_{\mathfrak{H}} \, d\sigma < \infty$$

It is clear from the definition that the exponentials $e^{-i\lambda(t)}$, $t \in {}^+\mathfrak{a}$, act as multiplicators in $H_2^-(\mathfrak{a}^*; \mathfrak{H})$.

Theorem 2.12. *The wave operator* $\mathbf{W}_+$ *maps the space* $\mathcal{D}_+$ *of outgoing initial data onto* $H_2^-(\mathfrak{a}^*; \mathfrak{H})$.

Recall that the integrals (2.11) defining the intertwining operators S^w_λ converge for $\operatorname{Re} i\lambda \in \mathfrak{a}^*_+$; it is easy to check that they too are acting as multiplicators in $H_2^-(\mathfrak{a}^*; \mathfrak{H})$. Moreover, they satisfy the following causality property. Let s_α, $\alpha \in P$, be the generators of the Weyl group associated with simple roots. The length $l(w)$ of $w \in W$ is the minimal number of factors in the decomposition $w = s_{\alpha_1} \cdots s_{\alpha_n}$, $\alpha_i \in P$. For any $\gamma \in P$ we write $w_1 \overset{\gamma}{\to} w_2$ if $w_1 = s_\gamma w_2$ and $l(w_1) = l(w_2) + 1$. We set $w \succcurlyeq v$ if there exist elements $\alpha_1, \dots, \alpha_m \in P$ such that $w \overset{\alpha_1}{\to} w_1 \overset{\alpha_2}{\to} w_2 \cdots \overset{\alpha_m}{\to} v$. This relation defines a partial ordering in W.

Proposition 2.13. *For any $w \in W$ $S^w H_2^-(\mathfrak{a}^*;\mathfrak{H}) \subset H_2^-(\mathfrak{a}^*;\mathfrak{H})$ is a closed subspace. Moreover, $w \succcurlyeq v$ implies that*

$$S^w H_2^-(\mathfrak{a}^*;\mathfrak{H}) \subset S^v H_2^-(\mathfrak{a}^*;\mathfrak{H}). \tag{2.28}$$

The causality property (2.28) is based on the Gindikin–Karpelevich factorization for the intertwining operators. (I would guess it is in fact *equivalent* to this factorization property, though I do not know the proof.) The scattering operators $\mathbf{S}^w$ which we computed in (2.26) differ from the normalized intertwining operators S^w only by a scalar multiplier $\frac{b(-w\lambda)}{b(-\lambda)}$; in general, however, this multiplier violates causality.

We are now able to give the technical formulation of the Huygens principle.

Definition 2.14. We shall say that system (2.1) has the Huygens property if all solution with compact support are eventually outgoing. In other words, for any $f \in \widetilde{\mathcal{H}}$ there exists $t \in \mathfrak{a}$ such that $U_t f \in \mathcal{D}_+$.

Theorem 2.15. *The following assertions are equivalent:*

(i) *Operators $\mathbf{S}^w$ are causal.*
(ii) *System (2.1) has the Huygens property.*
(iii) *All roots of $(\mathfrak{g},\mathfrak{a})$ have even multiplicities.*

Proof. It is well known that if all roots have even multiplicities, the root system of $(\mathfrak{g},\mathfrak{a})$ is reduced. Together with the Gindikin–Karpelevich formula (2.13) this implies that

(a) All roots of $(\mathfrak{g},\mathfrak{a})$ have even multiplicities if and only if the function $b(\lambda)^{-1} = \pi(\lambda)^{-1}c(\lambda)^{-1}$ is polynomial.
(b) Function $b(\lambda)^{-1}$ is regular in $\mathfrak{a}^* - i\mathfrak{a}_+^*$ if and only if it is polynomial.

To prove the implication (ii) $\Rightarrow$ (iii), note that for any $f \in \widetilde{\mathcal{H}}$ the function $b(\lambda)^{-1}F_+f(\lambda)$ is entire and lies in $\mathcal{K}_R$ for some $R > 0$. Hence the inequalities

$$\left|e^{-i\lambda(t)}b(\lambda)^{-1}F_+f(\lambda)\right| \le C_N(1+|\lambda|)^{-N}e^{-\eta(t)e^{R|\eta|}}, \tag{2.29}$$

which are valid for all $N \ge 0$, $\eta = \operatorname{Re} i\lambda \in \mathfrak{a}_+^*$; this shows that

$$e^{-i\lambda(t)}b(\lambda)^{-1}F_+f(\lambda) \in H_2^-(\mathfrak{a}^*;\mathfrak{H})$$

for sufficiently big $t \in \mathfrak{a}_+$. Conversely, suppose that the Huygens principle holds for (2.1); this means the validity of the implication

$$a \in \mathcal{K}_R \to e^{-i\lambda(t)}b(\lambda)^{-1}F_+f(\lambda) \in H_2^-(\mathfrak{a}^*;\mathfrak{H}) \quad \text{for some} \quad t \in \mathfrak{a}_+. \tag{2.30}$$

Without restricting the generality we may assume that a is constant on K/M; for such functions the condition $a \in \mathcal{K}_R$ means that a is an arbitrary entire function of exponential type which is rapidly decreasing on planes $\operatorname{Im}\lambda = \mathit{const}$. Hence (2.30) is valid if and only if $b(\lambda)^{-1}$ is holomorphic in $\mathfrak{a}^* - i\mathfrak{a}_+^*$.

In the next section, we shall establish another criterion: the Huygens principle is valid for system (2.1) if and only if it is valid for its analog in the Riemannian symmetric space of zero curvature.

3. Symmetric Spaces of Zero Curvature

For each symmetric space $X = G/K$ of non-positive curvature there exists a companion space of zero curvature. Let $\mathfrak{g} = \mathfrak{k} \bigoplus \mathfrak{p}$ be the Cartan decomposition associated with the Cartan involution θ. The subspace $\mathfrak{p} \subset \mathfrak{g}$ is invariant with respect to the representation $\mathrm{Ad}_G K$. We may regard $\mathfrak{p}$ as an additive abelian group; put $G_0 = K \ltimes \mathfrak{p}$. G_0 acts on $\mathfrak{p}$ by affine transformations and $\mathfrak{p} \simeq G_0/K$; it is called the Cartan motion group, and $\mathfrak{p}$ is the zero curvature symmetric space associated with (G, K).

Let $P^K(\mathfrak{p}^*)$ be the ring of K-invariant polynomials; we may regard these polynomials as symbols of K-invariant differential operators on $\mathfrak{p}$ with constant coefficients (or, equivalently, of G_0-invariant differential operators on $\mathfrak{p}$). The Killing form B allows to identify $\mathfrak{p}$ with its dual space. According to a classical Chevalley theorem, the restriction of polynomials on $\mathfrak{p}$ to the subspace $\mathfrak{a} \subset \mathfrak{p}$ induces a canonical isomorphism $P^K(\mathfrak{p}^*) \simeq P^W(\mathfrak{a}^*)$. Thus we get an analogue of the Harish-Chandra isomorphism Δ^0: $P^W(\mathfrak{a}^*) \to D^{G_0}(\mathfrak{p})$ defined by the ordinary Fourier transform on $\mathfrak{p}$; this isomorphism allows to define on $\mathfrak{a} \times \mathfrak{p}$ the multitemporal wave equation

$$'\sigma\left(-i\frac{\partial}{\partial t}\right)u(t,x) = \Delta^0_\sigma u(t,x), \quad \sigma \in P^W(\mathfrak{a}^*). \tag{3.1}$$

The construction of the initial value problem and of the energy form for this equation is completely parallel to the one given in Sec. 2. We keep the notation for the space of initial data introduced earlier.

Theorem 3.1. *The Huygens principle holds for the multitemporal wave equation in $X = G/K$ and in the associated space of zero curvature simultaneously.*

The Fourier theory for (3.1) is a straightforward generalization of the standard Fourier transform. For $\lambda \in \mathfrak{a}^*$, $k \in K$, $w \in W$ define the plane wave $\Phi_w(\cdot; \lambda, k)$ by

$$\Phi_w(t, x; \lambda, k) = \frac{1}{\pi(\lambda)} e^{-i\lambda(t)} e^{iB(x, k\cdot\lambda^w)}, \quad t \in \mathfrak{a}, x \in \mathfrak{p}. \tag{3.2}$$

Obviously, $\Phi_w(\cdot; \lambda, k)$ is a solution of (3.1). Define the Fourier transform of the initial data by

$$F_w f(\lambda, k) = \langle f, i_0(\Phi_w(\cdot; \lambda, k)\rangle_E. \tag{3.3}$$

Theorem 3.2. *For any $w \in W$ we have*

$$\|f\|_E^2 = \int_{\mathfrak{a}^* \times K/M} |F_w f(\lambda, k)|^2 \, d\mu(\lambda) dk, \quad \textit{where } d\mu(\lambda) = \prod_{\alpha \in \Sigma_+} |\langle \lambda, \alpha \rangle|^{m_\alpha} \, d\lambda. \tag{3.4}$$

F_w extends to an isometry of the space H of initial data onto $L_2(\mathfrak{a}^* \times K/M, d\mu\, dk)$.

This theorem is closely related to the Plancherel theorem for the group $G_0 = K \ltimes \mathfrak{p}$ [25].[b] The measure in (3.4) is of course the inverse image of the Euclidean measure on $\mathfrak{p}$ under the natural mapping

$$\beta\colon \mathfrak{a}_+ \times K/M \to \mathfrak{p}\colon\ (\lambda, k) \mapsto \mathrm{Ad}_G\, k \cdot \lambda. \tag{3.5}$$

We shall write F_+ instead of F_e; it is easy to see that the dependence of F_w on the Weyl group element is given by the simple formula $F_w = S^w \cdot F_+$, where S^w is the reflection operator in $L_2(K/M)$,

$$S^w \phi(k) = \phi(k^w), \tag{3.6}$$

induced by the natural right action of W on the K/M, $(kM, w) \mapsto km_w M$, where $m_w \in M'$ is a representative of $w \in M'/M = W$; clearly, this action commutes with the left action of K on K/M.

Each K-orbit in $\mathfrak{p}$ intersects the closure of the positive Weyl chamber $\mathfrak{a}_+$ in exactly one point. If $x = k \cdot t$ for some $k \in K$, $t \in \bar{\mathfrak{a}}_+$, we shall write $t = t(x)$. Put

$$B_R = \{x \in \mathfrak{p};\ |t(x)| \leq R\}. \tag{3.7}$$

Theorem 3.3. *For any solution u of the system* (3.1) *such that* $\mathrm{supp}\, i_0(u) \subset B_R$ *we have* $\mathrm{supp}\, i_t(u) \subset B_{R+|t|}$ *for any $t \in \mathfrak{a}$.*

This assertion follows from the classical Paley–Wiener theorem applied to the Fourier transform on $\mathfrak{p}$; the subtle point is that the Fourier transform of the initial data with compact support is a function defined not on $\mathfrak{p}^*_{\mathbb{C}}$, the complex hull of $\mathfrak{p}^*$, but rather on $\mathfrak{a}^*_{\mathbb{C}} \times K/M$.

Proposition 3.4. *Let ϕ be a W-invariant entire function on $\mathfrak{a}^*_{\mathbb{C}}$ such that*

$$|\phi(\lambda)| \leq C(1 + \|\lambda\|)^N e^{R\|\mathrm{Im}\,\lambda\|}.$$

There exists an entire $K_{\mathbb{C}}$-invariant function Φ on $\mathfrak{p}^*_{\mathbb{C}}$ such that $\phi = \beta^*\Phi$ and

$$|\Phi(\Lambda)| \leq C_1(1 + \|\Lambda\|)^N e^{R\|\mathrm{Im}\,\Lambda\|}. \tag{3.8}$$

The key point in the proof of this assertion is the following inequality:

Lemma 3.5. *For any $\lambda \in \mathfrak{a}^*_{\mathbb{C}}$ and $k \in K_{\mathbb{C}}$ we have*

$$\|\mathrm{Im}\,\lambda\| \leq \|\mathrm{Im}\, k \cdot \lambda\|. \tag{3.9}$$

---

[b] Let us note, however, the important difference between the Plancherel theorem for $L_2(\mathfrak{p})$ discussed in [25] and the harmonic analysis in the space of initial data. The principal series representations π_λ, π_μ for G_0 are again equivalent if and only if $\lambda, \mu \in \mathfrak{a}^*$ are conjugate with respect to the Weyl group. The natural intertwining operators coincide in this case with reflection operators S^w introduced in (3.6) Representation π_λ which decompose $L_2(\mathfrak{p})$ enter with multiplicity 1 (hence the symmetry of the Fourier amplitudes with respect to the action of the Weyl group); for the energy space $\mathcal{H}$ the multiplicity is equal to the order of W. Hence the integration in the Plancherel formula is over $\mathfrak{a}^*$ and not over $\mathfrak{a}^*_+$, like in the L_2 case.

The proof of Lemma 3.5 is based on the local study of the function $k \mapsto \|\mathrm{Im}\, k{\cdot}\lambda\|^2$ on the complex sphere $K_{\mathbb{C}}/M_{\mathbb{C}}$; one can show that the real slice $K/M \subset K_{\mathbb{C}}/M_{\mathbb{C}}$ is its critical manifold where it attains its minimum. Replacing λ with $i\lambda$, we get $\|\mathrm{Re}\,\lambda\| \le \|\mathrm{Re}\, k{\cdot}\lambda\|$ and hence also $\|\lambda\| \le \|k \cdot \lambda\|$ for any $k \in K_{\mathbb{C}}$. If ϕ is a polynomial, the assertion of Proposition 3.4 follows directly from the Chevalley isomorphism $P(\mathfrak{a}_{\mathbb{C}})^W \simeq P(\mathfrak{p}_{\mathbb{C}})^{K_{\mathbb{C}}}$. In general case, there exists a sequence of polynomials $\phi_n \in P(\mathfrak{a}_{\mathbb{C}})^W$ which converges to ϕ uniformly on each ball $B_m = \{\lambda;\ \|\lambda\| \le m\}$. Put $\Phi_n = (\beta^*)^{-1}\phi_n$, $\Phi_n \in P(\mathfrak{p}_{\mathbb{C}})^{K_{\mathbb{C}}}$. The inequality (3.9) implies that the sequence Φ_n converges uniformly in each ball $\{\Lambda \in \mathfrak{p}_{\mathbb{C}};\ \|\Lambda\| \le m\}$ and hence its limit is an entire function on $\mathfrak{p}_{\mathbb{C}}$; (3.9) implies then that the limit function $\Phi = (\beta^*)^{-1}\phi$ satisfies (3.8) on an open dense subset $K_{\mathbb{C}} \cdot \mathfrak{a}_{\mathbb{C}} \subset \mathfrak{p}_{\mathbb{C}}$ and hence everywhere on $\mathfrak{p}_{\mathbb{C}}$.

Functions satisfying the estimate (3.8) are Fourier transforms of distributions with compact support which have the singularity degree $\le N$. By duality, W-invariant entire function on $\mathfrak{a}^*$ which satisfy the estimates

$$\sup_{\lambda \in \mathfrak{a}^*} (1 + \|\mathrm{Re}\,\lambda\|)^N e^{-R\|\lambda\|} |\phi(\lambda)| < \infty \quad \text{for all} \quad N \ge 0 \tag{3.10}$$

extend to entire $K_{\mathbb{C}}$-invariant functions on $\mathfrak{p}_{\mathbb{C}}^*$ which are Fourier transforms of smooth functions on $\mathfrak{p}$ with compact support.

Let us now describe wave operators for the system (3.1). We must explore the asymptotics of the wave packets

$$u(t,x) = \int_{\mathfrak{a}^* \times K/M} e^{-i\lambda(t)} e^{iB(x,k\cdot\lambda)} a(\lambda,k)\, d\mu(\lambda)\, dk. \tag{3.11}$$

In contrast to the case of symmetric spaces of non-positive curvature described in Sec. 2 in order to compute this asymptotics we may use the stationary phase approximation.

Lemma 3.6. *Suppose that $x, H \in \mathfrak{a}$ are regular; then the function*

$$\phi_{x,H}\colon k \mapsto B(x, k \cdot H)$$

is a Morse function on K/M.

Proof. It is convenient to regard $\phi_{x,H}$ as a function on the K-orbit of $H \in \mathfrak{a}'$. Its differential is given by

$$\langle d\phi_{x,H}(k), Y\rangle = B([Y, k\cdot H], x) \quad \text{for all} \quad Y \in \mathfrak{k}. \tag{3.12}$$

Formula (3.12) immediately implies that the critical points of $\phi_{x,H}$ coincide with the intersection points of this orbit with the subspace $\mathfrak{a} \subset \mathfrak{p}$. Since by assumption $x, H \in \mathfrak{a}$ are regular, this intersection is transversal and all intersection points are conjugate with respect to the action of W; we shall parameterize them by the elements $w \in W$. The Hessian of $\phi_{x,H}$ in the point H^w is given by

$$\mathcal{H}_w(Y,Z) = -B([x,Y],\, [H^w, Z]), \quad Y, Z \in \mathfrak{k}. \tag{3.13}$$

Instead of bilinear form $\mathcal{H}_w$ on $\mathfrak{k} \times \mathfrak{k}$ it is convenient to consider its Hermitian complexification (clearly, the inertia indices of both forms coincide). Let $\tau\colon \mathfrak{g}_{\mathbb{C}} \to \mathfrak{g}_{\mathbb{C}}$ be the complex conjugation with respect to the real form $\mathfrak{u} = \mathfrak{k} \dot{+} i\mathfrak{p} \subset \mathfrak{g}_{\mathbb{C}}$. Put

$$\widetilde{\mathcal{H}}_w(Y,Z) = -B([x,Y],\,[H^w,\tau Z]), \quad Y,Z \in \mathfrak{k}_{\mathbb{C}}.$$

Choose a Cartan subalgebra $\mathfrak{h}_{\mathbb{C}} \supset \mathfrak{a}_{\mathbb{C}}$ in $\mathfrak{g}_{\mathbb{C}}$ and let $e_\alpha \in \mathfrak{g}_{\mathbb{C}}$ be the root space vectors which correspond to the roots α of $(\mathfrak{g}_{\mathbb{C}}, \mathfrak{h}_{\mathbb{C}})$ that are not identically zero on $\mathfrak{a}_{\mathbb{C}}$. Introduce an order the set Δ of such roots which is compatible with the order in the set Σ of restricted roots. We can choose e_α in such a way that $\tau \cdot e_\alpha = -e_{-\alpha}$ for all $\alpha \in \Delta_+$. The vectors $(1+\theta)e_\alpha, \alpha \in \Delta$, form a basis in $\mathfrak{k}_{\mathbb{C}} \ominus \mathfrak{m}_{\mathbb{C}}$, where the orthogonal complement is taken with respect to the positive definite Hermitian form $B_\tau := -B(\cdot, \tau\cdot)$ on $\mathfrak{g}_{\mathbb{C}}$. Using the well-known properties of root space vectors e_α and involutions τ and θ (see, for instance, [19]) it is easy to get an explicit expression for $\widetilde{\mathcal{H}}_w$. Put

$$Y = \sum_{\alpha\in\Delta_+} y_\alpha(1+\theta)e_\alpha, \quad Z = \sum_{\alpha\in\Delta_+} z_\alpha(1+\theta)e_\alpha.$$

Then

$$\widetilde{\mathcal{H}}_w = -\sum_{\alpha\in\Delta_+} \alpha(x)\alpha(H^w) y_\alpha \bar{z}_\alpha B_\tau((1-\theta)e_\alpha, (1-\theta)e_\alpha).$$

Hence in any basis of $\mathfrak{k}_{\mathbb{C}} \ominus \mathfrak{m}_{\mathbb{C}}$ which is orthonormal with respect to B_τ we have

$$|\det \widetilde{\mathcal{H}}_w| = \prod_{\alpha\in\Delta_+} |\alpha(x)|^{m_\alpha} |\alpha(H)|^{m_\alpha}.$$

The same expression is valid for the determinant of the real bilinear form $\mathcal{H}_w$ in any basis of $\mathfrak{k} \ominus \mathfrak{m}$ which is orthonormal with respect to $-B$. The positive inertia index of $\mathcal{H}_w$ is equal to

$$\mathrm{ind}_+ \mathcal{H}_w = \mathrm{card}\{\alpha \in \Delta_+\,;\, \alpha(H^w) < 0\}\,.$$

Let u be a solution of the system (3.1). We define the wave operator W_w by

$$W_w u(\tau,k) = \lim_{t\to\infty,\, t\in\mathfrak{a}_+} \prod_{\alpha\in\Delta_+} \alpha(t)^{m_\alpha/2} \cdot u(w^{-1}\cdot t + \tau, k\cdot t). \tag{3.14}$$

In order to prove the existence of this limit we shall compute it for wave packets (3.11) with the amplitude a satisfying the following conditions:

(a) a is a smooth function such that $\operatorname{supp} a \subset \mathfrak{a}^* \times K/M$ is compact;
(b) a is zero in some vicinity of $\mathfrak{a}^* \backslash (\mathfrak{a}^*)' \times K/M$.

Solutions with such amplitudes form an open dense set in the space of all solutions with finite energy.

Proposition 3.7. *Suppose that u is a solution of the form* (3.11) *with the amplitude satisfying the conditions* (a), (b) *above. Then the limit in* (3.14) *exists and is equal to*

$$W_w u(\tau, k) = (2\pi)^{n/2} e^{-i\pi n/4} \int_{\mathfrak{a}^*} \prod_{\alpha\in\Delta_+} |\alpha(\lambda)|^{m_\alpha/2} \epsilon_w(\lambda) a(\lambda, k^w) e^{-i\lambda(\tau)}\, d\lambda, \quad (3.15)$$

where $n = \dim K/M$ *and*

$$\epsilon_w(\lambda) = \prod_{\alpha\in\Delta_+,\, \alpha(\lambda^w)<0} e^{i\pi m_\alpha/2}. \quad (3.16)$$

Proof. We apply to the integral

$$I_{\lambda,x} = \int_{K/M} a(\lambda, k) e^{iB(x, k\cdot\lambda)} dk$$

with fixed $\lambda, x \in \mathfrak{a}$ the stationary phase method. All necessary calculations are contained in Lemma 3.6. We get:

$$I_\lambda(x) \sim (2\pi)^{n/2} \frac{e^{-\pi i n/4}}{\prod_{\alpha\in\Delta+} |\alpha(\lambda)\alpha(x)|^{m_\alpha/2}} \sum_{s\in W} \epsilon_s(\lambda) e^{i\lambda^s(x)} a(\lambda, k^s). \quad (3.17)$$

The correction term in (3.17) has the order

$$O\left(\frac{1}{\prod \alpha(x)^{3m_\alpha/2}}\right);$$

conditions (a), (b) guarantee that it is bounded uniformly in λ on the support of the amplitude a, which justifies the integration of the asymptotics over $\lambda \in \mathfrak{a}^*$. Substitute this asymptotics into the definition of the wave operator (3.14); the power of t in (3.14) is chosen to compensate exactly the decay of the solution along a light ray. The Riemann–Lebesgue lemma immediately implies that the only term in the sum over $s \in W$ which survives in the limit corresponds to $s = w$. A comparison with Theorem 3.2 implies that up a numerical factor we have

$$\|u\|_E^2 = \|W_w u\|^2_{L_2(\mathfrak{a}\times K/M)}. \quad (3.18)$$

Hence all wave operators (which are initially defined for wave packets with amplitudes satisfying conditions (a), (b)) admit an extension to isometric operators. Now finally we can lift these technical conditions and deal again with analytic amplitudes satisfying estimates (3.10). We put $\mathbf{W}_w = \Phi \circ W_w$.

Let us introduce two branches of the function $s \mapsto \sqrt{s}$ by the conditions

$$(s)_+^{1/2} = \begin{cases} \sqrt{s}, & s > 0, \\ i\sqrt{|s|}, & s < 0, \end{cases} \quad (s)_-^{1/2} = \begin{cases} \sqrt{s}, & s > 0, \\ -i\sqrt{|s|}, & s < 0. \end{cases} \quad (s)_+^{1/2} \cdot (s)_-^{1/2} = |s|.$$

Note that $(s)_\pm^{1/2}$ are boundary values of functions which are analytic in upper (respectively, lower) half-plane. Moreover,

$$\prod_{\alpha\in\Delta_+} |\alpha(\lambda)|^{m_\alpha/2} \epsilon_w(\lambda) = \prod_{\alpha\in\Delta_+} \left((\alpha(\lambda^w))_+^{1/2}\right)^{m_\alpha}.$$

Put

$$\tilde{a}(\lambda, k) = \prod_{\alpha \in \Delta_+} \left((\alpha(\lambda^w))_+^{1/2} \right)^{m_\alpha} \cdot a(\lambda, k).$$

Formulae (3.11), (3.15) take the form

$$u(t,x) = \int_{\mathfrak{a}^* \times K/M} e^{-i\lambda(t)} e^{iB(x, k\cdot\lambda)} \prod_{\alpha \in \Delta_+} \left((\alpha(\lambda))_-^{1/2} \right)^{m_\alpha} \tilde{a}(\lambda, k)\, d\lambda\, dk \tag{3.19}$$

and

$$W_+ u(\tau, k) = (2\pi)^{n/2} e^{-i\pi n/4} \int_{\mathfrak{a}^*} e^{-i\lambda(\tau)} \tilde{a}(\lambda, k)\, d\lambda. \tag{3.20}$$

A comparison of the asymptotics in different Weyl chambers given by (3.20), (3.15) immediately implies that the scattering operators $\mathbf{S}^w = \mathbf{W}_+ \mathbf{W}_w^{-1}$ are given by

$$\mathbf{S}^w = i^{l(w)} \prod_{\alpha \in \Delta_+ \cap w^{-1} \Delta_-} (\operatorname{sign} \alpha(\lambda))^{m_\alpha} (S^w)^{-1}, \tag{3.21}$$

where $l(w)$ is the length of w and S^w is the reflection operator acting in $\mathfrak{H} = L_2(K/M)$ which was defined in (3.6). The sign factors come from the mismatch of two branches of the square root. In particular, if all roots have even multiplicities, the sign factors disappear and the scattering operator is reduced to $i^{l(w)}(S^w)^{-1}$. When some of the multiplicities m_α are odd, the scattering operator contains the sign flips (or the Hilbert transforms) with respect to the variables $\lambda_\alpha := \alpha(\lambda)$. This situation is familiar for the ordinary wave equation in even-dimensional space and is in contrast with the case of spaces of non-positive curvature, where the scattering operators is meromorphic (albeit with poles in the wrong half-planes).

We define the light cone in complete analogy with (2.27),

$$C_+ = \left\{ (t,x) \in \mathfrak{a}_+ \times \mathfrak{p}; x \in B_{|t|} \right\}.$$

The outgoing subspace $\mathcal{D}_+ \subset \mathcal{H}$ is defined as the subspace of initial data of solution which vanish in C_+. In complete analogy with Theorem 2.12 we have

Theorem 3.8. *The wave operator* $\mathbf{W}_+$ *maps the space* $\mathcal{D}_+$ *of outgoing initial data onto* $H_2^-(\mathfrak{a}^*; \mathfrak{H})$.

The Huygens property for system (3.1) is defined in an exactly similar way: it holds if all solutions with compact support are eventually outgoing. Again in complete analogy with Theorem 2.15 we have

Theorem 3.9. (i) *If all roots of* $(\mathfrak{g}, \mathfrak{a})$ *have even multiplicities, system* (3.1) *has the Huygens property.*

(ii) *Conversely, if the Huygens principle holds true, all roots have even multiplicities.*

A comparison of the two criteria now yields our main result, Theorem 3.1 announced earlier.

Though the algebraic criteria given by Theorems 2.15 and 3.9 look very simple, the geometric understanding of the situation is still lacking. Recall that I. G. Petrovsky had linked the validity of the Huygens principle for the ordinary wave equation with the vanishing of a certain homology class associated with the wave operator, the *Petrovsky cycle* (see, for instance, [26]). I believe that the real key to the Huygens principle for the multitemporal (and may be also to the analytic properties of the scattering matrices, or intertwining operators in wave equation may be hidden in the geometry of the complex manifold $K_{\mathbb{C}}/M_{\mathbb{C}}$. After many years, this question still remains unresolved.

Let us finally return to our starting point, the automorphic wave equation on the upper half-plane. Here the only root of $\mathfrak{sl}(2)$ has multiplicity 1 and hence the Huygens principle does not hold in the whole upper half-plane. In a sense, the situation in the automorphic case looks much better. Recall that the main troubles which arise in the Lax–Phillips theory in the absence of Huygens principle are due to the fact that in this case the incoming and the outgoing subspaces are not orthogonal to each other [9]. This does not allow to define the contraction semigroup associated with the scattering. As observed by Faddeev and Pavlov in [1], in the automorphic case there exist natural incoming and outgoing subspaces which are orthogonal; hence the analytic properties of the scattering operator are much better than in the case of the whole upper half-plane. This curious fact brings us back to the exciting program launched in [1] which linked Scattering theory with the Riemann zeta. Once again, the story remains unfinished, and the fascinating program started by Faddeev and Pavlov should not be abandoned.

Acknowledgment

This work was partly supported by the RFBR grant 18-01-00271.

References

[1] B. S. Pavlov and L. D. Faddeev. Scattering theory and automorphic functions, *J. Soviet Math.* **3**(4) (1975) 522–548; Translated from *Zapiski Nauchnykh Seminarov Leningradskogo Otdeleniya Matematicheskogo Instituta im. V. A. Steklova Akademii Nauk SSSR* **27** (1972) 161–193.

[2] I. M. Gelfand, M. I. Graev and I. I. Pyatetskii-Shapiro, *Generalized Functions*, Vol. 6, Representation theory and automorphic functions (AMS Chelsea Publishing, Providence, RI, 2016); Reprint of the 1969 English translation.

[3] L. D. Faddeev, Expansion in eigenfunctions of the Laplace operator on the fundamental domain of a discrete group on the Lobacevskii plane, in *Fifty Years in Mathematical Physics* (World Scientific, 2016), pp. 45–74.

[4] A. B. Venkov, V. L. Kalinin and L. D. Faddeev, A nonarithmetic derivation of the Selberg trace formula, in *Differential Geometry, Lie Groups and Mechanics, Zap. Nauchn. Sem. LOMI*, Vol. 37 (Nauka, Leningrad, 1973), pp. 5–42; *J. Soviet Math.* **8** (1977) 171–199.

[5] P. D. Lax and R. S. Phillips, *Scattering Theory* (Academic Press, New York, 1967).

[6] M. A. Semenov-Tian-Shansky, Harmonic analysis on Riemannian symmetric spaces of nonpositive curvature and scattering theory, *Sov. Math. Izv.* **40** (1976) 562–592.

[7] P. D. Lax and R. S. Phillips, *Scattering Theory for Automorphic Functions*, Annals of Mathematics Studies, No. 87 (Princeton Univ. Press, Princeton, N.J., 1976).

[8] I. M. Gelfand, Automorphic functions and the theory of representations, in *Proc. Internat. Congr. Mathematicians* (*Stockholm, 1962*) (Inst. Mittag-Leffler, Djursholm, 1963), pp. 74–85.

[9] P. D. Lax and R. S. Phillips, Scattering theory for the acoustic equation in an even number of space dimensions, *Indiana Univ. Math J.* **22** (1972) 101–134.

[10] S. G. Gindikin and F. I. Karpelevich, An integral associated with Riemannian symmetric spaces of non-positive curvature, *Izv. Akad. Nauk SSSR Ser. Mat.* **30**(5) (1966) 1147–1156.

[11] Harish-Chandra, Spherical functions on a semisimple Lie group, I, II, *Amer. J. Math.* **80**(2) (1958) 241–310; **80**(3) (1958) 553–613.

[12] E. K. Sklyanin, Separation of variables — New trends, Quantum field theory, integrable models and beyond (Kyoto, 1994), *Progr. Theor. Phys. Suppl.* **118** (1995) 35–60.

[13] A. Gerasimov, S. Kharchev and D. Lebedev, Representation theory and quantum inverse scattering method: The open Toda chain and the hyperbolic Sutherland model, *Int. Math. Res. Not.* **2004**(17) (2004) 823–854.

[14] L. D. Faddeev, Quantum completely integrable models in field theory, in *Math. Physics Reviews, Sec. C*, Vol. 1 (Harwood, 1980), pp. 107–155; L. D. Faddeev, in *40 Years in Mathematical Physics*, World Scientific Series in 20th Century Mathematics, Vol. 2 (World Scientific, 1995), pp. 187–235.

[15] B. Kostant, Quantization and representation theory, in *Proc. of Symposium on Representations of Lie Groups* (*Oxford, 1977*), London Math. Soc. Lect. Notes Series, Vol. 34 (Cambridge Univ. Press, 1979), pp. 287–316.

[16] S. Helgason, Huygens' principle for wave equations on symmetric spaces, *J. Funct. Anal.* **107**(2) (1992) 279–288.

[17] S. Helgason, Integral geometry and multitemporal wave equations. Dedicated to the memory of Fritz John, *Comm. Pure Appl. Math.* **51**(9–10) (1998) 1035–1071.

[18] B. Kostant, On convexity, the Weyl group and the Iwasawa decomposition, *Ann. Sci. École Norm. Sup.* (*4*) **6** (1973) 413–455.

[19] S. Helgason, *Geometric Analysis on Symmetric Spaces*, Math. Surveys Monographs, Vol. 39 (Amer. Math. Soc., Providence, RI, 1994).

[20] J. J. Duistermaat and G. J. Heckman, On the variation in the cohomology of the symplectic form of the reduced phase space, *Invent. Math.* **69** (1982) 259–268.

[21] M. F. Atiyah and R. Bott, The moment map and equivariant cohomology, *Topology* **23**(1) (1984) 1–28.

[22] M. A. Semenov-Tian-Shansky, A certain property of the Kirillov integral, in Differential geometry, Lie groups and mechanics, *Zap. Nauchn. Sem. LOMI* **37** (1973) 53–65; *J. Sov. Math.* **8**(2) (1977) 208–218.

[23] G. Olafsson and H. Schlichtkrull, Wave propagation on Riemannian symmetric spaces, *J. Funct. Anal.* **107**(2) (1992) 270–278.

[24] T. Branson, G. Olafsson and H. Schlichtkrull, Huygens' principle in Riemannian symmetric spaces, *Math. Ann.* **301**(3) (1995) 445–462.

[25] S. G. Gindikin, Unitary representations of groups of automorphisms of Riemann symmetric spaces of null curvature, *Funct. Anal. Appl.* **1**(1) (1967) 28–32.

[26] M. F. Atiyah, R. Bott and L. Gårding, Lacunas for hyperbolic differential operators with constant coefficients I, *Acta Math.* **124**(1) (1970) 109–189.

Spectral Approach to Homogenization of Elliptic Operators in a Perforated Space

T. A. Suslina

St. Petersburg State University, Universitetskaya Nab. 7/9, St. Petersburg, 199034, Russia

t.suslina@spbu.ru

Let $\Gamma \subset \mathbb{R}^d$ be a lattice. For $\varepsilon > 0$, we consider the perforated space $\Pi_\varepsilon \subset \mathbb{R}^d$ which is an $(\varepsilon\Gamma)$-periodic open connected set with Lipschitz boundary. In $L_2(\Pi_\varepsilon; \mathbb{C}^n)$, we consider a self-adjoint strongly elliptic second-order differential operator $\mathcal{A}_\varepsilon$ with periodic coefficients depending on $\mathbf{x}/\varepsilon$. We study the behavior of the resolvent $(\mathcal{A}_\varepsilon + I)^{-1}$ for small ε. Approximations for this resolvent in the $(L_2 \to L_2)$ and $(L_2 \to H^1)$-operator norms with sharp order error estimates are found. The results are obtained by the operator-theoretic (spectral) approach. General results are applied to particular periodic operators of mathematical physics: the acoustics operator, the elasticity operator, and the Schrödinger operator with a singular potential.

To the memory of Ludvig Dmitrievich Faddeev

Keywords: Periodic differential operators; perforated space; homogenization; effective operator; operator error estimates.

Mathematics Subject Classification: 35B27

0. Introduction

The paper is concerned with homogenization for periodic differential operators (DO's). A broad literature is devoted to homogenization problems; first, we mention the books [2, 3, 17, 25]. For homogenization problems in $\mathbb{R}^d$, one of the methods is the spectral approach based on the Floquet–Bloch theory; see, e.g., [3, Chap. 4], [25, Chap. 2], and [12, 23].

Our main goal is to develop the spectral approach to homogenization of elliptic operators in a perforated space.

0.1. *The class of operators*

Let $\Gamma \subset \mathbb{R}^d$ be a lattice, and let $\Omega \subset \mathbb{R}^d$ be the cell of this lattice. We consider the perforated space $\Pi \subset \mathbb{R}^d$ which is a Γ-periodic open connected set with Lipschitz boundary. The set $\Omega^0 = \Pi \cap \Omega$ is the cell of the perforated space.

In $L_2(\Pi; \mathbb{C}^n)$, we consider a self-adjoint elliptic second-order DO given in a factorized form

$$\mathcal{A} = b(\mathbf{D})^* g(\mathbf{x}) b(\mathbf{D}) \tag{0.1}$$

with the Neumann (natural) boundary condition. Here $b(\mathbf{D}) = \sum_{l=1}^{d} b_l D_l$ is the $(m \times n)$-matrix first-order DO. We assume that $m \geq n$ and that the symbol $b(\boldsymbol{\xi})$ has maximal rank for any $0 \neq \boldsymbol{\xi} \in \mathbb{C}^d$. A matrix-valued function $g(\mathbf{x})$ of size $m \times m$ is assumed to be Γ-periodic and such that $g(\mathbf{x}) > 0$; $g, g^{-1} \in L_\infty$. The precise definition of the operator $\mathcal{A}$ is given in terms of the quadratic form; see Sec. 3.

We also study a more general operator of the form

$$\widetilde{\mathcal{A}} = f(\mathbf{x})^* \mathcal{A} f(\mathbf{x}) = f(\mathbf{x})^* b(\mathbf{D})^* g(\mathbf{x}) b(\mathbf{D}) f(\mathbf{x}). \tag{0.2}$$

Here a matrix-valued function $f(\mathbf{x})$ of size $n \times n$ is assumed to be Γ-periodic and such that $f, f^{-1} \in L_\infty$.

Many operators of mathematical physics can be represented in the form (0.1) or (0.2); see [5]. The simplest example is the acoustics operator $\mathcal{A} = -\operatorname{div} g(\mathbf{x}) \nabla = \mathbf{D}^* g(\mathbf{x}) \mathbf{D}$. The operator of elasticity theory also can be written in the required form.

Now we introduce the small parameter $\varepsilon > 0$ and consider the $(\varepsilon\Gamma)$-periodic perforated space $\Pi_\varepsilon = \varepsilon\Pi$. For any Γ-periodic function $\varphi(\mathbf{x})$ in Π, we denote $\varphi^\varepsilon(\mathbf{x}) := \varphi(\varepsilon^{-1}\mathbf{x})$, $\mathbf{x} \in \Pi_\varepsilon$. Our *main objects* are the operators $\mathcal{A}_\varepsilon$ and $\widetilde{\mathcal{A}}_\varepsilon$ acting in $L_2(\Pi_\varepsilon; \mathbb{C}^n)$ and given by

$$\mathcal{A}_\varepsilon = b(\mathbf{D})^* g^\varepsilon(\mathbf{x}) b(\mathbf{D}), \tag{0.3}$$

$$\widetilde{\mathcal{A}}_\varepsilon = (f^\varepsilon(\mathbf{x}))^* b(\mathbf{D})^* g^\varepsilon(\mathbf{x}) b(\mathbf{D}) f^\varepsilon(\mathbf{x}) \tag{0.4}$$

with the natural conditions on $\partial\Pi_\varepsilon$.

0.2. *Operator error estimates for elliptic problems in* $\mathbb{R}^d$

In a series of papers [5–8] by Birman and Suslina, an operator-theoretic approach to homogenization of elliptic equations in $\mathbb{R}^d$ was suggested and developed. This approach was based on the scaling transformation, the Floquet–Bloch theory, and the analytic perturbation theory.

Let A_ε be the operator in $L_2(\mathbb{R}^d; \mathbb{C}^n)$ given by the expression $b(\mathbf{D})^* g^\varepsilon(\mathbf{x}) b(\mathbf{D})$. In [5], it was proved that

$$\|(A_\varepsilon + I)^{-1} - (A^0 + I)^{-1}\|_{L_2(\mathbb{R}^d) \to L_2(\mathbb{R}^d)} \leq C\varepsilon. \tag{0.5}$$

Here $A^0 = b(\mathbf{D})^* g^0 b(\mathbf{D})$ is the *effective operator* with the constant *effective matrix* g^0. In [6, 7], a more accurate approximation for $(A_\varepsilon + I)^{-1}$ in the $(L_2 \to L_2)$-norm was found:

$$\|(A_\varepsilon + I)^{-1} - (A^0 + I)^{-1} - \varepsilon K_\varepsilon\|_{L_2(\mathbb{R}^d)\to L_2(\mathbb{R}^d)} \leq C\varepsilon^2. \tag{0.6}$$

Here K_ε is the so-called *corrector*; it contains rapidly oscillating factors and so depends on ε. We have $\|K_\varepsilon\|_{L_2\to L_2} = O(1)$. In [8], approximation of the resolvent in the "energy norm" (i.e., in the norm of operators acting from $L_2(\mathbb{R}^d;\mathbb{C}^n)$ to the Sobolev space $H^1(\mathbb{R}^d;\mathbb{C}^n)$) was obtained:

$$\|(A_\varepsilon + I)^{-1} - (A^0 + I)^{-1} - \varepsilon K_\varepsilon^{(1)}\|_{L_2(\mathbb{R}^d)\to H^1(\mathbb{R}^d)} \leq C\varepsilon. \tag{0.7}$$

Note that $\|K_\varepsilon^{(1)}\|_{L_2\to H^1} = O(\varepsilon^{-1})$ because the corrector $K_\varepsilon^{(1)}$ involves a rapidly oscillating factor. It turned out that the correctors in (0.6) and (0.7) are different. The corrector $K_\varepsilon^{(1)}$ is similar to the standard corrector of homogenization theory (but it involves an auxiliary smoothing operator), while the corrector K_ε was not known in classical homogenization. This operator is the sum of three terms: $K_\varepsilon = K_\varepsilon^{(1)} + (K_\varepsilon^{(1)})^* + K^{(3)}$, where $K^{(3)}$ does not depend on ε.

Estimates of the form (0.5)–(0.7) are called *operator error estimates* in homogenization theory. They are order-sharp. A different approach to operator error estimates (the *shift method*) was suggested by Zhikov; see [24, 26] and a recent survey [27].

Homogenization problems in a perforated space were actively studied by different methods; see, e.g., [1], [25, Sec. 3.1], and [9–11, 14, 17]. However, error estimates of operator type were not discussed in these works.

Operator error estimates for elliptic problems in a perforated space (and also in a bounded perforated domain) were obtained in [26] by the shift method; see also survey [27, Sec. 2.3]. In [26], estimates of the form (0.5) and (0.7) were proved for the acoustics operator and the elasticity operator.

0.3. *Main results of the paper*

In this paper, we study the behavior of the resolvents of the operators (0.3) and (0.4) for small ε. Our main results are approximations for the resolvents with sharp-order error estimates. Let us discuss the simpler operator (0.3). By $\mathcal{A}^0(\varepsilon)$ we denote the operator acting in $L_2(\Pi_\varepsilon;\mathbb{C}^n)$ given by $b(\mathbf{D})^* g^0 b(\mathbf{D})$ with the natural condition on $\partial\Pi_\varepsilon$. The effective matrix g^0 is different from the case of $\mathbb{R}^d$. As usual, it is defined in terms of the solution of the auxiliary problem on the cell Ω^0; see Sec. 4. We show that

$$\|(\mathcal{A}_\varepsilon + I)^{-1} - (\mathcal{A}^0(\varepsilon) + I)^{-1}\|_{L_2(\Pi_\varepsilon)\to L_2(\Pi_\varepsilon)} \leq C\varepsilon. \tag{0.8}$$

Also, we obtain more accurate approximation:

$$\|(\mathcal{A}_\varepsilon + I)^{-1} - (\mathcal{A}^0(\varepsilon) + I)^{-1} - \varepsilon\mathcal{K}_\varepsilon\|_{L_2(\Pi_\varepsilon)\to L_2(\Pi_\varepsilon)} \leq C\varepsilon^2. \tag{0.9}$$

Finally, we find approximation for the resolvent in the "energy norm":

$$\|(\mathcal{A}_\varepsilon + I)^{-1} - (\mathcal{A}^0(\varepsilon) + I)^{-1} - \varepsilon \mathcal{K}_\varepsilon^{(1)}\|_{L_2(\Pi_\varepsilon)\to H^1(\Pi_\varepsilon)} \le C\varepsilon. \tag{0.10}$$

As in the case of $\mathbb{R}^d$, the correctors in (0.9) and (0.10) are different. The corrector $\mathcal{K}_\varepsilon^{(1)}$ is similar to the standard corrector; it contains the rapidly oscillating factor and the auxiliary smoothing operator $\mathcal{P}_\varepsilon$. The corrector $\mathcal{K}_\varepsilon$ is the sum of three terms: $\mathcal{K}_\varepsilon = \mathcal{K}_\varepsilon^{(1)} + (\mathcal{K}_\varepsilon^{(1)})^* + \mathcal{K}_\varepsilon^{(3)}$. The operator $\mathcal{K}_\varepsilon^{(3)}$ does not contain oscillating factors, but involves $\mathcal{P}_\varepsilon$. Besides approximation (0.10), we find approximation for the "flux" $g^\varepsilon b(\mathbf{D})(\mathcal{A}_\varepsilon + I)^{-1}$ in the $(L_2 \to L_2)$-norm.

We find similar, but more complicated, approximations for the resolvent of the operator (0.4). In particular, the analog of estimate (0.8) looks as follows:

$$\|(\widetilde{\mathcal{A}}_\varepsilon + I)^{-1} - (f^\varepsilon)^{-1}(\mathcal{A}^0(\varepsilon) + \overline{Q})^{-1}((f^\varepsilon)^*)^{-1}\|_{L_2(\Pi_\varepsilon)\to L_2(\Pi_\varepsilon)} \le C\varepsilon.$$

Here $\overline{Q} = |\Omega^0|^{-1} \int_{\Omega^0} (f(\mathbf{x})f(\mathbf{x})^*)^{-1}\, d\mathbf{x}$.

We distinguish some additional conditions under which the smoothing operator $\mathcal{P}_\varepsilon$ in the correctors can be removed (i.e., replaced by the identity).

The general results are applied to specific operators of mathematical physics: the acoustics operator, the elasticity operator, and the Schrödinger operator with a singular periodic potential.

Let us compare our results with the results of [26]. As has been already mentioned, estimates of the form (0.8) and (0.10) were proved in [26] for the acoustics and elasticity operators by the shift method. In our work, the operator-theoretic (spectral) method is used, a more general class of operators is studied, and estimate of the form (0.9) is proved. Such estimate was not known before in the case of a perforated space.

0.4. *Method*

We apply the operator-theoretic approach which was suggested and developed in [5–8]. Let us comment on the proof of estimate (0.8). By the scaling transformation, this estimate is equivalent to

$$\|(\mathcal{A} + \varepsilon^2 I)^{-1} - (\mathcal{A}^0 + \varepsilon^2 I)^{-1}\|_{L_2(\Pi)\to L_2(\Pi)} \le C\varepsilon^{-1}. \tag{0.11}$$

Here $\mathcal{A}^0$ is the operator acting in $L_2(\Pi; \mathbb{C}^n)$ and given by $b(\mathbf{D})^* g^0 b(\mathbf{D})$ with the natural condition on $\partial\Pi$. Next, using the Gelfand transformation, we expand the operator $\mathcal{A}$ in the direct integral of the operators $\mathcal{A}(\mathbf{k})$ acting in $L_2(\Omega^0; \mathbb{C}^n)$ and depending on the parameter $\mathbf{k} \in \widetilde{\Omega}$ (the *quasimomentum*). Here $\widetilde{\Omega}$ is the Brillouin zone of the dual lattice. The operator $\mathcal{A}(\mathbf{k})$ is given by the differential expression $b(\mathbf{D}+\mathbf{k})^* g(\mathbf{x}) b(\mathbf{D}+\mathbf{k})$ with the periodic boundary conditions on $\partial\Omega \cap \partial\Omega^0$ and the Neumann condition on $\partial\Omega^0 \setminus \partial\Omega$. Estimate (0.11) is equivalent to the inequality

$$\|(\mathcal{A}(\mathbf{k}) + \varepsilon^2 I)^{-1} - (\mathcal{A}^0(\mathbf{k}) + \varepsilon^2 I)^{-1}\|_{L_2(\Omega^0)\to L_2(\Omega^0)} \le C\varepsilon^{-1}, \quad \mathbf{k} \in \widetilde{\Omega},$$

which must be uniform in $\mathbf{k}$.

The operator $\mathcal{A}(\mathbf{k})$ is an elliptic operator in a bounded domain; its resolvent is compact. It is an analytic self-adjoint operator family. However, the parameter $\mathbf{k}$ is multidimensional. As in [5], we put $\mathbf{k} = t\boldsymbol{\theta}$, where $t = |\mathbf{k}|$ and $\boldsymbol{\theta} \in \mathbb{S}^{d-1}$. The operator family $\mathcal{A}(\mathbf{k}) = \mathcal{A}(t\boldsymbol{\theta})$ is studied by means of the analytic perturbation theory with respect to the one-dimensional parameter t. For this, we apply the abstract operator scheme.

In the abstract setting, we consider the family of operators $A(t) = X(t)^*X(t)$ acting in some Hilbert space $\mathfrak{H}$. Here $X(t) = X_0 + tX_1$. (The family $A(t)$ models the operator family $\mathcal{A}(\mathbf{k}) = \mathcal{A}(t\boldsymbol{\theta})$, but the parameter $\boldsymbol{\theta}$ is absent in the abstract setting.) It is assumed that the point $\lambda_0 = 0$ is an eigenvalue of $A(0)$ of finite multiplicity n. Then for $|t| \leq t^0$ the perturbed operator $A(t)$ has exactly n eigenvalues (counted with multiplicities) on the interval $[0, \delta]$ (here δ and t^0 are controlled explicitly). These eigenvalues and the corresponding eigenvectors are real-analytic functions of t. The coefficients of the corresponding power series expansions are called *threshold characteristics* of the operator $A(t)$. We distinguish a finite rank operator S (the *spectral germ* of $A(t)$) which acts in the space $\mathfrak{N} = \operatorname{Ker} A(0)$. The germ contains information about the threshold characteristics of principal order. In terms of the germ, it is possible to approximate the resolvent $(A(t) + \varepsilon^2 I)^{-1}$. Application of the abstract results leads to the required estimates for DO's.

Thus, the main idea of our approach is that homogenization can be treated as a *spectral threshold effect* near the bottom of the spectrum of the operator $\mathcal{A}$.

0.5. *Plan of the paper*

The paper consists of Introduction and nine sections. Section 1 contains the necessary operator-theoretic material. Preliminaries concerning lattices, perforated space, and the Gelfand transformation are collected in Sec. 2. In Sec. 3, we describe the class of operators and the direct integral expansions for operators of the form (0.1) and (0.2). In Sec. 4, the family of operators $\mathcal{A}(\mathbf{k})$ is incorporated in the framework of the abstract scheme; the effective characteristics are described. In Sec. 5, using the abstract results, we obtain approximations for the resolvents $(\mathcal{A}(\mathbf{k}) + \varepsilon^2 I)^{-1}$ and $(\widetilde{\mathcal{A}}(\mathbf{k}) + \varepsilon^2 I)^{-1}$. In Sec. 6, by employing the direct integral expansions, we deduce approximations for the resolvents $(\mathcal{A}+\varepsilon^2 I)^{-1}$ and $(\widetilde{\mathcal{A}}+\varepsilon^2 I)^{-1}$. Section 7 is devoted to homogenization problems; here, by the scaling transformation, we deduce approximations for the resolvents $(\mathcal{A}_\varepsilon + I)^{-1}$ and $(\widetilde{\mathcal{A}}_\varepsilon + I)^{-1}$ from the results of Sec. 6. In Sec. 8, we show that the smoothing operator in the correctors can be removed under some additional assumptions; some special cases are considered. Section 9 is devoted to applications of the general results to the particular operators.

0.6. *Notation*

Let $\mathfrak{H}$ and $\mathfrak{H}_*$ be complex separable Hilbert spaces. The symbols $(\cdot,\cdot)_{\mathfrak{H}}$ and $\|\cdot\|_{\mathfrak{H}}$ stand for the inner product and the norm in $\mathfrak{H}$, respectively; the symbol $\|\cdot\|_{\mathfrak{H}\to\mathfrak{H}_*}$ stands for the norm of a linear continuous operator from $\mathfrak{H}$ to $\mathfrak{H}_*$. Sometimes we omit the indices. By $I = I_{\mathfrak{H}}$ we denote the identity operator in $\mathfrak{H}$. If $\mathfrak{N}$ is a subspace of $\mathfrak{H}$, then $\mathfrak{N}^\perp := \mathfrak{H} \ominus \mathfrak{N}$. If P is the orthogonal projection of $\mathfrak{H}$ onto $\mathfrak{N}$, then $P^\perp$ is the orthogonal projection onto $\mathfrak{N}^\perp$. If $A\colon \mathfrak{H} \to \mathfrak{H}_*$ is a linear operator, then $\operatorname{Dom} A$ and $\operatorname{Ker} A$ stand for its domain and its kernel.

The symbols $\langle\cdot,\cdot\rangle$ and $|\cdot|$ denote the inner product and the norm in $\mathbb{C}^n$; $\mathbf{1}_n$ is the unit $(n\times n)$-matrix. If a is an $(m \times n)$-matrix, then the symbol $|a|$ denotes the norm of a viewed as a linear operator from $\mathbb{C}^n$ to $\mathbb{C}^m$. Next, we use the notation $\mathbf{x} = (x_1, \dots, x_d) \in \mathbb{R}^d$, $iD_j = \partial_j = \partial/\partial x_j$, $j = 1, \dots, d$, $\mathbf{D} = -i\nabla = (D_1, \dots, D_d)$. We denote $B(r) := \{\mathbf{x} \in \mathbb{R}^d\colon\ |\mathbf{x}| < r\}$. If $\mathcal{O} \subset \mathbb{R}^d$ is a measurable set, denote $|\mathcal{O}| := \operatorname{meas} \mathcal{O}$. The L_p-classes (where $1 \le p \le \infty$) and the Sobolev classes of $\mathbb{C}^n$-valued functions in a domain $\mathcal{O} \subset \mathbb{R}^d$ are denoted by $L_p(\mathcal{O};\mathbb{C}^n)$ and $H^s(\mathcal{O};\mathbb{C}^n)$, respectively. If $n = 1$, we write simply $L_p(\mathcal{O})$, $H^s(\mathcal{O})$, but sometimes we use such notation also for the spaces of vector-valued or matrix-valued functions.

By $c, C, \mathcal{C}, \mathfrak{C}$ (with indices and marks) we denote various constants in estimates.

1. Abstract Operator-Theoretic Scheme

We use the operator-theoretic approach to homogenization problems. It is based on the abstract operator-theoretic scheme; see [5, Chap. 1], [6] and [8, Chap. 1].

1.1. *The operators $X(t)$ and $A(t)$*

Let $\mathfrak{H}$ and $\mathfrak{H}_*$ be complex separable Hilbert spaces. Suppose that $X_0\colon \mathfrak{H} \to \mathfrak{H}_*$ is a densely defined and closed operator, and $X_1\colon \mathfrak{H} \to \mathfrak{H}_*$ is a bounded operator. Then the operator $X(t) := X_0 + tX_1$, $t \in \mathbb{R}$, is closed on the domain $\operatorname{Dom} X(t) = \operatorname{Dom} X_0$. Consider the family of self-adjoint (and nonnegative) operators $A(t) := X(t)^* X(t)$ in $\mathfrak{H}$. The operator $A(t)$ is generated by the closed quadratic form $\|X(t)u\|^2_{\mathfrak{H}_*}$, $u \in \operatorname{Dom} X_0$. Denote $A(0) = X_0^* X_0 =: A_0$ and $\mathfrak{N} := \operatorname{Ker} A_0 = \operatorname{Ker} X_0$. We impose the following condition.

Condition 1.1. The point $\lambda_0 = 0$ is an isolated point of the spectrum of A_0, and $0 < n := \dim \mathfrak{N} < \infty$.

Let d^0 be the *distance from the point* $\lambda_0 = 0$ *to the rest of the spectrum of* A_0. We put $\mathfrak{N}_* := \operatorname{Ker} X_0^*$, $n_* := \dim \mathfrak{N}_*$, and *assume that* $n \le n_* \le \infty$. Let P and P_* be the orthogonal projections of $\mathfrak{H}$ onto $\mathfrak{N}$ and of $\mathfrak{H}_*$ onto $\mathfrak{N}_*$, respectively.

Denote by $F(t;\sigma)$, $\sigma > 0$, the spectral projection of $A(t)$ for the interval $[0,\sigma]$, and put $\mathfrak{F}(t;\sigma) := F(t;\sigma)\mathfrak{H}$. *We fix a number* $\delta > 0$ *such that* $8\delta < d^0$. Next, we choose a number $t^0 > 0$ such that

$$t^0 \le \delta^{1/2} \|X_1\|^{-1}. \tag{1.1}$$

According to [5, Chap. 1, (1.3)], $F(t;\delta) = F(t;3\delta)$ and $\operatorname{rank} F(t;\delta) = n$ for $|t| \le t^0$.

1.2. *The operators Z, R, and S*

Now we introduce some operators appearing in the analytic perturbation theory considerations; see [5, Chap. 1, Sec. 1] and [6, Sec. 1].

Let $\omega \in \mathfrak{N}$, and let $\psi = \psi(\omega) \in \operatorname{Dom} X_0 \cap \mathfrak{N}^\perp$ be a (weak) solution of the equation $X_0^*(X_0\psi + X_1\omega) = 0$. We define a bounded operator $Z\colon \mathfrak{H} \to \mathfrak{H}$ by the relation $Zu = \psi(Pu)$, $u \in \mathfrak{H}$. Note that Z takes $\mathfrak{N}$ to $\mathfrak{N}^\perp$ and $\mathfrak{N}^\perp$ to $\{0\}$. Let $R\colon \mathfrak{N} \to \mathfrak{N}_*$ be the operator defined by $R\omega = (X_0 Z + X_1)\omega$, $\omega \in \mathfrak{N}$. Another representation for R is given by $R = P_* X_1|_{\mathfrak{N}}$.

According to [5, Chap. 1], the operator $S := R^*R\colon \mathfrak{N} \to \mathfrak{N}$ is called the *spectral germ of the operator family $A(t)$ at $t = 0$.* The germ S can be represented as $S = PX_1^* P_* X_1|_{\mathfrak{N}}$. The spectral germ is said to be *nondegenerate* if $\operatorname{Ker} S = \{0\}$.

As shown in [5, Chap. 1] and [6], the operators defined above satisfy the estimates

$$\begin{aligned} &\|X_0 Z\|_{\mathfrak{H}\to\mathfrak{H}_*} \le \|X_1\|, \quad \|Z\|_{\mathfrak{H}\to\mathfrak{H}} \le (8\delta)^{-1/2}\|X_1\|, \\ &\|RP\|_{\mathfrak{H}\to\mathfrak{H}_*} \le \|X_1\|, \quad \|SP\|_{\mathfrak{H}\to\mathfrak{H}} \le \|X_1\|^2. \end{aligned} \tag{1.2}$$

1.3. *The analytic branches of eigenvalues and eigenvectors of $A(t)$*

According to the general analytic perturbation theory (see [15]), for $|t| \le t^0$ there exist real-analytic functions $\lambda_l(t)$ and real-analytic $\mathfrak{H}$-valued functions $\varphi_l(t)$ such that

$$A(t)\varphi_l(t) = \lambda_l(t)\varphi_l(t), \quad l = 1, \dots, n, \quad |t| \le t^0, \tag{1.3}$$

and the $\varphi_l(t)$, $l = 1, \dots, n$, form an *orthonormal basis* in $\mathfrak{F}(t;\delta)$. Moreover, for $|t| \le t_*$, where $0 < t_* \le t^0$ *is sufficiently small,* we have the following convergent power series expansions:

$$\lambda_l(t) = \gamma_l t^2 + \mu_l t^3 + \cdots, \quad \gamma_l \ge 0, \quad \mu_l \in \mathbb{R}, \quad l = 1, \dots, n, \tag{1.4}$$

$$\varphi_l(t) = \omega_l + t\psi_l^{(1)} + \cdots, \quad l = 1, \dots, n. \tag{1.5}$$

The elements $\omega_l = \varphi_l(0)$, $l = 1, \dots, n$, form an orthonormal basis in $\mathfrak{N}$.

Substituting expansions (1.4) and (1.5) in (1.3) and comparing the coefficients of powers t and t^2, we see that $\widehat{\omega}_l := \psi_l^{(1)} - Z\omega_l \in \mathfrak{N}$, $l = 1, \dots, n$, and

$$S\omega_l = \gamma_l \omega_l, \quad l = 1, \dots, n. \tag{1.6}$$

(Cf. [5, Chap. 1, Sec. 1].) Thus, the *numbers γ_l and the elements ω_l are eigenvalues and eigenvectors of the germ S.*

1.4. *The operator N*

In what follows, we will need the operator N acting in $\mathfrak{H}$ and defined by

$$N = Z^* X_1^* RP + (RP)^* X_1 Z. \tag{1.7}$$

Note that N takes $\mathfrak{N}$ to $\mathfrak{N}$ and $\mathfrak{N}^\perp$ to $\{0\}$. This operator can be also described in terms of the coefficients of the power series expansions (1.4), (1.5). According to [6, Sec. 4.2], $N = N_0 + N_*$, where

$$N_0 = \sum_{l=1}^{n} \mu_l(\cdot, \omega_l)\omega_l, \quad N_* = \sum_{l=1}^{n} \gamma_l \left((\cdot, \widehat{\omega}_l)\omega_l + (\cdot, \omega_l)\widehat{\omega}_l\right).$$

In the basis $\omega_1, \dots, \omega_n$ the operator N_0 is diagonal, while N_* has zero diagonal elements. Thus,

$$(N\omega_l, \omega_l)_{\mathfrak{H}} = \mu_l, \quad l = 1, \dots, n. \tag{1.8}$$

By (1.2) and (1.7), the operator N satisfies the following estimate

$$\|N\|_{\mathfrak{H}\to\mathfrak{H}} \leq (2\delta)^{-1/2}\|X_1\|^3. \tag{1.9}$$

1.5. *Nondegeneracy condition*

We impose the following additional assumption.

Condition 1.2. There exists a constant $c_* > 0$ such that $A(t) \geq c_* t^2 I$ for $|t| \leq t^0$.

From Condition 1.2 it follows that $\lambda_l(t) \geq c_* t^2$, $l = 1, \dots, n$, for $|t| \leq t^0$. By (1.4), this implies $\gamma_l \geq c_* > 0$, $l = 1, \dots, n$, i.e., the germ S is nondegenerate: $S \geq c_* I_{\mathfrak{N}}$.

1.6. *Approximations for the resolvent* $(A(t) + \varepsilon^2 I)^{-1}$

The following result was obtained in [5, Chap. 1, Sec. 5.2].

Theorem 1.3. *Suppose that the operator family $A(t)$ satisfies the assumptions of Sec. 1.1 and Condition 1.2. Let P be the orthogonal projection of $\mathfrak{H}$ onto $\mathfrak{N} = \operatorname{Ker} A(0)$, and let S be the spectral germ of $A(t)$ defined in Sec. 1.2. Denote $\Xi(t,\varepsilon) := (t^2 S + \varepsilon^2 I_{\mathfrak{N}})^{-1} P$. Then for $\varepsilon > 0$ and $|t| \leq t^0$*

$$\|(A(t) + \varepsilon^2 I)^{-1} - \Xi(t,\varepsilon)\|_{\mathfrak{H}\to\mathfrak{H}} \leq C_1 \varepsilon^{-1}. \tag{1.10}$$

A more accurate approximation of the resolvent was found in [6, Theorem 5.1].

Theorem 1.4. *Suppose that the assumptions of Theorem 1.3 are satisfied. Let Z be the operator defined in Sec. 1.2. Let N be given by (1.7). Denote*

$$K(t,\varepsilon) := t(Z\Xi(t,\varepsilon) + \Xi(t,\varepsilon)Z^*) - t^3 \Xi(t,\varepsilon) N \Xi(t,\varepsilon). \tag{1.11}$$

Then for $\varepsilon > 0$ and $|t| \leq t^0$ we have

$$\|(A(t) + \varepsilon^2 I)^{-1} - \Xi(t,\varepsilon) - K(t,\varepsilon)\|_{\mathfrak{H}\to\mathfrak{H}} \leq C_2. \tag{1.12}$$

Approximation of the resolvent $(A(t) + \varepsilon^2 I)^{-1}$ in the "energy norm" was found in [8, Theorem 2.1].

Theorem 1.5. *Suppose that the assumptions of Theorem* 1.3 *are satisfied. Let* Z *be the operator defined in Sec.* 1.2. *Then for* $\varepsilon > 0$ *and* $|t| \le t^0$ *we have*

$$\|A(t)^{1/2}\big((A(t)+\varepsilon^2 I)^{-1} - (I+tZ)\Xi(t,\varepsilon)\big)\|_{\mathfrak{H}\to\mathfrak{H}} \le C_3. \tag{1.13}$$

Remark 1.6. The constants C_1, C_2, C_3 are polynomials of the parameters $\delta^{-1/2}$, $\|X_1\|$, and $c_*^{-1/2}$ with absolute positive coefficients.

1.7. *The operator family* $\widetilde{A}(t) = M^*A(t)M$

Let $A(t)$: $\mathfrak{H} \to \mathfrak{H}$ be the operator family defined in Sec. 1.1. Let $\widetilde{\mathfrak{H}}$ be yet another separable Hilbert space. Let M: $\widetilde{\mathfrak{H}} \to \mathfrak{H}$ be an isomorphism. In the Hilbert space $\widetilde{\mathfrak{H}}$, we consider the operator family

$$\widetilde{A}(t) = M^*A(t)M. \tag{1.14}$$

Clearly, the operator (1.14) also satisfies the assumptions of Sec. 1.1. We have $\widetilde{A}(t) = \widetilde{X}(t)^*\widetilde{X}(t)$, where $\widetilde{X}(t) = X(t)M$. Then $\widetilde{X}(t) = \widetilde{X}_0 + t\widetilde{X}_1$, where $\widetilde{X}_0 = X_0M$ and $\widetilde{X}_1 = X_1M$. Suppose that $A(t)$ satisfies Condition 1.2. Then the family $\widetilde{A}(t)$ also satisfies such condition. The objects corresponding to the family (1.14) will be marked by tilde. For example, $\widetilde{\delta}$, $\widetilde{t}^0$, $\widetilde{c}_*$ are the analogs of δ, t^0, c_* for $\widetilde{A}(t)$.

Of course, the results of Secs. 1.1–1.6 can be directly applied to the operator family (1.14). But, for further applications, it is more convenient to formulate the results in terms of the objects associated with the family $A(t)$ and isomorphism M. In $\mathfrak{H}$, we consider the positive definite operator $Q := (MM^*)^{-1}$. Let $Q_{\mathfrak{N}} = PQ|_{\mathfrak{N}}$ be the block of Q in the subspace $\mathfrak{N}$. Obviously, $Q_{\mathfrak{N}}$ is an isomorphism in $\mathfrak{N}$.

Similarly to Sec. 1.3, for $|t| \le \widetilde{t}^0$ there exist real-analytic branches of the eigenvalues $\widetilde{\lambda}_l(t)$ and the eigenvectors $\widetilde{\varphi}_l(t)$ of $\widetilde{A}(t)$, $l = 1, \ldots, n$. The vectors $\widetilde{\varphi}_l(t)$, $l = 1, \ldots, n$, form an orthonormal basis in the eigenspace of $\widetilde{A}(t)$ corresponding to the interval $[0, \widetilde{\delta}]$. For small $|t|$ we have the power series expansions

$$\widetilde{\lambda}_l(t) = \widetilde{\gamma}_l t^2 + \widetilde{\mu}_l t^3 + \cdots, \quad \widetilde{\gamma}_l \ge 0, \quad \widetilde{\mu}_l \in \mathbb{R}, \quad l = 1, \ldots, n, \tag{1.15}$$

$$\widetilde{\varphi}_l(t) = \widetilde{\omega}_l + t\widetilde{\psi}_l^{(1)} + \cdots, \quad l = 1, \ldots, n. \tag{1.16}$$

As was shown in [6], the numbers $\widetilde{\gamma}_l$ and the vectors $\phi_l := M\widetilde{\omega}_l \in \mathfrak{N}$ are eigenvalues and eigenvectors of the problem

$$S\phi_l = \widetilde{\gamma}_l Q_{\mathfrak{N}}\phi_l, \quad l = 1, \ldots, n, \tag{1.17}$$

where S is the spectral germ of $A(t)$.

Next, we introduce the operator Z_Q in $\mathfrak{H}$ taking an element $u \in \mathfrak{H}$ to the solution ψ_Q of the problem $X_0^*(X_0\psi_Q + X_1\omega) = 0$, $Q\psi_Q \perp \mathfrak{N}$, where $\omega = Pu$. We put $N_Q := Z_Q^*X_1^*RP + (RP)^*X_1Z_Q$. As was shown in [6], we have

$$(N_Q\phi_l, \phi_l)_{\mathfrak{H}} = \widetilde{\mu}_l, \quad l = 1, \ldots, n, \tag{1.18}$$

where $\widetilde{\mu}_l$ are the coefficients at t^3 in (1.15).

Obviously, we have $(\widetilde{A}(t)+\varepsilon^2 I)^{-1} = M^{-1}(A(t)+\varepsilon^2 Q)^{-1}(M^*)^{-1}$. The following three theorems were proved in [5, Chap. 1, Sec. 5.3], [6, Theorems 6.7 and 6.8], and [8, Sec. 3], respectively.

Theorem 1.7. *Suppose that the operator family $A(t)$ in $\mathfrak{H}$ satisfies the assumptions of Sec.* 1.1 *and Condition* 1.2. *Suppose that the operator family $\widetilde{A}(t)$ in $\widetilde{\mathfrak{H}}$ is given by* (1.14), *where $M\colon \widetilde{\mathfrak{H}} \to \mathfrak{H}$ is an isomorphism. Let P be the orthogonal projection of $\mathfrak{H}$ onto $\mathfrak{N} = \operatorname{Ker} A(0)$. Let $S\colon \mathfrak{N} \to \mathfrak{N}$ be the spectral germ of the family $A(t)$. Let $Q := (MM^*)^{-1}$ and $Q_{\mathfrak{N}} = PQ|_{\mathfrak{N}}$. Denote $\Xi_Q(t,\varepsilon) := (t^2 S + \varepsilon^2 Q_{\mathfrak{N}})^{-1}P$. Then for $\varepsilon > 0$ and $|t| \le \widetilde{t}^0$ we have*

$$\|(\widetilde{A}(t)+\varepsilon^2 I)^{-1} - M^{-1}\Xi_Q(t,\varepsilon)(M^*)^{-1}\|_{\widetilde{\mathfrak{H}}\to\widetilde{\mathfrak{H}}} \le \widetilde{C}_1 \varepsilon^{-1},$$

$$\|(A(t)+\varepsilon^2 Q)^{-1} - \Xi_Q(t,\varepsilon)\|_{\mathfrak{H}\to\mathfrak{H}} \le \widetilde{C}_1 \|M\|^2 \varepsilon^{-1}.$$

Theorem 1.8. *Suppose that the assumptions of Theorem* 1.7 *are satisfied. Let*

$$K_Q(t,\varepsilon) := t\big(Z_Q \Xi_Q(t,\varepsilon) + \Xi_Q(t,\varepsilon) Z_Q^*\big) - t^3 \Xi_Q(t,\varepsilon) N_Q \Xi_Q(t,\varepsilon).$$

Then for $\varepsilon > 0$ and $|t| \le \widetilde{t}^0$ we have

$$\|(\widetilde{A}(t)+\varepsilon^2 I)^{-1} - M^{-1}(\Xi_Q(t,\varepsilon) + K_Q(t,\varepsilon))(M^*)^{-1}\|_{\widetilde{\mathfrak{H}}\to\widetilde{\mathfrak{H}}} \le \widetilde{C}_2,$$

$$\|(A(t)+\varepsilon^2 Q)^{-1} - \Xi_Q(t,\varepsilon) - K_Q(t,\varepsilon)\|_{\mathfrak{H}\to\mathfrak{H}} \le \widetilde{C}_2 \|M\|^2.$$

Theorem 1.9. *Under the assumptions of Theorem* 1.7, *for $\varepsilon > 0$ and $|t| \le \widetilde{t}^0$ we have*

$$\|\widetilde{A}(t)^{1/2}\big((\widetilde{A}(t)+\varepsilon^2 I)^{-1} - M^{-1}(I + tZ_Q)\Xi_Q(t,\varepsilon)(M^*)^{-1}\big)\|_{\widetilde{\mathfrak{H}}\to\widetilde{\mathfrak{H}}} \le \widetilde{C}_3,$$

$$\|A(t)^{1/2}\big((A(t)+\varepsilon^2 Q)^{-1} - (I + tZ_Q)\Xi_Q(t,\varepsilon)\big)\|_{\mathfrak{H}\to\mathfrak{H}} \le \widetilde{C}_3 \|M\|.$$

Remark 1.10. The constants $\widetilde{C}_1$, $\widetilde{C}_2$, $\widetilde{C}_3$ are polynomials of the parameters $\widetilde{\delta}^{-1/2}$, $\|\widetilde{X}_1\|$, and $\widetilde{c}_*^{-1/2}$ with absolute positive coefficients.

2. Lattices. Perforated Space. The Gelfand Transformation

2.1. *Lattices in* $\mathbb{R}^d$

Let Γ be a lattice in $\mathbb{R}^d$, $d \ge 2$, generated by the basis $\mathbf{a}_1, \dots, \mathbf{a}_d$, i.e., $\Gamma = \big\{\mathbf{a} \in \mathbb{R}^d\colon \mathbf{a} = \sum_{j=1}^d n_j \mathbf{a}_j,\ n_j \in \mathbb{Z}\big\}$, and let Ω be the (elementary) cell of this lattice: $\Omega := \big\{\mathbf{x} \in \mathbb{R}^d\colon \mathbf{x} = \sum_{j=1}^d \xi_j \mathbf{a}_j,\ 0 < \xi_j < 1\big\}$. The basis $\mathbf{b}_1, \dots, \mathbf{b}_d$ dual to $\mathbf{a}_1, \dots, \mathbf{a}_d$ is defined by the relations $\langle \mathbf{b}_l, \mathbf{a}_j \rangle = 2\pi\delta_{jl}$. This basis generates the *lattice* $\widetilde{\Gamma}$ *dual to* Γ. Let $\widetilde{\Omega}$ be the central *Brillouin zone* of the lattice $\widetilde{\Gamma}$: $\widetilde{\Omega} := \big\{\mathbf{k} \in \mathbb{R}^d\colon |\mathbf{k}| < |\mathbf{k} - \mathbf{b}|,\ 0 \ne \mathbf{b} \in \widetilde{\Gamma}\big\}$. Note that $|\Omega||\widetilde{\Omega}| = (2\pi)^d$. Let $2r_1 := \operatorname{diam} \widetilde{\Omega}$.

By $\widetilde{H}^1(\Omega;\mathbb{C}^n)$ we denote the subspace in $H^1(\Omega;\mathbb{C}^n)$ consisting of the functions whose Γ-periodic extension to $\mathbb{R}^d$ belongs to $H^1_{\mathrm{loc}}(\mathbb{R}^d;\mathbb{C}^n)$.

2.2. *The perforated space*

Let $\Pi \subset \mathbb{R}^d$ be a Γ-*periodic open connected set*; Π is called a *perforated space.* Periodicity of Π means that if $\mathbf{x} \in \Pi$, then $\mathbf{x}+\mathbf{a} \in \Pi$ for any $\mathbf{a} \in \Gamma$. We assume that Π *is Lipschitz domain.* Then for each point $\mathbf{x}^0 \in \partial\Pi$ there exists a neighborhood $\mathcal{O}$ and a Lipschitz diffeomorphism Ψ: $\mathcal{O} \to B(1)$ such that $\Psi(\mathbf{x}^0) = 0$, $\Psi(\mathcal{O} \cap \Pi) = \{\mathbf{y} \in B(1)\colon y_d > 0\}$ and $\Psi(\mathcal{O} \cap \partial\Pi) = \{\mathbf{y} \in B(1)\colon y_d = 0\}$.

Denote $\Pi_r := \Pi \cap B(r)$, $r > 0$. The domain $\Omega^0 := \Omega \cap \Pi$ is the cell of the perforated space Π. We assume that Ω^0 *is Lipschitz domain.* Note that $\partial\Omega$ is the parallelepipedal surface. Since Π is connected, it follows that the intersection of $\partial\Omega^0$ with each face of $\partial\Omega$ has positive ($(d-1)$-dimensional) measure.

By $\widetilde{H}^1(\Omega^0; \mathbb{C}^n)$ *we denote the space of restrictions of functions* $\mathbf{u} \in \widetilde{H}^1(\Omega; \mathbb{C}^n)$ *to* Ω^0. *In other words,* $\widetilde{H}^1(\Omega^0; \mathbb{C}^n)$ *consists of the functions whose* Γ*-periodic extension to* Π *belongs to* $H^1(\Pi_r; \mathbb{C}^n)$ *for any* $r > 0$.

2.3. *The Gelfand transformation*

Now we define the Gelfand transformation $\mathcal{U}$ on the perforated space Π. Let $\mathcal{S}(\Pi; \mathbb{C}^n)$ be the space of restrictions to Π of the functions belonging to the Schwartz class $\mathcal{S}(\mathbb{R}^d; \mathbb{C}^n)$. Initially, $\mathcal{U}$ is defined on the functions $\mathbf{v} \in \mathcal{S}(\Pi; \mathbb{C}^n)$ by the formula

$$\widetilde{\mathbf{v}}(\mathbf{k}, \mathbf{x}) = (\mathcal{U}\mathbf{v})(\mathbf{k}, \mathbf{x}) = |\widetilde{\Omega}|^{-1/2} \sum_{\mathbf{a} \in \Gamma} e^{-i\langle \mathbf{k}, \mathbf{x}+\mathbf{a} \rangle} \mathbf{v}(\mathbf{x} + \mathbf{a}), \quad \mathbf{x} \in \Omega^0, \quad \mathbf{k} \in \widetilde{\Omega}. \tag{2.1}$$

Since $\|\widetilde{\mathbf{v}}\|_{L_2(\widetilde{\Omega} \times \Omega^0)} = \|\mathbf{v}\|_{L_2(\Pi)}$, the transformation $\mathcal{U}$ extends by continuity up to a *unitary mapping*

$$\mathcal{U}\colon L_2(\Pi; \mathbb{C}^n) \to \int_{\widetilde{\Omega}} \oplus L_2(\Omega^0; \mathbb{C}^n)\, d\mathbf{k} =: \mathfrak{K}.$$

The relation $\mathbf{v} \in H^1(\Pi; \mathbb{C}^n)$ is equivalent to the fact that $\widetilde{\mathbf{v}}(\mathbf{k}, \cdot) \in \widetilde{H}^1(\Omega^0; \mathbb{C}^n)$ for almost every $\mathbf{k} \in \widetilde{\Omega}$ and

$$\int_{\widetilde{\Omega}} \int_{\Omega^0} \left(|(\mathbf{D} + \mathbf{k})\widetilde{\mathbf{v}}(\mathbf{k}, \mathbf{x})|^2 + |\widetilde{\mathbf{v}}(\mathbf{k}, \mathbf{x})|^2 \right) d\mathbf{x}\, d\mathbf{k} < \infty.$$

Under the Gelfand transformation $\mathcal{U}$, the operator of multiplication by a bounded Γ-periodic matrix-valued function in $L_2(\Pi; \mathbb{C}^n)$ turns into multiplication by the same function on the fibers of the direct integral $\mathfrak{K}$. The operator $\mathbf{D}$ applied to $\mathbf{v} \in H^1(\Pi; \mathbb{C}^n)$ turns into the operator $(\mathbf{D} + \mathbf{k})$ applied to $\widetilde{\mathbf{v}}(\mathbf{k}, \cdot) \in \widetilde{H}^1(\Omega^0; \mathbb{C}^n)$.

3. Periodic Differential Operators in $L_2(\Pi; \mathbb{C}^n)$

3.1. *The operator* $b(\mathbf{D})$

Consider the first-order matrix DO given by

$$b(\mathbf{D}) = \sum_{j=1}^{d} b_j D_j, \tag{3.1}$$

where b_j are constant $(m \times n)$-matrices, in general, with complex entries. Assume that $m \geq n$. Let $b(\boldsymbol{\xi}) = \sum_{j=1}^{d} b_j \xi_j$ be the symbol of the operator (3.1). We impose the following condition.

Condition 3.1. The symbol $b(\boldsymbol{\xi})$ satisfies

$$\operatorname{rank} b(\boldsymbol{\xi}) = n, \quad 0 \neq \boldsymbol{\xi} \in \mathbb{C}^d. \tag{3.2}$$

In particular, (3.2) implies that

$$\alpha_0 \mathbf{1}_n \leq b(\boldsymbol{\theta})^* b(\boldsymbol{\theta}) \leq \alpha_1 \mathbf{1}_n, \quad \boldsymbol{\theta} \in \mathbb{S}^{d-1}, \quad 0 < \alpha_0 \leq \alpha_1 < \infty, \tag{3.3}$$

for some positive constants α_0, α_1. From (3.3), it follows that

$$|b_j| \leq \alpha_1^{1/2}, \quad j = 1, \dots, d. \tag{3.4}$$

Condition 3.1 means that the operator $b(\mathbf{D})$ is *algebraically complete* (in the terminology of [22]). Then, by [22, Sec. 3.7.4, Theorems 7.7 and 7.8], the form $\|b(\mathbf{D})\mathbf{u}\|^2_{L_2(\Omega^0)}$ is coercive on $H^1(\Omega^0; \mathbb{C}^n)$: there exist constants $\mathfrak{C}_1 > 0$, $\mathfrak{C}_2 \geq 0$ such that

$$\|\mathbf{D}\mathbf{u}\|^2_{L_2(\Omega^0)} \leq \mathfrak{C}_1 \|b(\mathbf{D})\mathbf{u}\|^2_{L_2(\Omega^0)} + \mathfrak{C}_2 \|\mathbf{u}\|^2_{L_2(\Omega^0)}, \quad \mathbf{u} \in H^1(\Omega^0; \mathbb{C}^n). \tag{3.5}$$

Using (3.5) for the shifted cells $\Omega^0 + \mathbf{a}$ and summing over $\mathbf{a} \in \Gamma$, we obtain

$$\|\mathbf{D}\mathbf{u}\|^2_{L_2(\Pi)} \leq \mathfrak{C}_1 \|b(\mathbf{D})\mathbf{u}\|^2_{L_2(\Pi)} + \mathfrak{C}_2 \|\mathbf{u}\|^2_{L_2(\Pi)}, \quad \mathbf{u} \in H^1(\Pi; \mathbb{C}^n). \tag{3.6}$$

Below we will show that estimate of the form (3.6) with $\mathfrak{C}_2 = 0$ is valid; see Proposition 3.8.

According to [21, Sec. 1], the algebraically complete operator $b(\mathbf{D})$ has a "polynomial property": the kernel

$$\mathfrak{P} := \{\mathbf{u} \in H^1(\Omega^0; \mathbb{C}^n) : \ b(\mathbf{D})\mathbf{u} = 0\} \tag{3.7}$$

is a finite-dimensional subspace of $\mathbb{C}^n$-valued polynomials (of some degree). Obviously, $\mathfrak{P}$ contains the subspace of constant functions.

Remark 3.2. A polynomial $\mathbf{p}(\mathbf{x})$ belongs to the subspace $\mathfrak{P}$ if and only if its coefficients satisfy a certain system of linear algebraic equations (which is independent of the domain Ω^0). If we extend $\mathbf{p} \in \mathfrak{P}$ to Ω as the polynomial with the same coefficients, then the equation $b(\mathbf{D})\mathbf{p}(\mathbf{x}) = 0$ will hold for $\mathbf{x} \in \Omega$.

3.2. *The operator $\mathcal{A}$*

Let $g(\mathbf{x})$ be a Hermitian $(m \times m)$-matrix-valued function in Π (in general, with complex entries) such that $g, g^{-1} \in L_\infty(\Pi)$ and $g(\mathbf{x}) > 0$. Suppose that $g(\mathbf{x})$ is periodic with respect to the lattice Γ.

In $L_2(\Pi; \mathbb{C}^n)$, we consider the operator $\mathcal{A}$ given formally by the differential expression $b(\mathbf{D})^* g(\mathbf{x}) b(\mathbf{D})$ with the Neumann condition (natural condition) on the boundary $\partial\Pi$. The precise definition of $\mathcal{A}$ is given in terms of the quadratic form

$$\mathfrak{a}[\mathbf{u}, \mathbf{u}] = \int_{\Pi} \langle g(\mathbf{x}) b(\mathbf{D})\mathbf{u}(\mathbf{x}), b(\mathbf{D})\mathbf{u}(\mathbf{x})\rangle \, d\mathbf{x}, \quad \mathbf{u} \in H^1(\Pi; \mathbb{C}^n). \tag{3.8}$$

Obviously, the form (3.8) satisfies

$$\|g^{-1}\|_{L_\infty}^{-1} \|b(\mathbf{D})\mathbf{u}\|_{L_2(\Pi)}^2 \leq \mathfrak{a}[\mathbf{u}, \mathbf{u}] \leq \|g\|_{L_\infty} \|b(\mathbf{D})\mathbf{u}\|_{L_2(\Pi)}^2, \quad u \in H^1(\Pi; \mathbb{C}^n). \tag{3.9}$$

So, the form (3.8) is nonnegative. Let us check that it is closed. By (3.1), (3.4), and the upper estimate (3.9),

$$\mathfrak{a}[\mathbf{u}, \mathbf{u}] \leq c_1 \|\mathbf{D}\mathbf{u}\|_{L_2(\Pi)}^2, \quad \mathbf{u} \in H^1(\Pi; \mathbb{C}^n), \tag{3.10}$$

where $c_1 = d\alpha_1 \|g\|_{L_\infty}$. By (3.6) and the lower estimate (3.9), we have

$$\mathfrak{a}[\mathbf{u}, \mathbf{u}] \geq c_2 \big(\|\mathbf{D}\mathbf{u}\|_{L_2(\Pi)}^2 - \mathfrak{C}_2 \|\mathbf{u}\|_{L_2(\Pi)}^2\big), \quad \mathbf{u} \in H^1(\Pi; \mathbb{C}^n), \tag{3.11}$$

where $c_2 = \mathfrak{C}_1^{-1} \|g^{-1}\|_{L_\infty}^{-1}$. Thus, by (3.10) and (3.11), the form (3.8) is closed. A self-adjoint operator in $L_2(\Pi; \mathbb{C}^n)$ generated by this form is denoted by $\mathcal{A}$.

3.3. *The operators $\mathcal{A}(\mathbf{k})$*

Let $\mathbf{k} \in \mathbb{R}^d$. In $L_2(\Omega^0; \mathbb{C}^n)$, we consider the operator $\mathcal{A}(\mathbf{k})$ given formally by the differential expression $b(\mathbf{D}+\mathbf{k})^* g(\mathbf{x}) b(\mathbf{D}+\mathbf{k})$ with the Neumann condition (natural condition) on $\partial\Omega^0 \backslash \partial\Omega$ and the periodic boundary conditions on $\partial\Omega^0 \cap \partial\Omega$. The precise definition of $\mathcal{A}(\mathbf{k})$ is given in terms of the quadratic form

$$\mathfrak{a}(\mathbf{k})[\mathbf{u}, \mathbf{u}] = \int_{\Omega^0} \langle g(\mathbf{x}) b(\mathbf{D}+\mathbf{k})\mathbf{u}(\mathbf{x}), b(\mathbf{D}+\mathbf{k})\mathbf{u}(\mathbf{x})\rangle \, d\mathbf{x}, \quad \mathbf{u} \in \widetilde{H}^1(\Omega^0; \mathbb{C}^n). \tag{3.12}$$

Obviously, the form (3.12) satisfies

$$\|g^{-1}\|_{L_\infty}^{-1} \|b(\mathbf{D}+\mathbf{k})\mathbf{u}\|_{L_2(\Omega^0)}^2 \leq \mathfrak{a}(\mathbf{k})[\mathbf{u}, \mathbf{u}] \leq \|g\|_{L_\infty} \|b(\mathbf{D}+\mathbf{k})\mathbf{u}\|_{L_2(\Omega^0)}^2,$$
$$\mathbf{u} \in \widetilde{H}^1(\Omega^0; \mathbb{C}^n). \tag{3.13}$$

So, the form (3.12) is nonnegative. By (3.1), (3.4), and the upper estimate (3.13),

$$\mathfrak{a}(\mathbf{k})[\mathbf{u}, \mathbf{u}] \leq c_1 \|(\mathbf{D}+\mathbf{k})\mathbf{u}\|_{L_2(\Omega^0)}^2, \quad \mathbf{u} \in \widetilde{H}^1(\Omega^0; \mathbb{C}^n). \tag{3.14}$$

From (3.5) and (3.13), we deduce the lower estimate for the form (3.12):

$$\mathfrak{a}(\mathbf{k})[\mathbf{u}, \mathbf{u}] \geq \frac{c_2}{2} \|\mathbf{D}\mathbf{u}\|_{L_2(\Omega^0)}^2 - c_3(\mathbf{k}) \|\mathbf{u}\|_{L_2(\Omega^0)}^2, \quad \mathbf{u} \in \widetilde{H}^1(\Omega^0; \mathbb{C}^n).$$

Here $c_3(\mathbf{k}) = \|g^{-1}\|_{L_\infty}^{-1} (\mathfrak{C}_2 (2\mathfrak{C}_1)^{-1} + \alpha_1 |\mathbf{k}|^2)$. Thus, the form (3.12) is closed. A self-adjoint operator in $L_2(\Omega^0; \mathbb{C}^n)$ generated by this form is denoted by $\mathcal{A}(\mathbf{k})$.

Now, we consider the kernel of $\mathcal{A}(0)$:

$$\mathfrak{N} := \operatorname{Ker} \mathcal{A}(0) = \{\mathbf{u} \in \widetilde{H}^1(\Omega^0; \mathbb{C}^n) : \ b(\mathbf{D})\mathbf{u} = 0\}. \tag{3.15}$$

Lemma 3.3. *The subspace* (3.15) *consists of constant* $\mathbb{C}^n$*-valued functions:*

$$\mathfrak{N} = \{\mathbf{u} \in L_2(\Omega^0; \mathbb{C}^n) : \ \mathbf{u}(\mathbf{x}) = \mathbf{c}, \ \mathbf{c} \in \mathbb{C}^n\}. \tag{3.16}$$

Proof. Let $\mathfrak{P}$ be the subspace defined by (3.7). Obviously, $\mathfrak{N} = \mathfrak{P} \cap \widetilde{H}^1(\Omega^0; \mathbb{C}^n)$.

Let $\mathbf{p} \in \mathfrak{N}$. Then $\mathbf{p}(\mathbf{x})$ is a polynomial in Ω^0. We extend $\mathbf{p}(\mathbf{x})$ to Ω as the polynomial with the same coefficients. Then, by Remark 3.2, $b(\mathbf{D})\mathbf{p}(\mathbf{x}) = 0$ for $\mathbf{x} \in \Omega$. Let us check that $\mathbf{p} \in \widetilde{H}^1(\Omega; \mathbb{C}^n)$. Indeed, $\mathbf{p}(\mathbf{x})$ satisfies periodic boundary conditions on $\partial\Omega^0 \cap \partial\Omega$. This is equivalent to some algebraic conditions on the coefficients of the polynomial $\mathbf{p}(\mathbf{x})$ (which are independent of Ω^0). Then $\mathbf{p}(\mathbf{x})$ satisfies periodic boundary conditions on the whole $\partial\Omega$.

Using the Fourier series expansion for a function $\mathbf{u} \in \widetilde{H}^1(\Omega; \mathbb{C}^n)$ and the lower estimate (3.3), we see that

$$\|b(\mathbf{D})\mathbf{u}\|^2_{L_2(\Omega)} \geq \alpha_0 \|\mathbf{D}\mathbf{u}\|^2_{L_2(\Omega)}, \quad \mathbf{u} \in \widetilde{H}^1(\Omega; \mathbb{C}^n).$$

Hence, relations $\mathbf{p} \in \widetilde{H}^1(\Omega; \mathbb{C}^n)$ and $b(\mathbf{D})\mathbf{p}(\mathbf{x}) = 0$ in Ω imply that $\mathbf{D}\mathbf{p}(\mathbf{x}) = 0$ in Ω. Consequently, $\mathbf{p}(\mathbf{x})$ is a constant $\mathbb{C}^n$-valued function. □

Denote by $\mathcal{H}(\mathbf{k})$, $\mathbf{k} \in \mathbb{R}^d$, the operator in $L_2(\Omega^0; \mathbb{C}^n)$ generated by the closed nonnegative quadratic form $\|b(\mathbf{D} + \mathbf{k})\mathbf{u}\|^2_{L_2(\Omega^0)}$, $\mathbf{u} \in \widetilde{H}^1(\Omega^0; \mathbb{C}^n)$. Due to estimate (3.5) and the compactness of the embedding of $\widetilde{H}^1(\Omega^0; \mathbb{C}^n)$ into $L_2(\Omega^0; \mathbb{C}^n)$, its spectrum is discrete. Let $\nu_1(\mathbf{k}) \leq \nu_2(\mathbf{k}) \leq \cdots \leq \nu_j(\mathbf{k}) \leq \cdots$ be the consecutive eigenvalues of the operator $\mathcal{H}(\mathbf{k})$.

Lemma 3.3 shows that the point zero is an eigenvalue of $\mathcal{H}(0)$ of multiplicity n; the corresponding eigenspace is given by (3.16). Thus, $\nu_1(0) = \cdots = \nu_n(0) = 0$. So, $\nu := \nu_{n+1}(0) > 0$ is the first nonzero eigenvalue of $\mathcal{H}(0)$. We have

$$\|b(\mathbf{D})\mathbf{u}\|^2_{L_2(\Omega^0)} \geq \nu \|\mathbf{u}\|^2_{L_2(\Omega^0)}, \quad \mathbf{u} \in \widetilde{H}^1(\Omega^0; \mathbb{C}^n) \cap \mathfrak{N}^\perp. \tag{3.17}$$

The number ν *depends on the operator* $b(\mathbf{D})$ *and the domain* Ω^0.

Lemma 3.4. *Let* $r_0 := \frac{1}{4}\nu^{1/2}\alpha_1^{-1/2}$. *For* $|\mathbf{k}| \leq r_0$, *we have*

$$\mathcal{A}(\mathbf{k}) \geq c_* |\mathbf{k}|^2 I, \quad c_* = \alpha_0 \|g^{-1}\|^{-1}_{L_\infty}. \tag{3.18}$$

Proof. Obviously,

$$\|b(\mathbf{D} + \mathbf{k})\mathbf{u}\|^2_{L_2(\Omega^0)} \geq \frac{1}{2}\|b(\mathbf{D})\mathbf{u}\|^2_{L_2(\Omega^0)} - \|b(\mathbf{k})\mathbf{u}\|^2_{L_2(\Omega^0)}, \quad \mathbf{u} \in \widetilde{H}^1(\Omega^0; \mathbb{C}^n).$$

Next, by (3.3), $\|b(\mathbf{k})\mathbf{u}\|^2_{L_2(\Omega^0)} \leq \alpha_1 |\mathbf{k}|^2 \|\mathbf{u}\|^2_{L_2(\Omega^0)}$. Hence,

$$\|b(\mathbf{D} + \mathbf{k})\mathbf{u}\|^2_{L_2(\Omega^0)} \geq \frac{1}{2}\|b(\mathbf{D})\mathbf{u}\|^2_{L_2(\Omega^0)} - \alpha_1 |\mathbf{k}|^2 \|\mathbf{u}\|^2_{L_2(\Omega^0)}, \quad \mathbf{u} \in \widetilde{H}^1(\Omega^0; \mathbb{C}^n).$$

By the variational principle for eigenvalues, we obtain the lower estimate for the $(n+1)$th eigenvalue of $\mathcal{H}(\mathbf{k})$:

$$\nu_{n+1}(\mathbf{k}) \geq \frac{1}{2}\nu - \alpha_1|\mathbf{k}|^2 \geq \frac{1}{4}\nu, \quad |\mathbf{k}| \leq \frac{1}{2}\nu^{1/2}\alpha_1^{-1/2}. \tag{3.19}$$

Let $\mathbf{k} = t\boldsymbol{\theta}$, where $t = |\mathbf{k}|$, $\boldsymbol{\theta} \in \mathbb{S}^{d-1}$. We have $b(\mathbf{k})^* b(\mathbf{k}) = |\mathbf{k}|^2 b(\boldsymbol{\theta})^* b(\boldsymbol{\theta})$. Due to condition (3.3), the matrix $b(\boldsymbol{\theta})^* b(\boldsymbol{\theta})$ is positive definite. Denote its eigenvalues by $\nu_j^\circ(\boldsymbol{\theta})$, $j = 1, \ldots, n$. Then

$$\alpha_0 \leq \nu_j^\circ(\boldsymbol{\theta}) \leq \alpha_1, \quad j = 1, \ldots, n. \tag{3.20}$$

The corresponding eigenvectors are denoted by $\mathbf{c}_j(\boldsymbol{\theta}) \in \mathbb{C}^n$. Then the numbers $|\mathbf{k}|^2 \nu_j^\circ(\boldsymbol{\theta})$, $j = 1, \ldots, n$, are eigenvalues of the operator $\mathcal{H}(\mathbf{k})$; the corresponding eigenfunctions are constants $\mathbf{c}_j(\boldsymbol{\theta})$. By (3.20), $|\mathbf{k}|^2 \nu_j^\circ(\boldsymbol{\theta}) \leq \alpha_1 |\mathbf{k}|^2 < \frac{1}{4}\nu$ for $|\mathbf{k}| \leq r_0$. Consequently, by (3.19), for $|\mathbf{k}| \leq r_0$ the numbers $|\mathbf{k}|^2 \nu_j^\circ(\boldsymbol{\theta})$, $j = 1, \ldots, n$, are the first n eigenvalues of $\mathcal{H}(\mathbf{k})$: $\nu_j(\mathbf{k}) = |\mathbf{k}|^2 \nu_j^\circ(\boldsymbol{\theta})$. Hence, by (3.20),

$$\|b(\mathbf{D}+\mathbf{k})\mathbf{u}\|^2_{L_2(\Omega^0)} \geq \alpha_0 |\mathbf{k}|^2 \|\mathbf{u}\|^2_{L_2(\Omega^0)}, \quad \mathbf{u} \in \widetilde{H}^1(\Omega^0; \mathbb{C}^n), \quad |\mathbf{k}| \leq r_0.$$

It remains to use the lower estimate (3.13). □

Lemma 3.5. *Let $0 < r \leq r_1 = \frac{1}{2}\operatorname{diam}\widetilde{\Omega}$. For $\mathbf{k} \in \widetilde{\Omega}\backslash B(r)$ we have*

$$\mathcal{A}(\mathbf{k}) \geq C(r)I. \tag{3.21}$$

The constant $C(r)$ depends only on $\|g^{-1}\|_{L_\infty}$, the operator $b(\mathbf{D})$, r, and the domain Ω^0.

Proof. Let us check that $\nu_1(\mathbf{k}) > 0$ for $\mathbf{k} \in \widetilde{\Omega}\setminus\{0\}$. Indeed, suppose that $\|b(\mathbf{D}+\mathbf{k})\mathbf{u}\|^2_{L_2(\Omega^0)} = 0$ for some $\mathbf{u} \in \widetilde{H}^1(\Omega^0; \mathbb{C}^n)$. Let $\mathbf{v}(\mathbf{x}) := \mathbf{u}(\mathbf{x})e^{i\langle \mathbf{k},\mathbf{x}\rangle}$. Then $\mathbf{v} \in H^1(\Omega^0; \mathbb{C}^n)$ and $b(\mathbf{D})\mathbf{v} = 0$, i.e., $\mathbf{v} \in \mathfrak{P}$ (see (3.7)). Thus, $\mathbf{v}(\mathbf{x})$ is a polynomial in Ω^0. We extend $\mathbf{v}(\mathbf{x})$ to Ω as the polynomial with the same coefficients and define $\mathbf{u}(\mathbf{x})$ in Ω by the formula $\mathbf{u}(\mathbf{x}) = \mathbf{v}(\mathbf{x})e^{-i\langle \mathbf{k},\mathbf{x}\rangle}$. By Remark 3.2, $b(\mathbf{D})\mathbf{v}(\mathbf{x}) = 0$ in Ω, whence $b(\mathbf{D}+\mathbf{k})\mathbf{u}(\mathbf{x}) = 0$ in Ω. The polynomial $\mathbf{v}(\mathbf{x})$ satisfies $\mathbf{k}$-quasiperiodic boundary conditions on $\partial\Omega^0 \cap \partial\Omega$. This is equivalent to some algebraic conditions on the coefficients of this polynomial (which are independent of the domain Ω^0). Hence, $\mathbf{v}(\mathbf{x})$ satisfies $\mathbf{k}$-quasiperiodic boundary conditions on the whole $\partial\Omega$. It follows that $\mathbf{u} \in \widetilde{H}^1(\Omega; \mathbb{C}^n)$. Using the Fourier series expansion and the lower estimate (3.3), we see that

$$\|b(\mathbf{D}+\mathbf{k})\mathbf{u}\|^2_{L_2(\Omega)} \geq \alpha_0 \|(\mathbf{D}+\mathbf{k})\mathbf{u}\|^2_{L_2(\Omega)}, \quad \mathbf{u} \in \widetilde{H}^1(\Omega; \mathbb{C}^n). \tag{3.22}$$

Consequently, relation $b(\mathbf{D}+\mathbf{k})\mathbf{u}(\mathbf{x}) = 0$ in Ω implies that $(\mathbf{D}+\mathbf{k})\mathbf{u}(\mathbf{x}) = 0$, i.e., $\mathbf{u}(\mathbf{x}) = \mathbf{C}e^{-i\langle \mathbf{k},\mathbf{x}\rangle}$, $\mathbf{C} \in \mathbb{C}^n$. If $\mathbf{k} \in \widetilde{\Omega}\backslash\{0\}$, the function $e^{-i\langle \mathbf{k},\mathbf{x}\rangle}$ does not satisfy the periodic boundary conditions. Hence, $\mathbf{u}(\mathbf{x}) \equiv 0$. It means that $\operatorname{Ker}\mathcal{H}(\mathbf{k}) = \{0\}$ for $\mathbf{k} \in \widetilde{\Omega}\backslash\{0\}$.

Next, $\nu_1(\mathbf{k})$ is a continuous function of the parameter $\mathbf{k}$, and the set $K(r) := \operatorname{clos}\widetilde{\Omega}\backslash B(r)$ is compact. Therefore,

$$\check{C}(r) := \min_{\mathbf{k}\in K(r)} \nu_1(\mathbf{k}) > 0. \tag{3.23}$$

Clearly, $\check{C}(r)$ depends only on the operator $b(\mathbf{D})$, r and the domain Ω^0. Combining this with the lower estimate (3.13), we obtain inequality (3.21) with the constant $C(r) = \check{C}(r)\|g^{-1}\|^{-1}_{L_\infty}$. □

Now, we obtain the analog of estimate (3.22) for $\mathbf{u} \in \widetilde{H}^1(\Omega^0;\mathbb{C}^n)$.

Lemma 3.6. *For $\mathbf{k} \in \operatorname{clos}\widetilde{\Omega}$ we have*

$$\|(\mathbf{D}+\mathbf{k})\mathbf{u}\|^2_{L_2(\Omega^0)} \le \widetilde{\mathfrak{C}}_1\|b(\mathbf{D}+\mathbf{k})\mathbf{u}\|^2_{L_2(\Omega^0)}, \quad \mathbf{u} \in \widetilde{H}^1(\Omega^0;\mathbb{C}^n). \tag{3.24}$$

The constant $\widetilde{\mathfrak{C}}_1$ depends on the operator $b(\mathbf{D})$ and the domain Ω^0.

Proof. Let $\mathcal{H}_0(\mathbf{k})$ be the operator in $L_2(\Omega^0;\mathbb{C}^n)$ generated by the form $\|(\mathbf{D}+\mathbf{k})\mathbf{u}\|^2_{L_2(\Omega^0)}$, $\mathbf{u} \in \widetilde{H}^1(\Omega^0;\mathbb{C}^n)$. The subspace $\mathfrak{N}$ (see (3.16)) is the eigenspace of $\mathcal{H}_0(\mathbf{k})$ corresponding to the eigenvalue $|\mathbf{k}|^2$. At the same time, the subspace $\mathfrak{N}$ is the eigenspace of $\mathcal{H}(\mathbf{k})$ corresponding to the eigenvalues $|\mathbf{k}|^2\nu_j^\circ(\boldsymbol{\theta})$, $j = 1,\dots,n$; see the proof of Lemma 3.4. Hence, it suffices to check (3.24) for $\mathbf{u} \in \mathfrak{N}$ and for $\mathbf{u} \in \widetilde{H}^1(\Omega^0;\mathbb{C}^n)\cap\mathfrak{N}^\perp$.

For $\mathbf{u} = \mathbf{c} \in \mathfrak{N}$ the required inequality follows from (3.3). We have

$$|\mathbf{k}|^2\|\mathbf{c}\|^2_{L_2(\Omega^0)} \le \alpha_0^{-1}\|b(\mathbf{k})\mathbf{c}\|^2_{L_2(\Omega^0)}, \quad \mathbf{c} \in \mathbb{C}^n,$$

which is equivalent to

$$\|(\mathbf{D}+\mathbf{k})\mathbf{u}\|^2_{L_2(\Omega^0)} \le \alpha_0^{-1}\|b(\mathbf{D}+\mathbf{k})\mathbf{u}\|^2_{L_2(\Omega^0)}, \quad \mathbf{u} \in \mathfrak{N}.$$

Next, from the proof of Lemma 3.4 it follows that

$$\|b(\mathbf{D}+\mathbf{k})\mathbf{u}\|^2_{L_2(\Omega^0)} \ge \frac{1}{4}\nu\|\mathbf{u}\|^2_{L_2(\Omega^0)}, \quad \mathbf{u} \in \widetilde{H}^1(\Omega^0;\mathbb{C}^n)\cap\mathfrak{N}^\perp, \quad |\mathbf{k}| \le r_0. \tag{3.25}$$

By (3.23),

$$\|b(\mathbf{D}+\mathbf{k})\mathbf{u}\|^2_{L_2(\Omega^0)} \ge \check{C}(r_0)\|\mathbf{u}\|^2_{L_2(\Omega^0)}, \quad \mathbf{u} \in \widetilde{H}^1(\Omega^0;\mathbb{C}^n), \quad \mathbf{k} \in \operatorname{clos}\widetilde{\Omega}\backslash B(r_0). \tag{3.26}$$

Estimates (3.25) and (3.26) imply that

$$\|b(\mathbf{D}+\mathbf{k})\mathbf{u}\|^2_{L_2(\Omega^0)} \ge c_4\|\mathbf{u}\|^2_{L_2(\Omega^0)}, \quad \mathbf{u} \in \widetilde{H}^1(\Omega^0;\mathbb{C}^n)\cap\mathfrak{N}^\perp, \quad \mathbf{k} \in \operatorname{clos}\widetilde{\Omega}, \tag{3.27}$$

where $c_4 = \min\{\frac{1}{4}\nu, \check{C}(r_0)\}$. Using (3.5) and (3.27), it is easy to deduce estimate of the form (3.24) for $\mathbf{u} \in \widetilde{H}^1(\Omega^0;\mathbb{C}^n)\cap\mathfrak{N}^\perp$. □

Combining (3.13) and (3.24), we obtain the following statement.

Corollary 3.7. *For $\mathbf{k} \in \operatorname{clos}\widetilde{\Omega}$ we have*

$$\mathfrak{a}(\mathbf{k})[\mathbf{u},\mathbf{u}] \ge \widetilde{\mathfrak{C}}_1^{-1}\|g^{-1}\|^{-1}_{L_\infty}\|(\mathbf{D}+\mathbf{k})\mathbf{u}\|^2_{L_2(\Omega^0)}, \quad \mathbf{u} \in \widetilde{H}^1(\Omega^0;\mathbb{C}^n). \tag{3.28}$$

3.4. *Direct integral expansion for the operator $\mathcal{A}$*

With the help of the Gelfand transformation, the operator $\mathcal{A}$ is represented as

$$\mathcal{U}\mathcal{A}\mathcal{U}^{-1} = \int_{\widetilde{\Omega}} \oplus \mathcal{A}(\mathbf{k})\, d\mathbf{k}. \tag{3.29}$$

This means the following. If $\mathbf{v} \in H^1(\Pi;\mathbb{C}^n)$ and $\widetilde{\mathbf{v}} = \mathcal{U}\mathbf{v}$, then

$$\widetilde{\mathbf{v}}(\mathbf{k},\cdot) \in \widetilde{H}^1(\Omega^0;\mathbb{C}^n) \quad \text{for a.e. } \mathbf{k} \in \widetilde{\Omega}, \tag{3.30}$$

$$\mathfrak{a}[\mathbf{v},\mathbf{v}] = \int_{\widetilde{\Omega}} \mathfrak{a}(\mathbf{k})[\widetilde{\mathbf{v}}(\mathbf{k},\cdot),\widetilde{\mathbf{v}}(\mathbf{k},\cdot)]\, d\mathbf{k}. \tag{3.31}$$

Conversely, if $\widetilde{\mathbf{v}} \in \mathfrak{K}$ satisfies (3.30) and the integral in (3.31) is finite, then $\mathbf{v} \in H^1(\Pi;\mathbb{C}^n)$ and (3.31) is valid.

Using the direct integral expansion, from (3.24) and (3.28) we deduce the following statement which improves estimates (3.6) and (3.11).

Proposition 3.8. *We have*

$$\begin{aligned} \|b(\mathbf{D})\mathbf{u}\|^2_{L_2(\Pi)} &\geq \widetilde{\mathfrak{C}}_1^{-1}\|\mathbf{D}\mathbf{u}\|^2_{L_2(\Pi)}, \quad \mathbf{u} \in H^1(\Pi;\mathbb{C}^n), \\ \mathfrak{a}[\mathbf{u},\mathbf{u}] &\geq \widetilde{\mathfrak{C}}_1^{-1}\|g^{-1}\|^{-1}_{L_\infty}\|\mathbf{D}\mathbf{u}\|^2_{L_2(\Pi)}, \quad \mathbf{u} \in H^1(\Pi;\mathbb{C}^n). \end{aligned} \tag{3.32}$$

3.5. *The operator $\widetilde{\mathcal{A}}$*

Let $\mathcal{A}$ be the operator in $L_2(\Pi;\mathbb{C}^n)$ defined in Sec. 3.2. Let $f(\mathbf{x})$ be a Γ-periodic $(n \times n)$-matrix-valued function in Π such that $f, f^{-1} \in L_\infty(\Pi)$. In $L_2(\Pi;\mathbb{C}^n)$, we define the operator

$$\widetilde{\mathcal{A}} = f^*\mathcal{A}f. \tag{3.33}$$

Formally, $\widetilde{\mathcal{A}}$ is given by the differential expression $f(\mathbf{x})^*b(\mathbf{D})^*g(\mathbf{x})b(\mathbf{D})f(\mathbf{x})$ with the natural condition on $\partial\Pi$. The precise definition is given in terms of the quadratic form

$$\widetilde{\mathfrak{a}}[\mathbf{u},\mathbf{u}] = \int_\Pi \langle g(\mathbf{x})b(\mathbf{D})\mathbf{v}, b(\mathbf{D})\mathbf{v}\rangle\, d\mathbf{x}, \quad \mathbf{v} = f\mathbf{u} \in H^1(\Pi;\mathbb{C}^n). \tag{3.34}$$

Similarly to (3.29), we have

$$\mathcal{U}\widetilde{\mathcal{A}}\mathcal{U}^{-1} = \int_{\widetilde{\Omega}} \oplus \widetilde{\mathcal{A}}(\mathbf{k})\, d\mathbf{k}.$$

Here $\widetilde{\mathcal{A}}(\mathbf{k})$ is the operator in $L_2(\Omega^0;\mathbb{C}^n)$ generated by the quadratic form

$$\widetilde{\mathfrak{a}}(\mathbf{k})[\mathbf{u},\mathbf{u}] = \int_{\Omega^0} \langle g(\mathbf{x})b(\mathbf{D}+\mathbf{k})\mathbf{v}, b(\mathbf{D}+\mathbf{k})\mathbf{v}\rangle\, d\mathbf{x}, \quad \mathbf{v} = f\mathbf{u} \in \widetilde{H}^1(\Omega^0;\mathbb{C}^n). \tag{3.35}$$

4. The Effective Characteristics of the Operator $\mathcal{A}$

4.1. *Incorporation of the operators* $\mathcal{A}(\mathbf{k})$ *in the abstract scheme*

For $\mathbf{k} \in \mathbb{R}^d$, we put $\mathbf{k} = t\boldsymbol{\theta}$, $t = |\mathbf{k}|$, $\boldsymbol{\theta} \in \mathbb{S}^{d-1}$. We will apply the scheme of Sec. 1 to the operator family

$$A(t) = A(t;\boldsymbol{\theta}) := \mathcal{A}(t\boldsymbol{\theta}).$$

Then all constructions and estimates will depend on the additional parameter $\boldsymbol{\theta}$. We have to make our estimates uniform in $\boldsymbol{\theta}$.

Now, the spaces $\mathfrak{H}$ and $\mathfrak{H}_*$ are given by $\mathfrak{H} = L_2(\Omega^0;\mathbb{C}^n)$ and $\mathfrak{H}_* = L_2(\Omega^0;\mathbb{C}^m)$. The role of $X(t)$ is played by the operator $X(t;\boldsymbol{\theta})$ given by the differential expression $g^{1/2}b(\mathbf{D}+t\boldsymbol{\theta})$ on the domain $\widetilde{H}^1(\Omega^0;\mathbb{C}^n)$. We have $X(t;\boldsymbol{\theta}) = X_0 + tX_1(\boldsymbol{\theta})$, where $X_0 = g^{1/2}b(\mathbf{D})$, $\operatorname{Dom} X_0 = \widetilde{H}^1(\Omega^0;\mathbb{C}^n)$, and $X_1(\boldsymbol{\theta}) = g^{1/2}b(\boldsymbol{\theta})$. We have $A(t;\boldsymbol{\theta}) = X(t;\boldsymbol{\theta})^*X(t;\boldsymbol{\theta})$. By Lemma 3.3, the kernel $\mathfrak{N} = \operatorname{Ker} X_0$ consists of constant functions:

$$\mathfrak{N} = \{\mathbf{u} \in L_2(\Omega^0;\mathbb{C}^n)\colon\ \mathbf{u}(\mathbf{x}) = \mathbf{c},\ \mathbf{c} \in \mathbb{C}^n\}. \tag{4.1}$$

So, $\dim \mathfrak{N} = n$. The projection P of $L_2(\Omega^0;\mathbb{C}^n)$ onto $\mathfrak{N}$ is given by

$$P\mathbf{u} = |\Omega^0|^{-1}\int_{\Omega^0} \mathbf{u}(\mathbf{x})\,d\mathbf{x}. \tag{4.2}$$

Next, the kernel $\mathfrak{N}_* = \operatorname{Ker} X_0^*$ is described by

$$\mathfrak{N}_* = \{\mathbf{v} \in L_2(\Omega^0;\mathbb{C}^m)\colon\ g^{1/2}\mathbf{v} \in \widetilde{H}^1(\Omega^0;\mathbb{C}^m),\ b(\mathbf{D})^*(g^{1/2}\mathbf{v}) = 0\}.$$

We have $n_* = \dim \mathfrak{N}_* = \infty$ if $m > n$ and $n_* = n$ if $m = n$. So, the condition $n \le n_*$ is satisfied.

According to (3.17), the distance d^0 from the point $\lambda_0 = 0$ to the rest of the spectrum of $\mathcal{A}(0)$ satisfies the estimate

$$d^0 \ge \nu \|g^{-1}\|^{-1}_{L_\infty}. \tag{4.3}$$

In Sec. 1.1, it was required to fix a number $\delta \in (0, d^0/8)$. Using (4.3), we choose

$$\delta = \frac{1}{16}\nu\|g^{-1}\|^{-1}_{L_\infty}. \tag{4.4}$$

By (3.3),

$$\|X_1(\boldsymbol{\theta})\| \le \alpha_1^{1/2}\|g\|^{1/2}_{L_\infty}, \quad \boldsymbol{\theta} \in \mathbb{S}^{d-1}. \tag{4.5}$$

This allows us to choose t^0 (see (1.1)) to be independent of $\boldsymbol{\theta}$. We put

$$t^0 = \frac{1}{4}\nu^{1/2}\alpha_1^{-1/2}\|g\|^{-1/2}_{L_\infty}\|g^{-1}\|^{-1/2}_{L_\infty}. \tag{4.6}$$

Condition 1.2 is also satisfied due to Lemma 3.4:

$$A(t;\boldsymbol{\theta}) \ge c_* t^2 I, \quad c_* = \alpha_0 \|g^{-1}\|^{-1}_{L_\infty}, \quad t = |\mathbf{k}| \le t^0. \tag{4.7}$$

It is important that δ, t^0, and c_* do not depend on $\boldsymbol{\theta}$ (see (4.4), (4.6), (4.7)).

4.2. *The operators $Z(\boldsymbol{\theta})$ and $R(\boldsymbol{\theta})$*

According to Sec. 1.2, we construct the operator $Z(\boldsymbol{\theta})$ which plays the role of Z for the family $A(t;\boldsymbol{\theta})$. Let $\mathbf{u} \in L_2(\Omega^0;\mathbb{C}^n)$, $P\mathbf{u} = \mathbf{c}$, and let $\mathbf{w} \in \widetilde{H}^1(\Omega^0;\mathbb{C}^n)$ be the (weak) solution of the problem

$$b(\mathbf{D})^* g(\mathbf{x})(b(\mathbf{D})\mathbf{w}(\mathbf{x}) + b(\boldsymbol{\theta})\mathbf{c}) = 0, \quad \mathbf{x} \in \Omega^0; \quad \int_{\Omega^0} \mathbf{w}(\mathbf{x})\, d\mathbf{x} = 0. \tag{4.8}$$

Then $\mathbf{w} = Z(\boldsymbol{\theta})\mathbf{u}$. It is convenient to give the following representation for $Z(\boldsymbol{\theta})$. Suppose that an $(n \times m)$-matrix-valued function $\Lambda \in \widetilde{H}^1(\Omega^0)$ is the (weak) solution of the problem

$$b(\mathbf{D})^* g(\mathbf{x})(b(\mathbf{D})\Lambda(\mathbf{x}) + \mathbf{1}_m) = 0, \quad \mathbf{x} \in \Omega^0; \quad \int_{\Omega^0} \Lambda(\mathbf{x})\, d\mathbf{x} = 0. \tag{4.9}$$

Clearly, the operator $Z(\boldsymbol{\theta})$ can be represented as

$$Z(\boldsymbol{\theta}) = \Lambda b(\boldsymbol{\theta}) P. \tag{4.10}$$

By (1.2) and (4.5),

$$\begin{aligned} \|X_0 Z(\boldsymbol{\theta})\|_{L_2(\Omega^0)\to L_2(\Omega^0)} &\le \alpha_1^{1/2}\|g\|_{L_\infty}^{1/2}, \\ \|Z(\boldsymbol{\theta})\|_{L_2(\Omega^0)\to L_2(\Omega^0)} &\le (8\delta)^{-1/2}\alpha_1^{1/2}\|g\|_{L_\infty}^{1/2} \\ &= \sqrt{2}\nu^{-1/2}\alpha_1^{1/2}\|g\|_{L_\infty}^{1/2}\|g^{-1}\|_{L_\infty}^{1/2} =: C_Z. \end{aligned} \tag{4.11}$$

According to Sec. 1.2, the operator $R(\boldsymbol{\theta})\colon \mathfrak{N} \to \mathfrak{N}_*$, which plays the role of R for $A(t;\boldsymbol{\theta})$, is given by $R(\boldsymbol{\theta})\mathbf{c} = g^{1/2}(b(\mathbf{D})\mathbf{w} + b(\boldsymbol{\theta})\mathbf{c})$, $\mathbf{c} \in \mathfrak{N}$. In the matrix form, this can be written as

$$R(\boldsymbol{\theta}) = g^{1/2}(b(\mathbf{D})\Lambda + \mathbf{1}) b(\boldsymbol{\theta}). \tag{4.12}$$

4.3. *The spectral germ $S(\boldsymbol{\theta})$*

According to Sec. 1.2, the spectral germ $S(\boldsymbol{\theta})\colon \mathfrak{N} \to \mathfrak{N}$ of the operator family $A(t;\boldsymbol{\theta})$ at $t = 0$ is given by $S(\boldsymbol{\theta}) = R(\boldsymbol{\theta})^* R(\boldsymbol{\theta})$. Let us calculate the quadratic form of $S(\boldsymbol{\theta})$:

$$(S(\boldsymbol{\theta})\mathbf{c}, \mathbf{c})_{L_2(\Omega^0)} = \|R(\boldsymbol{\theta})\mathbf{c}\|^2_{L_2(\Omega^0)} = \|g^{1/2}(b(\mathbf{D})\mathbf{w} + b(\boldsymbol{\theta})\mathbf{c})\|^2_{L_2(\Omega^0)}, \quad \mathbf{c} \in \mathfrak{N}.$$

Hence, by (4.8), we obtain

$$\begin{aligned} (S(\boldsymbol{\theta})\mathbf{c}, \mathbf{c})_{L_2(\Omega^0)} &= \int_{\Omega^0} \langle g(\mathbf{x})(b(\mathbf{D})\mathbf{w}(\mathbf{x}) + b(\boldsymbol{\theta})\mathbf{c}), b(\mathbf{D})\mathbf{w}(\mathbf{x}) + b(\boldsymbol{\theta})\mathbf{c}\rangle\, d\mathbf{x} \\ &= \int_{\Omega^0} \langle g(\mathbf{x})(b(\mathbf{D})\mathbf{w}(\mathbf{x}) + b(\boldsymbol{\theta})\mathbf{c}), b(\boldsymbol{\theta})\mathbf{c}\rangle\, d\mathbf{x} = |\Omega^0| \langle g^0 b(\boldsymbol{\theta})\mathbf{c}, b(\boldsymbol{\theta})\mathbf{c}\rangle. \end{aligned} \tag{4.13}$$

Here the matrix g^0 is defined by

$$g^0 = |\Omega^0|^{-1} \int_{\Omega^0} \widetilde{g}(\mathbf{x})\, d\mathbf{x}, \tag{4.14}$$

$$\widetilde{g}(\mathbf{x}) := g(\mathbf{x})(b(\mathbf{D})\Lambda(\mathbf{x}) + \mathbf{1}). \tag{4.15}$$

By (4.13), the germ $S(\boldsymbol{\theta})$ acts as multiplication by the matrix

$$S(\boldsymbol{\theta}) = b(\boldsymbol{\theta})^* g^0 b(\boldsymbol{\theta}), \quad \boldsymbol{\theta} \in \mathbb{S}^{d-1}. \tag{4.16}$$

The matrix (4.14) is called the *effective matrix.*

According to Sec. 1.5, estimate (4.7) implies that the germ is nondegenerate:

$$S(\boldsymbol{\theta}) \geq c_* I_{\mathfrak{N}}, \quad \boldsymbol{\theta} \in \mathbb{S}^{d-1}. \tag{4.17}$$

4.4. *The analytic branches of eigenvalues and eigenvectors*

The analytic (in t) branches of the eigenvalues $\lambda_l(t;\boldsymbol{\theta})$ and the branches of the eigenvectors $\varphi_l(t;\boldsymbol{\theta})$ of the operator $A(t;\boldsymbol{\theta})$ depend on the parameter $\boldsymbol{\theta} \in \mathbb{S}^{d-1}$. We have

$$A(t;\boldsymbol{\theta})\varphi_l(t;\boldsymbol{\theta}) = \lambda_l(t;\boldsymbol{\theta})\varphi_l(t;\boldsymbol{\theta}), \quad t = |\mathbf{k}| \leq t^0, \quad l = 1, \dots, n.$$

According to (1.4) and (1.5), for small t we have the convergent power series expansions:

$$\begin{aligned} \lambda_l(t;\boldsymbol{\theta}) &= \gamma_l(\boldsymbol{\theta})t^2 + \mu_l(\boldsymbol{\theta})t^3 + \cdots, \quad & l = 1, \dots, n, \\ \varphi_l(t;\boldsymbol{\theta}) &= \boldsymbol{\omega}_l(\boldsymbol{\theta}) + t\boldsymbol{\psi}_l^{(1)}(\boldsymbol{\theta}) + \cdots, \quad & l = 1, \dots, n. \end{aligned} \tag{4.18}$$

The elements $\boldsymbol{\omega}_l(\boldsymbol{\theta})$, $l = 1, \dots, n$, form an orthonormal basis in the subspace (4.1). By (1.6) and (4.16), the numbers $\gamma_l(\boldsymbol{\theta})$ and the vectors $\boldsymbol{\omega}_l(\boldsymbol{\theta})$ are eigenvalues and eigenvectors of the spectral germ $S(\boldsymbol{\theta})$, i.e.,

$$b(\boldsymbol{\theta})^* g^0 b(\boldsymbol{\theta})\boldsymbol{\omega}_l(\boldsymbol{\theta}) = \gamma_l(\boldsymbol{\theta})\boldsymbol{\omega}_l(\boldsymbol{\theta}), \quad l = 1, \dots, n.$$

4.5. *The properties of the effective matrix* g^0

Denote

$$\overline{g} := |\Omega^0|^{-1} \int_{\Omega^0} g(\mathbf{x})\, d\mathbf{x}, \quad \underline{g} := \left(|\Omega^0|^{-1} \int_{\Omega^0} g(\mathbf{x})^{-1}\, d\mathbf{x} \right)^{-1}.$$

Proposition 4.1. *The effective matrix g^0 satisfies the estimates*

$$\underline{g} \leq g^0 \leq \overline{g}. \tag{4.19}$$

If $m = n$, then $g^0 = \underline{g}$.

In homogenization theory, estimates (4.19) are known as the Voigt–Reuss bracketing.

Proof. The proof is similar to that of Theorem 1.5 from [5, Chap. 3]. Let $\mathbf{C} \in \mathbb{C}^m$ and let $\mathbf{v}(\mathbf{x}) = \Lambda(\mathbf{x})\mathbf{C}$. Then $\mathbf{v} \in \widetilde{H}^1(\Omega^0; \mathbb{C}^n)$ is the solution of problem

$$b(\mathbf{D})^* g(\mathbf{x})(b(\mathbf{D})\mathbf{v}(\mathbf{x}) + \mathbf{C}) = 0, \quad \int_{\Omega^0} \mathbf{v}(\mathbf{x})\, d\mathbf{x} = 0. \tag{4.20}$$

Obviously,

$$g^{1/2}\mathbf{C} = g^{1/2}(b(\mathbf{D})\mathbf{v} + \mathbf{C}) - g^{1/2}b(\mathbf{D})\mathbf{v}. \tag{4.21}$$

The summands in the right-hand side are orthogonal to each other, because the first term belongs to $\mathfrak{N}_* = \operatorname{Ker} X_0^*$, while the second term belongs to $\operatorname{Ran} X_0$. Hence,

$$\|g^{1/2}(b(\mathbf{D})\mathbf{v} + \mathbf{C})\|^2_{L_2(\Omega^0)} + \|g^{1/2}b(\mathbf{D})\mathbf{v}\|^2_{L_2(\Omega^0)} = \|g^{1/2}\mathbf{C}\|^2_{L_2(\Omega^0)}. \tag{4.22}$$

It follows that

$$\|g^{1/2}(b(\mathbf{D})\mathbf{v} + \mathbf{C})\|^2_{L_2(\Omega^0)} \le \|g^{1/2}\mathbf{C}\|^2_{L_2(\Omega^0)}. \tag{4.23}$$

By (4.9), (4.14), and (4.15), the left-hand side is equal to $|\Omega^0|\langle g^0\mathbf{C}, \mathbf{C}\rangle$. Obviously, the right-hand side coincides with $|\Omega^0|\langle \overline{g}\mathbf{C}, \mathbf{C}\rangle$. Consequently, $\langle g^0\mathbf{C}, \mathbf{C}\rangle \le \langle \overline{g}\mathbf{C}, \mathbf{C}\rangle$, $\mathbf{C} \in \mathbb{C}^m$, which proves the upper estimate (4.19).

To prove the lower estimate, we consider the subspace $\mathfrak{M} := \{g^{-1/2}\mathbf{C}\colon \mathbf{C} \in \mathbb{C}^m\}$ of $\mathfrak{N}_*$. Next, we define the operator T in $L_2(\Omega^0; \mathbb{C}^m)$ by

$$T\mathbf{z} = |\Omega^0|^{-1} g^{-1/2} \underline{g} \int_{\Omega^0} g(\mathbf{x})^{-1/2}\mathbf{z}(\mathbf{x})\, d\mathbf{x}, \quad \mathbf{z} \in L_2(\Omega^0; \mathbb{C}^m).$$

It is easily seen that $T\mathbf{z} \in \mathfrak{M}$ for any $\mathbf{z} \in L_2(\Omega^0; \mathbb{C}^m)$; $T\mathbf{z} = \mathbf{z}$ for $\mathbf{z} \in \mathfrak{M}$; and

$$(T\mathbf{z}, \mathbf{z})_{L_2(\Omega^0)} = |\Omega^0|^{-1}\langle \underline{g}\mathbf{C}_{\mathbf{z}}, \mathbf{C}_{\mathbf{z}}\rangle, \quad \mathbf{C}_{\mathbf{z}} = \int_{\Omega^0} g(\mathbf{x})^{-1/2}\mathbf{z}(\mathbf{x})\, d\mathbf{x}, \quad \mathbf{z} \in L_2(\Omega^0; \mathbb{C}^m). \tag{4.24}$$

This means that T is the orthogonal projection of $L_2(\Omega^0; \mathbb{C}^m)$ onto $\mathfrak{M}$. Let us apply T to (4.21):

$$Tg^{1/2}\mathbf{C} = Tg^{1/2}(b(\mathbf{D})\mathbf{v} + \mathbf{C}) - Tg^{1/2}b(\mathbf{D})\mathbf{v}.$$

Since $g^{1/2}b(\mathbf{D})\mathbf{v} \in \operatorname{Ran} X_0 = \mathfrak{N}_*^{\perp}$ and $\mathfrak{M} \subset \mathfrak{N}_*$, then $Tg^{1/2}b(\mathbf{D})\mathbf{v} = 0$. Consequently, by (4.9), (4.14), and (4.15),

$$\begin{aligned}\|Tg^{1/2}\mathbf{C}\|^2_{L_2(\Omega^0)} &= \|Tg^{1/2}(b(\mathbf{D})\mathbf{v} + \mathbf{C})\|^2_{L_2(\Omega^0)} \le \|g^{1/2}(b(\mathbf{D})\mathbf{v} + \mathbf{C})\|^2_{L_2(\Omega^0)} \\ &= |\Omega^0|\langle g^0\mathbf{C}, \mathbf{C}\rangle.\end{aligned} \tag{4.25}$$

By (4.24) with $\mathbf{z} = g^{1/2}\mathbf{C}$, we have

$$\|Tg^{1/2}\mathbf{C}\|^2_{L_2(\Omega^0)} = |\Omega^0|\langle \underline{g}\mathbf{C}, \mathbf{C}\rangle. \tag{4.26}$$

It follows that $\langle \underline{g}\mathbf{C}, \mathbf{C}\rangle \le \langle g^0\mathbf{C}, \mathbf{C}\rangle$, $\mathbf{C} \in \mathbb{C}^m$, which is equivalent to the lower estimate (4.19).

Now, let $m = n$. Then $\dim \mathfrak{N}_* = n = m$. Since $\dim \mathfrak{M} = m$ and $\mathfrak{M} \subset \mathfrak{N}_*$, we have $\mathfrak{M} = \mathfrak{N}_*$. Since $g^{1/2}(b(\mathbf{D})\mathbf{v} + \mathbf{C}) \in \mathfrak{N}_*$, then the inequality in (4.25) becomes identity. Hence, $g^0 = \underline{g}$. $\square$

From (4.19) we deduce the following estimates for the norms of the matrices g^0 and $(g^0)^{-1}$:

$$|g^0| \le \|g\|_{L_\infty}, \quad |(g^0)^{-1}| \le \|g^{-1}\|_{L_\infty}. \tag{4.27}$$

Now, we distinguish the cases where one of the inequalities in (4.19) becomes equality.

Proposition 4.2. *The identity $g^0 = \overline{g}$ is equivalent to the relations*

$$b(\mathbf{D})^* \mathbf{g}_k(\mathbf{x}) = 0, \quad k = 1, \dots, m, \tag{4.28}$$

where $\mathbf{g}_k(\mathbf{x})$, $k = 1, \dots, m$, are the columns of the matrix $g(\mathbf{x})$.

Proof. The identity $g^0 = \overline{g}$ means that (4.23) becomes equality for any $\mathbf{C} \in \mathbb{C}^m$. By (4.22), this is equivalent to the relation $b(\mathbf{D})\mathbf{v} = 0$, where $\mathbf{v} \in \widetilde{H}^1(\Omega^0; \mathbb{C}^n)$ is the solution of problem (4.20). In this case, (4.20) means that $b(\mathbf{D})^* g(\mathbf{x})\mathbf{C} = 0$, $\mathbf{C} \in \mathbb{C}^m$, which is equivalent to (4.28). □

Proposition 4.3. *The identity $g^0 = \underline{g}$ is equivalent to the representations*

$$\mathbf{l}_k(\mathbf{x}) = \mathbf{l}_k^0 + b(\mathbf{D})\mathbf{v}_k(\mathbf{x}), \quad \mathbf{l}_k^0 \in \mathbb{C}^m, \quad \mathbf{v}_k \in \widetilde{H}^1(\Omega^0; \mathbb{C}^n), \quad k = 1, \dots, m, \tag{4.29}$$

where $\mathbf{l}_k(\mathbf{x})$, $k = 1, \dots, m$, are the columns of the matrix $g(\mathbf{x})^{-1}$.

Proof. By (4.25) and (4.26), the identity $g^0 = \underline{g}$ is equivalent to the relation $g^{1/2}(b(\mathbf{D})\mathbf{v} + \mathbf{C}) \in \mathfrak{M}$ for any $\mathbf{C} \in \mathbb{C}^m$. Here $\mathbf{v} \in \widetilde{H}^1(\Omega^0; \mathbb{C}^n)$ is the solution of problem (4.20). In other words, for any $\mathbf{C} \in \mathbb{C}^m$ there exists $\mathbf{C}_* \in \mathbb{C}^m$ such that $g^{1/2}(b(\mathbf{D})\mathbf{v} + \mathbf{C}) = g^{-1/2}\mathbf{C}_*$. Hence, $b(\mathbf{D})\mathbf{v} + \mathbf{C} = g^{-1}\mathbf{C}_*$. Integrating over Ω^0, we obtain $\mathbf{C} = \underline{g}^{-1}\mathbf{C}_*$. Let $\mathbf{e}_1, \dots, \mathbf{e}_m$ be the standard orthonormal basis in $\mathbb{C}^m$. Putting $\mathbf{C}_* = \mathbf{e}_k$, $k = 1, \dots, d$, we arrive at the representations (4.29). Automatically, $\mathbf{l}_k^0 = \underline{g}^{-1}\mathbf{e}_k$ and $\mathbf{v}_k$ is the solution of problem (4.20) with $\mathbf{C} = \mathbf{l}_k^0$. □

Remark 4.4. From the proof of Proposition 4.3 it follows that, under conditions (4.29), the matrix (4.15) is constant: $\widetilde{g}(\mathbf{x}) = g^0 = \underline{g}$.

4.6. *Estimates for Λ and $\widetilde{g}$*

Below we will need the estimates for Λ and $\widetilde{g}$.

Proposition 4.5. *Let $\Lambda \in \widetilde{H}^1(\Omega^0)$ be the solution of problem (4.9). Let $\widetilde{g}$ be given by (4.15). Then*

$$\|\Lambda\|_{L_2(\Omega^0)} \le C_\Lambda^{(1)} |\Omega^0|^{1/2} = \nu^{-1/2} \|g\|_{L_\infty}^{1/2} \|g^{-1}\|_{L_\infty}^{1/2} |\Omega^0|^{1/2}, \tag{4.30}$$

$$\|\Lambda\|_{H^1(\Omega^0)} \le C_\Lambda^{(2)} |\Omega^0|^{1/2} = (\nu^{-1} + \widetilde{\mathfrak{C}}_1)^{1/2} \|g\|_{L_\infty}^{1/2} \|g^{-1}\|_{L_\infty}^{1/2} |\Omega^0|^{1/2}, \tag{4.31}$$

$$\|\widetilde{g}\|_{L_2(\Omega^0)} \le \|g\|_{L_\infty} |\Omega^0|^{1/2}. \tag{4.32}$$

Proof. Let $\mathbf{C} \in \mathbb{C}^m$ and let $\mathbf{v}(\mathbf{x}) = \Lambda(\mathbf{x})\mathbf{C}$. Then $\mathbf{v} \in \widetilde{H}^1(\Omega^0; \mathbb{C}^n)$ satisfies

$$\int_{\Omega^0} \langle g(\mathbf{x})(b(\mathbf{D})\mathbf{v}(\mathbf{x}) + \mathbf{C}), b(\mathbf{D})\mathbf{v}(\mathbf{x})\rangle \, d\mathbf{x} = 0, \qquad \int_{\Omega^0} \mathbf{v}(\mathbf{x}) \, d\mathbf{x} = 0. \tag{4.33}$$

It follows that

$$\|g^{1/2} b(\mathbf{D})\mathbf{v}\|_{L_2(\Omega^0)} \leq \|g^{1/2}\mathbf{C}\|_{L_2(\Omega^0)} \leq \|g\|_{L_\infty}^{1/2} |\Omega^0|^{1/2} |\mathbf{C}|,$$

whence

$$\|b(\mathbf{D})\mathbf{v}\|_{L_2(\Omega^0)} \leq \|g\|_{L_\infty}^{1/2} \|g^{-1}\|_{L_\infty}^{1/2} |\Omega^0|^{1/2} |\mathbf{C}|. \tag{4.34}$$

Then, by (3.17), $\mathbf{v} \in \widetilde{H}^1(\Omega^0; \mathbb{C}^n) \cap \mathfrak{N}^\perp$ satisfies

$$\|\mathbf{v}\|_{L_2(\Omega^0)} \leq \nu^{-1/2} \|g\|_{L_\infty}^{1/2} \|g^{-1}\|_{L_\infty}^{1/2} |\Omega^0|^{1/2} |\mathbf{C}|. \tag{4.35}$$

This implies (4.30). Using (3.24) with $\mathbf{k} = 0$ and (4.34), we obtain

$$\|\mathbf{D}\mathbf{v}\|_{L_2(\Omega^0)} \leq \widetilde{\mathfrak{C}}_1^{1/2} \|b(\mathbf{D})\mathbf{v}\|_{L_2(\Omega^0)} \leq \widetilde{\mathfrak{C}}_1^{1/2} \|g\|_{L_\infty}^{1/2} \|g^{-1}\|_{L_\infty}^{1/2} |\Omega^0|^{1/2} |\mathbf{C}|. \tag{4.36}$$

Estimates (4.35) and (4.36) yield (4.31). Finally, by (4.33),

$$\int_{\Omega^0} \langle g(\mathbf{x})(b(\mathbf{D})\mathbf{v}(\mathbf{x}) + \mathbf{C}), b(\mathbf{D})\mathbf{v}(\mathbf{x}) + \mathbf{C}\rangle \, d\mathbf{x} = \int_{\Omega^0} \langle g(\mathbf{x})(b(\mathbf{D})\mathbf{v}(\mathbf{x}) + \mathbf{C}), \mathbf{C}\rangle \, d\mathbf{x}.$$

Consequently, $\|g^{1/2}(b(\mathbf{D})\mathbf{v}+\mathbf{C})\| \leq \|g\|_{L_\infty}^{1/2} |\Omega^0|^{1/2} |\mathbf{C}|$, which implies estimate (4.32). □

4.7. *The effective operator*

We put

$$S(\mathbf{k}) := t^2 S(\boldsymbol{\theta}) = b(\mathbf{k})^* g^0 b(\mathbf{k}), \quad \mathbf{k} = t\boldsymbol{\theta} \in \mathbb{R}^d. \tag{4.37}$$

Then $S(\mathbf{k})$ is the symbol of the DO $\mathcal{A}^0$ with constant coefficients generated in $L_2(\Pi; \mathbb{C}^n)$ by the quadratic form

$$\mathfrak{a}^0[\mathbf{u}, \mathbf{u}] = \int_\Pi \langle g^0 b(\mathbf{D})\mathbf{u}(\mathbf{x}), b(\mathbf{D})\mathbf{u}(\mathbf{x})\rangle \, d\mathbf{x}, \quad \mathbf{u} \in H^1(\Pi; \mathbb{C}^n). \tag{4.38}$$

By (4.27), we have

$$\|g^{-1}\|_{L_\infty}^{-1} \|b(\mathbf{D})\mathbf{u}\|_{L_2(\Pi)}^2 \leq \mathfrak{a}^0[\mathbf{u}, \mathbf{u}] \leq \|g\|_{L_\infty} \|b(\mathbf{D})\mathbf{u}\|_{L_2(\Pi)}^2, \quad \mathbf{u} \in H^1(\Pi; \mathbb{C}^n).$$

Then the form (4.38) satisfies analogs of estimates (3.10) and (3.32) with the same constants. Hence, this form is nonnegative and closed. The operator $\mathcal{A}^0$ is called the *effective operator* for $\mathcal{A}$. If the boundary $\partial\Pi$ is sufficiently smooth (it suffices to assume that $\partial\Pi \in C^{1,1}$), then $\mathcal{A}^0$ is given by $b(\mathbf{D})^* g^0 b(\mathbf{D})$ on the domain $\{\mathbf{u} \in H^2(\Pi; \mathbb{C}^n) \colon b(\mathbf{n}(\mathbf{x}))^* g^0 b(\mathbf{D})\mathbf{u}|_{\partial\Pi} = 0\}$. Here $\mathbf{n}(\mathbf{x})$ is the outer unit normal vector to $\partial\Pi$ at $\mathbf{x} \in \partial\Pi$.

Similarly to (3.29), the effective operator admits the following expansion:

$$\mathcal{U}\mathcal{A}^0\mathcal{U}^{-1} = \int_{\widetilde{\Omega}} \oplus \mathcal{A}^0(\mathbf{k}) \, d\mathbf{k}. \tag{4.39}$$

Here the operator $\mathcal{A}^0(\mathbf{k})$ acts in $L_2(\Omega^0;\mathbb{C}^n)$ and corresponds to the quadratic form

$$\mathfrak{a}^0(\mathbf{k})[\mathbf{u},\mathbf{u}] = \int_{\Omega^0} \langle g^0 b(\mathbf{D}+\mathbf{k})\mathbf{u}(\mathbf{x}), b(\mathbf{D}+\mathbf{k})\mathbf{u}(\mathbf{x})\rangle\, d\mathbf{x}, \quad \mathbf{u}\in \widetilde{H}^1(\Omega^0;\mathbb{C}^n). \tag{4.40}$$

By (4.27), we have

$$\|g^{-1}\|_{L_\infty}^{-1}\|b(\mathbf{D}+\mathbf{k})\mathbf{u}\|^2_{L_2(\Omega^0)} \le \mathfrak{a}^0(\mathbf{k})[\mathbf{u},\mathbf{u}] \le \|g\|_{L_\infty}\|b(\mathbf{D}+\mathbf{k})\mathbf{u}\|^2_{L_2(\Omega^0)}, \quad \mathbf{u}\in \widetilde{H}^1(\Omega^0;\mathbb{C}^n). \tag{4.41}$$

Then the form (4.40) satisfies analogs of estimates (3.14) and (3.28) with the same constants. Hence, this form is closed and nonnegative.

4.8. *The operator $N(\boldsymbol{\theta})$*

Now we construct the operator $N(\boldsymbol{\theta})$ that plays the role of N for the operator family $A(t;\boldsymbol{\theta})$. By (1.7), $N(\boldsymbol{\theta}) = Z(\boldsymbol{\theta})^* X_1(\boldsymbol{\theta})^* R(\boldsymbol{\theta})P + (R(\boldsymbol{\theta})P)^* X_1(\boldsymbol{\theta}) Z(\boldsymbol{\theta})$. According to (4.10) and (4.12), we obtain $N(\boldsymbol{\theta}) = Pb(\boldsymbol{\theta})^*(\Lambda^* b(\boldsymbol{\theta})^*\widetilde{g} + \widetilde{g}^* b(\boldsymbol{\theta})\Lambda)\, b(\boldsymbol{\theta})P$. Here $\widetilde{g}$ is given by (4.15). We introduce the notation

$$L(\boldsymbol{\theta}) := |\Omega^0|^{-1}\int_{\Omega^0} (\Lambda(\mathbf{x})^* b(\boldsymbol{\theta})^*\widetilde{g}(\mathbf{x}) + \widetilde{g}(\mathbf{x})^* b(\boldsymbol{\theta})\Lambda(\mathbf{x}))\, d\mathbf{x}. \tag{4.42}$$

Then

$$N(\boldsymbol{\theta}) = b(\boldsymbol{\theta})^* L(\boldsymbol{\theta}) b(\boldsymbol{\theta})P. \tag{4.43}$$

According to (1.8),

$$(N(\boldsymbol{\theta})\boldsymbol{\omega}_l(\boldsymbol{\theta}), \boldsymbol{\omega}_l(\boldsymbol{\theta}))_{L_2(\Omega^0)} = \mu_l(\boldsymbol{\theta}), \quad l = 1,\dots,n, \tag{4.44}$$

where $\mu_l(\boldsymbol{\theta})$ are the coefficients at t^3 in the expansions (4.18).

By (1.9), (4.4), and (4.5),

$$\|N(\boldsymbol{\theta})\|_{L_2(\Omega^0)\to L_2(\Omega^0)} \le (2\delta)^{-1/2}\alpha_1^{3/2}\|g\|^{3/2}_{L_\infty} = \sqrt{8}\nu^{-1/2}\alpha_1^{3/2}\|g\|^{3/2}_{L_\infty}\|g^{-1}\|^{1/2}_{L_\infty} =: C_N. \tag{4.45}$$

By (3.3), (4.30), and (4.32), the matrix (4.42) satisfies

$$|L(\boldsymbol{\theta})| \le 2\alpha_1^{1/2} C_\Lambda^{(1)}\|g\|_{L_\infty} =: C_L. \tag{4.46}$$

Observe that $L(\mathbf{k}) := tL(\boldsymbol{\theta})$ is a Hermitian $(m\times m)$-matrix-valued function first order homogeneous in $\mathbf{k}\in\mathbb{R}^d$. We put $N(\mathbf{k}) := t^3 N(\boldsymbol{\theta})$. Then

$$N(\mathbf{k}) = b(\mathbf{k})^* L(\mathbf{k}) b(\mathbf{k})P. \tag{4.47}$$

Let us distinguish some cases where $N(\boldsymbol{\theta}) = 0$. If $b(\boldsymbol{\theta})$ and $g(\mathbf{x})$ are matrices with *real entries*, then $\Lambda(\mathbf{x})$ has purely imaginary entries and $\widetilde{g}(\mathbf{x})$ has real entries. Hence, $L(\boldsymbol{\theta})$ and $b(\boldsymbol{\theta})^* L(\boldsymbol{\theta}) b(\boldsymbol{\theta})$ are Hermitian matrices with purely imaginary entries. In the case $n = 1$, it follows that $b(\boldsymbol{\theta})^* L(\boldsymbol{\theta}) b(\boldsymbol{\theta}) = 0$. We arrive at the following statement.

Proposition 4.6. *For the operator* $\mathcal{A} = \mathbf{D}^* g(\mathbf{x})\mathbf{D}$, *where the matrix* $g(\mathbf{x})$ *has real entries, we have* $N(\boldsymbol{\theta}) = 0$, $\boldsymbol{\theta} \in \mathbb{S}^{d-1}$. *The coefficient* $\mu_1(\boldsymbol{\theta})$ *at* t^3 *in the expansion* (4.18) *for the first eigenvalue* $\lambda_1(t;\boldsymbol{\theta})$ *is equal to zero.*

Some other cases where $N(\boldsymbol{\theta}) = 0$ are described by the following statement.

Proposition 4.7. (1) *Under conditions* (4.28) (*i.e., in the case where* $g^0 = \overline{g}$), *we have* $L(\boldsymbol{\theta}) = 0$ *and* $N(\boldsymbol{\theta}) = 0$, $\boldsymbol{\theta} \in \mathbb{S}^{d-1}$. (2) *Under conditions* (4.29) (*i.e., in the case where* $g^0 = \underline{g}$), *we have* $L(\boldsymbol{\theta}) = 0$ *and* $N(\boldsymbol{\theta}) = 0$, $\boldsymbol{\theta} \in \mathbb{S}^{d-1}$.

Proof. Obviously, under conditions (4.28), we have $\Lambda = 0$, whence $L(\boldsymbol{\theta}) = 0$ and $N(\boldsymbol{\theta}) = 0$.

By Remark 4.4, under conditions (4.29), the matrix $\widetilde{g}(\mathbf{x})$ is constant: $\widetilde{g}(\mathbf{x}) = g^0 = \underline{g}$. By (4.42) and the condition $\int_{\Omega^0} \Lambda(\mathbf{x})\, d\mathbf{x} = 0$, it follows that $L(\boldsymbol{\theta}) = 0$. Hence, $N(\boldsymbol{\theta}) = 0$. □

4.9. *The effective characteristics of the operator* $\widetilde{\mathcal{A}}$

Let $\widetilde{\mathcal{A}}$ be the operator (3.33), and let $\widetilde{\mathcal{A}}(\mathbf{k})$ be the corresponding operator family in $L_2(\Omega^0;\mathbb{C}^n)$ generated by the form (3.35). Then the operator $\widetilde{A}(t;\boldsymbol{\theta}) = \widetilde{\mathcal{A}}(t\boldsymbol{\theta})$ satisfies the assumptions of Sec. 1.7. Now $\widetilde{\mathfrak{H}} = \mathfrak{H} = L_2(\Omega^0;\mathbb{C}^n)$; the roles of $A(t)$ and M are played by the operator $A(t;\boldsymbol{\theta})$ and the operator of multiplication by the matrix-valued function f. We have $\widetilde{A}(t;\boldsymbol{\theta}) = f^* A(t;\boldsymbol{\theta}) f$.

By the variational arguments, the distance $\widetilde{d}^0$ from the point $\lambda_0 = 0$ to the rest of the spectrum of $\widetilde{\mathcal{A}}(0)$ satisfies $\widetilde{d}^0 \geq d^0 \|f^{-1}\|^{-2}_{L_\infty}$. Then, by (4.3), $\widetilde{d}^0 \geq \nu \|g^{-1}\|^{-1}_{L_\infty} \|f^{-1}\|^{-2}_{L_\infty}$. We fix a number $\widetilde{\delta} \in (0, \widetilde{d}^0/8)$ as follows: $\widetilde{\delta} = \frac{1}{16}\nu \|g^{-1}\|^{-1}_{L_\infty} \|f^{-1}\|^{-2}_{L_\infty}$. Next, the operator $\widetilde{X}_1(\boldsymbol{\theta}) = g^{1/2} b(\boldsymbol{\theta}) f$ satisfies $\|\widetilde{X}_1(\boldsymbol{\theta})\| \leq \alpha_1^{1/2} \|g\|^{1/2}_{L_\infty} \|f\|_{L_\infty}$. This allows us to choose the number $\widetilde{t}^0 \leq \widetilde{\delta}^{1/2} \|\widetilde{X}_1(\boldsymbol{\theta})\|^{-1}$ as follows:

$$\widetilde{t}^0 = \frac{1}{4} \nu^{1/2} \alpha_1^{-1/2} \|g\|^{-1/2}_{L_\infty} \|g^{-1}\|^{-1/2}_{L_\infty} \|f\|^{-1}_{L_\infty} \|f^{-1}\|^{-1}_{L_\infty}.$$

Note that $\widetilde{t}^0 \leq t^0$. Finally, from (4.7) it follows that

$$\widetilde{A}(t;\boldsymbol{\theta}) \geq \widetilde{c}_* t^2 I, \quad \widetilde{c}_* = \alpha_0 \|g^{-1}\|^{-1}_{L_\infty} \|f^{-1}\|^{-2}_{L_\infty}, \quad t = |\mathbf{k}| \leq \widetilde{t}^0.$$

The role of Q is played by the operator of multiplication by the matrix-valued function $Q(\mathbf{x}) := (f(\mathbf{x}) f(\mathbf{x})^*)^{-1}$. The block of Q in the subspace $\mathfrak{N}$ is multiplication by the constant matrix $\overline{Q} = |\Omega^0|^{-1} \int_{\Omega^0} Q(\mathbf{x})\, d\mathbf{x}$.

The analytic (in t) branches of the eigenvalues $\widetilde{\lambda}_l(t;\boldsymbol{\theta})$ and the branches of the eigenvectors $\widetilde{\varphi}_l(t;\boldsymbol{\theta})$ of $\widetilde{A}(t;\boldsymbol{\theta})$ depend on $\boldsymbol{\theta}$. We have

$$\widetilde{A}(t;\boldsymbol{\theta}) \widetilde{\varphi}_l(t;\boldsymbol{\theta}) = \widetilde{\lambda}_l(t;\boldsymbol{\theta}) \widetilde{\varphi}_l(t;\boldsymbol{\theta}), \quad t = |\mathbf{k}| \leq \widetilde{t}^0, \quad l = 1, \ldots, n.$$

According to (1.15) and (1.16), for small t we have the convergent power series expansions:

$$\widetilde{\lambda}_l(t;\boldsymbol{\theta}) = \widetilde{\gamma}_l(\boldsymbol{\theta})t^2 + \widetilde{\mu}_l(\boldsymbol{\theta})t^3 + \cdots, \quad l=1,\dots,n,$$
$$\widetilde{\varphi}_l(t;\boldsymbol{\theta}) = \widetilde{\omega}_l(\boldsymbol{\theta}) + t\widetilde{\psi}_l^{(1)}(\boldsymbol{\theta}) + \cdots, \quad l=1,\dots,n. \tag{4.48}$$

By (1.17), the numbers $\widetilde{\gamma}_l(\boldsymbol{\theta})$ and the vectors $\phi_l(\boldsymbol{\theta}) = f\widetilde{\omega}_l(\boldsymbol{\theta}) \in \mathfrak{N}$ satisfy

$$b(\boldsymbol{\theta})^* g^0 b(\boldsymbol{\theta})\phi_l(\boldsymbol{\theta}) = \widetilde{\gamma}_l(\boldsymbol{\theta})\overline{Q}\phi_l(\boldsymbol{\theta}), \quad l=1,\dots,n. \tag{4.49}$$

Similarly to (4.9), (4.10), the role of Z_Q is played by the operator $Z_Q(\boldsymbol{\theta}) = \Lambda_Q b(\boldsymbol{\theta})P$. Here $\Lambda_Q(\mathbf{x})$ is the $(n\times m)$-matrix-valued function such that $\Lambda_Q \in \widetilde{H}^1(\Omega^0)$ and

$$b(\mathbf{D})^* g(\mathbf{x})(b(\mathbf{D})\Lambda_Q(\mathbf{x}) + \mathbf{1}_m) = 0, \quad \mathbf{x}\in\Omega^0; \quad \int_{\Omega^0} Q(\mathbf{x})\Lambda_Q(\mathbf{x})\,d\mathbf{x} = 0.$$

Clearly, Λ_Q differs from Λ by a constant summand:

$$\Lambda_Q(\mathbf{x}) = \Lambda(\mathbf{x}) + \Lambda_Q^0, \quad \Lambda_Q^0 = -(\overline{Q})^{-1}\overline{Q\Lambda} = -|\Omega^0|^{-1}(\overline{Q})^{-1}\int_{\Omega^0} Q(\mathbf{x})\Lambda(\mathbf{x})\,d\mathbf{x}. \tag{4.50}$$

Next, the role of N_Q is played by the operator $N_Q(\boldsymbol{\theta})$ given by $N_Q(\boldsymbol{\theta}) = Z_Q(\boldsymbol{\theta})^* X_1(\boldsymbol{\theta})^* R(\boldsymbol{\theta})P + (R(\boldsymbol{\theta})P)^* X_1(\boldsymbol{\theta}) Z_Q(\boldsymbol{\theta})$. Similarly to (4.42) and (4.43), we obtain $N_Q(\boldsymbol{\theta}) = b(\boldsymbol{\theta})^* L_Q(\boldsymbol{\theta}) b(\boldsymbol{\theta})P$, where

$$L_Q(\boldsymbol{\theta}) := |\Omega^0|^{-1}\int_{\Omega^0} (\Lambda_Q(\mathbf{x})^* b(\boldsymbol{\theta})^* \widetilde{g}(\mathbf{x}) + \widetilde{g}(\mathbf{x})^* b(\boldsymbol{\theta})\Lambda_Q(\mathbf{x}))\,d\mathbf{x}.$$

Clearly, $L_Q(\boldsymbol{\theta}) = L(\boldsymbol{\theta}) + L_Q^0(\boldsymbol{\theta})$, where $L_Q^0(\boldsymbol{\theta}) = (\Lambda_Q^0)^* b(\boldsymbol{\theta})^* g^0 + g^0 b(\boldsymbol{\theta})\Lambda_Q^0$. According to (1.18),

$$(N_Q(\boldsymbol{\theta})\phi_l(\boldsymbol{\theta}), \phi_l(\boldsymbol{\theta}))_{L_2(\Omega^0)} = \widetilde{\mu}_l(\boldsymbol{\theta}), \quad l=1,\dots,n, \tag{4.51}$$

where $\widetilde{\mu}_l(\boldsymbol{\theta})$ are the coefficients at t^3 in the expansions (4.48).

The following statement can be checked similarly to Proposition 4.6.

Proposition 4.8. *For the operator $\widetilde{A} = f(\mathbf{x})^* \mathbf{D}^* g(\mathbf{x}) \mathbf{D} f(\mathbf{x})$, where the matrix $g(\mathbf{x})$ has real entries, we have $N_Q(\boldsymbol{\theta}) = 0$, $\boldsymbol{\theta} \in \mathbb{S}^{d-1}$.*

Remark 4.9. Under conditions (4.28) (i.e., in the case where $g^0 = \underline{g}$), we have $\Lambda_Q = 0$, whence $Z_Q(\boldsymbol{\theta}) = 0$ and $N_Q(\boldsymbol{\theta}) = 0$, $\boldsymbol{\theta} \in \mathbb{S}^{d-1}$.

5. Approximations for the Resolvent $(\mathcal{A}(\mathbf{k}) + \varepsilon^2 I)^{-1}$

5.1. *Approximation for the resolvent $(\mathcal{A}(\mathbf{k}) + \varepsilon^2 I)^{-1}$ in the $(L_2 \to L_2)$-norm*

We are going to apply Theorem 1.3 to the operator family $\mathcal{A}(\mathbf{k}) = A(t;\boldsymbol{\theta})$. The number t^0 is given by (4.6); it does not depend on $\boldsymbol{\theta}$. By (4.2) and (4.37), we have

$$t^2 S(\boldsymbol{\theta})P = S(\mathbf{k})P = b(\mathbf{k})^* g^0 b(\mathbf{k})P = b(\mathbf{D}+\mathbf{k})^* g^0 b(\mathbf{D}+\mathbf{k})P = \mathcal{A}^0(\mathbf{k})P. \tag{5.1}$$

Then the role of $\Xi(t,\varepsilon)$ is played by $(\mathcal{A}^0(\mathbf{k})+\varepsilon^2 I)^{-1}P$. Consequently, the operator in (1.10) takes the form $(\mathcal{A}(\mathbf{k})+\varepsilon^2 I)^{-1}-(\mathcal{A}^0(\mathbf{k})+\varepsilon^2 I)^{-1}P$. By Remark 1.6, the constant C_1 from (1.10) is a polynomial with absolute positive coefficients of $\delta^{-1/2}$, $\|X_1(\boldsymbol{\theta})\|$, and $c_*^{-1/2}$. Now the parameters c_* and δ are given by (3.18) and (4.4); they do not depend on $\boldsymbol{\theta}$. Due to (4.5), we can replace $\|X_1(\boldsymbol{\theta})\|$ by $\alpha_1^{1/2}\|g\|^{1/2}_{L_\infty}$. Then the constant C_1 becomes independent of $\boldsymbol{\theta}$; it depends only on α_0, α_1, $\|g\|_{L_\infty}$, $\|g^{-1}\|_{L_\infty}$, and ν (and so, on the operator $b(\mathbf{D})$ and the domain Ω^0).

By Theorem 1.3, we have

$$\|(\mathcal{A}(\mathbf{k})+\varepsilon^2 I)^{-1}-(\mathcal{A}^0(\mathbf{k})+\varepsilon^2 I)^{-1}P\|_{L_2(\Omega^0)\to L_2(\Omega^0)}\le C_1\varepsilon^{-1},\quad \varepsilon>0,\quad |\mathbf{k}|\le t^0. \tag{5.2}$$

Applying Lemma 3.5 to the operators $\mathcal{A}(\mathbf{k})$ and $\mathcal{A}^0(\mathbf{k})$, we obtain

$$\begin{aligned}\|(\mathcal{A}(\mathbf{k})+\varepsilon^2 I)^{-1}\|_{L_2(\Omega^0)\to L_2(\Omega^0)}&\le C(t^0)^{-1},\quad \varepsilon>0,\quad \mathbf{k}\in\widetilde{\Omega}\backslash B(t^0),\\ \|(\mathcal{A}^0(\mathbf{k})+\varepsilon^2 I)^{-1}\|_{L_2(\Omega^0)\to L_2(\Omega^0)}&\le C(t^0)^{-1},\quad \varepsilon>0,\quad \mathbf{k}\in\widetilde{\Omega}\backslash B(t^0).\end{aligned} \tag{5.3}$$

The constant $C(t^0)$ depends only on $\|g\|_{L_\infty}$, $\|g^{-1}\|_{L_\infty}$, the operator $b(\mathbf{D})$, and the domain Ω^0. Thus, by (5.2) and (5.3),

$$\|(\mathcal{A}(\mathbf{k})+\varepsilon^2 I)^{-1}-(\mathcal{A}^0(\mathbf{k})+\varepsilon^2 I)^{-1}P\|_{L_2(\Omega^0)\to L_2(\Omega^0)}\le C_4\varepsilon^{-1},\quad \varepsilon>0,\quad \mathbf{k}\in\widetilde{\Omega}, \tag{5.4}$$

where $C_4=\max\{C_1, 2C(t^0)^{-1/2}\}$. Next, (3.27) and (4.41) imply that

$$\|(\mathcal{A}^0(\mathbf{k})+\varepsilon^2 I)^{-1}P^\perp\|_{L_2(\Omega^0)\to L_2(\Omega^0)}\le c_4^{-1}\|g^{-1}\|_{L_\infty},\quad \varepsilon>0,\quad \mathbf{k}\in\widetilde{\Omega}. \tag{5.5}$$

Combining estimates (5.4) and (5.5), we arrive at the following result.

Theorem 5.1. *For $\varepsilon>0$ and $\mathbf{k}\in\widetilde{\Omega}$, we have*

$$\|(\mathcal{A}(\mathbf{k})+\varepsilon^2 I)^{-1}-(\mathcal{A}^0(\mathbf{k})+\varepsilon^2 I)^{-1}\|_{L_2(\Omega^0)\to L_2(\Omega^0)}\le \mathcal{C}_1\varepsilon^{-1}.$$

The constant $\mathcal{C}_1$ depends only on $\|g\|_{L_\infty}$, $\|g^{-1}\|_{L_\infty}$, the operator $b(\mathbf{D})$, and the domain Ω^0.

5.2. *More accurate approximation of the resolvent $(\mathcal{A}(\mathbf{k})+\varepsilon^2 I)^{-1}$ in the $(L_2\to L_2)$-norm*

Now, we will apply Theorem 1.4 to the operator family $\mathcal{A}(\mathbf{k})=A(t;\boldsymbol{\theta})$. By (4.10),

$$tZ(\boldsymbol{\theta})P=\Lambda b(\mathbf{k})P=\Lambda b(\mathbf{D}+\mathbf{k})P. \tag{5.6}$$

Next, by (4.47), $t^3N(\boldsymbol{\theta})=b(\mathbf{k})^*L(\mathbf{k})b(\mathbf{k})P=b(\mathbf{D}+\mathbf{k})^*L(\mathbf{D}+\mathbf{k})b(\mathbf{D}+\mathbf{k})P$. Consequently, the operator (1.11) takes the form

$$\begin{aligned}\mathcal{K}(\mathbf{k},\varepsilon)&=\Lambda b(\mathbf{D}+\mathbf{k})(\mathcal{A}^0(\mathbf{k})+\varepsilon^2 I)^{-1}P+(\mathcal{A}^0(\mathbf{k})+\varepsilon^2 I)^{-1}Pb(\mathbf{D}+\mathbf{k})^*\Lambda^*\\&\quad-(\mathcal{A}^0(\mathbf{k})+\varepsilon^2 I)^{-1}b(\mathbf{D}+\mathbf{k})^*L(\mathbf{D}+\mathbf{k})b(\mathbf{D}+\mathbf{k})(\mathcal{A}^0(\mathbf{k})+\varepsilon^2 I)^{-1}P.\end{aligned} \tag{5.7}$$

Note that, by (4.11) and (4.45),

$$\|\mathcal{K}(\mathbf{k},\varepsilon)\|_{L_2(\Omega^0)\to L_2(\Omega^0)} \le 2C_Z|\mathbf{k}|\|(\mathcal{A}^0(\mathbf{k})+\varepsilon^2 I)^{-1}\|_{L_2(\Omega^0)\to L_2(\Omega^0)} + C_N|\mathbf{k}|^3\|(\mathcal{A}^0(\mathbf{k})+\varepsilon^2 I)^{-1}\|^2_{L_2(\Omega^0)\to L_2(\Omega^0)}. \quad (5.8)$$

Applying (1.12), we obtain

$$\|(\mathcal{A}(\mathbf{k})+\varepsilon^2 I)^{-1} - (\mathcal{A}^0(\mathbf{k})+\varepsilon^2 I)^{-1}P - \mathcal{K}(\mathbf{k},\varepsilon)\|_{L_2(\Omega^0)\to L_2(\Omega^0)} \le C_2, \quad |\mathbf{k}|\le t^0, \quad \varepsilon>0. \quad (5.9)$$

Here the constant C_2 does not depend on $\boldsymbol{\theta}$; it depends only on α_0, α_1, $\|g\|_{L_\infty}$, $\|g^{-1}\|_{L_\infty}$, and ν. (The reasons are the same as for C_1; cf. Sec. 5.1.)

For $\mathbf{k}\in\widetilde{\Omega}\backslash B(t^0)$ estimates are trivial. The norms of the resolvents $(\mathcal{A}(\mathbf{k})+\varepsilon^2 I)^{-1}$ and $(\mathcal{A}^0(\mathbf{k})+\varepsilon^2 I)^{-1}$ satisfy (5.3). For the operator (5.7), from (5.3) and (5.8) we obtain

$$\|\mathcal{K}(\mathbf{k},\varepsilon)\|_{L_2(\Omega^0)\to L_2(\Omega^0)} \le 2C_Z r_1 C(t^0)^{-1} + C_N r_1^3 C(t^0)^{-2}, \quad \varepsilon>0, \quad \mathbf{k}\in\widetilde{\Omega}\backslash B(t^0).$$

Consequently,

$$\|(\mathcal{A}(\mathbf{k})+\varepsilon^2 I)^{-1} - (\mathcal{A}^0(\mathbf{k})+\varepsilon^2 I)^{-1}P - \mathcal{K}(\mathbf{k},\varepsilon)\|_{L_2(\Omega^0)\to L_2(\Omega^0)} \le C_5, \quad \varepsilon>0, \quad \mathbf{k}\in\widetilde{\Omega}\backslash B(t^0), \quad (5.10)$$

where $C_5 = 2C(t^0)^{-1} + 2C_Z r_1 C(t^0)^{-1} + C_N r_1^3 C(t^0)^{-2}$.

Finally, estimates (5.5), (5.9), and (5.10) imply the following result.

Theorem 5.2. *For $\varepsilon>0$ and $\mathbf{k}\in\widetilde{\Omega}$, we have*

$$\|(\mathcal{A}(\mathbf{k})+\varepsilon^2 I)^{-1} - (\mathcal{A}^0(\mathbf{k})+\varepsilon^2 I)^{-1} - \mathcal{K}(\mathbf{k},\varepsilon)\|_{L_2(\Omega^0)\to L_2(\Omega^0)} \le \mathcal{C}_2.$$

The constant $\mathcal{C}_2$ depends only on $\|g\|_{L_\infty}$, $\|g^{-1}\|_{L_\infty}$, the operator $b(\mathbf{D})$, and the domain Ω^0.

5.3. *Approximation of the resolvent $(\mathcal{A}(\mathbf{k})+\varepsilon^2 I)^{-1}$ in the "energy norm"*

Now, we will apply Theorem 1.5 to the operator family $\mathcal{A}(\mathbf{k}) = A(t;\boldsymbol{\theta})$. By (5.1) and (5.6), the operator in (1.13) takes the form

$$\mathcal{A}(\mathbf{k})^{1/2}((\mathcal{A}(\mathbf{k})+\varepsilon^2 I)^{-1} - (I+\Lambda b(\mathbf{D}+\mathbf{k}))(\mathcal{A}^0(\mathbf{k})+\varepsilon^2 I)^{-1}P).$$

By (1.13), we obtain

$$\|\mathcal{A}(\mathbf{k})^{1/2}((\mathcal{A}(\mathbf{k})+\varepsilon^2 I)^{-1} - (I+\Lambda b(\mathbf{D}+\mathbf{k}))(\mathcal{A}^0(\mathbf{k})+\varepsilon^2 I)^{-1}P)\|_{L_2(\Omega^0)\to L_2(\Omega^0)} \le C_3, \quad |\mathbf{k}|\le t^0, \quad (5.11)$$

for $\varepsilon>0$. Here the constant C_3 does not depend on $\boldsymbol{\theta}$; it depends only on α_0, α_1, $\|g\|_{L_\infty}$, $\|g^{-1}\|_{L_\infty}$, and ν. (The reasons are the same as for C_1; cf. Sec. 5.1.)

Now, we consider the case $\mathbf{k}\in\widetilde{\Omega}\backslash B(t^0)$. By (5.3),

$$\|\mathcal{A}(\mathbf{k})^{1/2}(\mathcal{A}(\mathbf{k})+\varepsilon^2 I)^{-1}\|_{L_2(\Omega^0)\to L_2(\Omega^0)} \le C(t^0)^{-1/2}, \quad \varepsilon>0, \quad \mathbf{k}\in\widetilde{\Omega}\backslash B(t^0). \quad (5.12)$$

By the lower estimate (4.41) and (5.3),

$$\begin{aligned}
&\|\mathcal{A}(\mathbf{k})^{1/2}(\mathcal{A}^0(\mathbf{k})+\varepsilon^2 I)^{-1}\|_{L_2(\Omega^0)\to L_2(\Omega^0)}\\
&\quad\le \|g\|_{L_\infty}^{1/2}\|b(\mathbf{D}+\mathbf{k})(\mathcal{A}^0(\mathbf{k})+\varepsilon^2 I)^{-1}\|_{L_2(\Omega^0)\to L_2(\Omega^0)}\\
&\quad\le \|g\|_{L_\infty}^{1/2}\|g^{-1}\|_{L_\infty}^{1/2}C(t^0)^{-1/2},\quad \varepsilon>0,\quad \mathbf{k}\in\widetilde{\Omega}\backslash B(t^0).
\end{aligned}\tag{5.13}$$

Note that

$$\begin{aligned}
&\|\mathcal{A}(\mathbf{k})^{1/2}\Lambda b(\mathbf{D}+\mathbf{k})(\mathcal{A}^0(\mathbf{k})+\varepsilon^2 I)^{-1}P\|_{L_2(\Omega^0)\to L_2(\Omega^0)}\\
&\quad\le \|\mathcal{A}(\mathbf{k})^{1/2}\Lambda b(\mathbf{k})P\|_{L_2(\Omega^0)\to L_2(\Omega^0)}\|(\mathcal{A}^0(\mathbf{k})+\varepsilon^2 I)^{-1}\|_{L_2(\Omega^0)\to L_2(\Omega^0)}.
\end{aligned}\tag{5.14}$$

By (4.5) and (4.11),

$$\begin{aligned}
\|\mathcal{A}(\mathbf{k})^{1/2}\Lambda b(\mathbf{k})P\|_{L_2(\Omega^0)\to L_2(\Omega^0)} &= \|(X_0+tX_1(\boldsymbol{\theta}))tZ(\boldsymbol{\theta})\|_{L_2(\Omega^0)\to L_2(\Omega^0)}\\
&\le \alpha_1^{1/2}\|g\|_{L_\infty}^{1/2}(r_1+r_1^2C_Z).
\end{aligned}$$

Together with (5.3) and (5.14), this yields

$$\begin{aligned}
\|\mathcal{A}(\mathbf{k})^{1/2}\Lambda b(\mathbf{D}+\mathbf{k})(\mathcal{A}^0(\mathbf{k})+\varepsilon^2 I)^{-1}P\|_{L_2(\Omega^0)\to L_2(\Omega^0)} &\le C_6,\\
\varepsilon>0,\quad \mathbf{k}\in\widetilde{\Omega}\backslash B(t^0),&
\end{aligned}\tag{5.15}$$

where $C_6=C(t^0)^{-1}\alpha_1^{1/2}\|g\|_{L_\infty}^{1/2}(r_1+r_1^2C_Z)$.

Relations (5.11)–(5.13) and (5.15) imply that

$$\begin{aligned}
&\|\mathcal{A}(\mathbf{k})^{1/2}((\mathcal{A}(\mathbf{k})+\varepsilon^2 I)^{-1}\\
&\quad-(I+\Lambda b(\mathbf{D}+\mathbf{k}))(\mathcal{A}^0(\mathbf{k})+\varepsilon^2 I)^{-1}P)\|_{L_2(\Omega^0)\to L_2(\Omega^0)}\le C_7,\quad \mathbf{k}\in\widetilde{\Omega},
\end{aligned}\tag{5.16}$$

where $C_7=\max\{C_3,(1+\|g\|_{L_\infty}^{1/2}\|g^{-1}\|_{L_\infty}^{1/2})C(t^0)^{-1/2}+C_6\}$.

Finally, by the lower estimate (4.41) and (5.5),

$$\|\mathcal{A}(\mathbf{k})^{1/2}(\mathcal{A}^0(\mathbf{k})+\varepsilon^2 I)^{-1}P^{\perp}\|_{L_2(\Omega^0)\to L_2(\Omega^0)}\le \|g\|_{L_\infty}^{1/2}\|g^{-1}\|_{L_\infty}c_4^{-1/2},\quad \mathbf{k}\in\widetilde{\Omega}.\tag{5.17}$$

Estimates (5.16) and (5.17) imply the following result.

Theorem 5.3. *For $\varepsilon>0$ and $\mathbf{k}\in\widetilde{\Omega}$, we have*

$$\|\mathcal{A}(\mathbf{k})^{1/2}((\mathcal{A}(\mathbf{k})+\varepsilon^2 I)^{-1}-(I+\Lambda b(\mathbf{D}+\mathbf{k})P)(\mathcal{A}^0(\mathbf{k})+\varepsilon^2 I)^{-1})\|_{L_2(\Omega^0)\to L_2(\Omega^0)}\le \mathcal{C}_3.$$

The constant $\mathcal{C}_3$ depends only on $\|g\|_{L_\infty}$, $\|g^{-1}\|_{L_\infty}$, the operator $b(\mathbf{D})$, and the domain Ω^0.

5.4. *Approximations for the resolvent $(\widetilde{\mathcal{A}}(\mathbf{k})+\varepsilon^2 I)^{-1}$ and the generalized resolvent $(\mathcal{A}(\mathbf{k})+\varepsilon^2 Q)^{-1}$*

Now, we consider the operator family $\widetilde{\mathcal{A}}(\mathbf{k})$ in $L_2(\Omega^0;\mathbb{C}^n)$ generated by the form (3.35). The corresponding effective characteristics were described in Sec. 4.9.

Similarly to the proof of Theorems 5.1 and 5.2, using the results of Sec. 4.9, we deduce the following results from Theorems 1.7 and 1.8. The details are omitted.

Theorem 5.4. *For $\varepsilon > 0$ and $\mathbf{k} \in \widetilde{\Omega}$, we have*

$$\|(\widetilde{\mathcal{A}}(\mathbf{k}) + \varepsilon^2 I)^{-1} - f^{-1}(\mathcal{A}^0(\mathbf{k}) + \varepsilon^2 \overline{Q})^{-1}(f^*)^{-1}\|_{L_2(\Omega^0)\to L_2(\Omega^0)} \leq \widetilde{C}_1 \varepsilon^{-1},$$

$$\|(\mathcal{A}(\mathbf{k}) + \varepsilon^2 Q)^{-1} - (\mathcal{A}^0(\mathbf{k}) + \varepsilon^2 \overline{Q})^{-1}\|_{L_2(\Omega^0)\to L_2(\Omega^0)} \leq \widetilde{C}_1 \|f\|^2_{L_\infty} \varepsilon^{-1}.$$

The constant $\widetilde{C}_1$ depends on $\|g\|_{L_\infty}$, $\|g^{-1}\|_{L_\infty}$, $\|f\|_{L_\infty}$, $\|f^{-1}\|_{L_\infty}$, the operator $b(\mathbf{D})$, and Ω^0.

Theorem 5.5. *Let*

$$\mathcal{K}_Q(\mathbf{k}, \varepsilon) = \Lambda_Q b(\mathbf{D} + \mathbf{k})(\mathcal{A}^0(\mathbf{k}) + \varepsilon^2 \overline{Q})^{-1} P + (\mathcal{A}^0(\mathbf{k}) + \varepsilon^2 \overline{Q})^{-1} P b(\mathbf{D} + \mathbf{k})^* \Lambda_Q^* \\ - (\mathcal{A}^0(\mathbf{k}) + \varepsilon^2 \overline{Q})^{-1} b(\mathbf{D} + \mathbf{k})^* L_Q(\mathbf{D} + \mathbf{k}) b(\mathbf{D} + \mathbf{k})(\mathcal{A}^0(\mathbf{k}) + \varepsilon^2 \overline{Q})^{-1} P.$$

For $\varepsilon > 0$ and $\mathbf{k} \in \widetilde{\Omega}$, we have

$$\|(\widetilde{\mathcal{A}}(\mathbf{k}) + \varepsilon^2 I)^{-1} - f^{-1}((\mathcal{A}^0(\mathbf{k}) + \varepsilon^2 \overline{Q})^{-1} + \mathcal{K}_Q(\mathbf{k}, \varepsilon))(f^*)^{-1}\|_{L_2(\Omega^0)\to L_2(\Omega^0)} \leq \widetilde{C}_2,$$

$$\|(\mathcal{A}(\mathbf{k}) + \varepsilon^2 Q)^{-1} - (\mathcal{A}^0(\mathbf{k}) + \varepsilon^2 \overline{Q})^{-1} - \mathcal{K}_Q(\mathbf{k}, \varepsilon)\|_{L_2(\Omega^0)\to L_2(\Omega^0)} \leq \widetilde{C}_2 \|f\|^2_{L_\infty}.$$

The constant $\widetilde{C}_2$ depends on $\|g\|_{L_\infty}$, $\|g^{-1}\|_{L_\infty}$, $\|f\|_{L_\infty}$, $\|f^{-1}\|_{L_\infty}$, the operator $b(\mathbf{D})$, and Ω^0.

Similarly to the proof of Theorem 5.3, using Theorem 1.9, we obtain

$$\|\widetilde{\mathcal{A}}(\mathbf{k})^{1/2}((\widetilde{\mathcal{A}}(\mathbf{k}) + \varepsilon^2 I)^{-1} \\ - f^{-1}(I + \Lambda_Q b(\mathbf{D} + \mathbf{k}) P)(\mathcal{A}^0(\mathbf{k}) + \varepsilon^2 \overline{Q})^{-1}(f^*)^{-1})\|_{L_2\to L_2} \leq \check{C}_3 \tag{5.18}$$

for $\varepsilon > 0$ and $\mathbf{k} \in \widetilde{\Omega}$. Recall that $\Lambda_Q(\mathbf{x}) = \Lambda(\mathbf{x}) + \Lambda_Q^0$, where Λ_Q^0 is constant matrix. It is easily seen that, in (5.18), Λ_Q can be replaced by Λ, because the term with Λ_Q^0 is of order $O(1)$. We arrive at the following result.

Theorem 5.6. *For $\varepsilon > 0$ and $\mathbf{k} \in \widetilde{\Omega}$, we have*

$$\|\widetilde{\mathcal{A}}(\mathbf{k})^{1/2}((\widetilde{\mathcal{A}}(\mathbf{k}) + \varepsilon^2 I)^{-1} \\ - f^{-1}(I + \Lambda b(\mathbf{D} + \mathbf{k}) P)(\mathcal{A}^0(\mathbf{k}) + \varepsilon^2 \overline{Q})^{-1}(f^*)^{-1})\|_{L_2(\Omega^0)\to L_2(\Omega^0)} \leq \widetilde{C}_3,$$

$$\|\mathcal{A}(\mathbf{k})^{1/2}((\mathcal{A}(\mathbf{k}) + \varepsilon^2 Q)^{-1} \\ - (I + \Lambda b(\mathbf{D} + \mathbf{k}) P)\,(\mathcal{A}^0(\mathbf{k}) + \varepsilon^2 \overline{Q})^{-1})\|_{L_2(\Omega^0)\to L_2(\Omega^0)} \leq \widetilde{C}_3 \|f\|_{L_\infty}.$$

The constant $\widetilde{C}_3$ depends on $\|g\|_{L_\infty}$, $\|g^{-1}\|_{L_\infty}$, $\|f\|_{L_\infty}$, $\|f^{-1}\|_{L_\infty}$, the operator $b(\mathbf{D})$, and Ω^0.

6. Approximations for the Resolvent $(\mathcal{A}+\varepsilon^2 I)^{-1}$

6.1. *Approximation for the resolvent $(\mathcal{A}+\varepsilon^2 I)^{-1}$ in the $(L_2 \to L_2)$-norm*

Let $\mathcal{A}$ be the operator in $L_2(\Pi;\mathbb{C}^n)$ corresponding to the quadratic form (3.8); see Sec. 3.2. Let $\mathcal{A}^0$ be the effective operator corresponding to the form (4.38); see Sec. 4.7. Using (3.29) and (4.39), we see that

$$\mathcal{U}\big((\mathcal{A}+\varepsilon^2 I)^{-1}-(\mathcal{A}^0+\varepsilon^2 I)^{-1}\big)\mathcal{U}^{-1}$$
$$=\int_{\widetilde{\Omega}}\oplus\big((\mathcal{A}(\mathbf{k})+\varepsilon^2 I)^{-1}-(\mathcal{A}^0(\mathbf{k})+\varepsilon^2 I)^{-1}\big)\,d\mathbf{k}. \tag{6.1}$$

Consequently,

$$\|(\mathcal{A}+\varepsilon^2 I)^{-1}-(\mathcal{A}^0+\varepsilon^2 I)^{-1}\|_{L_2(\Pi)\to L_2(\Pi)}$$
$$=\operatorname*{ess\,sup}_{\mathbf{k}\in\widetilde{\Omega}}\|(\mathcal{A}(\mathbf{k})+\varepsilon^2 I)^{-1}-(\mathcal{A}^0(\mathbf{k})+\varepsilon^2 I)^{-1}\|_{L_2(\Omega^0)\to L_2(\Omega^0)}.$$

Applying Theorem 5.1, we arrive at the following result.

Theorem 6.1. *For $\varepsilon>0$, we have*

$$\|(\mathcal{A}+\varepsilon^2 I)^{-1}-(\mathcal{A}^0+\varepsilon^2 I)^{-1}\|_{L_2(\Omega^0)\to L_2(\Omega^0)}\leq \mathcal{C}_1\varepsilon^{-1}.$$

The constant $\mathcal{C}_1$ depends only on $\|g\|_{L_\infty}$, $\|g^{-1}\|_{L_\infty}$, the operator $b(\mathbf{D})$, and the domain Ω^0.

6.2. *The operator $\mathcal{P}$*

In $L_2(\Pi;\mathbb{C}^n)$, we define the operator

$$\mathcal{P}:=\mathcal{U}^{-1}\Big(\int_{\widetilde{\Omega}}\oplus P\,d\mathbf{k}\Big)\mathcal{U}=\mathcal{U}^{-1}[P]\mathcal{U}. \tag{6.2}$$

Here $[P]$ is the orthogonal projection in $\mathfrak{K}=L_2(\widetilde{\Omega}\times\Omega^0;\mathbb{C}^n)$ that acts on the fibers as the operator P (the operator of averaging over Ω^0). Since $\mathcal{U}$ is a unitary mapping of $L_2(\Pi;\mathbb{C}^n)$ onto $\mathfrak{K}$, the operator $\mathcal{P}$ is an orthogonal projection in $L_2(\Pi;\mathbb{C}^n)$.

Let us calculate the operator $\mathcal{P}$. Let $\mathbf{v}\in\mathcal{S}(\Pi;\mathbb{C}^n)$ and $\widetilde{\mathbf{v}}=\mathcal{U}\mathbf{v}$. Then, by (2.1),

$$([P]\mathcal{U}\mathbf{v})(\mathbf{k})=|\Omega^0|^{-1}\int_{\Omega^0}\widetilde{\mathbf{v}}(\mathbf{k},\mathbf{x})\,d\mathbf{x}=|\Omega^0|^{-1}|\widetilde{\Omega}|^{-1/2}\sum_{\mathbf{a}\in\Gamma}\int_{\Omega^0}e^{-i\langle\mathbf{k},\mathbf{x}+\mathbf{a}\rangle}\mathbf{v}(\mathbf{x}+\mathbf{a})\,d\mathbf{x}$$
$$=|\Omega^0|^{-1}|\widetilde{\Omega}|^{-1/2}\int_{\Pi}e^{-i\langle\mathbf{k},\mathbf{x}\rangle}\mathbf{v}(\mathbf{x})\,d\mathbf{x}=:|\Omega^0|^{-1/2}(\mathcal{F}\mathbf{v})(\mathbf{k}),\quad \mathbf{k}\in\widetilde{\Omega}. \tag{6.3}$$

Here we introduce the operator $\mathcal{F}$ defined on the Schwartz class $\mathcal{S}(\Pi;\mathbb{C}^n)$ by the relation

$$(\mathcal{F}\mathbf{v})(\mathbf{k})=|\Omega^0|^{-1/2}|\widetilde{\Omega}|^{-1/2}\int_{\Pi}e^{-i\langle\mathbf{k},\mathbf{x}\rangle}\mathbf{v}(\mathbf{x})\,d\mathbf{x},\quad \mathbf{k}\in\widetilde{\Omega}.$$

Since $[P]$ is an orthogonal projection and $\mathcal{U}$ is unitary, we have

$$\|[P]\mathcal{U}\mathbf{v}\|_{L_2(\widetilde{\Omega}\times\Omega^0)} \leq \|\mathbf{v}\|_{L_2(\Pi)}, \quad \mathbf{v} \in \mathcal{S}(\Pi;\mathbb{C}^n).$$

By (6.3), it follows that

$$\|\mathcal{F}\mathbf{v}\|_{L_2(\widetilde{\Omega})} \leq \|\mathbf{v}\|_{L_2(\Pi)}, \quad \mathbf{v} \in \mathcal{S}(\Pi;\mathbb{C}^n).$$

By continuity, we extend $\mathcal{F}$ to a bounded mapping from $L_2(\Pi;\mathbb{C}^n)$ to $L_2(\widetilde{\Omega};\mathbb{C}^n)$.

We calculate the quadratic form of $\mathcal{P}$, taking into account that $\mathcal{P}$ is an orthogonal projection:

$$(\mathcal{P}\mathbf{v},\mathbf{v})_{L_2(\Pi)} = \|\mathcal{P}\mathbf{v}\|^2_{L_2(\Pi)} = \|[P]\mathcal{U}\mathbf{v}\|^2_{L_2(\widetilde{\Omega}\times\Omega^0)} = \|\mathcal{F}\mathbf{v}\|^2_{L_2(\widetilde{\Omega})} = (\mathcal{F}^*\mathcal{F}\mathbf{v},\mathbf{v})_{L_2(\Pi)}.$$

Thus, $\mathcal{P} = \mathcal{F}^*\mathcal{F}$. The operator $\mathcal{P}$ can be represented as the integral operator

$$(\mathcal{P}\mathbf{v})(\mathbf{x}) = \int_\Pi p(\mathbf{x},\mathbf{y})\mathbf{v}(\mathbf{y})\,d\mathbf{y}, \quad p(\mathbf{x},\mathbf{y}) = |\Omega^0|^{-1}|\widetilde{\Omega}|^{-1}\int_{\widetilde{\Omega}} e^{i\langle\mathbf{k},\mathbf{x}-\mathbf{y}\rangle}\,d\mathbf{k}. \tag{6.4}$$

Finally, one more interpretation of $\mathcal{P}$ is possible. Let $\widetilde{\mathcal{P}}$ be the pseudodifferential operator acting in $L_2(\mathbb{R}^d;\mathbb{C}^n)$ with the symbol $c\chi_{\widetilde{\Omega}}(\boldsymbol{\xi})$, where $\chi_{\widetilde{\Omega}}(\boldsymbol{\xi})$ is the characteristic function of the set $\widetilde{\Omega}$ and $c = (2\pi)^d|\Omega^0|^{-1}|\widetilde{\Omega}|^{-1}$. If the space $L_2(\Pi;\mathbb{C}^n)$ is identified with the subspace $\{\mathbf{u} \in L_2(\mathbb{R}^d;\mathbb{C}^n)\colon \mathbf{u}|_{\mathbb{R}^d\setminus\Pi} = 0\}$, then $\mathcal{P}\mathbf{v} = (\widetilde{\mathcal{P}}\mathbf{v})|_\Pi$, $\mathbf{v} \in L_2(\Pi;\mathbb{C}^n)$. It follows that the operator $\mathcal{P}$ is infinitely smoothing: it is continuous from $L_2(\Pi;\mathbb{C}^n)$ to $H^s(\Pi;\mathbb{C}^n)$ for any $s > 0$.

6.3. *More accurate approximation of the resolvent $(\mathcal{A}+\varepsilon^2 I)^{-1}$ in the $(L_2 \to L_2)$-norm*

In $L_2(\Pi;\mathbb{C}^n)$, we define the operator

$$\begin{aligned}\mathcal{K}(\varepsilon) &= \Lambda b(\mathbf{D})(\mathcal{A}^0+\varepsilon^2 I)^{-1}\mathcal{P} + (\mathcal{A}^0+\varepsilon^2 I)^{-1}\mathcal{P}b(\mathbf{D})^*\Lambda^* \\ &\quad - (\mathcal{A}^0+\varepsilon^2 I)^{-1}b(\mathbf{D})^*L(\mathbf{D})b(\mathbf{D})(\mathcal{A}^0+\varepsilon^2 I)^{-1}\mathcal{P}.\end{aligned} \tag{6.5}$$

Similarly to (6.1), the operator $(\mathcal{A}+\varepsilon^2 I)^{-1} - (\mathcal{A}^0+\varepsilon^2 I)^{-1} - \mathcal{K}(\varepsilon)$ is unitarily equivalent to the direct integral of the operators $(\mathcal{A}(\mathbf{k})+\varepsilon^2 I)^{-1} - (\mathcal{A}^0(\mathbf{k})+\varepsilon^2 I)^{-1} - \mathcal{K}(\mathbf{k},\varepsilon)$. Applying Theorem 5.2, we arrive at the following result.

Theorem 6.2. *For $\varepsilon > 0$, we have*

$$\|(\mathcal{A}+\varepsilon^2 I)^{-1} - (\mathcal{A}^0+\varepsilon^2 I)^{-1} - \mathcal{K}(\varepsilon)\|_{L_2(\Pi)\to L_2(\Pi)} \leq C_2, \quad \varepsilon > 0. \tag{6.6}$$

The constant C_2 depends only on $\|g\|_{L_\infty}$, $\|g^{-1}\|_{L_\infty}$, the operator $b(\mathbf{D})$, and the domain Ω^0.

6.4. *Approximation of the resolvent $(\mathcal{A}+\varepsilon^2 I)^{-1}$ in the "energy norm"*

Similarly to (6.1), the operator $\mathcal{A}^{1/2}\big((\mathcal{A}+\varepsilon^2 I)^{-1} - (I+\Lambda b(\mathbf{D})\mathcal{P})(\mathcal{A}^0+\varepsilon^2 I)^{-1}\big)$ is unitarily equivalent to the direct integral of the operators

$$\mathcal{A}(\mathbf{k})^{1/2}((\mathcal{A}(\mathbf{k})+\varepsilon^2 I)^{-1} - (I+\Lambda b(\mathbf{D}+\mathbf{k})P)(\mathcal{A}^0(\mathbf{k})+\varepsilon^2 I)^{-1}).$$

Applying Theorem 5.3, we arrive at the following result.

Theorem 6.3. *For $\varepsilon>0$, we have*

$$\|\mathcal{A}^{1/2}\big((\mathcal{A}+\varepsilon^2 I)^{-1} - (I+\Lambda b(\mathbf{D})\mathcal{P})(\mathcal{A}^0+\varepsilon^2 I)^{-1}\big)\|_{L_2(\Pi)\to L_2(\Pi)} \le \mathcal{C}_3, \quad \varepsilon>0. \tag{6.7}$$

The constant $\mathcal{C}_3$ depends only on $\|g\|_{L_\infty}$, $\|g^{-1}\|_{L_\infty}$, the operator $b(\mathbf{D})$, and the domain Ω^0.

6.5. *Approximations for the resolvent $(\widetilde{\mathcal{A}}+\varepsilon^2 I)^{-1}$ and the generalized resolvent $(\mathcal{A}+\varepsilon^2 Q)^{-1}$*

Now, we consider the operator $\widetilde{\mathcal{A}}$ in $L_2(\Pi;\mathbb{C}^n)$ defined by (3.33). Using the direct integral expansions, we deduce the following results from Theorems 5.4–5.6.

Theorem 6.4. *For $\varepsilon>0$, we have*

$$\|(\widetilde{\mathcal{A}}+\varepsilon^2 I)^{-1} - f^{-1}(\mathcal{A}^0+\varepsilon^2\overline{Q})^{-1}(f^*)^{-1}\|_{L_2(\Pi)\to L_2(\Pi)} \le \widetilde{\mathcal{C}}_1\varepsilon^{-1},$$

$$\|(\mathcal{A}+\varepsilon^2 Q)^{-1} - (\mathcal{A}^0+\varepsilon^2\overline{Q})^{-1}\|_{L_2(\Pi)\to L_2(\Pi)} \le \widetilde{\mathcal{C}}_1\|f\|^2_{L_\infty}\varepsilon^{-1}.$$

The constant $\widetilde{\mathcal{C}}_1$ depends on $\|g\|_{L_\infty}$, $\|g^{-1}\|_{L_\infty}$, $\|f\|_{L_\infty}$, $\|f^{-1}\|_{L_\infty}$, the operator $b(\mathbf{D})$, and Ω^0.

Theorem 6.5. *Let*

$$\begin{aligned}\mathcal{K}_Q(\varepsilon) &= \Lambda_Q b(\mathbf{D})(\mathcal{A}^0+\varepsilon^2\overline{Q})^{-1}\mathcal{P} + (\mathcal{A}^0+\varepsilon^2\overline{Q})^{-1}\mathcal{P}b(\mathbf{D})^*\Lambda_Q^* \\ &\quad - (\mathcal{A}^0+\varepsilon^2\overline{Q})^{-1}b(\mathbf{D})^*L_Q(\mathbf{D})b(\mathbf{D})(\mathcal{A}^0+\varepsilon^2\overline{Q})^{-1}\mathcal{P}.\end{aligned}$$

For $\varepsilon>0$, we have

$$\|(\widetilde{\mathcal{A}}+\varepsilon^2 I)^{-1} - f^{-1}((\mathcal{A}^0+\varepsilon^2\overline{Q})^{-1} + \mathcal{K}_Q(\varepsilon))(f^*)^{-1}\|_{L_2(\Pi)\to L_2(\Pi)} \le \widetilde{\mathcal{C}}_2,$$

$$\|(\mathcal{A}+\varepsilon^2 Q)^{-1} - (\mathcal{A}^0+\varepsilon^2\overline{Q})^{-1} - \mathcal{K}_Q(\varepsilon)\|_{L_2(\Pi)\to L_2(\Pi)} \le \widetilde{\mathcal{C}}_2\|f\|^2_{L_\infty}.$$

The constant $\widetilde{\mathcal{C}}_2$ depends on $\|g\|_{L_\infty}$, $\|g^{-1}\|_{L_\infty}$, $\|f\|_{L_\infty}$, $\|f^{-1}\|_{L_\infty}$, the operator $b(\mathbf{D})$, and Ω^0.

Theorem 6.6. *For $\varepsilon > 0$, we have*

$$\|\widetilde{\mathcal{A}}^{1/2}((\widetilde{\mathcal{A}}+\varepsilon^2 I)^{-1} - f^{-1}(I+\Lambda b(\mathbf{D})\mathcal{P})(\mathcal{A}^0+\varepsilon^2\overline{Q})^{-1}(f^*)^{-1})\|_{L_2(\Pi)\to L_2(\Pi)} \le \widetilde{C}_3,$$

$$\|\mathcal{A}^{1/2}((\mathcal{A}+\varepsilon^2 Q)^{-1} - (I+\Lambda b(\mathbf{D})\mathcal{P})(\mathcal{A}^0+\varepsilon^2\overline{Q})^{-1})\|_{L_2(\Pi)\to L_2(\Pi)} \le \widetilde{C}_3\|f\|_{L_\infty}.$$

The constant $\widetilde{C}_3$ depends on $\|g\|_{L_\infty}$, $\|g^{-1}\|_{L_\infty}$, $\|f\|_{L_\infty}$, $\|f^{-1}\|_{L_\infty}$, the operator $b(\mathbf{D})$, and Ω^0.

7. Homogenization of the Operator $\mathcal{A}_\varepsilon$

7.1. *The operator $\mathcal{A}_\varepsilon$. The scaling transformation*

Let $\Pi_\varepsilon := \varepsilon\Pi$. Then the perforated space Π_ε is periodic with respect to the lattice $\varepsilon\Gamma$. If $\psi(\mathbf{x})$ is a Γ-periodic measurable function in Π, we define an $(\varepsilon\Gamma)$-periodic function $\psi^\varepsilon(\mathbf{x})$ on Π_ε by $\psi^\varepsilon(\mathbf{x}) := \psi(\varepsilon^{-1}\mathbf{x})$, $\varepsilon > 0$.

Our main object is the operator $\mathcal{A}_\varepsilon$ acting in $L_2(\Pi_\varepsilon;\mathbb{C}^n)$ and formally given by

$$\mathcal{A}_\varepsilon := b(\mathbf{D})^* g^\varepsilon(\mathbf{x}) b(\mathbf{D})$$

with the Neumann condition (natural condition) on the boundary $\partial\Pi_\varepsilon$. The matrix-valued function $g(\mathbf{x})$ and the operator $b(\mathbf{D})$ satisfy the assumptions of Secs. 3.1 and 3.2. The precise definition of $\mathcal{A}_\varepsilon$ is given in terms of the quadratic form

$$\mathfrak{a}_\varepsilon[\mathbf{u},\mathbf{u}] = \int_{\Pi_\varepsilon} \langle g^\varepsilon(\mathbf{x}) b(\mathbf{D})\mathbf{u}(\mathbf{x}), b(\mathbf{D})\mathbf{u}(\mathbf{x})\rangle\, d\mathbf{x}, \quad \mathbf{u} \in H^1(\Pi_\varepsilon;\mathbb{C}^n). \tag{7.1}$$

Our goal is to approximate the resolvent $(\mathcal{A}_\varepsilon + I)^{-1}$ in the $(L_2 \to L_2)$ and $(L_2 \to H^1)$-norms.

Let T_ε be the *scaling transformation acting from $L_2(\Pi_\varepsilon;\mathbb{C}^n)$ to $L_2(\Pi;\mathbb{C}^n)$ and defined by*

$$(T_\varepsilon \mathbf{u})(\mathbf{x}) = \varepsilon^{d/2}\mathbf{u}(\varepsilon\mathbf{x}), \quad \mathbf{x} \in \Pi, \quad \varepsilon > 0.$$

Then T_ε is unitary. Obviously,

$$\mathfrak{a}_\varepsilon[\mathbf{u},\mathbf{u}] = \varepsilon^{-2}\mathfrak{a}[T_\varepsilon\mathbf{u}, T_\varepsilon\mathbf{u}], \quad \mathbf{u} \in H^1(\Pi_\varepsilon;\mathbb{C}^n). \tag{7.2}$$

Combining (3.10), (3.32), and (7.2), we see that

$$\widetilde{\mathfrak{C}}_1^{-1}\|g^{-1}\|_{L_\infty}^{-1}\|\mathbf{D}\mathbf{u}\|^2_{L_2(\Pi_\varepsilon)} \le \mathfrak{a}_\varepsilon[\mathbf{u},\mathbf{u}] \le c_1\|\mathbf{D}\mathbf{u}\|^2_{L_2(\Pi_\varepsilon)}, \quad \mathbf{u} \in H^1(\Pi_\varepsilon;\mathbb{C}^n). \tag{7.3}$$

Relation (7.2) means that $\mathcal{A}_\varepsilon = \varepsilon^{-2}T_\varepsilon^* \mathcal{A} T_\varepsilon$. Hence,

$$(\mathcal{A}_\varepsilon + I)^{-1} = \varepsilon^2 T_\varepsilon^*(\mathcal{A}+\varepsilon^2 I)^{-1}T_\varepsilon. \tag{7.4}$$

Using this identity, we deduce approximations for the resolvent $(\mathcal{A}_\varepsilon + I)^{-1}$ from the results of Sec. 6.

7.2. Approximation for the resolvent $(\mathcal{A}_\varepsilon + I)^{-1}$ in the $(L_2 \to L_2)$-norm

Let $\mathcal{A}^0(\varepsilon)$ be the effective operator acting in $L_2(\Pi_\varepsilon; \mathbb{C}^n)$ and given by

$$\mathcal{A}^0(\varepsilon) := b(\mathbf{D})^* g^0 b(\mathbf{D})$$

with the Neumann condition (natural condition) on the boundary $\partial\Pi_\varepsilon$. The precise definition of $\mathcal{A}^0(\varepsilon)$ is given in terms of the quadratic form

$$\mathfrak{a}^0(\varepsilon)[\mathbf{u}, \mathbf{u}] = \int_{\Pi_\varepsilon} \langle g^0 b(\mathbf{D})\mathbf{u}(\mathbf{x}), b(\mathbf{D})\mathbf{u}(\mathbf{x})\rangle \, d\mathbf{x}, \quad \mathbf{u} \in H^1(\Pi_\varepsilon; \mathbb{C}^n).$$

Note that $\mathcal{A}^0(\varepsilon)$ has constant coefficients; it depends on ε because Π_ε depends on ε. Similarly to (7.4), we have

$$(\mathcal{A}^0(\varepsilon) + I)^{-1} - \varepsilon^2 T_\varepsilon^* (\mathcal{A}^0 + \varepsilon^2 I)^{-1} T_\varepsilon. \tag{7.5}$$

Using (7.4), (7.5), and taking into account that T_ε is unitary, we obtain

$$\begin{aligned} &\|(\mathcal{A}_\varepsilon + I)^{-1} - (\mathcal{A}^0(\varepsilon) + I)^{-1}\|_{L_2(\Pi_\varepsilon) \to L_2(\Pi_\varepsilon)} \\ &\quad = \varepsilon^2 \|(\mathcal{A} + \varepsilon^2 I)^{-1} - (\mathcal{A}^0 + \varepsilon^2 I)^{-1}\|_{L_2(\Pi) \to L_2(\Pi)}. \end{aligned}$$

Then, applying Theorem 6.1, we arrive at the following result.

Theorem 7.1. *For $\varepsilon > 0$, we have*

$$\|(\mathcal{A}_\varepsilon + I)^{-1} - (\mathcal{A}^0(\varepsilon) + I)^{-1}\|_{L_2(\Pi_\varepsilon) \to L_2(\Pi_\varepsilon)} \leq C_1 \varepsilon. \tag{7.6}$$

The constant C_1 depends only on $\|g\|_{L_\infty}$, $\|g^{-1}\|_{L_\infty}$, the operator $b(\mathbf{D})$, and the domain Ω^0.

7.3. The operator $\mathcal{P}_\varepsilon$

In $L_2(\Pi_\varepsilon; \mathbb{C}^n)$ we define the operator

$$\mathcal{P}_\varepsilon := T_\varepsilon^* \mathcal{P} T_\varepsilon. \tag{7.7}$$

Here the operator $\mathcal{P}$ is given by (6.2). Then, by (6.4), $\mathcal{P}_\varepsilon$ can be represented as the integral operator

$$(\mathcal{P}_\varepsilon \mathbf{v})(\mathbf{x}) = \int_{\Pi_\varepsilon} p_\varepsilon(\mathbf{x}, \mathbf{y}) \mathbf{v}(\mathbf{y}) \, d\mathbf{y}, \quad p_\varepsilon(\mathbf{x}, \mathbf{y}) = |\Omega^0|^{-1} |\widetilde{\Omega}|^{-1} \int_{\widetilde{\Omega}/\varepsilon} e^{i\langle \mathbf{k}, \mathbf{x}-\mathbf{y}\rangle} \, d\mathbf{k}.$$

Let us give one more interpretation of $\mathcal{P}_\varepsilon$. Let $\widetilde{\mathcal{P}}_\varepsilon$ be the pseudodifferential operator acting in $L_2(\mathbb{R}^d; \mathbb{C}^n)$ with the symbol $c\chi_{\widetilde{\Omega}/\varepsilon}(\boldsymbol{\xi})$, where $\chi_{\widetilde{\Omega}/\varepsilon}(\boldsymbol{\xi})$ is the characteristic function of the set $\widetilde{\Omega}/\varepsilon$ and $c = (2\pi)^d |\Omega^0|^{-1} |\widetilde{\Omega}|^{-1}$. If the space $L_2(\Pi_\varepsilon; \mathbb{C}^n)$ is identified with the subspace $\{\mathbf{u} \in L_2(\mathbb{R}^d; \mathbb{C}^n)\colon \mathbf{u}|_{\mathbb{R}^d \setminus \Pi_\varepsilon} = 0\}$, then $\mathcal{P}_\varepsilon \mathbf{v} = (\widetilde{\mathcal{P}}_\varepsilon \mathbf{v})|_{\Pi_\varepsilon}$, $\mathbf{v} \in L_2(\Pi_\varepsilon; \mathbb{C}^n)$. It follows that the operator $\mathcal{P}_\varepsilon$ is infinitely smoothing: it is continuous from $L_2(\Pi_\varepsilon; \mathbb{C}^n)$ to $H^s(\Pi_\varepsilon; \mathbb{C}^n)$ for any $s > 0$ (but the $(L_2 \to H^s)$-norm of $\mathcal{P}_\varepsilon$ depends on ε).

7.4. *More accurate approximation for the resolvent* $(\mathcal{A}_\varepsilon + I)^{-1}$ *in the* $(L_2 \to L_2)$*-norm*

Now, we introduce a *corrector*

$$\begin{aligned}\mathcal{K}_\varepsilon &= \Lambda^\varepsilon b(\mathbf{D})(\mathcal{A}^0(\varepsilon)+I)^{-1}\mathcal{P}_\varepsilon + (\mathcal{A}^0(\varepsilon)+I)^{-1}\mathcal{P}_\varepsilon b(\mathbf{D})^*(\Lambda^\varepsilon)^* \\ &\quad - (\mathcal{A}^0(\varepsilon)+I)^{-1}b(\mathbf{D})^*L(\mathbf{D})b(\mathbf{D})(\mathcal{A}^0(\varepsilon)+I)^{-1}\mathcal{P}_\varepsilon. \end{aligned} \tag{7.8}$$

It is a bounded operator in $L_2(\Pi_\varepsilon;\mathbb{C}^n)$. The operators (6.5) and (7.8) satisfy

$$\mathcal{K}_\varepsilon := \varepsilon T_\varepsilon^* \mathcal{K}(\varepsilon) T_\varepsilon. \tag{7.9}$$

Theorem 6.2 and relations (7.4), (7.5), and (7.9) imply the following result.

Theorem 7.2. *For* $\varepsilon > 0$, *we have*

$$\|(\mathcal{A}_\varepsilon + I)^{-1} - (\mathcal{A}^0(\varepsilon)+I)^{-1} - \varepsilon\mathcal{K}_\varepsilon\|_{L_2(\Pi_\varepsilon)\to L_2(\Pi_\varepsilon)} \le \mathcal{C}_2\varepsilon^2.$$

The constant $\mathcal{C}_2$ *depends only on* $\|g\|_{L_\infty}$, $\|g^{-1}\|_{L_\infty}$, *the operator* $b(\mathbf{D})$, *and the domain* Ω^0.

7.5. *Approximation of the resolvent* $(\mathcal{A}_\varepsilon + I)^{-1}$ *in the* $(L_2 \to H^1)$*-operator norm*

We have

$$\begin{aligned}&\mathcal{A}_\varepsilon^{1/2}((\mathcal{A}_\varepsilon+I)^{-1} - (I+\varepsilon\Lambda^\varepsilon b(\mathbf{D})\mathcal{P}_\varepsilon)(\mathcal{A}^0(\varepsilon)+I)^{-1}) \\ &\quad = \varepsilon T_\varepsilon^* \mathcal{A}^{1/2}((\mathcal{A}+\varepsilon^2 I)^{-1} - (I+\Lambda b(\mathbf{D})\mathcal{P})(\mathcal{A}^0+\varepsilon^2 I)^{-1})T_\varepsilon.\end{aligned}$$

Then, applying Theorem 6.3, we obtain

$$\|\mathcal{A}_\varepsilon^{1/2}((\mathcal{A}_\varepsilon+I)^{-1} - (I+\varepsilon\Lambda^\varepsilon b(\mathbf{D})\mathcal{P}_\varepsilon)(\mathcal{A}^0(\varepsilon)+I)^{-1})\|_{L_2(\Pi_\varepsilon)\to L_2(\Pi_\varepsilon)} \le \mathcal{C}_3\varepsilon,\ \varepsilon>0.$$

Together with the lower estimate (7.3), this implies

$$\|\mathbf{D}((\mathcal{A}_\varepsilon+I)^{-1} - (I+\varepsilon\Lambda^\varepsilon b(\mathbf{D})\mathcal{P}_\varepsilon)(\mathcal{A}^0(\varepsilon)+I)^{-1})\|_{L_2(\Pi_\varepsilon)\to L_2(\Pi_\varepsilon)} \le \mathcal{C}_4\varepsilon, \quad \varepsilon>0, \tag{7.10}$$

where $\mathcal{C}_4 = \mathcal{C}_3\widetilde{\mathfrak{C}}_1^{1/2}\|g^{-1}\|_{L_\infty}^{1/2}$. Let us check that

$$\|(\mathcal{A}_\varepsilon+I)^{-1} - (I+\varepsilon\Lambda^\varepsilon b(\mathbf{D})\mathcal{P}_\varepsilon)(\mathcal{A}^0(\varepsilon)+I)^{-1}\|_{L_2(\Pi_\varepsilon)\to L_2(\Pi_\varepsilon)} \le \mathcal{C}_5\varepsilon, \quad \varepsilon>0, \tag{7.11}$$

where the constant $\mathcal{C}_5$ depends only on $\|g\|_{L_\infty}$, $\|g^{-1}\|_{L_\infty}$, the operator $b(\mathbf{D})$, and the domain Ω^0. The difference of the resolvents satisfies estimate (7.6). Next, by (7.5), (7.7), and the direct integral expansions, we see that

$$\begin{aligned}&\|\Lambda^\varepsilon b(\mathbf{D})\mathcal{P}_\varepsilon(\mathcal{A}^0(\varepsilon)+I)^{-1}\|_{L_2(\Pi_\varepsilon)\to L_2(\Pi_\varepsilon)} \\ &\quad = \varepsilon\|\Lambda b(\mathbf{D})\mathcal{P}(\mathcal{A}^0+\varepsilon^2 I)^{-1}\|_{L_2(\Pi)\to L_2(\Pi)} \\ &\quad = \varepsilon \operatorname*{ess\,sup}_{\mathbf{k}\in\widetilde{\Omega}} \|\Lambda b(\mathbf{D}+\mathbf{k})P(\mathcal{A}^0(\mathbf{k})+\varepsilon^2 I)^{-1}\|_{L_2(\Omega^0)\to L_2(\Omega^0)}.\end{aligned} \tag{7.12}$$

Recalling (5.1) and (5.6), we arrive at

$$\begin{aligned}&\|\Lambda^\varepsilon b(\mathbf{D})\mathcal{P}_\varepsilon(\mathcal{A}^0(\varepsilon)+I)^{-1}\|_{L_2(\Pi_\varepsilon)\to L_2(\Pi_\varepsilon)}\\&\quad=\varepsilon \operatorname*{ess\,sup}_{\mathbf{k}\in\widetilde{\Omega}}\|tZ(\boldsymbol{\theta})(t^2S(\boldsymbol{\theta})+\varepsilon^2I_{\mathfrak{N}})^{-1}P\|_{L_2(\Omega^0)\to L_2(\Omega^0)}.\end{aligned}\tag{7.13}$$

Then estimates (4.11) and (4.17) imply that

$$\|\Lambda^\varepsilon b(\mathbf{D})\mathcal{P}_\varepsilon(\mathcal{A}^0(\varepsilon)+I)^{-1}\|_{L_2(\Pi_\varepsilon)\to L_2(\Pi_\varepsilon)}\le C_Zc_*^{-1/2},\quad \varepsilon>0.\tag{7.14}$$

Combining this with (7.6), we obtain estimate (7.11) with the constant $\mathcal{C}_5=\mathcal{C}_1+C_Zc_*^{-1/2}$.

Finally, estimates (7.10) and (7.11) imply the following result.

Theorem 7.3. *For $\varepsilon>0$, we have*

$$\|(\mathcal{A}_\varepsilon+I)^{-1}-(I+\varepsilon\Lambda^\varepsilon b(\mathbf{D})\mathcal{P}_\varepsilon)(\mathcal{A}^0(\varepsilon)+I)^{-1}\|_{L_2(\Pi_\varepsilon)\to H^1(\Pi_\varepsilon)}\le\mathcal{C}_6\varepsilon.\tag{7.15}$$

The constant $\mathcal{C}_6$ depends only on $\|g\|_{L_\infty}$, $\|g^{-1}\|_{L_\infty}$, the operator $b(\mathbf{D})$, and the domain Ω^0.

7.6. *Approximation of the "flux"*

Now, we deduce approximation for the operator $g^\varepsilon b(\mathbf{D})(\mathcal{A}_\varepsilon+I)^{-1}$ corresponding to the "flux". By the direct integral expansion and the scaling transformation, from (5.16) it follows that

$$\begin{aligned}&\|g^\varepsilon b(\mathbf{D})(\mathcal{A}_\varepsilon+I)^{-1}\\&\quad-g^\varepsilon b(\mathbf{D})(I+\varepsilon\Lambda^\varepsilon b(\mathbf{D}))(\mathcal{A}^0(\varepsilon)+I)^{-1}\mathcal{P}_\varepsilon\|_{L_2(\Pi_\varepsilon)\to L_2(\Pi_\varepsilon)}\le C_7\|g\|_{L_\infty}^{1/2}\varepsilon,\quad\varepsilon>0.\end{aligned}\tag{7.16}$$

By (3.1), we have

$$\begin{aligned}\varepsilon g^\varepsilon b(\mathbf{D})\Lambda^\varepsilon b(\mathbf{D})(\mathcal{A}^0(\varepsilon)+I)^{-1}\mathcal{P}_\varepsilon&-g^\varepsilon(b(\mathbf{D})\Lambda)^\varepsilon b(\mathbf{D})(\mathcal{A}^0(\varepsilon)+I)^{-1}\mathcal{P}_\varepsilon\\&+\varepsilon\sum_{l=1}^d g^\varepsilon b_l\Lambda^\varepsilon b(\mathbf{D})D_l(\mathcal{A}^0(\varepsilon)+I)^{-1}\mathcal{P}_\varepsilon.\end{aligned}\tag{7.17}$$

Let us estimate the second term in the right-hand side of (7.17). Similarly to (7.12)–(7.14),

$$\begin{aligned}&\varepsilon\left\|\sum_{l=1}^d g^\varepsilon b_l\Lambda^\varepsilon b(\mathbf{D})D_l(\mathcal{A}^0(\varepsilon)+I)^{-1}\mathcal{P}_\varepsilon\right\|_{L_2(\Pi_\varepsilon)\to L_2(\Pi_\varepsilon)}\\&\quad=\varepsilon\operatorname*{ess\,sup}_{\mathbf{k}\in\widetilde{\Omega}}\left\|\sum_{l=1}^d gb_lt^2Z(\boldsymbol{\theta})\theta_l(t^2S(\boldsymbol{\theta})+\varepsilon^2I_{\mathfrak{N}})^{-1}P\right\|_{L_2(\Omega^0)\to L_2(\Omega^0)}\le C_8\varepsilon,\end{aligned}\tag{7.18}$$

where $C_8=d\|g\|_{L_\infty}\alpha_1^{1/2}C_Zc_*^{-1}$.

Recalling the notation (4.15) and using (7.16)–(7.18), we arrive at the following statement.

Theorem 7.4. *For $\varepsilon > 0$, we have*

$$\|g^\varepsilon b(\mathbf{D})(\mathcal{A}_\varepsilon + I)^{-1} - \widetilde{g}^\varepsilon b(\mathbf{D})(\mathcal{A}^0(\varepsilon) + I)^{-1}\mathcal{P}_\varepsilon\|_{L_2(\Pi_\varepsilon)\to L_2(\Pi_\varepsilon)} \leq \mathcal{C}_7\varepsilon, \quad \varepsilon > 0. \tag{7.19}$$

The constant $\mathcal{C}_7$ depends only on $\|g\|_{L_\infty}$, $\|g^{-1}\|_{L_\infty}$, the operator $b(\mathbf{D})$, and the domain Ω^0.

7.7. *The operator $\widetilde{\mathcal{A}}_\varepsilon$. Approximations for the resolvent $(\widetilde{\mathcal{A}}_\varepsilon + I)^{-1}$ and the generalized resolvent $(\mathcal{A}_\varepsilon + Q^\varepsilon)^{-1}$*

Let $\mathcal{A}_\varepsilon$ be the operator in $L_2(\Pi_\varepsilon; \mathbb{C}^n)$ generated by the quadratic form (7.1). Let $f(\mathbf{x})$ be a Γ-periodic $(n\times n)$-matrix-valued function in Π such that $f, f^{-1} \in L_\infty(\Pi)$. In $L_2(\Pi_\varepsilon; \mathbb{C}^n)$, we define the operator

$$\widetilde{\mathcal{A}}_\varepsilon = (f^\varepsilon)^* \mathcal{A}_\varepsilon f^\varepsilon.$$

Formally, $\widetilde{\mathcal{A}}_\varepsilon$ is given by the differential expression $(f^\varepsilon(\mathbf{x}))^* b(\mathbf{D})^* g^\varepsilon(\mathbf{x}) b(\mathbf{D}) f^\varepsilon(\mathbf{x})$ with the natural condition on $\partial\Pi_\varepsilon$. The precise definition of $\widetilde{\mathcal{A}}_\varepsilon$ is given in terms of the quadratic form

$$\widetilde{\mathfrak{a}}_\varepsilon[\mathbf{u},\mathbf{u}] = \int_{\Pi_\varepsilon} \langle g^\varepsilon(\mathbf{x}) b(\mathbf{D})(f^\varepsilon(\mathbf{x})\mathbf{u}(\mathbf{x})), b(\mathbf{D})(f^\varepsilon(\mathbf{x})\mathbf{u}(\mathbf{x}))\rangle\, d\mathbf{x}, \quad f^\varepsilon \mathbf{u} \in H^1(\Pi_\varepsilon; \mathbb{C}^n).$$

Let $Q(\mathbf{x}) = (f(\mathbf{x})f(\mathbf{x})^*)^{-1}$. Obviously, $(\widetilde{\mathcal{A}}_\varepsilon + I)^{-1} = (f^\varepsilon)^{-1}(\mathcal{A}_\varepsilon + Q^\varepsilon)^{-1}((f^\varepsilon)^*)^{-1}$. We deduce approximations for the resolvent $(\widetilde{\mathcal{A}}_\varepsilon + I)^{-1}$ and the generalized resolvent $(\mathcal{A}_\varepsilon + Q^\varepsilon)^{-1}$ from the theorems of Sec. 6.5.

Let $\widetilde{\mathcal{A}}$ be the operator in $L_2(\Pi; \mathbb{C}^n)$ generated by the quadratic form (3.34). We have

$$(\widetilde{\mathcal{A}}_\varepsilon + I)^{-1} = \varepsilon^2 T_\varepsilon^* (\widetilde{\mathcal{A}} + \varepsilon^2 I)^{-1} T_\varepsilon, \tag{7.20}$$

$$(\mathcal{A}_\varepsilon + Q^\varepsilon)^{-1} = \varepsilon^2 T_\varepsilon^* (\mathcal{A} + \varepsilon^2 Q)^{-1} T_\varepsilon. \tag{7.21}$$

Theorems 6.4, 6.5 and relations (7.20), (7.21) directly imply the following results.

Theorem 7.5. *For $\varepsilon > 0$, we have*

$$\|(\widetilde{\mathcal{A}}_\varepsilon + I)^{-1} - (f^\varepsilon)^{-1}(\mathcal{A}^0(\varepsilon) + \overline{Q})^{-1}((f^\varepsilon)^*)^{-1}\|_{L_2(\Pi_\varepsilon)\to L_2(\Pi_\varepsilon)} \leq \widetilde{\mathcal{C}}_1\varepsilon,$$

$$\|(\mathcal{A}_\varepsilon + Q^\varepsilon)^{-1} - (\mathcal{A}^0(\varepsilon) + \overline{Q})^{-1}\|_{L_2(\Pi_\varepsilon)\to L_2(\Pi_\varepsilon)} \leq \widetilde{\mathcal{C}}_1 \|f\|^2_{L_\infty}\varepsilon.$$

The constant $\widetilde{\mathcal{C}}_1$ depends on $\|g\|_{L_\infty}$, $\|g^{-1}\|_{L_\infty}$, $\|f\|_{L_\infty}$, $\|f^{-1}\|_{L_\infty}$, the operator $b(\mathbf{D})$, and Ω^0.

Theorem 7.6. *Let*

$$\mathcal{K}_{Q,\varepsilon} = \Lambda_Q^\varepsilon b(\mathbf{D})(\mathcal{A}^0(\varepsilon) + \overline{Q})^{-1}\mathcal{P}_\varepsilon + (\mathcal{A}^0(\varepsilon) + \overline{Q})^{-1}\mathcal{P}_\varepsilon b(\mathbf{D})^*(\Lambda_Q^\varepsilon)^*$$
$$- (\mathcal{A}^0(\varepsilon) + \overline{Q})^{-1} b(\mathbf{D})^* L_Q(\mathbf{D}) b(\mathbf{D})(\mathcal{A}^0(\varepsilon) + \overline{Q})^{-1}\mathcal{P}_\varepsilon.$$

For $\varepsilon > 0$, we have

$$\|(\widetilde{\mathcal{A}}_\varepsilon + I)^{-1} - (f^\varepsilon)^{-1}((\mathcal{A}^0(\varepsilon) + \overline{Q})^{-1} + \varepsilon\mathcal{K}_{Q,\varepsilon})((f^\varepsilon)^*)^{-1}\|_{L_2(\Pi_\varepsilon)\to L_2(\Pi_\varepsilon)} \leq \widetilde{C}_2\varepsilon^2,$$

$$\|(\mathcal{A}_\varepsilon + Q^\varepsilon)^{-1} - (\mathcal{A}^0(\varepsilon) + \overline{Q})^{-1} - \varepsilon\mathcal{K}_{Q,\varepsilon}\|_{L_2(\Pi_\varepsilon)\to L_2(\Pi_\varepsilon)} \leq \widetilde{C}_2\|f\|^2_{L_\infty}\varepsilon^2.$$

The constant $\widetilde{C}_2$ depends on $\|g\|_{L_\infty}$, $\|g^{-1}\|_{L_\infty}$, $\|f\|_{L_\infty}$, $\|f^{-1}\|_{L_\infty}$, the operator $b(\mathbf{D})$, and Ω^0.

Similarly to the proof of Theorems 7.3 and 7.4, we deduce the following result from Theorem 6.6 (the details are omitted).

Theorem 7.7. *For $\varepsilon > 0$, we have*

$$\|f^\varepsilon(\widetilde{\mathcal{A}}_\varepsilon + I)^{-1} - (I + \varepsilon\Lambda^\varepsilon b(\mathbf{D})\mathcal{P}_\varepsilon)(\mathcal{A}^0(\varepsilon) + \overline{Q})^{-1}((f^\varepsilon)^*)^{-1}\|_{L_2(\Pi_\varepsilon)\to H^1(\Pi_\varepsilon)} \leq \widetilde{C}_6\varepsilon,$$

$$\|(\mathcal{A}_\varepsilon + Q^\varepsilon)^{-1} - (I + \varepsilon\Lambda^\varepsilon b(\mathbf{D})\mathcal{P}_\varepsilon)(\mathcal{A}^0(\varepsilon) + \overline{Q})^{-1}\|_{L_2(\Pi_\varepsilon)\to H^1(\Pi_\varepsilon)} \leq \widetilde{C}_6\|f\|_{L_\infty}\varepsilon,$$

$$\|g^\varepsilon b(\mathbf{D})f^\varepsilon(\widetilde{\mathcal{A}}_\varepsilon + I)^{-1} - \widetilde{g}^\varepsilon b(\mathbf{D})(\mathcal{A}^0(\varepsilon) + \overline{Q})^{-1}\mathcal{P}_\varepsilon((f^\varepsilon)^*)^{-1}\|_{L_2(\Pi_\varepsilon)\to L_2(\Pi_\varepsilon)} \leq \widetilde{C}_7\varepsilon,$$

$$\|g^\varepsilon b(\mathbf{D})(\mathcal{A}_\varepsilon + Q^\varepsilon)^{-1} - \widetilde{g}^\varepsilon b(\mathbf{D})(\mathcal{A}^0(\varepsilon) + \overline{Q})^{-1}\mathcal{P}_\varepsilon\|_{L_2(\Pi_\varepsilon)\to L_2(\Pi_\varepsilon)} \leq \widetilde{C}_7\|f\|_{L_\infty}\varepsilon.$$

The constants $\widetilde{C}_6$ and $\widetilde{C}_7$ depend only on $\|g\|_{L_\infty}$, $\|g^{-1}\|_{L_\infty}$, $\|f\|_{L_\infty}$, $\|f^{-1}\|_{L_\infty}$, the operator $b(\mathbf{D})$, and the domain Ω^0.

8. Removal of the Smoothing Operator. Special Cases

8.1. *More accurate approximation of the resolvent $(\mathcal{A}_\varepsilon + I)^{-1}$ in the $(L_2 \to L_2)$-norm: Removal of the operator $\mathcal{P}_\varepsilon$*

Now, we show that, under the additional assumptions, it is possible to remove the smoothing operator from the corrector $\mathcal{K}_\varepsilon$.

First, we check that, if $\Lambda \in L_\infty$, then $\mathcal{P}_\varepsilon$ can be replaced by I in the first two terms of $\mathcal{K}_\varepsilon$.

Lemma 8.1. *Suppose that $\Lambda \in L_\infty$. Then for $\varepsilon > 0$, we have*

$$\|\Lambda^\varepsilon b(\mathbf{D})(\mathcal{A}^0(\varepsilon) + I)^{-1}(\mathcal{P}_\varepsilon - I)\|_{L_2(\Pi_\varepsilon)\to L_2(\Pi_\varepsilon)} \leq \mathcal{C}_8\varepsilon. \tag{8.1}$$

The constant $\mathcal{C}_8$ depends on $\|g^{-1}\|_{L_\infty}$, the operator $b(\mathbf{D})$, the domain Ω^0, and the norm $\|\Lambda\|_{L_\infty}$.

Proof. By the scaling transformation, estimate (8.1) is equivalent to the inequality

$$\|\Lambda b(\mathbf{D})(\mathcal{A}^0 + \varepsilon^2 I)^{-1}(\mathcal{P} - I)\|_{L_2(\Pi)\to L_2(\Pi)} \leq \mathcal{C}_8. \tag{8.2}$$

By (5.5) and the direct integral expansions (4.39) and (6.2), we have

$$\|(\mathcal{A}^0 + \varepsilon^2 I)^{-1}(\mathcal{P} - I)\|_{L_2(\Pi)\to L_2(\Pi)} \leq c_4^{-1}\|g^{-1}\|_{L_\infty}. \tag{8.3}$$

By (4.27),

$$\|b(\mathbf{D})(\mathcal{A}^0+\varepsilon^2 I)^{-1/2}\|_{L_2(\Pi)\to L_2(\Pi)} \le |(g^0)^{-1/2}| \|(\mathcal{A}^0)^{1/2}(\mathcal{A}^0+\varepsilon^2 I)^{-1/2}\|_{L_2(\Pi)\to L_2(\Pi)} \le \|g^{-1}\|^{1/2}_{L_\infty}. \tag{8.4}$$

Estimates (8.3), (8.4), and the condition $\Lambda \in L_\infty$ imply the required estimate (8.2) with $\mathcal{C}_8 = c_4^{-1/2}\|g^{-1}\|_{L_\infty}\|\Lambda\|_{L_\infty}$. □

Theorem 7.2 and Lemma 8.1 imply the following result.

Theorem 8.2. *Let $\Lambda \in \widetilde{H}^1(\Omega^0)$ be the solution of problem (4.9). Suppose that $\Lambda \in L_\infty$. Let*

$$\mathcal{K}'_\varepsilon = \Lambda^\varepsilon b(\mathbf{D})(\mathcal{A}^0(\varepsilon)+I)^{-1} + (\mathcal{A}^0(\varepsilon)+I)^{-1}b(\mathbf{D})^*(\Lambda^\varepsilon)^* - (\mathcal{A}^0(\varepsilon)+I)^{-1}b(\mathbf{D})^*L(\mathbf{D})b(\mathbf{D})(\mathcal{A}^0(\varepsilon)+I)^{-1}\mathcal{P}_\varepsilon.$$

Then for $\varepsilon > 0$, we have

$$\|(\mathcal{A}_\varepsilon+I)^{-1} - (\mathcal{A}^0(\varepsilon)+I)^{-1} - \varepsilon\mathcal{K}'_\varepsilon\|_{L_2(\Pi_\varepsilon)\to L_2(\Pi_\varepsilon)} \le \mathcal{C}_9\varepsilon^2.$$

The constant $\mathcal{C}_9 = \mathcal{C}_2 + 2\mathcal{C}_8$ depends only on $\|g\|_{L_\infty}$, $\|g^{-1}\|_{L_\infty}$, the operator $b(\mathbf{D})$, the domain Ω^0, and the norm $\|\Lambda\|_{L_\infty}$.

Now, we show that, if the boundary $\partial\Pi$ is sufficiently smooth, then $\mathcal{P}_\varepsilon$ can be replaced by I in the third term of $\mathcal{K}_\varepsilon$.

Lemma 8.3. *Suppose that $\partial\Pi \in C^{1,1}$. Then for $\varepsilon > 0$, we have*

$$\|(\mathcal{A}^0(\varepsilon)+I)^{-1}b(\mathbf{D})^*L(\mathbf{D})b(\mathbf{D})(\mathcal{A}^0(\varepsilon)+I)^{-1}(\mathcal{P}_\varepsilon - I)\|_{L_2(\Pi_\varepsilon)\to L_2(\Pi_\varepsilon)} \le \mathcal{C}_{10}\varepsilon. \tag{8.5}$$

The constant $\mathcal{C}_{10}$ depends only on $\|g\|_{L_\infty}$, $\|g^{-1}\|_{L_\infty}$, the operator $b(\mathbf{D})$, and the domain Ω^0.

Proof. By the scaling transformation, the required estimate (8.5) is equivalent to

$$\|(\mathcal{A}^0+\varepsilon^2 I)^{-1}b(\mathbf{D})^*L(\mathbf{D})b(\mathbf{D})(\mathcal{A}^0+\varepsilon^2 I)^{-1}(\mathcal{P} - I)\|_{L_2(\Pi)\to L_2(\Pi)} \le \mathcal{C}_{10}. \tag{8.6}$$

Condition $\partial\Pi \in C^{1,1}$ ensures that the operator $(\mathcal{A}^0+I)^{-1}$ is continuous from $L_2(\Pi;\mathbb{C}^n)$ to $H^2(\Pi;\mathbb{C}^n)$. We have

$$\|(\mathcal{A}^0+I)^{-1}\|_{L_2(\Pi)\to H^2(\Pi)} \le \widehat{c}, \tag{8.7}$$

where the constant $\widehat{c}$ depends only on $\|g\|_{L_\infty}$, $\|g^{-1}\|_{L_\infty}$, the operator $b(\mathbf{D})$, and the domain Ω^0. This follows from the theorems about regularity of solutions of the strongly elliptic systems; see, e.g., [19, Chap. 4]. Next, by (8.3),

$$\begin{aligned}&\|(\mathcal{A}^0+I)(\mathcal{A}^0+\varepsilon^2 I)^{-1}(\mathcal{P} - I)\|_{L_2(\Pi)\to L_2(\Pi)} \\ &\le \|\mathcal{A}^0(\mathcal{A}^0+\varepsilon^2 I)^{-1}\|_{L_2(\Pi)\to L_2(\Pi)} + \|(\mathcal{A}^0+\varepsilon^2 I)^{-1}(\mathcal{P} - I)\|_{L_2(\Pi)\to L_2(\Pi)} \\ &\le 1 + c_4^{-1}\|g^{-1}\|_{L_\infty}.\end{aligned} \tag{8.8}$$

Since the operator $\mathcal{P}$ is an orthogonal projection and commutes with differential operators with constant coefficients, we obtain

$$\begin{aligned}&\|(\mathcal{A}^0+\varepsilon^2 I)^{-1}b(\mathbf{D})^*L(\mathbf{D})b(\mathbf{D})(\mathcal{A}^0+\varepsilon^2 I)^{-1}(\mathcal{P}-I)\|_{L_2(\Pi)\to L_2(\Pi)}\\&\le \|b(\mathbf{D})(\mathcal{A}^0+\varepsilon^2 I)^{-1}(\mathcal{P}-I)\|_{L_2(\Pi)\to L_2(\Pi)}\\&\times \|L(\mathbf{D})b(\mathbf{D})(\mathcal{A}^0+\varepsilon^2 I)^{-1}(\mathcal{P}-I)\|_{L_2(\Pi)\to L_2(\Pi)}.\end{aligned} \tag{8.9}$$

By (8.3) and (8.4),

$$\|b(\mathbf{D})(\mathcal{A}^0+\varepsilon^2 I)^{-1}(\mathcal{P}-I)\|_{L_2(\Pi)\to L_2(\Pi)}\le c_4^{-1/2}\|g^{-1}\|_{L_\infty}. \tag{8.10}$$

Next, relations (3.1), (3.4), (4.46), (8.7), and (8.8) imply that

$$\|L(\mathbf{D})b(\mathbf{D})(\mathcal{A}^0+\varepsilon^2 I)^{-1}(\mathcal{P}-I)\|_{L_2(\Pi)\to L_2(\Pi)}\le C_L\alpha_1^{1/2}d\widehat{c}(1+c_4^{-1}\|g^{-1}\|_{L_\infty}). \tag{8.11}$$

Finally, estimates (8.9)–(8.11) imply the required estimate (8.6) with the constant $\mathcal{C}_{10}=C_L\alpha_1^{1/2}d\widehat{c}c_4^{-1/2}\|g^{-1}\|_{L_\infty}(1+c_4^{-1}\|g^{-1}\|_{L_\infty})$. □

Theorem 7.2 and Lemma 8.3 imply the following result.

Theorem 8.4. *Suppose that $\partial\Pi\in C^{1,1}$. Let*

$$\begin{aligned}\mathcal{K}''_\varepsilon&=\Lambda^\varepsilon b(\mathbf{D})(\mathcal{A}^0(\varepsilon)+I)^{-1}\mathcal{P}_\varepsilon+(\mathcal{A}^0(\varepsilon)+I)^{-1}\mathcal{P}_\varepsilon b(\mathbf{D})^*(\Lambda^\varepsilon)^*\\&\quad-(\mathcal{A}^0(\varepsilon)+I)^{-1}b(\mathbf{D})^*L(\mathbf{D})b(\mathbf{D})(\mathcal{A}^0(\varepsilon)+I)^{-1}.\end{aligned}$$

Then for $\varepsilon>0$, we have

$$\|(\mathcal{A}_\varepsilon+I)^{-1}-(\mathcal{A}^0(\varepsilon)+I)^{-1}-\varepsilon\mathcal{K}''_\varepsilon\|_{L_2(\Pi_\varepsilon)\to L_2(\Pi_\varepsilon)}\le\mathcal{C}_{11}\varepsilon^2.$$

The constant $\mathcal{C}_{11}=\mathcal{C}_2+\mathcal{C}_{10}$ depends only on $\|g\|_{L_\infty}$, $\|g^{-1}\|_{L_\infty}$, the operator $b(\mathbf{D})$, and the domain Ω^0.

Now, we impose the following condition, under which it is possible to remove the smoothing operator in all terms of $\mathcal{K}_\varepsilon$.

Condition 8.5. 1°. Suppose that $\partial\Pi\in C^{1,1}$. 2°. Suppose that the solution $\Lambda\in\widetilde{H}^1(\Omega^0)$ of problem (4.9) is a multiplier from $H^1(\Pi;\mathbb{C}^m)$ to $L_2(\Pi;\mathbb{C}^n)$.

The norm of the operator $[\Lambda]$ of multiplication by the matrix-valued function $\Lambda(\mathbf{x})$ is denoted by $M_\Lambda:=\|[\Lambda]\|_{H^1(\Pi)\to L_2(\Pi)}$.

Under Condition 8.5, instead of the corrector (7.8), we will use the simpler operator

$$\begin{aligned}\mathcal{K}^0_\varepsilon&=\Lambda^\varepsilon b(\mathbf{D})(\mathcal{A}^0(\varepsilon)+I)^{-1}+(\mathcal{A}^0(\varepsilon)+I)^{-1}b(\mathbf{D})^*(\Lambda^\varepsilon)^*\\&\quad-(\mathcal{A}^0(\varepsilon)+I)^{-1}b(\mathbf{D})^*L(\mathbf{D})b(\mathbf{D})(\mathcal{A}^0(\varepsilon)+I)^{-1}.\end{aligned} \tag{8.12}$$

Condition 8.5(1°) ensures that the operator $(\mathcal{A}^0(\varepsilon)+I)^{-1}$ is continuous from $L_2(\Pi_\varepsilon;\mathbb{C}^n)$ to $H^2(\Pi_\varepsilon;\mathbb{C}^n)$. Then the operator $b(\mathbf{D})(\mathcal{A}^0(\varepsilon)+I)^{-1}$ is a continuous mapping of $L_2(\Pi_\varepsilon;\mathbb{C}^n)$ to $H^1(\Pi_\varepsilon;\mathbb{C}^m)$. Hence, Condition 8.5(2°) implies that the operator $\Lambda^\varepsilon b(\mathbf{D})(\mathcal{A}^0(\varepsilon)+I)^{-1}$ (the first term in (8.12)) is continuous in $L_2(\Pi_\varepsilon;\mathbb{C}^n)$.

The second term in (8.12) is adjoint to the first one; so, it is also continuous. Finally, Condition 8.5(1°) implies that the third term in (8.12) is continuous too.

Theorem 8.6. *Under Condition* 8.5, *for* $\varepsilon > 0$, *we have*

$$\|(\mathcal{A}_\varepsilon + I)^{-1} - (\mathcal{A}^0(\varepsilon) + I)^{-1} - \varepsilon \mathcal{K}_\varepsilon^0\|_{L_2(\Pi_\varepsilon)\to L_2(\Pi_\varepsilon)} \le \mathcal{C}_{12}\varepsilon^2. \tag{8.13}$$

The constant $\mathcal{C}_{12}$ *depends only on* $\|g\|_{L_\infty}$, $\|g^{-1}\|_{L_\infty}$, *the operator* $b(\mathbf{D})$, *the domain* Ω^0, *and the multiplier norm* M_Λ.

Proof. By the scaling transformation, estimate (8.13) is equivalent to

$$\|(\mathcal{A} + \varepsilon^2 I)^{-1} - (\mathcal{A}^0 + \varepsilon^2 I)^{-1} - \mathcal{K}^0(\varepsilon)\|_{L_2(\Pi)\to L_2(\Pi)} \le \mathcal{C}_{12}, \tag{8.14}$$

where

$$\begin{aligned}\mathcal{K}^0(\varepsilon) = \Lambda b(\mathbf{D})(\mathcal{A}^0 + \varepsilon^2 I)^{-1} + (\mathcal{A}^0 + \varepsilon^2 I)^{-1} b(\mathbf{D})^* \Lambda^* \\ - (\mathcal{A}^0 + \varepsilon^2)^{-1} b(\mathbf{D})^* L(\mathbf{D}) b(\mathbf{D})(\mathcal{A}^0 + \varepsilon^2 I)^{-1}.\end{aligned}$$

In order to deduce (8.14) from (6.6), we need to estimate the operator

$$\begin{aligned}\mathcal{K}(\varepsilon) - \mathcal{K}^0(\varepsilon) = \Lambda b(\mathbf{D})(\mathcal{A}^0 + \varepsilon^2 I)^{-1}(\mathcal{P} - I) + (\mathcal{A}^0 + \varepsilon^2 I)^{-1}(\mathcal{P} - I) b(\mathbf{D})^* \Lambda^* \\ - (\mathcal{A}^0 + \varepsilon^2 I)^{-1} b(\mathbf{D})^* L(\mathbf{D}) b(\mathbf{D})(\mathcal{A}^0 + \varepsilon^2 I)^{-1}(\mathcal{P} - I).\end{aligned} \tag{8.15}$$

Condition 8.5(1°) ensures estimate (8.7). Hence, by (3.1) and (3.4),

$$\|b(\mathbf{D})(\mathcal{A}^0 + I)^{-1}\|_{L_2(\Pi)\to H^1(\Pi)} \le (d\alpha_1)^{1/2}\widehat{c}.$$

Then Condition 8.5(2°) implies that

$$\|\Lambda b(\mathbf{D})(\mathcal{A}^0 + I)^{-1}\|_{L_2(\Pi)\to L_2(\Pi)} \le (d\alpha_1)^{1/2}\widehat{c} M_\Lambda. \tag{8.16}$$

Estimates (8.8) and (8.16) imply that

$$\|\Lambda b(\mathbf{D})(\mathcal{A}^0 + \varepsilon^2 I)^{-1}(I - \mathcal{P})\|_{L_2(\Pi)\to L_2(\Pi)} \le \mathcal{C}_{13}, \tag{8.17}$$

where $\mathcal{C}_{13} = (1 + c_4^{-1}\|g^{-1}\|_{L_\infty})(d\alpha_1)^{1/2}\widehat{c} M_\Lambda$.

From (8.6), (8.15), and (8.17), it follows that

$$\|\mathcal{K}(\varepsilon) - \mathcal{K}^0(\varepsilon)\|_{L_2(\Pi)\to L_2(\Pi)} \le 2\mathcal{C}_{13} + \mathcal{C}_{10}.$$

Together with (6.6), this implies the required estimate (8.14) with

$$\mathcal{C}_{12} = \mathcal{C}_2 + 2\mathcal{C}_{13} + \mathcal{C}_{10}.$$

□

Similarly to Lemma 6.7 from [7], using the Sobolev embedding theorem, it is easy to check the following statement.

Proposition 8.7. *If* $\Lambda \in L_p(\Omega^0)$, *where* $p > 2$ *for* $d = 2$ *and* $p = d$ *for* $d \ge 3$, *then the operator* $[\Lambda]$ *is continuous from* $H^1(\Pi; \mathbb{C}^m)$ *to* $L_2(\Pi; \mathbb{C}^n)$, *and* $M_\Lambda \le c(\Omega^0)\|\Lambda\|_{L_p(\Omega^0)}$.

By the Sobolev embedding theorem and estimate (4.31), we deduce the following corollary from Proposition 8.7; cf. [7, Proposition 6.8].

Proposition 8.8. *If $d \leq 4$, then the operator $[\Lambda]$ is continuous from $H^1(\Pi;\mathbb{C}^m)$ to $L_2(\Pi;\mathbb{C}^n)$, and the norm M_Λ depends only on $\|g\|_{L_\infty}$, $\|g^{-1}\|_{L_\infty}$, the operator $b(\mathbf{D})$, and the domain Ω^0.*

8.2. *Approximation of the resolvent $(\mathcal{A}_\varepsilon + I)^{-1}$ in the $(L_2 \to H^1)$-norm: Removal of the operator $\mathcal{P}_\varepsilon$*

Now, we show that, under the additional assumptions, it is possible to remove the smoothing operator in estimate (7.15).

Condition 8.9. 1°. Suppose that $\partial\Pi \in C^{1,1}$. 2°. Suppose that the solution $\Lambda \in \widetilde{H}^1(\Omega^0)$ of problem (4.9) is bounded, i.e., $\Lambda \in L_\infty$.

Similarly to Lemma 8.3 from [8], it is easy to check the following statement.

Lemma 8.10. *Suppose that the solution $\Lambda \in \widetilde{H}^1(\Omega^0)$ of problem (4.9) is bounded. Then for any $u \in H^1(\Pi)$, we have*

$$\int_\Pi |g(\mathbf{x})^{1/2} b(\mathbf{D})\Lambda(\mathbf{x})|^2 |u(\mathbf{x})|^2 \, d\mathbf{x} \leq 4\|g\|_{L_\infty} \int_\Pi |u(\mathbf{x})|^2 \, d\mathbf{x} + 10 d\alpha_1 \|g\|_{L_\infty} \|\Lambda\|^2_{L_\infty} \int_\Pi |\mathbf{D}u(\mathbf{x})|^2 \, d\mathbf{x}.$$

Lemma 8.11. *Under Condition 8.9, for $\varepsilon > 0$, we have*

$$\|\mathcal{A}^{1/2}\big(\Lambda b(\mathbf{D})(\mathcal{A}^0 + \varepsilon^2 I)^{-1}(\mathcal{P} - I)\big)\|_{L_2(\Pi)\to L_2(\Pi)} \leq \mathcal{C}_{14}, \tag{8.18}$$

where the constant $\mathcal{C}_{14}$ depends only on $\|g\|_{L_\infty}$, $\|g^{-1}\|_{L_\infty}$, the operator $b(\mathbf{D})$, the domain Ω^0, and the norm $\|\Lambda\|_{L_\infty}$.

Proof. By (3.1), we have

$$\begin{aligned}
&\|\mathcal{A}^{1/2}\big(\Lambda b(\mathbf{D})(\mathcal{A}^0 + \varepsilon^2 I)^{-1}(\mathcal{P} - I)\big)\|_{L_2(\Pi)\to L_2(\Pi)} \\
&\quad = \|g^{1/2} b(\mathbf{D}) \big(\Lambda b(\mathbf{D})(\mathcal{A}^0 + \varepsilon^2 I)^{-1}(\mathcal{P} - I)\big)\|_{L_2(\Pi)\to L_2(\Pi)} \\
&\quad \leq \|g^{1/2}(b(\mathbf{D})\Lambda) b(\mathbf{D})(\mathcal{A}^0 + \varepsilon^2 I)^{-1}(\mathcal{P} - I)\|_{L_2(\Pi)\to L_2(\Pi)} \\
&\qquad + \sum_{l=1}^d \|g^{1/2} b_l \Lambda D_l b(\mathbf{D})(\mathcal{A}^0 + \varepsilon^2 I)^{-1}(\mathcal{P} - I)\|_{L_2(\Pi)\to L_2(\Pi)}.
\end{aligned}$$

Denote the terms on the right by J_1 and J_2. By Lemma 8.10,

$$\begin{aligned}
J_1^2 &\leq 4\|g\|_{L_\infty} \|b(\mathbf{D})(\mathcal{A}^0 + \varepsilon^2 I)^{-1}(\mathcal{P} - I)\|^2_{L_2(\Pi)\to L_2(\Pi)} \\
&\quad + 10 d\alpha_1 \|g\|_{L_\infty} \|\Lambda\|^2_{L_\infty} \|\mathbf{D} b(\mathbf{D})(\mathcal{A}^0 + \varepsilon^2 I)^{-1}(\mathcal{P} - I)\|^2_{L_2(\Pi)\to L_2(\Pi)}.
\end{aligned}$$

The first term is estimated by (8.10). To estimate the second term, we use (3.1), (3.4), (8.7), and (8.8):

$$\|\mathbf{D} b(\mathbf{D})(\mathcal{A}^0 + \varepsilon^2 I)^{-1}(\mathcal{P} - I)\|_{L_2(\Pi)\to L_2(\Pi)} \leq d^{1/2}\alpha_1^{1/2}\widehat{c}(1 + c_4^{-1}\|g^{-1}\|_{L_\infty}). \tag{8.19}$$

Hence, the term J_1 satisfies estimate $J_1 \le \mathcal{C}'_{14}$ with the constant $\mathcal{C}'_{14}$ depending on $\|g\|_{L_\infty}$, $\|g^{-1}\|_{L_\infty}$, the operator $b(\mathbf{D})$, the domain Ω^0, and the norm $\|\Lambda\|_{L_\infty}$.

By (3.4), the term J_2 satisfies

$$J_2 \le d^{1/2}\alpha_1^{1/2}\|g\|_{L_\infty}^{1/2}\|\Lambda\|_{L_\infty}\|\mathbf{D}b(\mathbf{D})(\mathcal{A}^0+\varepsilon^2 I)^{-1}(\mathcal{P}-I)\|_{L_2(\Pi)\to L_2(\Pi)}.$$

Hence, by (8.19), we see that $J_2 \le \mathcal{C}''_{14}$ with the constant $\mathcal{C}''_{14}$ depending on $\|g\|_{L_\infty}$, $\|g^{-1}\|_{L_\infty}$, the operator $b(\mathbf{D})$, the domain Ω^0, and the norm $\|\Lambda\|_{L_\infty}$.

Thus, we arrive at the required estimate (8.18) with $\mathcal{C}_{14} = \mathcal{C}'_{14} + \mathcal{C}''_{14}$. □

Combining (6.7) and (8.18), we obtain

$$\|\mathcal{A}^{1/2}((\mathcal{A}+\varepsilon^2 I)^{-1} - (I+\Lambda b(\mathbf{D}))(\mathcal{A}^0+\varepsilon^2 I)^{-1})\|_{L_2(\Pi)\to L_2(\Pi)} \le \mathcal{C}_3 + \mathcal{C}_{14}, \quad \varepsilon > 0.$$

By the scaling transformation, it follows that

$$\|\mathcal{A}_\varepsilon^{1/2}((\mathcal{A}_\varepsilon + I)^{-1} - (I+\varepsilon\Lambda^\varepsilon b(\mathbf{D}))(\mathcal{A}^0(\varepsilon)+I)^{-1})\|_{L_2(\Pi_\varepsilon)\to L_2(\Pi_\varepsilon)} \le (\mathcal{C}_3+\mathcal{C}_{14})\varepsilon, \quad \varepsilon > 0.$$

Together with the lower estimate (7.3), this yields

$$\begin{aligned}&\|\mathbf{D}((\mathcal{A}_\varepsilon + I)^{-1} - (I+\varepsilon\Lambda^\varepsilon b(\mathbf{D}))(\mathcal{A}^0(\varepsilon)+I)^{-1})\|_{L_2(\Pi_\varepsilon)\to L_2(\Pi_\varepsilon)}\\ &\quad \le (\mathcal{C}_3+\mathcal{C}_{14})\widetilde{\mathfrak{C}}_1^{1/2}\|g^{-1}\|_{L_\infty}^{1/2}\varepsilon. \end{aligned} \tag{8.20}$$

Using (7.6) and the condition $\Lambda \in L_\infty$, we see that

$$\begin{aligned}&\|(\mathcal{A}_\varepsilon + I)^{-1} - (I+\varepsilon\Lambda^\varepsilon b(\mathbf{D}))(\mathcal{A}^0(\varepsilon)+I)^{-1}\|_{L_2(\Pi_\varepsilon)\to L_2(\Pi_\varepsilon)}\\ &\quad \le (\mathcal{C}_1 + \|\Lambda\|_{L_\infty}\|g^{-1}\|_{L_\infty}^{1/2})\varepsilon. \end{aligned} \tag{8.21}$$

Finally, estimates (8.20) and (8.21) imply the following result.

Theorem 8.12. *Under Condition* 8.9, *for* $\varepsilon > 0$, *we have*

$$\|(\mathcal{A}_\varepsilon + I)^{-1} - (I+\varepsilon\Lambda^\varepsilon b(\mathbf{D}))(\mathcal{A}^0(\varepsilon)+I)^{-1}\|_{L_2(\Pi_\varepsilon)\to H^1(\Pi_\varepsilon)} \le \mathcal{C}_{15}\varepsilon. \tag{8.22}$$

The constant $\mathcal{C}_{15}$ *depends only on* $\|g\|_{L_\infty}$, $\|g^{-1}\|_{L_\infty}$, *the operator* $b(\mathbf{D})$, *the domain* Ω^0, *and the norm* $\|\Lambda\|_{L_\infty}$.

Similarly, the smoothing operator in approximation for the "flux" can be removed under Condition 8.9.

Theorem 8.13. *Under Condition* 8.9, *for* $\varepsilon > 0$, *we have*

$$\|g^\varepsilon b(\mathbf{D})(\mathcal{A}_\varepsilon + I)^{-1} - \widetilde{g}^\varepsilon b(\mathbf{D})(\mathcal{A}^0(\varepsilon)+I)^{-1}\|_{L_2(\Pi_\varepsilon)\to L_2(\Pi_\varepsilon)} \le \mathcal{C}_{16}\varepsilon. \tag{8.23}$$

The constant $\mathcal{C}_{16}$ *depends only on* $\|g\|_{L_\infty}$, $\|g^{-1}\|_{L_\infty}$, *the operator* $b(\mathbf{D})$, *the domain* Ω^0, *and the norm* $\|\Lambda\|_{L_\infty}$.

Proof. In order to deduce estimate (8.23) from Theorem 7.4, we need to check the estimate

$$\|\widetilde{g}^\varepsilon b(\mathbf{D})(\mathcal{A}^0(\varepsilon)+I)^{-1}(\mathcal{P}_\varepsilon - I)\|_{L_2(\Pi_\varepsilon)\to L_2(\Pi_\varepsilon)} \le \mathcal{C}_{17}\varepsilon. \tag{8.24}$$

By the scaling transformation, (8.24) is equivalent to

$$\|\widetilde{g}b(\mathbf{D})(\mathcal{A}^0+\varepsilon^2 I)^{-1}(\mathcal{P} - I)\|_{L_2(\Pi)\to L_2(\Pi)} \le \mathcal{C}_{17}. \tag{8.25}$$

Recalling that $\widetilde{g} = g(b(\mathbf{D})\Lambda + \mathbf{1})$ and applying Lemma 8.10, we obtain

$$\begin{aligned}
&\|\widetilde{g}b(\mathbf{D})(\mathcal{A}^0+\varepsilon^2 I)^{-1}(\mathcal{P} - I)\|_{L_2(\Pi)\to L_2(\Pi)}\\
&\quad\le 3\|g\|_{L_\infty}\|b(\mathbf{D})(\mathcal{A}^0+\varepsilon^2 I)^{-1}(\mathcal{P} - I)\|_{L_2(\Pi)\to L_2(\Pi)}\\
&\qquad + (10d\alpha_1)^{1/2}\|g\|_{L_\infty}\|\Lambda\|_{L_\infty}\|\mathbf{D}b(\mathbf{D})(\mathcal{A}^0+\varepsilon^2 I)^{-1}(\mathcal{P} - I)\|_{L_2(\Pi)\to L_2(\Pi)}.
\end{aligned}$$

Estimating the first term by (8.10) and the second one by (8.19), we arrive at the required estimate (8.25) with the constant $\mathcal{C}_{17}$ depending on $\|g\|_{L_\infty}$, $\|g^{-1}\|_{L_\infty}$, the operator $b(\mathbf{D})$, the domain Ω^0, and the norm $\|\Lambda\|_{L_\infty}$.

Finally, estimates (7.19) and (8.24) imply (8.23) with the constant $\mathcal{C}_{16} = \mathcal{C}_7 + \mathcal{C}_{17}$. □

Corollary 8.14. *Suppose that* $d = 2$ *and* $\partial\Pi \in C^{1,1}$. *Then estimates* (8.22) *and* (8.23) *are valid.*

Proof. The solution $\Lambda \in \widetilde{H}^1(\Omega^0)$ of problem (4.9) satisfies $\Lambda \in W_p^1(\Omega^0)$ with some $p > 2$ (see [13]). By the Sobolev embedding theorem for $d = 2$, $W_p^1(\Omega^0) \subset L_\infty$. Hence, Condition 8.9 is satisfied. Theorems 8.12 and 8.13 can be applied. □

Another condition ensuring estimate of the form (8.23) is given by the following theorem.

Theorem 8.15. *Suppose that the matrix-valued function* (4.15) *is bounded, i.e.,* $\widetilde{g} \in L_\infty$. *Then for* $\varepsilon > 0$, *we have*

$$\|g^\varepsilon b(\mathbf{D})(\mathcal{A}_\varepsilon + I)^{-1} - \widetilde{g}^\varepsilon b(\mathbf{D})(\mathcal{A}^0(\varepsilon)+I)^{-1}\|_{L_2(\Pi_\varepsilon)\to L_2(\Pi_\varepsilon)} \le \mathcal{C}'_{16}\varepsilon.$$

The constant $\mathcal{C}'_{16}$ *depends only on* $\|g\|_{L_\infty}$, $\|g^{-1}\|_{L_\infty}$, *the operator* $b(\mathbf{D})$, *the domain* Ω^0, *and the norm* $\|\widetilde{g}\|_{L_\infty}$.

Proof. It suffices to check estimate of the form (8.25). Such estimate follows from (8.10) and the condition $\widetilde{g} \in L_\infty$. □

8.3. *Approximations for the operators* $(\widetilde{\mathcal{A}}_\varepsilon + I)^{-1}$ *and* $(\mathcal{A}_\varepsilon + Q^\varepsilon)^{-1}$*: Removal of the smoothing operator*

Similarly to Secs. 8.1 and 8.2, under some additional assumptions, it is possible to remove the smoothing operator $\mathcal{P}_\varepsilon$ in approximations for the operators $(\widetilde{\mathcal{A}}_\varepsilon + I)^{-1}$ and $(\mathcal{A}_\varepsilon + Q^\varepsilon)^{-1}$. We formulate the corresponding results (details of the proofs are omitted).

Theorem 8.16. *Let $\Lambda \in \widetilde{H}^1(\Omega^0)$ be the solution of problem (4.9). Suppose that $\Lambda \in L_\infty$. Let*

$$\mathcal{K}'_{Q,\varepsilon} = \Lambda_Q^\varepsilon b(\mathbf{D})(\mathcal{A}^0(\varepsilon)+\overline{Q})^{-1} + (\mathcal{A}^0(\varepsilon)+\overline{Q})^{-1} b(\mathbf{D})^*(\Lambda_Q^\varepsilon)^* \\ - (\mathcal{A}^0(\varepsilon)+\overline{Q})^{-1} b(\mathbf{D})^* L_Q(\mathbf{D}) b(\mathbf{D})(\mathcal{A}^0(\varepsilon)+\overline{Q})^{-1}\mathcal{P}_\varepsilon. \tag{8.26}$$

Then for $\varepsilon > 0$, we have

$$\|(\widetilde{\mathcal{A}}_\varepsilon + I)^{-1} - (f^\varepsilon)^{-1}((\mathcal{A}^0(\varepsilon)+\overline{Q})^{-1} + \varepsilon\mathcal{K}'_{Q,\varepsilon})((f^\varepsilon)^*)^{-1}\|_{L_2(\Pi_\varepsilon)\to L_2(\Pi_\varepsilon)} \le \widetilde{C}_9\varepsilon^2,$$

$$\|(\mathcal{A}_\varepsilon + Q^\varepsilon)^{-1} - (\mathcal{A}^0(\varepsilon)+\overline{Q})^{-1} - \varepsilon\mathcal{K}'_{Q,\varepsilon}\|_{L_2(\Pi_\varepsilon)\to L_2(\Pi_\varepsilon)} \le \widetilde{C}_9\|f\|^2_{L_\infty}\varepsilon^2.$$

The constant $\widetilde{C}_9$ depends only on $\|g\|_{L_\infty}$, $\|g^{-1}\|_{L_\infty}$, $\|f\|_{L_\infty}$, $\|f^{-1}\|_{L_\infty}$, the operator $b(\mathbf{D})$, the domain Ω^0, and the norm $\|\Lambda\|_{L_\infty}$.

Theorem 8.17. *Suppose that $\partial\Pi \in C^{1,1}$. Let*

$$\mathcal{K}''_{Q,\varepsilon} = \Lambda_Q^\varepsilon b(\mathbf{D})(\mathcal{A}^0(\varepsilon)+\overline{Q})^{-1}\mathcal{P}_\varepsilon + (\mathcal{A}^0(\varepsilon)+\overline{Q})^{-1}\mathcal{P}_\varepsilon b(\mathbf{D})^*(\Lambda_Q^\varepsilon)^* \\ - (\mathcal{A}^0(\varepsilon)+\overline{Q})^{-1} b(\mathbf{D})^* L_Q(\mathbf{D}) b(\mathbf{D})(\mathcal{A}^0(\varepsilon)+\overline{Q})^{-1}.$$

Then for $\varepsilon > 0$, we have

$$\|(\widetilde{\mathcal{A}}_\varepsilon + I)^{-1} - (f^\varepsilon)^{-1}((\mathcal{A}^0(\varepsilon)+\overline{Q})^{-1} + \varepsilon\mathcal{K}''_{Q,\varepsilon})((f^\varepsilon)^*)^{-1}\|_{L_2(\Pi_\varepsilon)\to L_2(\Pi_\varepsilon)} \le \widetilde{C}_{11}\varepsilon^2,$$

$$\|(\mathcal{A}_\varepsilon + Q^\varepsilon)^{-1} - (\mathcal{A}^0(\varepsilon)+\overline{Q})^{-1} - \varepsilon\mathcal{K}''_{Q,\varepsilon}\|_{L_2(\Pi_\varepsilon)\to L_2(\Pi_\varepsilon)} \le \widetilde{C}_{11}\|f\|^2_{L_\infty}\varepsilon^2.$$

The constant $\widetilde{C}_{11}$ depends only on $\|g\|_{L_\infty}$, $\|g^{-1}\|_{L_\infty}$, $\|f\|_{L_\infty}$, $\|f^{-1}\|_{L_\infty}$, the operator $b(\mathbf{D})$, and the domain Ω^0.

Theorem 8.18. *Suppose that Condition 8.5 is satisfied. Let*

$$\mathcal{K}^0_{Q,\varepsilon} = \Lambda_Q^\varepsilon b(\mathbf{D})(\mathcal{A}^0(\varepsilon)+\overline{Q})^{-1} + (\mathcal{A}^0(\varepsilon)+\overline{Q})^{-1} b(\mathbf{D})^*(\Lambda_Q^\varepsilon)^* \\ - (\mathcal{A}^0(\varepsilon)+\overline{Q})^{-1} b(\mathbf{D})^* L_Q(\mathbf{D}) b(\mathbf{D})(\mathcal{A}^0(\varepsilon)+\overline{Q})^{-1}.$$

Then for $\varepsilon > 0$, we have

$$\|(\widetilde{\mathcal{A}}_\varepsilon + I)^{-1} - (f^\varepsilon)^{-1}((\mathcal{A}^0(\varepsilon)+\overline{Q})^{-1} + \varepsilon\mathcal{K}^0_{Q,\varepsilon})((f^\varepsilon)^*)^{-1}\|_{L_2(\Pi_\varepsilon)\to L_2(\Pi_\varepsilon)} \le \widetilde{C}_{12}\varepsilon^2,$$

$$\|(\mathcal{A}_\varepsilon + Q^\varepsilon)^{-1} - (\mathcal{A}^0(\varepsilon)+\overline{Q})^{-1} - \varepsilon\mathcal{K}^0_{Q,\varepsilon}\|_{L_2(\Pi_\varepsilon)\to L_2(\Pi_\varepsilon)} \le \widetilde{C}_{12}\|f\|^2_{L_\infty}\varepsilon^2.$$

The constant $\widetilde{C}_{12}$ depends only on $\|g\|_{L_\infty}$, $\|g^{-1}\|_{L_\infty}$, $\|f\|_{L_\infty}$, $\|f^{-1}\|_{L_\infty}$, the operator $b(\mathbf{D})$, the domain Ω^0, and the multiplier norm M_Λ.

Theorem 8.19. *Under Condition* 8.9, *for* $\varepsilon > 0$, *we have*

$$\|f^\varepsilon(\widetilde{\mathcal{A}}_\varepsilon + I)^{-1} - (I + \varepsilon\Lambda^\varepsilon b(\mathbf{D}))(\mathcal{A}^0(\varepsilon) + \overline{Q})^{-1}((f^\varepsilon)^*)^{-1}\|_{L_2(\Pi_\varepsilon)\to H^1(\Pi_\varepsilon)} \leq \widetilde{\mathcal{C}}_{15}\varepsilon, \tag{8.27}$$

$$\|(\mathcal{A}_\varepsilon + Q^\varepsilon)^{-1} - (I + \varepsilon\Lambda^\varepsilon b(\mathbf{D}))(\mathcal{A}^0(\varepsilon) + \overline{Q})^{-1}\|_{L_2(\Pi_\varepsilon)\to H^1(\Pi_\varepsilon)} \leq \widetilde{\mathcal{C}}_{15}\|f\|_{L_\infty}\varepsilon, \tag{8.28}$$

$$\|g^\varepsilon b(\mathbf{D})f^\varepsilon(\widetilde{\mathcal{A}}_\varepsilon + I)^{-1} - \widetilde{g}^\varepsilon b(\mathbf{D})(\mathcal{A}^0(\varepsilon) + \overline{Q})^{-1}((f^\varepsilon)^*)^{-1}\|_{L_2(\Pi_\varepsilon)\to L_2(\Pi_\varepsilon)} \leq \widetilde{\mathcal{C}}_{16}\varepsilon, \tag{8.29}$$

$$\|g^\varepsilon b(\mathbf{D})(\mathcal{A}_\varepsilon + Q^\varepsilon)^{-1} - \widetilde{g}^\varepsilon b(\mathbf{D})(\mathcal{A}^0(\varepsilon) + \overline{Q})^{-1}\|_{L_2(\Pi_\varepsilon)\to L_2(\Pi_\varepsilon)} \leq \widetilde{\mathcal{C}}_{16}\|f\|_{L_\infty}\varepsilon. \tag{8.30}$$

The constants $\widetilde{\mathcal{C}}_{15}$ *and* $\widetilde{\mathcal{C}}_{16}$ *depend only on* $\|g\|_{L_\infty}$, $\|g^{-1}\|_{L_\infty}$, $\|f\|_{L_\infty}$, $\|f^{-1}\|_{L_\infty}$, *the operator* $b(\mathbf{D})$, *the domain* Ω^0, *and the norm* $\|\Lambda\|_{L_\infty}$.

Corollary 8.20. *Suppose that* $d = 2$ *and* $\partial\Pi \in C^{1,1}$. *Then estimates* (8.27)–(8.30) *are valid.*

Theorem 8.21. *Suppose that the matrix-valued function* (4.15) *is bounded, i.e.,* $\widetilde{g} \in L_\infty$. *Then for* $\varepsilon > 0$, *we have*

$$\|g^\varepsilon b(\mathbf{D})f^\varepsilon(\widetilde{\mathcal{A}}_\varepsilon + I)^{-1} - \widetilde{g}^\varepsilon b(\mathbf{D})(\mathcal{A}^0(\varepsilon) + \overline{Q})^{-1}((f^\varepsilon)^*)^{-1}\|_{L_2(\Pi_\varepsilon)\to L_2(\Pi_\varepsilon)} \leq \widetilde{\mathcal{C}}'_{16}\varepsilon,$$

$$\|g^\varepsilon b(\mathbf{D})(\mathcal{A}_\varepsilon + Q^\varepsilon)^{-1} - \widetilde{g}^\varepsilon b(\mathbf{D})(\mathcal{A}^0(\varepsilon) + \overline{Q})^{-1}\|_{L_2(\Pi_\varepsilon)\to L_2(\Pi_\varepsilon)} \leq \widetilde{\mathcal{C}}'_{16}\|f\|_{L_\infty}\varepsilon.$$

The constant $\widetilde{\mathcal{C}}'_{16}$ *depends only on* $\|g\|_{L_\infty}$, $\|g^{-1}\|_{L_\infty}$, $\|f\|_{L_\infty}$, $\|f^{-1}\|_{L_\infty}$, *the operator* $b(\mathbf{D})$, *the domain* Ω^0, *and the norm* $\|\widetilde{g}\|_{L_\infty}$.

8.4. *Smoothness conditions on the boundary*

Remark 8.22. Condition $\partial\Pi \in C^{1,1}$ can be replaced by the following implicit condition*:* suppose that the operator $(\mathcal{A}^0 + I)^{-1}$ is continuous from $L_2(\Pi;\mathbb{C}^n)$ to $H^2(\Pi;\mathbb{C}^n)$ and estimate (8.7) holds. In the case of scalar elliptic equations, wide conditions ensuring (8.7) are known; see, e.g., [18, Chap. 7]. (In particular, it suffices that $\partial\Pi \in C^\alpha$, $\alpha > 3/2$.) The statements of Lemmas 8.3, 8.11, Theorems 8.4, 8.6, 8.12, 8.13, 8.17–8.19, and Corollaries 8.14, 8.20 remain true if the condition $\partial\Pi \in C^{1,1}$ is replaced by the implicit condition (8.7).

8.5. *Special cases*

Suppose that $g^0 = \overline{g}$, i.e., relations (4.28) are satisfied. Then the solution $\Lambda \in \widetilde{H}^1(\Omega^0)$ of problem (4.9) is equal to zero, i.e., $\Lambda(\mathbf{x}) = 0$. Consequently, $L(\mathbf{D}) = 0$ and $\mathcal{K}_\varepsilon = 0$; see (4.42) and (7.8). Also, we have $\widetilde{g} = g \in L_\infty$. So, the assumptions of Theorem 8.15 are satisfied. Applying Theorems 7.2, 7.3, and 8.15, we arrive at the following statement.

Proposition 8.23. *Suppose that $g^0 = \overline{g}$, i.e., relations (4.28) are satisfied. Then for $\varepsilon > 0$*

$$\|(\mathcal{A}_\varepsilon + I)^{-1} - (\mathcal{A}^0(\varepsilon) + I)^{-1}\|_{L_2(\Pi_\varepsilon)\to L_2(\Pi_\varepsilon)} \le \mathcal{C}_2\varepsilon^2,$$

$$\|(\mathcal{A}_\varepsilon + I)^{-1} - (\mathcal{A}^0(\varepsilon) + I)^{-1}\|_{L_2(\Pi_\varepsilon)\to H^1(\Pi_\varepsilon)} \le \mathcal{C}_6\varepsilon,$$

$$\|g^\varepsilon b(\mathbf{D})(\mathcal{A}_\varepsilon + I)^{-1} - g^\varepsilon b(\mathbf{D})(\mathcal{A}^0(\varepsilon) + I)^{-1}\|_{L_2(\Pi_\varepsilon)\to L_2(\Pi_\varepsilon)} \le \mathcal{C}'_{16}\varepsilon.$$

Similarly, if $g^0 = \overline{g}$, then $\Lambda_Q(\mathbf{x}) = 0$ and $\mathcal{K}_{Q,\varepsilon} = 0$. Since $\widetilde{g} = g \in L_\infty$, the assumptions of Theorem 8.21 are satisfied. Applying Theorems 7.6, 7.7, and 8.21, we arrive at the following statement.

Proposition 8.24. *Suppose that $g^0 = \overline{g}$, i.e., relations (4.28) are satisfied. Then for $\varepsilon > 0$*

$$\|(\widetilde{\mathcal{A}}_\varepsilon + I)^{-1} - (f^\varepsilon)^{-1}(\mathcal{A}^0(\varepsilon) + \overline{Q})^{-1}((f^\varepsilon)^*)^{-1}\|_{L_2(\Pi_\varepsilon)\to L_2(\Pi_\varepsilon)} \le \widetilde{\mathcal{C}}_2\varepsilon^2,$$

$$\|(\mathcal{A}_\varepsilon + Q^\varepsilon)^{-1} - (\mathcal{A}^0(\varepsilon) + \overline{Q})^{-1}\|_{L_2(\Pi_\varepsilon)\to L_2(\Pi_\varepsilon)} \le \widetilde{\mathcal{C}}_2\|f\|^2_{L_\infty}\varepsilon^2,$$

$$\|f^\varepsilon(\widetilde{\mathcal{A}}_\varepsilon + I)^{-1} - (\mathcal{A}^0(\varepsilon) + \overline{Q})^{-1}((f^\varepsilon)^*)^{-1}\|_{L_2(\Pi_\varepsilon)\to H^1(\Pi_\varepsilon)} \le \widetilde{\mathcal{C}}_6\varepsilon,$$

$$\|(\mathcal{A}_\varepsilon + Q^\varepsilon)^{-1} - (\mathcal{A}^0(\varepsilon) + \overline{Q})^{-1}\|_{L_2(\Pi_\varepsilon)\to H^1(\Pi_\varepsilon)} \le \widetilde{\mathcal{C}}_6\|f\|_{L_\infty}\varepsilon,$$

$$\|g^\varepsilon b(\mathbf{D})f^\varepsilon(\widetilde{\mathcal{A}}_\varepsilon + I)^{-1} - g^\varepsilon b(\mathbf{D})(\mathcal{A}^0(\varepsilon) + \overline{Q})^{-1}((f^\varepsilon)^*)^{-1}\|_{L_2(\Pi_\varepsilon)\to L_2(\Pi_\varepsilon)} \le \widetilde{\mathcal{C}}'_{16}\varepsilon,$$

$$\|g^\varepsilon b(\mathbf{D})(\mathcal{A}_\varepsilon + Q^\varepsilon)^{-1} - g^\varepsilon b(\mathbf{D})(\mathcal{A}^0(\varepsilon) + \overline{Q})^{-1}\|_{L_2(\Pi_\varepsilon)\to L_2(\Pi_\varepsilon)} \le \widetilde{\mathcal{C}}'_{16}\|f\|_{L_\infty}\varepsilon.$$

Now, suppose that $g^0 = \underline{g}$, i.e., relations (4.29) are satisfied. Then, by Remark 4.4, $\widetilde{g}(\mathbf{x}) = g^0 = \underline{g}$. Next, by Proposition 4.7(2), we have $L(\mathbf{D}) = 0$. Consequently, the third term of the corrector (7.8) is equal to zero. Since $\widetilde{g} = g^0$, the assumptions of Theorem 8.15 are satisfied. Applying Theorems 7.2 and 8.15, we obtain the following statement.

Proposition 8.25. *Suppose that $g^0 = \underline{g}$, i.e., relations (4.29) are satisfied. Let*

$$\check{\mathcal{K}}_\varepsilon = \Lambda^\varepsilon b(\mathbf{D})(\mathcal{A}^0(\varepsilon) + I)^{-1}\mathcal{P}_\varepsilon + (\mathcal{A}^0(\varepsilon) + I)^{-1}\mathcal{P}_\varepsilon b(\mathbf{D})^*(\Lambda^\varepsilon)^*.$$

Then for $\varepsilon > 0$, we have

$$\|(\mathcal{A}_\varepsilon + I)^{-1} - (\mathcal{A}^0(\varepsilon) + I)^{-1} - \varepsilon\check{\mathcal{K}}_\varepsilon\|_{L_2(\Pi_\varepsilon)\to L_2(\Pi_\varepsilon)} \le \mathcal{C}_2\varepsilon^2,$$

$$\|g^\varepsilon b(\mathbf{D})(\mathcal{A}_\varepsilon + I)^{-1} - g^0 b(\mathbf{D})(\mathcal{A}^0(\varepsilon) + I)^{-1}\|_{L_2(\Pi_\varepsilon)\to L_2(\Pi_\varepsilon)} \le \mathcal{C}'_{16}\varepsilon. \qquad (8.31)$$

Applying Theorem 8.21 in the case where $g^0 = \underline{g}$, we obtain analogs of estimate (8.31) for the operators $(\widetilde{\mathcal{A}}_\varepsilon + I)^{-1}$ and $(\mathcal{A}_\varepsilon + Q^\varepsilon)^{-1}$.

Proposition 8.26. *Suppose that $g^0 = \underline{g}$, i.e., relations (4.29) are satisfied. Then for $\varepsilon > 0$*

$$\|g^\varepsilon b(\mathbf{D})f^\varepsilon(\widetilde{\mathcal{A}}_\varepsilon + I)^{-1} - g^0 b(\mathbf{D})(\mathcal{A}^0(\varepsilon) + \overline{Q})^{-1}((f^\varepsilon)^*)^{-1}\|_{L_2(\Pi_\varepsilon)\to L_2(\Pi_\varepsilon)} \le \widetilde{\mathcal{C}}'_{16}\varepsilon,$$

$$\|g^\varepsilon b(\mathbf{D})(\mathcal{A}_\varepsilon + Q^\varepsilon)^{-1} - g^0 b(\mathbf{D})(\mathcal{A}^0(\varepsilon) + \overline{Q})^{-1}\|_{L_2(\Pi_\varepsilon)\to L_2(\Pi_\varepsilon)} \le \widetilde{\mathcal{C}}'_{16}\|f\|_{L_\infty}\varepsilon.$$

9. Applications of General Results

9.1. *The acoustics operator*

In $L_2(\Pi)$, $d \geq 2$, we consider the operator given formally by the expression

$$\mathcal{A} = \mathbf{D}^* g(\mathbf{x})\mathbf{D} = -\operatorname{div} g(\mathbf{x})\nabla \tag{9.1}$$

with the Neumann condition on $\partial\Pi$. Here $g(\mathbf{x})$ is a Γ-periodic Hermitian $(d \times d)$-matrix-valued function such that

$$g(\mathbf{x}) > 0; \qquad g, g^{-1} \in L_\infty. \tag{9.2}$$

If $g(\mathbf{x})$ is a symmetric matrix with real entries, the operator (9.1) *is called the acoustics operator.* The operator (9.1) is a particular case of the operator $\mathcal{A}$ defined in Sec. 3.2. Now we have $n = 1$, $m = d$, and $b(\mathbf{D}) = \mathbf{D}$.

Let $\mathbf{e}_1, \ldots, \mathbf{e}_d$ be the standard orthonormal basis in $\mathbb{R}^d$. Let $\Phi_j \in \widetilde{H}^1(\Omega^0)$ be the weak solutions of the problems

$$\operatorname{div} g(\mathbf{x})(\nabla\Phi_j(\mathbf{x}) + \mathbf{e}_j) = 0, \qquad \int_{\Omega^0} \Phi_j(\mathbf{x})\, d\mathbf{x} = 0, \tag{9.3}$$

$j = 1, \ldots, d$. Then the solution $\Lambda \in \widetilde{H}^1(\Omega^0)$ of problem (4.9) is the row: $\Lambda(\mathbf{x}) = i(\Phi_1(\mathbf{x}), \ldots, \Phi_d(\mathbf{x}))$. According to (4.15), the matrix $\widetilde{g}(\mathbf{x})$ is the $(d\times d)$-matrix with the columns $g(\mathbf{x})(\nabla\Phi_j(\mathbf{x}) + \mathbf{e}_j)$, $j = 1, \ldots, d$. By (4.14), the effective matrix g^0 is given by $g^0 = |\Omega^0|^{-1} \int_{\Omega^0} \widetilde{g}(\mathbf{x})\, d\mathbf{x}$. The subspace $\mathfrak{N}$ (see (4.1)) is one-dimensional; it consists of constant functions. According to (4.16), the spectral germ $S(\boldsymbol{\theta})$ acts as multiplication by the number $\gamma(\boldsymbol{\theta}) = \langle g^0\boldsymbol{\theta}, \boldsymbol{\theta}\rangle$. The expansion (4.18) for the first eigenvalue $\lambda_1(t;\boldsymbol{\theta})$ of $\mathcal{A}(\mathbf{k}) = A(t;\boldsymbol{\theta})$ starts with the term $\gamma(\boldsymbol{\theta})t^2 = \langle g^0\mathbf{k}, \mathbf{k}\rangle$:

$$\lambda_1(t;\boldsymbol{\theta}) = \langle g^0\boldsymbol{\theta}, \boldsymbol{\theta}\rangle t^2 + \mu(\boldsymbol{\theta})t^3 + \cdots. \tag{9.4}$$

If $g(\mathbf{x})$ is a symmetric matrix with real entries, then, by Proposition 4.6, $N(\boldsymbol{\theta}) = 0$ for any $\boldsymbol{\theta} \in \mathbb{S}^{d-1}$. Hence, by (4.44), $\mu(\boldsymbol{\theta}) = 0$ in the expansion (9.4).

If $g(\mathbf{x})$ is a Hermitian matrix with complex entries, then, in general, $N(\boldsymbol{\theta})$ is not zero. A calculation (cf. [7, Sec. 10.3]) shows that

$$N(\boldsymbol{\theta}) = \mu(\boldsymbol{\theta}) = -i \sum_{j,l,k=1}^{d} (a_{jlk} - a_{jlk}^*)\theta_j\theta_l\theta_k, \quad \boldsymbol{\theta} \in \mathbb{S}^{d-1},$$

$$a_{jlk} = |\Omega^0|^{-1} \int_{\Omega^0} \Phi_j(\mathbf{x})^* \langle g(\mathbf{x})(\nabla\Phi_l(\mathbf{x}) + \mathbf{e}_l), \mathbf{e}_k\rangle\, d\mathbf{x}, \quad j,l,k = 1, \ldots, d.$$

Now, let $\mathcal{A}_\varepsilon$ be the operator in $L_2(\Pi_\varepsilon)$ given formally by the expression

$$\mathcal{A}_\varepsilon = \mathbf{D}^* g^\varepsilon(\mathbf{x})\mathbf{D} = -\operatorname{div} g^\varepsilon(\mathbf{x})\nabla$$

with the Neumann condition on $\partial\Pi_\varepsilon$. The effective operator $\mathcal{A}^0(\varepsilon)$ is given by $\mathcal{A}^0(\varepsilon) = \mathbf{D}^* g^0\mathbf{D} = -\operatorname{div} g^0\nabla$ with the Neumann condition on $\partial\Pi_\varepsilon$. The corrector

(7.8) takes the form

$$\mathcal{K}_\varepsilon = \sum_{j=1}^{d} \left(\Phi_j^\varepsilon \partial_j (\mathcal{A}^0(\varepsilon) + I)^{-1} \mathcal{P}_\varepsilon - (\mathcal{A}^0(\varepsilon) + I)^{-1} \mathcal{P}_\varepsilon \partial_j (\Phi_j^\varepsilon)^* \right)$$

$$- \sum_{j,l,k=1}^{d} (\mathcal{A}^0(\varepsilon) + I)^{-1} (a_{jlk} - a_{jlk}^*) \partial_j \partial_l \partial_k (\mathcal{A}^0(\varepsilon) + I)^{-1} \mathcal{P}_\varepsilon.$$

Applying Theorems 7.1–7.4, we obtain the following result.

Proposition 9.1. *Under the assumptions of Sec.* 9.1, *for* $\varepsilon > 0$, *we have*

$$\|(\mathcal{A}_\varepsilon + I)^{-1} - (\mathcal{A}^0(\varepsilon) + I)^{-1}\|_{L_2(\Pi_\varepsilon) \to L_2(\Pi_\varepsilon)} \leq C_1 \varepsilon,$$

$$\|(\mathcal{A}_\varepsilon + I)^{-1} - (\mathcal{A}^0(\varepsilon) + I)^{-1} - \varepsilon \mathcal{K}_\varepsilon\|_{L_2(\Pi_\varepsilon) \to L_2(\Pi_\varepsilon)} \leq C_2 \varepsilon^2,$$

$$\left\| (\mathcal{A}_\varepsilon + I)^{-1} - \Big(I + \varepsilon \sum_{j=1}^{d} \Phi_j^\varepsilon \partial_j \mathcal{P}_\varepsilon \Big) (\mathcal{A}^0(\varepsilon) + I)^{-1} \right\|_{L_2(\Pi_\varepsilon) \to H^1(\Pi_\varepsilon)} \leq C_6 \varepsilon,$$

$$\|g^\varepsilon \nabla (\mathcal{A}_\varepsilon + I)^{-1} - \widetilde{g}^\varepsilon \nabla (\mathcal{A}^0(\varepsilon) + I)^{-1} \mathcal{P}_\varepsilon)\|_{L_2(\Pi_\varepsilon) \to L_2(\Pi_\varepsilon)} \leq C_7 \varepsilon.$$

The constants C_1, C_2, C_6, *and* C_7 *depend on* $\|g\|_{L_\infty}$, $\|g^{-1}\|_{L_\infty}$, *and the domain* Ω^0.

Of course, as in the general case, we can apply Theorem 8.6 if $d \leq 4$ and $\partial\Pi \in C^{1,1}$; see Proposition 8.8. We can apply Corollary 8.14 if $d = 2$ and $\partial\Pi \in C^{1,1}$.

Now, we consider the acoustics operator, i.e., the case where $g(\mathbf{x})$ has real entries.

Lemma 9.2. *Suppose that* Π *is Lipschitz domain and* $g(\mathbf{x})$ *is a* Γ*-periodic symmetric* $(d \times d)$*-matrix-valued function with real entries satisfying* (9.2). *Then the solution* $\Phi_j \in \widetilde{H}^1(\Omega^0)$ *of problem* (9.3) *is bounded and the norm* $\|\Phi_j\|_{L_\infty}$ *is controlled in terms of* $\|g\|_{L_\infty}$, $\|g^{-1}\|_{L_\infty}$, *and the domain* Ω^0.

Proof. We extend Φ_j to Π as a Γ-periodic function. Then Φ_j is the weak solution of the equation $-\operatorname{div} g(\mathbf{x}) \nabla \Phi_j(\mathbf{x}) = \operatorname{div} g(\mathbf{x}) \mathbf{e}_j$ in Π with the natural boundary condition. By [20, Theorem 13.1, Chap. 3], for any bounded strictly interior subdomain $\mathcal{O} \subset \Pi$ we have $\Phi_j \in L_\infty(\mathcal{O})$ (with appropriate control of the norm).

It remains to show that Φ_j is bounded near the boundary $\partial\Pi$. Since Π is Lipschitz domain, for any $\mathbf{x}^0 \in \partial\Pi$ there exists a neighborhood $\mathcal{O}$ and a Lipschitz diffeomorphism $\Psi\colon \mathcal{O} \to B(1)$ that rectifies the boundary. Then the function $\widetilde{\Phi}_j = \Phi_j \circ \Psi^{-1}$ satisfies (in the weak sense) equation of the same type in $B_+(1) = \{\mathbf{y} \in B(1)\colon y_d > 0\}$: $-\operatorname{div} G(\mathbf{y}) \nabla \widetilde{\Phi}_j(\mathbf{y}) = \operatorname{div} \mathbf{F}(\mathbf{y})$, with $\mathbf{F} \in L_\infty$, and the natural condition on $\{\mathbf{y} \in B(1)\colon y_d = 0\}$. Next, we extend $\widetilde{\Phi}_j(\mathbf{y})$ to $B(1)$ as an even function in y_d (keeping the same notation). It is standard to show that $\widetilde{\Phi}_j(\mathbf{y})$ satisfies equation of the same type in $B(1)$. Applying [20, Theorem 13.1, Chap. 3] once more, we conclude that $\widetilde{\Phi}_j \in L_\infty(B(1/2))$. Hence, Φ_j is also bounded in some neighborhood of the point $\mathbf{x}^0$. Since $\partial\Omega^0$ is compact, it follows that $\Phi_j \in L_\infty(\Omega^0)$. Estimates for the norm $\|\Phi_j\|_{L_\infty(\Omega^0)}$ are derived in a standard way. □

Recall that, by Proposition 4.6, $\mathbf{D}^*L(\mathbf{D})\mathbf{D} = 0$ if $g(\mathbf{x})$ has real entries. Using Theorems 8.2, 8.12, 8.13, and Lemma 9.2, and taking Remark 8.22 into account, we improve the results of Proposition 9.1 in this case.

Proposition 9.3. *Under the assumptions of Sec.* 9.1, *suppose that the matrix* $g(\mathbf{x})$ *is a* Γ*-periodic symmetric* $(d \times d)$*-matrix-valued function with real entries satisfying* (9.2). *Then the following statements hold.*
1°. *Let*

$$\mathcal{K}_\varepsilon^0 = \sum_{j=1}^{d} \left(\Phi_j^\varepsilon \partial_j (\mathcal{A}^0(\varepsilon) + I)^{-1} - (\mathcal{A}^0(\varepsilon) + I)^{-1} \partial_j \Phi_j^\varepsilon\right). \tag{9.5}$$

Then for $\varepsilon > 0$, *we have*

$$\|(\mathcal{A}_\varepsilon + I)^{-1} - (\mathcal{A}^0(\varepsilon) + I)^{-1} - \varepsilon \mathcal{K}_\varepsilon^0\|_{L_2(\Pi_\varepsilon) \to L_2(\Pi_\varepsilon)} \leq \mathcal{C}_9 \varepsilon^2.$$

2°. *Suppose that* $\partial\Pi \in C^\alpha$, $\alpha > 3/2$. *For* $\varepsilon > 0$, *we have*

$$\left\| (\mathcal{A}_\varepsilon + I)^{-1} - \Big(I + \varepsilon \sum_{j=1}^{d} \Phi_j^\varepsilon \partial_j\Big)(\mathcal{A}^0(\varepsilon) + I)^{-1} \right\|_{L_2(\Pi_\varepsilon) \to H^1(\Pi_\varepsilon)} \leq \mathcal{C}_{15} \varepsilon,$$

$$\|g^\varepsilon \nabla (\mathcal{A}_\varepsilon + I)^{-1} - \widetilde{g}^\varepsilon \nabla (\mathcal{A}^0(\varepsilon) + I)^{-1})\|_{L_2(\Pi_\varepsilon) \to L_2(\Pi_\varepsilon)} \leq \mathcal{C}_{16} \varepsilon.$$

The constants $\mathcal{C}_9$, $\mathcal{C}_{15}$, *and* $\mathcal{C}_{16}$ *depend on* $\|g\|_{L_\infty}$, $\|g^{-1}\|_{L_\infty}$, *and the domain* Ω^0.

9.2. *The elasticity operator*

Let $d \geq 2$. We represent the operator of elasticity theory as in [5, Chap. 5, Sec. 2]. Let ζ be an orthogonal second rank tensor in $\mathbb{R}^d$; in the standard orthonormal basis in $\mathbb{R}^d$, it can be represented by a matrix $\zeta = \{\zeta_{jl}\}_{j,l=1}^d$. We shall consider *symmetric* tensors ζ, which will be identified with vectors $\zeta_* \in \mathbb{C}^m$, $2m = d(d+1)$, by the following rule. The vector ζ_* is formed by all components ζ_{jl}, $j \leq l$, and the pairs (j, l) are put in order in some fixed way.

For the *displacement vector* $\mathbf{u} \in H^1(\Pi; \mathbb{C}^n)$, we introduce the deformation tensor $e(\mathbf{u}) = \frac{1}{2}\left\{\frac{\partial u_j}{\partial x_l} + \frac{\partial u_l}{\partial x_j}\right\}$. Let $e_*(\mathbf{u})$ be the vector corresponding to the tensor $e(\mathbf{u})$ in accordance with the rule described above. The relation $b(\mathbf{D})\mathbf{u} = -ie_*(\mathbf{u})$ determines an $(m \times d)$-matrix homogeneous DO $b(\mathbf{D})$ uniquely; the symbol of this DO is a matrix with real entries. For instance, with an appropriate ordering, we have

$$b(\boldsymbol{\xi}) = \begin{pmatrix} \xi_1 & 0 \\ \frac{1}{2}\xi_2 & \frac{1}{2}\xi_1 \\ 0 & \xi_2 \end{pmatrix}, \quad d = 2.$$

Let $\sigma(\mathbf{u})$ be the *stress tensor*, and let $\sigma_*(\mathbf{u})$ be the corresponding vector. In the accepted way of writing, the *Hooke law* about proportionality of stresses and deformations can be expressed by the relation $\sigma_*(\mathbf{u}) = g(\mathbf{x})e_*(\mathbf{u})$, where $g(\mathbf{x})$ is an

$(m \times m)$-matrix-valued function with real entries (which gives a "concise" description of the Hooke tensor). The matrix $g(\mathbf{x})$ characterizes the parameters of the elastic (in general, anisotropic) medium under study. We assume that the matrix-valued function $g(\mathbf{x})$ is Γ-periodic in Π and such that $g(\mathbf{x}) > 0$, and $g, g^{-1} \in L_\infty$.

The energy of elastic deformations is given by the quadratic form

$$w[\mathbf{u},\mathbf{u}] = \frac{1}{2}\int_\Pi \langle \sigma_*(\mathbf{u}), e_*(\mathbf{u})\rangle_{\mathbb{C}^m}\, d\mathbf{x} = \frac{1}{2}\int_\Pi \langle g(\mathbf{x})b(\mathbf{D})\mathbf{u}, b(\mathbf{D})\mathbf{u}\rangle_{\mathbb{C}^m}\, d\mathbf{x}, \qquad \mathbf{u} \in H^1(\Pi;\mathbb{C}^d).$$

The operator $\mathcal{W}$ in $L_2(\Pi;\mathbb{C}^d)$ generated by this form is the *elasticity operator.* Thus, the operator $2\mathcal{W} = b(\mathbf{D})^* g b(\mathbf{D}) = \mathcal{A}$ is a particular case of the operator $\mathcal{A}$ defined in Sec. 3.2. In this case, $n = d$ and $m = d(d+1)/2$.

In the case of isotropic medium, the matrix $g(\mathbf{x})$ depends only on two functional *Lame parameters* $\lambda(\mathbf{x})$ and $\mu(\mathbf{x})$. The parameter $\mu(\mathbf{x})$ is the *shear modulus.* Often, another parameter $K(\mathbf{x}) = \lambda(\mathbf{x}) + \frac{2\mu(\mathbf{x})}{d}$ is introduced instead of $\lambda(\mathbf{x})$; $K(\mathbf{x})$ is called the *modulus of volume compression.* The modulus $\lambda(\mathbf{x})$ may be negative. In the isotropic case, the conditions that ensure the positive definiteness of the matrix $g(\mathbf{x})$ are as follows: $\mu(\mathbf{x}) \geq \mu_0 > 0$, $K(\mathbf{x}) \geq K_0 > 0$. As an example, we write down the matrix $g(\mathbf{x})$ in the isotropic case for $d = 2$:

$$g(\mathbf{x}) = \begin{pmatrix} K(\mathbf{x}) + \mu(\mathbf{x}) & 0 & K(\mathbf{x}) - \mu(\mathbf{x}) \\ 0 & 4\mu(\mathbf{x}) & 0 \\ K(\mathbf{x}) - \mu(\mathbf{x}) & 0 & K(\mathbf{x}) + \mu(\mathbf{x}) \end{pmatrix}, \qquad d = 2.$$

The effective matrix g^0 is defined by the general rules (see (4.14), (4.15)); the effective operator is given by $\mathcal{W}^0 = \frac{1}{2}\mathcal{A}^0 = \frac{1}{2}b(\mathbf{D})^* g^0 b(\mathbf{D})$ (with the natural condition on $\partial\Pi$). Even in the isotropic case, in general, the matrix g^0 is not isotropic.

Now, we consider the operator $\mathcal{W}_\varepsilon = \frac{1}{2}\mathcal{A}_\varepsilon$ with rapidly oscillating coefficients. This operator acts in $L_2(\Pi_\varepsilon;\mathbb{C}^d)$ and corresponds to the quadratic form

$$\frac{1}{2}\int_{\Pi_\varepsilon} \langle g^\varepsilon(\mathbf{x})b(\mathbf{D})\mathbf{u}, b(\mathbf{D})\mathbf{u}\rangle_{\mathbb{C}^m}\, d\mathbf{x}, \quad \mathbf{u} \in H^1(\Pi_\varepsilon;\mathbb{C}^d).$$

All general results apply to the operator $\mathcal{W}_\varepsilon$: Theorems 7.1–7.3 are applicable. We can apply Theorem 8.6 if $d \leq 4$ and $\partial\Pi \in C^{1,1}$; see Proposition 8.8. We can apply Corollary 8.14 if $d = 2$ and $\partial\Pi \in C^{1,1}$. No simplification in general formulations occurs (even in the isotropic case with variable periodic coefficients $\lambda(\mathbf{x})$ and $\mu(\mathbf{x})$).

9.3. *The Schrödinger operator. Factorization*

In this subsection, we describe a factorization for the Schrödinger operator; in the case of $\mathbb{R}^d$, such factorization is well known; see [16] and [4, Sec. 4].

In $L_2(\Pi)$, we consider the Schrödinger operator (with metric) formally given by

$$\mathcal{H} = \mathbf{D}^* \breve{g}(\mathbf{x})\mathbf{D} + V(\mathbf{x}), \quad \mathbf{x} \in \Pi, \tag{9.6}$$

with the Neumann condition on $\partial\Pi$. Here $\breve{g}(\mathbf{x})$ is a Γ-periodic $(d\times d)$-matrix-valued function with real entries such that $\breve{g}(\mathbf{x})>0$ and $\breve{g},\breve{g}^{-1}\in L_\infty$; $V(\mathbf{x})$ is a Γ-periodic real-valued function such that $V\in L_s(\Omega^0)$, $s>\frac{d}{2}$. Precisely, $\mathcal{H}$ is a self-adjoint operator in $L_2(\Pi)$ generated by the quadratic form

$$\mathfrak{h}[u,u]=\int_\Pi\left(\langle\breve{g}(\mathbf{x})\mathbf{D}u(\mathbf{x}),\mathbf{D}u(\mathbf{x})\rangle+V(\mathbf{x})|u(\mathbf{x})|^2\right)d\mathbf{x},\quad u\in H^1(\Pi). \tag{9.7}$$

Under our assumptions, this form is closed and semibounded from below.

With the help of the Gelfand transformation, the operator $\mathcal{H}$ admits the following direct integral expansion (cf. (3.29))

$$\mathcal{U}\mathcal{H}\mathcal{U}^{-1}=\int_{\widetilde{\Omega}}\oplus\mathcal{H}(\mathbf{k})d\mathbf{k},$$

where $\mathcal{H}(\mathbf{k})$ is a selfadjoint operator in $L_2(\Omega^0)$ generated by the quadratic form

$$\mathfrak{h}(\mathbf{k})[u,u]=\int_{\Omega^0}\left(\langle\breve{g}(\mathbf{x})(\mathbf{D}+\mathbf{k})u(\mathbf{x}),(\mathbf{D}+\mathbf{k})u(\mathbf{x})\rangle+V(\mathbf{x})|u(\mathbf{x})|^2\right)d\mathbf{x},\quad u\in\widetilde{H}^1(\Omega^0). \tag{9.8}$$

Adding an appropriate constant to $V(\mathbf{x})$, we *assume that*

$$\inf_{0\neq u\in\widetilde{H}^1(\Omega^0)}\frac{\mathfrak{h}(0)[u,u]}{\|u\|^2_{L_2(\Omega^0)}}=0. \tag{9.9}$$

Since the embedding of $\widetilde{H}^1(\Omega^0)$ in $L_2(\Omega^0)$ is compact, the infimum in (9.9) is reached on some (real-valued) function $\psi(\mathbf{x})$ and

$$\int_{\Omega^0}\left(\langle\breve{g}(\mathbf{x})\mathbf{D}\psi(\mathbf{x}),\mathbf{D}\eta(\mathbf{x})\rangle+V(\mathbf{x})\psi(\mathbf{x})\eta(\mathbf{x})\right)d\mathbf{x}=0,\quad\forall\eta\in\widetilde{H}^1(\Omega^0). \tag{9.10}$$

Then the infimum in (9.9) is also reached on the function $\omega(\mathbf{x})=|\psi(\mathbf{x})|$ and this function also satisfies the identity (9.10).

Moreover, $\omega\in C^\tau$ with some (small) $\tau>0$ and, by the Harnak inequality, $\omega(\mathbf{x})>0$. Indeed, we consider the periodic extension of ω to Π. Then ω is the weak solution of the equation $-\operatorname{div}\breve{g}\nabla\omega+V\omega=0$ in Π with the natural boundary condition. Then, under our assumptions on $\breve{g}$ and V, the required properties of ω in any bounded strictly interior subdomain of Π are known; see [20]. It is easy to check these properties near the boundary $\partial\Pi$ by using standard arguments, as in the proof of Lemma 9.2. It can be also shown that the functions ω and ω^{-1} are multipliers on $H^1(\Pi)$ and $\widetilde{H}^1(\Omega^0)$.

Thus, the equation $-\operatorname{div}\breve{g}\nabla\omega+V\omega=0$ in Π with the natural boundary condition has a positive periodic solution $\omega(\mathbf{x})$. It is convenient to normalize ω by the condition

$$\int_{\Omega^0}\omega^2(\mathbf{x})\,d\mathbf{x}=|\Omega^0|. \tag{9.11}$$

By substitution $u = \omega v$, the forms (9.7) and (9.8) turn into

$$\mathfrak{h}[u,u] = \int_\Pi \omega^2(\mathbf{x})\langle \check{g}(\mathbf{x})\mathbf{D}v(\mathbf{x}), \mathbf{D}v(\mathbf{x})\rangle \, d\mathbf{x}, \quad u = \omega v, \ v \in H^1(\Pi); \tag{9.12}$$

$$\mathfrak{h}(\mathbf{k})[u,u] = \int_{\Omega^0} \omega^2(\mathbf{x})\langle \check{g}(\mathbf{x})(\mathbf{D}+\mathbf{k})v(\mathbf{x}), (\mathbf{D}+\mathbf{k})v(\mathbf{x})\rangle \, d\mathbf{x},$$

$$u = \omega v, \ v \in \widetilde{H}^1(\Omega^0). \tag{9.13}$$

From (9.13) it directly follows that $\operatorname{Ker}\mathcal{H}(0) = \{c\omega\colon c \in \mathbb{C}\}$. Let $g = \omega^2\check{g}$. We see that $\mathfrak{h}(\mathbf{k})[u,u] = \mathfrak{a}(\mathbf{k})[v,v]$, $u = \omega v$, where the form

$$\mathfrak{a}(\mathbf{k})[v,v] = \int_{\Omega^0} \langle g(\mathbf{x})(\mathbf{D}+\mathbf{k})v(\mathbf{x}), (\mathbf{D}+\mathbf{k})v(\mathbf{x})\rangle \, d\mathbf{x}, \quad v \in \widetilde{H}^1(\Omega^0),$$

corresponds to the acoustics operator. Relation (9.12) means that the operator $\mathcal{H}$ admits a factorization of the form

$$\mathcal{H} = \omega^{-1}\mathbf{D}^* g \mathbf{D}\omega^{-1}, \quad g := \omega^2\check{g}. \tag{9.14}$$

Remark 9.4. We can take (9.14) as the definition of the operator $\mathcal{H}$ assuming that ω is an arbitrary Γ-periodic function in Π such that

$$\omega(\mathbf{x}) > 0; \quad \omega, \omega^{-1} \in L_\infty. \tag{9.15}$$

The initial form (9.6) can be recovered, but under conditions (9.15) the potential $V = \omega^{-1}(\operatorname{div}\check{g}\nabla\omega)$ may be strongly singular.

The operator (9.14) is a particular case of the operator $\widetilde{\mathcal{A}}$ defined by (3.33). Now we have $\mathcal{A} = \mathbf{D}^* g(\mathbf{x})\mathbf{D}$, $f(\mathbf{x}) = \omega(\mathbf{x})^{-1}$. Hence, $Q(\mathbf{x}) = \omega^2(\mathbf{x})$ and $\overline{Q} = 1$ (due to (9.11)).

Let $\widetilde{\lambda}_1(\mathbf{k}) = \widetilde{\lambda}_1(t;\boldsymbol{\theta})$ be the first eigenvalue of the operator $\mathcal{H}(\mathbf{k})$. According to (4.49), the coefficient $\widetilde{\gamma}_1(\boldsymbol{\theta})$ in the expansion (4.48) for $\widetilde{\lambda}_1(t;\boldsymbol{\theta})$ is given by $\widetilde{\gamma}_1(\boldsymbol{\theta}) = \langle g^0\boldsymbol{\theta}, \boldsymbol{\theta}\rangle$. By Proposition 4.8, we have $N_Q(\boldsymbol{\theta}) = 0$. Hence, by (4.51), $\widetilde{\mu}_1(\boldsymbol{\theta}) = 0$. Then (4.48) implies that

$$\widetilde{\lambda}_1(\mathbf{k}) = \widetilde{\lambda}_1(t;\boldsymbol{\theta}) = \langle g^0\boldsymbol{\theta}, \boldsymbol{\theta}\rangle t^2 + O(t^4).$$

In quantum mechanics, the matrix $(g^0)^{-1}$ is associated with the *tensor of the effective masses.* We see that *the tensors of the effective masses for the Schrödinger operator* $\mathcal{H} = \mathbf{D}^*\check{g}\mathbf{D} + V$ *and the acoustics operator* $\mathcal{A} = \mathbf{D}^* g\mathbf{D}$ (where $g = \check{g}\omega^2$) *coincide.*

9.4. *The periodic Schrödinger operator with a singular potential. Homogenization*

Now, we consider the operator $\mathcal{H}_\varepsilon$ acting in $L_2(\Pi_\varepsilon)$ and given by

$$\mathcal{H}_\varepsilon = (\omega^\varepsilon)^{-1}\mathbf{D}^* g^\varepsilon \mathbf{D}(\omega^\varepsilon)^{-1}, \quad \varepsilon > 0, \tag{9.16}$$

with the natural boundary condition. In the initial terms, we have

$$\mathcal{H}_\varepsilon = \mathbf{D}^*\check{g}^\varepsilon\mathbf{D} + \varepsilon^{-2}V^\varepsilon.$$

The potential $\varepsilon^{-2}V^\varepsilon$ is singular in the sense that it involves a large factor ε^{-2}.

In $L_2(\Pi_\varepsilon)$, we consider the operator $\mathcal{A}^0(\varepsilon)$ given by the differential expression $\mathbf{D}^* g^0 \mathbf{D}$ with the Neumann condition on $\partial\Pi_\varepsilon$, where g^0 is the effective matrix for the acoustics operator (9.1) (with $g = \check{g}\omega^2$). We can apply Theorem 7.5, taking into account that $\overline{Q} = 1$.

Proposition 9.5. *Let $\mathcal{H}_\varepsilon$ be the operator* (9.16). *Then for $\varepsilon > 0$, we have*

$$\|(\mathcal{H}_\varepsilon + I)^{-1} - \omega^\varepsilon(\mathcal{A}^0(\varepsilon) + I)^{-1}\omega^\varepsilon\|_{L_2(\Pi_\varepsilon)\to L_2(\Pi_\varepsilon)} \le \widetilde{\mathcal{C}}_1\varepsilon.$$

The constant $\widetilde{\mathcal{C}}_1$ depends on $\|\check{g}\|_{L_\infty}$, $\|\check{g}^{-1}\|_{L_\infty}$, $\|\omega\|_{L_\infty}$, $\|\omega^{-1}\|_{L_\infty}$, and the domain Ω^0.

Next, the matrix Λ_Q (see (4.50)) is the row: $\Lambda_Q = i(\Phi_{Q,1}, \dots, \Phi_{Q,d})$, where $\Phi_{Q,j} = \Phi_j + \Phi_j^0$, $\Phi_j^0 = -|\Omega^0|^{-1}\int_{\Omega^0}\omega^2\Phi_j\,d\mathbf{x}$. By Lemma 9.2, $\Phi_j \in L_\infty$, $j = 1, \dots, d$, i.e., $\Lambda \in L_\infty$. Then the assumptions of Theorem 8.16 are satisfied. By Proposition 4.8, we have $N_Q(\boldsymbol{\theta}) = 0$. Thus, the third term of the corrector is equal to zero: $b(\mathbf{D})^* L_Q(\mathbf{D}) b(\mathbf{D}) = 0$. Consequently, the corrector (8.26) takes the form

$$\mathcal{K}'_{Q,\varepsilon} = \sum_{j=1}^d \left(\Phi_j^\varepsilon \partial_j (\mathcal{A}^0(\varepsilon) + I)^{-1} - (\mathcal{A}^0(\varepsilon) + I)^{-1}\partial_j \Phi_j^\varepsilon\right)$$

and coincides with $\mathcal{K}_\varepsilon^0$ (see (9.5)). We have taken into account the obvious identity $\sum_{j=1}^d \left(\Phi_j^0 \partial_j (\mathcal{A}^0(\varepsilon) + I)^{-1} - (\mathcal{A}^0(\varepsilon) + I)^{-1}\partial_j \Phi_j^0\right) = 0$. Theorem 8.16 leads to the following result.

Proposition 9.6. *Let $\mathcal{K}_\varepsilon^0$ be given by* (9.5). *Then for $\varepsilon > 0$, we have*

$$\|(\mathcal{H}_\varepsilon + I)^{-1} - \omega^\varepsilon((\mathcal{A}^0(\varepsilon) + I)^{-1} + \varepsilon\mathcal{K}_\varepsilon^0)\omega^\varepsilon\|_{L_2(\Pi_\varepsilon)\to L_2(\Pi_\varepsilon)} \le \widetilde{\mathcal{C}}_9\varepsilon^2.$$

The constant $\widetilde{\mathcal{C}}_9$ depends on $\|\check{g}\|_{L_\infty}$, $\|\check{g}^{-1}\|_{L_\infty}$, $\|\omega\|_{L_\infty}$, $\|\omega^{-1}\|_{L_\infty}$, and the domain Ω^0.

Finally, applying Theorem 7.7 in the general case and Theorem 8.19 in the case of smooth boundary, and taking Remark 8.22 into account, we obtain the following statement.

Proposition 9.7. 1°. *For $\varepsilon > 0$, we have*

$$\left\|(\omega^\varepsilon)^{-1}(\mathcal{H}_\varepsilon + I)^{-1} - \left(I + \varepsilon\sum_{j=1}^d \Phi_j^\varepsilon \partial_j \mathcal{P}_\varepsilon\right)(\mathcal{A}^0(\varepsilon) + I)^{-1}\omega^\varepsilon\right\|_{L_2(\Pi_\varepsilon)\to H^1(\Pi_\varepsilon)} \le \widetilde{\mathcal{C}}_6\varepsilon,$$

$$\|g^\varepsilon\nabla(\omega^\varepsilon)^{-1}(\mathcal{H}_\varepsilon + I)^{-1} - \widetilde{g}^\varepsilon\nabla(\mathcal{A}^0(\varepsilon) + I)^{-1}\mathcal{P}_\varepsilon\omega^\varepsilon\|_{L_2(\Pi_\varepsilon)\to L_2(\Pi_\varepsilon)} \le \widetilde{\mathcal{C}}_7\varepsilon.$$

2°. *Suppose that* $\partial\Pi \in C^\alpha$, $\alpha > 3/2$. *Then for* $\varepsilon > 0$, *we have*

$$\left\|(\omega^\varepsilon)^{-1}(\mathcal{H}_\varepsilon + I)^{-1} - \left(I + \varepsilon \sum_{j=1}^{d} \Phi_j^\varepsilon \partial_j\right)(\mathcal{A}^0(\varepsilon) + I)^{-1}\omega^\varepsilon\right\|_{L_2(\Pi_\varepsilon)\to H^1(\Pi_\varepsilon)} \leq \widetilde{\mathcal{C}}_{15}\varepsilon,$$

$$\|g^\varepsilon \nabla(\omega^\varepsilon)^{-1}(\mathcal{H}_\varepsilon + I)^{-1} - \widetilde{g}^\varepsilon \nabla(\mathcal{A}^0(\varepsilon) + I)^{-1}\omega^\varepsilon\|_{L_2(\Pi_\varepsilon)\to L_2(\Pi_\varepsilon)} \leq \widetilde{\mathcal{C}}_{16}\varepsilon.$$

The constants $\widetilde{\mathcal{C}}_6$, $\widetilde{\mathcal{C}}_7$, $\widetilde{\mathcal{C}}_{15}$, $\widetilde{\mathcal{C}}_{16}$ *depend on* $\|\check{g}\|_{L_\infty}$, $\|\check{g}^{-1}\|_{L_\infty}$, $\|\omega\|_{L_\infty}$, $\|\omega^{-1}\|_{L_\infty}$, *and the domain* Ω^0.

Acknowledgments

The author is grateful to Nikolai Filonov and Alexander Nazarov for fruitful discussions. This work was supported by Russian Foundation for Basic Research (grant no. 16-01-00087).

References

[1] G. Allaire, Homogenization and two-scale convergence, *SIAM J. Math. Anal.* **23**(6) (1992) 1482–1518.

[2] N. S. Bakhvalov and G. P. Panasenko, *Homogenization: Averaging Processes in Periodic Media. Mathematical Problems in Mechanics of Composite Materials* ("Nauka", Moscow, 1984); English transl., Math. Appl. (Soviet Ser.), Vol. 36 (Kluwer Acad. Publ. Group, Dordrecht, 1989).

[3] A. Bensoussan, J.-L. Lions and G. Papanicolaou, *Asymptotic Analysis for Periodic Structures*, Stud. Math. Appl., Vol. 5 (North-Holland Publishing Co., Amsterdam-New York, 1978).

[4] M. Sh. Birman and T. A. Suslina, Two-dimensional periodic Pauli operator. The effective masses at the lower edge of the spectrum, *Oper. Theory: Adv. and Appl.* **108** (1999) 13–31.

[5] M. Sh. Birman and T. A. Suslina, Second order periodic differential operators. Threshold properties and homogenization, *Algebra i Analiz* **15**(5) (2003) 1–108; English transl., *St. Petersburg Math. J.* **15**(5) (2004) 639–714.

[6] M. Sh. Birman and T. A. Suslina, Threshold approximations with corrector for the resolvent of a factorized operator family, *Algebra i Analiz* **17**(5) (2005) 69–90; English transl., *St. Petersburg Math. J.* **17**(5) (2006) 745–762.

[7] M. Sh. Birman and T. A. Suslina, Homogenization with corrector term for periodic elliptic differential operators, *Algebra i Analiz* **17**(6) (2005) 1–104; English transl., *St. Petersburg Math. J.* **17**(6) (2006) 897–973.

[8] M. Sh. Birman and T. A. Suslina, Homogenization with corrector term for periodic differential operators. Approximation of solutions in the Sobolev class $H^1(\mathbb{R}^d)$, *Algebra i Analiz* **18**(6) (2006) 1–130; English transl., *St. Petersburg Math. J.* **18**(6) (2007) 857–955.

[9] D. Cioranescu and J. S. J. Paulin, Homogenization in perforated media, in *Homogenization of Reticulated Structures*, Applied Mathematical Sciences, Vol. 136 (Springer, New York, NY, 1999), pp. 1–70.

[10] G. A. Chechkin, Homogenization in perforated domains, in *Topics on Concentration Phenomena and Problems with Multiple Scales*, eds. A. Braides and V. Chiado Piat,

Lecture Notes of the Unione Matematica Italiana, Vol. 2 (Springer-Verlag, Berlin, 2006), pp. 189–208.

[11] G. A. Chechkin and A. L. Piatnitski, Homogenization of boundary-value problem in a locally periodic perforated domain, *Appl. Anal.* **71**(1–4) (1999) 215–235.

[12] C. Conca, R. Orive and M. Vanninathan, Bloch approximation in homogenization and applications, *SIAM J. Math. Anal.* **33**(5) (2002) 1166–1198.

[13] M. Giaquinta, *Multiple Integrals in the Calculus of Variations and Nonlinear Elliptic Systems*, Ann. of Math. Stud., Vol. 105 (Princeton Univ. Press, NJ, 1983).

[14] G. Griso and M. T. Sanchez, Homogenization of an elasticity problem for a catalyst support by using the unfolding method, *Asymptot. Anal.* **67**(1) (2010) 45–84.

[15] T. Kato, *Perturbation Theory for Linear Operators* (Springer-Verlag, Berlin, 1995).

[16] W. Kirsch and B. Simon, Comparison theorems for the gap of Schrödinger operators, *J. Funct. Anal.* **75**(2) (1987) 396–410.

[17] V. A. Marchenko and E. Ya. Khruslov, *Homogenization of Partial Differential Equations*, Progress in Mathematical Physics, Vol. 46 (Birkhäuser Boston, 2006).

[18] V. G. Maz'ya and T. O. Shaposhnikova, *Theory of Multipliers in Spaces of Differentiable Functions*, Monographs and Studies in Mathematica, Vol. 23 (Pitman (Advanced Publishing Program), 1985).

[19] W. McLean, *Strongly Elliptic Systems and Boundary Integral Equations* (Cambridge Univ. Press, Cambridge, 2000).

[20] O. A. Ladyzhenskaya and N. N. Ural'tseva, *Linear and Quasilinear Elliptic Equations* (Academic Press, 1968).

[21] S. A. Nazarov, The polynomial property of self-adjoint elliptic boundary-value problems and an algebraic description of their attributes, *Uspekhi Matem. Nauk* **54**(5) (1999) 77–142; English transl., *Russian Math. Surveys* **54**(5) (1999) 947–1014.

[22] J. Nečas, *Direct Methods in the Theory of Elliptic Equations*, Springer Monographs in Mathematics (Springer, 2012).

[23] E. V. Sevost'yanova, Asymptotic expansion of the solution of a second-order elliptic equation with periodic rapidly oscillating coefficients, *Matem. Sbornik* **115**(2) (1981) 204–222; English transl., *Math. USSR-Sb.* **43**(2) (1982) 181–198.

[24] V. V. Zhikov, On some estimates of homogenization theory, *Dokl. Ros. Akad. Nauk* **406**(5) (2006) 597–601; English transl., *Dokl. Math.* **73** (2006) 96–99.

[25] V. V. Zhikov, S. M. Kozlov and O. A. Olejnik, *Homogenization of Differential Operators* ("Nauka", Moscow, 1993); English transl. (Springer-Verlag, Berlin, 1994).

[26] V. V. Zhikov and S. E. Pastukhova, On operator estimates for some problems in homogenization theory, *Russ. J. Math. Phys.* **12**(4) (2005) 515–524.

[27] V. V. Zhikov and S. E. Pastukhova, Operator estimates in homogenization theory, *Uspekhi Matem. Nauk* **71**(3) (2016) 27–122; English transl., *Russian Math. Surveys* **71**(3) (2016) 417–511.

Hopfions

Paul Sutcliffe
Department of Mathematical Sciences, Durham University,
Durham DH1 3LE, UK
p.m.sutcliffe@durham.ac.uk

More than 40 years ago, Faddeev proposed the existence of three-dimensional topological solitons classified by the integer-valued Hopf invariant. These solitons are now known as hopfions and have been investigated in a range of systems, including the original model suggested by Faddeev, where a variety of stable knot and link solutions have been computed numerically. Very recently, numerical computations have predicted the existence of nanoscale hopfions in frustrated magnets and experiments have realized micrometer-sized hopfions in chiral ferromagnetic fluids. All these examples of hopfions will be described and their similarities and differences discussed.

1. Introduction

Topological solitons are stable particle-like configurations of nonlinear fields (for a review, see [1]). In one space dimension, sine-Gordon kinks provide an example that is integrable in both the static and time-dependent situations. Examples of two-dimensional topological solitons of physical importance include magnetic vortices in superconductors and magnetic skyrmions in thin chiral ferromagnets. In three-dimensional space, there are non-abelian magnetic monopoles in Yang–Mills–Higgs gauge theories and skyrmions that model nuclei within an effective nonlinear pion theory. In all these examples, the topological charge that counts the number of solitons is a generalized winding number. In systems without gauge fields, the topological charge counts the winding of compactified space around the target space of fields, and in gauge theories, it counts the winding around the space of vacua of the fields at a large distance from the soliton. In both cases, there is a natural assignment of a point in space that corresponds to the position of a single soliton.

In 1975, Faddeev [2] proposed a theory in which the topological charge is not a generalized winding number but instead is a linking number of field lines, given by the integer-valued Hopf invariant. Faddeev's model is obtained as a target space reduction of the two-flavor Skyrme model [3] by restricting the $SU(2)$ field to lie on an equatorial two-sphere, and is therefore often known as the Skyrme–Faddeev model. Although the model is a simple restriction of the Skyrme model, the topology is very different and means that even a single soliton has a position that is a closed curve in space rather than a single point. In this sense, these solitons are string-like, even though they are localized in all three space dimensions. These solitons are known as hopfions and it is their string-like nature that distinguishes them from other topological solitons and results in new phenomena, such as the formation of stable knot solitons. The Skyrme–Faddeev model and its hopfion solutions, including knots and links, is the subject of Sec. 2.

Although the Skyrme–Faddeev model has interesting hopfion solutions, to date there is no convincing evidence that there is a physical system that is well-described by this theory. However, recent numerical computations suggest that similar hopfions may exist at the nanoscale in frustrated magnets [4], as discussed in Sec. 3. Furthermore, micrometer-sized hopfions have recently been created and observed experimentally in chiral ferromagnetic fluids [5] and the results agree well with those obtained from numerical simulations, as described in Sec. 4. Research into hopfions has therefore reached an exciting phase where comparisons with experimental results are now possible, 40 years after Faddeev proposed their existence.

2. Hopfions in the Skyrme–Faddeev Model

As static hopfion solutions are the subject of this article, time-dependence will not be discussed in what follows. In this case, the Skyrme–Faddeev model involves a map $\mathbf{m}\colon \mathbb{R}^3 \mapsto S^2$, which is realized as a real three-component unit vector $\mathbf{m} = (m_1, m_2, m_3)$, where $\mathbf{m}\cdot\mathbf{m} = 1$. The static energy of the Skyrme–Faddeev model [2] is defined to be

$$E = \frac{1}{32\pi^2\sqrt{2}} \int \partial_i\mathbf{m}\cdot\partial_i\mathbf{m} + \frac{1}{2}(\partial_i\mathbf{m}\times\partial_j\mathbf{m})\cdot(\partial_i\mathbf{m}\times\partial_j\mathbf{m})\, d^3x, \tag{1}$$

where the normalization is chosen for later convenience. The first term in the energy is that of the usual $O(3)$ sigma model and the second is a Skyrme term. As these two terms scale in opposite directions under a spatial rescaling, they balance each other and allow solutions with a finite non-zero size, in agreement with the requirements of Derrick's theorem [6].

For a field configuration to have finite energy, it must tend to a constant value at spatial infinity. Without loss of generality, this far field value may be chosen to be $\mathbf{m}(\infty) = (0,0,1) = \mathbf{e}_3$. This boundary condition compactifies space from $\mathbb{R}^3$ to S^3, therefore the field becomes a map $\mathbf{m}\colon S^3 \mapsto S^2$. Such maps are classified by $\pi_3(S^2) = \mathbb{Z}$, hence there is an integer-valued topological charge Q, the Hopf charge, which gives the soliton number. As mentioned above, this topological charge is not a winding number or degree of a mapping, but rather it is a linking number of field lines, as follows. Generically, the preimage of a point on the target two-sphere is a closed loop (or a collection of closed loops) in the three-sphere domain obtained from the compactification of $\mathbb{R}^3$. Two loops obtained as the preimages of any two distinct points on the target two-sphere are linked exactly Q times, where Q is the Hopf charge.

A lower bound on the energy

$$E \geq c\,Q^{3/4}, \tag{2}$$

has been proved [7, 8] with $c = (3/16)^{3/8} \approx 0.534$ and it is known that the fractional power is optimal [9]. However, it is expected that the above value of the constant c is not optimal and numerical results support Ward's conjecture [10] that the above energy bound holds with the value $c = 1$.

Hopfion solutions can only be obtained numerically but there is a useful analytic approximation [11] that captures the topological features of hopfions and can be used to generate good initial conditions for energy minimization computations. The first step for this approximation is to represent compactified space as the unit three-sphere in $\mathbb{C}^2$ via the introduction of the complex coordinates

$$(Z_1, Z_2) = \frac{1}{r^2 + \mu^2}\left(2\mu(x_1 + ix_2), r^2 - \mu^2 + 2i\mu x_3\right), \tag{3}$$

where $r^2 = x_1^2 + x_2^2 + x_3^2$ and the real constant $\mu > 0$ sets the scale of the configuration but has no impact on the topology. The field is then given by the formula

$$\mathbf{m} = \frac{1}{|p|^2 + |q|^2}\left(\bar{p}q + p\bar{q}, i\bar{p}q - ip\bar{q}, |q|^2 - |p|^2\right), \tag{4}$$

where $p(Z_1, Z_2)$ and $q(Z_1, Z_2)$ are polynomials in Z_1 and Z_2 with $p(0, 1) = 0$ to satisfy the boundary condition $\mathbf{m}(\infty) = \mathbf{e}_3$. The Hopf charge of the field (4) is equal to the number of preimages of a regular point close to the origin in $\mathbb{C}^2$ of the map (p, q): $\mathbb{B}^4 \mapsto \mathbb{C}^2$, where $\mathbb{B}^4$ is the unit 4-ball in $\mathbb{C}^2$ given by $|Z_1|^2 + |Z_2|^2 \leq 1$.

The position of a hopfion is the closed curve in space where $\mathbf{m} = -\mathbf{e}_3$, this being antipodal to the far field value. The position of the field (4) is therefore obtained by solving the equation $q = 0$. As an example, for any pair of positive integers n_1, n_2 the choice $(p, q) = (Z_1^{n_1}, Z_2^{n_2})$ yields an axially symmetric field with Hopf charge $Q = n_1 n_2$ and position given by the circle $x_1^2 + x_2^2 = \mu^2$ in the plane $x_3 = 0$. This field is axially symmetric in the sense that a spatial rotation $x_1 + ix_2 \mapsto e^{i\theta}(x_1 + ix_2)$ is equivalent to the rotation on target space $m_1 + im_2 \mapsto e^{in_1\theta}(m_1 + im_2)$, which is a symmetry of the theory and the far field boundary condition. The simplest case, $n_1 = n_2 = 1$, is illustrated in Fig. 1 by plotting several preimages, including the position preimage given by $\mathbf{m} = -\mathbf{e}_3$. The preimage of a given point $\mathbf{s} \in S^2$ is displayed by plotting an isosurface close to zero of the function $|\mathbf{m} - \mathbf{s}|^2$. It is clear from Fig. 1 that any two preimages are linked exactly once, in agreement with the fact that $Q = 1$.

Torus knots are classified by a pair (a, b) of coprime positive integers with $a > b$, where these integers correspond to the number of times that the knot winds around the poloidal and toroidal directions of the torus, respectively. The method described above can be used to create fields in which the position curve is an (a, b)-torus knot by setting $(p, q) = (Z_1^\alpha Z_2^\beta, Z_1^a + Z_2^b)$, where α is any positive integer and β is any

Fig. 1. Preimages of the field with unit Hopf charge obtained from $(p, q) = (Z_1, Z_2)$. The preimage associated with a given color is indicated by the position of the corresponding colored cone on the sphere in the inset. All pairs of preimages are linked exactly once. The yellow circle is the position preimage where $\mathbf{m} = -\mathbf{e}_3$.

non-negative integer. In this case, the Hopf charge is given by the cross ratio $Q = a\beta + b\alpha$. Links can also be obtained by choosing q to be reducible, for example, by taking the above form for an (a, b)-torus knot but with a and b no longer coprime. A simple case with Hopf charge $Q = 2\alpha$ is obtained taking $a = b = 2$ with $\beta = 0$ and $\alpha > 1$ to produce a two-component link with unit linking number between the two components.

It turns out that for all known numerically computed hopfions the topology of the field can be captured by a suitable choice of polynomials (p, q), and furthermore, by starting with these initial fields the required computational time to obtain the solution is dramatically reduced. Hopfions with $Q = 1$ and $Q = 2$ are axially symmetric and were first computed by imposing this symmetry directly within the numerical scheme [12, 13]. For both $Q = 1$ and $Q = 2$ the hopfion has the topology of the polynomials $(p, q) = (Z_1^Q, Z_2)$ and, as for all higher values of Q, the energy is around 20% above Ward's conjectured bound. These numerical solutions are shown in Figs. 2.1 and 2.2, where the position curve is displayed in yellow and a second preimage is displayed in red to confirm that the linking number is indeed one and two, respectively.

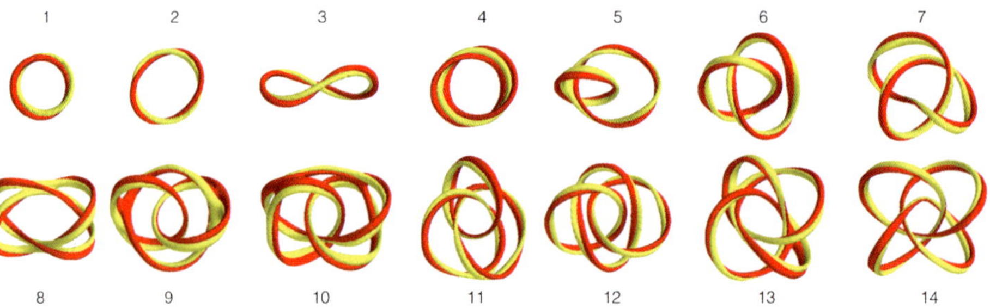

Fig. 2. The position curve (yellow) of hopfions in the Skyrme–Faddeev model for $1 \leq Q \leq 14$. Another preimage curve (red) is also shown to illustrate that the linking number is Q.

Numerical computations over a number of years [11, 14–18] have produced results for all minimal energy hopfions with $Q \leq 36$ and these are displayed in Fig. 2 for $Q \leq 14$. A highlight is the first hopfion knot, Fig. 2.7, originally computed in [14] following the prediction by Faddeev and Niemi [13] that stable knots might exist in this model as minimal energy hopfions. The position curve of the $Q = 7$ hopfion is a trefoil knot, which is also the (3, 2)-torus knot, and the hopfion has the topology of the polynomials $(p, q) = (Z_1^2 Z_2, Z_1^3 + Z_2^2)$. The minimal energy hopfions with $Q = 13$ and $Q = 14$ (Figs. 2.13 and 2.14) are also knots, but this time both are the (4, 3)-torus knot.

Hopfion links, in which the position curve has more than one component, are more prevalent than hopfion knots. Hopfions with $Q = 5, 6, 8$ are all two-component links and hopfions with $Q = 9, 10, 11, 12$ are all three-component links. Their topology can also be described by appropriate polynomials, for example the $Q = 6$ hopfion is described by $(p, q) = (Z_1^3, Z_1^2 + Z_2^2)$ and the $Q = 9$ hopfion by $(p, q) = (Z_1^3, Z_1^3 + Z_2^3)$. As Q increases, the hopfions become more complicated and include links in which one of the components is a knot, plus cable knots and links [18]. Some understanding of the structure of hopfions and the values of Q at which various knots and links appear can be obtained by using an approximate description in terms of elastic rods, with the general form of the elastic rod energy derived directly from the field theory [19].

The Skyrme–Faddeev energy (1) can be modified by the addition of a potential term (involving the field but not its derivative) and retains similar hopfion solutions [20, 21]. Theories that have more symmetries than the Skyrme–Faddeev model also support hopfions. The Nicole model [22] takes only the first term in the energy density in (1) but raised to a fractional power to endow the model with conformal symmetry. This symmetry allows the application of an axially symmetric ansatz in toroidal coordinates that reduces the partial differential equation to a single ordinary differential equation for a profile function [23] that yields solutions with the topology of the polynomials $(p, q) = (Z_1^{n_1}, Z_2^{n_2})$. However, numerical computations [24] reveal that the minimal energy hopfions are not axially symmetric for $Q > 2$ and instead are similar to those in the Skyrme–Faddeev model, forming links and knots.

The Aratyn–Ferreira–Zimerman model [25] is another conformal field theory with hopfions, but this model also has an additional infinite-dimensional symmetry group acting on target space. The energy of this model takes only the second term in the energy density in (1) but raised to one half of the fractional power appearing in the Nicole model. Again the conformal symmetry allows the use of a toroidal ansatz to reduce to an ordinary differential equation for a profile function but in this case the extra symmetry allows exact solutions in closed form [26] for all topologies of the type $(p, q) = (Z_1^{n_1}, Z_2^{n_2})$. The minimal energy hopfions are more often of this axial form than in other models, but again for some values of Q the minimal energy hopfion is a knot [27].

3. Hopfions in Frustrated Magnets

As the field of a hopfion is a three-component unit vector, then magnetic materials provide a natural system to search for a physical realization of hopfions, as the direction of the local magnetization in a magnetic medium is a three-component unit vector $\mathbf{m}$. Two-dimensional topological solitons in the form of nanoscale magnetic skyrmions [28] were first obtained experimentally [29, 30] within the last decade in thin samples of chiral ferromagnets. More recently, it has been suggested [31] that frustrated magnets, in which nearest-neighbor ferromagnetic exchange interactions compete with higher-neighbor anti-ferromagnetic exchange interactions, should also support nanoscale magnetic skyrmions. The possibility of hopfions in these materials was first mentioned in [32].

Within the continuum Ginzburg–Landau theory for frustrated magnets the energy contains the leading order terms [33]

$$E_{\text{frus}} = \int \left(\frac{J}{2} \partial_i \mathbf{m} \cdot \partial_i \mathbf{m} + \frac{\widetilde{J}}{2} |\partial_i \partial_i \mathbf{m}|^2 + H(1 - \mathbf{e}_3 \cdot \mathbf{m}) \right) d^3x, \tag{5}$$

where the constants J and $\widetilde{J}$ are linear combinations (with coefficients that depend upon the geometry of the atomic lattice) of the nearest-neighbor ferromagnetic and higher-neighbor antiferromagnetic exchange constants. The final term in (5) is the Zeeman interaction with an external magnetic field of magnitude H that is taken to lie along the $\mathbf{e}_3$ direction.

For solitons in frustrated magnets, a crucial feature is that $J < 0$ and $\widetilde{J} > 0$, which implies that the first term in the energy (5) favors an increase in scale when considering Derrick scaling for a three-dimensional soliton, rather than the usual decrease in scale for a ferromagnet with $J > 0$. Using scaled dimensionless units in which $J = -1$ and $\widetilde{J} = 1$ sets the scale of a nanosized soliton to be of order one and means that the saturation value for the external magnetic field is $H_s = \frac{1}{4}$ in these units. For $H \geq H_s$, the ground state is the zero energy fully polarized ferromagnetic state $\mathbf{m} = \mathbf{e}_3$, aligned with the external magnetic field. To consider isolated solitons embedded in this constant far field, the external magnetic field is required to be beyond the saturation value, which is satisfied by setting $H = 2H_s$.

Numerical computations of hopfions in frustrated magnets have been performed [4] by minimization of the energy (5), with initial conditions provided by the approximation described in the previous section using a pair of polynomials (p, q). The results show a surprising similarity to those of the Skyrme–Faddeev model, despite the fact that there are significant differences in the form of the energy for the two systems. In particular, although the first term in the energy in both systems is the standard continuum exchange energy that is quadratic in derivatives, because $J < 0$ this term comes with the opposite sign in the energy E_{frus} of the frustrated magnet as compared to the energy E of the Skyrme–Faddeev model given by (1).

The $Q = 1$ hopfion is displayed as the left image in Fig. 3, using the same method as earlier, where the yellow curve is the position curve given by the preimage

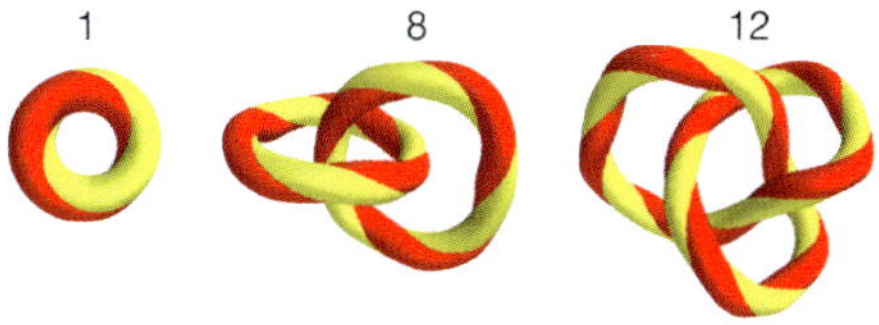

Fig. 3. The position curve (yellow) of minimal energy hopfions with $Q = 1, 8, 12$ in a frustrated magnet. Another preimage curve (red) is also shown to illustrate that the linking number is Q.

$\mathbf{m} = -\mathbf{e}_3$ antipodal to the far field direction, and the red curve is another preimage to illustrate that the linking number is the Hopf charge Q. This unit charge hopfion takes the standard axially symmetric form and has the topology of the polynomials $(p, q) = (Z_1, Z_2)$, as in the Skyrme–Faddeev model. The first minimal energy hopfion that is a link appears at $Q = 8$ and is shown as the middle image in Fig. 3. This hopfion is a two-component link and has the topology of the polynomials $(p, q) = (Z_1^4, Z_1^2 + Z_2^2)$. The first minimal energy hopfion that is a knot appears at $Q = 12$ and is shown as the right image in Fig. 3. It is a trefoil knot and has the topology of the polynomials $(p, q) = (Z_1^3 Z_2^2, Z_1^3 + Z_2^2)$.

As the above examples demonstrate, the general structure of hopfions in frustrated magnets is similar to hopfions in the Skyrme–Faddeev model, but the details of the knots and links that appear at specific charges are different. In particular, a larger value of the Hopf charge Q is required in frustrated magnets to form minimal energy links and knots than in the Skyrme–Faddeev model, with the minimal energy trefoil knot appearing at $Q = 12$ and $Q = 7$, respectively, in the two systems.

The numerical computations of hopfions in frustrated magnets predict the existence of stable nanoknots and nanolinks, providing impetus for future experimental work to engineer these hopfions in magnetic materials. These stable three-dimensional nanoparticles may be interpreted as closed loops of twisted strings of magnetic skyrmions, and given the immense interest in technological spintronic applications of magnetic skyrmions [34] there may be potential technological applications of three-dimensional hopfions encoding knotted magnetization. It is amusing to note that encoding data in knotted magnetization would provide a futuristic analogue of the quipu used by the Incas to record information in knotted strings.

4. Hopfions in Chiral Ferromagnetic Fluids

A physical medium in which hopfions have recently been realized experimentally [5] is a chiral ferromagnetic fluid. The medium consists of chiral ferromagnetic liquid crystal colloids formed from magnetic nanoplates with spontaneous long-range ferromagnetic ordering in a nematic fluid host. Ackerman and Smalyukh [5] were able to create micrometer-sized hopfions in their experiments using optical tweezers and to observe the complicated topology of hopfions using fluorescence polarizing microscopy to yield data in excellent agreement with their numerical simulations.

Within the one-constant approximation of the Frank free energy of the chiral liquid crystal, the direction of the coupled molecular alignment and magnetization fields is given by a three-component unit vector $\mathbf{m}$ with an associated energy

$$E_{\text{chiral}} = K \int \left(\frac{1}{2} \partial_i \mathbf{m} \cdot \partial_i \mathbf{m} + D\varepsilon_{ijk} m_i \partial_j m_k \right) d^3x. \tag{6}$$

Here K is the Frank elastic constant of the chiral ferromagnetic liquid crystal colloids and $2\pi/D$ is the intrinsic helical pitch.

The first term in the chiral energy (6) is symmetric under both the spatial rotation $x_1+ix_2 \mapsto e^{i\theta}(x_1+ix_2)$ and the internal rotation $m_1+im_2 \mapsto e^{i\psi}(m_1+im_2)$, for arbitrary angles θ and ψ. However, the second term in (6) breaks this symmetry to the diagonal subgroup where $\theta = \psi$. This means that a hopfion field can have axial symmetry only if a spatial rotation is equivalent to an internal rotation through the same angle. In particular, this implies that the field given by the polynomials $(p, q) = (Z_1^Q, Z_2)$ is axially symmetric only in the case $Q = 1$, in contrast to the other systems described in the previous sections, where such a field is axially symmetric for any value of Q.

For chiral hopfions, the dimensions of the confining medium and the associated boundary conditions play a vital role. In the experiments [5], the sample is thin in the third dimension with a thickness approximately equal to the intrinsic helical pitch, typically around 30 μm, and the sample surfaces were treated to enforce the boundary condition $\mathbf{m} = \mathbf{e}_3$. The $Q = 1$ numerical hopfion solution displayed in Fig. 4 models this situation by minimization of the chiral energy (6) in a thin slab with the smallest dimension equal to the intrinsic helical pitch and $\mathbf{m}$ fixed to the value $\mathbf{e}_3$ on the boundary of the slab. The initial condition uses the axially symmetric polynomials $(p, q) = (Z_1, Z_2)$ and this topology is preserved in the final energy minimizing hopfion. Similar numerical solutions are presented in the original work of Ackerman and Smalyukh [5], together with simulations that relax the one-constant approximation. Both models provide a good description of the experimental results and capture the essential features of hopfions in chiral ferromagnetic fluids.

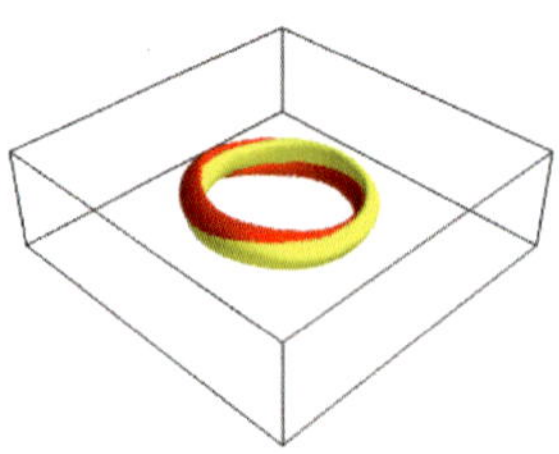

Fig. 4. The position curve (yellow) of the $Q = 1$ chiral hopfion together with another preimage curve (red) to illustrate that the linking number is one. The box represents the size of the thin sample, with the smallest dimension equal to the intrinsic helical pitch.

5. Conclusion

Faddeev introduced the concept of the hopfion as a new type of three-dimensional topological soliton with novel string-like properties. Numerical computations have provided a detailed description of hopfions in his original model and have confirmed his prediction of the existence of stable knots in field theory. Recent experiments have realized micrometer-sized hopfions in chiral ferromagnetic fluids and nanoscale hopfions have been predicted by simulations of frustrated magnets. These latest developments mean that hopfions may very well become an important soliton in a wide range of areas and could even have future technological applications.

References

[1] N. S. Manton and P. M. Sutcliffe, *Topological Solitons* (CUP, Cambridge 2004).
[2] L. D. Faddeev, Quantization of solitons, Princeton preprint IAS-75-QS70 (1975).
[3] T. H. R. Skyrme, *Proc. R. Soc. A* **260** (1961) 127.
[4] P. M. Sutcliffe, *Phys. Rev. Lett.* **118** (2017) 247203.
[5] P. J. Ackerman and I. I. Smalyukh, *Nat. Mater.* **16** (2017) 426.
[6] G. H. Derrick, *J. Math. Phys.* **5** (1964) 1252.
[7] A. F. Vakulenko and L. V. Kapitanski, *Dokl. Akad. Nauk USSR* **246** (1979) 840.
[8] A. Kundu and Yu. P. Rybakov, *J. Phys. A* **15** (1982) 269.
[9] F. Lin and Y. Yang, *Commun. Math. Phys.* **249** (2004) 273.
[10] R. S. Ward, *Nonlinearity* **12** (1999) 241.
[11] P. M. Sutcliffe, *Proc. R. Soc. A* **463** (2007) 3001.
[12] J. Gladikowski and M. Hellmund, *Phys. Rev. D* **56** (1997) 5194.
[13] L. Faddeev and A. J. Niemi, *Nature* **387** (1997) 58.
[14] R. A. Battye and P. M. Sutcliffe, *Phys. Rev. Lett.* **81** (1998) 4798.
[15] R. A. Battye and P. M. Sutcliffe, *Proc. R. Soc. Lond. A* **455** (1999) 4305.
[16] J. Hietarinta and P. Salo, *Phys. Lett. B* **451** (1999) 60.
[17] J. Hietarinta and P. Salo, *Phys. Rev. D* **62** (2000) 081701(R).
[18] P. Jennings, *J. Phys. A* **48** (2015) 315401.
[19] D. Harland, M. Speight and P. M. Sutcliffe, *Phys. Rev. D* **83** (2011) 065008.
[20] D. Foster, *Phys. Rev. D* **83** (2011) 085026.
[21] A. Samoilenka and Y. Shnir, *JHEP* **09** (2017) 29.
[22] D. A. Nicole, *J. Phys. G* **4** (1978) 1363.
[23] O. Babelon and L. A. Ferreira, *JHEP* **11** (2002) 20.
[24] M. Gillard and P. M. Sutcliffe, *J. Math. Phys.* **51** (2010) 122305.
[25] H. Aratyn, L. A. Ferreira and A. H. Zimerman, *Phys. Lett. B* **456** (1999) 162.
[26] H. Aratyn, L. A. Ferreira and A. H. Zimerman, *Phys. Rev. Lett.* **83** (1999) 1723.
[27] M. Gillard, *Nonlinearity* **24** (2011) 2729.
[28] A. N. Bogdanov and D. A Yablonskii, *Sov. Phys. JETP* **68** (1989) 101.
[29] S. Mühlbauer *et al*, *Science* **323** (2009) 915.
[30] X. Z. Yu *et al.*, *Nature* **465** (2010) 901.
[31] A. O. Leonov and M. Mostovoy, *Nat. Commun.* **6** (2015) 8275.
[32] P. M. Sutcliffe, *Nat. Mater.* **16** (2017) 392.
[33] S.-Z. Lin and S. Hayami, *Phys. Rev. B.* **93** (2016) 064430.
[34] A. Fert, V. Cros, V. and J. Sampaio, *Nat. Nanotechnology* **8** (2013) 152.

Completeness of the Bethe Ansatz for the Periodic Isotropic Heisenberg Model

V. Tarasov

Department of Mathematical Sciences,
Indiana University – Purdue University Indianapolis,
402 North Blackford St., Indianapolis, IN 46202-3216, USA
vtarasov@iupui.edu
and
St. Petersburg Branch of Steklov Mathematical Institute,
Fontanka 27, St. Petersburg, 191023, Russia
vt@pdmi.ras.ru

For the periodic isotropic Heisenberg model with arbitrary spins and inhomogeneities, we describe the system of algebraic equations whose solutions are in bijection with eigenvalues of the transfer-matrix. The system describes pairs of polynomials with the given discrete Wronskian (Casorati determinant) and additional divisibility conditions on discrete Wronskians with multiple steps. If the polynomial of the smaller degree in the pair is coprime with the Wronskian, this system turns into the standard Bethe ansatz equations. Moreover, if the transfer-matrix is diagonalizable, then its spectrum is necessarily simple modulo natural degeneration.

To the memory of Ludwig Dmitrievich Faddeev

1. Introduction

The problem to find eigenvalues and eigenvectors of the Hamiltonian of the Heisenberg model was first addressed by Bethe [4] who looked for the eigenvectors in a certain specific form. His method and its various further extensions are nowadays called the Bethe ansatz. The current literature on the Heisenberg model and its counterparts in statistical mechanics, the six- and eight-vertex models, is enormous, so we limit our references here just to two classical books [2, 17].

The Hamiltonian of the Heisenberg model can be included into a one-parameter family of commuting linear operators called the transfer-matrix, see [2, 6, 17], and the actual problem is to find eigenvalues and eigenvectors of the transfer-matrix. This can be done by the algebraic Bethe ansatz, see [6, 7], in the framework of the quantum inverse scattering method (QISM). Moreover, the QISM produces an integrable higher spin generalization of the Heisenberg model [19, 18], that can be solved by the algebraic Bethe ansatz as well.

The Bethe ansatz approach reduces the problem of diagonalizing the transfer-matrix of the Heisenberg model to solving a system of algebraic equations, called the Bethe ansatz equations, and gives a recipe to construct an eigenvector and an eigenvalue of the transfer-matrix given a solution of the Bethe ansatz equations. In fact, not all solutions of the Bethe ansatz equations are equally good, and there are also selection rules on how to pick up acceptable solutions, that is, those that are better than others. It is desirable that all eigenvectors of the transfer-matrix can be obtained from acceptable solutions of the Bethe ansatz equations, and whether it is possible to achieve, say, by tuning the selection rules, is the question of completeness of the Bethe ansatz for the Heisenberg model on a finite lattice.

The completeness question was discussed in [4] by Bethe himself and many times since then by various authors, see for instance, [12, 13, 20, 36, 3] and references therein. However, the question is not considered to be closed yet with more papers dealing with different aspects of the problem, see for example, [8–10, 14].

In this paper, we revisit the question of completeness of the Bethe ansatz and present the results that were obtained in the author's previous works with Mukhin and Varchenko, together with recent progress for the higher spin model. For the already established statements, the goal is to translate them from the language of commutative algebra to a more familiar and accessible language of linear algebra, making the results easier to appreciate. To this end, the assertions might be slightly weakened for the sake of simplicity of presentation. Also, because of the change of language, the statements need not look the same as in the original papers referred to, or even it there may not be any straightforward counterparts there at all. To keep the paper shorter, details of translation are mostly omitted and the proofs of new results will appear elsewhere.

It turns out that the key to achieving the goal of completeness of the Bethe ansatz for the Heisenberg model is an appropriate modification of the standard Bethe ansatz equations. For a periodic spin-1/2 chain of length n with inhomogeneities $z_1, \dots, z_n$, the right system of algebraic equations describes the problem of finding a pair of polynomials $p(u)$, $q(u)$ in one variable with the prescribed discrete Wronskian (Casorati determinant),

$$p\Big(u+\frac{i}{2}\Big)q\Big(u-\frac{i}{2}\Big)-p\Big(u-\frac{i}{2}\Big)q\Big(u+\frac{i}{2}\Big)=\prod_{s=1}^{n}(u-z_s)\,, \tag{1.1}$$

see [24, 26]. For a periodic model, the transfer-matrix is SL(2)-invariant, that determines a natural degeneration of its spectrum. And modulo this degeneration, eigenvectors and eigenvalues of the transfer-matrix are in bijection with classes of pairs of polynomials p, q satisfying (1.1), provided the pairs p_1, q_1 and p_2, q_2 are in the same class if they span the same two-dimensional vector space. In particular, the spectrum of the transfer-matrix is simple modulo the natural degeneration induced by the SL(2)-invariance of the model.

Similar result holds for the periodic higher spin chains. In this case, the spectrum of the transfer-matrix is also simple modulo the natural degeneration, and eigenvalues of the transfer-matrix are in bijection with the classes of pairs of polynomials $p(u)$, $q(u)$ that have the prescribed discrete Wronskian $p(u+\frac{i}{2})q(u-\frac{i}{2}) - p(u-\frac{i}{2})q(u+\frac{i}{2})$ and whose discrete Wronskians with multiple steps

$$p\left(u+m\frac{i}{2}\right)q\left(u-m\frac{i}{2}\right) - p\left(u-m\frac{i}{2}\right)q\left(u+m\frac{i}{2}\right) \tag{1.2}$$

are divisible by certain given polynomials. For the spin chain with the largest spin ℓ, the number m in (1.2) runs through integers between 1 and 2ℓ.

This result has a grain of surprise. One might expect that additional conditions on polynomials $p(u)$, $q(u)$ for higher spin models could be formulated in terms of the Baxter equation

$$a(u)p(u-i) + d(u)p(u+i) = E(u)p(u)\,.$$

However, what one can get this way is only the divisibility condition for the discrete Wronskian (1.2) with $m = 2$, and it is not sufficient for most of spin chains that contain spins higher than 1, see an example in Sec. 6.

The Bethe ansatz for the Gaudin model is understood nowadays much better than for lattice integrable models. The completeness of the Bethe ansatz is proved for the Gaudin model associated to any semisimple Lie algebra and any tensor product of its finite-dimensional representations, see [22, 32]. The proof in [22] relies upon the reformulation of the Bethe ansatz equations as the intersection equations for Schubert cells in the Grassmannians relative to osculating flags. For lattice integrable models, the Schubert calculus approach is applicable only partially and does not cover the most interesting cases, in particular, the homogeneous model. The divisibility conditions for discrete Wronskians with multiple steps do describe the intersection of Schubert cells in Grassmannians relative to secant flags once this picture is applicable. And it is an interesting question to find an appropriate geometric picture for the divisibility conditions that works for all cases.

In this note, we consider the most rigid and degenerate case of the isotropic Heisenberg model (*XXX* model) with periodic boundary conditions. The results extend almost literally with due modifications to the case of the *XXX* model with quasiperiodic (a.k.a. twisted) boundary conditions. An interesting question is to work out the case of the *XXX* model with open boundary conditions solvable by the Bethe ansatz. It turns out that solving the Bethe ansatz equations for a model with open boundary conditions is equivalent to imposing an additional symmetry on solutions of the Bethe ansatz equations for the "doubled" model with (quasi)periodic boundary conditions.

The case of the *XXZ* model, which is mostly discussed in the literature for completeness of the Bethe ansatz, is studied less than the case of the *XXX* model in the direction described in the paper. For generic anisotropy, the case of the *XXZ* model is likely to have only technical subtleties rather than principal difficulties compared to the *XXX* case. For generic quasiperiodic boundary conditions, including the pure periodic case, the results are expected to be similar to the case of the quasiperiodic XXX model, while for certain specific quasiperiodic boundary conditions, degeneration of the spectrum of the transfer-matrix can occur resembling the case of the periodic XXX model. On the other hand, the case of the XXZ model at roots of unity is more sophisticated due to the known large degeneracy of eigenvalues of the transfer-matrix. Nevertheless, it is plausible for this case too that the appropriate form of the Bethe ansatz equations yielding completeness of the Bethe ansatz describes pairs of functions with the given discrete Wronskian.

The plan of the paper is as follows. Section 2 describes the higher spin XXX model. In Sec. 3, we review the algebraic Bethe ansatz method and start to discuss the completeness question. In Sec. 4, we remind reformulation of the Bethe ansatz equations via discrete Wronskians of pairs of polynomials and the second-order difference equations with a basis of polynomial solutions. The results on completeness of the Bethe ansatz for spin-1/2 models are given in Sec. 5, and for higher spin models in Sec. 6.

2. Higher Spin XXX Model

Let S^+, S^-, S^3 be the generators of the Lie algebra $\mathfrak{sl}_2$:

$$[S^3, S^\pm] = \pm S^\pm, \qquad [S^+, S^-] = 2S^3.$$

Let V_ℓ be the irreducible $(2\ell+1)$-dimensional representation of $\mathfrak{sl}_2$, that is, the representation of spin ℓ. Denote by w_ℓ the highest weight vector of V_ℓ: $S^+ w_\ell = 0$, $S^3 w_\ell = \ell w_\ell$.

The space of states of the higher spin XXX model with spins $\ell_1, \ldots, \ell_n$ is the tensor product $\mathcal{H} = V_{\ell_1} \otimes \cdots \otimes V_{\ell_n}$. The local L-operator of the model is

$$L_a(u) = \begin{pmatrix} u + iS_a^3 & iS_a^- \\ iS_a^+ & u - iS_a^3 \end{pmatrix} \in \operatorname{End}(\mathbb{C}^2 \otimes \mathcal{H}),$$

where $S_a^\pm$, S_a^3 are linear operators acting as $S^\pm$, S^3, respectively, on the ath tensor factor and as the identity operator on the other tensor factors.

The monodromy matrix of the XXX model with inhomogeneities $z_1, \ldots, z_n$ equals

$$T(u) = L_n(u - z_n) \cdots L_1(u - z_1) = \begin{pmatrix} A(u) & B(u) \\ C(u) & D(u) \end{pmatrix},$$

and for the model with periodic boundary conditions, the transfer-matrix is

$$\mathcal{T}(u) = \operatorname{tr} T(u) = A(u) + D(u). \tag{2.1}$$

It is known that the transfer-matrix forms a commutative family of operators on $\mathcal{H}$:

$$\mathcal{T}(u)\mathcal{T}(v) = \mathcal{T}(v)\mathcal{T}(u)$$

for any u, v. The question we are interested in is to find eigenvectors and eigenvalues of the transfer-matrix.

The Lie algebra $\mathfrak{sl}_2$ acts naturally on the space $\mathcal{H}$, the generator S^ν acting as the sum $S_1^\nu + \cdots + S_n^\nu$. The transfer-matrix $\mathcal{T}(u)$ commutes with this action,

$$[\mathcal{T}(u), S^\nu] = 0\,, \qquad \nu = +, -, 3\,, \tag{2.2}$$

hence, every eigenspace of $\mathcal{T}(u)$ is invariant under the action of $\mathfrak{sl}_2$. To eliminate this degeneracy of the spectrum of the transfer-matrix, we consider the action of $\mathcal{T}(u)$ on the multiplicity spaces of irreducible representations of $\mathfrak{sl}_2$ in the representation $\mathcal{H}$. The multiplicity spaces are conveniently visualized as the subspaces of weight singular vectors

$$\mathcal{H}_k = \{v \in \mathcal{H} | S^+ v = 0,\ S^3 v = (\ell_1 + \cdots + \ell_n - k)v\}\,,$$

that are invariant under the action of $\mathcal{T}(u)$ due to (2.2). The subspace $\mathcal{H}_k$ is not trivial if only if k is an integer such that $0 \leqslant k \leqslant \ell_1 + \cdots + \ell_n$.

We say that the spectrum of $\mathcal{T}(u)$ is simple modulo natural degeneration if $\mathcal{T}(u)$ is diagonalizable and all its eigenspaces are irreducible representations of $\mathfrak{sl}_2$.

We will assume that the inhomogeneities $z_1, \ldots, z_n$ satisfy the following condition:

Let $X_s = \{z_s - i\ell_s, z_s - i(\ell_s - 1), \ldots, z_s + i\ell_s\}$, $s = 1, \ldots, n$. Then for any r, $s = 1, \ldots, n$, the sets X_r and X_s either do not intersect or one of them is a subset of the other. (2.3)

This condition means that the space $\mathcal{H}$ has no invariant subspaces under the action of the entries $A(u)$, $B(u)$, $C(u)$, $D(u)$ of $T(u)$, see [35]. In other words, $T(u)$ defines an irreducible representation of the Yangian $Y(\mathfrak{gl}_2)$ on $\mathcal{H}$. It is known that under condition (2.3), the equivalence class of this representation does not depend on the order of pairs $(\ell_1, z_1), \ldots, (\ell_n, z_n)$. In particular, the spectrum of $\mathcal{T}(u)$ remains unchanged.

We will not assume that $z_1, \ldots, z_n$ satisfy some kind of reality conditions. So the transfer-matrix $\mathcal{T}(u)$ might be not diagonalizable. The results on completeness of the Bethe ansatz extends to such cases as well, but for the sake of simplicity, they are not completely covered in this paper.

It is relatively easy to show that $\mathcal{T}(u)$ is diagonalizable if all $z_1, \ldots, z_n$ are sufficiently large and after a permutation satisfy inequalities $|z_1 - z_2| \ll |z_2 - z_3| \ll \cdots \ll |z_{n-1} - z_n|$. Thus $\mathcal{T}(u)$ is diagonalizable for generic $z_1, \ldots, z_n$. The description of the exceptional locus of $z_1, \ldots, z_n$ is a very interesting problem, but it is out of reach for the moment. However, there are two cases when $\mathcal{T}(u)$ is known to be diagonalizable because its coefficients are normal operators relative to some positive definite scalar product on $\mathcal{H}$, see [27, 28]. The cases are as follows:

There exists a number c such that the polynomial $W(u - c)$ has real coefficients and $|\operatorname{Im}(z_s - c)| < 1/2$ for all $s = 1, \ldots, n$. (2.4)

All $z_1, \ldots, z_n$ have the same real part and there is a subset $I \subset \{1, \ldots, n\}$ such that $|z_r - z_s| > \ell_r + \ell_s$ for all $r, s \in I$ and all $r, s \notin I$. (2.5)

Condition (2.4) covers the most physically interesting and, perhaps, the most challenging case of the homogeneous XXX model, $z_1 = \cdots = z_n = 0$.

3. Algebraic Bethe Ansatz

The algebraic Bethe ansatz gives the following recipe to obtain eigenvectors and eigenvalues of the transfer-matrix $\mathcal{T}(u)$ on the spaces $\mathcal{H}_k$. Consider the reference state

$$|\emptyset\rangle = w_{\ell_1} \otimes \cdots \otimes w_{\ell_n} \in \mathcal{H}_0 .$$

It is an eigenvector of the entries $A(u)$ and $D(u)$, and is annihilated by the entry $C(u)$:

$$A(u)|\emptyset\rangle = a(u)|\emptyset\rangle , \qquad D(u)|\emptyset\rangle = d(u)|\emptyset\rangle , \qquad C(u)|\emptyset\rangle = 0 , \tag{3.1}$$

where

$$a(u) = \prod_{s=1}^{n} (u - z_s + i\ell_s) , \qquad d(u) = \prod_{s=1}^{n} (u - z_s - i\ell_s) . \tag{3.2}$$

Let k be an integer such that $0 \leqslant k \leqslant \ell_1 + \cdots + \ell_n$. Set

$$|t_1, \ldots, t_k\rangle = B(t_1) \cdots B(t_k)|\emptyset\rangle .$$

It is known that the vector $|t_1, \ldots, t_k\rangle$ is a symmetric function of $t_1, \ldots, t_k$.

The *Bethe ansatz equations* is the following system of algebraic equations on $t_1, \ldots, t_k$,

$$a(t_s) \prod_{\substack{r=1 \\ r \neq s}}^{k} (t_s - t_r - i) = d(t_s) \prod_{\substack{r=1 \\ r \neq s}}^{k} (t_s - t_r + i) , \quad s = 1, \ldots, k . \tag{3.3}$$

Given a solution of the Bethe ansatz equations, the algebraic Bethe ansatz suggests a presumable eigenvector of the transfer-matrix.

Theorem 3.1. *Let $t_1, \ldots, t_k$ be a solution of system* (3.3) *such that*

$$t_r \neq t_s , \qquad 1 \leqslant r < s \leqslant k . \tag{3.4}$$

Then $|t_1, \ldots, t_k\rangle \in \mathcal{H}_k$ and

$$\mathcal{T}(u)|t_1, \ldots, t_k\rangle = E(u; t_1, \ldots, t_k)|t_1, \ldots, t_k\rangle ,$$

where

$$E(u; t_1, \ldots, t_k) = a(u) \prod_{s=1}^{k} \frac{u - t_s - i}{u - t_s} + d(u) \prod_{s=1}^{k} \frac{u - t_s + i}{u - t_s} .$$

For a solution $t_1, \dots, t_k$ of the Bethe ansatz equations, the vector $|t_1, \dots, t_k\rangle$ is called the *Bethe vector*. The question whether the construction of Theorem 3.1 produces all eigenvectors of the transfer-matrix $\mathcal{T}(u)$ in $\mathcal{H}_k$ is called the question of completeness of the Bethe ansatz. A part of discussion of the completeness is counting solutions of the Bethe ansatz equations. Since both the vector $|t_1, \dots, t_k\rangle$ and $E(u, t_1, \dots, t_k)$ are symmetric functions of $t_1, \dots, t_k$, this counting should not distinguish solutions of the Bethe ansatz equations obtained from each other by permutation of coordinates.

Introduce a polynomial $p(u)$ with the roots $t_1, \dots, t_k$:

$$p(u) = \prod_{s=1}^{k} (u - t_s) . \tag{3.5}$$

We will always assume that $p(u)$ is related to $t_1, \dots, t_k$ via (3.5).

It turns out that assumption (3.4) that the roots of $p(u)$ are distinct can be eliminated.

Theorem 3.2. *Assume that the ratio*

$$E(u; p) = \frac{a(u)p(u-i) + d(u)p(u+i)}{p(u)} \quad \textit{is a polynomial.} \tag{3.6}$$

Then $|t_1, \dots, t_k\rangle \in \mathcal{H}_k$ *and*

$$\mathcal{T}(u)|t_1, \dots, t_k\rangle = E(u; p)|t_1, \dots, t_k\rangle . \tag{3.7}$$

The proof of this statement is similar to that of Theorem 3.1 and relies upon the following well known formulae that can be proved by induction on k:

$$\mathcal{T}(u)|t_1, \dots, t_k\rangle = \frac{a(u)p(u-i) + d(u)p(u+i)}{p(u)}|t_1, \dots, t_k\rangle$$

$$+ i\sum_{s=1}^{k} \frac{a(t_s)p(t_s - i) + d(t_s)p(t_s + i)}{(u - t_s)p'(t_s)}|u, t_1, \dots, t_{s-1}, t_{s+1}, \dots, t_n\rangle ,$$

$$S^+|t_1, \dots, t_k\rangle = i\sum_{s=1}^{k} \frac{a(t_s)p(t_s - i) + d(t_s)p(t_s + i)}{(u - t_s)p'(t_s)}|t_1, \dots, t_{s-1}, t_{s+1}, \dots, t_n\rangle .$$

However, instead of canceling the "unwanted terms" individually, all those terms in each formula are summed up, and the obtained expressions are simplified using identities for rational functions.

Condition (3.6) gives a system of algebraic equations on the coefficients of the polynomial $p(u)$ explicitly written as follows. Denote by $y(u; p)$ be the remainder of division of the polynomial $a(u)p(u - i) + d(u)p(u + i)$ by $p(u)$, and consider its coefficients: $y(u; p) = \sum_{s=1}^{k} y_s(p)u^{k-s}$. Then the system of equations defined by (3.6) is

$$y_s(u; p) = 0 , \qquad s = 1, \dots, k . \tag{3.8}$$

Condition (3.6), hence system (3.8) clearly implies Eqs. (3.3), while under assumption (3.4), Eqs. (3.3) and (3.8) are equivalent.

Notice that the eigenvalue $E(u,p)$ in formula (3.7) uniquely determines the polynomial $p(u)$ via relation (3.6), see Proposition 3.3 below. This observation implies that nonzero Bethe vectors $|t_1,\dots,t_k\rangle$ are linearly independent as eigenvectors of the transfer-matrix $\mathcal{T}(u)$ with different eigenvalues.

Proposition 3.3. *Let $f(u)$ be a monic polynomial such that* $\deg f \leqslant \ell_1+\dots+\ell_n$ *and*

$$a(u)f(u-i)+d(u)f(u+i)=E(u;p)f(u)\,. \tag{3.9}$$

Then $f(u)=p(u)$.

Proof. By definition of $E(u,p)$, the polynomial $p(u)$ solves Eq. (3.9). And if $f(u)$ is a polynomial solution of (3.9) not proportional to $p(u)$, then the polynomial

$$Q(u)=p\Bigl(u+\frac{i}{2}\Bigr)f\Bigl(u-\frac{i}{2}\Bigr)-p\Bigl(u-\frac{i}{2}\Bigr)f\Bigl(u+\frac{i}{2}\Bigr)$$

is nonzero and satisfy $a(u)Q(u-\frac{i}{2})=d(u)Q(u+\frac{i}{2})$. Taking the subleading coefficient in u in this equality yields $\deg Q=2(\ell_1+\dots+\ell_n)$. Hence,

$$\deg f=\deg Q-\deg p+1\geqslant \ell_1+\dots+\ell_n+1$$

since $\deg p\leqslant \ell_1+\dots+\ell_n$. □

One more requirement usually imposed on solutions of the Bethe ansatz equations is that

$$\begin{aligned} t_r &\neq t_s+i\,, && r,s=1,\dots,k\,,\\ a(t_s)&\neq 0\,, && s=1,\dots,k\,,\\ d(t_s)&\neq 0\,, && s=1,\dots,k\,. \end{aligned} \tag{3.10}$$

In other words, it is assumed that the polynomial $p(u)$ is coprime with each of the polynomials $p(u-i)$, $a(u)$, and $d(u)$. It is easy to show that if $p(u)$ obeys (3.6), then each line in (3.10) implies the others, and there exist $r,s=1,\dots,n$, and $m=0,\dots,k-1$, such that $m=\ell_r+\ell_s+i(z_r-z_s)$ and $p(z_r-i\ell_r+il)=0$ for any $l=0,\dots,m$. However, under condition (2.3), one has

$$|z_r-i\ell_r,z_r-i\ell_r+i,\dots,z_s+i\ell_s,\tilde t_{m+2},\dots,\tilde t_k\rangle=0$$

for any $\tilde t_{m+2},\dots,\tilde t_k$. Thus if assumption (3.10) fails, the Bethe vector $|t_1,\dots,t_k\rangle$ vanishes and equality (3.7) becomes trivial.

Solutions of the Bethe ansatz equations satisfying conditions (3.10) are called *admissible*. It turns out that for an admissible solution of the Bethe ansatz equations, Theorem 3.2 does produce an eigenvector of the transfer-matrix.

Theorem 3.4. *Let $p(u)$ obey conditions* (3.6) *and* (3.10). *Then* $|t_1,\dots,t_k\rangle\neq 0$.

If $p(u)$ is a nondegenerate solution of system (3.8), then the result follows from Theorem 3.6 below, which is a minor generalization of the Gaudin–Korepin determinant formula (3.11) for the norm of a Bethe vector. For degenerate solutions of system (3.8), the result of Theorem 3.4 is new, and its proof is similar to the proof of the analogous statement for the Gaudin model given in [29].

Define a function $N(t_1, \ldots, t_k)$ by the rule

$$C(t_1) \cdots C(t_k)|t_1, \ldots, t_k\rangle = N(t_1, \ldots, t_k)|\emptyset\rangle .$$

This function is usually called the norm of vector $|t_1, \ldots, t_k\rangle$. The Gaudin–Korepin formula (3.11) shows that the norm of a Bethe vector is essentially given by the Jacobian of the Bethe ansatz equations.

Theorem 3.5 ([16]). *Let $t_1, \ldots, t_k$ be a solution of system* (3.3) *such that $t_r \neq t_s$ for all $1 \leqslant r < s \leqslant k$. Then*

$$N(t_1, \ldots, t_k) = (-i)^{k^2} \prod_{s=1}^{k} d(t_s) \prod_{r=1}^{k} \prod_{1 \leqslant r < s \leqslant k} \frac{1}{(t_r - t_s)^2} \tag{3.11}$$

$$\times \det \left[\frac{\partial}{\partial t_s} \left(a(t_s) \prod_{\substack{r=1 \\ r \neq s}}^{k} (t_s - t_r - i) - d(t_s) \prod_{\substack{r=1 \\ r \neq s}}^{k} (t_s - t_r + i) \right) \right]_{r,s=1}^{k} .$$

Alternative proofs of formula (3.11) are given in [33, 34, 36].

Using the Vandermonde determinant, the Jacobian of system (3.3) in formula (3.11) can be converted into the Jacobian of system (3.8), see formula (3.12). This allows us to eliminate assumption (3.4) in Theorem 3.5.

Theorem 3.6. *Let $p(u)$ be a solution of system* (3.8). *Then*

$$N(t_1, \ldots, t_k) = (-1)^k \prod_{s=1}^{k} d(t_s) \, \det \left(\frac{\partial y_r}{\partial p_s} \right)_{r,s=1}^{k} , \tag{3.12}$$

where $p_1, \ldots, p_k$ are the coefficients of $p(u)$: $p(u) = u^k + \sum_{s=1}^{k} p_s \, u^{k-s}$.

Formula (3.12) follows from (3.11) by the deformation arguments using Proposition 3.7 below. Also, it is plausible that the proof of formula (3.11) given in [33, 34] can be modified to work without assumption (3.4).

The next proposition established in [24] is an important technically tool.

Proposition 3.7. *For any $z_1, \ldots, z_n$, system* (3.8) *has finitely many solutions, and their number counted with multiplicities does not depend on $z_1, \ldots, z_n$. For generic $z_1, \ldots, z_n$, all solutions of system* (3.8) *are nondegenerate and satisfy conditions* (3.4) *and* (3.10).

All statements in this section, except the assertion of Theorem 3.4 for degenerate solutions of the Bethe ansatz equations, are obtained by a "direct" analysis of the algebraic Bethe ansatz formulae. These statements show that the Bethe ansatz equations (3.8) can have at most $\dim \mathcal{H}_k$ admissible solutions. And if system (3.8) does have $\dim \mathcal{H}_k$ nondegenerate admissible solutions, then the corresponding Bethe vectors form a basis of $\mathcal{H}_k$, that is, the Bethe ansatz is complete. Moreover, the restriction of the transfer-matrix $\mathcal{T}(u)$ on $\mathcal{H}_k$ in this case is diagonalizable and has simple spectrum.

It is widely believed that the Bethe ansatz is complete for generic $z_1, \ldots, z_n$. It is indeed true, but the proof is not easy at all. Theorem 3.8 follows from the results of [24], though the details are not straightforward.

Theorem 3.8. *Let $z_1, \ldots, z_n$ be generic. Then for any integer k such that $0 \leqslant k \leqslant \ell_1 + \cdots + \ell_n$, system (3.3) has exactly $\dim \mathcal{H}_k$ nondegenerate admissible solutions and the corresponding Bethe vectors form a basis of $\mathcal{H}_k$. Moreover, the spectrum of the transfer-matrix $\mathcal{T}(u)$ is simple modulo natural degeneration.*

However, for particular $z_1, \ldots, z_n$, it can happen that the number of admissible solution of the Bethe ansatz equations is strictly less than required for completeness of the Bethe ansatz. The simplest example is given by the homogeneous spin-1/2 model with $n = 4$, and goes back to [1].

Example 3.9. Let $\ell_s = 1/2$, $z_s = 0$, $s = 1, \ldots, 4$, and $k = 2$. Then $\dim \mathcal{H}_2 = 2$, but there is only one quadratic polynomial $p(u)$ satisfying conditions (3.6) and (3.10):

$$p(u) = u^2 - \frac{1}{12}, \qquad E(u) = 2u^4 + 3u^2 + \frac{13}{8}.$$

It follows from Theorem 5.1 below that for the homogeneous spin-1/2 model with even n, the number of admissible solution of the Bethe ansatz equations for any $k \geqslant 2$ is strictly less than $\dim \mathcal{H}_k$.

4. Pairs of Polynomials

To deal with the question of completeness of the Bethe ansatz for arbitrary $z_1, \ldots, z_n$, one has to take into account solutions of the Bethe ansatz equations that do not satisfy additional conditions (3.10). And there are two natural questions coming out: to figure out selection rules picking up the solutions that give rise to eigenvalues of the transfer-matrix, and to show that all eigenvalues can be eventually obtained this way. The key observation to answer the first question is described in Theorem 4.1. The second question is worked out in Theorem 4.2.

Define

$$W(u) = \prod_{s=1}^{n} \prod_{r=1}^{2\ell_s} \left(u - z_s + i\left(\ell_s - r + \frac{1}{2}\right)\right), \tag{4.1}$$

and for $m = 1, 2, \ldots,$

$$W_m(u) = \prod_{\substack{s=1 \\ 2\ell_s+1\geqslant m}}^{n} \prod_{r=0}^{2\ell_s-m} \left(u - z_s + i\left(\ell_s - r - \frac{m}{2}\right)\right). \tag{4.2}$$

Clearly, $W_1(u) = W(u)$, and $W_m(u) = 1$ if $m > 2\max\{\ell_1, \ldots, \ell_n\}$. Notice that

$$a(u) = \frac{W(u+\frac{i}{2})}{W_2(u)}, \qquad d(u) = \frac{W(u-\frac{i}{2})}{W_2(u)}. \tag{4.3}$$

Theorem 4.1. *Let the polynomial $p(u)$ satisfy conditions (3.6) and (3.10). Then there exists a polynomial $q(u)$ such that*

$$p\left(u+\frac{i}{2}\right)q\left(u-\frac{i}{2}\right) - p\left(u-\frac{i}{2}\right)q\left(u+\frac{i}{2}\right) = W(u) \tag{4.4}$$

and

for any $m = 2, 3, \ldots,$ the polynomial (4.5)

$$Q_m(u) = p\left(u+m\frac{i}{2}\right)q\left(u-m\frac{i}{2}\right) - p\left(u-m\frac{i}{2}\right)q\left(u+m\frac{i}{2}\right)$$

is divisible by $W_m(u)$. (Recall that $W_m(u) = 1$ if $m > 2\max\{\ell_1, \ldots, \ell_n\}$.)

The eigenvalue $E(u, p)$ in Theorem 3.2 is expressed as

$$E(u,p) = \frac{p(u+i)q(u-i) - p(u-i)q(u+i)}{W_2(u)}. \tag{4.6}$$

The statement of Theorem 4.1 under assumption (3.4) goes back to [31]. It was independently obtained in [30] in more general form. The proof of Theorem 4.1 avoiding (3.4) is similar to that of [21, Proposition 4.1]. The part (4.5) for $m \geqslant 3$ is new. It follows by induction from the equality $Q_1(u) = W(u)$, see (4.4), and the recurrence relation

$$\begin{aligned} &Q_m(u)p\left(u+(m-2)\frac{i}{2}\right) \\ &= Q_{m-1}\left(u-\frac{i}{2}\right)p\left(u+m\frac{i}{2}\right) + Q_1\left(u+(m-1)\frac{i}{2}\right)p\left(u-m\frac{i}{2}\right). \end{aligned} \tag{4.7}$$

The next statement shows that under conditions (4.4), (4.5), formula (4.6) produce all eigenvalues of the transfer-matrix. The statement follows from the results of [23], see Corollaries 7.4 and 7.5 therein.

Theorem 4.2. *Let a polynomial $E(u)$ be an eigenvalue of the restriction of $\mathcal{T}(u)$ on $\mathcal{H}_k$. Then there are a unique monic polynomial $p(u)$ of degree k and a polynomial $q(u)$ such that they satisfy conditions (4.4) and*

$$E(u) = \frac{p(u+i)q(u-i) - p(u-i)q(u+i)}{W_2(u)}, \tag{4.8}$$

cf. (4.6). Furthermore, the polynomials $p(u)$, $q(u)$ satisfy condition (4.5).

Lemma 4.3. *For a polynomial $E(u)$, the difference equation*

$$a(u)f(u-i)+d(u)f(u+i)=E(u)f(u) \tag{4.9}$$

has two linearly independent polynomial solutions if and only if there are polynomials $p(u)$, $q(u)$ satisfying conditions (4.4) *and* (4.8). *Moreover, if Eq.* (4.9) *has a polynomial solution that obey condition* (3.10), *then the polynomials $p(u)$, $q(u)$ satisfy condition* (4.5).

Proof. Clearly, if polynomials $p(u)$, $q(u)$ satisfy conditions (4.4), (4.8), then they are linearly independent solutions of Eq. (4.9).

On the other hand, if $p(u)$, $q(u)$ are linearly independent solutions of Eqs. (4.9), then the polynomial

$$Q_1(u)=p\left(u+\frac{i}{2}\right)q\left(u-\frac{i}{2}\right)-p\left(u-\frac{i}{2}\right)q\left(u+\frac{i}{2}\right)$$

is nonzero and satisfy

$$\frac{Q_1(u+\frac{i}{2})}{Q_1(u-\frac{i}{2})}=\frac{a(u)}{d(u)}=\frac{W(u+\frac{i}{2})}{W(u-\frac{i}{2})},$$

see (4.3). Normalizing $p(u)$, $q(u)$ such that $Q_1(u)$ is monic yields equality (4.4), while formula (4.6) follows from (4.9) with $f(u)=p(u)$.

To prove the second sentence of Lemma 4.3, notice that the polynomials $Q_m(u)$ do not depend up to proportionality on a choice of linearly independent solutions $p(u)$, $q(u)$ of Eq. (4.9). Thus one may assume that $p(u)$ obeys (3.10), so the claim follows by induction from the equality $Q_1(u)=W(u)$ and relation (4.7). □

5. Completeness for Spin-1/2 Model

In the next two sections, we formulate the final results on completeness of the Bethe ansatz for the *XXX* model. We will describe eigenvectors and eigenvalues of the transfer-matrix in terms of pairs of polynomials $p(u)$, $q(u)$ such that $\deg p<\deg q$ and $p(u)$ is a monic polynomial. We call such pairs *ordered pairs of polynomials*. By a minor abuse of notation, we will assume that in the ordered pair of polynomials $p(u)$, $q(u)$, the polynomial $q(u)$ of the higher degree is determined only up to adding a multiple of the polynomial $p(u)$ of the lower degree.

Clearly, there is a bijection between ordered pairs of polynomials and two-dimensional subspaces of $\mathbb{C}[u]$ defined by assigning to an ordered pair of polynomials $p(u)$, $q(u)$ the span of $p(u)$, $q(u)$. Nevertheless, we prefer to talk about ordered pairs of polynomials rather than two-dimensional spaces of polynomials since the former has closer psychological association with the original Bethe ansatz equations.

Clearly, for an ordered pair of polynomials $p(u)$, $q(u)$, any determinant $p(u_1)q(u_2)-p(u_2)q(u_1)$ is not affected by the ambiguity in $q(u)$.

In this section, we consider spin-1/2 models. That is, we take $\ell_s = 1/2$ for all $s = 1, \dots, n$, so that $\mathcal{H} = (\mathbb{C}^2)^{\otimes n}$,

$$a(u) = W\left(u + \frac{i}{2}\right), \quad d(u) = W\left(u - \frac{i}{2}\right), \qquad W(u) = \prod_{s=1}^{n} (u - z_s)\,. \tag{5.1}$$

We replace condition (2.3) by a weaker condition

$$z_s - z_r \neq i\,, \qquad 1 \leqslant r < s \leqslant n\,. \tag{5.2}$$

This condition means that vectors $|t_1, \dots, t_k\rangle$ with various $k = 0, \dots, n$, and $t_1, \dots, t_k$ span the space $\mathcal{H}$, see [35].

Notice that for $\mathcal{H} = (\mathbb{C}^2)^{\otimes n}$, the subspace $\mathcal{H}_k$ is not trivial if and only if k is an integer such that $0 \leqslant 2k \leqslant n$. Farther on in this section, we will assume that $2k \leqslant n$.

The next theorem is proved in [24]. Similar result for the XXX-type integrable model associated with the Lie algebra $\mathfrak{gl}_N$, is obtained in [26], see also [25].

Theorem 5.1. *Let $\ell_s = 1/2$ for all $s = 1, \dots, n$, and $z_1, \dots, z_n$ satisfy condition* (5.2). *A polynomial $E(u)$ is an eigenvalue of the restriction of $\mathcal{T}(u)$ on $\mathcal{H}_k$ if and only if there exists an ordered pair of polynomials $p(u)$, $q(u)$ such that* $\deg p = k$,

$$p\left(u + \frac{i}{2}\right) q\left(u - \frac{i}{2}\right) - p\left(u - \frac{i}{2}\right) q\left(u + \frac{i}{2}\right) = W(u)\,, \tag{5.3}$$

$$p(u + i) q(u - i) - p(u - i) q(u + i) = E(u)\,. \tag{5.4}$$

Moreover, the corresponding eigenvector is unique up to proportionality. In particular, if the restriction of $\mathcal{T}(u)$ on $\mathcal{H}_k$ is diagonalizable, then there exist exactly $\dim \mathcal{H}_k$ *ordered pairs of polynomials $p(u)$, $q(u)$ such that* $\deg p = k$ *and equality* (5.3) *holds.*

Applying Lemma 4.3 gives another version of Theorem 5.1.

Theorem 5.2. *Let $\ell_s = 1/2$ for all $s = 1, \dots, n$, and $z_1, \dots, z_n$ satisfy condition* (5.2). *A polynomial $E(u)$ is an eigenvalue of the restriction of $\mathcal{T}(u)$ on $\mathcal{H}_k$ if and only if the difference equation*

$$a(u) f(u - i) + d(u) f(u + i) = E(u) f(u) \tag{5.5}$$

has polynomial solutions $p(u)$, $q(u)$ such that $\deg p = k < \deg q$. *Moreover the corresponding eigenvector is unique up to proportionality. In particular, if the restriction of $\mathcal{T}(u)$ on $\mathcal{H}_k$ is diagonalizable, then there exist exactly* $\dim \mathcal{H}_k$ *polynomials $E(u)$ such that Eq.* (5.5) *has polynomial solutions $p(u)$, $q(u)$ with* $\deg p = k < \deg q$.

Suppose there exists an ordered pair of polynomials $p(u)$, $q(u)$ such that $\deg p = k$, equality (5.3) holds, and the roots of $p(u)$ do not satisfy condition (3.10). Then by Theorems 5.1, 4.1, and 4.2, the number of admissible solutions of system (3.8) in this case is strictly less than the dimension of the space $\mathcal{H}_k$. The simplest example is the same as at the end of Sec. 3.

Example 5.3. Let $n = 4$, $\ell_s = 1/2$, $z_s = 0$, $s = 1, \dots, 4$, and $k = 2$. Then $\dim \mathcal{H}_2 = 2$, and there are two ordered pairs of polynomials $p(u)$, $q(u)$ such that $\deg p = 2$ and equality (5.3) holds: $p(u) = u^2 + 1/4$, $q(u) = iu(u^2 + 5/4)$, and $p(u) = u^2 - 1/12$, $q(u) = iu(u^2 + 1/4)$. However, only for the second pair, the roots of $p(u)$ satisfy condition (3.10) and give an admissible solution of system (3.8).

This example can be generalized. One can show that for the homogeneous spin-1/2 model, $z_s = 0$, $s = 1, \dots, n$, with even n, for any $k \geqslant 2$, there exist ordered pairs of polynomials $p(u)$, $q(u)$ such that $\deg p = k$, equality (5.3) holds, and $p(i/2) = p(-i/2) = 0$, so the roots of $p(u)$ do not satisfy condition (3.10).

6. Completeness for Higher Spin Model

In this section, we return to the general case. Recall formulae (4.1)–(4.3):

$$W(u) = \prod_{s=1}^{n} \prod_{r=1}^{2\ell_s} \left(u - z_s + i\left(\ell_s - r + \frac{1}{2}\right)\right),$$

$$W_m(u) = \prod_{\substack{s=1 \\ 2\ell_s+1 \geqslant m}}^{n} \prod_{r=0}^{2\ell_s - m} \left(u - z_s + i\left(\ell_s - r - \frac{m}{2}\right)\right).$$

The next theorem is the main result of the paper.

Theorem 6.1. *Let $z_1, \dots, z_n$ satisfy condition (2.3). A polynomial $E(u)$ is an eigenvalue of the restriction of $\mathcal{T}(u)$ on $\mathcal{H}_k$ if and only if there exists an ordered pair of polynomials $p(u)$, $q(u)$ such that $\deg p = k$,*

$$p\left(u + \frac{i}{2}\right) q\left(u - \frac{i}{2}\right) - p\left(u - \frac{i}{2}\right) q\left(u + \frac{i}{2}\right) = W(u), \tag{6.1}$$

for any $m = 2, 3, \dots,$ the polynomial

$$Q_m(u) = p\left(u + m\frac{i}{2}\right) q\left(u - m\frac{i}{2}\right) - p\left(u - m\frac{i}{2}\right) q\left(u + m\frac{i}{2}\right) \tag{6.2}$$

is divisible by $W_m(u)$, and

$$\frac{p(u+i)q(u-i) - p(u-i)q(u+i)}{W_2(u)} = E(u). \tag{6.3}$$

(Recall that $W_m(u) = 1$ if $m > 2\max\{\ell_1, \dots, \ell_n\}$.) Moreover, the corresponding eigenvector is unique up to proportionality. In particular, if the restriction of $\mathcal{T}(u)$ on $\mathcal{H}_k$ is diagonalizable, then there exist exactly $\dim \mathcal{H}_k$ ordered pairs of polynomials $p(u)$, $q(u)$ such that $\deg p = k$, equality (5.3) holds, and for any $m = 2, 3, \dots,$ the polynomial $Q_m(u)$ is divisible by $W_m(u)$.

Theorem 6.1 is similar to Theorem 5.1. However, there is no counterpart of Theorem 5.2 for the higher spin case. If the difference equation

$$a(u)f(u-i)+d(u)f(u+i)=E(u)f(u)$$

has two linearly independent polynomial solutions $p(u)$, $q(u)$, then they should obey relations (6.1) and (6.3). But the next illuminating example shows that the divisibility of $Q_2(u)$ by $W_2(u)$ given by (6.3) need not imply the divisibility of $Q_m(u)$ by $W_m(u)$ for $m \geqslant 3$.

Example 6.2. Let $n=4$, $\ell_1=\ell_2=1/2$, $\ell_3=\ell_4=3/2$, $z_s=0$, $s=1,\dots,4$, and $k=4$, so that

$$W(u)=u^4(u^2+1)^2\,, \qquad W_2(u)=\left(u^2+\frac{1}{4}\right)^2, \qquad W_3(u)-u^2.$$

Then $\dim \mathcal{H}_4=2$, but there are three ordered pairs of polynomials $p(u)$, $q(u)$ such that $\deg p=4$, equality (6.1) holds, and the polynomial $Q_2(u)$ is divisible by $W_2(u)$:

$$p(u)=u^4+\frac{7}{6}u^2-\frac{3}{80}\,, \qquad q(u)=iu\left(u^2+\frac{1}{4}\right)\left(u^2+\frac{9}{4}\right), \tag{6.4}$$

$$p(u)=\left(u^2+\frac{1}{4}\right)\left(u^2+\frac{9}{4}\right), \qquad q(u)=iu\left(u^4+\frac{13}{2}u^2+\frac{89}{16}\right), \tag{6.5}$$

$$p(u)=\left(u^2+\frac{1}{4}\right)^2, \qquad q(u)=iu\left(u^2+\frac{1}{4}\right)^2. \tag{6.6}$$

Among these three pairs, only for one of them, (6.4), the roots of $p(u)$ satisfy condition (3.10), and exactly for two of them, (6.4) and (6.5), the polynomial $Q_3(u)$ is divisible by $W_3(u)$. Therefore, the eigenvalues of the restriction of $\mathcal{T}(u)$ on $\mathcal{H}_4$ are given by the polynomials $2u^4+9u^2-15/8$ and $2u^4+9u^2+49/8$ corresponding by formula (6.3) to pairs (6.4), (6.5), respectively, while the polynomial $2u^4+9u^2+81/8$ corresponding to pair (6.6) is not an eigenvalue of $\mathcal{T}(u)$.

Nevertheless, Theorem 5.2 survives almost literally for higher spin models satisfying the following condition:

Let $X_s=\{z_s-i\ell_s,\, z_s-i(\ell_s-1),\dots,z_s+i\ell_s\}$, $s=1,\dots,n$. Then for any $r,s=1,\dots,n$, the sets X_r and X_s either do not intersect or coincide. (6.7)

cf. (2.3). In particular, condition (6.7) holds for homogeneous models. Notice that if $X_r=X_s$ in (6.7), then $z_r=z_s$ and $\ell_r=\ell_s$.

Theorem 6.3. *Let $z_1,\dots,z_n$ satisfy condition (6.7). A polynomial $E(u)$ is an eigenvalue of the restriction of $\mathcal{T}(u)$ on $\mathcal{H}_k$ if and only if the difference equation*

$$a(u)f(u-i)+d(u)f(u+i)=E(u)f(u) \tag{6.8}$$

has polynomial solutions $p(u)$, $q(u)$ such that $\deg p = k < \deg q$. *Moreover the corresponding eigenvector is unique up to proportionality. In particular, if the restriction of $\mathcal{T}(u)$ on $\mathcal{H}_k$ is diagonalizable, then there exist exactly* $\dim \mathcal{H}_k$ *polynomials $E(u)$ such that Eq.* (5.5) *has polynomial solutions $p(u)$, $q(u)$ with* $\deg p = k < \deg q$.

Theorem 6.3 follows from Theorem 6.1 by Lemma 4.3 and the next statement.

Lemma 6.4. *Let $z_1, \dots, z_n$ satisfy condition* (6.7). *Assume that the polynomials $p(u)$, $q(u)$ satisfy condition* (6.1), *and the polynomial $Q_2(u)$ is divisible by $W_2(u)$. Then the polynomial $Q_m(u)$ is divisible by $W_m(u)$ for all $m \geqslant 3$.*

Proof. The statement follows by induction on m from the next Plücker relation

$$Q_m(u)Q_1\left(u + (m-3)\frac{i}{2}\right)$$
$$= Q_{m-1}\left(u - \frac{i}{2}\right)Q_2\left(u + (m-2)\frac{i}{2}\right) - Q_{m-2}(u-i)Q_1\left(u + (m-1)\frac{i}{2}\right). \qquad \square$$

Acknowledgment

This work was supported in part by the Simons Foundation grant #430235.

References

[1] L. V. Avdeev and A. A. Vladimirov, Exceptional solutions of the Bethe ansatz equations, *Teor. Matem. Fiz.* **69**(2) (1986) 163–174 (in Russian); *Theor. Math. Phys.* **69**(2) (1987) 1071–1079.

[2] R. Baxter, *Exactly Solved Models in Statistical Mechanics* (Academic Press, Inc., London, 1982).

[3] R. Baxter, Completeness of the Bethe ansatz for the six- and eight-vertex models, *J. Stat. Phys.* **108**(1–2) (2002) 1–48.

[4] H. Bethe, Zur Theorie der Metalle: I. Eigenwerte und Eigenfunktionen der linearen Atomkette, *Z. Phys.* **71** (1931) 205–226.

[5] B. Feigin, E. Frenkel and L. Rybnikov, Opers with irregular singularity and spectra of the shift of argument subalgebra, *Duke Math. J.* **155**(2) (2010) 337–363.

[6] L. D. Faddeev and L. A. Takhtajan, The quantum method for the inverse problem and the *XYZ* Heisenberg model, *Uspekhi Matem. Nauk* **34**(5) (1979) 13–63 (in Russian); *Russian Math. Surveys* **34**(5) (1979) 11–68.

[7] L. D. Faddeev and L. A. Takhtajan, The spectrum and scattering of excitations in the one-dimensional isotropic Heisenberg model, *Zap. Nauch. Semin. LOMI* **109** (1981) 134–178 (in Russian); *J. Soviet Math.* **24** (1984) 241–267.

[8] P. R. Giri and T. Deguchi, Singular eigenstates in the even (odd) length Heisenberg spin chain, *J. Phys. A* **48**(17) (2015) 175207, 26 pp.

[9] T. Deguchi and P. R. Giri, Exact quantum numbers of collapsed and non-collapsed 2-string solutions in the Heisenberg spin chain, *J. Phys. A* **49**(17) (2016) 174001, 24 pp.

[10] A. Gainutdinov and R. Nepomechie, Algebraic Bethe ansatz for the quantum group invariant open XXZ chain at roots of unity, *Nucl. Phys. B* **909** (2016) 796–839.
[11] W. Hao, R. Nepomechie and A. Sommes, Singular solutions, repeated roots and completeness for higher-spin chains, *J. Stat. Mech. Theory Exp.* (2014)(3) P03024, 20 pp.
[12] A. N. Kirillov, Combinatorial identities and the completeness of states for Heisenberg magnet, *Zap. Nauch. Semin. LOMI* **131** (1983) 88–105 (in Russian); *J. Soviet Math.* **30** (1985) 2298–3310.
[13] A. N. Kirillov, Completeness of the states for the generalized Heisenberg model, *Zap. Nauch. Semin. LOMI* **134** (1985) 169–189 (in Russian); *J. Soviet Math.* **36** (1987) 115–128.
[14] A. N. Kirillov and R. Sakamoto, Singular solutions to the Bethe ansatz equations and rigged configurations, *J. Phys. A* **47**(20) (2014) 205207, 20 pp.
[15] A. N. Kirillov and R. Sakamoto, Bethe's quantum numbers and rigged configurations, *Nucl. Phys. B* **905** (2016) 359–372.
[16] V. E. Korepin, Calculation of norms of Bethe wave functions, *Commun. Math. Phys.* **86**(3) (1982) 391–418.
[17] V. E. Korepin, N. M. Bogoliubov and A. G. Izergin, *Quantum Inverse Scattering Method and Correlation Functions* (Cambridge University Press, 1993).
[18] P. P. Kulish and N. Yu. Reshetikhin, Quantum linear problem for the sine-Gordon equation and higher representations, *Zap. Nauch. Semin. LOMI* **101** (1981) 101–110 (in Russian); *J. Soviet Math.* **23**(4) (1983) 2435–2441.
[19] P. P. Kulish, N. Yu. Reshetikhin and E. K. Sklyanin, Yang–Baxter equations and representation theory I, *Lett. Math. Phys.* **5**(5) (1981) 393–403.
[20] R. P. Langlands and Y. Saint-Aubin, Combinational aspects of the Bethe equations, in *Advances in Mathematical Sciences*, CRM Proc. Lect. Notes, Vol. 11 (Amer. Math. Soc., 1997), pp. 231–301.
[21] J. R. Li and V. Tarasov, *XXZ*-type Bethe ansatz equations and quasi-polynomials, *J. Phys.: Conf. Ser.* **411** (2013) 012020, 20 pp.
[22] E. Mukhin, V. Tarasov and A. Varchenko, Schubert calculus and representations of the general linear group, *J. Amer. Math. Soc.* **22**(4) (2009) 909–940.
[23] E. Mukhin, V. Tarasov and A. Varchenko, Generating operator of *XXX* or Gaudin transfer matrices has quasi-exponential kernel, *SIGMA* **6** (2007) 060, 31 pp.
[24] E. Mukhin, V. Tarasov and A. Varchenko, Bethe algebra of homogeneous *XXX* Heisenberg model has simple spectrum, *Comm. Math. Phys.* **288**(1) (2009) 1–42.
[25] E. Mukhin, V. Tarasov and A. Varchenko, Bethe subalgebras of the group algebra of the symmetric group, *Transform. Groups* **18**(3) (2013) 767–801.
[26] E. Mukhin, V. Tarasov and A. Varchenko, Spaces of quasi-exponentials and representations of the Yangian $Y(\mathfrak{gl}_N)$, *Transform. Groups* **19**(3) (2014) 861–885.
[27] E. Mukhin, V. Tarasov and A. Varchenko, On reality property of Wronski maps, *Confluentes Math.* **1**(2) (2009) 225–247.
[28] E. Mukhin, V. Tarasov and A. Varchenko, Reality property of discrete Wronski map with imaginary step, *Lett. Math. Phys.* **100**(2) (2012) 151–160.
[29] E. Mukhin, V. Tarasov and A. Varchenko, Bethe algebra of the $\mathfrak{gl}_{N+1}$ Gaudin model and algebra of functions on the critical set of the master function, in *New Trends in Quantum Integrable Systems*, eds. B. Feigin, M. Jimbo and M. Okado (World Scientific, 2011), pp. 307–324.
[30] E. Mukhin and A. Varchenko, Solutions to the *XXX* type Bethe ansatz equations and flag varieties, *Cent. Eur. J. Math.* **1**(2) (2003) 238–271.

[31] G. P. Pronko and Yu. G. Stroganov, Bethe equations "on the wrong side of equator", *J. Phys. A* **32**(12) (1999) 2333–2340.

[32] L. Rybnikov, A proof of the Gaudin Bethe ansatz conjecture, Preprint (2016) 15 pp; arXiv:1608.04625.

[33] N. A. Slavnov, Calculation of scalar products of wave functions and form-factors in the framework of the algebraic Bethe ansatz, *Teor. Matem. Fiz.* **79**(2) (1989) 232–240 (in Russian); *Theor. Math. Phys.* **79**(2) (1989) 502–508.

[34] N. A. Slavnov, Algebraic Bethe ansatz and quantum integrable systems, *Russian Math. Surveys* **62**(4) (2007) 727–766.

[35] V. Tarasov, Irreducible monodromy matrices for the R-matrix of the XXZ-model and lattice local quantum Hamiltonians, *Teor. Matem. Fiz.* **63**(2) (1985) 175–196 (in Russian); *Theor. Math. Phys.* **63**(2) (1985) 440–454.

[36] V. Tarasov and A. Varchenko, Completeness of Bethe vectors and difference equations with regular singular points, *Internat. Math. Res. Notices* (1995)(13) 637–669.

Analytic Scattering Theory for Jacobi Operators and Bernstein–Szegö Asymptotics of Orthogonal Polynomials

D. R. Yafaev

IRMAR, Université de Rennes I,
Campus de Beaulieu, Rennes, 35042, France
and
SPGU, Univ. Nab. 7/9, Saint Petersburg, 199034, Russia
yafaev@univ-rennes1.fr

We study semi-infinite Jacobi matrices $H = H_0 + V$ corresponding to trace class perturbations V of the "free" discrete Schrödinger operator H_0. Our goal is to construct various spectral quantities of the operator H, such as the weight function, eigenfunctions of its continuous spectrum, the wave operators for the pair H_0, H, the scattering matrix, the spectral shift function, etc. This allows us to find the asymptotic behavior of the orthonormal polynomials $P_n(z)$ associated to the Jacobi matrix H as $n \to \infty$. In particular, we consider the case of z inside the spectrum $[-1, 1]$ of H_0 when this asymptotic has an oscillating character of the Bernstein–Szegö type and the case of z at the end points ± 1.

To the memory of Lyudvig Dmitrievich Faddeev

Keywords: Jacobi matrices; discrete Schrödinger operator; orthogonal polynomials; asymptotics for large numbers; Szegö function.

Mathematics Subject Classification: 33C45, 39A70, 47A40, 47B39

1. Introduction

1.1. *Jacobi versus differential operators*

The theory of the Schrödinger operator $D^2 + b(x)$, $D = -id/dx$, with a short-range potential $b(x)$ is to a large extent due to L. D. Faddeev. We note, in particular, his classical papers [11, 12] on the direct and inverse quantum scattering problems (both in the one- and multi-dimensional cases) and [10] on the trace formulas for the operator $D^2 + b(x)$. Later, the paper [10] was significantly generalized by him jointly with Buslaev in [4]. Some of these results were exposed in the book [13] by Faddeev and Takhtajan.

As is well known, the theories of Jacobi operators given by three-diagonal matrices and of differential operators $Da(x)D + b(x)$ are to a large extent similar. This is true for Jacobi operators acting in the space $\ell^2(\mathbb{Z})$ and differential operators acting in the space $L^2(\mathbb{R})$ as well as for the corresponding operators acting in the spaces $\ell^2(\mathbb{Z}_+)$ and $L^2(\mathbb{R}_+)$, respectively. We refer to the book [27] where this analogy is described in a sufficiently detailed way. Both classes of the operators are very important in physical applications. Moreover, Jacobi operators in the space $\ell^2(\mathbb{Z}_+)$ are intimately related (see, e.g., the classical book [1]) to the theory of orthogonal polynomials. Necessary information on orthogonal polynomials can be found in the other classical book [26]; the theory of Jacobi operators is carefully presented in the books [1, 2].

Our goal is to develop the spectral and scattering theory for Jacobi operators in the space $\ell^2(\mathbb{Z}_+)$ using basically the same approach as for the differential operator $Da(x)D + b(x)$ in the space $L^2(\mathbb{R}_+)$. This leads to new results for orthogonal polynomials.

1.2. *Orthogonal polynomials*

Jacobi operators are defined in the space $\ell^2(\mathbb{Z}_+)$ by matrices

$$H = \begin{pmatrix} b_0 & a_0 & 0 & 0 & 0 & \cdots \\ a_0 & b_1 & a_1 & 0 & 0 & \cdots \\ 0 & a_1 & b_2 & a_2 & 0 & \cdots \\ 0 & 0 & a_2 & b_3 & a_3 & \cdots \\ \vdots & \vdots & \vdots & \ddots & \ddots & \ddots \end{pmatrix}. \tag{1.1}$$

The entries a_n here are arbitrary positive numbers, and b_n are arbitrary real numbers. We denote by e_n, $n \in \mathbb{Z}_+$, the canonical basis in $\ell^2(\mathbb{Z}_+)$, that is, all components of the vector e_n are zeros, except the nth component which equals 1. If the sequences $\{a_n\}$ and $\{b_n\}$ are bounded, then the Jacobi operator H is bounded in $\ell^2(\mathbb{Z}_+)$. Let us denote by $dE(\lambda)$ the spectral family of the self-adjoint operator H and define the corresponding spectral measure $d\rho(\lambda) = d(E(\lambda)e_0, e_0)$.

Given matrix (1.1), one constructs polynomials $P_n(z)$ by the recurrent relation

$$a_{n-1}P_{n-1}(z) + b_n P_n(z) + a_n P_{n+1}(z) = zP_n(z), \quad n \in \mathbb{Z}_+, \tag{1.2}$$

and the boundary conditions $P_{-1}(z) = 0$, $P_0(z) = 1$. Then $P_n(z)$ is a polynomial of degree n, that is,

$$P_n(z) = k_n(z^n + r_n z^{n-1} + \cdots). \tag{1.3}$$

Comparing the coefficients at z^{n+1} and z^n in the left- and right-hand sides of (1.2), we see that $k_n = (a_0 a_1 \cdots a_{n-1})^{-1} > 0$ and

$$a_n = \frac{k_n}{k_{n+1}}, \quad b_n = r_n - r_{n+1}. \tag{1.4}$$

Obviously, $P(z) = \{P_n(z)\}_{n=0}^{\infty}$ satisfies the equation $HP(z) = zP(z)$, that is, it is an "eigenvector" of the operator H.

Putting together (1.1) and (1.2), we find that

$$e_n = P_n(H)e_0 \tag{1.5}$$

for all $n \in \mathbb{Z}_+$. It follows that

$$d(E(\lambda)e_n, e_m) = P_n(\lambda)P_m(\lambda)d\rho(\lambda),$$

whence

$$\int_{-\infty}^{\infty} P_n(\lambda)P_m(\lambda)d\rho(\lambda) = \delta_{n,m}; \tag{1.6}$$

as usual, $\delta_{n,n} = 1$ and $\delta_{n,m} = 0$ for $n \neq m$. Thus, the polynomials $P_n(\lambda)$ are orthogonal and normalized in the space $L^2(\mathbb{R}; d\rho)$. Formula (1.5) also shows that the set of all vectors $H^n e_0$, $n \in \mathbb{Z}_+$, is dense in the space $\ell^2(\mathbb{Z}_+)$, and hence the spectrum of the Jacobi operator H is simple with the generating vector e_0.

Alternatively, $\{P_0(\lambda), P_1(\lambda), \ldots, P_n(\lambda), \ldots\}$ can be obtained by the Gram–Schmidt orthogonalization of the monomials $\{1, \lambda, \ldots, \lambda^n, \ldots\}$ in the space $L^2(\mathbb{R}_+; d\rho)$; one also has to additionally require that $k_n > 0$ in (1.3). This fact is known as the Favard theorem. In contract to the continuous case (see the paper [15] where an integral equation was used), the inverse problem of reconstructing the Jacobi operator H given its spectral measure $d\rho(\lambda)$ admits a quite explicit solution. Indeed, let a measure $d\rho(\lambda)$ have a bounded and infinite support, and let $P_n(z)$ be the corresponding polynomials satisfying (1.3) and (1.6). Then the coefficients of the operator H can be recovered by formulas (1.4).

1.3. *The discrete Schrödinger operator and its perturbations*

In the particular case $a_n = 1/2$ and $b_n = 0$, the operator (1.1) is denoted H_0. It plays the role of the "free" differential operator D^2 in the space $L^2(\mathbb{R}_+)$ with the boundary condition $u(0) = 0$. The operator H_0 can be diagonalized explicitly.

In this paper, we suppose that the perturbation $V = H - H_0$ satisfies a "short-range" assumption

$$\sum_{n=0}^{\infty} (|a_n - 1/2| + |b_n|) < \infty. \tag{1.7}$$

Then V belongs to the trace class $\mathfrak{S}_1$, and its trace norm is equivalent to the sum (1.7). Under assumption (1.7) the spectrum $\sigma(H)$ of the operator H is absolutely continuous on the interval $(-1, 1)$, but the operator H may have discrete spectrum $\sigma_d(H)$ (possibly, infinite) in $\mathbb{R}\setminus[-1, 1]$; moreover, the points $+1$ or -1 may be its eigenvalues.

We construct various spectral quantities of the operator H, such as the weight function $w(\lambda)$ defined by the equation $d\rho(\lambda) = w(\lambda)d\lambda$ for $\lambda \in (-1, 1)$, eigenfunctions of its continuous spectrum, the perturbation determinant $\Delta(z)$ and the wave operators $W_\pm(H, H_0)$ for the pair H_0, H, the scattering matrix $S(\lambda)$, the spectral shift function $\xi(\lambda)$, etc. Under assumption (1.7) the functions $w(\lambda)$, $S(\lambda)$ and $\xi(\lambda)$

are continuous in $\lambda \in (-1,1)$. Moreover, $w(\lambda) \neq 0$ and, as shown in [20], the so-called Szegö condition

$$\int_{-1}^{1} \ln w(\lambda)(1-\lambda^2)^{-1/2} d\lambda > -\infty \tag{1.8}$$

is satisfied.

Spectral results on Jacobi operators lead to the corresponding assertions for the polynomials $P_n(z)$ defined by (1.2). For example, we show (see Theorem 2.5) that, for $\lambda \in (-1,1)$,

$$P_n(\lambda) = \sqrt{\frac{2}{\pi}} w(\lambda)^{-1/2}(1-\lambda^2)^{-1/4} \sin((n+1)\arccos\lambda + \pi\xi(\lambda)) + o(1) \tag{1.9}$$

as $n \to \infty$. This asymptotic relation is the classical result of Bernstein [3] (see also formula (12.1.8) in the Szegö book [26]). It is required in [3, 26] that $\operatorname{supp}\rho \subset [-1,1]$ and Lipschitz–Dini conditions are imposed on the weight function $w(\lambda)$. Under assumption (1.7) formula (1.9) is probably new. In particular, we do no assume that $\operatorname{supp}\rho \subset [-1,1]$. Note that the spectral shift function $\xi(\lambda)$ in formula (1.9) is usually replaced in the theory of orthogonal polynomials by the so-called Szego function.

As shows the example of Pollaczek polynomials (see formula (7.14), below), assumption (1.7) cannot be significantly relaxed. For Pollaczek polynomials, $a_n - 1/2$ and b_n have order n^{-1} as $n \to \infty$, and the phase in formula (1.9) is essentially changed (see Sec. 5 in the Appendix to [26]). This resembles the modification of the phase function for the Schrödinger operator with the Coulomb potential (see, e.g., formula (36,23) in the book [22]).

As shown in the paper [4] by Buslaev and Faddeev (see Sec. 4.6 in [30], for a detailed presentation), the expansion as $|z| \to \infty$ of the perturbation determinant $\Delta(z)$ for a pair H_0, H of Schrödinger operators contains both integer and half-integer powers of z^{-1}. This leads to two series of trace identities: of integer and of half-integer orders. The first of them is stated in terms of integer powers of eigenvalues and moments of the spectral shift function. The second series is stated in terms of half-integer powers of eigenvalues and moments of the modulus of the perturbation determinant. For Jacobi operators H_0, H, the expansion of the perturbation determinant $\Delta(z)$ as $|z| \to \infty$ contains integer powers of z^{-1} only. So there are no identities of half-integer orders. On the other hand, there is a version of trace identities of integer order (known as Case sum rules) stated in terms of the weight function, that is, of the modulus of the perturbation determinant. We note that the Case sum rules involve Chebyshev polynomials of the operators H_0 and H while the identities in spirit of [4] yield expressions for the traces $\operatorname{Tr}(H^n - H_0^n)$.

This paper contains relatively few new results. However, we hope that a consistent analogy between Jacobi and differential operators might shed a new light on some aspects of the theory of orthogonal polynomials. A similar point of view was adopted in the paper [20] devoted to Hilbert–Schmidt perturbations V of the operator H_0.

Note that, in [20], necessary and sufficient conditions in terms of spectral data of H were found for V to be in the Hilbert–Schmidt class. The problem of characterization of spectral data was also solved in [24] for perturbations V in the class (it is known now as the Ryckman class) relatively close to the trace class $\mathfrak{S}_1$. It looks tempting to obtain exhaustive results of such type for $V \in \mathfrak{S}_1$.

1.4. *The structure of the paper*

The paper is organized as follows. In Sec. 2, we consider Eq. (1.2) and, in addition to its polynomial solutions $P_n(z)$, we introduce so-called Jost solutions $f_n(z)$ of this equation. The Jost function $\Omega(z)$ basically coincides with $f_{-1}(z)$. The solutions $f_n(z)$ exponentially decay as $n \to \infty$ for $z \notin \sigma(H_0) = [-1, 1]$ and oscillate for $z = \lambda \pm i0$ where $\lambda \in (-1, 1)$. Then a link between $f_n(\lambda \pm i0)$ and $P_n(\lambda)$ leads to asymptotic formulas for polynomials $P_n(\lambda)$. Thus, we obtain formula (1.9) (with the argument of $\Omega(\lambda \pm i0)$ playing the role of $\xi(\lambda)$); see Theorem 2.5. Polynomial solutions $P_n(z)$ and their asymptotics as $n \to \infty$ for $z \notin [-1, 1]$ are studied in Sec. 3. Our proofs of the existence of $f_n(z)$ and uniform bounds on $P_n(z)$ rely on discrete "Volterra integral" equations. Their study is postponed until Sec. 9.

Section 4 is devoted to an investigation of the Jost solutions $f_n(z)$ and the polynomials $P_n(z)$ as $z \to \pm 1$. This corresponds to the low energy scattering for the Schrödinger equation. In particular, we exhibit here examples of Jacobi operators with eigenvalues at the points 1 or -1.

In Sec. 5, we construct the perturbation determinant $\Delta(z)$ and the spectral shift function for the pair H_0, H and derive formulas for the traces $\operatorname{Tr}(H^n - H_0^n)$. We also show in Theorem 5.6 that the Jost function $\Omega(z)$ differs from $\Delta(z)$ by a numerical factor only. Scattering theory for the pair H_0, H is developed in Sec. 6.

A link of the Szegö function with the perturbation determinant (or the Jost function) is established in Sec. 7. We here also show (again on the example of Pollaczek polynomials) that the weight function $w(\lambda)$ may tend to zero exponentially if assumption (1.7) is slightly relaxed. Therefore the Szegö condition (1.8) is sharp for trace class perturbations. In Sec. 8, the results of the preceding sections are illustrated on the example of the Jacobi polynomials when $a_n - 1/2$ and b_n have order n^{-2} as $n \to \infty$.

Some parts of this paper have non-trivial intersections with Sec. 2 of the paper [20] where, however, specific features of the Jacobi operators were extensively used. The analogy with the continuous case exploited in the present paper allows us to obtain some results for free.

The modern approach based on a Riemann–Hilbert problem for matrix-valued functions of [14] combined with the steepest descent method of [7] is out of the scope of the present paper.

2. Jost Solutions of the Jacobi Equations and Bernstein–Szegö Asymptotics

2.1. *Jost solutions*

In the canonical basis e_n, $n \in \mathbb{Z}_+$, the Jacobi operator H is defined by the formula

$$He_n = a_{n-1}e_{n-1} + b_n e_n + a_n e_{n+1}, \quad n \in \mathbb{Z}_+, \tag{2.1}$$

where we accept that $e_{-1} = 0$. Relations (1.1) and (2.1) are of course equivalent. The sequences $a_n > 0$ and $b_n = \bar{b}_n$ where $n = 0, 1, \dots$ are assumed to be bounded, so that H is a bounded self-adjoint operator in the space $\ell^2(\mathbb{Z}_+)$.

Consider now the equation $Hu = zu$, that is

$$a_{n-1}u_{n-1} + b_n u_n + a_n u_{n+1} = zu_n, \quad n \in \mathbb{Z}_+, \tag{2.2}$$

for a sequence $u = \{u_n\}_{n=-1}^{\infty}$. The number $a_{-1} \neq 0$ can be chosen at our convenience; for definiteness, we put $a_{-1} = 1/2$. Obviously, the values of u_{k-1} and u_k for some $k \in \mathbb{Z}_+$ determine the whole sequence u_n satisfying Eq. (2.2).

Let $f = \{f_n\}_{n=-1}^{\infty}$ and $g = \{g_n\}_{n=-1}^{\infty}$ be two solutions of Eq. (2.2). A direct calculation shows that their Wronskian

$$\{f, g\} := a_n(f_n g_{n+1} - f_{n+1} g_n) \tag{2.3}$$

does not depend on $n = -1, 0, 1, \dots$. In particular, for $n = -1$ and $n = 0$, we have

$$\{f, g\} = 2^{-1}(f_{-1}g_0 - f_0 g_{-1}) \quad \text{and} \quad \{f, g\} = a_0(f_0 g_1 - f_1 g_0). \tag{2.4}$$

Calculating the Wronskian (2.3) for $n \to \infty$, we see that Eq. (2.2) may have at most one (up to a multiplicative constant) solution u_n such that $u_n \to 0$ as $n \to \infty$.

In the case $a_n = 1/2$, $b_n = 0$, the operator (2.1) is known as the free discrete Schrödinger operator; it will be denoted H_0. The spectrum of the operator H_0 is simple, absolutely continuous and coincides with the interval $[-1, 1]$. The corresponding spectral measure $d\rho_0(\lambda) = d(E_0(\lambda)e_0, e_0)$ is given by the formula

$$d\rho_0(\lambda) = 2\pi^{-1}\sqrt{1 - \lambda^2}\, d\lambda, \quad \lambda \in (-1, 1).$$

Below we fix the branch of the analytic function $\sqrt{z^2 - 1}$ of $z \in \mathbb{C} \setminus [-1, 1] =: \Pi$ by the condition $\sqrt{z^2 - 1} > 0$ for $z > 1$. Obviously, this function is continuous on the closure $\operatorname{clos}\Pi$ of Π, it equals $\pm i\sqrt{1 - \lambda^2}$ for $z = \lambda \pm i0$, $\lambda \in (-1, 1)$, and $\sqrt{z^2 - 1} < 0$ for $z < -1$. Put

$$\zeta(z) = z - \sqrt{z^2 - 1} = (z + \sqrt{z^2 - 1})^{-1}; \tag{2.5}$$

then $|\zeta(z)| < 1$ for $z \in \Pi$. Since

$$2z = \zeta(z) + \zeta(z)^{-1}, \tag{2.6}$$

the sequence $\{\zeta(z)^n\}_{n=-1}^{\infty}$ satisfies the "free" equation (2.2): $\zeta(z)^{n-1} + \zeta(z)^{n+1} = 2z\zeta(z)^n$. For $\lambda \in [-1, 1]$, it is common to set $\lambda = \cos\theta$ where $\theta \in [0, \pi]$. Then

$$\zeta(\lambda \pm i0) = e^{\mp i\theta}. \tag{2.7}$$

It is well known (see, e.g., [31] for a detailed proof) that the matrix elements of the resolvent $R_0(z) = (H_0 - zI)^{-1}$ (here and below I is the identity operator) of the operator H_0 are given by the formula

$$(R_0(z)e_n, e_m) = \frac{\zeta(z)^{n+m+2} - \zeta(z)^{|n-m|}}{\sqrt{z^2-1}}$$

for $z \in \Pi$ and all $n, m \in \mathbb{Z}_+$. In particular, $(R_0(z)e_0, e_0) = -2\zeta(z)$. The polynomials $P_n^{(0)}(z)$ defined by the recurrent equation (1.2) where $a_n = 1/2$, $b_n = 0$ and obeying the conditions $P_{-1}^{(0)}(z) = 0$, $P_0^{(0)}(z) = 1$ are normalized Chebyshev polynomials of the second kind. They satisfy the equation

$$P_n^{(0)}(z) = \frac{1}{2\sqrt{z^2-1}}\left(\zeta(z)^{-n-1} - \zeta(z)^{n+1}\right) \tag{2.8}$$

and, in particular,

$$P_n^{(0)}(\lambda) = \frac{\sin((n+1)\theta)}{\sqrt{1-\lambda^2}}, \quad \lambda = \cos\theta \in (-1,1). \tag{2.9}$$

Evidently, the "perturbation" $V = H - H_0$ is given by the equality

$$Ve_n = (a_{n-1} - 1/2)e_{n-1} + b_n e_n + (a_n - 1/2)e_{n+1}. \tag{2.10}$$

If $a_n \to 1/2$ and $b_n \to 0$ as $n \to \infty$, then the operator V is compact. In this case the essential spectrum of the operator H coincides with the interval $[-1,1]$, and its discrete spectrum $\sigma_d(H)$ consists of simple eigenvalues accumulating, possibly, to the points 1 and -1 only.

Under assumption (1.7), Eq. (2.2) has the so-called Jost solution $f(z) = \{f_n(z)\}_{n=-1}^{\infty}$ distinguished by its asymptotics as $n \to \infty$. Our proof of this fact is similar to the continuous case, but we give it for the completeness of our presentation in Sec. 9.1. It is based on the discrete Volterra integral equation

$$f_n(z) = \zeta(z)^n - \frac{1}{\sqrt{z^2-1}} \sum_{m=n+1}^{\infty} (\zeta(z)^{n-m} - \zeta(z)^{m-n})(Vf(z))_m \tag{2.11}$$

where $z \in \Pi$ and V is defined by formula (2.10). A somewhat different proof can be found in [20] (see also the book [27]). Put

$$\rho_n = \sum_{m=n}^{\infty} (|a_m - 1/2| + |b_m|). \tag{2.12}$$

Then $\rho_n \to 0$ as $n \to \infty$.

Theorem 2.1. *Let assumption* (1.7) *be satisfied, and let* $z \in \operatorname{clos}\Pi$, $z \neq \pm 1$. *Then Eq.* (2.2) *has a solution satisfying the condition*

$$f_n(z) = \zeta(z)^n(1 + O(\rho_n)) \tag{2.13}$$

as $n \to \infty$. *Every function* $f_n(z)$, $n = -1, 0, 1, \ldots$, *depends analytically on* $z \in \Pi$, *and it is continuous in* z *up to the cut along* $[-1,1]$ *except, possibly, the points* ± 1.

Since $|\zeta(z)| < 1$, it follows from (2.13) that $f_n(z) \to 0$ exponentially as $n \to \infty$ for $z \in \Pi$. In particular, Eq. (2.2) has only one solution satisfying (2.13). Note that for the operator H_0, the error term in (2.13) disappears and the Jost function is $f_n(z) = \zeta(z)^n$.

Recall that the polynomials $P_n(z)$ are defined by Eq. (2.2) and the conditions $P_{-1}(z) = 0$, $P_0(z) = 1$. Put $P(z) = \{P_n(z)\}_{n=-1}^{\infty}$, $f(z) = \{f_n(z)\}_{n=-1}^{\infty}$,

$$\omega(z) := \{P(z), f(z)\} = -2^{-1} f_{-1}(z), \tag{2.14}$$

where the first formula (2.4) has been used. In particular, for the operator H_0 we have $\omega_0(z) = -(2\zeta(z))^{-1}$. By analogy with the continuous case, we define the Jost function

$$\Omega(z) := \omega(z)/\omega_0(z) = -2\zeta(z)\omega(z). \tag{2.15}$$

The following result is a direct consequence of Theorem 2.1.

Corollary 2.2. *Under assumption* (1.7) *the Jost function* $\Omega(z)$ *depends analytically on* $z \in \Pi$, *and it is continuous in* z *up to the cut along* $[-1, 1]$ *except, possibly, the points* ± 1. *A point* $z \in \Pi$ *is an eigenvalue of the operator* H *if and only if* $\Omega(z) = 0$.

Theorem 2.1 can be supplemented by the following assertion.

Theorem 2.3. *Under the assumptions of Theorem* 2.1 *put* $a_n^{(N)} = a_n$, $b_n^{(N)} = b_n$ *for* $n \le N$ *and* $a_n^{(N)} = 1/2$, $b_n^{(N)} = 0$ *for* $n > N$. *Let* $f_n^{(N)}(z)$ *be the Jost solution of Eq.* (2.2) *with the coefficients* $a_n^{(N)}$, $b_n^{(N)}$. *Then for each* $n \ge -1$, *we have*

$$\lim_{N\to\infty} f_n^{(N)}(z) = f_n(z), \quad z \in \operatorname{clos}\Pi, \quad z \neq \pm 1. \tag{2.16}$$

2.2. *The Bernstein–Szegö asymptotics*

For $\lambda \in (-1, 1)$, Eq. (2.2) where $z = \lambda$ has two solutions $f(\lambda \pm i0) = \{f_n(\lambda \pm i0)\}_{n=-1}^{\infty}$; of course $f_n(\lambda - i0) = \overline{f_n(\lambda + i0)}$. It follows from Theorem 2.1 and formula (2.7) that asymptotics of $f_n(\lambda \pm i0)$ as $n \to \infty$ is oscillating:

$$f_n(\lambda \pm i0) = e^{\mp i\theta n}(1 + O(\rho_n)), \quad \theta = \arccos\lambda \in (0, \pi). \tag{2.17}$$

Calculating their Wronskian (2.3) for $n \to \infty$, we find that

$$\{f(\lambda + i0), f(\lambda - i0)\} = i\sin\theta \neq 0, \quad \sin\theta = \sqrt{1 - \lambda^2}, \tag{2.18}$$

so that these solutions are linearly independent. Thus,

$$P_n(\lambda) = \overline{c(\lambda)} f_n(\lambda + i0) + c(\lambda) f_n(\lambda - i0) \tag{2.19}$$

for some constant $c(\lambda)$. Taking the Wronskians of this equation with $f(\lambda + i0)$ and using (2.14), (2.18), we find that

$$\sin\theta\, c(\lambda) = i\omega(\lambda + i0).$$

Of course $\omega(\lambda - i0) = \overline{\omega(\lambda + i0)}$. Thus (2.19) leads to an intermediary result.

Lemma 2.4. *Under assumption* (1.7) *the representation*

$$P_n(\lambda) = \frac{\omega(\lambda - i0) f_n(\lambda + i0) - \omega(\lambda + i0) f_n(\lambda - i0)}{i\sqrt{1-\lambda^2}}, \quad \lambda \in (-1,1), \quad n = 0,1,2,\ldots, \tag{2.20}$$

holds true.

In particular, representation (2.20) implies that

$$\omega(\lambda \pm i0) \neq 0, \quad \lambda \in (-1,1). \tag{2.21}$$

Indeed, otherwise we would have $P_n(\lambda) = 0$ for some $\lambda \in (-1,1)$ and all $n \in \mathbb{Z}_+$. However, $P_0(\lambda) = 1$ for all λ.

Let us set

$$\varkappa(\theta) = |\Omega(\cos\theta + i0)|, \quad \Omega(\cos\theta + i0) = \varkappa(\theta) e^{i\eta(\theta)}, \quad \theta \in (0,\pi). \tag{2.22}$$

In the theory of the Schrödinger operator, the functions $\varkappa(\theta)$ and $\eta(\theta)$ are known as the limit amplitude and the limit phase, respectively; the function $\eta(\theta)$ is also known as the scattering phase or the phase shift. Definition (2.22) fixes $\eta(\theta)$ only up to a term $2\pi k$ where $k \in \mathbb{Z}$.

It follows from (2.7) and (2.15) that

$$2\omega(\lambda \pm i0) = -\Omega(\lambda \pm i0)\zeta(\lambda \pm i0)^{-1} = -\varkappa(\theta) e^{\pm i(\eta(\theta)+\theta)}.$$

Therefore, combined together relations (2.17) and (2.20) yield the Bernstein–Szegö asymptotics of the polynomials $P_n(\lambda)$.

Theorem 2.5. *Under assumption* (1.7) *for* $\lambda \in (-1,1)$ *the polynomials* $P_n(\lambda)$ *have asymptotics*

$$P_n(\lambda) = \varkappa(\theta)(\sin\theta)^{-1}\sin((n+1)\theta + \eta(\theta)) + O(\rho_n), \quad \theta = \arccos\lambda, \tag{2.23}$$

as $n \to \infty$. *Relation* (2.23) *is uniform in* λ *on compact subintervals of* $(-1,1)$.

2.3. *The resolvent and spectral measure*

Let us construct the resolvent $R(z) = (H - zI)^{-1}$ of the operator H. Recall that $\omega(z)$ is the Wronskian (2.14).

Lemma 2.6. *For all* $n, m \in \mathbb{Z}_+$, *we have*

$$(R(z)e_n, e_m) = \omega(z)^{-1} P_n(z) f_m(z), \quad z \in \Pi, \tag{2.24}$$

if $n \leq m$ *and* $(R(z)e_n, e_m) = (R(z)e_m, e_n)$.

Proof. We will show that the operator $R(z)$ defined by (2.24) is the resolvent of H. We have

$$\omega(z)(R(z)u)_n = f_n(z)A_n(z) + P_n(z)B_n(z) \tag{2.25}$$

where

$$A_n(z) = \sum_{m=0}^{n} P_m(z)u_m, \quad B_n(z) = \sum_{m=n+1}^{\infty} f_m(z)u_m, \tag{2.26}$$

at least for all sequences $u = \{u_n\}$ with a finite number of non-zero components u_n. In this case $R(z)u \in \ell^2(\mathbb{Z}_+)$ because $f_n(z) \in \ell^2(\mathbb{Z}_+)$ for all $z \in \Pi$.

Our goal is to check that $(H - z)R(z)u = u$. It follows from definition (1.1) of the Jacobi operator H and formula (2.25) that

$$\begin{aligned}\omega((H-z)Ru)_n &= a_{n-1}\big(f_{n-1}A_{n-1} + P_{n-1}B_{n-1}\big) + (b_n - z)\big(f_nA_n + P_nB_n\big)\\ &\quad + a_n\big(f_{n+1}A_{n+1} + P_{n+1}B_{n+1}\big).\end{aligned} \tag{2.27}$$

According to (2.26) we have

$$f_{n-1}A_{n-1} + P_{n-1}B_{n-1} = f_{n-1}(A_n - P_nu_n) + P_{n-1}(B_n + f_nu_n)$$

and

$$f_{n+1}A_{n+1} + P_{n+1}B_{n+1} = f_{n+1}A_n + P_{n+1}B_n.$$

Let us substitute these expressions into the right-hand side of (2.27) and observe that the coefficients at A_n and B_n equal zero by virtue of Eq. (2.2) for $\{f_n\}$ and $\{P_n\}$, respectively. It follows that

$$((H-z)Ru)_n = \omega^{-1}a_{n-1}(-P_nf_{n-1} + f_nP_{n-1})u_n = u_n$$

whence $R(z) = (H - z)^{-1}$; in particular, the operator $R(z)$ defined by (2.24) is bounded in $\ell^2(\mathbb{Z}_+)$. □

In view of Theorem 2.1, $f_n(z)$ and $\omega(z)$ are continuous functions of $z \in \mathbb{C}\setminus[-1,1]$ up to the cut along $[-1,1]$ with possible exception of the points $z = \pm1$. Therefore using (2.21), we obtain the following result.

Theorem 2.7. *Let assumption* (1.7) *hold. Then, for all $n, m \in \mathbb{Z}_+$, the functions $(R(z)e_n, e_m)$ are continuous as $z \in \Pi$ approaches the interval $(-1,1)$ from above or below, and the spectrum of the operator H is absolutely continuous on $(-1,1)$.*

We emphasize that the points 1 and -1 may be eigenvalues of H; see Example 4.15 below.

Let us now calculate the spectral family $dE(\lambda)$ of the operator H. We proceed from the identity

$$2\pi i\frac{d(E(\lambda)e_n, e_m)}{d\lambda} = (R(\lambda + i0)e_n, e_m) - (R(\lambda - i0)e_n, e_m). \tag{2.28}$$

It follows from formula (2.24) that

$$(R(\lambda \pm i0)e_n, e_m) = \omega(\lambda \pm i0)^{-1}P_n(\lambda)f_m(\lambda \pm i0), \quad n \le m.$$

Substituting this expression into (2.28), we find that

$$2\pi i\frac{d(E(\lambda)e_n, e_m)}{d\lambda} = P_n(\lambda)\frac{\omega(\lambda - i0)f_m(\lambda + i0) - \omega(\lambda + i0)f_m(\lambda - i0)}{|\omega(\lambda + i0)|^2}.$$

Let us combine this representation with formula (2.20) for $P_m(\lambda)$. Since $2|\omega(\lambda + i0)| = |\Omega(\lambda + i0)|$, we obtain the following result.

Theorem 2.8. *Let assumption* (1.7) *hold. Then, for all* $n, m \in \mathbb{Z}_+$ *and* $\lambda \in (-1, 1)$, *we have the representation*

$$\frac{d(E(\lambda)e_n, e_m)}{d\lambda} = 2\pi^{-1}\sqrt{1-\lambda^2}|\Omega(\lambda + i0)|^{-2}P_n(\lambda)P_m(\lambda).$$

Corollary 2.9. *For* $\lambda \in (-1, 1)$, *the spectral measure of the operator* H *equals*

$$d\rho(\lambda) := d(E(\lambda)e_0, e_0) = w(\lambda)d\lambda, \tag{2.29}$$

where the weight function

$$w(\lambda) = 2\pi^{-1}\sqrt{1-\lambda^2}|\Omega(\lambda + i0)|^{-2}. \tag{2.30}$$

In particular, for the operator H_0, *we have*

$$w_0(\lambda) = 2\pi^{-1}\sqrt{1-\lambda^2}. \tag{2.31}$$

In view of (2.22), (2.30) the amplitude in (2.23) can be written as

$$\varkappa(\theta)(\sin\theta)^{-1} = 2^{1/2}\pi^{-1/2}(1-\lambda^2)^{-1/4}w(\lambda)^{-1/2} \tag{2.32}$$

which is more common in the orthogonal polynomials literature.

Note that Theorem 2.1 does not give any information on the behavior of the Jost function $\Omega(z)$ as $z \to \pm 1$. However relation (2.30) implies that

$$\int_{-1}^{1}\sqrt{1-\lambda^2}|\Omega(\lambda + i0)|^{-2}d\lambda = \frac{\pi}{2}\int_{-1}^{1} w(\lambda)d\lambda \le \pi/2, \tag{2.33}$$

and hence $\Omega(\lambda + i0)$ cannot vanish too rapidly as $\lambda \to 1 - 0$ and $\lambda \to -1 + 0$ (even if 1 or -1 are eigenvalues of H). For example, the behavior $\Omega(\lambda + i0) \sim c_\pm(\lambda \mp 1)$ with $c_\pm \ne 0$ is excluded.

3. Regular Solutions of the Jacobi Equations

3.1. *Regular solutions*

The regular solution $P(z) = \{P_n(z)\}$ of the Jacobi equation (1.2) is determined by the boundary conditions $P_{-1}(z) = 0$, $P_0(z) = 1$. Then $P_n(z)$ is a polynomial of degree n and relation (1.6) is satisfied. By analogy with the continuous case (see, e.g., part 1 of Sec. 4.1 in [30]), we here obtain bounds on $P_n(z)$ for large n. As usual, the variable $\zeta(z)$ is defined by (2.5).

Theorem 3.1. *If assumption* (1.7) *is satisfied, then*

$$|P_n(z)| \le C|\zeta(z)|^{-n} \tag{3.1}$$

with some positive constant C *not depending on* n *and on* z *in compact subsets of* $\mathbb{C}\setminus\{-1, 1\}$.

Theorem 3.1 will be proven in Sec. 9.2. The proof relies on the equation (cf. Eq. (2.11))

$$P_n(z) = P_n^{(0)}(z) + \frac{1}{\sqrt{z^2-1}} \sum_{m=0}^{n-1} (\zeta(z)^{n-m} - \zeta(z)^{m-n})(VP(z))_m, \quad n \geq 1, \tag{3.2}$$

where $z \in \Pi$, $P_n^{(0)}(z)$ (normalized Chebyshev polynomials of the second kind) are given by Eq. (2.8) and V is defined by formula (2.10).

Recall now formula (2.24) for the resolvent $R(z)$ of the operator H. Putting together estimates (2.13) and (3.1), we obtain the following result.

Theorem 3.2. *Let assumption* (1.7) *be satisfied. For all* $n, m \in \mathbb{Z}_+$, *the functions* $(R(z)e_n, e_m)$ *are continuous in* z *up to the cut along* $[-1, 1]$ *except, possibly, the points* ± 1. *Moreover,*

$$|(R(z)e_n, e_m)| \leq C|\omega(z)|^{-1}|\zeta(z)|^{|n-m|}$$

with some positive constant C *that does not depend on* n, m *and on* z *in compact subsets of* $\operatorname{clos}\Pi$ *and away from the points* ± 1.

We need also a representation of the Jost function $\Omega(z)$ in terms of the orthogonal polynomials $P_n(z)$. It plays the role of the representation (see, e.g., formula (1.38) in Chap. 4 of [30]) of the Jost function via the regular solution of the Schrödinger equation in the continuous case.

Proposition 3.3. *Let assumption* (1.7) *be satisfied, and let the Jost function* $\Omega(z)$ *be defined by* (2.14). *Then*

$$\Omega(z) = 1 - 2\sum_{n=0}^{\infty} \zeta(z)^{n+1}(VP(z))_n \tag{3.3}$$

for all $z \in \operatorname{clos}\Pi$ *except, possibly, the points* ± 1.

Proof. Suppose first that $z = \lambda + i0$ where $\lambda \in (-1, 1)$. Substituting expressions (2.9) and (2.20) into Eq. (3.2) we see that

$$\begin{aligned} &\omega(\lambda - i0)f_n(\lambda + i0) - \omega(\lambda + i0)f_n(\lambda - i0) \\ &\quad = 2^{-1}(e^{i(n+1)\theta} - e^{-i(n+1)\theta}) \\ &\qquad + \sum_{m=0}^{n-1}(e^{-i(n-m)\theta} - e^{i(n-m)\theta})(VP(\lambda + i0))_m. \end{aligned}$$

Let us consider the asymptotics of both sides of this equation as $n \to \infty$ and use relation (2.17). Then comparing coefficients at $e^{in\theta}$, we get (3.3) for $z = \lambda + i0$. Both sides of (3.3) are analytic in $z \in \Pi$ and continuous up to the cut along $[-1, 1]$ according to Theorems 2.1 and 3.1. Therefore relation (3.3) extends to all z. □

3.2. *Exponentially growing solutions*

Let us find asymptotics of the polynomials $P_n(z)$ for $z \notin [-1,1]$. We follow here closely the scheme exposed in Sec. 4.1 (see, in particular, Lemma 1.11) of [30].

We start by introducing solutions $g_n(z)$ of Eq. (2.2) exponentially growing as $n \to \infty$. Perhaps this construction is of interest in its own sake. For $z \in \Pi$, fix $n_0 = n_0(z)$ such that $f_n(z) \neq 0$ for $n \geq n_0 - 1$. Note that, for $\operatorname{Im} z \neq 0$, one can set $n_0 = 0$ because the equality $f_{n_0-1}(z) = 0$ implies that the Jacobi operator $H^{(n_0)}$ with the matrix elements $a_n^{(n_0)} = a_{n+n_0}$, $b_n^{(n_0)} = b_{n+n_0}$ has the eigenvalue z. Put

$$g_n(z) = f_n(z)\Theta_n(z) \quad \text{where} \quad \Theta_n(z) = \sum_{m=n_0}^{n} (a_{m-1}f_{m-1}(z)f_m(z))^{-1}. \tag{3.4}$$

An elementary calculation shows that this sequence satisfies Eq. (2.2). It is also easy to find the asymptotics of $g_n(z)$ as $n \to \infty$:

$$\begin{aligned}\zeta^n g_n(z) &= 2\zeta^{2n} \sum_{m=n_0}^{n} \zeta^{-2m+1}(1+o(1)) = 2\zeta^{2n+1}\frac{\zeta^{-2n-2}-1}{\zeta^{-2}-1}(1+o(1)) \\ &= 2\frac{1-\zeta^{2n+2}}{\zeta^{-1}-\zeta}(1+o(1)) = \frac{1+o(1)}{\sqrt{z^2-1}}\end{aligned}$$

where $\zeta = \zeta(z)$. This yields the following result.

Lemma 3.4. *Let $z \in \Pi$. Under assumption (1.7) the sequence $g_n(z)$ defined by (3.4) satisfies Eq. (2.2) and*

$$\lim_{n\to\infty} \zeta(z)^n g_n(z) = \frac{1}{\sqrt{z^2-1}}.$$

The Wronskian (2.3) of $f(z) = \{f_n(z)\}$, $g(z) = \{g_n(z)\}$ equals

$$\{f(z), g(z)\} = a_n f_n(z) f_{n+1}(z)(\Theta_{n+1}(z) - \Theta_n(z)) = 1,$$

and hence solutions $f(z)$ and $g(z)$ are linearly independent. It follows that

$$P_n(z) = d_+(z)f_n(z) + d_-(z)g_n(z) \tag{3.5}$$

where

$$\{P(z), f(z)\} = d_-(z)\{g(z), f(z)\} = -d_-(z)$$

and

$$\{P(z), g(z)\} = d_+(z)\{f(z), g(z)\} = d_+(z).$$

According to (2.14), (2.15) we have

$$d_-(z) = (2\zeta(z))^{-1}\Omega(z). \tag{3.6}$$

Obviously, $d_+(z) \neq 0$ if $d_-(z) = 0$. Therefore, Lemma 3.4 implies the following result.

Theorem 3.5. *Under assumption* (1.7) *for all* $z \in \Pi$, *we have the relation*

$$\lim_{n\to\infty} \zeta(z)^n P_n(z) = \frac{\Omega(z)}{1-\zeta(z)^2} \tag{3.7}$$

with convergence uniform on compact subsets of Π. *Moreover, if* $\Omega(z) = 0$, *then*

$$\lim_{n\to\infty} \zeta(z)^{-n} P_n(z) = \{P(z), g(z)\} \neq 0. \tag{3.8}$$

The existence of the limit in (3.7) is the classical result of the Szegö theory. It is stated as Theorem 12.1.2 in the book [26] where the assumptions are imposed on the measure $d\rho(\lambda)$; in particular, it is assumed that $\operatorname{supp}\rho \subset [-1,1]$. Under assumption (1.7) asymptotic relation (3.7) was established in Theorem 2.20 of [20]; our proof is rather different from that in [20]. Relation (3.8) is perhaps new.

4. Edge Points of the Spectrum

4.1. *The Jost function at edge points*

A study of the Jost function $\Omega(z)$ as $z \to 1$ and $z \to -1$ requires an additional assumption on the coefficients a_n, b_n. This is quite similar to the Schrödinger operator in the space $L^2(\mathbb{R}_+)$ that has a threshold at the point $z = 0$. The presentation here follows very closely Sec. 4.3 of [30] where the continuous case was considered; so some technical details will be omitted.

Theorem 4.1. *Suppose that*

$$\sum_{n=0}^{\infty} n(|a_n - 1/2| + |b_n|) < \infty. \tag{4.1}$$

Then all functions $f_n(z)$ *and, in particular, the Jost function* $\Omega(z)$ *are continuous in* z *up to the cut along* $[-1,1]$. *Moreover,* $f_n(\pm 1)$ *satisfies Eq.* (2.2) *for* $z = \pm 1$, *that is*

$$a_{n-1} f_{n-1}(\pm 1) + b_n f_n(\pm 1) + a_n f_{n+1}(\pm 1) = \pm f_n(\pm 1), \quad n \in \mathbb{Z}_+, \tag{4.2}$$

and $f_n(\pm 1) = (\pm 1)^n + o(1)$ *as* $n \to \infty$.

Theorem 3.1 can be supplemented by the following result.

Theorem 4.2. *If assumption* (4.1) *is satisfied, then*

$$|P_n(z)| \leq C(n+1)|\zeta(z)|^{-n} \tag{4.3}$$

with some positive constant C *not depending on* n *and on* z *in compact subsets of* $\mathbb{C}$.

The proofs of Theorems 4.1 and 4.2 will be discussed in Sec. 9.3. We emphasize that assumption (4.1) cannot be relaxed; see Remark 4.13 below.

Passing in (2.11) to the limit $z \to \pm 1$ and taking into account the relation

$$\zeta(z)^k = (\pm 1)^k (1 \mp k\sqrt{z^2-1}) + O(|z^2-1|), \quad z \to \pm 1, \tag{4.4}$$

we obtain an equation for the sequences $f_n(\pm 1)$:

$$f_n(\pm 1) = (\pm 1)^n + 2 \sum_{m=n+1}^{\infty} (\pm 1)^{n-m+1}(n-m)(Vf(\pm 1))_m.$$

Similarly, since $P_n^{(0)}(\pm 1) = (n+1)(\pm 1)^n$, Eq. (3.2) yields

$$P_n(\pm 1) = (n+1)(\pm 1)^n - 2\sum_{m=0}^{n-1} (\pm 1)^{n-m+1}(n-m)(VP(\pm 1))_m. \tag{4.5}$$

Estimate (4.3) allows us to pass to the limit $z \to \pm 1$ in (3.3) which leads to the following result.

Lemma 4.3. *Under assumption* (4.1), *the representation*

$$\Omega(\pm 1) = 1 - 2\sum_{n=0}^{\infty} (\pm 1)^{n+1}(VP(\pm 1))_n \tag{4.6}$$

holds.

Similarly to Lemma 3.4, we can introduce a solution $g_n(\pm 1)$ of Eq. (4.2) linearly independent with $f_n(\pm 1)$.

Lemma 4.4. *Let assumption* (4.1) *hold. Define the sequence* $g_n(\pm 1)$ *by formula* (3.4). *Then* $g_n(\pm 1)$ *satisfies Eq.* (4.2), $g_n(\pm 1) = 2(\pm 1)^{n+1} n(1+o(1))$ *as* $n \to \infty$ *and* $\{f(\pm 1), g(\pm 1)\} = 1$.

Since neither of solutions $f_n(\pm 1)$, $g_n(\pm 1)$ nor their linear combinations tend to zero as $n \to \infty$, we obtain

Theorem 4.5. *Under assumption* (4.1), *Eq.* (4.2) *do not have solutions tending to zero as* $n \to \infty$. *In particular, the operator* H *cannot have eigenvalues* 1 *and* -1.

To find the asymptotics of the polynomials $P_n(z)$ at critical points $z = \pm 1$, we use equalities (3.5) and (3.6) where $z = \pm 1$. The next result is a direct consequence of Lemma 4.4.

Theorem 4.6. *Let assumption* (4.1), *and let* $\Omega(\pm 1) \neq 0$. *Then*

$$P_n(\pm 1) = \Omega(\pm 1)(\pm 1)^n n(1+o(1)). \tag{4.7}$$

Of course, if $\Omega(\pm 1) \neq 0$, then according to (2.24) the resolvent kernel $(R(z)e_n, e_m)$ is continuous as $z \to \pm 1$.

4.2. *Resonances at edge points*

Our next goal is to find the asymptotic behavior as $z \to \pm 1$ of $\Omega(z)$ and hence of $(R(z)e_n, e_m)$ in the exceptional case $\Omega(\pm 1) = 0$. Let us use the same terminology as for Schrödinger operators.

Definition 4.7. Let assumption (4.1) hold. If $\Omega(\pm 1) = 0$, we say that the operator H has a resonance at $z = \pm 1$.

Clearly, the condition $\Omega(\pm 1) = 0$ is equivalent to the linear dependence of the solutions $P_n(\pm 1)$ and $f_n(\pm 1)$ of Eq. (4.2). In this case the sequence $P_n(\pm 1)$ is bounded as $n \to \infty$. The next result follows from Eqs. (4.5) and (4.6).

Lemma 4.8. *Suppose that* $\Omega(\pm 1) = 0$ *and put*

$$\gamma_\pm = 1 + 2 \sum_{m=0}^{\infty} (\pm 1)^{m-1} m (VP(\pm 1))_m. \tag{4.8}$$

Then there exists

$$\lim_{n \to \infty} (\pm 1)^n P_n(\pm 1) = \gamma_\pm \neq 0. \tag{4.9}$$

In particular, $P_n(\pm 1) = (\pm 1)^n \gamma_\pm f_n(\pm 1)$.

Now we are in a position to find the asymptotic behavior of the Jost function $\Omega(z)$ as $z \to \pm 1$ in the case $\Omega(\pm 1) = 0$. Let us proceed from representation (3.3). According to (4.4), for each n, we have

$$\zeta(z)^{n+1} (VP(z))_n = \big((\pm 1)^{n+1} - (\pm 1)^n (n+1)\sqrt{z^2-1}\big)(VP(\pm 1))_n + O(z^2 - 1).$$

Substituting this expression into (3.3) and taking into account Eq. (4.6), we see that

$$\Omega(z) = \Omega(\pm 1) + 2 \sum_{n=0}^{\infty} (\pm 1)^n (n+1)(VP(\pm 1))_n \sqrt{z^2-1} + o(\sqrt{|z^2-1|}).$$

If $\Omega(\pm 1) = 0$, then the coefficient at $\sqrt{z^2-1}$ here equals $\pm\gamma_\pm$. Let us state the result obtained.

Theorem 4.9. *Under assumption* (4.1) *suppose that* $\Omega(\pm 1) = 0$ *and define the number* $\gamma_\pm$ *by* (4.8) *or, equivalently, by* (4.9). *Then*

$$\Omega(z) = \pm \gamma_\pm \sqrt{z^2-1} + o(\sqrt{|z^2-1|})$$

as $z \to \pm 1$.

Using formula (2.30), we obtain the following consequence for the weight function.

Corollary 4.10. *Under the assumptions of Theorem* 4.9, *we have*

$$w(\lambda) = 2\pi^{-1} \gamma_\pm^{-2} (1 - \lambda^2)^{-1/2} (1 + o(1)), \quad \lambda \in (-1, 1),$$

as $\lambda \to \pm 1$.

In view of Lemmas 2.6 and 4.8, Theorem 4.9 implies also the following result.

Corollary 4.11. *Under the assumptions of Theorem* 4.9 *for all* $n, m \in \mathbb{Z}_+$, *the representation*

$$(R(z)e_n, e_m) = -2\frac{(\pm 1)^{\min\{n,m\}} f_n(\pm 1) f_m(\pm 1) + o(1)}{\sqrt{z^2 - 1}}$$

as $z \to \pm 1$ *is satisfied.*

Remark 4.12. Combined together, Theorems 4.1 and 4.9 ensure that under assumption (4.1) the discrete spectrum of the operator H is finite.

Remark 4.13. Under assumption (4.1) the asymptotics of the orthogonal polynomials $P_n(\pm 1)$ is given by formulas (4.7) and (4.9). An example of the Jacobi polynomials shows that the asymptotics of $P_n(\pm 1)$ is significantly changed (see formula (8.18), below) if the condition (4.1) is even slightly relaxed. In particular, estimate (4.3) is violated in this case.

Remark 4.14. The problem of characterization of spectral data corresponding to assumption (4.1) was considered in [17].

4.3. *Eigenvalues at edge points*

The condition (4.1) of Theorem 4.5 guaranteeing, in particular, the absence of eigenvalues ± 1 is optimal. Indeed, there exist Jacobi operators satisfying condition (1.7) but not (4.1) with an eigenvalue at the points $+1$ or -1. In the example below, we suppose that $a_n = 1/2$. Set

$$\psi_n^{(\pm)} = (\pm 1)^n (n+1)^{-l} \tag{4.10}$$

for all $n \in \mathbb{Z}_+$ and $\psi_{-1}^{(\pm)} = 0$. Then $\psi^{(\pm)} = \{\psi_n^{(\pm)}\} \subset \ell^2(\mathbb{Z}_+)$ if $l > 1/2$. Let us consider Eq. (2.2) where $u_n = \psi_n^{(\pm)}$ and $z = \pm 1$ as an equation for $b_n^{(\pm)}$:

$$b_n^{(\pm)} = \pm 1 - (2\psi_n^{(\pm)})^{-1}(\psi_{n-1}^{(\pm)} + \psi_{n+1}^{(\pm)}). \tag{4.11}$$

It follows from (4.10) that

$$b_n^{(\pm)} = \mp 2^{-1} l(l+3) n^{-2} + O(n^{-3})$$

as $n \to \infty$. In particular, condition (1.7) holds. Moreover, up to a finite rank operator, the perturbation $V^{(\pm)} = H^{(\pm)} - H_0$ of the operator H_0 is negative for the upper sign and positive for the lower sign. Thus, we have

Example 4.15. Let $a_n = 1/2$ and $b_n^{(\pm)}$ be given by formulas (4.10) where $l > 1/2$ and (4.11). Then condition (1.7) is satisfied. The Jacobi operator $H^{(\pm)}$ has the eigenvalue ± 1 with the eigenvector $\psi_n^{(\pm)}$ defined by (4.10). The spectrum of $H^{(+)}$ is finite above the point 1, and spectrum of $H^{(-)}$ is finite below the point -1.

Note that the operators $H^{(\pm)}$ have infinite number of eigenvalues. For instance, this fact follows from a general result of [6, Theorem 3], stating that the spectrum of a Jacobi operator is purely absolutely continuous on $[-1,1]$ if it has only a finite number of eigenvalues outside $[-1,1]$. In our case $H^{(\pm)}$ has the eigenvalue ± 1.

Example 4.15 exhibits an asymptotic behavior of orthogonal polynomials $P_n(z)$ intermediary between exponential (for $z \notin [-1,1]$) and oscillating (for $z \in (-1,1)$) regimes. Indeed, if $a_n = 1/2$ and $b_n^{(\pm)}$ are defined by (4.10) and (4.11), then $P_n(\pm 1) = \psi_n^{(\pm)}$ are given by (4.10). So, this sequence behaves as some power of n (negative or positive) as $n \to \infty$.

4.4. *Schrödinger operators with zero eigenvalues*

A similar result is true for the Schrödinger operator $\mathbf{H} = -d^2/dx^2 + b(x)$ with the boundary condition $u(0) = 0$ in the space $L^2(\mathbb{R}_+)$. Let $\psi \in C^\infty(\mathbb{R}_+)$, $\psi(x) > 0$ for all $x > 0$, $\psi(x) = x$ in a neighborhood of the point $x = 0$ and

$$\psi(x) = x^{-l} \quad \text{where} \quad l > 1/2 \tag{4.12}$$

for sufficiently large x. Then $\psi \in L^2(\mathbb{R}_+)$. Put

$$b(x) = \psi''(x)\psi(x)^{-1} \tag{4.13}$$

so that $b(x) = l(l+1)x^{-2}$ if x is large enough. This yields

Example 4.16. Let $b(x)$ be given by formulas (4.12) and (4.13). Then the operator $\mathbf{H}$ has the eigenvalue 0 with the eigenfunction $\psi(x)$ defined by (4.12), and its negative spectrum is finite.

The decay of $b(x)$ as $|x|^{-2}$ at infinity is critical. Actually, it is known (see part 1 in Sec. 4.3 of [30]) that $\mathbf{H}$ cannot have the zero eigenvalue provided

$$\int_0^\infty (1+x)|b(x)|dx < \infty.$$

5. The Perturbation Determinant and the Spectral Shift Function. Trace Identities

5.1. *Abstract background*

Let H_0 and H be arbitrary self-adjoint operators with a trace class difference $V = H - H_0$. Then the perturbation determinant

$$\Delta(z) = \operatorname{Det}\big(I + VR_0(z)\big) \tag{5.1}$$

for the pair H_0, H is well defined and is an analytic function of $z \in \mathbb{C}\backslash\sigma(H_0)$. Obviously, $\Delta(\bar{z}) = \overline{\Delta(z)}$ and

$$\Delta(z) \to 1 \quad \text{as} \quad \operatorname{dist}\{z, \sigma(H_0)\} \to \infty. \tag{5.2}$$

Note also the general formula

$$\operatorname{Tr}\big(R(z) - R_0(z)\big) = -\frac{\Delta'(z)}{\Delta(z)}. \tag{5.3}$$

In view of our applications to Jacobi operators, below we suppose that the operators H_0 and H are bounded.

The Kreĭn spectral shift function $\xi(\lambda)$ is defined in terms of the perturbation determinant (5.1). According to (5.2) we can fix the branch of the function $\ln \Delta(z)$ for $\operatorname{Im} z \neq 0$ by the condition

$$\arg \Delta(z) \to 0 \quad \text{as} \quad \operatorname{dist}\{z, \sigma(H_0)\} \to \infty. \tag{5.4}$$

Then

$$\xi(\lambda) := \pi^{-1} \lim_{\varepsilon \to +0} \arg \Delta(\lambda + i\varepsilon). \tag{5.5}$$

This limit exists for a.e. $\lambda \in \mathbb{R}$,

$$\int_{-\infty}^{\infty} |\xi(\lambda)| d\lambda \leq \|V\|_1 \tag{5.6}$$

and the representation

$$\ln \Delta(z) = \int_{-\infty}^{\infty} \xi(\lambda)(\lambda - z)^{-1} d\lambda, \quad \operatorname{Im} z \neq 0, \tag{5.7}$$

holds. The function $\xi(\lambda)$ is constant on intervals not containing points of $\sigma(H_0) \cup \sigma(H)$. In particular, $\xi(\lambda) = 0$ for $\lambda < \inf\big(\sigma(H_0) \cup \sigma(H)\big)$ and $\lambda > \sup\big(\sigma(H_0) \cup \sigma(H)\big)$. If λ_1 is an isolated eigenvalue of finite multiplicity k_0 of the operator H_0 and multiplicity k of the operator H, then

$$\xi(\lambda_1 + 0) - \xi(\lambda_1 - 0) = k_0 - k. \tag{5.8}$$

We refer to the books [16, 25, 29] for a detailed presentation of these notions.

5.2. *A trace formula*

Let us come back to Jacobi operators. Under the assumption (1.7), the perturbation V belongs to the trace class $\mathfrak{S}_1$ so that all results mentioned in the previous subsection are true.

Let us find a link of the perturbation determinant $\Delta(z)$ with the Jost function (2.15). We follow here closely the scheme of the paper [4] where the operator $-D^2 + b(x)$ was considered; for its detailed presentation, see Sec. 4.1 of [30]. Note that there exists also a different to proofs of such assertions relying on the Fredholm expansion of determinants; this approach was developed in [19].

Let us find an expression of the left-hand side of (5.3) in terms of the Jost function $\Omega(z)$. The first assertion is true without any assumptions on a_n and b_n. Recall that $\{f_n(z)\}_{n=-1}^{\infty}$ is the Jost solution of Eq. (2.2) and $\omega(z)$ is defined by (2.14).

Lemma 5.1. *For all $z \in \Pi$ and all $N \geq 0$, we have the identity*

$$\sum_{n=0}^{N} f_n(z)P_n(z) = -\omega'(z) + a_N\big(P_N(z)f'_{N+1}(z) - P_{N+1}(z)f'_N(z)\big). \tag{5.9}$$

Proof. Let us differentiate (in z) Eq. (2.2) for $f_n(z)$ and then multiply it by $P_n(z)$:

$$a_{n-1}f'_{n-1}(z)P_n(z) + a_n f'_{n+1}(z)P_n(z) = (z - b_n)f'_n(z)P_n(z) + f_n(z)P_n(z). \tag{5.10}$$

Multiplying Eq. (1.2) for $P_n(z)$ by $f'_n(z)$, we see that

$$a_{n-1}P_{n-1}(z)f'_n(z) + a_n P_{n+1}(z)f'_n(z) = (z - b_n)P_n(z)f'_n(z). \tag{5.11}$$

Then we subtract (5.11) from (5.10):

$$\begin{aligned} f_n(z)P_n(z) = {} & \big(a_{n-1}f'_{n-1}(z)P_n(z) - a_n f'_n(z)P_{n+1}(z)\big) \\ & + \big(a_n f'_{n+1}(z)P_n(z) - a_{n-1}f'_n(z)P_{n-1}(z)\big). \end{aligned} \tag{5.12}$$

Let us calculate the sums of the right-hand sides over $n = 0, 1, \ldots, N$:

$$\sum_{n=0}^{N} \big(a_{n-1}f'_{n-1}(z)P_n(z) - a_n f'_n(z)P_{n+1}(z)\big) = 2^{-1}f'_{-1}(z)P_0(z) - a_N f'_N(z)P_{N+1}(z)$$

and

$$\sum_{n=0}^{N} \big(a_n f'_{n+1}(z)P_n(z) - a_{n-1}f'_n(z)P_{n-1}(z)\big) = a_N f'_{N+1}(z)P_N(z) - 2^{-1}f'_0(z)P_{-1}(z).$$

Since $P_{-1}(z) = 0$, $P_0(z) = 1$ and $f_{-1}(z) = -2\omega(z)$, taking the sum of Eq. (5.12), we obtain the identity (5.9). □

Let us calculate the asymptotics of the right-hand side of (5.9) as $n \to \infty$ supposing first that the perturbation V has finite support, that is, $a_n = 1/2$ and $b_n = 0$ for sufficiently large n. In view of (2.6), Eq. (2.2) are satisfied in this case for all $z \in \Pi$ and n large enough if $u_n(z)$ is a arbitrary linear combination of the functions $\zeta(z)^n$ and $\zeta(z)^{-n}$. In particular, we have

$$f_n(z) = \zeta(z)^n \quad \text{and} \quad P_n(z) = \gamma_+(z)\zeta(z)^n + \gamma_-(z)\zeta(z)^{-n} \tag{5.13}$$

for some numbers $\gamma_\pm(z)$. Recall that $|\zeta(z)| < 1$ for $z \in \Pi$.

Lemma 5.2. *For all $z \in \Pi$, $z \notin \sigma_d(H)$ and $n \to \infty$, we have the relation*

$$\omega(z)^{-1}\big(P_n(z)f'_{n+1}(z) - P_{n+1}(z)f'_n(z)\big) = -\frac{2n}{\sqrt{z^2-1}} + \frac{\zeta(z)}{z^2-1} + O(\zeta(z)^{2n}). \tag{5.14}$$

Proof. It follows from formulas (5.13) that the Wronskian (2.14) calculated for large n is given by the equation

$$2\omega = (\gamma_+\zeta^n + \gamma_-\zeta^{-n})\zeta^{n+1} - (\gamma_+\zeta^{n+1} + \gamma_-\zeta^{-n-1})\zeta^n = \gamma_-(\zeta - \zeta^{-1}). \tag{5.15}$$

Formulas (5.13) also imply that

$$\begin{aligned} P_n f'_{n+1} - P_{n+1} f'_n &= (\gamma_+\zeta^n + \gamma_-\zeta^{-n})(\zeta^{n+1})' - (\gamma_+\zeta^{n+1} + \gamma_-\zeta^{-n-1})(\zeta^n)' \\ &= -\frac{1}{\sqrt{z^2-1}}\Big(\gamma_-\big((n+1)\zeta - n\zeta^{-1}\big) + \gamma_+\zeta^{2n+1}\Big) \end{aligned} \tag{5.16}$$

where we used that

$$\zeta(z)' = -\frac{\zeta(z)}{\sqrt{z^2-1}}.$$

Dividing now (5.16) by (5.15), we arrive at (5.14). ◻

Using expression (2.24) for $(R(z)e_n, e_n)$ and putting together Lemmas 5.1 and 5.2, we get the following result.

Theorem 5.3. *Suppose that $a_n = 1/2$ and $b_n = 0$ for sufficiently large n. Then for all $z \in \Pi$, $z \notin \sigma_d(H)$ and $N \to \infty$, we have the relation*

$$\sum_{n=0}^{N} (R(z)e_n, e_n) = -\omega'(z)\omega(z)^{-1} - \frac{N}{\sqrt{z^2-1}} + \frac{\zeta(z)}{2(z^2-1)} + O(\zeta(z)^{2N}). \tag{5.17}$$

Of course the same formula is true for the operator H_0 when $\omega_0(z) = -2^{-1}\zeta(z)^{-1}$. Comparing formulas (5.17) for H and H_0, we find that

$$\sum_{n=0}^{N} \big((R(z)e_n, e_n) - (R_0(z)e_n, e_n)\big) = -\omega'(z)\omega(z)^{-1} + \omega_0'(z)\omega_0(z)^{-1} + O(\zeta(z)^{2N}).$$

Passing here to the limit $N \to \infty$, we see that

$$\operatorname{Tr}\big(R(z) - R_0(z)\big) = -\frac{\Omega'(z)}{\Omega(z)}, \quad z \in \Pi, \tag{5.18}$$

where the function $\Omega(z)$ is defined by (2.15).

Theorem 2.3 allows us to extend this result to arbitrary operators satisfying condition (1.7). Indeed, set $a_n^{(N)} = a_n$, $b_n^{(N)} = b_n$ for $n \le N$ and $a_n^{(N)} = 1/2$, $b_n^{(N)} = 0$ for $n > N$. Let $H^{(N)}$ be the Jacobi operator with the matrix elements $a_n^{(N)}$, $b_n^{(N)}$, and let $\Omega^{(N)}(z) = -2\zeta(z)\{P^{(N)}(z), f^{(N)}(z)\}$. Write formula (5.18) for the pair $H_0, H^{(N)}$ and pass to the limit $N \to \infty$. Since $V^{(N)} := H^{(N)} - H_0 \to V$ in the trace norm, we see that

$$\operatorname{Tr}\big(R^{(N)}(z) - R_0(z)\big) \to \operatorname{Tr}\big(R(z) - R_0(z)\big)$$

as $N \to \infty$. According to Theorem 2.3 we also have $\Omega^{(N)}(z) \to \Omega(z)$ and hence $d\Omega^{(N)}(z)/dz \to d\Omega(z)/dz$ as $N \to \infty$. This leads to the desired result.

Theorem 5.4. *The relation* (5.18) *holds true under assumption* (1.7).

5.3. *The Wronskian and the perturbation determinant*

Putting together relations (5.3) and (5.18), we see that

$$\Delta(z) = A\,\Omega(z), \quad z \in \Pi, \tag{5.19}$$

for some constant $A \in \mathbb{C}$. Our goal is to check that

$$A = \prod_{k=0}^{\infty}(2a_k). \tag{5.20}$$

Note that under assumption (1.7) the infinite product here converges and $A > 0$.

Let us compare the asymptotics of $\Delta(z)$ and $\Omega(z)$ as $|z| \to \infty$. The first of them is given by (5.2). So we only have to consider the Jost function $\Omega(z)$.

Lemma 5.5. *Suppose that $a_n = 1/2$ and $b_n = 0$ for all sufficiently large n. Then for all $n = -1, 0, 1, \dots$ and $|z| \to \infty$, we have the asymptotic relation*

$$f_n(z) = \Bigl(\prod_{k=n}^{\infty}(2a_k)\Bigr)^{-1} \zeta(z)^n(1 + O(\zeta(z))). \tag{5.21}$$

In particular,

$$\Omega(z) = A^{-1} + O(\zeta(z)). \tag{5.22}$$

Proof. Since $f_n(z) = \zeta^n$ for n large enough, relations (5.21) are certainly satisfied for such n. Suppose that, for some n_0, relations (5.21) are true for all $n \geq n_0$. According to Eq. (2.2) we have

$$a_{n_0-1}f_{n_0-1} = \bigl(2^{-1}(\zeta + \zeta^{-1}) - b_{n_0}\bigr)f_{n_0} - a_{n_0}f_{n_0+1}.$$

By our assumption on f_{n_0} and f_{n_0+1}, this implies

$$a_{n_0-1}f_{n_0-1} = 2^{-1}\zeta^{-1}\Bigl(\prod_{k=n_0}^{\infty}(2a_k)\Bigr)^{-1}\zeta^{n_0}(1 + O(\zeta)) + O(\zeta^{n_0})$$

which yields (5.21) for f_{n_0-1}. In view of (2.14), (2.15), relation (5.22) is a consequence of (5.21) for $n = -1$. □

Thus relations (5.19) and (5.20) are verified for perturbations of finite support. It remains to extend them to arbitrary perturbations satisfying assumption (1.7). Let us use the arguments and the notation $H^{(N)} = H_0 + V^{(N)}$, $\Omega^{(N)}(z)$ already used by the proof of Theorem 5.4. Let $\Delta^{(N)}(z)$ be the perturbation determinant for the pair H_0, $H^{(N)}$. For all N, we have the relation

$$\Delta^{(N)}(z) = A^{(N)}\Omega^{(N)}(z) \quad \text{where} \quad A^{(N)} = \prod_{k=0}^{N-1}(2a_k).$$

Let us pass here to the limit $N \to \infty$. Since $V^{(N)} \to V$ in the trace norm, $\Delta^{(N)}(z) \to \Delta(z)$ as $N \to \infty$. We also see that $\Omega^{(N)}(z) \to \Omega(z)$ by Theorem 2.3 and $A^{(N)} \to A$ under assumption (1.7). This yields the desired result.

Theorem 5.6. *Under assumption* (1.7), *the equality* (5.19) *is true with constant* (5.20).

Since $\Delta(z)$ satisfies (5.2), this gives us the asymptotics of the Jost function at infinity.

Corollary 5.7. *Under assumption* (1.7), *we have*

$$\lim_{|z|\to\infty} \Omega(z) = A^{-1}.$$

Thus, we can fix the branch of the function $\arg \Omega(z)$ for $\operatorname{Im} z \neq 0$ by the condition

$$\lim_{|z|\to\infty} \arg \Omega(z) = 0.$$

Then in view of normalization (5.4), we have

$$\arg \Delta(z) = \arg \Omega(z), \quad \operatorname{Im} z \neq 0.$$

5.4. *The spectral shift function and the scattering phase*

Next, we discuss the spectral shift function $\xi(\lambda)$ for a pair of Jacobi operators H_0, H satisfying assumption (1.7). According to Theorem 2.1 and formula (5.19), the perturbation determinant $\Delta(z)$ is a continuous function of $z \in \operatorname{clos} \Pi$ except, possibly, the points $z = \pm 1$. Moreover, according to (2.21), we have $\Omega(\lambda \pm i0) \neq 0$. Therefore $\xi(\lambda)$ is also a continuous function of $\lambda \in (-1, 1)$.

A link between $\xi(\lambda)$ and the scattering phase $\eta(\theta)$ follows from definitions (2.22) and (5.5).

Theorem 5.8. *Under assumption* (1.7), *the relation*

$$\xi(\lambda) = \pi^{-1} \eta(\arccos \lambda) \tag{5.23}$$

holds for all $\lambda \in (-1, \lambda)$.

Substituting (5.23) into (2.23) and taking into account (2.32), we can reformulate Theorem 2.5 in terms of the weight function $w(\lambda)$ and the spectral shift function $\xi(\lambda)$. This yields asymptotic relation (1.9); note that the remainder $o(1)$ in the right-hand side can be replaced by $O(\rho_n)$.

We emphasize that $\eta(\theta)$ is a continuous function of $\theta \in (0, \pi)$, but Theorem 2.1 yields no information about its behavior as $\theta \to 0$ and $\theta \to \pi$. Comparing relations (5.6) and (5.23), we however see that

$$\int_0^\pi |\eta(\theta)| \sin\theta \, d\theta \leq \pi \|V\|_1.$$

5.5. *Trace identities*

To find expressions of $\operatorname{Tr}(H^n - H_0^n)$ in terms of the spectral shift function, we only have to compare asymptotic expansions as $|z| \to \infty$ of both sides of the representation (5.7). It follows from relation (5.3) that

$$\ln \Delta(z) = -\sum_{n=1}^{\infty} n^{-1} \operatorname{Tr}(H^n - H_0^n) z^{-n}.$$

Since

$$\int_{-\infty}^{\infty} \xi(\lambda)(\lambda - z)^{-1} d\lambda = -\sum_{n=1}^{\infty} \int_{-\infty}^{\infty} \xi(\lambda)\lambda^{n-1} d\lambda\, z^{-n},$$

equating the coefficients at z^{-n}, we see that

$$\operatorname{Tr}(H^n - H_0^n) = n \int_{-\infty}^{\infty} \xi(\lambda)\lambda^{n-1} d\lambda. \tag{5.24}$$

On the discrete spectrum, the spectral shift function can be explicitly calculated. Indeed, let $\lambda_1^{(+)} > \lambda_2^{(+)} > \cdots > 1$ (and $\lambda_1^{(-)} < \lambda_2^{(-)} < \cdots < -1$) be eigenvalues of H lying above the point 1 (respectively, below the point -1). It follows from formula (5.8) that $\xi(\lambda) = n$ for $\lambda \in (\lambda_{n+1}^{(+)}, \lambda_n^{(+)})$ and $\xi(\lambda) = -n$ for $\lambda \in (\lambda_n^{(-)}, \lambda_{n+1}^{(-)})$.

Using formula (5.8), we find that

$$n \int_{1}^{\infty} \xi(\lambda)\lambda^{n-1} d\lambda = \sum_{k=1}^{\infty} k\big((\lambda_k^{(+)})^n - (\lambda_{k+1}^{(+)})^n\big)$$

and, similarly, for the integral over $(-\infty, -1)$. The series here is convergent by virtue of estimate (5.6). Putting together this relation with (5.24), we obtain the following result.

Theorem 5.9. *Let assumption* (1.7) *be satisfied. Then*

$$\operatorname{Tr}(H^n - H_0^n) = n \int_{-1}^{1} \xi(\lambda)\lambda^{n-1} d\lambda + \sum_{k=1}^{\infty} k\big((\lambda_k^{(+)})^n - (\lambda_{k+1}^{(+)})^n\big)$$
$$+ \sum_{k=1}^{\infty} k\big((\lambda_k^{(-)})^n - (\lambda_{k+1}^{(-)})^n\big). \tag{5.25}$$

In view of (5.23), the integral on the right can be expressed in terms of the phase function:

$$\int_{-1}^{1} \xi(\lambda)\lambda^{n-1} d\lambda = \frac{1}{\pi} \int_{0}^{\pi} \eta(\theta) \cos^{n-1}\theta \sin\theta\, d\lambda.$$

In a very general framework, formulas of type (5.25) were studied in the book [27, Chap. 6].

5.6. *The Levinson formula*

The trace formula of zero order (the Levinson theorem) requires a special discussion. Now we assume a stronger condition (4.1) on the coefficients of the operator H. Then according to Theorem 4.1, the corresponding perturbation determinant $\Delta(z)$ is continuous as $z \to \pm 1$. One has to distinguish the cases $\Delta(1) = 0$ and $\Delta(-1) = 0$ when the operator H has threshold resonances at the points $\lambda = 1$ or $\lambda = -1$. Note also (see Remark 4.12) that under assumption (4.1) the operator H has only a finite number N of discrete eigenvalues.

Theorem 5.10. *Let assumption* (4.1) *be satisfied. Then the limits* $\xi(1-0)$ *and* $\xi(-1+0)$ *exist and*

$$\xi(1-0) - \xi(-1+0) = N + \varkappa \tag{5.26}$$

where $\varkappa = 0$ *if* $\Delta(\pm 1) \neq 0$ *for both signs,* $\varkappa = 1/2$ *if* $\Delta(\pm 1) = 0$ *for one of the signs and* $\varkappa = 1$ *if* $\Delta(\pm 1) = 0$ *for both signs.*

Proof. Let us consider a contour which consists of a part

$$\begin{aligned} \mathcal{C}_{r_-,r_+} &= [-1 + r_- + i0, 1 - r_+ + i0] \cup [-1 + r_- - i0, 1 - r_+ - i0] \\ &\quad \cup \{|z-1] = r_+\} \cup \{|z+1] = r_-\} \end{aligned}$$

encircling the cut along $[-1, 1]$ and of the circle $|z| = R$. We suppose that r_- and r_+ are sufficiently small, R is sufficiently large, and we go in the clockwise over $\mathcal{C}_{r_-,r_+}$ and in the counterclockwise direction over $|z| = R$. Since $\Delta(z)$ has simple zeros at the eigenvalues of H, the argument principle implies that

$$\operatorname{var}_{\mathcal{C}_{r_-,r_+}} \arg \Delta(z) + \operatorname{var}_{|z|=R} \arg \Delta(z) = 2\pi N. \tag{5.27}$$

According to (5.4) we have $\operatorname{var}_{|z|=R} \arg \Delta(z) \to 0$ as $R \to \infty$. It follows from (5.5) that

$$\begin{aligned} \operatorname{var}_{\mathcal{C}_{r_-,r_+}} \arg \Delta(z) &= 2\pi(\xi(1 - r_+) - \xi(1 + r_-)) \\ &\quad + \operatorname{var}_{|z-1|=r_+} \arg \Delta(z) + \operatorname{var}_{|z+1|=r_-} \arg \Delta(z). \end{aligned} \tag{5.28}$$

Let us now pass here to the limit $r_-, r_+ \to 0$. If $\Delta(\pm 1) \neq 0$, then $\operatorname{var}_{|z\mp 1|=r_\pm} \arg \Delta(z) \to 0$. In the case $\Delta(\pm 1) = 0$ it follows from Theorem 4.9 that

$$\lim_{r_\pm \to 0} \operatorname{var}_{|z\mp 1|=r_\pm} \arg \Delta(z) = \pi.$$

Thus relation (5.26) is a direct consequence of (5.27) and (5.28). These arguments prove also the existence of the limits $\xi(1-0)$ and $\xi(-1+0)$. □

6. Scattering Theory

6.1. *Wave and scattering operators*

First, we briefly recall basic notions of scattering theory. We refer, for example, to the book [29] for a more complete presentation of this material.

The wave operators $W_\pm(H, H_0)$ for a pair of self-adjoint operators H_0, H in a Hilbert space $\mathcal{H}$ are defined as strong limits

$$W_\pm(H, H_0) = \operatorname*{s-lim}_{t\to\pm\infty} e^{iHt}e^{-iH_0t}; \tag{6.1}$$

here H_0 is supposed to be absolutely continuous. Under the assumption of the existence of limits (6.1), $W_\pm(H, H_0)$ are isometric operators and enjoy the intertwining property $HW_\pm(H, H_0) = W_\pm(H, H_0)H_0$. The wave operator $W_\pm(H, H_0)$ is called complete if its range coincides with the absolutely continuous subspace of the operator H. The scattering operator

$$\mathbf{S} = W_+(H, H_0)^*W_-(H, H_0) \tag{6.2}$$

commutes with H_0, $\mathbf{S}H_0 = H_0\mathbf{S}$, and it is unitary if both wave operators $W_+(H, H_0)$ and $W_-(H, H_0)$ are complete.

To define the scattering matrix, we suppose for definiteness that the spectrum of H_0 is simple and coincides with $[-1, 1]$. Let F_0: $\mathcal{H} \to L^2(-1, 1)$ be unitary and $F_0H = \mathsf{A}F_0$ where A is the operator of multiplication by λ in the space $L^2(-1, 1)$. Then the scattering matrix $S(\lambda) \in \mathbb{C}$ is defined by the relation

$$(F_0\mathbf{S}f)(\lambda) = S(\lambda)(F_0f)(\lambda), \quad \lambda \in (-1, 1); \tag{6.3}$$

obviously, $|S(\lambda)| = 1$ if $\mathbf{S}$ is unitary. Note that the scattering matrix does not depend on the diagonalization F_0 of H_0 if it has simple spectrum.

Let us come back to Jacobi operators. Under condition (1.7) which is always assumed in this section, the perturbation $V = H - H_0$ is trace class. Therefore the wave operators $W_\pm(H, H_0)$ exist and are complete by the classical Kato–Rosenblum theorem.

Our goal here is to obtain representations of the wave operators and of the scattering matrix in terms of the polynomials $P_n(\lambda)$ and of the Jost function $\Omega(\lambda + i0)$ for $\lambda \in (-1, 1)$. Such expressions are known as stationary representations. As a by-product of our considerations, we will also give a direct proof of the existence and completeness of the wave operators.

6.2. *Diagonalization of the Jacobi operators*

Let $d\rho(\lambda)$ be the spectral measure of the operator H. We define a mapping U: $\ell^2(\mathbb{Z}_+) \to L^2(\mathbb{R}; d\rho)$ by the formula

$$(Ue_n)(\lambda) = P_n(\lambda).$$

This mapping is isometric according to (1.6). It is also unitary because the set of all polynomials $P_n(\lambda)$, $n \in \mathbb{Z}_+$, is dense in $L^2(\mathbb{R}; d\rho)$. Finally, the intertwining property

$$(UHf)(\lambda) = \lambda(Uf)(\lambda) \tag{6.4}$$

holds. Indeed, it suffices to check it for $f = e_n$ when according to definition (2.1) of the operator H, $(UHe_n)(\lambda)$ coincides with the left-hand side of (1.2) (where $z = \lambda$) while $\lambda(Ue_n)(\lambda)$ equals its right-hand side.

Next, we reduce the absolutely continuous part of the operator H to the operator A of multiplication by λ in $L^2(-1, 1)$. To that end, we put

$$\psi_n(\lambda) = \sqrt{w(\lambda)} P_n(\lambda), \quad \lambda \in (-1, 1), \tag{6.5}$$

where $w(\lambda)$ is the weight function defined by (2.30) and introduce a mapping F: $\ell^2(\mathbb{Z}_+) \to L^2(-1, 1)$ by the formula

$$(Fe_n)(\lambda) = \psi_n(\lambda), \quad \lambda \in (-1, 1). \tag{6.6}$$

The operator F^*: $L^2(-1, 1) \to \ell^2(\mathbb{Z}_+)$ adjoint to F is given by the equality

$$(F^* g)_n = \int_{-1}^{1} \psi_n(\lambda) g(\lambda) d\lambda, \quad n \in \mathbb{Z}_+. \tag{6.7}$$

According to (6.5) it follows from the unitarity of the operator U that

$$FF^* = I, \quad F^* F = E(-1, 1), \tag{6.8}$$

where $E(-1, 1)$ is the spectral projection of the operator H corresponding to the interval $(-1, 1)$. Thus the operator F^* is isometric, and it is unitary if H has no point spectrum. By virtue of Eq. (6.4), the intertwining property

$$HF^* = F^* \mathsf{A} \tag{6.9}$$

holds.

For the free operator H_0, the corresponding unitary mapping F_0: $\ell^2(\mathbb{Z}_+) \to L^2(-1, 1)$ is defined by the formula

$$(F_0 e_n)(\lambda) = \psi_n^{(0)}(\lambda) \quad \text{where} \quad \psi_n^{(0)}(\lambda) = \sqrt{w_0(\lambda)} P_n^{(0)}(\lambda), \quad \lambda \in (-1, 1). \tag{6.10}$$

Here $w_0(\lambda)$ is defined by (2.31) and $P_n^{(0)}(\lambda)$ are normalized Chebyshev polynomials of the second kind.

6.3. *Stationary representations*

The presentation below is very close to Sec. 4.2 of the book [30] where the continuous case was considered. We start with two elementary facts. Recall that $\zeta(\lambda \pm i0)$ and ρ_n are defined by formulas (2.7) and (2.12), respectively.

Lemma 6.1. *Let* $\int_{-1}^{1} |g(\lambda)|^2 \sqrt{1 - \lambda^2} d\lambda < \infty$. *Then*

$$\lim_{t \to \pm\infty} \sum_{n=0}^{\infty} \left| \int_{-1}^{1} \zeta(\lambda \mp i0)^n e^{-i\lambda t} g(\lambda) d\lambda \right|^2 = 0. \tag{6.11}$$

Proof. It is more convenient to work in the variable θ using that

$$\int_{-1}^{1} \zeta(\lambda \mp i0)^n e^{-i\lambda t} g(\lambda) d\lambda = \int_0^\pi e^{\pm in\theta} e^{-it\cos\theta} g(\cos\theta)\sin\theta\, d\theta. \tag{6.12}$$

The Parseval identity implies the uniform in t estimate

$$\sum_{n=0}^{\infty} \left| \int_{-1}^{1} \zeta(\lambda \mp i0)^n e^{-i\lambda t} g(\lambda) d\lambda \right|^2$$

$$\leq 2\pi \int_0^\pi |g(\cos\theta)|^2 \sin^2\theta\, d\theta = 2\pi \int_{-1}^{1} |g(\lambda)|^2 \sqrt{1-\lambda^2} d\lambda.$$

Therefore it suffices to check (6.11) for $g \in C_0^\infty((-1,1))$.

Integrating in (6.12) by parts and observing that

$$\big|\pm n + t\sin\theta\big| \geq c\big|n + |t|\big| \quad \text{for } \pm t > 0$$

where $\theta \in [\varepsilon_0, \pi - \varepsilon_0]$, $\varepsilon_0 > 0$ and $c = c(\varepsilon_0) > 0$, we find an estimate

$$\left| \int_0^\pi e^{\pm in\theta} e^{-it\cos\theta} g(\cos\theta)\sin\theta\, d\theta \right| \leq C(n + |t|)^{-1}, \quad n \in \mathbb{Z}_+, \quad \pm t > 0.$$

It follows that

$$\sum_{n=0}^{\infty} \left| \int_{-1}^{1} \zeta(\lambda \mp i0)^n e^{-i\lambda t} g(\lambda) d\lambda \right|^2 \leq C \sum_{n=0}^{\infty} (n + |t|)^{-2} \leq C_1 |t|^{-1}, \quad \pm t > 0,$$

which implies (6.11). □

Lemma 6.2. *Under assumption* (1.7), *suppose additionally that*

$$\sum_{n=0}^{\infty} \rho_n^2 < \infty. \tag{6.13}$$

Let $g \in L^1(-1,1)$ *and* $\operatorname{supp} g \subset (-1,1)$. *Then*

$$\lim_{|t|\to\infty} \sum_{n=0}^{\infty} \left| \int_{-1}^{1} (f_n(\lambda \pm i0) - \zeta(\lambda \pm i0)^n) e^{-i\lambda t} g(\lambda) d\lambda \right|^2 = 0. \tag{6.14}$$

Proof. Theorem 2.1 yields the estimate

$$\left| \int_{-1}^{1} (f_n(\lambda \pm i0) - \zeta(\lambda \pm i0)^n) e^{-i\lambda t} g(\lambda) d\lambda \right| \leq C\rho_n$$

where C does not depend on t. For every n, the integrals in (6.14) tend to zero as $|t| \to \infty$ by the Riemann–Lebesgue lemma. So, under assumption (6.13), the dominated convergence theorem implies (6.14). □

Let us now set

$$\sigma_\pm(\lambda) = \frac{\Omega(\lambda \pm i0)}{|\Omega(\lambda \pm i0)|} = e^{\pm i\eta(\arccos\lambda)}, \quad \lambda \in (-1,1). \tag{6.15}$$

The operator $\Sigma_\pm(\lambda)$ of multiplication by $\sigma_\pm(\lambda)$ is of course unitary in the space $L^2(-1,1)$.

Using definitions (2.30) and (6.5) of the functions $w(\lambda)$ and $\psi_n(\lambda)$, we rewrite relation (2.20) as

$$\psi_n(\lambda) = \frac{1}{i\sqrt{2\pi}\sqrt[4]{1-\lambda^2}}\Big(-\zeta(\lambda+i0)\sigma_-(\lambda)f_n(\lambda+i0) + \zeta(\lambda-i0)\sigma_+(\lambda)f_n(\lambda-i0)\Big). \tag{6.16}$$

For the operator H_0 and the function $\psi_n^{(0)}(\lambda)$ defined in (6.10), this reduces to the formula

$$\psi_n^{(0)}(\lambda) = \frac{1}{i\sqrt{2\pi}\sqrt[4]{1-\lambda^2}}\Big(-\zeta(\lambda+i0)^{n+1} + \zeta(\lambda-i0)^{n+1}\Big). \tag{6.17}$$

Recall that the operators F and F_0 are defined by relations (6.6) and (6.10). Then F^* is given by relation (6.7) and similarly for F_0^*. The operator A acts as multiplication by λ in the space $L^2(-1,1)$.

Lemma 6.3. *For all $g \in L^2(-1,1)$, we have*

$$\lim_{t\to\pm\infty} \|(F^*\Sigma_\pm - F_0^*)e^{-i\mathsf{A}t}g\| = 0. \tag{6.18}$$

Proof. Since $\sigma_+(\lambda)\sigma_-(\lambda) = 1$, it follows from relations (6.16) and (6.17) that, for both signs "$\pm$",

$$\begin{aligned} &i\sqrt{2\pi}\sqrt[4]{1-\lambda^2}\big(\psi_n(\lambda)\sigma_\pm(\lambda) - \psi_n^{(0)}(\lambda)\big) \\ &\qquad = \pm(\sigma_\pm(\lambda)^2 - 1)\zeta(\lambda \mp i0)^{n+1} + \sigma_\pm(\lambda)r_n(\lambda) \end{aligned} \tag{6.19}$$

where $r_n(\lambda) = -r_n^{(+)}(\lambda) + r_n^{(-)}(\lambda)$ and

$$r_n^{(\pm)}(\lambda) = \zeta(\lambda \pm i0)\sigma_\pm(\lambda)\big(f_n(\lambda \pm i0) - \zeta(\lambda \pm i0)^n\big).$$

According to (6.19) we have

$$\begin{aligned} (F^*\Sigma_\pm - F_0^*)e^{-i\mathsf{A}t}g)_n &= \pm \int_{-1}^1 (\sigma_\pm(\lambda)^2 - 1)\zeta(\lambda \mp i0)^{n+1}e^{-i\lambda t}\tilde{g}(\lambda)d\lambda \\ &\quad + \int_{-1}^1 r_n(\lambda)\sigma_\pm(\lambda)e^{-i\lambda t}\tilde{g}(\lambda)d\lambda \end{aligned} \tag{6.20}$$

where

$$\tilde{g}(\lambda) = \frac{g(\lambda)}{i\sqrt{2\pi}\sqrt[4]{1-\lambda^2}}.$$

Let us come back to relation (6.18) where we may assume that $\operatorname{supp} g \subset (-1,1)$. If $t \to +\infty$ (if $t \to -\infty$), we use (6.20) for the upper (lower) sign. Then the contribution to the norm in $\ell^2(\mathbb{Z}_+)$ of the first term in the right-hand side of (6.20) tends to zero according to Lemma 6.1. The contribution of the second term tends to zero according to Lemma 6.2. □

Theorem 6.4. *Let assumptions* (1.7) *and* (6.13) *be satisfied. Then the strong limits* (6.1) *exist and*

$$W_\pm(H, H_0) = F^* \Sigma_\pm F_0. \tag{6.21}$$

Proof. We have to check that

$$\lim_{t\to\pm\infty} \| e^{iHt} e^{-iH_0 t} f - F^* \Sigma_\pm F_0 f \| = 0$$

for all $f \in \ell^2(\mathbb{Z}_+)$. In view of the intertwining property (6.9) this relation can be rewritten as

$$\lim_{t\to\pm\infty} \| e^{-iH_0 t} f - F^* \Sigma_\pm e^{-i\mathsf{A}t} F_0 f \| = 0. \tag{6.22}$$

Set now $g = F_0 f$. Then again in view of the intertwining property (6.9) for H_0, we see that relations (6.18) and (6.22) are equivalent. □

It follows from relations (6.8) and (6.21) that the scattering operator (6.2) is given by the equality

$$\mathbf{S} = F_0^* \Sigma_+^* \Sigma_- F_0.$$

Putting together this relation with the definition (6.3) of the scattering matrix and definition (6.15) of $\sigma_\pm(\lambda)$, we can state the following result. Recall also that the scattering phase $\eta(\theta)$ was defined by formula (2.22).

Theorem 6.5. *Under assumptions of Theorem* 6.4, *the scattering matrix for the pair* H_0, H *satisfies the equality*

$$S(\lambda) = \frac{\Omega(\lambda - i0)}{\Omega(\lambda + i0)} = e^{-2i\eta(\arccos \lambda)}, \quad \lambda \in (-1, 1). \tag{6.23}$$

In view of relation (5.23), formula (6.23) can be rewritten as the Birman–Kreĭn formula

$$S(\lambda) = e^{-2i\pi\xi(\lambda)}.$$

6.4. *Weakening the assumptions*

Theorems 6.4 and 6.5 remain true without additional assumption (6.13). The proof of this fact requires some tools of abstract scattering theory. We give only basic ideas of one of its possible proofs.

The following result is a direct consequence of Theorem 3.2.

Lemma 6.6. *Let an operator Q in the space $\ell^2(\mathbb{Z}_+)$ be defined by the equality $(Qf)_n = q_n f_n$ where $q_n = \bar{q}_n$ and $\sum_{n=0}^{\infty} q_n^2 < \infty$. Under assumption* (1.7) *the operator-valued function $QR(z)Q$ defined for* $\operatorname{Im} z \neq 0$ *is continuous in the Hilbert–Schmidt norm up to the cut along $[-1,1]$ with possible exception of the points $z = \pm 1$. In particular, the operator Q is H-smooth in the sense of Kato on every compact subinterval of $(-1,1)$.*

Under assumption (1.7) Lemma 6.6 is true with

$$q_n = (|a_n - 1/2| + |b_n|)^{1/2}.$$

Note that the perturbation $V = QGQ$ where G is a bounded operator in the space $\ell^2(\mathbb{Z}_+)$. Using Theorems 5.3.4 and 5.6.1 in [29], we can now deduce from Lemma 6.6 the following assertion.

Theorem 6.7. *The wave operators $W_\pm(H, H_0)$ exist, are complete and representations* (6.21), (6.23) *are true under the only assumption* (1.7).

7. The Szegö Function and the Case Sum Rules

Here we establish a link between the perturbation determinant $\Delta(z)$ and the Szegö function $D(\zeta)$. We also show that the Case sum rules are direct consequences of this link. Since the Jost function $\Omega(z)$ is related to $\Delta(z)$ by simple formula (5.19), we do not discuss it here. As usual, $|\zeta| < 1$ and $2z = \zeta + \zeta^{-1}$.

7.1. *The Szegö function and the perturbation determinant*

We define the Szegö function $D(\zeta)$ by the formula

$$D(\zeta) = \exp\left(\frac{1}{4\pi}\int_{-\pi}^{\pi} \frac{e^{i\theta} + \zeta}{e^{i\theta} - \zeta} \ln\bigl(w(\cos\theta)|\sin\theta|\bigr)d\theta\right), \quad |\zeta| < 1, \tag{7.1}$$

where $w(\lambda)$ is the weight function (2.29). This is exactly formula (10.2.10) (see also Theorem 12.1.2) in [26], but in contrast to [26], we do not suppose that H has no eigenvalues. Of course definition (7.1) requires that

$$\int_{-\pi}^{\pi} \left|\ln\bigl(w(\cos\theta)|\sin\theta|\bigr)\right| d\theta < \infty$$

or, equivalently, (1.8) be satisfied. This is known as the Szegö condition.

Recall the standard Jensen–Poisson representation of functions $f(\zeta)$ analytic in the unit disc $|\zeta| < 1$. If $f(\zeta)$ is continuous in the closed disc $|\zeta| \leq 1$ and $\operatorname{Im} f(0) = 0$, then

$$f(\zeta) = \frac{1}{2\pi}\int_{-\pi}^{\pi} \frac{e^{i\theta} + \zeta}{e^{i\theta} - \zeta} \operatorname{Re} f(e^{i\theta})d\theta. \tag{7.2}$$

In particular, we see that

$$\begin{aligned}\ln(1+\zeta) &= \frac{1}{2\pi}\int_{-\pi}^{\pi}\frac{e^{i\theta}+\zeta}{e^{i\theta}-\zeta}\ln\cos(\theta/2)d\theta,\\ \ln(1-\zeta) &= \frac{1}{2\pi}\int_{-\pi}^{\pi}\frac{e^{i\theta}+\zeta}{e^{i\theta}-\zeta}\ln|\sin(\theta/2)|d\theta.\end{aligned}\tag{7.3}$$

We here fix $\arg(1+\zeta)$ and $\arg(1-\zeta)$ by the condition $\arg 1 = 0$.

Set $\widetilde{\Delta}(\zeta) = \Delta(z)$. It follows from Corollary 2.2 that the function $\widetilde{\Delta}(\zeta)$ is analytic in the unit disc and is continuous up to the unit circle with a possible exception of the points ± 1. Moreover, according to (2.21), $\widetilde{\Delta}(\zeta) \neq 0$ if $|\zeta| = 1$ but $\zeta \neq \pm 1$. Under assumption (1.7), an information on the behavior of $\widetilde{\Delta}(\zeta)$ as $\zeta \to \pm 1$ follows from Theorem 9.14 in [20] where it is shown that the function $\widetilde{\Delta}(\zeta)$ belongs to the Nevanlinna class N and does not have a singular inner component. We refer to the book [8], Chap. 2, for precise definitions of these notions. We also note that $\widetilde{\Delta}(0) = 1$ according to (5.2).

Let λ_k be eigenvalues (lying on $(-\infty,-1)\cup(1,\infty)$) of the operator H. In contrast to Sec. 5.5, we do not distinguish positive and negative eigenvalues in notation and suppose that $|\lambda_1| \geq |\lambda_2| \geq \cdots > 1$. The numbers $\mu_k \in (-1,1)$ defined by $\mu_k + \mu_k^{-1} = 2\lambda_k$ are zeros of the function $\widetilde{\Delta}(\zeta)$. Since $\widetilde{\Delta} \in N$, we have

$$\sum_{k=1}^{\infty}(1-|\mu_k|) < \infty, \tag{7.4}$$

so that the Blaschke product

$$B(\zeta) = \prod_{k=1}^{\infty}\frac{\mu_k}{|\mu_k|}\frac{\mu_k-\zeta}{1-\mu_k\zeta}, \quad |\zeta| < 1, \tag{7.5}$$

is well defined. The function $B(\zeta)$ is continuous in the closed disc $|\zeta| \leq 1$ except, possibly, the points ± 1 and $|B(\zeta)| = 1$ for $|\zeta| = 1$. Note that the proof of the inclusion $\widetilde{\Delta} \in N$ in [20] relied on the condition (7.4) established under assumption (1.7) earlier in [18].

Let us define the outer function

$$G(\zeta) = \exp\left(\frac{1}{2\pi}\int_{-\pi}^{\pi}\frac{e^{i\theta}+\zeta}{e^{i\theta}-\zeta}\ln|\widetilde{\Delta}(e^{i\theta})|d\theta\right) \tag{7.6}$$

where the function $\ln|\widetilde{\Delta}(e^{i\theta})|$ belongs to $L^1(-\pi,\pi)$ because $\widetilde{\Delta} \in N$. This is equivalent to the Szegö condition (1.8) since, by relations (2.30) and (5.19),

$$|\widetilde{\Delta}(e^{i\theta})|^2 = |\Delta(\cos\theta)|^2 = A^2\frac{2}{\pi}\frac{|\sin\theta|}{w(\cos\theta)},$$

whence

$$2\ln|\widetilde{\Delta}(e^{i\theta})| = -\ln(w(\cos\theta)|\sin\theta|) + \ln(2\pi^{-1}A^2\sin^2\theta). \tag{7.7}$$

It follows from Theorem 2.9 in [8] that the factorization

$$\widetilde{\Delta}(\zeta) = B(\zeta)G(\zeta) \tag{7.8}$$

holds. Of course (7.6) is the Jensen–Poisson representation (7.2) of the function $f(\zeta) = \ln\big(\widetilde{\Delta}(\zeta)/B(\zeta)\big)$, but, since this function is not continuous in the disc $|\zeta| \leq 1$, its justification relies on the properties of $\widetilde{\Delta}$ stated above.

Let us now compare definitions (7.1) and (7.6). We substitute expression (7.7) into (7.1) and take into account formulas (7.3). Thus, using factorization (7.8), we arrive at the following result.

Theorem 7.1. *Let assumption* (1.7) *be satisfied, and let* $|\zeta| < 1$. *Set* $\widetilde{\Delta}(\zeta) = \Delta(2^{-1}(\zeta + \zeta^{-1}))$ *where* $\Delta(z)$ *is the perturbation determinant* (5.1). *Define the Blaschke product* $B(\zeta)$ *by formula* (7.5), *the Szegö function* $D(\zeta)$ — *by* (7.1), *and the product* A — *by* (5.20). *Then the identity*

$$\widetilde{\Delta}(\zeta) = AB(\zeta)\frac{1-\zeta^2}{\sqrt{2\pi}D(\zeta)} \tag{7.9}$$

is true.

We emphasize that Theorem 7.1 is a direct consequence of classical results on the factorization of functions in the Nevanlinna class combined with the analytical results of [20].

According to (7.9), formulas (5.7) and (7.1) provide two different representations of an essentially the same object named the perturbation determinant $\Delta(z)$ or the Szegö function $D(\zeta)$. In view of (5.5), the first of them is given in terms of $\arg\Delta(\lambda+i0)$ while according to (2.30) and (5.19) the second representation is stated in terms of $\ln|\Delta(\lambda + i0)|$. Obviously, these two functions are harmonic conjugate.

It is of course possible to rewrite representation (7.1) in terms of the variable $z = 2^{-1}(\zeta + \zeta^{-1}) \in \Pi$. Let us also introduce the weight function $w_0(\lambda)$ of the operator H_0; it is given by formula (2.31). Taking into account that the function $w(\cos\theta)|\sin\theta|$ is even, making the change of variables $\lambda = \cos\theta$ and using formulas (7.3), we see that

$$D(\zeta) = \frac{1-\zeta^2}{\sqrt{2\pi}}\exp\left(\frac{1-\zeta^2}{2\pi}\int_{-1}^{1}\frac{\ln\big(w(\lambda)/w_0(\lambda)\big)}{1-2\zeta\lambda+\zeta^2}\frac{d\lambda}{\sqrt{1-\lambda^2}}\right) \tag{7.10}$$

or, equivalently,

$$D(\zeta(z)) = \zeta(z)\sqrt{\frac{2(z^2-1)}{\pi}}\exp\left(-\frac{\sqrt{z^2-1}}{2\pi}\int_{-1}^{1}\frac{\ln\big(w(\lambda)/w_0(\lambda)\big)}{\lambda - z}\frac{d\lambda}{\sqrt{1-\lambda^2}}\right).$$

7.2. *The Case sum rules*

Let us, finally, obtain the Case sum rules for the pair H_0, H. Putting together (7.9) and (7.10), we see that

$$\ln \widetilde{\Delta}(\zeta) - \ln B(\zeta) - \ln A = \frac{\zeta^2 - 1}{2\pi} \int_{-1}^{1} \frac{\ln\bigl(w(\lambda)/w_0(\lambda)\bigr)}{1 - 2\zeta\lambda + \zeta^2} \frac{d\lambda}{\sqrt{1-\lambda^2}}. \tag{7.11}$$

Let us compare the behavior of both sides of this equation as $\zeta \to 0$. Since $\widetilde{\Delta}(0) = 1$, setting $\zeta = 0$ and using definition (7.5), we obtain the identity

$$\sum_{k=1}^{\infty} \ln(2a_k) + \sum_{k=1}^{\infty} \ln|\mu_k| = \frac{1}{2\pi} \int_{-1}^{1} \ln\bigl(w(\lambda)/w_0(\lambda)\bigr) \frac{d\lambda}{\sqrt{1-\lambda^2}}. \tag{7.12}$$

This Case sum rule of zero order is of course different from the Levinson theorem (5.26).

More generally, let us consider the asymptotic expansions of both sides of (7.11) as $\zeta \to 0$ and compare the coefficients at the same powers of ζ. According to Theorem 2.13 in [20] we have

$$\ln \widetilde{\Delta}(\zeta) = -2 \sum_{n=1}^{\infty} n^{-1} \operatorname{Tr}\bigl(T_n(H) - T_n(H_0)\bigr)\zeta^n$$

where $T_n(\lambda) = \cos(n \arccos \lambda)$ are the Chebyshev polynomials of the first kind. It directly follows from definition (7.5) that

$$\ln B(\zeta) = \sum_{k=1}^{\infty} \ln|\mu_k| + \sum_{n=1}^{\infty} n^{-1} \left(\sum_{k=1}^{\infty} (\mu_k^n - \mu_k^{-n}) \right) \zeta^n$$

where the series over k are convergent due to the condition (7.4). Finally, we use formula (10.11.29) in [9]:

$$\frac{1-\zeta^2}{1 - 2\zeta\lambda + \zeta^2} = 1 + 2 \sum_{n=1}^{\infty} T_n(\lambda)\zeta^n.$$

Thus the equality of the coefficients at ζ^n in the left- and right-hand sides of (7.11) yields the identity

$$\operatorname{Tr}\bigl(T_n(H) - T_n(H_0)\bigr) = -\frac{1}{2} \sum_{k=1}^{\infty} (\mu_k^n - \mu_k^{-n}) + \frac{n}{2\pi} \int_{-1}^{1} \ln\bigl(w(\lambda)/w_0(\lambda)\bigr) T_n(\lambda) \frac{d\lambda}{\sqrt{1-\lambda^2}}, \quad n = 1, 2, \ldots. \tag{7.13}$$

The trace identities (7.12) and (7.13) known as the Case sum rules are not new. They were obtained by him in [5] and rigorously proven in [20]. We note that, in the paper [20], the identities (7.12) and (7.13) were first checked for finite rank perturbations $H - H_0$ and then (7.12) and (7.13) (for $n = 1$) were used for the proof of the inclusion $\widetilde{\Delta} \in N$.

7.3. *The Szegö condition and the Pollaczek polynomials*

The Szegö condition (1.8) implies that the weight function $w(\lambda)$ does not vanish too rapidly as $\lambda \to -1$ and $\lambda \to 1$ or, to put it differently, $|\Omega(\lambda + i0)|$ does not tend to infinity too rapidly. The example of the Pollaczek polynomials shows that condition (1.8) may be violated without assumption (1.7).

Recall that the normalized Pollaczek polynomials are defined (see, e.g., Appendix to [26]) by recurrent relations (1.2) with

$$a_n = \frac{n+1}{\sqrt{(2n+2\alpha+1)(2n+2\alpha+3)}}, \quad b_n = -\frac{2\beta}{2n+2\alpha+1}; \tag{7.14}$$

here the parameters $\alpha, \beta \in \mathbb{R}$ and $\alpha > |\beta|$. It follows that

$$a_n = 2^{-1} - \alpha(2n)^{-1} + O(n^{-2}), \quad b_n = -\beta n^{-1} + O(n^{-2}). \tag{7.15}$$

The corresponding normalized weight function is given by the formula

$$w(\lambda) = (\alpha + 1/2)e^{(2\theta-\pi)h(\theta)}\big(\cosh(\pi h(\theta))\big)^{-1} \quad \text{where} \quad h(\theta) = (\alpha\cos\theta + \beta)(\sin\theta)^{-1}$$

and as usual $\lambda = \cos\theta$. It is easy to see that

$$\ln w(\lambda) = -\pi(\alpha+\beta)\theta^{-1} + O(1) \tag{7.16}$$

as $\lambda \to 1 - 0$ and a similar formula is true as $\lambda \to -1 + 0$.

It follows from (7.15) that $V = H - H_0$ is Hilbert–Schmidt, but the series $\sum_n (a_n - 1/2)$ and $\sum_n b_n$ are divergent; in particular, assumption (1.7) is not satisfied. According to (7.16) the Szegö condition (1.8) is violated for Pollaczek polynomials. This is consistent with the classical theorem of Szegö, Shohat, Geronimus, Kreĭn and Kolmogorov; see, e.g., Theorem 4 in [20]. On the other hand, relation (7.16) implies that

$$\int_{-1}^{1} \ln w(\lambda)(1-\lambda^2)^{1/2} d\lambda > -\infty.$$

This is consistent with Theorem 1 in [20].

7.4. *The continuous case*

For the continuous operator $\mathbf{H} = -D^2 + b(x)$, an analogue of the Szegö condition means that the corresponding weight function $\mathbf{w}(\lambda)$ (for its definition, see, e.g., subsection 6 of §4.1 in [30]) does not vanish too rapidly as $\lambda \to +0$, that is,

$$\int_0^1 \ln \mathbf{w}(\lambda)\lambda^{-\alpha} d\lambda > -\infty \tag{7.17}$$

where $\alpha = 1/2$ if $b \in L^1(\mathbb{R}_+)$. Note however that we were unable to find this assertion in the literature.

Condition (7.17) appears to be sufficiently sharp. Indeed, suppose that b is a non-negative smooth function such that

$$b(x) = b_0 x^{-\rho}, \quad b_0 > 0, \quad \rho \in (0,2),$$

for large x. It is shown in [28] that for such potentials

$$\ln \mathbf{w}(\lambda) \sim -w_0 \lambda^{-(2-\rho)/(2\rho)}, \quad w_0 = w_0(b_0, \rho) > 0,$$

as $\lambda \to +0$. In this case condition (7.17) is satisfied if and only if $\alpha < 3/2 - 1/\rho$. Observe that formula (7.16) for Pollaczek polynomials corresponds to the exponential decay as $\lambda \to 0$ of the weight function $\mathbf{w}(\lambda)$ for Coulomb repulsive potentials.

The trace class condition for the operator $\mathbf{H}$ requires that $\rho > 1$ which yields (7.17) for any $\alpha < 1/2$. Similarly, the Hilbert–Schmidt condition for the operator $\mathbf{H}$ requires that $\rho > 1/2$ which yields (7.17) for any $\alpha < -1/2$. This is very close to the so-called quasi-Szegö condition established in [20] for the Jacobi operator $H = H_0 + V$ with a Hilbert–Schmidt perturbation V.

8. Example. The Jacobi Polynomials

8.1. *Definitions*

In this section, we suppose that the measure $d\rho(\lambda)$ is absolutely continuous and supported on the interval $[-1,1]$ so that

$$d\rho(\lambda) = w(\lambda) d\lambda, \quad \lambda \in (-1,1), \tag{8.1}$$

and

$$w(\lambda) = \kappa (1-\lambda)^{\alpha} (1+\lambda)^{\beta}, \quad \alpha, \beta > -1. \tag{8.2}$$

The weight function $w(\lambda) = w_{\alpha,\beta}(\lambda)$ (as well as all other objects discussed below) depends on α and β, but these parameters are often omitted in notation. The constant

$$\kappa = \kappa_{\alpha,\beta} = \frac{\Gamma(\alpha+\beta+2)}{2^{\alpha+\beta+1}\Gamma(\alpha+1)\Gamma(\beta+1)} \tag{8.3}$$

is chosen in such a way that the measure (8.1) is normalized, i.e., $\rho(\mathbb{R}) = \rho((-1,1)) = 1$. The orthonormal polynomials $P_n(\lambda) = P_n^{(\alpha,\beta)}(\lambda)$ constructed by the measure (8.1), (8.2) are known as the Jacobi polynomials.

Usually the Jacobi polynomials $\mathsf{P}_n(\lambda) = \mathsf{P}_n^{(\alpha,\beta)}(\lambda)$ are normalized (see, e.g., the books [9, 26]) by the condition

$$\mathsf{P}_n(1) = \frac{\Gamma(\alpha+1+n)}{\Gamma(\alpha+1)\Gamma(n+1)}. \tag{8.4}$$

According to formulas (10.8.4) and (10.8.5) in the book [9] we have

$$\int_{-1}^{1} |\mathsf{P}_n(\lambda)|^2 (1-\lambda)^\alpha (1+\lambda)^\beta d\lambda$$

$$= \frac{2^{\alpha+\beta+1}\Gamma(\alpha+1+n)\Gamma(\beta+1+n)}{n!(2n+\alpha+\beta+1)\Gamma(\alpha+\beta+n+1)} =: \mathsf{h}_n \tag{8.5}$$

and

$$\mathsf{P}_n(z) = \mathsf{k}_n\left(z^n + \frac{n(\alpha-\beta)}{2n+\alpha+\beta} z^{n-1} + \cdots\right)$$

where

$$\mathsf{k}_n = \frac{\Gamma(\alpha+\beta+2n+1)}{2^n n! \Gamma(\alpha+\beta+n+1)}. \tag{8.6}$$

For $\alpha+\beta \le -1$ and $n=0$, we have to set

$$\mathsf{h}_0 = 2^{\alpha+\beta+1}\frac{\Gamma(\alpha+1)\Gamma(\beta+1)}{\Gamma(\alpha+\beta+2)} = \kappa^{-1} \quad \text{and} \quad \mathsf{k}_0 = 1. \tag{8.7}$$

It follows from equalities (8.2) and (8.5) that

$$P_n(z) = (\kappa h_n)^{-1/2}\mathsf{P}_n(z).$$

Let H be the Jacobi operator corresponding to the measure (8.1), (8.2). Its coefficients are given by formulas (1.4) which yields

$$a_n^2 = \frac{\mathsf{k}_n^2}{\mathsf{k}_{n+1}^2}\frac{\mathsf{h}_{n+1}}{\mathsf{h}_n} = \frac{4(n+1)(n+\alpha+1)(n+\beta+1)}{(2n+\alpha+\beta+3)(2n+\alpha+\beta+2)^2}\frac{n+\alpha+\beta+1}{2n+\alpha+\beta+1} \tag{8.8}$$

(if $n=0$ and $\alpha+\beta=-1$ the last factor here should be replaced by 1) and

$$b_n = (\alpha-\beta)\left(\frac{n}{2n+\alpha+\beta} - \frac{n+1}{2n+2+\alpha+\beta}\right) \tag{8.9}$$

(if $n=0$ and $\alpha+\beta=0$ this formula should be replaced by $b_0 = (\beta-\alpha)/2$). Obviously, for a symmetric weight function (8.2) when $\alpha=\beta$, we have $b_n=0$. In the general case, we have the asymptotic formulas

$$a_n = 1/2 + 2^{-4}(1-2\alpha^2-2\beta^2)n^{-2} + O(n^{-3}), \quad b_n = 2^{-2}(\beta^2-\alpha^2)n^{-2} + O(n^{-3}). \tag{8.10}$$

Assumption (1.7) is of course true in this case, and the sequence (2.12) satisfies the condition $\rho_n = O(n^{-1})$ as $n\to\infty$. Note that the coefficients at n^{-2} in (8.10) are not zeros unless $|\alpha|=|\beta|=1/2$.

Next, we calculate the product (5.20). It follows from (8.8) that

$$A^{(n)} := \prod_{k=0}^{n-1}(2a_k) = \frac{\mathsf{k}_0}{\sqrt{\mathsf{h}_0}}\frac{\sqrt{\mathsf{h}_n}}{\mathsf{k}_n}2^n.$$

Let us find the limit of the right-hand side as $n\to\infty$. According to formula (1.18.4) in the book [9], for all $p\in\mathbb{R}$ we have

$$\Gamma(n+p) = n^p\,\Gamma(n)(1+O(n^{-1})). \tag{8.11}$$

Therefore the sequence h_n defined by (8.5) satisfies

$$\mathsf{h}_n = 2^{\alpha+\beta} n^{-1} + O(n^{-2}). \tag{8.12}$$

Let us use additionally (see formula (1.3.11) in [9]) that

$$\Gamma(2n+\alpha+\beta+1) = \pi^{-1/2} 2^{2n+\alpha+\beta} \Gamma(n+(\alpha+\beta+1)/2)\Gamma(n+(\alpha+\beta+2)/2).$$

So, we see that the sequence k_n defined by (8.6) satisfies $\mathsf{k}_n = 2^{\alpha+\beta} 2^n (\pi n)^{-1/2}(1+O(n^{-1}))$. It follows that

$$\lim_{n\to\infty} 2^n \frac{\sqrt{\mathsf{h}_n}}{\mathsf{k}_n} = \sqrt{\pi} 2^{-(\alpha+\beta)/2}.$$

Taking also into account formulas (8.7), we finally find that

$$A = \lim_{n\to\infty} A^{(n)} = \sqrt{\pi\kappa}\, 2^{-(\alpha+\beta)/2} \tag{8.13}$$

where κ is defined by (8.3).

8.2. *The perturbation determinant and the spectral shift function*

Let us calculate the Szegö function defined by (7.1) for the measure (8.1), (8.2). Now we have

$$w(\cos\theta)|\sin\theta| = \kappa\, 2^{\alpha+\beta+1} |\sin(\theta/2)|^{2\alpha+1} (\cos(\theta/2))^{2\beta+1}.$$

Substituting this expression into (7.1) and using relations (7.3), we find that

$$D(\zeta) = \sqrt{\kappa} 2^{-(\alpha+\beta+1)/2} (1-\zeta)^{\alpha+1/2} (1+\zeta)^{\beta+1/2}. \tag{8.14}$$

In particular, for the operator H_0 when $\alpha = \beta = 1/2$, we have $D_0(\zeta) = (2\pi)^{-1/2}(1-\zeta^2)$. Setting in (8.14) $\zeta = 0$ and taking into account that, by (7.9), $\sqrt{2\pi} D(0) = A$, we recover expression (8.13).

Now the perturbation determinant (5.1) for the pair H_0, H can be found from formulas (7.9) where $B(\zeta) = 1$ and (8.13), (8.14). Let us state the results obtained.

Theorem 8.1. *Let H be the Jacobi operator with the coefficients* (8.8), (8.9) *or, equivalently, corresponding to the spectral measure* (8.1), (8.2). *Then the Szegö function $D(\zeta)$ and the perturbation determinant $\Delta(z)$ are given by formulas* (8.14) *and*

$$\Delta(z) = (1-\zeta(z))^{-\alpha+1/2} (1+\zeta(z))^{-\beta+1/2}. \tag{8.15}$$

Recall that the Jost function $\Omega(z)$ is related to $\Delta(z)$ by formula (5.19). So it follows from Theorem 3.5 that

$$\lim_{n\to\infty} \zeta(z)^n P_n(z) = \kappa^{-1/2} \pi^{-1/2} 2^{(\alpha+\beta)/2} (1-\zeta(z))^{-\alpha-1/2} (1+\zeta(z))^{-\beta-1/2}, \quad z \in \Pi.$$

Thus we recover the well known asymptotic formula (see, for example, formula (8.21.9) in the book [26]) for the Jacobi polynomials.

Let us now calculate $\Omega(\lambda + i0)$ for $\lambda \in (-1, 1)$; as usual, we set $\lambda = \cos\theta$ where $\theta \in (0, \pi)$. Then $\zeta = \zeta(\lambda + i0) = e^{-i\theta}$ belongs to the lower half-circle. It is easy to see that

$$1 - \zeta = 1 - \cos\theta + i\sin\theta = 2\sin(\theta/2)e^{i(\pi-\theta)/2} = \sqrt{2(1-\lambda)}e^{i(\pi-\theta)/2}$$

and

$$1 + \zeta = 1 + \cos\theta - i\sin\theta = 2\cos(\theta/2)e^{-i\theta/2} = \sqrt{2(1+\lambda)}e^{-i\theta/2}.$$

Substituting these expressions into formula (8.15) we obtain the following result.

Theorem 8.2. *Under the assumptions of Theorem* 8.1, *the perturbation determinant is given on the cut along* $[-1, 1]$ *by the formula*

$$\Delta(\lambda + i0) = 2^{(1-\alpha-\beta)/2}(1-\lambda)^{(1-2\alpha)/4}(1+\lambda)^{(1-2\beta)/4}e^{i\pi\xi(\lambda)}, \quad \lambda \in (-1, 1), \tag{8.16}$$

where the spectral shift function

$$\xi(\lambda) = (2\pi)^{-1}(\alpha + \beta - 1)\arccos\lambda - 4^{-1}(2\alpha - 1). \tag{8.17}$$

Using (5.19), (8.13) and substituting expression (8.16) for $|\Delta(\lambda + i0)|$ into formula (2.30), we recover relations (8.2), (8.3) for $w(\lambda)$. In view of definition (2.22), Theorem 8.2 yields expressions for the limit amplitude and the limit phase. Thus, formula (2.23) means that

$$P_n(\lambda) = 2^{1/2}(\pi\kappa)^{-1/2}(1-\lambda)^{-(1+2\alpha)/4}(1+\lambda)^{-(1+2\beta)/4}$$
$$\times \sin\left(\left(n + \frac{\alpha+\beta+1}{2}\right)\arccos\lambda - \frac{2\alpha-1}{4}\pi\right) + O(n^{-1})$$

where κ is given by (8.3). This coincides of course with asymptotics (8.21.10) of the Jacobi polynomials in the book [26].

Substituting expressions (8.15) for the perturbation determinant and (8.17) for the spectral shift function into formula (5.7) we obtain a curious identity

$$\int_{-1}^{1} \frac{\arccos\lambda}{\lambda - z}d\lambda = -2\pi\ln(1 + \zeta(z)).$$

The asymptotics of Jacobi polynomials at the edge points $z = \pm 1$ is different from general results (4.7) and (4.9) of Sec. 4 because in view of formulas (8.9) and (8.10) the condition (4.1) is not satisfied unless $|\alpha| = |\beta| = 1/2$. Indeed, putting together formulas (8.4), (8.11) and (8.12), we see that

$$P_n(1) = \kappa^{-1/2}\Gamma(\alpha + 1)^{-1}2^{-(\alpha+\beta)/2}n^{\alpha+1/2}(1 + O(n^{-1})) \tag{8.18}$$

where κ is defined by (8.3). This is consistent with asymptotics (4.7) if $\alpha = 1/2$ and with asymptotics (4.9) if $\alpha = -1/2$. The results for $P_n(-1)$ are quite similar.

8.3. *The scattering matrix*

Let us now discuss stationary representations for the wave operators $W_\pm(H, H_0)$ and the corresponding scattering matrix $S(\lambda)$. Recall that the operator F was defined by Eqs. (6.5) and (6.6). Using formulas (6.15) and (6.23), we can state the following result.

Theorem 8.3. *Under the assumptions of Theorem* 8.1, *the wave operator* $W_\pm(H, H_0)$ *is given by the equality* (6.21) *where* $\Sigma_\pm$ *is the operator of multiplication by the function*

$$\sigma_\pm(\lambda) = e^{\pm i((\alpha+\beta-1)\arccos\lambda - \pi(\alpha-1/2))/2}.$$

The scattering matrix satisfies the equation

$$S(\lambda) = e^{i\pi(\alpha-1/2)} e^{-i(\alpha+\beta-1)\arccos\lambda}.$$

In view of asymptotics (8.10) the condition (4.1) is not satisfied (unless $|\alpha| = |\beta| = 1/2$), and hence the results of Secs. 4.1 and 4.2, as well as Theorem 5.10, are not applicable. Nevertheless representation (8.15) allows us to find the singularity of the perturbation determinant $\Delta(z)$ as $z \to \pm 1$. If $z \to 1$, then

$$\Delta(z) = 2^{-\beta+1/2}(1-\zeta(z))^{-\alpha+1/2}(1+O(|\zeta(z)-1|)).$$

If $z \to -1$, we have

$$\Delta(z) = 2^{-\alpha+1/2}(1+\zeta(z))^{-\beta+1/2}(1+O(|\zeta(z)+1|)).$$

It follows from (8.17) that the limits of $\xi(\lambda)$ as $\lambda \to -1+0$ and as $\lambda \to 1-0$ exist and

$$\xi(-1+0) = (1-2\beta)/4, \quad \xi(1-0) = (1-2\alpha)/4.$$

Recall that $\xi(\lambda) = 0$ for $|\lambda| > 1$. Thus, the spectral shift function is continuous at the point -1 (at the point 1) if and only if $\beta = 1/2$ ($\alpha = 1/2$). We are not aware of the results of this type for the differential operator $Da(x)D + b(x)$ in the space $L^2(\mathbb{R}_+)$.

Let us, finally, discuss the exceptional case $|\alpha| = |\beta| = 1/2$:

1^0. If $\alpha = \beta = 1/2$, then $a_n = 1/2$, $b_n = 0$ for all $n \in \mathbb{Z}_+$, and hence $H = H_0$.

2^0. If $\alpha = \beta = -1/2$, then $a_0 = 1/\sqrt{2}$, $a_n = 1/2$ for all $n \geq 1$, $b_n = 0$ for all $n \in \mathbb{Z}_+$ and the perturbation V has rank 2. In this case, $\Delta(z) = 1 - \zeta(z)^2$ so that the corresponding Jacobi operator has resonances at both points $z = 1$ and $z = -1$.

3^0. If $\alpha = -\beta = \pm 1/2$, then $a_n = 1/2$ for all $n \in \mathbb{Z}_+$, $b_0 = \mp 1$, $b_n = 0$ for all $n \geq 1$ and the perturbation V has rank 1. In this case, $\Delta(z) = 1 \pm \zeta(z)$ so that the corresponding Jacobi operator has a resonance at the point $z = \mp 1$.

Observe that in cases 1^0 and 2^0, $P_n(\lambda)$ are the normalized Chebyshev polynomials of the second and first kind, respectively.

8.4. *Related results*

Note that very detailed results on the asymptotics of the orthogonal polynomials were obtained in [21] for the weight function

$$\tilde{w}(\lambda) = (1-\lambda)^{\alpha}(1+\lambda)^{\beta}h(\lambda);$$

here $h(\lambda)$ is an arbitrary real analytic function such that $h(\lambda) > 0$ for $\lambda \in [-1, 1]$. On the contrary, it was shown in [23] that, for the weight function $\tilde{w}(\lambda)\chi_c(\lambda)$ where $\chi_c(\lambda) = 1$ for $\lambda \in [-1, 0)$ and $\chi_c(\lambda) = c \neq 1$, $c > 0$, for $\lambda \in (0, 1]$, the classical asymptotics (1.9) of the orthogonal polynomials is significantly changed. For such weight functions, $a_n - 1/2 = O(n^{-1})$ and $b_n = O(n^{-1})$ so that condition (1.7) is not satisfied.

Both papers [21] and [23] rely on the approach of [7, 14].

9. Discrete Volterra Equations

9.1. *Equation for the Jost solution*

Here we prove Theorems 2.1 and 2.3. We follow rather closely the scheme exposed for the Schrödinger equation in a detailed way in Sec. 4.1 of [30]. Note however that the operator H is obtained by a second order perturbation of the operator H_0. This circumstance should be taken into account.

Let us reduce Eq. (2.2) to a discrete Volterra integral equation.

Lemma 9.1. *Let assumption* (1.7) *be satisfied, and let* $z \in \operatorname{clos}\Pi$, $z \neq \pm 1$. *Suppose that* $f(z) = \{f_n(z)\} \in \ell^\infty(\mathbb{Z}_+)$ *and define* V *by formula* (2.10). *Then a solution of Eq.* (2.11) *satisfies also Eq.* (2.2).

Proof. Under assumption (1.7), the sequence $\phi(z) := Vf(z) \in \ell^1(\mathbb{Z}_+)$. Put

$$\Sigma_n^{(+)}(z) = \sum_{m=n+1}^{\infty} \zeta(z)^m \phi_m(z), \quad \Sigma_n^{(-)}(z) = \sum_{m=n+1}^{\infty} \zeta(z)^{-m} \phi_m(z).$$

Then Eq. (2.11) can be written as

$$f_n = \zeta^n - \frac{\zeta^n}{\sqrt{z^2-1}}\Sigma_n^{(-)} + \frac{\zeta^{-n}}{\sqrt{z^2-1}}\Sigma_n^{(+)},$$

whence

$$f_{n-1} = \zeta^{n-1} - \frac{\zeta^{n-1}}{\sqrt{z^2-1}}\Sigma_n^{(-)} + \frac{\zeta^{-n+1}}{\sqrt{z^2-1}}\Sigma_n^{(+)} + \frac{\zeta-\zeta^{-1}}{\sqrt{z^2-1}}\phi_n,$$

$$f_{n+1} = \zeta^{n+1} - \frac{\zeta^{n+1}}{\sqrt{z^2-1}}\Sigma_n^{(-)} + \frac{\zeta^{-n-1}}{\sqrt{z^2-1}}\Sigma_n^{(+)}.$$

It follows that

$$\begin{aligned}2\sqrt{z^2-1}((H_0-z)f)_n &= \sqrt{z^2-1}\big(f_{n-1}+f_{n+1}-2zf_n\big)\\ &= (\zeta-\zeta^{-1})\phi_n + (\zeta^{-n+1}+\zeta^{-n-1})\Sigma_n^{(+)} - (\zeta^{n+1}+\zeta^{n-1})\Sigma_n^{(-)}\\ &\quad - 2z(\zeta^{-n}\Sigma_n^{(+)} - \zeta^n\Sigma_n^{(-)}). \qquad (9.1)\end{aligned}$$

The sums of the terms containing $\Sigma_n^{(+)}$ and $\Sigma_n^{(-)}$ equal zero because $\zeta+\zeta^{-1}=2z$. Since $\zeta-\zeta^{-1}=-2\sqrt{z^2-1}$, equality (9.1) yields the desired equation $((H_0-z)f)_n = -(Vf)_n$. □

Next, we study Eq. (2.11). Let us distinguish the term corresponding to $m=n+1$ in its right-hand side. Then Eq. (2.11) reads as

$$\begin{aligned}2a_nf_n(z) &= \zeta(z)^n - 2b_{n+1}f_{n+1}(z) + (1-2a_{n+1})f_{n+2}(z)\\ &\quad - \frac{1}{\sqrt{z^2-1}}\sum_{m=n+2}^{\infty}(\zeta(z)^{n-m}-\zeta(z)^{m-n})(Vf(z))_m.\end{aligned}$$

It is convenient to rewrite this equation in terms of the sequence

$$g_n(z) = 2a_n\zeta(z)^{-n}f_n(z) \qquad (9.2)$$

as

$$g_n(z) = 1 + \sum_{m=n+1}^{\infty} G_{n,m}(z)g_m(z) \qquad (9.3)$$

where

$$\begin{aligned}2a_m\sqrt{z^2-1}G_{n,m}(z) &= (\zeta(z)^{2m-2n}-1)b_m\\ &\quad + (\zeta(z)^{2m-2n-1}-\zeta(z))(a_{m-1}-1/2)\\ &\quad + (\zeta(z)^{2m-2n+1}-\zeta(z)^{-1})(a_m-1/2). \qquad (9.4)\end{aligned}$$

It is important that

$$|G_{n,m}(z)| \le C\alpha_m \qquad (9.5)$$

where

$$\alpha_m = |a_{m-1}-1/2| + |a_m-1/2| + |b_m|$$

and the constant $C=C(z)$ does not depend on n, m and on z in compact subsets of $\operatorname{clos}\Pi$ away from the points ±1.

Let us solve Eq. (9.3) by iterations. Put $g_n^{(0)}=1$ for all $n\in\mathbb{Z}_+$ and

$$g_n^{(k+1)}(z) = \sum_{m=n+1}^{\infty} G_{n,m}(z)g_m^{(k)}(z), \quad k\ge 0. \qquad (9.6)$$

Lemma 9.2. *The estimates*

$$|g_n^{(k)}(z)| \leq \frac{C^k}{k!}\left(\sum_{m=n+1}^{\infty} \alpha_m\right)^k, \quad \forall n \in \mathbb{Z}_+, \tag{9.7}$$

are true for all $k \in \mathbb{Z}_+$.

Proof. Suppose that (9.7) is satisfied for some $k \in \mathbb{Z}_+$. We have to check that the same estimate (with k replaced by $k+1$ in the right-hand side) holds for $g_n^{(k+1)}(z)$. It follows from (9.5) and (9.7) that

$$|g_n^{(k+1)}(z)| \leq \frac{C^{k+1}}{k!} \sum_{m=n+1}^{\infty} \alpha_m \left(\sum_{p=m+1}^{\infty} \alpha_p\right)^k. \tag{9.8}$$

Observe that

$$(k+1)\alpha_m \left(\sum_{p=m+1}^{\infty} \alpha_p\right)^k + \left(\sum_{p=m+1}^{\infty} \alpha_p\right)^{k+1} \leq \left(\sum_{p=m}^{\infty} \alpha_p\right)^{k+1},$$

and hence, for all $N \in \mathbb{Z}_+$,

$$\begin{aligned}(k+1) &\sum_{m=n+1}^{N} \alpha_m \left(\sum_{p=m+1}^{\infty} \alpha_p\right)^k \\ &\leq \sum_{m=n+1}^{N} \left(\left(\sum_{p=m}^{\infty} \alpha_p\right)^{k+1} - \left(\sum_{p=m+1}^{\infty} \alpha_p\right)^{k+1}\right) \leq \left(\sum_{p=n+1}^{\infty} \alpha_p\right)^{k+1}.\end{aligned}$$

Substituting this estimate into (9.8), we obtain (9.7) for $g_n^{(k+1)}(z)$. $\square$

It is now easy to conclude the proof of Theorem 2.1. Set

$$g_n(z) = \sum_{k=0}^{\infty} g_n^{(k)}(z).$$

Estimate (9.7) shows that this series converges and, by the recurrent definition (9.6), the functions $g_n(z)$ satisfy Eq. (9.3) equivalent to (2.11). Since every function $g_n^{(k)}(z)$ is analytic in $z \in \Pi$ and is continuous up to the cut $[-1, 1]$ (away from the points ± 1), estimate (9.7) guarantees that $g_n(z)$ and hence $f_n(z)$ possess the same properties.

It also follows from (9.3) and (9.5) that

$$|g_n(z) - 1| \leq C \sum_{m=n+1}^{\infty} \alpha_m$$

which in view of (9.2) implies (2.13). $\square$

As far as Theorem 2.3 is concerned, we note that all functions $g_n^{(k)}(z)$ are continuous with respect to the cut-off parameter $N \to \infty$, and therefore estimates obtained in the proof of Theorem 2.1 imply relation (2.16).

9.2. *Equation for the regular solution*

Next, we discuss Theorem 3.1. The first assertion is quite similar to Lemma 9.1, and so its proof will be omitted.

Lemma 9.3. *Let* $P(z) = \{P_n(z)\}_{n=-1}^{\infty}$ *where* $P_{-1}(z) = 0$, $P_0(z) = 1$ *satisfy Eq.* (3.2). *Then* $P_n(z)$ *satisfy also Eq.* (1.2); *in particular,* $P_n(z)$ *is a polynomial of degree* n.

We proceed from Eq. (3.2) and, following the scheme of the previous subsection, distinguish the term corresponding to $m = n-1$ in the sum (3.2). Then Eq. (3.2) reads as

$$2a_{n-1}P_n(z) = P_n^{(0)}(z) - 2a_{n-2}P_{n-2}(z) - 2b_{n-1}P_{n-1}(z) + \frac{1}{\sqrt{z^2-1}}\sum_{m=0}^{n-2}(\zeta(z)^{n-m} - \zeta(z)^{m-n})(VP(z))_m, \quad n \geq 2.$$

It is convenient to rewrite this equation in terms of the sequence

$$Q_n(z) = 2a_{n-1}\zeta(z)^n P_n(z)$$

as

$$Q_n(z) = Q_n^{(0)}(z) - \sum_{m=0}^{n-1} \mathsf{G}_{n,m}(z)Q_m(z). \tag{9.9}$$

Here $Q_n^{(0)}(z) = \zeta(z)^n P_n^{(0)}(z)$ are bounded uniformly in n in view of (2.8). The matrix elements $\mathsf{G}_{n,m}(z)$ are quite similar to $G_{n,m}(z)$ defined by (9.4). It is only important that $\mathsf{G}_{n,m}(z)$ satisfy estimate (9.5).

We again solve Eq. (9.9) by iterations. Put $Q_0^{(k)} = 0$ for $k \geq 1$ and

$$Q_n^{(k+1)}(z) = \sum_{m=0}^{n-1} \mathsf{G}_{n,m}(z)Q_m^{(k)}(z), \quad k \geq 0, \quad n \geq 1. \tag{9.10}$$

Lemma 9.4. *The estimates*

$$|Q_n^{(k)}(z)| \leq \frac{C^k}{k!}\left(\sum_{m=0}^{n-1}\alpha_m\right)^k, \quad \forall n \geq 1, \tag{9.11}$$

are true for all $k \in \mathbb{Z}_+$.

Proof. Suppose that (9.11) is satisfied for some $k \in \mathbb{Z}_+$. We have to check that the same estimate (with k replaced by $k+1$ in the right-hand side) holds for $Q_n^{(k+1)}(z)$. It follows from (9.10), (9.11) that

$$|Q_n^{(k+1)}(z)| \leq \frac{C^{k+1}}{k!}\sum_{m=0}^{n-1}\alpha_m\left(\sum_{p=0}^{m-1}\alpha_p\right)^k. \tag{9.12}$$

Observe that

$$(k+1)\alpha_m\left(\sum_{p=0}^{m-1}\alpha_p\right)^k+\left(\sum_{p=0}^{m-1}\alpha_p\right)^{k+1}\leq\left(\sum_{p=0}^{m}\alpha_p\right)^{k+1},$$

and hence

$$(k+1)\sum_{m=0}^{n-1}\alpha_m\left(\sum_{p=0}^{m-1}\alpha_p\right)^k\leq\sum_{m=0}^{n-1}\left(\left(\sum_{p=0}^{m}\alpha_p\right)^{k+1}-\left(\sum_{p=0}^{m-1}\alpha_p\right)^{k+1}\right)\leq\left(\sum_{p=0}^{n-1}\alpha_p\right)^{k+1}.$$

Substituting this estimate into (9.12), we obtain (9.11) for $k+1$. □

Let us now set

$$Q_n(z)=\sum_{k=0}^{\infty}Q_n^{(k)}(z).$$

Estimate (9.11) shows that this series converges and

$$Q_n(z)\leq\sum_{k=0}^{\infty}\frac{C^k}{k!}\left(\sum_{m=0}^{n-1}\alpha_m\right)^k=\exp\left(C\sum_{m=0}^{n-1}\alpha_m\right).$$

By the recurrent definition (9.10), the functions $Q_n(z)$ satisfy Eq. (9.9) equivalent to (3.2). This proves estimate (3.1). □

9.3. *Jost solutions at the edge points*

The proof of Theorem 4.1 can be obtained similarly to that of Theorem 2.1 if one observes that

$$\left|\frac{\zeta^{2k}-1}{\sqrt{z^2-1}}\right|=2\left|\zeta\frac{\zeta^{2k}-1}{\zeta^2-1}\right|=2|\zeta||\zeta^{2(k-1)}+\zeta^{2(k-2)}+\cdots+\zeta^2+1|\leq 2k. \tag{9.13}$$

Thus instead of (9.5), we now have a bound

$$|G_{n,m}(z)|\leq C(m+1)\alpha_m \tag{9.14}$$

with a constant C not depending on ζ in the unit disc. Under assumption (4.1) this bound allows us to repeat the arguments of Theorem 2.1.

The proof of Theorem 4.2 follows the scheme of proof of Theorem 3.1, but now we again have to use estimate (9.13). This estimate implies that the kernels $\mathsf{G}_{n,m}(z)$ in (9.9) obey bound (9.14).

Acknowledgments

This work was supported by the Russian Science Foundation project 17-11-01126.

References

[1] N. Akhiezer, *The Classical Moment Problem and Some Related Questions in Analysis* (Oliver and Boyd, Edinburgh and London, 1965).
[2] Yu. M. Berezanskii, *Expansion in Eigenfunctions of Selfadjoint Operators* (Amer. Math. Soc., Providence, R.I., 1968).
[3] S. Bernstein, Sur les polynômes orthogonaux relatifs à un segment fini, *J. Mathé.* **9** (1930) 127–177; **10** (1931) 219–286.
[4] V. S. Buslaev and L. D. Faddeev, Formulas for traces for a singular Sturm–Liouville differential operator, *Soviet Math. Dokl.* **1** (1960) 451–454.
[5] K. M. Case, Orthogonal polynomials. II, *J. Math. Phys.* **16** (1975) 1435–1440.
[6] D. Damanik and R. Killip, Half-line Schrödinger operators with no bound states, *Acta Math.* **193** (2004) 31–72.
[7] P. Deift and X. Zhou, A steepest descent method for oscillatory Riemann–Hilbert problem, *Ann. Math.* **137** (1993) 295–368.
[8] P. L. Duren, *Theory of H^p Spaces* (Academic Press, New York and London, 1970).
[9] A. Erdélyi, W. Magnus, F. Oberhettinger and F. G. Tricomi, *Higher Transcendental Functions*, Vols. 1 and 2 (McGraw-Hill, New York-Toronto-London, 1953).
[10] L. D. Faddeev, An expression for the trace of the difference between two singular differential operators of Sturm–Liouville type, *Dokl. AN SSSR* **115**(5) (1957) 878–881 (in Russian).
[11] L. D. Faddeev, Properties of the S-matrix of the one-dimensional Schrödinger equation, *Amer. Math. Soc. Transl.* (*Ser. 2*) **65** (1967) 139–166.
[12] L. D. Faddeev, Inverse problem of quantum scattering theory. II, *J. Soviet. Math.* **5** (1976) 334–396.
[13] L. Faddeev and L. Takhtajan, *Hamiltonian Methods in the Theory of Solitons* (Springer-Verlag, Berlin, Heidelberg, 2007).
[14] A. Fokas, A. Its and A. Kitaev, The isomonodromy approach to matrix models in $2D$ quantum gravity, *Comm. Math. Phys.* **147** (1992) 395–430.
[15] I. M. Gel'fand and B. M. Levitan, On the determination of a differential equation from its spectral function, *Izv. Akad. Nauk SSSR, Ser. Mat.* **15** (1951) 309–361; *Amer. Math. Soc. Transl.* (*Ser. 2*) **1** (1955) 253–304.
[16] I. C. Gokhberg and M. G. Kreĭn, *Introduction to the Theory of Linear Nonselfadjoint Operators in Hilbert Space* (Amer. Math. Soc., Providence, Rhode Island, 1970).
[17] G. Sh. Guseinov, The scattering problem for an infinite Jacobi matrix, *Izv. Akad. Nauk Armyan. SSR Ser. Mat.* **12** (1977) 365–379 (in Russian).
[18] D. Hundertmark and B. Simon, Lieb–Thirring inequalities for Jacobi matrices, *J. Approx. Theory* **118** (2002) 106–130.
[19] P. Jost and A. Pais, On the scattering of a particle by a static potential, *Phys. Rev.* **82** (1951) 840–851.
[20] R. Killip and B. Simon, Sum rules for Jacobi matrices and their applications to spectral theory, *Ann. Math.* **158** (2003) 253–321.
[21] A. B. J. Kuijlaars, K. T.-R. McLaughlin, W. Van Assche and M. Vanlessen, The Riemann–Hilbert approach to strong asymptotics for orthogonal polynomials on $[-1, 1]$, *Adv. Math.* **188** (2004) 337–398.
[22] L. D. Landau and E. M. Lifshitz, *Quantum Mechanics* (Pergamon Press, 1965).
[23] F. Moreno, A. Martinez-Finkelshtein and V. L. Sousa, Asymptotics of orthogonal polynomials for a weight with a jump on $[-1, 1]$, *Constr. Approx.* **33**(2) (2011) 219–263.
[24] E. Ryckman, A strong Szegö theorem for Jacobi matrices, *Comm. Math. Phys.* **271** (2007) 791–820; Erratum, *Comm. Math. Phys.* **275** (2007) 581–585.

[25] B. Simon, *Trace Ideal Methods*, London Math. Soc. Lecture Notes (Cambridge Univ. Press, London and New York, 1979).

[26] G. Szegö, *Orthogonal Polynomials* (Amer. Math. Soc., Providence, R.I., 1978).

[27] G. Teschl, *Jacobi Operators and Completely Integrable Nonlinear Lattices* (Amer. Math. Soc., Providence, R.I., 2000).

[28] D. R. Yafaev, The low energy scattering for slowly decreasing potentials, *Comm. Math. Phys.* **85** (1982) 177–196.

[29] D. R. Yafaev, *Mathematical Scattering Theory: General Theory* (Amer. Math. Soc., Providence, R.I., 1992).

[30] D. R. Yafaev, *Mathematical Scattering Theory: Analytic Theory* (Amer. Math. Soc., Providence, R.I., 2010).

[31] D. R. Yafaev, A point interaction for the discrete Schrödinger operator and generalized Chebyshev polynomials, *J. Math. Phys.* **58** (2017) 063511.